Nanotechnology Applications for Clean Water

Solutions for Improving Water Quality

Second Edition

Edited by

Anita Street

US Department of Energy

Richard Sustich

Center of Advanced Materials for the Purification of Water with Systems, University of Illinois at Urbana-Champaign

Jeremiah Duncan

Department of Atmospheric Science and Chemistry, Plymouth State University

Nora Savage

Office of Research and Development, US Environmental Protection Agency

Foreword by

George Gray

ELSEVIER

AMSTERDAM • BOSTON • HEIDELBERG • LONDON
NEW YORK • OXFORD • PARIS • SAN DIEGO
SAN FRANCISCO • SINGAPORE • SYDNEY • TOKYO
William Andrew is an imprint of Elsevier

William Andrew is an imprint of Elsevier
The Boulevard, Langford Lane, Kidlington, Oxford, OX5 1GB
225 Wyman Street, Waltham, MA 02451, USA

First published 2009
Second edition 2014

British Library Cataloguing-in-Publication Data
A catalogue record for this book is available from the British Library

Library of Congress Cataloging-in-Publication Data
A catalog record for this book is availabe from the Library of Congress

ISBN: 978-1-4557-3116-9

14 15 16 17 18 10 9 8 7 6 5 4 3 2 1

Contents

PART 6 OUTLOOK

List of Contributors

Shirish Agarwal
Department of Civil and Environmental Engineering, University of Cincinnati, Cincinnati, OH, USA

Souhail R. Al-Abed
National Risk Management Research Laboratory, US Environmental Protection Agency, Cincinnati, OH, USA

Pedro J.J. Alvarez
Department of Civil and Environmental Engineering, Rice University, Houston, TX, USA

Anshup
InnoNano Research Private Limited, An IIT Madras-Incubated Company Focusing on Technologies for Clean Drinking Water, Chennai, India

Abdullah Mohamed Asiri
Chemistry Department, King Abdulaziz University, Jeddah, Saudi Arabia

Chaoyi Ba
Center of Advanced Materials for the Purification of Water with Systems, Department of Materials Science and Engineering, University of Illinois at Urbana-Champaign, Urbana, IL, USA

Leonidas Bachas
University of Miami, Miami, FL, USA

Olgica Bakajin
Chief Executive Officer at Porifera, Inc., San Francisco, CA, USA

David M. Berube
North Carolina State University, Raleigh, NC, USA

Madhuleena Bhadra
Department of Chemistry and Environmental Science, New Jersey Institute of Technology, University Heights, Newark, NJ, USA

Dibakar Bhattacharyya
Department of Chemical and Materials Engineering, University of Kentucky, Lexington, KY, USA

Jagan Bontha
Pacific Northwest National Laboratory, Richland, WA, USA

M.S. Bootharaju
Department of Chemistry, Indian Institute of Technology Madras, Chennai, India

Dick Brown
ERM Corporation, Ewing, NJ, USA

Lena Brunet
Institut Francais du Petrole Lyon, France

So-Ryong Chae
Department of Civil and Environmental Engineering, School of Engineering, Duke University, Durham, NC, USA

Daiwon Choi
Pacific Northwest National Laboratory, Richland, WA, USA

Hyeok Choi
Department of Civil Engineering, The University of Texas at Arlington, Arlington, TX, USA

Kimberly M. Cross
Department of Chemical & Biomolecular Engineering, University of California at Los Angeles, Los Angeles, CA, USA

David Culpepper
NanoFex LLC., New Orleans, LA, USA

Mamadou S. Diallo
Department of Civil Engineering, Howard University, Washington, DC, USA; Materials and Process Simulation Center, Division of Chemistry and Chemical Engineering, California Institute of Technology, Pasadena, CA, USA

Dionysios D. Dionysiou
Environmental Engineering and Science Program, University of Cincinnati, Cincinnati, OH, USA; Nireas – International Water Research Centre, University of Cyprus, Nicosia, Cyprus

Dan Du
Key Laboratory of Pesticide and Chemical Biology of Ministry of Education, College of Chemistry, Central China Normal University, Wuhan, PR China; School of Mechanical and Materials Engineering, Washington State University, Pullman, WA, USA

Nian Du
Department of Chemistry, Center for Advanced Sensors and Environmental Systems, State University of New York at Binghamton, Binghamton, NY, USA

Jeremiah S. Duncan
Department of Atmospheric Science and Chemistry, Plymouth State University, Plymouth, NH, USA

James Economy
Center of Advanced Materials for Purification of Water with Systems, Department of Materials Science and Engineering, University of Illinois at Urbana-Champaign, Urbana, IL, USA

Lisa Farmen
Crystal Clear Technologies Inc., Portland, OR, USA

Emma Fauss
Department of Electrical and Computer Engineering, University of Virginia, Charlottesville, VA, USA

Francesco Fornasiero
Molecular Biophysics and Functional Nanostructures Group, Chemistry, Materials, Earth, and Life Sciences Directorate, Lawrence Livermore National Laboratory, Livermore, CA, USA

Asim K. Ghosh
Desalination Division, Bhabha Atomic Research Centre, Trombay, Mumbai, India

Michael E. Gorman
Department of Science, Technology and Society, University of Virginia, Charlottesville, VA, USA

Costas P. Grigoropoulos
Department of Mechanical Engineering, University of California at Berkeley, Berkeley, CA, USA

David J. Grimshaw
Royal Holloway, University of London, Egham Hill, UK

Lawrence D. Gudza
Practical Action, Bourton-on-Dunsmore, Rugby, UK

Changseok Han
Environmental Engineering and Science Program, University of Cincinnati, Cincinnati, OH, USA; Nireas, International Water Research Centre, University of Cyprus, Nicosia, Cyprus

Feng He
Oak Ridge National Laboratory, Oak Ridge, TN, USA

Thembela Hillie
National Centre for Nano-Structured Materials, DST/CSIR Nanotechnology Innovation Centre, Pretoria, South Africa; Physics Department, University of the Free State, Bloemfontein, South Africa

Mbhuti Hlophe
Department of Chemistry, North-West University (Mafikeng Campus), Mmabatho, South Africa

Eric M.V. Hoek
Department of Civil and Environmental Engineering and California NanoSystems Institute, University of California, Los Angeles, CA, USA

Jason K. Holt
Molecular Biophysics and Functional Nanostructures Group, Chemistry, Materials, Earth, and Life Sciences Directorate, Lawrence Livermore National Laboratory, Livermore, CA, USA

Ernest M. Hotze
Department of Civil and Environmental Engineering, School of Engineering, Duke University, Durham, NC, USA

Jung Bin In
Department of Mechanical Engineering, University of California at Berkeley, Berkeley, CA, USA; Molecular Biophysics and Functional Nanostructures Group, Chemistry, Materials, Earth, and Life Sciences Directorate, Lawrence Livermore National Laboratory, Livermore, CA, USA

Vijay T. John
Department of Chemical and Biomolecular Engineering, Tulane University, New Orleans, LA, USA

Isaac K'Owino
Department of Chemistry, Center for Advanced Sensors and Environmental Systems, State University of New York at Binghamton, Binghamton, NY, USA

Sangil Kim
Molecular Biophysics and Functional Nanostructures Group, Chemistry, Materials, Earth, and Life Sciences Directorate, Lawrence Livermore National Laboratory, Livermore, CA, USA

Jian Ku Shang
Center of Advanced Materials for Purification of Water with Systems, Department of Materials Science and Engineering, University of Illinois at Urbana-Champaign, Urbana, IL, USA

Qi Li
Center of Advanced Materials for Purification of Water with Systems, Department of Materials Science and Engineering, University of Illinois at Urbana-Champaign, Urbana, IL, USA

Qilin Li
Department of Civil and Environmental Engineering, Rice University, Houston, TX, USA

Yuehe Lin
Chemistry Department, King Abdulaziz University, Jeddah, Saudi Arabia; School of Mechanical and Materials Engineering, Washington State University, Pullman, WA, USA

Juewen Liu
University of Waterloo, Waterloo, Ontario, Canada

Ruiqiang Liu
Environmental Engineering Program, Department of Civil Engineering, Auburn University, Auburn, AL, USA

Gregory V. Lowry
Department of Civil and Environmental Engineering, Carnegie Mellon University, Pittsburgh, PA, USA

Yi Lu
Center of Advanced Materials for the Purification of Water with Systems, Department of Chemistry, University of Illinois at Urbana – Champaign, Urbana, IL, USA

Yunfeng Lu
Department of Chemical & Biomolecular Engineering, University of California at Los Angeles, Los Angeles, CA, USA

Delina Y. Lyon
Shell Oil Company, Houston, TX, USA

Shaily Mahendra
Department of Civil and Environmental Engineering, University of California at Los Angeles, Los Angeles, CA, USA

Debapriya Mazumdar
ANDalyze, Inc. Chicago, IL, USA

Gary L. McPherson
Department of Chemical and Biomolecular Engineering, Tulane University, New Orleans, LA

Evan S. Michelson
Alfred P. Sloan Foundation, New York, NY, USA

Somenath Mitra
Department of Chemistry and Environmental Science, New Jersey Institute of Technology, University Heights, Newark, NJ, USA

Gordon Nameni
Accelerate DBA, Milwaukee, WI, USA

Aleksandr Noy
Molecular Biophysics and Functional Nanostructures Group, Chemistry, Materials, Earth, and Life Sciences Directorate, Lawrence Livermore National Laboratory, Livermore, CA, USA

Denis M. O'Carroll
Department of Civil and Environmental Engineering, The University of Western Ontario, London, Ontario, Canada

Veronica Okello
Department of Chemistry, Center for Advanced Sensors and Environmental Systems, State University of New York at Binghamton, Binghamton, NY, USA

Marcells A. Omole
Department of Chemistry, Center for Advanced Sensors and Environmental Systems, State University of New York at Binghamton, Binghamton, NY, USA

Francis Osonga
Department of Chemistry, Center for Advanced Sensors and Environmental Systems, State University of New York at Binghamton, Binghamton, NY, USA

Sehinde Owoseni
Department of Chemical and Biomolecular Engineering, Tulane University, New Orleans, LA, USA

Hyung Gyu Park
Molecular Biophysics and Functional Nanostructures Group, Chemistry, Materials, Earth, and Life Sciences Directorate, Lawrence Livermore National Laboratory, Livermore, CA, USA

Mary Theresa M. Pendergast
Department of Civil and Environmental Engineering and California NanoSystems Institute, University of California, Los Angeles, CA, USA

Tanapon Phenrat
Department of Civil Engineering, Naresuan University, Phitsanulok, Thailand

Brian Pianfetti
Center of Advanced Materials for Purification of Water with Systems, University of Illinois at Urbana-Champaign, Urbana, IL, USA

Saik Choon Poh
CH2MHILL, Los Angeles, CA, USA

T. Pradeep
DST Unit of Nanoscience and Thematic Unit of Excellence, Department of Chemistry, Indian Institute of Technology Madras, Chennai, India

David Rejeski
Science and Technology Innovation Program, Woodrow Wilson International Center for Scholars, Washington, DC, USA

S.M.C. Ritchie
Department of Chemical & Biological Engineering, University of Alabama, Tuscaloosa, AL, USA

Chris Roberts
Department of Chemical Engineering, Auburn University, Auburn, AL, USA

Curtis D. Roth
DXV Water Technologies, LLC, Tustin, CA, USA

Omowunmi A. Sadik
Department of Chemistry, Center for Advanced Sensors and Environmental Systems, State University of New York at Binghamton, Binghamton, NY, USA

Nora Savage
Office of Research and Development, US Environmental Protection Agency, Washington, DC, USA

Hatice Sengül
Department of Environmental Engineering, Hacettepe University, Ankara, Turkey

Mark Shannon
Center of Advanced Materials for Purification of Water with Systems, University of Illinois at Urbana-Champaign, Urbana, IL, USA

John R. Shapley
Department of Chemistry and Center of Advanced Materials for the Purification of Water with Systems, University of Illinois at Urbana–Champaign, Urbana, IL, USA

Jack Stilgoe
University College London, London, UK

Timothy J. Strathmann
Department of Civil and Environmental Engineering and Center of Advanced Materials for the Purification of Water with Systems, University of Illinois at Urbana–Champaign, Urbana, IL, USA

Anita Street
US Department of Energy, Washington, DC, USA

Bhanukiran Sunkara
Department of Chemical and Biomolecular Engineering, Tulane University, New Orleans, LA, USA

Richard C. Sustich
Center of Advanced Materials for Purification of Water with Systems, University of Illinois at Urbana-Champaign, Urbana, IL, USA

Nathan Swami
Department of Electrical and Computer Engineering, University of Virginia, Charlottesville, VA, USA

Yukiko Takahashi
Department of Civil and Environmental Engineering, Nagaoka University of Technology, Nagaoka, Niigata, Japan

Volodymyr V. Tarabara
Department of Civil and Environmental Engineering, Michigan State University, East Lansing, Michigan, USA

Thomas L. Theis
Institute for Environmental Science and Policy, Department of Civil and Materials Engineering, University of Illinois at Chicago, Chicago, IL, USA

Pradeep Venkataraman
Department of Chemical and Biomolecular Engineering, Tulane University, New Orleans, LA, USA

Diem X. Vuong
DXV Water Technologies, LLC, Tustin, CA, USA

Jinwen Wang
Center of Advanced Materials for the Purification of Water with Systems, Department of Materials Science and Engineering, University of Illinois at Urbana-Champaign, Urbana, IL, USA

Jun Wang
Pacific Northwest National Laboratory, Richland, WA, USA

Ahson Wardak
Department of Systems and Information Engineering, University of Virginia, Charlottesville, VA, USA

Charles J. Werth
Department of Civil and Environmental Engineering and Center of Advanced Materials for the Purification of Water with Systems, University of Illinois at Urbana–Champaign, Urbana, IL, USA

Mark R. Wiesner
Department of Civil and Environmental Engineering, School of Engineering, Duke University, Durham, NC, USA

Pinggui Wu
Superior Graphite Company, Chicago, IL, USA

Jian Xu
Coskata, Warrenville, IL, USA

Yinhui Xu
Environmental Engineering Program, Department of Civil Engineering, Auburn University, Auburn, AL, USA

Idris Yazgan
Department of Chemistry, Center for Advanced Sensors and Environmental Systems, State University of New York at Binghamton, Binghamton, NY, USA

Abolfazl Zakersalehi
Department of Civil Engineering, The University of Texas at Arlington, Arlington, TX, USA

Jingjing Zhan
Department of Chemical and Biomolecular Engineering Tulane University, New Orleans, LA, USA

Weiying Zhang
Key Laboratory of Pesticide and Chemical Biology of Ministry of Education, College of Chemistry, Central China Normal University, Wuhan, PR China; School of Mechanical and Materials Engineering, Washington State University, Pullman, WA, United States

Dongye Zhao
Environmental Engineering Program, Department of Civil Engineering, Auburn University, Auburn, AL, USA

Rubo Zheng
Department of Chemical and Biomolecular Engineering, Tulane University, New Orleans, LA, USA

Tonghua Zheng
Department of Chemical and Biomolecular Engineering, Tulane University, New Orleans, LA, USA

Foreword

Imagine brushing your teeth with reclaimed wastewater. Would you be comfortable doing so? Would testing data indicating that this water was cleaner than the public water supply be enough to put aside any trepidation?

On a recent visit, I learned about Singapore's NEWater[1]. Surrounded by the sea, with limited surface area and reservoir capacity and a growing population, Singapore has embraced advanced technology – including nanotechnology – to meet its water needs. The program uses advanced filtration, reverse osmosis and disinfection technologies to allow wastewater to supplement their potable water supply. Singapore's NEWater has been tested and found to be cleaner than its very high quality public drinking water but it has not (yet) been put directly into the public drinking supply. It is used to recharge reservoirs (indirect potable use) or used in industries such as chip fabrication, which requires ultra-clean water. This illustrates the challenges of new technologies – not just having the tools to improve on what came before but also being aware of and addressing societal issues, including public acceptance. Yet advances are necessary if we are to provide safe water to the 7 billion (and growing) population of Earth. This new 2nd Edition of Nanotechnology Applications for Clean Water: Solutions for Improving Water Quality brings us the very latest thinking on both applications and implications of nanotechnology for enhancing water quality around the world.

Humans have a very long history of using technology to improve water. Thousands of years ago filtration, exposure to sunlight and other techniques were used to purify water. Tomb paintings suggest the ancient Egyptians may have invented flocculation with alum, still practiced in modern water treatment facilities. The Greeks and Romans made links between water and health and sought ways to enhance water quality. Over the centuries, while advances to these technologies improved water quality, pathogens remained a concern. The use of disinfectants like chlorine and ozone was the next big leap in water safety. It is estimated that clean water technologies, especially chlorination of drinking water (which started in the US in 1908) was responsible for 50% of the decrease in mortality, and 75% of the decrease in infant mortality, in American cities between 1900 and 1936[2]. Improvements in water quality have resulted in remarkable advances in public health. But there is more to do. Many parts of the world lack appropriate and effective water quality technology. Estimates vary but waterborne disease still causes at least 700,000 deaths per year. Contamination of groundwater puts aquifers out of bounds for drinking water and water shortages threaten to bring suffering and conflict.

---

[1]http://www.pub.gov.sg/about/historyfuture/Pages/NEWater.aspx.

[2]Source: Cutler, D. and Miller, G. (2005) The Role of Public Health Improvements in Health Advances: The Twentieth-Century United States. Demography 42:1–22.

We now stand at the cusp of another technological leap – the use of nanotechnology to monitor, decontaminate and disinfect water. In this 2nd Edition of Nanotechnology Applications for Clean Water: Solutions for Improving Water Quality Jeremiah Duncan, Nora Savage, Anita Street and Richard Sustich bring together in one place the knowledge and the vision to guide this leap. This new version brings the latest information, new topics and a format that demonstrates the wide applicability of nanotechnology to water quality. The book also addresses the key societal issues that must be addressed to realize the promise of nanotechnology.

With the advent of the National Nanotechnology Initiative (NNI), the US embarked on a program of research and development focused on both the many potential applications of nanoscale technology and the range of possible implications of the technology for health, safety and the environment. This is one of the first times that those important issues around safety and social acceptance were addressed as a technology develops. When I co-chaired the Nanotechnology Health and Safety Initiative, part of the NNI, federal agencies worked to coordinate research to identify and manage the potential risks of nanomaterials, and these efforts continue. During my tenure leading the US Environmental Protection Agency (EPA) Office of Research and Development, our budget for studying both applications and implications of nanotechnology more than doubled. These initiatives, together with research in academia and the private sector have contributed to the advances described in this volume.

Each of this book's editors has a link to the US EPA, whether as agency scientists, fellows or advisors. They deeply understand the promise of nanotechnology for improving water quality and the need to ensure that any risks from these tools are properly managed. Everyone involved with this book knows the importance of water quality for the future of the human population.

We have come to a point where real progress is evident and new technologies hold tremendous promise for enhancing water quality. The cutting edge thinking and scientific developments described in this volume demonstrate the ingenuity and commitment of these scientists to improving water quality for all. If the contributors to Nanotechnology Applications for Clean Water: Solutions for Improving Water Quality have their way, then very soon everyone on earth will be able to drink their water (and brush their teeth) with confidence.

George Gray
Professor of Environmental and Occupational Health;
Director, Center for Risk Science and Public Health;
George Washington University;
School of Public Health and Health Services

Preface

Introduction

The US National Nanotechnology Initiative (NNI) defines "nanoscience" as "research to discover new behaviors and properties of materials with dimensions at the nanoscale, which ranges roughly from 1 to 100 nanometers (nm)." A key objective of nanotechnology is to develop materials, devices and systems with fundamentally different properties by exploiting the unique properties of molecular and supramolecular systems at the nanoscale. Nearly every field of science has been affected by the tools and ideas of nanotechnology, and breakthroughs continue to be made in computing, medicine, sensing, energy production, and environmental protection.

Four years after the publication of the first edition of this book, the use of nanotechnology and nanomaterials continues to mature. The first edition introduced to many the use of nanotechnology to improve water quality and the societal implications therein that may affect acceptance or widespread applications. Its primary objectives were to:

1. Provide a summary of the state of the field to interested scientists, engineers and policy makers;
2. Consider the technological advances in the context of society's interests and needs and;
3. Identify grand challenges and directions for future research in the field.

This second edition renews these objectives, with the goal of providing an update on the field.

Organization and content

This book consists of 41 contributions or chapters. In this second edition, the chapters in the first four parts have been reorganized around the essential functions of the nanotechnologies presented rather than their anticipated applications. Continuing in the spirit of the first edition, the last section (Part 5) addresses societal implications.

- Part 1 presents nanotechnologies for sensing impurities in water at concentration levels of significance to human health and regulatory compliance.
- Part 2 presents separation technologies that exclude contaminants while allowing water to pass through the barrier. Separation technologies can include both size and charge exclusion.

- Part 3 presents transformation technologies that can reduce the toxic effects of contaminants and destroy or reduce the infectivity of pathogens in water. Transformation technologies include catalytic and photocatalytic particles, fibers and films.
- Part 4 presents sequestration and immobilization technologies that bind to contaminants either for extraction and recovery, or for in situ stabilization to prevent contaminants from migrating into groundwater from the in situ treatment site.
- Part 5 presents societal implications of implementing new technologies, including cultural, ethical, and environmental concerns.

Through a reorganization, updating of most chapters, and addition of new material, we hope this new edition will continue to inspire the development and implementation of nanotechnology-based solutions to critical water issues.

Disclaimer

The views expressed on the parts of Nora Savage and Anita Street do not reflect the views of the Environmental Protection Agency or the Department of Energy, and no official endorsement should be inferred.

Anita Street
Richard Sustich
Jeremiah Duncan
Nora Savage

Acknowledgment

The editors would like to extend a special acknowledgment to Dr. Mamadou S. Diallo. As a co-editor of the first edition, on which this volume heavily relies, he provided an indelible, singular insight and perspective that continues to resonate with us all.

Introduction: Water Purification in the Twenty-First Century—Challenges and Opportunities

Richard C. Sustich, Mark Shannon and Brian Pianfetti
Center of Advanced Materials for Purification of Water with Systems, University of Illinois at Urbana-Champaign, Urbana, IL, USA

I.1 Current water issues

Ensuring the availability of clean, abundant fresh water for human use is among the most pressing issues facing the United States and the world, as depicted in Fig. I.1 [1]. *Nature* framed the issues by stating, "More than one billion people in the world lack access to clean water, and things are getting worse. Over the next two decades, the average supply of water per person will drop by a third, possibly condemning millions of people to an avoidable premature death" [2]. In the United States as in the rest of the world, water has a broad impact on health, food, energy, and economy. While it might seem that the United States is blessed with abundant fresh water, regional scarcities and competing demands leave little room for growth. Recent environmental catastrophes (e.g., the Indian Ocean Tsunami (2004), Hurricane Katrina (2005)) demonstrate how fragile our municipal water supplies and infrastructure are, as well as supplies needed to provide energy and irrigation for food. Environmental stress resulting from climate change and population growth and migration is expected to increase over next two decades (Fig. I.2) [3].

The challenges facing water production for human use in the United States and elsewhere are not fully defined nor do their solutions fit into a neat box. Rather, they are a complex and interrelated set of problems requiring a suite of individual, local, regional, national, and even international solutions incorporating an integrated information base and efficient, cost-effective, and reliable analytical and treatment technologies. Aquifers throughout the United States (e.g., Ogallala, Mahomet) and around the world (eg., Central China) are suffering from declining water levels, saltwater intrusion, contamination from surface waters, and inadequately replenished fresh groundwater. Major rivers and watersheds are also being overdrawn (e.g., Colorado River), while return flows are contributing to downstream nutrient loading and salinity problems (e.g., Gulf of Mexico). Lakes and wetlands are also experiencing increased salting in many areas. In much of the developed world, this is an opportune time to address water problems, because the aging water infrastructure will need to be upgraded over the next 40 years.

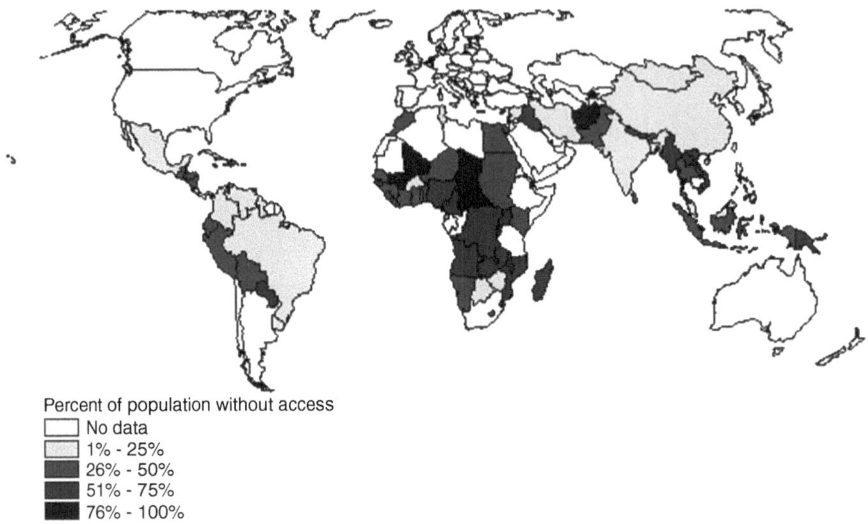

FIGURE I.1

Populations without access to safe drinking water.

(From The World's Water *1998–1999 by Peter H. Gleick. Copyright © 1998 Island Press. Reproduced by permission of Island Press, Washington, D.C.)*

Estimated costs through 2019 for infrastructure replacement in the United States alone are estimated to range from $485 billion to $896 billion (excluding operations and maintenance using current technologies) [4]. As staggering as this sum appears, it is most likely only a fraction of the true cost needed to create new water supplies, which is an additional cost. Although conservation has allowed substantive reduction in per capita water use in developed countries over the past 30 years, conservation alone cannot halt the demand an expanding global population and developing economies place on water. Redistributing current supplies to meet future needs carries additional costs in infrastructure (pumps, canals, and pipelines) and energy. This approach leads to collateral economic losses from rationing and lost opportunities to expand energy, agriculture, and business activities. Such confounding challenges, however, also offer substantial opportunities to create new water-related businesses, products, and industries that can be marketed all over the world, which is, as a whole, experiencing similar issues and problems. There is little doubt that satisfying humankind's demand for water in a sustainable manner requires visionary new approaches to management and conservation of water resources augmented by new technologies capable of dramatically reducing the cost of supplying clean fresh water—technologies that can best be derived from tightly coupled basic and applied research.

There is a nearly one-to-one correlation throughout the world between national economic output and per capita water use. The United States has the highest Gross National Product (GNP) and the highest fresh water usage in the world at

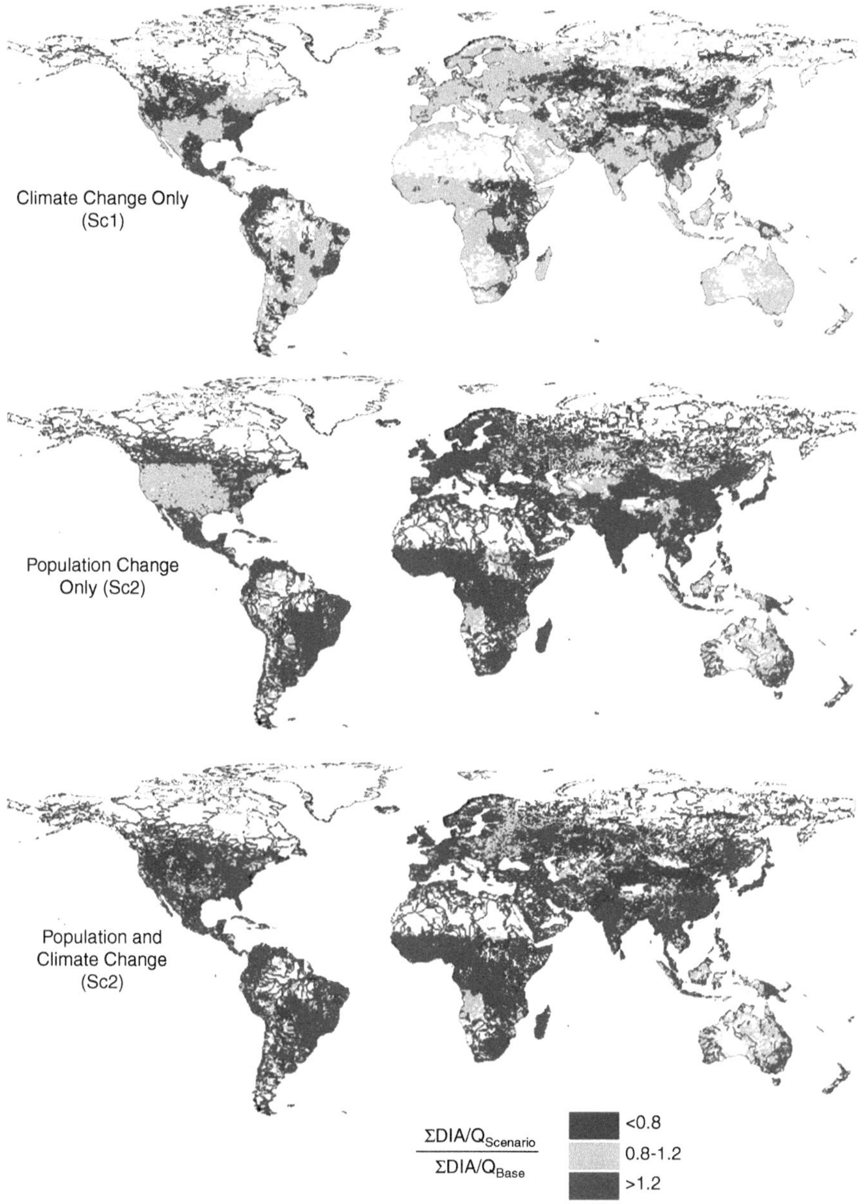

FIGURE I.2

Relative change in demand per discharge.

(From Vörösmarty et al., Science *289: 284 (July 14, 2000). Reprinted with permission from AAAS.)*

approximately 2,000 m^3 per person per year, whereas sub-Saharan Africa has the lowest GNP and water usage (approximately 100 m^3 per person per year). The two notable exceptions to this correlation—Singapore and Israel—are important lessons for developed and developing nations alike. Both have significant fresh-water resource issues and both spend a much larger percentage of their GNP for water production and water research than the United States and other developed nations. The Congressional Budget Office estimated in 2001 that average annual costs for US water systems, including operations and maintenance, over the period 2000–2019 will range from $70.7 billion to $98 billion (in 2001 dollars). Cumulatively, costs for the 20-year period are expected to range from $1.4 trillion to $1.96 trillion (in 2001 dollars) [5]. These estimates assume that water resources and water availability remain stable at current levels. Under this scenario, many American households could see water rates increase by an order of magnitude, bringing residential water bills on a par with energy bills. Energy production (coal/gas/nuclear generation, mining, and refining) presently accounts for the largest fresh water withdrawals in the United States, and while more than 97 percent of water withdrawn for energy production is returned to the environment, this return does not replenish highest quality sourcewaters and aquifers [6].With projected increasing energy demands, consumptive water loss for energy production and detrimental water quality impacts could increase dramatically with increased production of alternative biofuels [7]. (Chapter 35, "Competition for Water," includes a discussion of the impacts of the production of biofuels on water availability and competition.) As increases in acreage and irrigation are needed for the production of biofuels, further dramatic increases in demand will strain existing water supplies and new water sources will be needed. If lack of water limits growth in new energy supplies, every aspect of the US and global economy will be affected, increasing costs to the consumer. Clearly, the costs to upgrade infrastructure and create new supplies likely cannot be met without revolutionary improvements in the science and technology of water purification.

On top of the impending supply crisis, the United States faces a host of water quality issues that demand improved treatment methods to resolve. These problems include toxic organic compounds, heavy metals, and pathogens, many of which pose health risks at very low (less than parts per million) levels. The ability to robustly sense and remove low-level contaminants can better protect the health of Americans, but at what cost? The costs of contaminant removal and residuals management currently make regulating many trace contaminants prohibitive. However, if we can selectively and efficiently remove trace toxic and problematic contaminants, through filtering or transformation, without removing a large amount of nontoxic and even beneficial constituents, the costs can be greatly reduced. We also need to improve disinfection to inactivate pathogens that are resistant to current methods and to prevent introduced or emerging pathogens from causing large-scale harm, all without creating disinfection byproducts that are themselves highly toxic. Improved disinfection will also help the rest of the world, where estimated annual water- related mortality rates range from 2.2 to

12 million [8].There is a clear and urgent need for new, more effective methods to purify water for the people of the United States and the world.

I.2 Water purification: impacts and opportunities

I.2.1 Water-environment nexus

While ensuring adequate supplies of potable water for human use must continue to be a research priority, the environmental impacts of increasing human water consumption will also demand attention. Portions of the midwestern, southeastern, and southwestern United States are already seeing increased water stress. The US Environmental Protection Agency now has more than 90 contaminants on their Candidate Contaminate List that are, or soon need to be, addressed by water treatment plants. Additionally, relatively few aquifers in the United States have been adequately surveyed to determine actual available groundwater reserves. Better research on the availability, detection of contaminants, and strategies for remediation can increase utilization of currently available sources as well as facilitate development of new water sources such as brackish aquifers.

I.2.2 Water–energy nexus

In 2000, thermoelectric power production in the United States accounted for 132.4 billion gallons per day, or 39 percent, of all fresh water withdrawals, and 3 billion gallons, or 3 percent, of fresh water consumption [6]. By 2030, generating capacity is expected to expand by 22 percent, and whereas projected changes in withdrawals vary from − 21 to + 6 percent depending on the new generation and cooling technology deployed, fresh water consumption is expected to increase by 28–49 percent [9]. Concurrent efforts to reduce dependence on imported oil through new fuels such as biomass, syngas, and hydrogen are expected to expand the overall water footprint of the energy sector even further. Of the nation's 104 nuclear reactors, 24 are located in areas experiencing severe drought, such as the southeastern United States, and face potential shutdowns due to water scarcity and concerns over environmental impacts [10]. New sources of water, and new purification technologies to enhance water reuse, will be needed to keep pace with energy demands.

I.2.3 Water–food nexus

Food production is even more closely linked to the availability of water than energy production. Although most Americans know that they are recommended to consume one gallon of water per day for proper hydration, most generally don't recognize the amount of watcr used to produce the food that they consume (water footprint). Across the United States, 41 percent of all water withdrawn,

139,189 million gallons per day, is used for irrigation and livestock [6]. Additionally, 85 percent, or 84,956 million gallons of water per day, is lost (not returned) to the local water source by irrigation and livestock production [6]. New technologies that can dramatically enhance agricultural water conservation and increase the recovery and reuse of irrigation runoff and livestock wastewater could have the largest impact on future water availability.

I.2.4 Water–health nexus

Lack of access to potable water is a leading cause of death worldwide. Dehydration, diarrheal diseases, contaminated source waters, waterborne pathogens, water needed for food production (starvation), and water for sanitation are just some of the factors that impact health. Around the world, 1.1 billion people (17 percent) lack access to improved water and 2.4 billion (42 percent) lack access to improved sanitation. Every year 1.8 million people die from diarrheal diseases (including cholera). Between 28 and 35 million people in the Bangalore region in India suffer from arsenic poisoning from their water supplies. Every day, an estimated 3,900 children die from water-related disease or poor hygiene [11]. As our water sources become more stressed, these numbers could increase substantially. The water–health nexus is crucial for the survival of humanity. Creating better disinfection and purification technologies could significantly reduce these problems that much of the world currently faces, and, equally importantly, some regions of the developed world may soon face.

I.2.5 Water–economy nexus

In the past 20 years, over one trillion dollars have been spent in the United States alone on drinking water treatment and distribution, wastewater treatment, and residuals disposal. An estimate by the US EPA claims that simply to maintain current water distribution and wastewater collection systems through 2019, $485–$896 billion (with a point estimate of $662 million) will be needed for infrastructure and $72–$724 billion (with a point estimate of $309 billion) will be needed for operation and maintenance [4]. The Congressional Budget Office estimated that in the late 1990s, the average cost of water and wastewater services represented 0.5 percent of household income nationwide. By 2019, water and wastewater service costs are projected to account for 0.6–0.9 percent of national household income [5].The subsequent increase in the cost of water will be reflected in increases across the economy and especially in increased costs of food and energy, which are already strained. The increase in the percentage of a household's income spent on water, energy, and food will result in less discretionary funds for retail markets. The water–economy nexus demonstrates how the availability of fresh water impacts a country's potential for prosperity. Acting now to enhance reclamation and reuse and to develop new waters will ensure continued economic prosperity of developed nations and create new economic opportunities for the developing world.

I.2.6 Water–security nexus

Water supply and transport logistics have been a challenge for military campaigns since the time of ancient Rome, when water was transported great distances by pack-animals [12]. Recently, water supply has also emerged as a high-value target for terrorism. Disabling and/or contamination of urban water distribution systems can impact thousands, and in some cases millions of customers, and the inherently open nature of wastewater collection systems renders them vulnerable to the weaponization of a myriad of commercial and industrial chemicals. In response the US EPA has initiated the WaterSentinel Program for the design, development, and deployment of a robust, integrated water surveillance system that will incorporate real-time system-wide water quality monitoring, critical contaminant sampling arid analysis, and public health surveillance. The extent and complexity of the WaterSentinel Program is expected to create a powerful demand for nanoscale contaminant sensors, distributed signal acquisition and transmission, and decision support technology [13].

I.3 Critical problems to be addressed in water research

All over the world, people face profound threats to the availability of sufficient, safe and clean water, affecting their health and economic well-being. The problems with economically providing clean water are growing so quickly that incremental improvements in current methods of water purification could leave much of the world with inadequate supplies of clean water in mere decades. The challenges to overcome in science, technology, and society require a long-term vision of what needs to be solved. The critical problems that will have to be addressed over the next 20 years with regard to clean water are summarized briefly in the following sections.

I.3.1 Availability and sourcewater protection

At present, the United States and much of the world lack sufficient knowledge regarding the actual amount of water stored and recharged in currently utilized freshwater aquifers. Current data indicate that levels in some monitored aquifers are dropping rapidly. Regions of the High Plains Aquifer south of the Canadian River in New Mexico and Texas experienced water level declines of more than 60 feet between 1980 and 1999 [14]. While there are some regional efforts to look at these issues, a global effort to inventory and quantify the existing fixed and recharging supplies of fresh, brackish, and saline water is critical not only for projecting water availability and sustainable withdrawal capacities, but also for helping scientists and engineers choose solutions that will be viable. The effects of salting on lands and lakes, as well as contamination rates of aquifers also need to be quantified.

I.3.2 New water supplies

Meaningful increases in potable water supplies can only be achieved through reuse of existing wastewater and development of brackish arid saline sources—from the "sea to sink to the sea again." This effort will need to focus on augmenting water supplies via desalination of seawater and brackish aquifers, as well as through direct reuse of municipal and agricultural wastewaters. Brackish aquifers and wastewaters indeed present greater challenges than seawater desalination. Critical issues to utilizing inland brackish lakes and aquifers include developing methods and materials that can separate dissolved solids in hard water with minimal fouling, and minimizing residuals created during desalination and reclamation of contaminated and brackish source waters.

I.3.3 Contaminant detection and selective decontamination/removal

Efficient removal of contaminants from all types of water sources is needed to get the "drop of poison out of an ocean of water." Current treatment technologies are typically not contaminant-specific, resulting in excessive reagent use and removal of benign constituents and excessive generation of residuals requiring further processing and disposal. Efforts to develop more marginal water sources, due to increasing demand and depletion of existing sources, will likely become prohibitively expensive using conventional approaches. A major cost factor in removing trace amounts of critical contaminants from source waters is that large quantities of benign, potable constituents are also removed. Additionally, real-time, in situ detection, adsorption, and/or catalytic destruction of potential warfare/terrorism agents are major challenges for the water industry.

I.3.4 Pathogen deactivation and removal

Disinfection technologies that effectively deactivate known and emerging pathogens without producing toxic substances are needed to "beat chlorination." New and affordable materials, methods, and systems are necessary to provide drinking water free of harmful viral, bacterial, and protozoan pathogens, while avoiding the formation of toxic by-products or impairing the treatment of other contaminants. A key unsolved problem is the detection and removal of new and/or evolving infective viruses, and resistant pathogens.

I.3.5 Conservation and reuse

While projections show that conservation alone will not be enough to solve the problems, reduction in per capita water consumption remains an important part of the solution to the problem. Conservation via improved efficiencies and reduction in waste can dramatically reduce overall costs of providing clean water. Research efforts that focus on minimizing the withdrawal of water and on the conversion of direct draw applications to reuse systems have the potential to substantially reduce projected water needs, particularly for specific watersheds and aquifers.

I.3.6 Scalability, ramp-up, and technology diffusion

Researchers can make the greatest discoveries and solutions to our problems, but unless a means to move these advances to full production is possible, these innovations will remain in the laboratory. Further, many novel approaches to problems, although scientifically intriguing, may not take into consideration the costs of mass production or implementation. The scalability component focuses on capacity for researchers to incorporate benchmarking and manufacturing scale-up considerations as well as facilitating the testing and movement of new materials and procedures to industry. For a technology to be a success, the total cost cycle must be favorable and it must win in the marketplace. Moreover, with respect to potable water systems, a history of performance efficacy and costs of installation and operation must be available for water managers to select one technology over another with confidence.

I.4 Conclusion

There are clearly many aspects to the broad problems of water quality and many technology and policy components to the effort to ensure the sustainability of water for human use. This book focuses on the impacts of and opportunities for the application of nanotechnology to enhance water quality, and the societal concerns about the widespread use of nanotechnology in the water arena.

References

[1] P. Gleick, The World's Water—The Biennial Report on Freshwater Resources, Pacific Institute, Oakland, California, 1998, p. 40.

[2] "Global water crisis," Nature, <http://www.nature.com/nature/focus/water/> (accessed 25.01.08).

[3] C.J. Vörösmarty, P. Green, J. Salisbury, R.B. Lammers, Global water resources: vulnerability from climate change and population growth, Science, 289 (5477) (2000) 284.

[4] Office of Water, The Clean Water and Drinking Water Infrastructure Gap Analysis, United States Environmental Protection Agency, Washington, D.C., 2002, p. 5.

[5] United States Congressional Budget Office, Future Investment in Drinking Water and Wastewater Infrastructure, Washington, D.C., 2002, p. 17.

[6] P. Torcellini, N. Long, R. Judkoff, Consumptive Water Use for US Power Production, National Renewable Energy Laboratory, United States Department of Energy, Golden, Colorado, 2003, p. 8.

[7] Committee on Water Implications of Biofuels Production in the United States, Water Implications of Biofuels Production in the United States, National Research Council, The National Academies Press, Washington, D.C., 2007, pp. 6–7.

[8] P. Gleick, Dirty Water. Estimated Deaths from Water-Related Diseases 2000–2020, Pacific Institute, Oakland, California, 2002, p. 2.

[9] E. Shuster, A. McNemar, G.J. Stiegel Jr., and J. Murphy, Estimating Freshwater Needs to Meet Future Thermoelectric Generation Requirements—2007 Update, National Energy Technology Laboratory, United States Department of Energy, Pittsburgh, PA, 2007, pp. 1–2.

[10] M. Weiss, Drought could force nuke-plant shutdowns, Associated Press, <http://ap.google.com/article/ALeqM5isS6MZRQ2X9OAGwbMWhPcdsA1E_AD8UBQIU00> (accessed 24.01.08).

[11] World Health Organization, Water sanitation and hygiene links to health, <http://www.who.int/water_sanitation_health/publications/facts2004/en/index.html> (26.01.08).

[12] J. Roth, The Logistics of the Roman Army at War (264 B.C.–A.D. 235), Brill Publishing, Leiden, Netherlands, 1999, p. 201.

[13] M. Royer, Emerging Challenges in Water Security and Infrastructure Resilience, US Israeli Workshop on Nanotechnology for Water Purification, Arlington, VA, 2006.

[14] V. McGuire, Water-Level Changes in the High Plains Aquifer, 1980–1999, United States Geological Service, Lincoln, Nebraska, 2001, p. 1.

PART

Contaminant Sensing Technologies

1

CHAPTER

1

Sensors Based on Carbon Nanotube Arrays and Graphene for Water Monitoring

Dan Du[1,3], Weiying Zhang[1,3], Abdullah Mohamed Asiri[2] and Yuehe Lin[3,4]

[1]*Key Laboratory of Pesticide and Chemical Biology of Ministry of Education, College of Chemistry, Central China Normal University, Wuhan, PR China* [2]*Chemistry Department, King Abdulaziz University, Jeddah, Saudi Arabia* [3]*School of Mechanical and Materials Engineering, Washington State University, Pullman, WA, USA* [4]*Pacific Northwest National Laboratory, Richland, WA, USA*

1.1 Introduction

Environmental monitoring of toxic pollutants in surface and subsurface water sources and wastewaters presently relies on the collection of discrete liquid samples for subsequent laboratory analysis. Sensors that are field-deployable and able to measure part-per-billion (ppb) or nanomolar levels of toxic pollutants will reduce time and costs associated with environmental monitoring of hazardous

Street, Sustich, Duncan and Savage. Nanotechnology Applications for Clean Water, 2nd Edition.
 DOI: http://dx.doi.org/10.1016/B978-1-4557-3116-9.00001-9

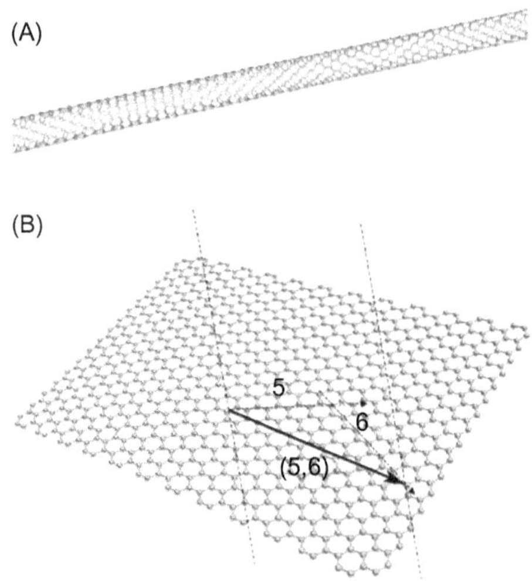

FIGURE 1.1

(A) The ideal structures of an SWCNT and (B) a graphene sheet.

From Ref. [4], reproduced by permission of Wiley-VCH.

chemicals. Electrochemical sensors appear to be a very promising technique that offers desired characteristics such as field-deployability, selectivity, sensitivity, robustness, and inexpensiveness [1–3].

Nanomaterials, particularly carbon nanomaterials, have a significant role to play in new developments in each of the biosensor size domains [4–6]. They have shown great promise in many applications, such as bioscience and biotechnology, energy storage and conversion, environmental and biomedical applications. In this chapter, we focus only on the use of two carbon nanomaterials, carbon nanotubes (CNTs) and graphene-based sensors for water monitoring and environmental applications.

CNTs are well-ordered, hollow graphitic nanomaterials made of cylinders of sp^2-hybridized carbon atoms. They have high aspect ratios, high mechanical strength, high surface areas, excellent chemical and thermal stabilities, and rich electronic and optical properties [7]. The latter properties make CNTs important transducer materials in biosensors: high conductivity along their length means they are excellent nanoscale electrode materials [8–10]. These materials are classed as single-walled carbon nanotubes (SWCNTs) (Figure 1.1A), which are single sheets of graphene "rolled" into tubes, or multiwalled carbon nanotubes (MWCNTs), each of which contains several concentric tubes that share a common longitudinal axis [11]. As one-dimensional (1D) carbon allotropes, CNTs have lengths that can vary from several hundred nanometers to several millimeters, but their diameters depend on their class: SWCNTs are 0.4–2 nm in diameter and

MWCNTs are 2–100 nm in diameter. An important part of the success of CNTs for these applications is their ability to promote electron transfer in electrochemical reactions.

Graphene is a 2D sheet of carbon atoms in a hexagonal configuration with atoms bonded by sp^2 bonds. These bonds and this electron configuration are the reasons for the extraordinary properties of graphene, which include a very large surface area (at 2630 m^2/g, it is double that of SWCNTs), a tunable band gap, room-temperature Hall effect, high mechanical strength (200 times greater than steel), and high elasticity and thermal conductivity [12]. Graphene is the most recent member of the multidimensional carbon-nanomaterial family, starting with fullerenes as a 0D material, SWCNTs as 1D nanomaterials, and ending with graphite as a 3D material. Graphene fills the gap for 2D carbon nanomaterials (Figure 1.1B). Isolation of individual graphene sheets was long sought, but only in 2004, it was achieved by a surprisingly simple technique [13]. Since then, fundamental research and research on applications have increased rapidly. Graphene is an ideal material for electrochemistry [14–17] because of its very large 2D electrical conductivity, large surface area, and low cost. The use of graphene in electrochemical sensors and biosensors is particularly interesting, with the first articles emerging in 2008. Since then, their number has grown explosively.

1.2 CNT-based electrochemical sensors

1.2.1 Various methods for preparation of CNT-based sensors

Although CNTs are relatively new in analytical fields, their unique electronic (electron transfer rate similar to edge-plane graphite), chemical (biocompatibility and ability to be covalently functionalized), and mechanical properties (3 times stronger than steel) make them extremely attractive for chemical and biochemical sensors [18,19]. Various methods have been used to prepare CNT-based sensors: (a) casting of CNT thin films, from the suspensions of CNTs in solvents [20–28], such as sulfuric acid [21], Nafion [25], dihexadecyl hydrogen phosphate (DHP) [26], DMF [27], and acetone [28], prior to being coated on electrode surfaces, (b) using CNTs as paste electrodes or electrode composites [29–31], and (c) using aligned CNT as electrode substrates [32–40]. With specific to metal ion sensors, DHP [26] and Nafion [25] have been used to disperse MWCNTs under ultrasonication prior to being drop-coated on glassy carbon electrodes. The CNT film enables the development of mercury-free electrodes that can detect from 10^{-9} to 10^{-6} M of Cd and Pb.

Most CNT-based sensors take advantage of the bulk properties of CNTs, including increased electrode surface area [41], fast electron transfer rate [42], and good electrocatalytivity in promoting electron transfer reactions of many important species [43,44]. Using CNTs as nanoelectrodes have been increasing explored since conventional macroelectrodes (having diameter in millimeters), such as glassy carbon and carbon paste, are known to have slow mass transport. Using nanoelectrodes (with diameter in nanometers) can enhance mass transport [37,38,45]; as electrodes decrease in size, radial (3D) diffusion becomes dominant and results in fast mass transport and fast electron transfer, and promotes rapid electrochemical detection. Nanoelectrodes also have higher responsiveness (or higher mass sensitivity) than macroelectrodes, attributed to their lower background (charging) currents [46]. Additionally, they are less influenced by solution resistance due to lower ohmic drop [38]. The high diffusion rate at nanoelectrodes enables the study of fast electrochemical and chemical reactions [47]. Despite its advantages, a single nanoelectrode offers extremely low capacitive current (in pico-amperes), thereby requiring expensive signal amplifiers. To solve this issue, nanoelectrode arrays (NEA) consisting of millions of nanoelectrodes have been developed at Pacific Northwest National Laboratory (PNNL) in collaboration with the Boston College, in order to provide magnified signals without the need for a signal amplifier.

High-density aligned CNT electrode arrays (i.e., dense CNT forest) have been reported to show fast electron transfer and electrocatalytic characteristics [48], but they do not maintain the properties of individual nanoelectrodes due to the overlapping of their diffusion layers [49–53]. To make each CNT on the array work as an individual nanoelectrode, the spacing must be sufficiently larger than the diameter of the nanotubes. A nanoscale in size of the electrode and the million numbers of electrodes will potentially result in an improved signal-to-noise ratio (and hence improved detection limits) by three to four orders of magnitude compared with that of single nanoelectrode.

1.2.2 Fabrication of aligned CNT NEA

The fabrication procedure for low-site density aligned CNT NEA has been refined [54,55]. Figure 1.2 shows the scheme of the fabrication procedure. After the electrochemical deposition of Ni nanoparticles (shown in Figure 1.2A), an aligned CNT array was grown (shown in Figure 1.2B), and then a thin layer of SiO_2 was coated on the surface by magnetron sputtering to insulate the Cr layer. After that, 5 m thick M-Bond 610 (two-component, solvent-thinned, epoxy-phenolic adhesive from Vishay Intertechnology, Inc.) was coated and cured at 170°C for 2 hours, which further insulates the Cr and also provides mechanical support to the CNTs. After these steps, the CNTs were half embedded in the polymer resin as shown in Figure 1.2C. In the next step, a fiber-free cloth was used to polish the surface that mechanically breaks the top part of the CNTs and exposes the tip of the CNTs as shown in Figure 1.2D. Finally, the sample surface was rinsed in deionized water, and an insulated copper wire (0.5 mm in radius) was attached to the corner of the substrate by applying a drop of conductive silver epoxy followed by insulating epoxy. The copper wire NEAs assembly

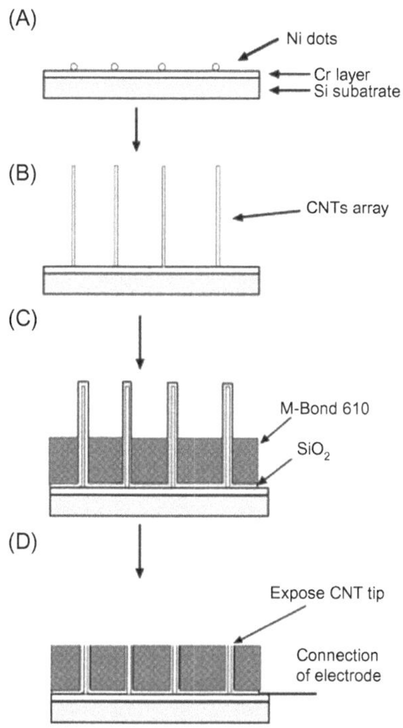

FIGURE 1.2

Fabrication scheme of the NEAs. (A) Ni nanoparticles electrodeposition; (B) aligned CNT growth; (C) coating of SiO_2 and M-Bond; and (D) polishing to expose CNTs.

From Ref. [54], reproduced by permission of American Chemical Society.

was left to cure in air at room temperature for several hours. Figure 1.3 shows the structure of the final CNT NEAs.

1.2.3 Applications of CNT-based sensors for metal ion monitoring

The CNT NEAs have been applied for detection of toxic metal ions. With Hg-coated CNT NEA, linear relationship between the Pb signals and the Pb concentration in the solutions ranging from 1 to 100 ppb (g/L) has been obtained [55]. Because Hg is highly toxic and not suitable for field-deployable use, the relatively benign bismuth, Bi(III), has been evaluated as a Hg substitute [56,57]. Bi-based electrodes performed as well as Hg-based electrode for Cr(VI) quantitation after the accumulation of Cr (VI)-diethylenetriammine pentaacetic acid (DTPA) chelate [56]. Highly responsive voltammetric analysis of Cd and Pb at the Bi-based CNT NEA was obtained [34]. Figure 1.4A shows well-defined peaks of 5 g/L of Cd and Pb obtained after only 2 min of accumulation, in which Bi was accumulated *in situ* with the target metals

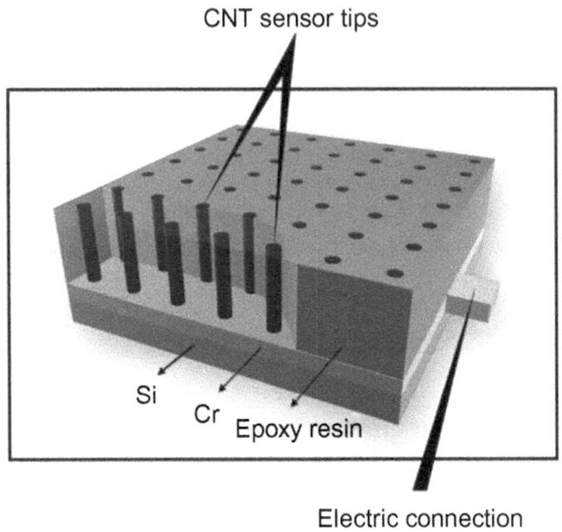

FIGURE 1.3

Structure of CNT NEAs.

at −1.2 V. Figure 1.4B shows a linear calibration curve of Cd achieved at a very useful concentration range of 0.5–8 g/L of Cd. The detection limit of 0.04 g/L was obtained under optimum experimental conditions.

The attractive behavior of the new CNT NEA sensing platform holds great promise for on-site environmental monitoring and biomonitoring of toxic metals.

1.3 Graphene-based sensors

1.3.1 Graphene-based electrochemical sensors

The excellent electrochemical behaviors of graphene indicate that graphene is a promising electrode material in electroanalysis. Graphene, together with its various derivatives, such as graphene oxide (GO), graphene nanoribbon, chemically reduced graphene oxide (rGO), or nitrogen doped graphene [58–62], has shown fascinating advantages in electrochemistry such as electrochemical devices, capacitors, or transistors due to its remarkable electrochemical properties [63–66]. The graphene-based materials have been used to construct various biosensors based on different sensing mechanisms. Because their 2D structure provides a large area for the immobilization of enzyme, and enriched oxygen-containing groups are apt to react with amino-acid residues, graphene derivatives are ideal substrates for immobilization of certain biomolecules. Enzymes can be immobilized on graphene surfaces through the following three methods: (1) electrostatic interaction between positively charged graphene and a negatively charged enzyme, (2) covalent linkage of the enzyme molecule to the graphene surface,

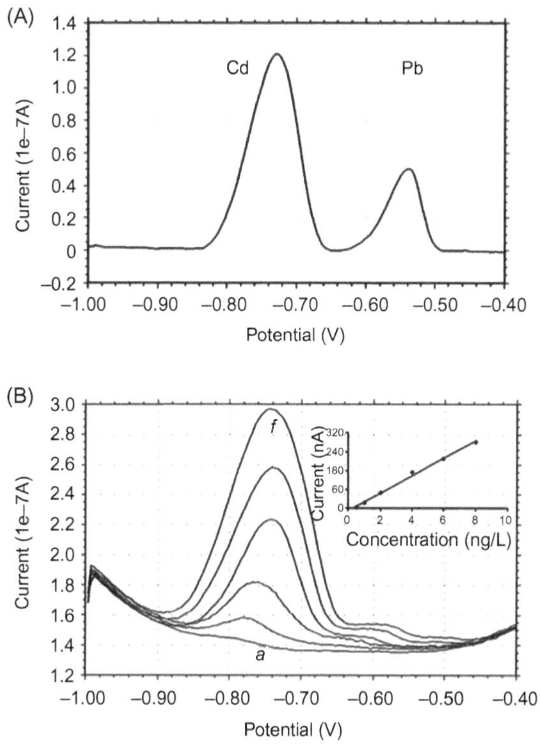

FIGURE 1.4

(A) Typical square wave voltammogram of 5 μg/L cadmium and lead on the CNT NEAs in 0.1 M acetate buffer (pH 4.5) in the presence of 500 μg/L bismuth. (B) SWV response for increasing the concentration of cadmium (0.5, 1, 2, 4, 6, 8SWV μg/L). Also shown (inset), the resulting calibration plot. Other conditions include *in situ* plated bismuth-coated CNT-NEA with 1 min pretreatment at 0.3 V; 2 min accumulation at −1.2 V; 10 s rest period (without stirring); square-wave voltammetric scan with a step potential of 5 mV; amplitude, 20 mV; frequency, 25 Hz.

From Ref. [34], reproduced by permission of The Royal Society of Chemistry.

and (3) entrapment of the enzyme using the polysaccharide chitosan or another polymer. Electrochemical approach is considered as one of the best methods for biomolecule detection due to its high sensitivity, low cost, quick response, and easy operation. The excellent electrochemical behaviors of graphene make it a promising electrode material to improve the detection of biomolecules [67]. Graphene-based materials enhance the direct electron transfer that is characteristic of some redox enzymes and retain bioactivity because of their favorable electronic properties and biocompatibility [68–70].

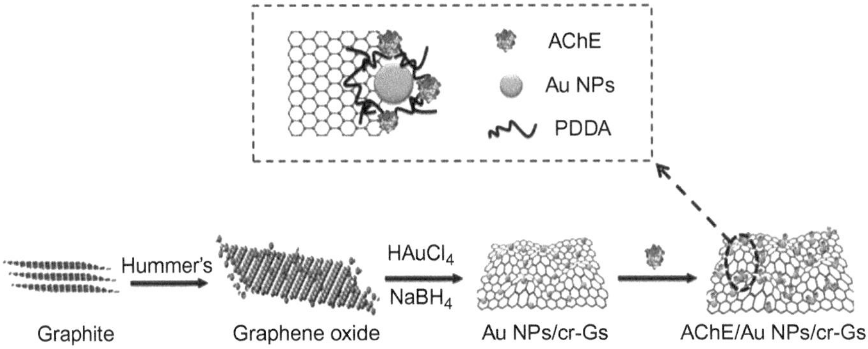

FIGURE 1.5

Schematic illustration of Au NP/cr-Gs hybrid synthesis and AChE/Au NP/cr-Gs nanoassembly generation by using PDDA. Graphite was used for producing GO with Hummer's method and Au NPs/cr-Gs was obtained by reducing $HAuCl_4$ on GO. AChE was stabilized on the surface of Au NP/cr-Gs hybrid by self-assembling.

From Ref. [75], reproduced by permission of The Royal Society of Chemistry.

1.3.2 Graphene sensors for pesticides

Because of the high toxicity of organophosphorus pesticides (OPs), they still remain a serious health threat to humans across the world. Rapid detection of these toxic agents has become increasingly important for health protection [71,72]. Although gas or liquid chromatography and mass spectrometry are accurate and routinely performed for analyzing OPs [73,74], these methods have a number of disadvantages, limiting their applications for rapid analyses under field conditions. Development of a simple, rapid, sensitive, and inexpensive method with real-time output to detect OPs is of considerable interest. Graphene has been developed as an enhanced sensing platform for on-site detection of OPs, based on immobilization of acetylcholinesterase (AChE) or OP hydrolase (OPH). Our group has reported a self-assembly of AChE on a gold nanoparticles (Au NPs)–graphene nanosheet hybrid using a long-chain polyelectrolyte, poly-(diallyldimethylammonium chloride) (PDDA) as a linker for detection of OP [75]. As shown in Figure 1.5, GO was used as the precursor to produce chemically reduced graphene oxide nanosheets (cr-Gs) and PDDA strands could terminate at the cr-Gs surface due to the electrostatic interaction between positive charged PDDA and negative charged GO. By adding $NaBH_4$, chemical reduction and Au NPs decoration were realized in one step. By using PDDA as a linker, AChE could be enriched and stabilized on the nanohybrid. Self-assembling was easy and fast to be realized between negatively charged AChE and positively charged PDDA. As a result, nanoassembly AChE/Au NPs/cr-Gs was formed in the presence of PDDA and was utilized as the catalyst platform of the electrochemical biosensor.

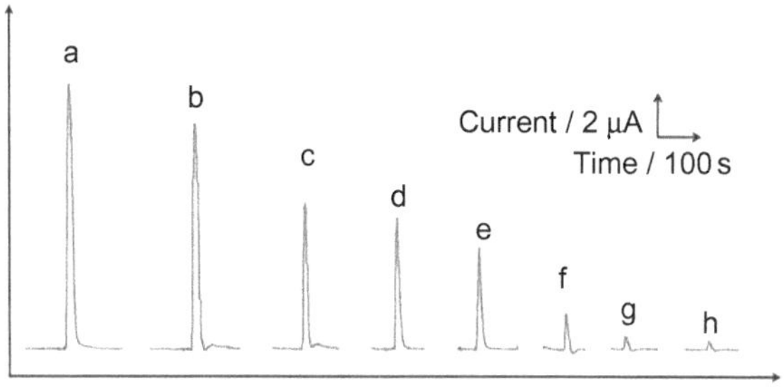

FIGURE 1.6

Typical amperometric responses of AChE/Au NP/cr-Gs-based SPEs to paraoxon by using flow injection analysis system. Signal (a) is from 2 mM ATCh, signals (b–h) are from 2 mM ATCh after the AChE/Au NPs/cr-Gs-based SPEs were incubated with paraoxon for 15 min with concentrations of 0.1 pM, 0.25 pM, 2.5 pM, 25 pM, 0.25 nM, 2.5 nM, and 5 nM, respectively. The data of amperometric responses are collected and shown together. Working potential: 0.4 V.

From Ref. [75], reproduced by permission of The Royal Society of Chemistry.

Under the optimized condition, ultrasensitive detection of paraoxon (a model compound of organophosphate pesticides) was realized with lab-built flow injection analysis system. Figure 1.6 is a typical *i–t* curve acquired at AChE/Au NPs/cr-Gs incubated with different concentrations of paraoxon. Signal (a) is from AChE/Au NPs/cr-Gs for 2 mM acetylthiocholine (ATCh), signals (b to h) are generated from 2 mM ATCh after the AChE/Au NPs/cr-Gs incubated with different concentrations of paraoxon: 0.1 pM, 0.25 pM, 2.5 pM, 25 pM, 0.25 nM, 2.5 nM, and 5 nM paraoxon for 15 min, respectively. Catalysis activity of AChE to ATCh is inhibited apparently after exposure to paraoxon, and even a very low concentration of paraoxon (0.1 pM) could cause the apparent inhibition. Oxidation current of enzymatic product decreased sharply as paraoxon concentration increased. The results demonstrate that the developed approach provides a promising strategy to improve the sensitivity and enzyme activity of electrochemical biosensors.

In another approach, we prepared a graphene–ZrO_2 nanocomposite (GZN) used for both enrichment and detection of methyl parathion (MP) [76]. GZN was synthesized by one-step electrodeposition onto a glass carbon electrode (GCE) and served as a selective sorbent to enrich MP in garlic samples due to the strong affinity between ZrO_2 and phosphoric group. The determinations included two main steps: (a) MP adsorption and (b) electrochemical stripping detection of adsorbed MP. Fast solid-phase extraction (SPE) was shown by the effective extraction of MP to the GZN modified electrode within 5 min without taking any special measures. Compared to plain graphene or a ZrO_2 nanoparticle modified electrode, GZN is more sensitive and selective to MP due to a synergistic effect. The improved performance combines the advantages of a large surface area and the fast electron transfer

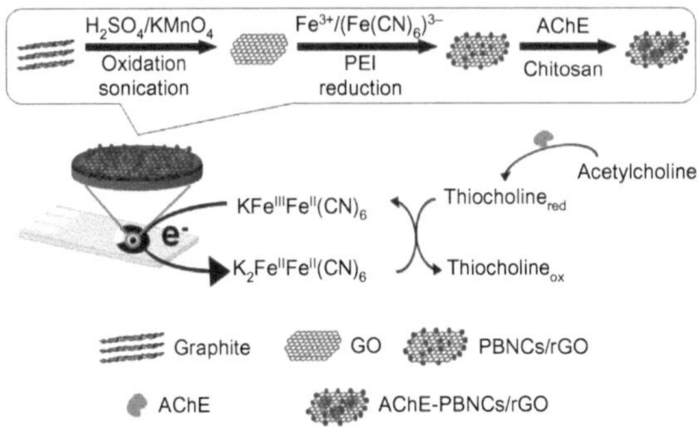

FIGURE 1.7

Preparation of PBNCs/rGO nanocomposite-based AChE biosensor and the electrocatalytic mechanism for ATCl oxidation.

From Ref. [82], reproduced by permission of The Royal Society of Chemistry.

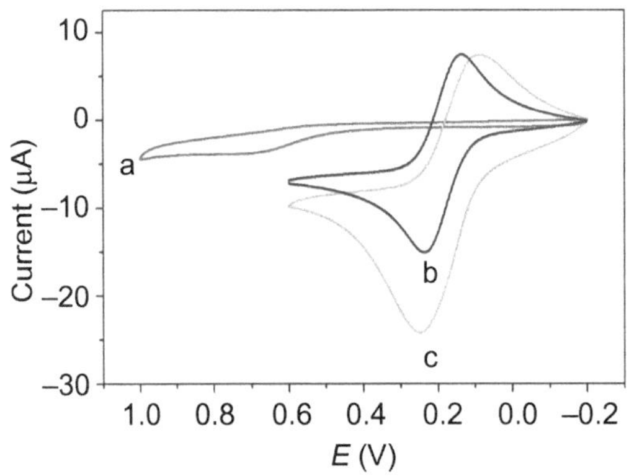

FIGURE 1.8

(Curve a) Cyclic voltammograms of bare SPE in PBS; (Curve b) AChE-PBNCs/rGO/SPE in PBS in the absence of ATCl; and (Curve c) AChE-PBNCs/rGO/SPE in PBS in the presence of ATCl.

From Ref. [82], reproduced by permission of The Royal Society of Chemistry.

of graphene with the strong affinity of ZrO_2 to the phosphate group to open new opportunities for a simple, sensitive, selective, and fast analysis of OPs.

Different kinds of metal nanoparticles have been introduced into graphene, such as Au [77,78], TiO_2 [79,80], and SnO_2 [81], to strengthen the activity. Graphene-based nanocomposites have been developed as an enhanced sensing platform for biosensors, because these kinds of nanocomposite films may generate synergistic effect to enhance the sensitivity. As shown in Figure 1.7, we demonstrated a facile procedure to efficiently prepare Prussian blue nanocubes/reduced graphene oxide (PBNCs/rGO) nanocomposites by directly mixing Fe^{3+} and $(Fe(CN)_6)^{3-}$ in the presence of GO in polyethyleneimine aqueous solution, resulting in a novel AChE biosensor for detection of monocrotophos [82]. Figure 1.8 compares the cyclic voltammograms of bare electrode and AChE–PBNCs/rGO/SPE in the presence of acetylthiocholine chloride (ATCl). The oxidation of thiocholine, a hydrolysis product of ATCl by AChE, reaches maximum oxidative current at 680 mV obtained at AChE modified SPE (curve a). In the case of AChE–PBNCs/rGO/SPE in PBS buffer, a pair of reversible peaks typifying the oxidation and reduction of PB is clearly observed (curve b). The small distance between peak potentials indicates its adsorption on the electrode surface. When adding 3.0 mM ATCl to the PBS, the oxidation current obtained at AChE-PBNCs/rGO/SPE (curve c) obviously increases compared to that observed in the absence of ATCl (curve b) and the peak potential shifts negatively to 220 mV (curve c). The over-potential decreases ~460 mV compared to curve a, suggesting that PBNCs/rGO has an electrocatalytic activity toward the oxidation of thiocholine. We further studied the amperometric response of ATCl by i–t measurements at the potential of 0.2 V obtained at different AChE modified electrodes including bare SPE, PB/SPE, and PBNCs/rGO/SPE. The oxidation current of thiocholine at bare SPE reached the maximum steady-state current (1.4 mA) with 26 s, while it increased to 10.7 mA at PB/SPE with response time of 20.6 s due to the good electron conductivity of PB and its high electrocatalytic activity. However, when PBNCs/rGO nanocomposite was introduced onto the electrode, further increased current (15.2 mA) and decreased response time (10.9 s) were observed. The results demonstrate that the largest catalytic current is observed at the PBNCs/rGO electrode because of the synergistic effects between the reduced GO and the PB nanocubes on thiocholine oxidation. Other researchers explained the observed synergy by the hypothesis that introducing CNTs into the polymer matrix containing redox mediators improved its electronic and ionic transport capacity and electron self-exchange in the polymer film [83,84]. This hypothesis can also be applied to our system of a PBNCs/rGO nanocomposite modified electrode. The reduced GO in the nanocomposite improves the electronic and potassium ionic transport capacity, resulting in an enhancement of the electrocatalytic activity. The general reactions on the electrode surface could be as follows:

$$\text{Acetylthiocholine chloride} \xrightarrow{\text{AChE}} \text{Thiocholine} + \text{HA (acetic acid)} \tag{1.1}$$

$$PB_{ox} + \text{Thiocholine}_{red} \rightarrow PB_{red} + \text{Thiocholine}_{ox} \tag{1.2}$$

$$PB_{red} \overset{e^-}{\rightleftarrows} PB_{ox} \tag{1.3}$$

The catalytic thiocholine oxidation in the case of PB can be achieved with the redox center active at ~220 mV, which greatly lower the oxidation potential to thiocholine detection. The AChE biosensor shows rapid response and high sensitivity for detection of monocrotophos with a linear range from 1.0 to 600 ng/mL and a detection limit of 0.1 ng/mL. These results suggest that the PBNCs/rGO hybrids nanocomposite exhibited high electrocatalytic activity toward the oxidation of thiocholine, which lead to the sensitive detection of OP pesticides.

Table 1.1 Graphene-Based Sensors for Environmental Applications

Target Analyte	Linear Range	Detection Limit	Modified Electrode	References
Pb^+	0.5–50 μg/L; 20–120 μg/L	0.02 μg/L 0.55 μg/L	Nafion–graphene; Bi-graphene	[85,86]
Cd^+	1.5–30 μg/L 20–120 μg/L 0.2–15 μg/L	0.02 μg/L 5 ng/L 2.8 μg/L	Nafion–graphene; Bi-graphene	[85–87]
Zn^{2+}		17 μg/L	Bi-graphene	[86]
Cu^{2+}		26 μg/L	Bi-graphene	[86]
Ag^+	1 μM to 10 mM	10 μM	Aptamer–graphene	[88]
Hg^{2+}	10 pM to 5 μM 8–100 nM	1 nM 5.0 nM	Aptamer–grapheme DNA–graphene	[88,89]
Bisphenol A	60 nM to 80 μM	42 nM	Pt/graphene–CNTs	[90]
Catechol	0.03–1.0 mM 0.01–22 μM	10.0 nM 6 nM	MnO_2–graphene; PAMAM–graphene	[91,92]
Hydroquinone	0.01–0.7 μM	7.0 nM	MnO_2–graphene	[93]
1-Aminopyrene	10–260 nM	3.6 nM	®-CD/Poly (*N*-acetylaniline)/ graphene	[94]
Chlorophenol	4–800 μM	1.5 μM	Pd–graphene	[95]
Hydrazine	1 nM to 0.05 μM	0.2 nM	Graphene	[96]

1.3.3 Graphene sensors for other pollutants

Graphene-based electrochemical sensors have been developed for environmental analysis. Table 1.1 collected some quantitative detection of heavy metal ions and environmental pollutants based on graphene sensors [85–96]. For example, Li et al. [85,87] report that Nafion–graphene composite film-based electrochemical sensors not only exhibits improved sensitivity for the metal ion (Pb^{2+} and Cd^{2+}) detections, but also alleviates the interferences due to the synergistic effect of graphene nanosheets and Nafion [85]. The stripping current signal is greatly enhanced on graphene electrodes. As shown in the differential pulse anodic stripping voltammograms (DPASV) in Figure 1.9, the stripping current signal is well distinguished. The linear range for the detection of Pb^{2+} and Cd^{2+} is wide (0.5–50 and 1.5–30 g/L for Pb^{2+} and Cd^{2+}, respectively). The detection limits ($S/N = 3$) are 0.02 mg/L for both Cd^{2+} and Pb^{2+}, which are more sensitive than those of Nafion film modified bismuth electrode [97], and ordered mesoporous carbon-coated GCE [98], and comparable to Nafion/CNT-coated bismuth film electrode [99]. The enhanced performance is attributed to the unique properties of the graphene (nanosized graphene sheet, nanoscale thickness of these sheets, and

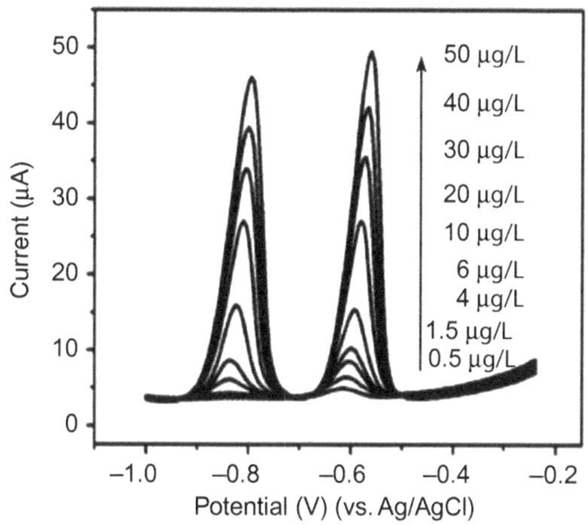

FIGURE 1.9

Stripping voltammograms for the different concentrations of Cd^{2+} and Pb^{2+} on an *in situ* plated Nafion-G-BFE (bismuth film electrode) in solution containing 0.4 mg/LBi^{3+}.

From Ref. [85], reproduced by permission of Elsevier.

high conductivity), which endowed the capability to strongly adsorb target ions, enhanced the surface concentration, improved the sensitivity, and alleviated the fouling effect of surfactants [85].

1.4 Conclusions and future work

This chapter review work is relevant to the two fastest growing carbon nanomaterials: CNTs and graphene, in water monitoring and environmental applications. Due to their excellent electrochemical behaviors, CNTs and graphene have shown excellent performance in electrochemical detection of metal ions, pesticides, and other pollutants.

Since their introduction in 1991, CNTs have been widely investigated for electrochemical sensors of many important biomolecules because of their electrocatalytic and antifouling properties, biocompatibility, high surface, and mechanical strength. For trace metal analysis, CNT thin film created by drop-coating of CNT−solvent suspensions on electrode surfaces has been explored in order to develop mercury-free sensors by exploiting the bulk properties of the CNTs. Array of low-site-density aligned CNTs has been grown on metal substrates by a nonlithographic method. Each CNT serves as a nanoelectrode which normally has greater mass transfer rate and higher mass sensitivity than conventional macroelectrodes. The array of millions of CNT nanoelectrodes provides magnified voltammetric signals for trace metal ions without the need for a signal amplifier.

As a new nanocarbon material, graphene has exhibited excellent performance in electrochemistry applications. Graphene has been synthesized with various strategies. Novel methods for well-controlled synthesis and processing of graphene should be developed. For electrochemical applications, the approach with chemical/thermal reduction of GO looks promising. Better understanding of physics and chemistry at the surface of graphene and interaction of chemicals and biomolecules at the interface of graphene will play an important role in applying graphene as nanoscaffold in water and other environmental pollutants monitoring.

Acknowledgments

This work was supported by the National Natural Science Foundation of China (21275062, 21075047) and the Program for New Century Excellent Talents in University (NCET-12-0871). PNNL is operated by Battelle for DOE under Contract DE-AC05-76RL01830.

References

[1] J. Wang, Y. Lin, Functionalized carbon nanotubes and nanofibers for biosensing applications, TrAC, Trends Anal. Chem. 27 (7) (2008) 619–626.

[2] G. Liu, Y. Lin, Nanomaterial labels in electrochemical immunosensors and immunoassays, Talanta 74 (3) (2007) 308–317.

[3] W. Yantasee, Y. Lin, K. Hongsirikarn, G.E. Fryxell, R.S. Addleman, C. Timchalk, Electrochemical sensors for the detection of lead and other toxic heavy metals: the next generation of personal exposure biomonitors, Environ. Health Perspect. 115 (12) (2007) 1683–1690.

[4] W. Yang, K.R. Ratinac, S.P. Ringer, P. Thordarson, J. Justin Gooding, F. Braet, Angew. Chem. Int. Ed. 49 (2010) 2114–2138.

[5] M. Pumera, A. Ambrosi, A. Bonanni, E.L.K. Chng, H. Poh, Trend. Anal. Chem. 29 (2010) 954–965.

[6] Y. Shao, J. Wang, H. Wu, J. Liu, I.A. Aksay, Y. Lin, Electroanalysis 22 (2010) 1027–1036.

[7] P.M. Ajayan, Chem. Rev. 99 (1999) 1787–1799.

[8] I. Heller, J. Kong, H.A. Heering, K.A. Williams, S.G. Lemay, C. Dekker, Nano Lett. 5 (2005) 137–142.

[9] D. Krapf, B.M. Quinn, M.Y. Wu, H.W. Zandbergen, C. Dekker, S.G. Lemay, Nano Lett. 6 (2006) 2531–2535.

[10] J. Justin Gooding, A. Chou, J.Q. Liu, D. Losic, J.G. Shapter, D.B. Hibbert, Electrochem. Commun. 9 (2007) 1677–1683.

[11] M.C. Hersam, Nat. Nanotechnol. 3 (2008) 387–394.

[12] A.K. Geim, K.S. Novoselov, Nat. Mater. 6 (2007) 183–191.

[13] K.S. Novoselov, A.K. Geim, S.V. Morozov, D. Jiang, Y. Zhang, S.V. Dubonos, et al., Science 306 (2004) 666–669.

[14] M. Pumera, Chem. Rec. 9 (2009) 211–223.

[15] M. Liang, L. Zhi, J. Mater. Chem. 19 (2009) 5871–5878.

[16] T. Seyller, A. Bostwick, K.V. Emtsev, K. Horn, L. Ley, J.L. McChesney, et al., Phys. Stat. Sol. B. 245 (2008) 1436–1446.
[17] C.N.R. Rao, A.K. Sood, K.S. Subrahmanyam, A. Govindaraj, Angew. Chem. Int. Ed. 48 (2009) 7752–7777.
[18] R.H. Baughman, A. Zakhidov, W.A. De Heer, Science 297 (2002) 787.
[19] Q. Zhao, Z. Gan, Q. Zhuang, Electroanalysis 14 (2002) 1609.
[20] J. Wang, M. Musameh, Y.J. Lin, Am. Chem. Soc. 125 (2003) 2408–2409.
[21] M. Musameh, J. Wang, A. Merkoci, Y. Lin, Electrochem. Commun. 4 (2002) 743–746.
[22] J. Wang, M. Li, Z. Shi, N. Li, Z. Gu, Anal. Chem. 74 (2002) 1993–1997.
[23] J. Wang, M. Li, Z. Shi, N. Li, Z. Gu, Electroanalysis 14 (2002) 225–230.
[24] Z. Wang, J. Liu, Q. Liang, Y. Wang, G. Luo, Analyst 127 (2002) 653–658.
[25] Y.-C. Tsai, J.-M. Chen, S.-C. Li, F. Marken, Electrochem. Commun. 6 (2004) 917–922.
[26] K. Wu, S. Hu, J. Fei, W. Bai, Analytica Chim. Acta 489 (2003) 215–221.
[27] H.X. Luo, Z.J. Shi, N.Q. Li, Z.N. Gu, Q.K. Zhuang, Anal. Chem. Commun. (2001) 915.
[28] F.H. Wu, G.C. Zhao, X.W. Wei, Electrochem. Commun. 4 (2002) 690.
[29] V.G. Gavalas, S.A. Law, J.C. Ball, R. Andrews, L.G. Bachas, Anal. Biochem. 329 (2004) 247–252.
[30] M.D. Rubianes, G.A. Rivas, Electrochem. Commun. 5 (2003) 689–694.
[31] F. Valentini, S. Orlanducci, M.L. Terranova, A. Amine, G. Palleschi, Sens. Actuat. B. 100 (2004) 117–125.
[32] Y. Lin, F. Lu, Y. Tu, Z.F. Ren, Nano Lett. 4 (2004) 191–195.
[33] M. Gao, L.M.G. Dai, G. Wallace, Electroanalysis 15 (2003) 1089–1094.
[34] G. Liu, Y. Lin, Y. Tu, Z. Ren, Analyst 130 (2005) 1098–1101.
[35] K.P. Loh, S.L. Zhao, W.D. Zhang. 13 (2004) 1075–1079.
[36] W.C. Poh, K.P. Loh, W.D. Zhang, T. Sudhiranjan, J.S. Ye, F.S. Sheu, Langmuir 20 (2004) 5484–5492.
[37] Y. Tu, Y. Lin, Z.F. Ren, Nano Lett. 3 (2003) 107–109.
[38] Y. Tu, Y. Lin, W. Yantasee, Z. Ren, Electroanalysis 17 (2005) 79–84.
[39] S.G. Wang, Q. Zhang, R. Wang, S.F. Yoon, Biochem. Biophys. Res. Commun. 311 (2003) 572–576.
[40] S.G. Wang, Q. Zhang, R. Wang, S.F. Yoon, J. Ahn, D.J. Yang, et al., Electrochem. Commun. 5 (2003) 800–803.
[41] J. Wang, G. Liu, R.J. Jan, Am. Chem. Soc. 126 (2004) 3010.
[42] J. Gooding, R. Wibowo, J. Liu, W. Yang, D. Losic, S. Orbons, et al., Am. Chem. Soc. 125 (2003) 9006.
[43] Y. Lin, W. Yantasee, F. Lu, J. Wang, M. Musameh, Y. Tu, et al., Biosensors based on carbon nanotubes, Dekker Encyclopedia of Nanoscience and Nanotechnology, Marcel Dekker, New York, 2004, pp. 361–374
[44] Y. Lin, W. Yantasee, J. Wang, Frontiers in Bioscience 10 (2005) 492–505.
[45] Y. Tu, Z.P. Huang, D.Z. Wang, J.G. Wen, Z.F. Ren, Appl. Phys. Lett. 80 (2002) 4018–4020.
[46] V. Menon, C. Martin, Anal. Chem. 67 (1995) 1920.
[47] D. Arrigan, Analyst 129 (2004) 1157–1165.
[48] X. Yu, D. Chattopadhyay, I. Galeska, F. Papadimitrakopoulos, J. Rusling, Electrochem. Commun. 5 (2003) 408.

[49] R. Feeney, S.P. Kounaves, Electroanalysis 12 (2000) 677.
[50] S. Fletcher, M.D. Horne, Electrochem. Commun. 1 (1999) 502.
[51] H.J. Lee, C. Beriet, R. Ferrigno, H.H.J. Girault, Electroanal, Chem. Commun. 502 (2001) 138.
[52] J. Li, H.T. Ng, A. Cassell, W. Fan, H. Chen, Q. Ye, et al., Nano Lett. 3 (2003) 597.
[53] M.E. Sandison, N. Anicet, A. Glidle, J.M. Cooper, Anal. Chem. 74 (2002) 5717.
[54] Y. Tu, Y. Lin, Z.F. Ren, Nano Lett. 3 (2003) 107–109.
[55] Y. Tu, Y. Lin, W. Yantasee, Z. Ren, Electroanalysis 17 (2005) 79–84.
[56] L. Lin, N.S. Lawrence, S. Thongngamdee, J. Wang, Y. Lin, Talanta 65 (2005) 144–148.
[57] J. Wang, J. Lu, S.B. Hocevar, B. Ogorevc, Electroanalysis 13 (2001) 13.
[58] K.S. Novoselov, A.K. Geim, S.V. Morozov, D. Jiang, M.I. Katsnelson, I.V. Grigorieva, et al., Nature 438 (2005) 197–200.
[59] S. Stankovich, D.A. Dikin, G.H.B. Dommett, K.M. Kohlhaas, E.J. Zimney, E.A. Stach, et al., Nature 442 (2006) 282–286.
[60] A.K. Geim, Science 324 (2009) 1530–1534.
[61] X.R. Wang, X.L. Li, L. Zhang, Y. Yoon, P.K. Weber, H.L. Wang, et al., Science 324 (2009) 768–771.
[62] Y.W. Son, M.L. Cohen, S.G. Louie, Nature 444 (2006) 347–349.
[63] W.Y. Kim, K.S. Kim, Nat. Nanotechnology 3 (2008) 408–412.
[64] K.P. Loh, Q.L. Bao, P.K. Ang, J.X. Yang, J. Mater. Chem. 20 (2010) 2277–2289.
[65] Y.M. Shi, W.J. Fang, K.K. Zhang, W.J. Zhang, L.J. Li, Small 5 (2009) 2005–2011.
[66] M.J. Allen, V.C. Tung, R.B. Kaner, Chem. Rev. 110 (2010) 132–145.
[67] D. Du, Y. Yang, Y. Lin, MRS. Bull. (2012).
[68] X. Kang, J. Wang, H. Wu, I.A. Aksay, J. Liu, Y. Lin, Biosens. Bioelectron. 25 (2009) 901–905.
[69] C. Shan, H. Yang, J. Song, D. Han, A. Ivaska, L. Niu, Anal. Chem. 81 (2009) 2378.
[70] Y. Wang, Y. Shao, D.W. Matson, J. Li, Y. Lin, ACS Nano. 4 (2010) 1790.
[71] M. Jokanovic, M. Kosanovic, Environ. Toxicol. Pharmacol. 29 (2010) 195–201.
[72] J.E. Casida, J. Agric. Food Chem. 59 (2011) 2762–2769.
[73] N. Fidalgo-Used, E. Blanco-Gonzalez, A. Sanz-Medel, Talanta 70 (2006) 1057–1063.
[74] K.D. Lin, S.S. Zhou, C. Xu, W.P. Liu, J. Agric. Food Chem. 54 (2006) 8134–8138.
[75] J. Mater, Chem. 21 (2011) 5319.
[76] J. Mater, Chem. 21 (2011) 8032.
[77] Y.X. Fang, S.J. Guo, C.Z. Zhu, Y.M. Zhai, E.K. Wang, Langmuir 26 (2010) 11277–11282.
[78] S. Zhang, Y.Y. Shao, H.G. Liao, J. Liu, I.A. Aksay, G.P. Yin, et al., Chem. Mater. 23 (2011) 1079–1081.
[79] Y.C. Qiu, K.Y. Yan, S.H. Yang, L.M. Jin, H. Deng, W.S. Li, ACS Nano. 4 (2010) 6515–6526.
[80] O. Akhavan, M. Abdolahad, A. Esfandiar, M. Mohatashamifar, J. Phys. Chem. C. 114 (2010) 12955–12959.
[81] S.M. Paek, E.J. Yoo, I. Honma, Nano Lett. 9 (2009) 72–75.
[82] Nanoscale (2012) 4 4674.
[83] M.G. Zhang, W. Gorski, J. Am. Chem. Soc. 127 (2005) 2058–2059.
[84] M.G. Zhang, W. Gorski, Anal. Chem. 77 (2005) 3960–3965.

[85] J. Li, S. Guo, Y. Zhai, E. Wang, Anal. Chim. Acta 649 (2009) 196.
[86] P.K. Sahoo, B. Panigrahy, S. Sahoo, A.K. Satpati, D. Li, D. Bahadur, Biosens. Bioelectron. 43 (2013) 293–296.
[87] J. Li, S. Guo, Y. Zhai, E.K. Wang, Electrochem. Commun. 11 (2009) 1085–1088.
[88] F. Liu, H.D. Ha, D.J. Han, T.S. Seo, Small. http://dx.doi:10.1002/smll.201300499.
[89] Y. Zhang, H. Zhao, Z. Wu, Y. Xue, X. Zhang, Y. He, et al., Biosens. Bioelectron. 48 (2013) 180–187.
[90] Z.X. Zheng, Y.L. Du, Z.H. Wang, Q.L. Feng, C.M. Wang, Analyst 138 (2013) 693–701.
[91] T. Gan, J. Sun, K. Huang, L. Song, Y. Li, Sens, Actuators B. Chem. 177 (2013) 412–418.
[92] E. Araque, R. Villalonga, M. Gamella, P.M. Ruiz, J. Reviejo, J.M. Pingarron, et al., Chem. B. 1 (2013) 2289–2296.
[93] T. Tan, J. Sun, K. Huang, L. Song, Y. Li, Sens. Actuators B. Chem. 177 (2013) 412–418.
[94] G. Zhu, L. Wu, X. Zhang. W. Liu, X. Zhang, J. Chen. A new dual-signalling electrochemical sensing strategy based on competitive hostGuest interaction of a β-cyclodextrin/Poly(*N*-acetylaniline)/graphene-modified electrode: sensitive electrochemical determination of organic pollutants. Chem. Eur. J. 19 (2013) 6368–6373.
[95] J.J. Shi, J.J. Zhu, Electrochimica Acta 56 (2011) 6008–6013.
[96] Y. Wang, Y. Wan, D. Zhang, Electrochem. Commun. 2 (2010) 187–190.
[97] G. Kefala, A. Economou, A. Voulgaropoulos, Analyst 129 (2004) 1082.
[98] L.D. Zhu, C.Y. Tian, R.L. Yang, J.L. Zhai, Electroanalysis 20 (2008) 527.
[99] H. Xu, L.P. Zeng, S.J. Xing, Y.Z. Xian, G.Y. Shi, Electroanalysis 20 (2008) 2655.

CHAPTER

Advanced Nanosensors for Environmental Monitoring

2

Omowunmi A. Sadik

Department of Chemistry, Center for Advanced Sensors and Environmental Systems, State University of New York at Binghamton, Binghamton, NY, USA

2.1 Introduction

Rapid and precise sensors capable of detecting pollutants at the molecular level could greatly enhance our ability to protect human health and the environment. Of particular interests are remote, *in situ*, and continuous monitoring devices capable of yielding real-time information, and also those that can detect pollutants at very low levels. The ability to functionalize metal particles, control the size, shape, and stability at the nanoscale level is important for fundamental and industrial applications [1,2]. This can be achieved by using polymeric membranes that

Street, Sustich, Duncan and Savage. Nanotechnology Applications for Clean Water, 2nd Edition.
 DOI: http://dx.doi.org/10.1016/B978-1-4557-3116-9.00002-0

provide an excellent dispersion and sequestering environment for the metal particles, thus generating polymer metal nanocomposites with the desired morphological, electrochemical, and structural properties.

Recently, an emerging area of research is the implementation of nanomaterials for the development of a novel class of biosensors, biochips, nanosensors, and other transducers. In this respect, metal nanoparticles have been used for the development of chemical and biological sensors because of their unique optical, magnetic, and electrical properties [3]. By adding complementary molecules to a bioconjugate probe, a systematic interaction of the target molecules with the probe creates aggregation of the nanoparticles and this can be monitored by extinction or scattering. Alternatively, the binding of target molecules results in the formation of a secondary polymer shell, as is commonly employed for the interaction between the conjugates of nanoparticles with monoclonal antibodies and the corresponding target antigen, or between the conjugates of nanoparticles with polyclonal antibodies and low molecular antigens (haptens) [4]. Moreover, the application of gold nanoparticles for signal amplification has attracted a lot of interest in the biosensor development field. This is not only attributed to their unique properties in the conjugation with biological recognition elements but also in the signal transduction with optical, electrical microgravimetric, and electrochemical methods [5]. The sensitive detection with these metal–nanoparticle-based bio-tags is attributed to the signal amplification with silver deposition also known as silver enhancement. Techniques that incorporate silver enhancement have been able to achieve 2- to 30-fold amplifications [6].

Many efforts are currently focused on finding the ideal material to ensure the greatest characteristics of the nanostructured sensor system. Basically, in the construction of biosensors there are several requirements for selecting an electrode material [7]: (i) biocompatibility with the biological element, (ii) absence of diffusion barriers, (iii) stability with changes in temperature, pH, ionic strength, or macroenvironment, (iv) sufficient sensitivity and selectivity for the analyte of interest, as well as (v) low cost and ease of mass production. Moreover, these materials should either possess the necessary functional groups needed for the attachment of biomolecules or be able to be easily functionalized. Some sensors require operation in harsh conditions and, in this case, materials with special mechanical and chemical resistance, almost inert, are needed [7]. Additionally, an ideal electrode material must be characterized by a good conductivity to ensure a rapid electron transfer.

Environmental monitoring typically involves several steps such as sampling, sample handling, and sample transportation to a specialized laboratory to determine the chemical composition and to establish the toxic effects. These conventional approaches are expensive, time consuming, and require highly trained personnel. Thus the multiple steps involved frequently prevent rapid information about the composition and/or the toxicity of the sample to be obtained in an efficient manner. Thus risk assessment and remediation efforts could take months or years. Consequently, the need for fast, sensitive, selective, and cheap alarm systems is becoming more apparent. In this chapter, we will focus on some of these novel approaches as well as the fundamental scientific questions involved in this emerging area of research. Advanced nanosensors will be described with respect to the mechanism of molecular recognition, material design and characterization, sensing procedure and efficiency, as well as potential application for improving water quality. We will describe specific evaluations of these sensors in terms of speed, simplicity, and cost, for rapid and reliable measurements of some environmental contaminants such as organophosphates (OPs), volatile organic chemicals, pathogenic bacteria, and DNA.

2.2 Nanostructured sensing materials developed

We have explored the feasibility of designing advanced conducting polymeric materials for sensing and remediation applications [8–34]. These include the synthesis of (i) polyamic acid–silver nanoparticle composite membranes [10], (ii) polyoxy-dianiline films [7–9], and (iii) electrochemical deposition of gold nanoparticle films onto functionalized conducting polymer substrates [8,9]. These were characterized using electrochemical and surface morphology techniques including transmission electron micrograph (TEM), atomic force microscopy (AFM), cyclic voltammetry (CV), and Fourier transform infrared (FTIR) spectroscopy. We have utilized the unique reactivity of polyamic acid (PAA) to design polymer-assisted nanostructured materials. This approach works by preventing the cyclization of the reactive soluble intermediate into polyimides at low temperature [9]. The ability to prevent the cyclization process has enabled the design of a new class of electrode materials. Three different materials that incorporate nanoparticles of varying sizes have been successfully synthesized and characterized using this approach. We have explored these as precursors for environmental sensing and remediation.

2.2.1 Incorporation of metal nanoparticles in photopolymerized organic conducting polymers

Electroactive conducting polymers (ECPs) can be patterned by inexpensive methods (e.g., chemical, electrochemical, hot embossing, and imprinting techniques). Imprint method, which can allow submicrometer patterning with dimensions smaller than 100 nm, has been demonstrated using hard imprint materials [32–37]. Electrochemical methods can enable the preparation of thin films on electrodes, but are limited by the conducting substrates to be used during deposition. Moreover, the polymers suffer from poor stoichiometry, insufficient conductivity, and reproducibility. However, photopolymerization technique allows the use of insulating substrates such that several metals and alloys can be deposited from an electrolyte onto the nonconducting surface. Photochemical preparation of ECP films utilizes a mixture of monomer, electron acceptor, and photoactive species exposed to ultraviolet (UV)/visible light [38–41]. We have found that regularly arranged metal nanoparticles could be spontaneously prepared using photochemical polymerization of ECPs [7,8]. The oxidation of a Π-conjugated ECP with gold trichloride, silver nitrate, palladium ions, and copper sulfate following photochemical reaction was found to produce conducting films having metal clusters in 5–100 nm range [7,8]. The incorporated metal clusters enhanced the conductivity of the resulting polymers. We also found that the temperature dependence of ECP materials upon exposure to different compounds produced changes in surface conductivity, and this can be used to monitor the compounds. These metal clusters could also serve as novel catalysts for the removal of heavy metals in aqueous media [42,43].

Alternative photochemical routes to polypyrrole (PPy or Ppy) that have advantages in some applications over the conventional chemical and electrochemical syntheses have been described. Segawa and coworkers reported that visible light irradiation of an aqueous pyrrole solution in the presence of $[Ru(bipy)_3]^{2+}$ (bipy = 2,2′-bipyridine) as the photosensitizer and $[CoCl(NH_3)_5]^{2+}$ as a sacrificial oxidant led to the deposition [41]. This powdery product exhibited a relatively low conductivity (3×10^{-4} S/cm^2) compared with the polypyrrole chloride [PpyCl] prepared via standard chemical or electrochemical methods. The photochemically initiated polymerization is believed to proceed via the mechanism shown in Figure 2.1, where the oxidation of the pyrrole monomer is performed by the strong oxidant $[Ru(bipy)_3]^{3+}$ generated by oxidative quenching of the photo-excited $^*[Ru(bipy)_3]^{2+}$ species by the sacrificial Co(III) complex.

In our laboratories, novel conducting polymers were prepared through a systematic photopolymerization of PPy with the incorporation of Group 1B metal particles [7,8]. This was achieved through a systematic photopolymerization of PPy with the incorporation of Group 1B metal particles (Eqs. (2.1) and (2.2)). The structural and morphological properties of the incorporated metal nanoparticles were examined with respect to the role of the metals ions, nature of substrates, exposure time, and monomer counterion ratios. Elemental analysis confirmed the identity of the particles with sizes of 100 nm diameter.

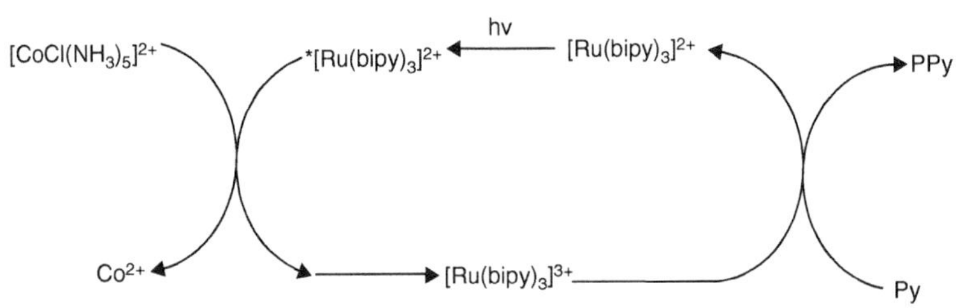

FIGURE 2.1

Photoelectrochemical generation of polypyrrole.

$$3Py + 6AuCl_4^- + hv \rightarrow [Py - Py - Py]^+ AuCl_4^- + 3HAuCl_4 + 2Au^0 + HCl + 7Cl^- \quad (2.1)$$

$$3Py + 4AuCl_3 + NaDS + hv \rightarrow [Py - Py - Py]^+ DS^- + HAuCl_4 + NaCl + 3Au^0 + 7HCl \quad (2.2)$$

2.2.1.1 Polymerization mechanism

A mechanism is proposed for the formation of metal nanostructure during photopolymerization [7,8]. The reaction schemes in Eqs. (2.1) and (2.2) are offered to describe our findings for the photoreaction where gold trichloride is incorporated into the film as $AuCl_4^-$ (Eq. (2.1)) and the condition where the additional anionic dopant, dodecylsulfate (DS^-), excludes the $AuCl_4^-$ (Eq. (2.2)). The reaction starts with the absorption of a photon and spontaneous reduction of $AuCl_3$ salt from solution to form gold clusters. This leads to the oxidation of pyrrole and the formation of pyrrole radical cation through excitation of electrons in the π-orbital. Next is the coupling of two radical cations to form a dimer and a deprotonation process where two protons are lost per coupling. Upon excitation of the monomer, the dimers or higher oligomers are also oxidized and this can further react with the radical cation of the monomer to build up the chain. Coupling of oxidized oligomers may also occur.

When each pyrrole monomer loses just two protons, the polymer is not conductive. The polymer becomes conductive when it is further oxidized by an anion in the electrolyte. For each proton lost, an anion must associate with the polymer for charge neutrality. A necessary condition for the reaction is that the polymerization steps should proceed with 100% efficiency. In this case, all electrons produced during the reduction of AuCl3 ions must result in the oxidation of the pyrrole monomer. If, indeed, the metal ion is an active participant in the initiation step by facilitating pyrrole oxidation, then it is likely that the same effect will be observed using copper or silver solutions, given the similarities in electronic configuration and electron affinity. Thus the incorporation of other Gp 1B metal ions was investigated.

The rate of film formation was found to be dependant upon the metal salt utilized. Subsequent analysis of the ECP films shows the incorporation of zero valent metal particles and the resulting film is conductive and electroactive. By varying the metal salt to monomer ratios, stoichiometric relationships were determined. Addition of anionic dopant sources, such as sodium dodecylsulfate, are shown to alter the reaction stoichiometry, indicating that competition between the metal salt anion and surfactant ion for inclusion in the p-doped film may be a crucial factor in the mechanism of photonic polymerization. The photopolymerization techniques do not require a conductive substrate to deposit the films, and may be used in conjunction with silicon technology, the integration of which could lead to hybrid modules or integrated sensors, where signal processing can be carried out on the same chip (Figure 2.2).

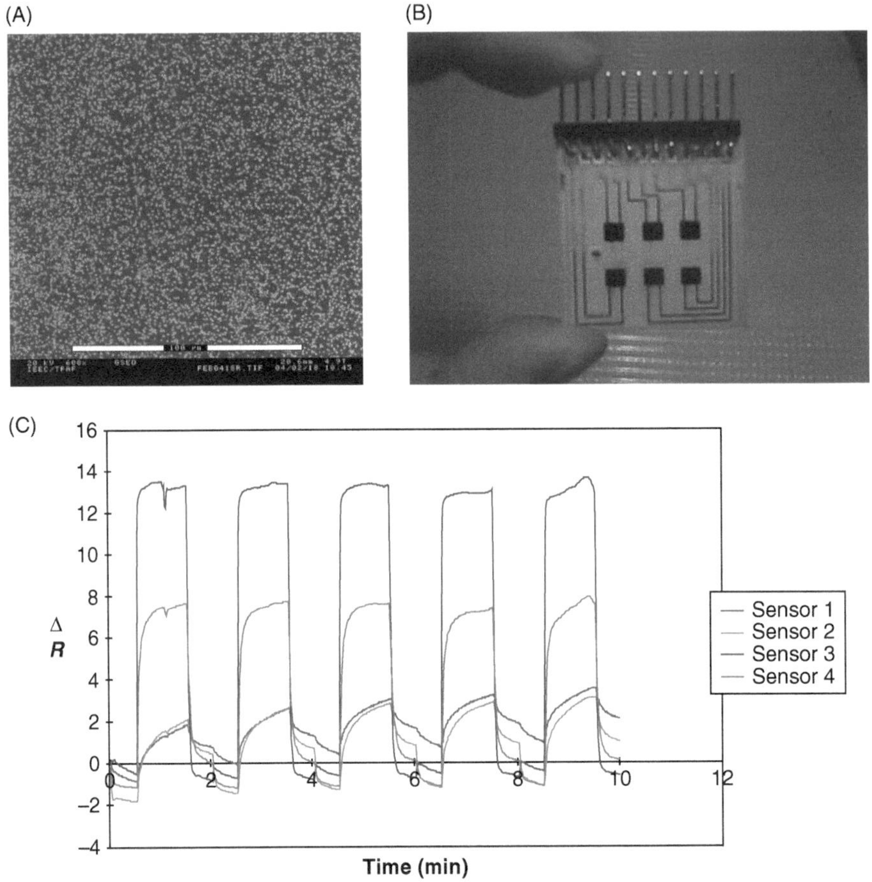

FIGURE 2.2

(A) Synthesis of nanostructured materials and hybrid polymers containing 5 nm Ag particles to enhance polymer conductivity; (B) arrays of polymers electrochemically deposited onto copper substrates; (C) measured electrical resistance changes of polymer arrays when exposed to different concentrations of OP nerve agents. Response time is about 2 min.

2.2.1.2 Polypyrrole films for vapor sensing

Additionally, we tested the possibility of photopolymerized [PPy] for vapor sensing (Figure 2.2). When the PP arrays are presented with pulses of organic vapors, they generate characteristic temporal response for each vapor. The resistance change was reversible upon the removal of the organic vapor and sensor signals returned to baseline. The reproducible signals obtained could be attributed to that caused by defects contained in the conjugated structure as well as the energy of sorption and desorption processes, and solubility consideration [8,38–42].

2.2.2 Nanostructured PAA membranes as novel electrode materials

We have also reported a new approach for the preparation of PAA composites containing Ag and Au nanoparticles. The composite film of PAA and metal particles were obtained upon electrodeposition of a PAA solution containing gold or silver salts with subsequent thermal treatment, whereas imidization to polyimide is prevented [9]. The composites of poly (amic acid)–metallic gold and silver nanoparticles were prepared using pyromellitic dianhydride (PMDA) and 4,4′-oxydianiline (ODA) solutions containing the salts of the metal ions [9]. The resulting PAA composites were thermally cured between 40°C and 105°C and then electrochemically reduced by galvanostatic deposition. Thermal curing at relatively low temperature was sufficient to eliminate all residual solvents after 2 h with concomitant reduction of the metal ions into metallic nanoparticles. FTIR spectroscopy, scanning electron microscopy (SEM), and CV and NMR analysis were used for structural and morphological characterization. The molecular weight of the solid PAA was estimated to be 10,000 dalton using gel permeation chromatography. Elemental analysis and TEM confirmed that electrochemical or thermal reduction of PAA-metal composites produced homogeneous metallic nanoparticles (Figure 2.3). In our system, we prepared novel PAA metal nano-composites onto RVC electrodes in which the polymer will retain the carboxyl groups and exhibit high stability. In this approach, salts of the desired metal particles were introduced into the PAA solution so that subsequent electrochemical reduction or thermal curing at low temperature would produce monodispersed nanoparticles. For this purpose, we used two solvent systems: first was in tetrahydrofuran with triethylamine (TEA) as surfactant and methanol as precipitant agent. The other was developed in acetonitrile using similar surfactant and precipitating agent. PAA was obtained as a viscous solution in the first solvent system, whereas it was obtained as solid using acetonitrile as solvent. The uniqueness of this approach lies in the incorporation of metal nanoparticles within the electropolymerized PAA at low temperature and the ability to prevent the cyclization process at low temperature. The low temperature ensured that the thermal reduction occurred and imidization process was avoided [9].

Figure 2.4 shows the chemistry of the synthetic process including the electrodeposition of silver–composite PAA film from PMDA and ODA. The rationale is that when constant current is applied, the PAA salt is dissociated. Under the influence of electric field, the negatively charged poly- or monocarboxylates migrate to the anode, accept protons, and precipitate as continuous polymer film [9]. The presence of free carboxylic groups in the PAA–silver-modified electrode will allow the utilization of this material as immobilization matrices in sensors and biosensor devices [9]. Structural analysis of PAA and Ag–PAA were achieved by using FTIR spectroscopy to confirm the presence of specific functional groups [9]. The strong peak around 1225 cm^{-1} that appeared in both spectra was associated to stretching vibration of the ether group [33]. All these bands

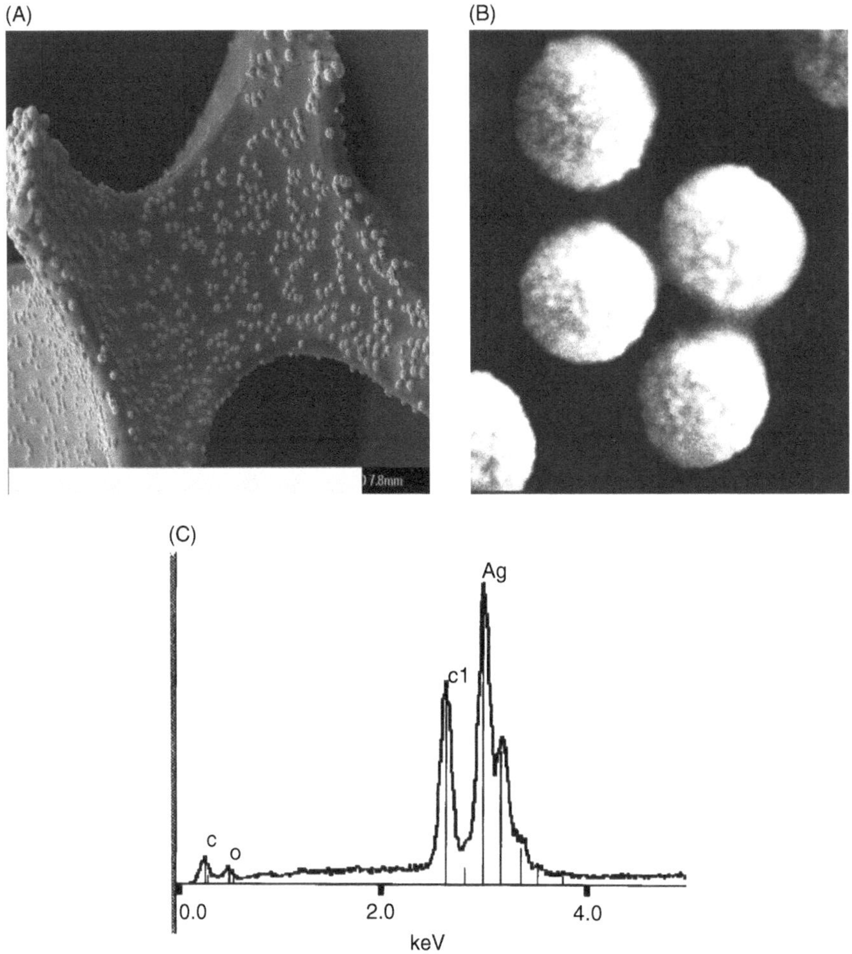

FIGURE 2.3

Preparation of nanostructured hybrid organic–polymeric materials—TEM characterization of sequestered PAA silver nanoparticles. (A) SEM of PAA with silver nanoparticles onto reticulous vitreous carbon (RVCs); (B) SEM picture of gold agglomerates stabilized with PAA membranes; and (C) EDX spectrum of PAA–AgNP film.

were still present in the Ag–PAA spectrum after thermal treatment. The absence of typical imide functional bands in the cured PAA and Ag–PAA films at 1778 cm^{-1} (for symmetric C = O) and 1374 cm^{-1} (C–N–C axial) [9] confirmed that the imidization process had been avoided at 105°C even after 5 days of thermal treatment. Moreover, the similarity in both spectra (spectra not shown) suggests that the characteristic functional groups present in PAA basic molecule retained their integrity after electrodeposition and drying.

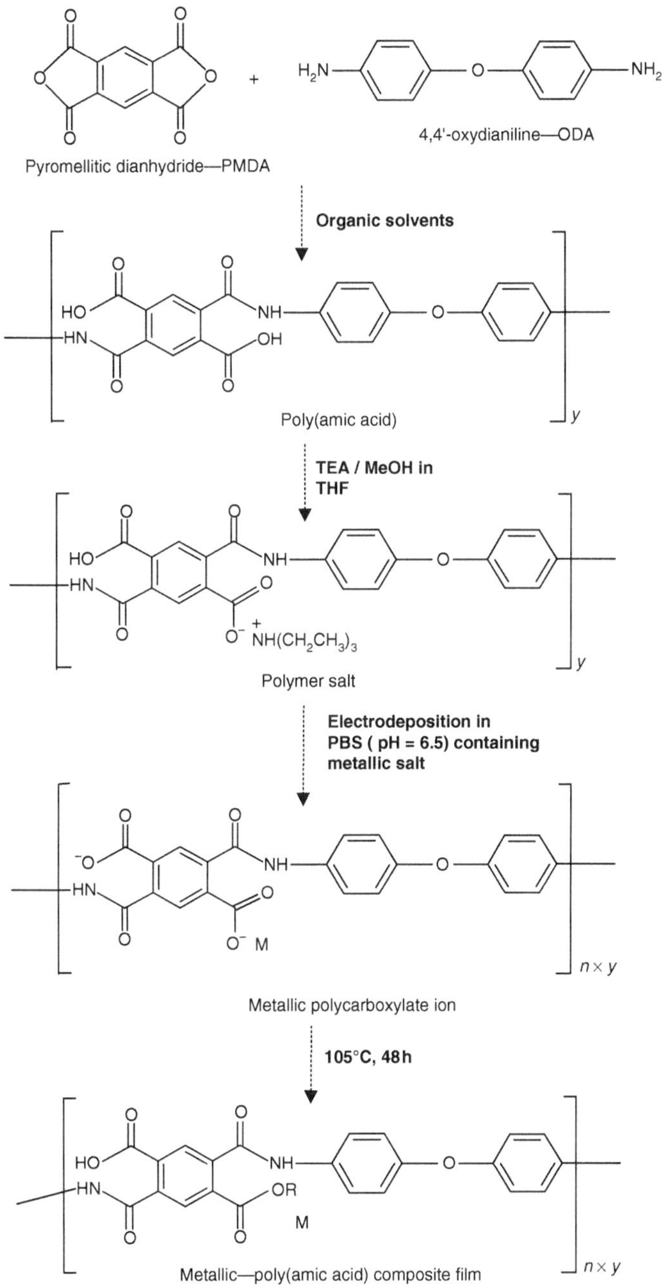

FIGURE 2.4

Schematic representation of metallic nanoparticles.

2.3 Chemical sensor arrays and pattern recognition

The nanostructured materials synthesized were subsequently incorporated as sensing elements in an array system. A sensor array system (SAS) consists of three functional components that operate serially on a sample containing the analyte, a sample handler, an array of polymer sensors, and a signal processing system (Figure 2.5). The output of the SAS can identify an environmental toxin or biological molecule, estimate the concentration, or determine the characteristic properties of the compound present in air or other samples.

Fundamental to the SAS is the idea that each sensor in the array has different sensitivity. For example, Compound No. 1 may produce a high response in one sensor and lower responses in the others, whereas Compound No. 2 may produce high readings for sensors other than the one that exhibit significant response to Compound No. 1. What is important is that the pattern of response across the sensors is distinct for different compound. This allows the system to identify an unknown agent from the pattern of sensor responses or database. Thus, each sensor in the array has a unique response profile to the spectrum of chemicals under test. The pattern of response across all sensors in the array is used to identify and/or characterize the analyte.

We have further demonstrated the effective use of our nanostructured polymer arrays coupled with machine learning for the detection and classification of OP nerve agents' stimulants. For OPs and volatile organics, we showed a significant 168% specificity improvement and a 40.5% positive predictive value improvement using the s2000 kernels at 100% and 98% sensitivities when compared to commercial system [30,31]. The OP molecules that are dissociated at the surface of the polymer electrode cause the change in sensor resistance. This chemisorption-driven process can change the electrical resistance considerably, and can make the sensor array sensitive to OP concentrations over a wide range of concentrations. Since the sensor completely regains its original resistance with OP cycling, it appears that the OP diffusion is confined to the surface layer.

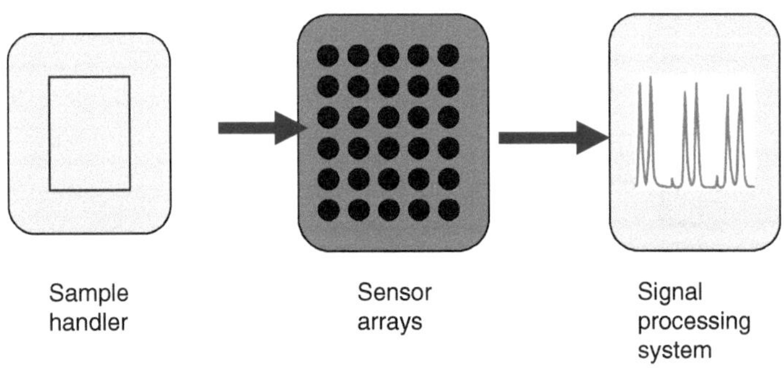

FIGURE 2.5

Schematic diagram of a standard multiarray sensor system.

2.3.1 Data processing, pattern recognition, and support vector machines

The task of a sensor array system is to identify the presence of toxic environmental chemicals (TECs) in the sample and perhaps to estimate its concentration. This is achieved by means of signal processing and pattern recognition (Figure 2.5). These two steps may be subdivided into four sequential stages: preprocessing, feature extraction, classification, and decision-making. Preprocessing compensates for sensor drift, compresses the transient response of the sensor array, and reduces sample-to-sample variations. Typical techniques include manipulation of the sensor baseline; normalization of sensor response ranges for all sensors in an array and compression of sensor transients. Feature extraction has two purposes: to reduce the dimensionality of the measurement space and to extract information relevant for pattern recognition. Feature extraction is generally performed with classical principal component analysis (PCA) and linear discriminant analysis (LDA). PCA finds projections of maximum variance and is the most widely used linear feature extraction techniques. However, it is not optimal for classification since it ignores the identity (class label) of the analyte examples in the database. LDA on the other hand looks at the class label of each example. Its goal is to find projections and maximize the distance between examples of the same TEC agent. As an example, PCA may do better with a projection that contains high variance random noise whereas LDA may do better with a projection that contains subtle, but maybe crucial, agent-discriminatory information. LDA is more appropriate for classification purposes.

Once the sensor signal is projected on an appropriate low-dimensional space, the classification stage can be trained to identify the patterns that are representative of each class of compound. The classical methods of performing the classification task are *k*-nearest neighbors (*k*-NN), Bayesian classifiers, and artificial neural networks (ANN). *k*-NN classifiers find examples in the TEC database that are closest to the unidentified agents and will assign the nature of the agent represented by a majority of those examples. Bayesian classifiers first build a probability density function of each agent class on the low-dimensional space. When presented with an identified compound, the Bayesian classifier will pick the class that maximizes the precompiled probability distribution. The classifier produces an estimate of the class for an unknown sample along with an estimate of the confidence placed on the class assignment. However, the inability to present error-free detection, requirement for large training data set, as well as the inability to minimize true risk has led to the development of other processing approaches such as the support vector machines (SVMs).

SVMs are a new and radically different type of classifiers or "learning machines" that use a hypothesis space of linear functions in a high-dimensional feature/space. SVMs are generally trained with learning algorithms originating from optimization theory that implements a learning bias derived from statistical learning theory. The use of SVMs for computational intelligence is a recent development, and certainly unknown for analytical monitoring of TECs. Several reviews provide extensive backgrounds to develop the mathematical foundation of SVMs ([30,31] and cited references). In the context of classifying TEC, the objective of SVMs is to construct an "optimal hyperplane" as the decision surface such that the margin of separation between two different chemical substances is maximized. SVMs are based on the fundamental ideas of: (i) structural/empirical risk minimization (SRM/ERM); (ii) the Vapnik–Chervonenkis (VC) dimension; (iii) the constrained optimization problem; and (iv) the SVM decision rule. Properly designed SVMs should have a good performance on untested data because of their ability to generalize and scale up to problems that are more complex. The fact that the margin does not depend on input dimensionality means it is immune to the curse of dimensionality. SVMs have been successfully applied to a variety of classification problems including text categorization, handwritten digit recognition, gene expression analysis, and simple chemical and mixtures recognition.

We have studied the integration of our sensor network with SVM for the detection of TEC. This approach reduced the number of false negative errors by 173%, while making no false positive errors when compared to the baseline performance [30,31]. The reader may recall that obtaining larger and larger sets of valid training data would sometimes produce (with a great deal of training experience) a better performing neural network (NN), which resulted from classical training methods. This restriction is not incumbent on the SRM principles and is the fundamental difference between training NNs and training SVMs. Finally, because SVMs minimize the true risk, they provide a global minimum.

2.3.2 Integration of sensor array with chromatographic systems

We have previously shown that combining ECP with conventional instrumental techniques (such as chromatographic and flow injection analysis (FIA)), provides a suitable approach to the development of sensitive and reproducible analytical signals [44,45]. Results obtained by the use of integrated conducting polymer sensors with chromatographic analysis and FIA revealed a significant improvement in sensor performance and the overall analysis with respect to time and selectivity [44–47]. In Section 2.3, we saw how sensor arrays and SVMs were utilized in the detection and classification of OP nerve agent simulants. However, sensor arrays cannot quantitatively detect mixtures of compounds. This is because SAS systems analyze mixtures of compounds as a one-component solution. To overcome this limitation, we have coupled sensor arrays to a gas chromatograph. In this scenario, the analyte mixture is first separated by the gas chromatography (GC) column based on their relative retention times and detected using the multiarray polymer sensors. This new system showed significant improvement over the existing multisensor array systems in the analysis of mixtures [46]. We have also demonstrated how the hyphenated combination of GC and conducting polymer sensor arrays could effectively be used in the separation and detection of mixtures of volatile organic compounds [47].

Other workers have also reported their research findings on GC–sensor arrays. A good example is a prototype portable GC that combines a multiadsorbent preconcentrator, a tandem-column separation stage, and a detection consisting of an integrated array of polymer-coated surface acoustic wave (SAW) devices [48–52]. In that report the determination of vapor mixtures of common indoor air contaminants including 2-propanol, 3-methyl butanol, and 1 octen-3-ol was demonstrated. The advantages of sensor arrays include the fact that they do not require auxiliary gases and hazardous materials commonly required in classical GC detectors (e.g., flammable hydrogen gas in FID) and hazardous radioactive material (Ni-63 in ECD) for operation. Most sensor arrays are small thus leading to small dead volumes and increased sensitivity. The small size also has inherent advantages in miniaturization of chromatography–sensor array systems. Furthermore, the sensor arrays can also be used in conjunction with machine learning programs.

2.4 Biosensing applications of nanostructured materials

Our laboratories and others have utilized the uniqueness of nanoparticles to develop biosensors for different environmental, clinical, and security applications [15,16]. Some of these are discussed in the following sections.

2.4.1 Biosensors for polychlorinated biphenyls

The persistence of polychlorinated biphenyls (PCBs) in the environment and their extensive usage in numerous industrial and commercial applications are currently of great concern. The limited information on the effect of PCBs indicated that these compounds might produce immunological abnormalities, reproductive dysfunction or an increased thyroid volume, increased prevalence of thyroid and liver disorders [53,54]. Nowadays, there is an increased interest to detect these environmental chemicals from both scientific and regulatory communities. Consequently, significant efforts are focused on developing a fast and reliable method for their determination.

Immunosensors have been reported to exhibit considerable potential for PCB detection [54–57]. For instance, we have developed a PCB immunosensor, constructed by immobilizing anti-PCB antibody (Ab) into a conducting PPy membrane. Pulsed-accelerated immunoassay for signal generation in stationary cell or FIA is obtained by applying a pulsed waveform between +0.60 and −0.60 V and a pulse frequency of 120 and 480 ms. With the optimized sensor, a linearity of 0.3–100 μg/L and a detection limit of 3.3, 1.56, 0.39, and 1.66 ng/mL, respectively, were obtained for Aroclors 1242, 1248, 1245, and 1016 [54,57].

Most studies with immunosensors are carried out in aqueous solutions in which large molecules as Ab function ideally. However, PCBs, as well as other important environmental analytes are poorly soluble in this medium. In addition, extraction and concentration of the sample are commonly carried out in organic solvents. In this context, a detection method for these compounds in the presence of organic medium is also required. In most cases, the immunological activity of immobilized Ab is generally lower in organic solvents compared to water. Significant improvement in this direction was achieved when Ab was encapsulated in reversed micelles [58]. Detection of PCBs was also achieved using DNA biosensors designed for environmental monitoring. In a recent work developed in our group we reported a detection limit of 10 nM PCB using supramolecular dsDNA sensor on Ag–Au-coated quartz crystal electrode with impedance spectroscopy [59]. Using an electrochemical system, Marrazza et al. detected as low as 0.2 mg/L PCBs using screen-printed disposable DNA biosensor [60]. In this case, determination was achieved by measuring changes of the electrochemical signal of guanine in calf thymus DNA extract immobilized onto the electrode surface.

2.4.2 Endocrine disrupting chemicals, chlorinated organics, and other analytes

2.4.2.1 Biosensors for polyphenols and other analytes

A novel label-free electrochemical scheme for probing the electronic properties of DNA binding with small molecules was recently described by our group [61]. The strategy, called metal-enhanced electrochemical detection (MED) involved the electrooxidation of silver ions, deposited as a monolayer onto a gold electrode surface in the presence of dsDNA molecules. The results indicated that by oxidizing the silver monolayer, highly reactive oxides of silver ions are generated, causing a change in the electronic properties of the immobilized dsDNA. Thus, in the presence of low molecular weight organic DNA-binding molecules, structural change in the DNA occurs, generating corresponding change in the redox properties of the silver monolayer. These variations are proportional to the concentration of DNA-binding molecules and can be easily quantified by voltammetric techniques. The approach offers an alternative route for the detection of DNA hybridization reactions, DNA–protein interactions, and gene detection. It was demonstrated that the performance of the DNA biosensors are strongly dependent on hybridization reactions at the transducer–solution interface. Thus, control of the hybridization conditions such as temperature and hybridization time can considerably extend the dynamic range and lower the sensitivity [61].

The strategy for MED using the reactivity of silver ions is illustrated in Figure 2.6. A monolayer of silver is deposited on a gold electrode or other conducting substrates (e.g., platinum or glassy carbon). The silver deposition can also be achieved by incubating silver compounds for approximately 5 min in the dark at room temperature. The electrochemical oxidation of silver produces silver ions and electrons accompanied by a reversible redox signal (Figure 2.6A). If dsDNA is present at the surface, the silver ions are dispersed and are held electrostatically. Upon reduction, the silver ions return to the surface and a reduction current

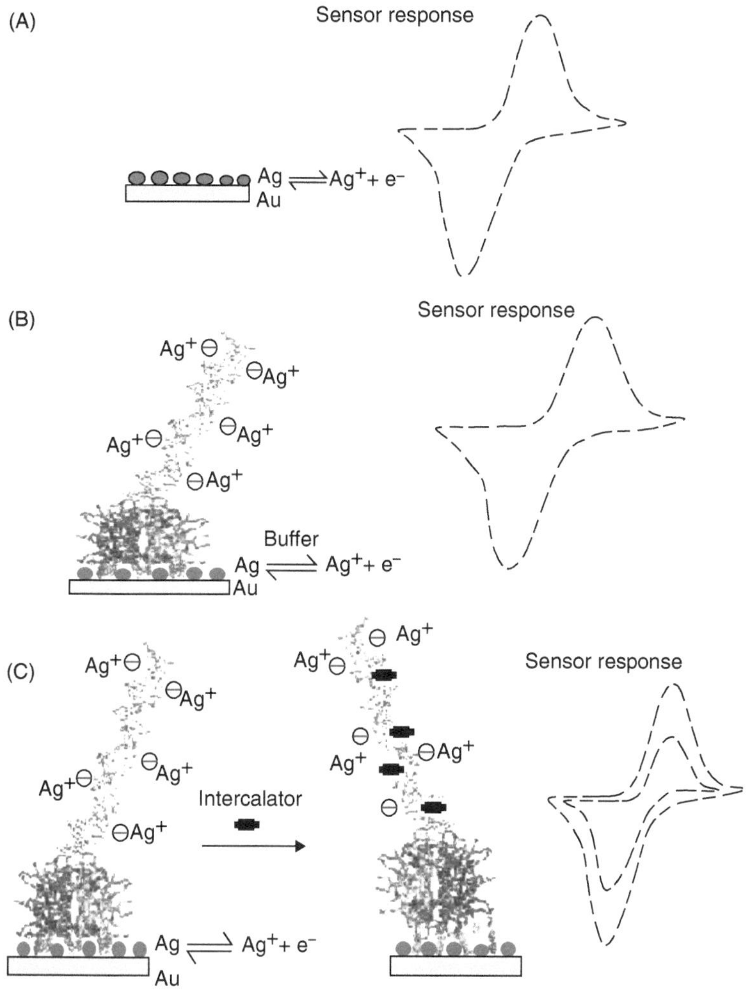

FIGURE 2.6

MED concept for electrochemical amplification using Ag^0/Ag^+ couple (A) in buffer only, (B) at an immobilized dsDNA electrode in buffer, or (C) at an immobilized dsDNA electrode in buffer and analytes.

is measured (Figure 2.6B). If a DNA-binding low molecular weight organic molecule is introduced into the solution, structural change in the DNA molecule occurs, which is signified by a simultaneous change in the redox currents from silver, and a decrease in current is measured (Figure 2.6C). This decrease is proportional to the concentration of the DNA-binding molecule. The underlying signal transformations produced here could result from the DNA conformational or structural changes in the presence of the analyte molecules, which ultimately hinder the flow of electrons. Typical results for MED are shown in Figure 2.7 using differential pulse voltammetry.

Using the proposed MED concept and un-optimized assay format, a detection limit of 4 ppt has been recorded for Bisphenol A (BPA), an endocrine disrupting chemical (EDC) (Table 2.1), whereas the limit of detection for the same analyte using ELISA was 228 ppb. This represents a 1000-fold improvement over the gold standard. In addition, the time required for the MED biosensor was about 20 min, compared to between 24 h and 3 days necessary for the ELISA techniques. Apart from small organics, we have tested the utility of the MED

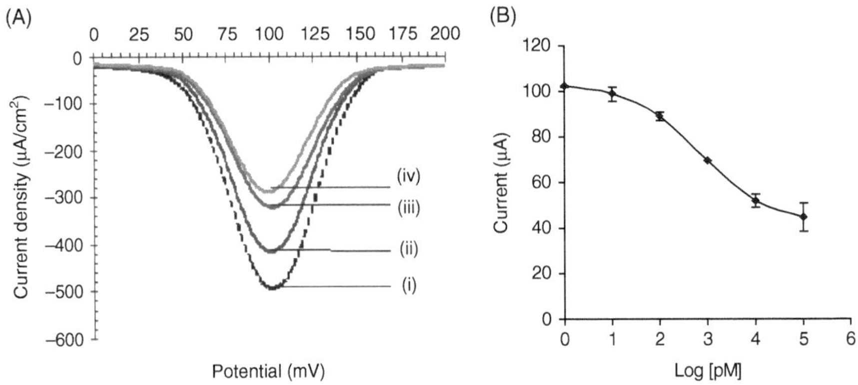

FIGURE 2.7

Typical MED signals at (A) different analyte concentrations of DNA (B) calibration curve for the MED-based assays.

Table 2.1 Comparison of MED-Based Biosensor Performance with Conventional Techniques

	Limit of Detection[a,b,c]		
Analytes	**ELISA**	**Electrochemical Fe^{2+}/Fe^{3+}**	**Detection Ag^{0}/Ag^{+} Couple**
BPA	228 ppb	10 nM	4 ppt
NP	37.5 ppb	ND	14 ppt
Cisplatin		1 nM	10 pM
PCBs		ND	5 ppt

BPA: Bisphenol A; ND: None Detected; nM: Nanomolar; PCB: Polychlorinated biphenyls; pM: Picomolar; ppt: Parts per trillion.
[a]Sadik et al, JACS, 123, 2001, 11335
[b]Ngundi M., PhD Binghamton University, 2003
[c]Kowino I., PhD Binghamton University 2006.

concept for other molecules with molar masses ranging from 200 to 150,000 amu. The range of analytes tested include nucleic acids, PCBs, nonylphenols, cisplatin, and other naturally occurring isoflavonoids [58,59].

2.4.2.2 Biosensors for EDCs

In recent years, it has become evident that many environmental chemicals, including synthetic and endogenous estrogens, can mimic, block, or alter the action of endogenous steroid hormones and can interfere with hormone-regulated physiological processes [13,62–66]. These industrial and environmental chemicals are known as EDCs and structurally resemble endogenous estrogens. They have been analyzed for many decades in numerous biological and medical investigations. Screening and confirmatory strategies for these steroids involve chemical or immunochemical methods followed by the complete instrumental confirmation of steroids by mass spectrometry (MS). Until recently, the standard technique for analyses has been GC/MS. Moreover, the limits of detection were not sufficient to analyze steroids at low levels in urine and environmental samples. Also, these techniques typically require sample pretreatment, expensive apparatus, and skilled personnel.

Synthetic estrogens are characterized by the presence of phenolic functional groups, a common structural feature that is also found in natural estrogens. This structural feature could facilitate binding to estrogen receptor [62–66] and possibly generate receptor-induced transformations. Sadik et al. presented a summary of different approaches reported for EDCs [13] and also demonstrated the feasibility of *in situ* monitoring of the interaction between BPA and dsDNA [59]. In another report, Ngundi et al. demonstrated the comparative electrochemical behavior of β-estradiol and selected EDCs, specifically alkylphenols, and proposed a possible link between the structure and their estrogenic activity [66]. More recently, we have reported the isolation and complete structural characterization of quercetin (QRC) and other isoflavonoids [67]. We have also developed and optimized a fully autonomous electrochemical biosensor for studying the role of alkylphenols on A549 lung adenocarcinoma cell line. This advanced biosensor uses a prototype 96-electrode (DOX-96) well-type device that allows the measurement of cell respiratory activity via the consumption of dissolved oxygen. The system provides a continuous, real-time monitoring of cell activity upon exposure to naturally occurring polyphenols, specifically resveratrol (RES), genistein (GEN), and QRC. The system is equipped with a multipotentiostat, a 96-electrode well for measurements and cell culturing with three disposable electrodes fitted into each well. A comparison with classical "cell-culture" techniques indicates that the biosensor provides real-time measurement with no added reagents. A detection limit of 1×10^4 was recorded versus 200 and 6×10^3 cells/well for MTT and fluorescence assays, respectively. This method was optimized with respect to cell stability, reproducibility, applied potential, cell density per well, volume/composition of cell culture medium per well, and incubation. Others include total measuring time, temperature, and sterilization procedure. This study represents a basic research tool that may allow researchers to study the type, level, and specific influence of isoflavonoids on cells.

The DOX approach utilizes a prototype multichannel system that enables simultaneous, quantitative, and continuous measurement of dissolved oxygen using a 96-well electrode biosensor prototype shown in Figure 2.8 known as the DOX device. DOX is fully automated, portable, equipped with a multipotentiostat, and can be connected to a computer. The latter enables external control of the instrument, online recording of experimental parameters, graphical presentation and data storage. The instrument software plots current intensity versus time for each well while data are simultaneously processed for 12 channels, each corresponding to 8 sensors. Experimental setup involves the use of 96 disposable electrodes in a three-electrode format (reference, working, and auxiliary). These 96 sensors are placed in a conventional 96-well plate in which the cells are cultured. Respiration of cells generates a reduction in the concentration of dissolved oxygen, which is determined using electrical current produced [19,20] according to the following reaction:

$$O_2 + 4H^+ + 4e^- \rightarrow 2H_2O \tag{2.3}$$

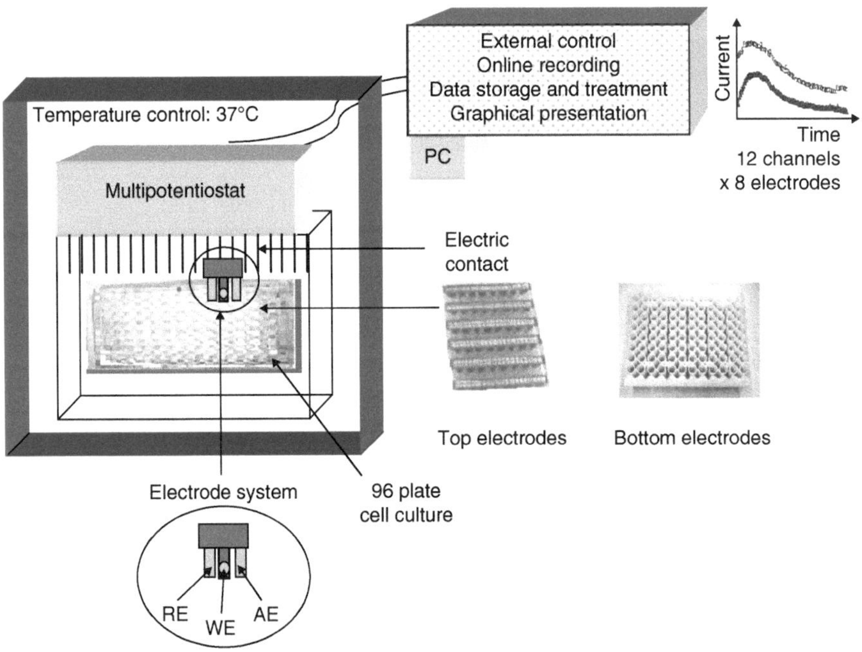

FIGURE 2.8

Schematic diagram of the multichannel DOX oxygen sensor system used for measuring cytotoxicity. Two configurations of the 96 electrodes are available disposed in top or bottom of the well plate. Each sensor consists of three electrodes: reference (RE), auxiliary (CE), and working (WE).

Cytotoxicity measurements involve the application of cells into the optimum growth medium containing a selected isoflavonoid. This is followed by a continuous monitoring of the oxygen consumed by the cells with time. Basically, if an inhibition of cell proliferation occurred in the presence of PPh this will generate higher oxygen content, proportional with the inactivation level induced by PPh. This could be indirectly correlated to the concentration of PPh introduced in the medium.

2.4.3 Multiarray electrochemical sensors for monitoring pathogenic bacteria, cell viability, and antibiotic susceptibility

Bacterial pathogens are found not only in soil, food, marine, and estuarine waters but also in intestinal tracts of humans and animals. Microbial diseases constitute one of the major causes of death in many developing countries. For this reason, significant efforts are needed to develop new systems for rapid detection of pathogenic bacteria in a variety of fields. Among these, enzyme, DNA or immunosensors coupled with electrochemical, piezoelectric, optical, acoustic, and thermal detection represent viable alternatives. We reported the use of amperometric signals of the DOX coupled with PCA for continuous monitoring, identification, and differentiation of bacteria. Two types of differentiation mechanisms were tested: (1) direct monitoring of respiratory activity via oxygen consumption and (2) quantification of the effect of three broad-spectrum antibiotics on bacteria growth and respiration over time (Figure 2.8). Five species of bacteria were examined including: *Escherichia coli*, *Escherichia adecarboxylata*, *Comamonas acidovorans*, *Corynebacterium glutamicum*, and *Staphylococcus epidermidis*. The addition of small concentrations of antibiotics to the growth medium alters the oxygen consumption of the cells and a unique fingerprint is created for a specific cell. This fingerprint is shown to evolve over a specific concentration range that is dependant of instrumental constraints of the DOX system. The application PCA to classify the data was also examined (Figure 2.9). It was shown that bacteria could be classified simply by their oxygen consumption rates over a varying concentration range. Discrimination between species can also be increased by the effects of the antibiotics on the oxygen consumption of varying concentrations of cells (Table 2.2). The proposed DOX–PCA system illustrates a generic template that can be tailored to meet specific research goals by the selection of specific cell/antibiotic combinations and concentrations.

Another prototype variation uses a bottom electrode design in which the electrodes are manufactured directly into the bottom of each well in the plate. All experiments in this project were performed with the top electrode system. Each element in the disposable electrode array is composed of a working, reference, and auxiliary electrode for individual wells in the plate. The gold working electrode has dimensions of $0.19\ mm^2 \times 5\ mm^2 \times 5\ mm^2$. After the electrodes are placed into the sample, the plate is inserted into the DOX-96 instrument, which resembles a small optical microplate reader. The instrument contains a

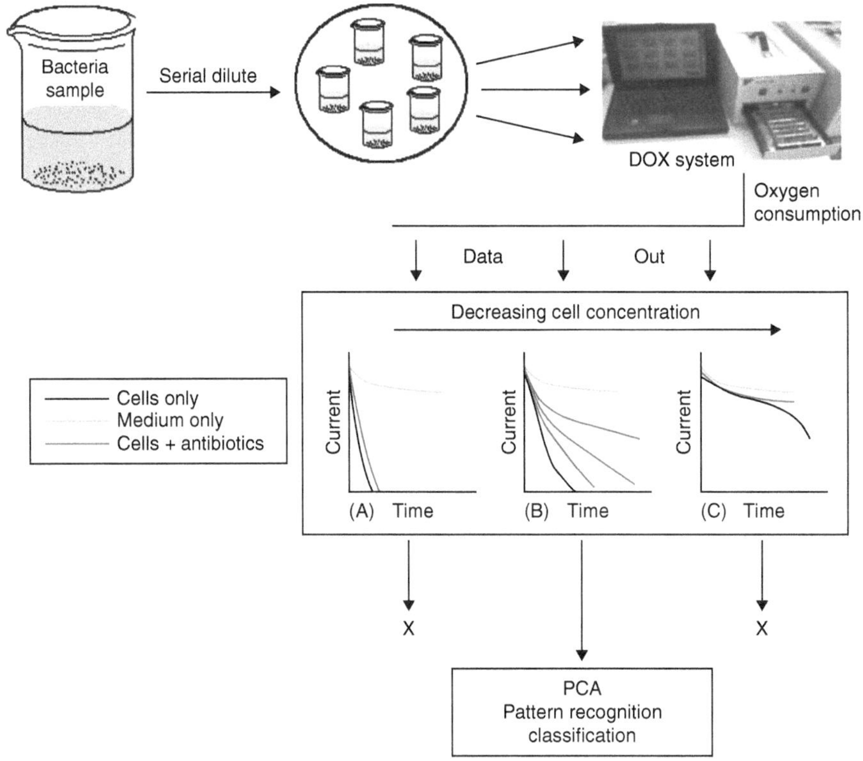

FIGURE 2.9

Schematic of the DOX–PCA concept. A, B, and C represent DOX responses for high, medium, and low cell concentrations, respectively.

Table 2.2 Bacteria Concentrations Used in the DOX–PCA Experiments

	Concentration × 10^6 cfu/mL				
Species of Bacteria	**C1**	**C2**	**C3**	**C4**	**C5**
C. acidovarans (ATCC# 51340)	112	56[a]	28	14	7.5
C. glutamicum (ATCC# 13032)	79.3	39.65	19.83[a]	9.91	4.96
S. epidermidis (ATCC# 12228)	81.7	40.85	20.43[a]	10.21	5.10
E. adecarboxylata (ATCC# 23216)	45	22.5	11.25[a]	5.63	2.81
E. coli (ATCC# 25922)	150	75	37.5[a]	18.8	9.4

[a]*Maximum cell concentration that can produce a DOX fingerprint.*

multipotentiostat, and the current is measured for each of the 96 wells automatically according to user-set parameters. Options include voltage (negative range only), total time, and measurement intervals (e.g., 1 measurement per minute). Data is processed by the software and represented graphically in a time versus current plot and simultaneously tabulated in Microsoft excel.

2.5 Conclusions and future perspectives

We have developed a number of nanostructured materials as candidates for assessing the occurrence of water contamination. These include PAA–metal nanoparticle composite membranes, polyoxy-dianiline membranes, sequestered gold, silver or palladium nanoparticles within electroconducting polymers and underpotential deposition of metal films, membranes, and colloids onto solid electrodes. We have also reported the application of these materials for the design of advanced sensors. In that respect, sensors have been developed for EDCs, volatile organic compounds, the detection and identification of bacteria based on antibiotic susceptibility, multiarray electrochemical sensor with pattern recognition techniques, as well as for metal-enhanced detection of cyanobacteria microcystis (M)[75] spp based on DNA base-pair mismatches. Most of these sensors have shown good-to-excellent pathogen recovery efficiencies as well as a reasonable efficacy for sensing contaminants from water in controlled laboratory experiments.

Chemical and biosensors will continue to provide important monitoring, diagnostic and mechanistic solutions to many environmental, clinical, food, and security applications; and may open new areas of modern analysis. Despite important progress already achieved, the biosensor market is still relatively small, requiring important fundamental and mechanistic studies in order to fully explore their real potentials. Today, more than 90% of commercial biosensors are designated to glucose analysis and few analytes could still be detected. This is mainly a consequence of insufficient reliability associated with poor stability of the sensing material, multiple matrix effect, and also a dependence upon the physico/chemical parameters and interferences within the transducers. Nevertheless, with respect to rapid environmental analysis, sensitive and selective low-cost determination of a great variety of analytes, no suitable alternative exists for biosensors.

Acknowledgments

The author acknowledges the United States Environmental Protection Agency through the STAR program (RD-83090601 and R825323) and the National Science Foundation (CHE-0513470) for funding. Also acknowledged are the contributions of members of the author's research group (past and present) at SUNY-Binghamton including Adam Wanekaya, Austin Aluoch, Isaac K'Owino,

Samuel Kikandi, Jason Karasinski, Sharin Benda, Syeda Begum, Miriam Masila, Fei Yan, Marc Briemer, Hongwu Xu, Sydney Sheldon, Eugen Yevgheny, and Sean Falvey.

References

[1] K.R. Shull, A.J.J. Kellock, Polym. Sci. B: Polym. Phys. 33 (1995) 1417.
[2] D.H. Cole, K.R. Shull, P. Baldo, L. Rehn, Macromolecules 32 (1999) 771.
[3] G. Khlebtsov, J. Quant, Spectro. Radiat. Transfer 89 (2004) 143.
[4] T.M. Lee, H. Cai, I. Hsing, Analyst 130 (2005) 364–369.
[5] X. Su, S.F.Y. Li, S.J. O'Shea, Chem. Commun. (2001) 755.
[6] P. Langer, A. Konan, M. Tajtakova, J. Petrik, J. Chovancova, B. Drobna, et al., J. Occup. Environ. Med. 45 (5) (2003) 526–532.
[7] D. Andreescu, S. Andreescu, O.A. Sadik, New materials for biosensors, biochips and molecular bioelectronics (A review-invited contribution), in: L. Gorton (Ed.), Comprehensive Analytical Chemistry, vol. 44, Elsevier, 2005.
[8] M.A. Breimer, G. Yevgeny, S. Sy, O.A. Sadik, Nano Lett. 1 (2001) 305.
[9] D. Andreescu, A. Wanekaya, O.A. Sadik, J. Wang, Langmuir 21 (15) (2005) 6891–6899.
[10] D. Andreescu, O.A. Sadik, J. Electrochem. Soc. 152 (10) (2005) E299–E307.
[11] J. Karasinski, S. Andreescu, O.A. Sadik, Advanced electrochemical sensors for cell cancer monitoring, Methods 37 (1) (2005) 84–93.
[12] S. Andreescu, J. Karasinski, O.A. Sadik, Multiarray biosensors for toxicity monitoring and bacterial pathogens, in: G.K. Knopf, A.S. Bassi (Eds.), Smart Biosensors, CRC Press, USA, 2007, pp. 521–538.
[13] O.A. Sadik, D. Witt, Novel monitoring strategies for environmental endocrine disruptors, Environ. Sci. Technol. 33 (17) (1999) A368–A375.
[14] O.A. Sadik, J.M. Van Emons, Applications of electrochemical immunosensors for environmental monitoring, Biosens. Bioelectronic. 11 (8) (1996) p1–11.
[15] S. Andreescu, O.A. Sadik, Trends and challenges in biochemical sensors for clinical and environmental monitoring, Pure Appl. Chem. 76 (4) (2004) 861–878.
[16] O.A. Sadik, S. Andreescu, A. Wanekaya, Advanced environmental methodologies for agent detection, J. Environ. Monit. 6 (2004) 513–522.
[17] O.A. Sadik, W. Land, J. Wang, Targeting chemical and biological warfare agents at the molecular level, Electroanalysis 15 (4) (2003) 1149–1159.
[18] (a)S. Andreescu, O.A. Sadik, Correlation of analyte structures with biosensor responses using the detection of phenolic estrogens as a model example, Anal. Chem. 76 (3) (2004) 552–560.(b)S. Andreescu, O.A. Sadik, D. McGee, Autonomous multielectrode system for monitoring the interactions of isoflavonoids with lung cancer cells, Anal. Chem. 76 (2004) 2321–2330.
[19] S. Andreescu, O.A. Sadik, N. McGee, Effect of natural and synthetic estrogens on A549 lung cancer cells: correlation of chemical structures with cytotoxic effects, Chem. Res. Toxicol. 18 (3) (2005) 466–474.

[20] J. Karasinski, S. Andreescu, O.A. Sadik, B. Lavine, Multiarray sensors with pattern recognition for the detection, classification and differentiation of bacteria at species and subspecies levels, Anal. Chem. 77 (24) (2005) 7941–7949.

[21] J. Karasinski, S. Andreescu, L. White, Y. Zliang, O.A. Sadik, B.K. Lavirre, et al., Detection, and identification of bacteria using antibiotic susceptibility and a multiarray electrochemical sensor with pattern recognition, Biosens. Bioelectron. 22 (2007) 2646–2649.

[22] M. Breimer, G. Yevgheny, O.A. Sadik, Integrated fluorescence capillary DNA biosensor, Biosens. Bioelectron. 18 (2003) 1135–1147.

[23] F.S. Ligler, M. Breimer, J.P. Golden, D.A. Nivens, J.P. Dobson, O.A. Sadik, Integrating waveguide biosensor, Anal. Chem. 74 (2002) 713–719.

[24] A. Fatah, R. Arcilesi, T. Chekol, Chalotte Latin, O.A. Sadik, and A. Aluoch, Guide for the Selection of Biological Agent Detection Equipment for Emergency First Responders, Second Edition, Guide 101-104, 2007, US Department of Homeland Security, Preparedness Directorate, Office of Grants and Training Systems Support Division, Washington, D.C.

[25] I. Kowino, S. Mwilu, O.A. Sadik, Metal-enhanced biosensor for mismatch detection, Anal. Biochem. 369 (2007) 8–17.

[26] F. Yan, A. Erdem, B. Meric, K. Kerman, M. Ozsoz, O.A. Sadik, Electrochemical DNA biosensors for gene related microcystis species, Electrochem. Commun. 3 (2001) 224–228.

[27] O.A. Sadik, F. Yan, Novel biosensor for pathogenic toxins using fluorescent cyclic polypeptide conjugate, Chem. Commun. (2004) 1136–1137.

[28] F. Yan, M. Ozsoz, O.A. Sadik, Electrochemical and conformational studies of microcystin-LR, Anal. Chim. Acta 409 (2000) 247–255.

[29] F. Yan, O.A. Sadik., Enzyme modulated cleavage of dsDNA for studying the interfacial biomolecular interactions, J. Am. Chem. Soc. 23 (2001) 11335–11340.

[30] O.A. Sadik, A. Wanekaya, L. Walker, M. Uematsu, L. Wong, M. Embrechts, Detection and classification of organophosphate nerve agent simulants using support vector machines with multiarray sensors, J. Chem. Inf. Model. 44 (2004) 499–507.

[31] W. Land, W.O. Sadik, D. Leibensperger, and M. Breimer, Using support vector machines for development and testing intelligent sensors to combat terrorism, Artificial Neural Networks, Smart Engineering System Design, Neural networks, Fuzzy Logic, Evolutionary Programming, Data Mining, and Complex Systems (ANNIE), Cihan H. Dagli, Anna L. Buczak, Joydeep Ghosh, Mark J. Embrechts, Okan Ersoy, Stephen L. Kernel, eds., vol. 2, pp. 811–816, 2002.

[32] S.Y. Chou, P.R. Krauss, P.J. Renstrom, Appl. Phys. Lett. 67 (1995) 3114.

[33] D.T. Chiu, et al., Proc. Natl. Acad Sci. USA 97 (2000) 2408. (a) E.W. Jager, E. Smela, O. Inganas, Issues in nanotechnology, Science 290 (2000) 1540. (b) N. Seeman, Trends Biotech. 17 (1999) 437. (c) L. Adleman, Science 266 (1994) 1021. (d) R.P. Falhman, D.J. Sen, JACS 121 (1999) 11079.

[34] R. Murray (Ed.), Molecular Design of Electrode Surfaces, Techniques of Chemistry, XXII, Wiley, 1992.

[35] A.J. Heeger, in: T.A. Skotheim (Ed.), Handbook of Conducting Polymers II, Marcel Dekker, New York, 1986, p 729, and references therein.

[36] J.I. Kroschwitz, in: J.I. Kroschwitz (Ed.), Electrical and Electronic Properties of Polymers, Wiley, New York, 1988.
[37] M.J. Croissant, T. Napporn, J. Leger, C. Lamy, Electrochim. Acta 43 (1998) 2447.
[38] O.J. Murphy, G.D. Hitchens, D. Hodko, E.T. Clarke, D.L. Miller, and D.L. Park. U.S. Patent #5,545,308 of 8/13/1996.
[39] J.M. Kern, J.P. Sauvage, J. Chem. Soc. Chem. Commun. (1989) 657.
[40] H. Segawa, T. Shimidzu, K. Honda, J. Chem. Soc., Chem. Commun. (1989) 132.
[41] Marcells A. Omole, Isaac O. K'Owino, Omowunmi A. Sadik, Appl. Catal., B 76 (2007) 158–176.
[42] I. Kowino, M. Omole, O.A. Sadik., J. Environ. Monit. 9 (2007) 657–665.
[43] O.A. Sadik, G.G. Wallace, Detection of electroinactive ions using conducting polymer microelectrode, Electroanalysis 6 (1994) 860.
[44] (a) O.A. Sadik, G.G. Wallace, Effect of electroinactive species on detection, Electroanalysis 5 (1993) 555. (b) O.A. Sadik, M.J. John, G.G. Wallace, D. Barnet, D. Clarke, D.G. Laing, Analyst 119 (1994) 1997. (c) D. Barnett, D.G. Laing, S. Skopec, O.A. Sadik, G.G. Wallace, Anal. Lett. 27 (1994) 2417.
[45] A. Wanekaya, O.A. Sadik, Multicomponent analysis of alcohol vapors using integrated gas chromatography with sensor array, Sens. Act. B. 110 (1) (2005) 41–48.
[46] A. Wanekaya., O.A. Sadik, Development of tandem chromatography sensor arrays technologies, in: G.A. Grimes, E.C. Dickey, M.V. Pishko (Eds.), Encyclopedia of Sensors, vol. 2, American Scientific Publishers, Valencia, CA, 2006, pp. 333–348.
[47] C.-J. Lu, J. Whiting, R.D. Sacks, E.T. Zellers, Anal. Chem. 75 (2003) 1400.
[48] A.J. Hoffman, G. Mills, H. Yee, M.R. Hoffman, J. Phys. Chem. 96 (1992) 5546.
[49] C. Lu, E. Zellers, Anal. Chem. 73 (2001) 3449.
[50] J.J. Whiting, C. Lu, E.T. Zellers, R.D. Sacks, Anal. Chem. 73 (2001) 4668.
[51] G. Watson, E. Staples, S. Viswanathan, Environ. Prog. 22 (2003) 215.
[52] A.C. Templeton, W.P. Wuelfung, R.W. Murray, Acc. Chem. Res. 33 (2000) 27.
[53] F.E. Ahmed, Trends Anal. Chem. 22 (3) (2003) 170–185.
[54] (a) S. Bender, O.A. Sadik, Environ. Sci. Technol. 32 (1998) 788–797. (b) M. Masila, F. Yan, O.A. Sadik, Environmental biosensors for organochlorines, cyanobacteria toxins and endocrine disrupting chemicals, Biotechnol. Bioprocess Eng. 5 (2000) 407–412.
[55] J. Sherry, Chemosphere 34 (5–7) (1997) 1011–1025.
[56] M. Sisak, M. Franek, K. Hruska, Anal. Chim. Acta 311 (1995) 415–422.
[57] J. Horacek, P. Skladal, Anal. Chim. Acta 412 (2000) 37–45.
[58] F. Yan, O.A. Sadik, Anal. Chem. 73 (2001) 5272–5280.
[59] G. Marrazza, I. Chianella, M. Mascini, Biosens. Bioelectron. 14 (1) (1999) 43–51.
[60] I. Kowino, O.A. Sadik, Novel electrochemical detection scheme for probing the interactions of small molecules with DNA, Langmuir 19 (2003) 4344–4350. A. Mantovani, Toxicology, 181–182, pp. 367–370, 2002.
[61] (a) M.J. Lopez de Alda, D. Barcelo Fres, J. Anal. Chem. 371 (2001) 437–447. (b) P. Tundo, P. Anastas, D. StC. Black, J. Breen, T. Collins, S. Memoli, et al., Pure Appl. Chem. 72 (7) (2000) 1207–1228.
[62] R. Steinmetz, N.G. Brown, D.L. Allen, R.M. Bigsby, N. Ben-Jonathan, Endocrinology 138 (1997) 1780–1786.

[63] S.F. Arnold, D.M. Klotz, B.M. Collins, P.M. Vonier, L.J. Guillette, J.A. McLachlan, Science 272 (1996) 1489–1492.
[64] M.N. Jacobs, D.F.V. Lewis, P. Nutr. Soc. 61 (2002) 105–122.
[65] M. Ngundi., O.A. Sadik, S. Suye, Y. Takashi, First comparative reaction mechanisms for estrogens and environmental hormones, Electrochem. Commun. 5 (2003) 61–67.
[66] A. Zhou, S. Kikandi, O.A. Sadik, Electrochemical degradation of quercetin: isolation and structural elucidation of the degradation products, Electrochem. Commun. 9 (2007) 2247–2256.
[67] C. Alocilja, S.M. Radke, Biosens. Bioelectron. 18 (2003) 841–846.

CHAPTER

3 Electrochemical Biosensors Based on Nanomaterials for Detection of Pesticides and Explosives

Dan Du[1,3], Abdullah Mohamed Asiri[2], Jun Wang[4] and Yuehe Lin[2,3]

[1]Key Laboratory of Pesticide and Chemical Biology of Ministry of Education, College of Chemistry, Central China Normal University, Wuhan, PR China [2]Chemistry Department, King Abdulaziz University, Jeddah, Saudi Arabia [3]School of Mechanical and Materials Engineering, Washington State University, Pullman, WA, USA [4]Pacific Northwest National Laboratory, Richland, WA, USA

3.1 Introduction

Both organophosphate (OP) pesticides and explosives are toxic compounds and arouse great environmental concerns because of their intensive and extensive use in agriculture and industries across the world [1–3]. The toxicity of OPs is due to their inhibition of enzymes (e.g., acetylcholinesterase (AChE) activities), which will lead to various diseases or even death. Serious health problems, including anemia and

Street, Sustich, Duncan and Savage. Nanotechnology Applications for Clean Water, 2nd Edition.
 DOI: http://dx.doi.org/10.1016/B978-1-4557-3116-9.00003-2

abnormal liver function, cataract development, and skin irritation, can result when people are exposed to an explosives-contaminated environment (such as 2,4,6-trinitrotoluene (TNT)) or when they drink water contaminated with explosive materials [4–6]. Therefore, simple, rapid, sensitive, and field-deployable sensors for detecting OPs or TNT are highly desirable for on-site monitoring of these toxic pollutants. This will provide time for proper procedures to be followed to mitigate dangers or assess and mitigate TNT-contaminated sites. On-site TNT analysis can also reduce both time and cost for cleanup and permit real-time decision-making.

Various analytical methods have been developed to detect OPs and TNT in environmental samples. Both TNT and OPs are small organic compounds and are routinely analyzed using analytical techniques, such as gas chromatography–mass spectrometry (GC–MS) or liquid chromatography (LC)–MS [7,8]. However, these analyses for OPs and TNT are generally performed at centralized laboratories, requiring extensive labor and analytical resources with a lengthy turnaround time. Such disadvantages of these methods limit their applications for in-field rapid analyses. Some new, simple, and sensitive analytical methods with real-time output and a low cost are urgently needed.

Electrochemical techniques offer a simple and inexpensive approach for rapid and on-site monitoring of OPs and TNT [9–12]. They have been widely used in various chem/biosensors because of their simplicity, low cost, high sensitivity, and miniaturization [1,9–18]. The sensitivity of these types of chem/biosensors can be further enhanced by using various nanotechnology-based amplifications [2,7]. Carbon nanotubes (CNTs) represent a new class of nanomaterials and are composed of graphitic carbons with one or several concentric tubules. Recent studies have demonstrated that CNTs can improve the direct electron transfer reaction of some biomolecules, including cytochrome C [2], catalase [2], or nicotinamide adenine dinucleotide [19], on an electrode surface. This is attributed to their unique electronic structure, high electrical conductivity, and redox-active sites. CNTs can greatly decrease the overpotential of oxidizing or reducing some enzymatic products, e.g., hydrogen peroxide [17,20] and thiocholine [1,21], because of their unique structure and electrochemical properties. Nanoparticles (NPs), such as functionalized silica NPs, carbon nanosphere and graphene, have also been developed for sensitive assays of tumor biomarkers [22–25].

In this chapter, we describe nanomaterial-based biosensors for detecting OP pesticides and explosives. Two kinds of enzymes biosensors, AChE and organophosphorus hydrolase (OPH), are summarized for OP pesticides detection. CNTs and functionalized silica NPs have been chosen for this study. The biosensors were combined with the flow-injection system, providing great advantages for on-site, real-time, and continuous detection of environmental pollutants such as OPs and TNT. The sensors take advantage of the unique electrochemical properties of CNTs, which makes it feasible to achieve a sensitive electrochemical detection of the products from enzymatic reactions at low potential. These methods offer a new environmental monitoring tool for rapid, inexpensive, and highly sensitive detection of OPs or TNT compounds.

3.2 Nanomaterials-based biosensors for pesticides

3.2.1 Biosensor based on AChE

3.2.1.1 Principle of electrochemical biosensor for OPs

The OP biosensors based on enzyme activity have been widely studied. Generally, two approaches are involved: either AChE alone or AChE combined with choline oxidase (ChO). The inhibition of AChE activity in the single and bienzyme system is monitored by measuring the oxidation or reduction current of the product of the enzymatic reaction. In the following bienzyme system (Eqs. (3.1) and (3.2)), AChE catalyzes the hydrolysis of the neurotransmitter acetylcholine into acetate and choline. The choline is subsequently converted by ChO, producing hydrogen peroxide in the presence of oxygen. Hydrogen peroxide can be detected amperometrically with different electrochemical transducers.

$$\text{Acetylcholine} + H_2O \xrightarrow{\text{AChE}} \text{Choline} + \text{Acetate acid} \tag{3.1}$$

$$\text{Choline} + O_2 \xrightarrow{\text{ChO}} \text{Betaine aldehyde} + H_2O_2 \tag{3.2}$$

In a single enzyme system (Eq. (3.3)), acetylthiocholine (ATCh) can be enzymatically hydrolyzed by AChE to thiocholine:

$$\text{Acetylthiocholine} + H_2O \xrightarrow{\text{AChE}} \text{Thiocholine} + \text{Acetate acid} \tag{3.3}$$

Thiocholine is electroactive and can be oxidized at the electrochemical transducer. The inhibition extent of AchE activity is well correlated with the amount of OPs. So the inhibition of AChE activity can be used for quantitation of OPs.

We have developed two types of CNT electrodes for sensitive detection of OPs. First, we developed an AChE/ChO-modified CNT/screen-printed electrode (SPE) as disposable sensor for on-site detection of OPs. Second, we employed the layer-by-layer (LBL) technique to assemble a multilayer of AChE on CNT for enhancing the sensitivity of the electrochemical biosensors and combined the biosensor with a flow injection system for automatic analysis of OP pesticides.

3.2.2 Biosensor based on ChO/AChE bienzyme

3.2.2.1 The preparation of OP biosensors based on immobilization of CHO/AChE bienzymes on CNTs thin-film electrode

ChO/AChE enzymes were co-immobilized onto the surface of multiwall carbon nanotubes (MWCNTs) modified SPE with the following procedure: (1) CNTs were oxidized to generate carboxylic acid groups by electrochemical techniques, e.g., applied with a potential of 1.4 V and (2) two enzymes were immobilized on the CNTs with a coupling agent such as 1-ethyl-3-(3-dimethylaminopropyl) carbodiimide (EDC). Amperometric and chronoamperometric techniques were used to measure enzyme inhibition in this study [2].

3.2.2.2 The kinetic of the inhibition of the enzyme activity

The kinetics of the enzyme inhibition by the OP insecticide, methyl parathion (MP), was studied using the AChE/ChO biosensor. Figure 3.1A shows the amperometric response of the biosensor from adding MP. It was found that the electrochemical response was rapid, and the signal reached a steady state within 30 s. The fast diffusion of the enzyme substrate and the product led to a fast response of the biosensor, which can be attributed to the membrane-free porous CNTs. When an OP insecticide, MP, was successively added to the sensing surface, the enzyme activity decreased rapidly in the first 10 min and continued to decrease slowly thereafter. An enzyme activity is used as an indicator of inhibition. Each reported value of enzyme activity is calculated as:

$$\text{Enzyme activity}(\%) = (I_i/I_0) \times 100 \tag{3.4}$$

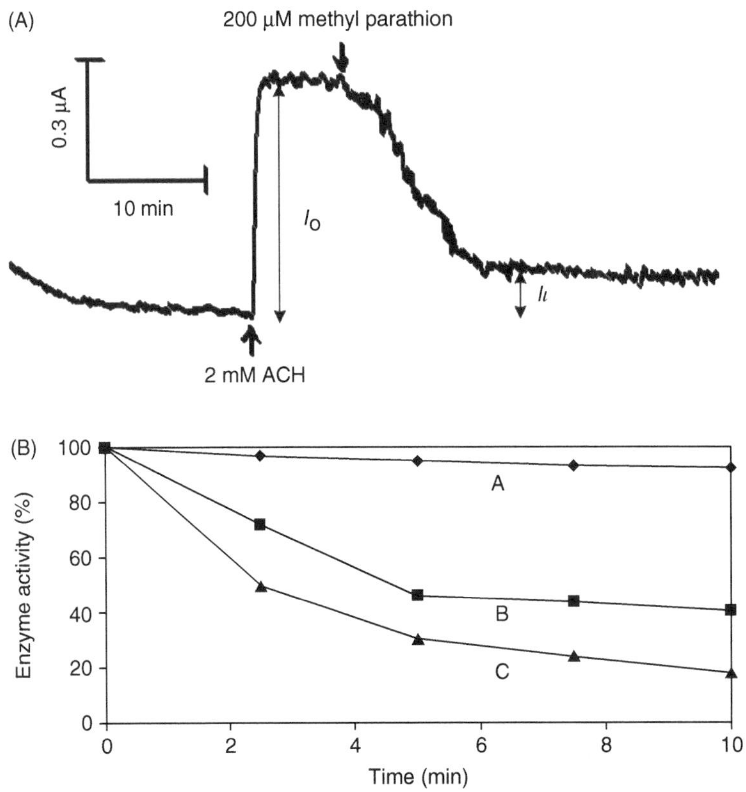

FIGURE 3.1

(A) Amperometric response of CNT-modified screen-printed biosensor for MP. Experimental conditions: 0.1 M phosphate buffer containing 0.1 M NaCl (pH 7.4); applied potential of 0.50 V. (B) Effect of inhibition reaction time on enzyme activity, measured with three OPs: A, chlorpyrifos; B, fenitrothion; and C, MP.

Source: From Ref. [2], with permission.

where I_0 is the current value at a steady state obtained in the absence of OP compounds (blank), and I_i is the current value obtained in the presence of OP compounds (sample).

Additional inhibition experiments have been performed with the biosensor using the OP insecticides chlopyrifos, fenitrothion, or MP (200 μM of each). A comparison of enzyme activity as a function of time after spiking with the three OP compounds is presented in Figure 3.1B. For all the OP compounds tested, the enzyme activity decreases with time, and the inhibition effect of OP compounds to the enzyme activity of AChE is in the following ascending order: MP > fenitrothion > chlorpyrifos.

3.2.2.3 Performance of the biosensor

The performance of the biosensor (including detection limit, dynamic range, and reproducibility) were investigated with the biosensor exposed to 1 mL of the incubation solution containing MP for 10 min before the enzyme activity change was determined. A successive incubation measurement was used for calibration experiments, and the relative inhibition percentage, RI (%), was calculated according to the following equation:

$$\mathrm{RI}(\%) = (I_{n-1} - I_n)/I_n \times 100 \tag{3.5}$$

where I is the steady-state current of the chronoamperometric response, and the subscript n represents the number of successive incubations in an MP solution. The biosensor shows good analytical characteristics for MP, including a broad dynamic linear range (up to 200 μM), high sensitivity (0.48% inhibition/μM, $r2 = 0.96$), and a low detection limit (0.05 μM). These improved characteristics are attributed to the catalytic activity of CNTs to hydrogen peroxide and the large surface area of the CNT material. Due to the low cost and ease of use of SPE, the SPE-based OP biosensors can be used for on-site rapid screening of potential pesticides contamination in water.

3.2.3 Biosensor based on LBL assembly of AChE on CNT

3.2.3.1 The preparation of multilayers of AChE on CNT-modified electrode

AChE was immobilized on the negatively charged CNT surface by alternatively assembling a cationic poly(diallyldimethylammonium chloride) (PDDA) layer and a negatively charged AChE layer (Figure 3.2). The positively charged polycation was adsorbed onto the surface of negatively charged CNT by dipping the CNT/GC electrode (Figure 3.2A) in an aqueous solution containing 1 mg/mL PDDA and 0.5 M NaCl for 20 min (Figure 3.2B). Then, the PDDA/CNT/GC electrode was rinsed with distilled water and dried in nitrogen. Using the same procedure, a layer of negatively charged AChE was adsorbed at a Tris−HCl buffer solution (pH 8.0) containing 0.2 unit/mL AChE (Figure 3.2C). Another PDDA layer was adsorbed on the top of the AChE layer using the same procedure to prevent AChE leaking from the electrode surface (Figure 3.2D) [1].

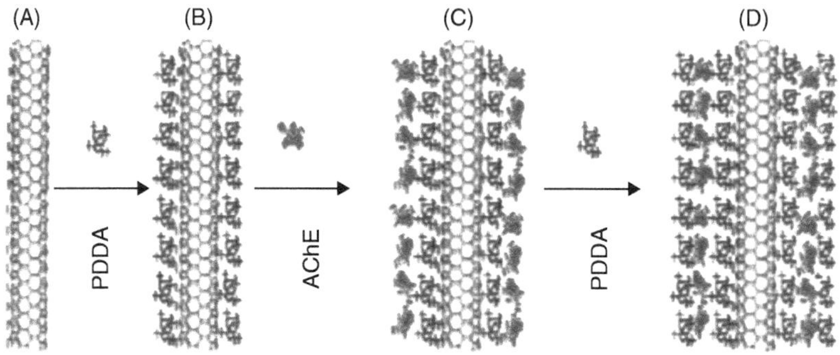

FIGURE 3.2

Schematics of LBL electrostatic self-assembly of AChE on CNT: (A) CNT/GC; (B) assembling positively charged PDDA on negatively charged CNT; (C) assembling negatively charged AChE; (D) assembling the second PDDA layer.

Source: From Ref. [1], with permission.

3.2.3.2 Flow injection amperometric detection of paraoxon with LBL-assembled biosensor

A laboratory-built flow-injection system, which consists of a carrier, a syringe pump (Model 1001, BAS), a sample injection valve (Valco Cheminert VIGI C2XL, Houston, TX), and a laboratory-built wall-jet-based electrochemical cell, was used for the amperometric measurement of OPs. The amperometric measurements were conducted at 0.15 V. All potentials are referred to the Ag/AgCl reference. The laboratory-built microelectrochemical cell based on a wall-jet (flow-onto) design (Figure 3.3A) integrates three electrodes. A laser-cut Teflon gasket is sandwiched between two acrylic blocks to form a flow-cell. The working electrode is placed into the bottom piece and is sealed with an O-ring; Ag/AgCl reference electrode and platinum wire counter electrode are placed into the upper piece inside a groove (Figure 3.3B). The solution flows onto the working electrode surface and exits through the groove to the outlet. This cell design allows quick installation of enzyme-modified working electrode.

Figure 3.4A shows the typical amperometric responses versus time during the inhibition and regeneration process of the biosensor: The amperometric response of the biosensor before and after exposure of the AChE to an OP model compound paraoxon is shown. First, two successive injections of 10 μL of 2 mM ATCh show the initial enzyme activity (peaks 1 and 2). Then 10 μL of 10^{-8} M paraoxon was injected into the cell, and the flow was stopped for 6 min. Reinjection of the 10 μL of 2 mM ATCh shows a significant decrease in enzyme activity (peaks 3 and 4). Successive incubation with 0.1 mM pyridine 2-aldoxime methiodide (PAM) and 10 mM ATCh for 2 min, respectively, resulted in recovery of enzyme activity (peaks 5 and 6). Reincubation of the biosensor with 10 μL of 10^{-10} M paraoxon caused a decrease again in the activity (peaks 7 and 8). The relative decrease in the activity

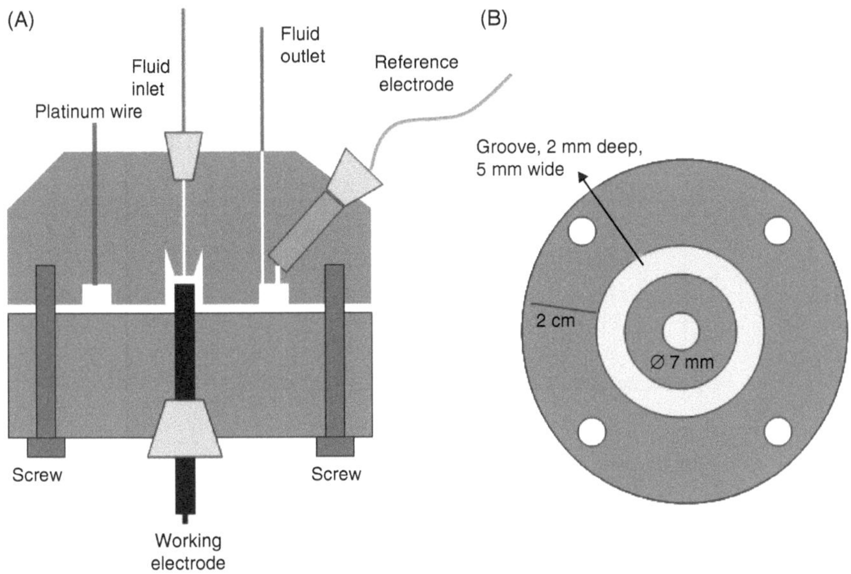

FIGURE 3.3

Electrochemical flow cell.

at the second inhibition (area of peak 7 to peak 5) is less than that obtained in the first inhibition (area of peaks 3 to 1). After the second regeneration step, the activity of AChE was recovered again (peaks 9 and 10).

3.2.3.3 Performance of the biosensor

The performance of this biosensor was studied under the optimum conditions established (Figure 3.4B). It was found that the RI of AChE activity increased with the concentration of paraoxon, ranging from 10^{-13} to 10^{-7} M, and is linearly with -log[paraoxon] at the concentration range 1×10^{-12}–10^{-8} M with a detection limit of 0.4 pM (calculated for 20% inhibition). This detection limit is three orders of magnitude lower than the covalent binding or adsorbing AChE on the CNT-modified SPE under batch conditions. The reproducibility of the biosensor for paraoxon detection was examined, and a relative standard deviation (RSD) of less than 5.6 ($n = 6$) was obtained. The biosensor can be reused as long as residual activity is at least 50% of the original value.

3.2.4 Biosensor based on OPH

A growing class of bacterial enzymes known as phosphotriesterases has recently been characterized. Unlike AChE, these enzymes catalyze the hydrolysis of OPs with a high turnover rate. The typical one is OPH, which exhibits the capability to hydrolyze a large variety of OP compounds by producing less toxic products,

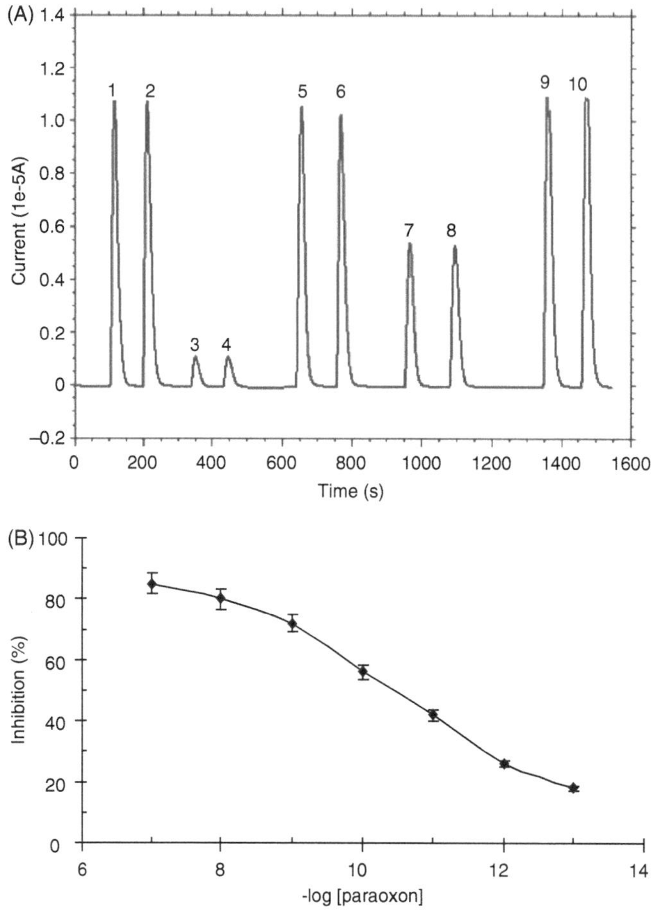

FIGURE 3.4

(A) Typical amperometric responses of PDDA/AChE/PDDA/CNT/GC biosensor during the flow injection analysis of paraoxon. Signals 1 and 2, initial enzyme activity; signals 3 and 4, enzyme activity after incubating 6 min with 10 μL of 1×10^{-8} M paraoxon; signals 5 and 6, enzyme activity after regeneration with 1 mM PAM and 10 mM ATCh; signals 7 and 8, enzyme activity after incubating 6 min with 10 μL of 1×10^{-10} M paraoxon; signals 9 and 10, enzyme activity after regeneration with 1 mM PAM and 10 mM ATCh. Note that the current versus time record was paused during the inhibition and regeneration. Flow rate, 0.25 mL/min; working potential, 150 mV. (B) Inhibition curve of the biosensor to different concentrations of paraoxon.

Source: From Ref. [1], with permission.

such as *p*-nitrophenol and diethyl phosphate [26]. Biosensors based on a catalytic reaction are superior to the inhibition mode since they can be potentially reused and are also suitable for continuous monitoring. Several types of OPH-based amperometric, potentiometric, or optical biosensors have been described [27,28].

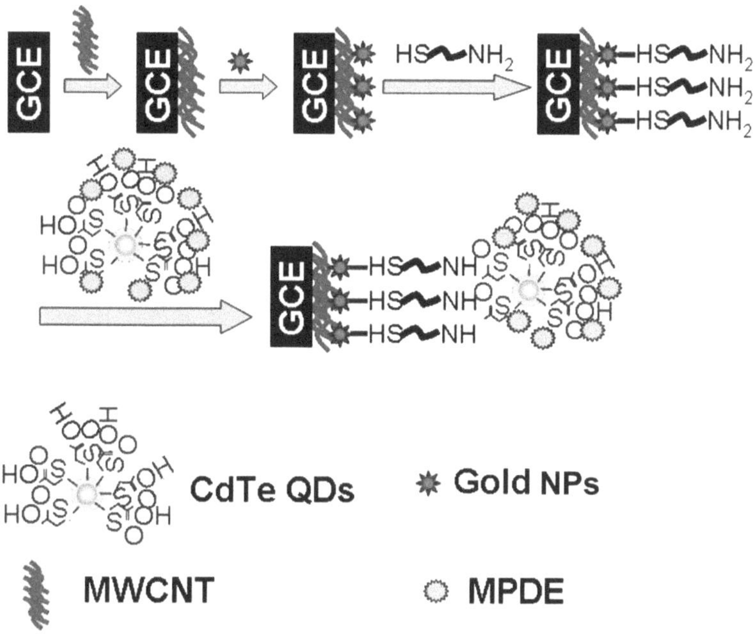

SCHEME 3.1

Preparation procedure for MP degrading enzyme biosensor.

Source: *From Ref. [29], with permission.*

We have developed an amperometric OPH biosensor for highly selective and sensitive determination of MP [29]. OPH can hydrolyze pesticides containing the P–S bond. One kind of OPH, MP degrading enzyme (MPDE), catalyzes the hydrolysis of MP to generate an equimolar amount of *p*-nitrophenol, which, in turn, is electrochemically oxidized at the working electrode poised at a fixed potential to generate a current that is proportional to the pesticide concentration. The sensitivity of the electrochemical signal can be further enhanced by using nanomaterial-based amplifications. As shown in Scheme 3.1, gold NPs were introduced by a multipotential step technique with CNTs precast on a glassy carbon electrode (GCE) to immobilize MPDE through a CdTe quantum dots (CdTe QDs) covalent attachment. It is easy to electrodeposit gold NPs, and the layer thickness can be controlled. The introduced CNTs and gold NPs significantly increased the surface area and exhibited synergistic effects toward enzymatic catalysis. CdTe QDs are further used as carriers to load a large amount of enzyme.

Figure 3.5 showed the linear scan voltammograms of MP with different concentrations at electrodes. No apparent peak was observed at Cys/Au_{nano}/MWCNT/GCE (curve a) and CdTe/Cys/Au_{nano}/MWCNT/GCE (curve b). However, when the MPDE-QD conjugate was coupled to the electrode, the MPDE-CdTe/Cys/Au_{nano}/MWCNT/GCE displayed an oxidation peak with Epa = 0.75 V (curve c).

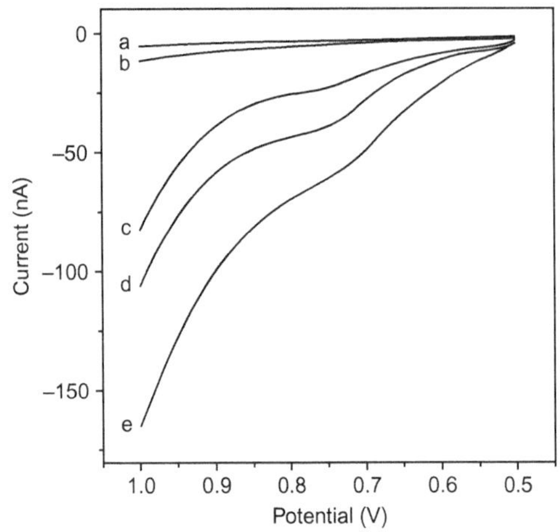

FIGURE 3.5

The linear scan voltammograms of 100 ng/mL MP at (a) Cys/Au_{nano}/MWCNT/GCE, (b) QD/Cys/Au_{nano}/MWCNT/GCE, (c) MPDE-CdTe/Cys/Au_{nano}/MWCNT/GCE, (d) 200 ng/mL MP, and (e) 400 ng/mL MP at MPDE-CdTe/Cys/Au_{nano}/MWCNT/GCE.

Source: From Ref. [29], with permission.

Obviously this peak came from the oxidation of *p*-nitrophenol, a hydrolysis product of MP, catalyzed by the immobilized MPDE. The oxidation current increased with increasing MP concentration (curves d and e). The calibration curve under the optimized experimental conditions was shown in Figure 3.6. The linear response range of the biosensor to MP concentration was described with two segments. One was from 5.0 to 200 ng/mL, and the other one was from 200 to 1000 ng/mL with correlation coefficients of 0.994 and 0.996, respectively. The detection limit was 1.0 ng/mL. Unlike AChE biosensor, MPDE is not poisoned by OP pesticides and it can be potentially reused and is suitable for continuous monitoring.

3.3 NP-based electrochemical immunoassay of TNT

3.3.1 The principle of NP-based TNT sensor

A NP label-based electrochemical immunoassay for TNT has been developed at our laboratory recently [3]. The poly(guanine) functionalized silica NPs were used as electroactive reporters for this method. The fundamental principle of the sensor relies on the specificity of the recognition element (e.g., anti-TNT antibody) to recognize and bind the explosive molecule (TNT) with the subsequent releasing of reporter-labeled analog complexes, which can be electrochemically detected. Therefore, the reporter-functionalized NP enhances the sensitivity of the method.

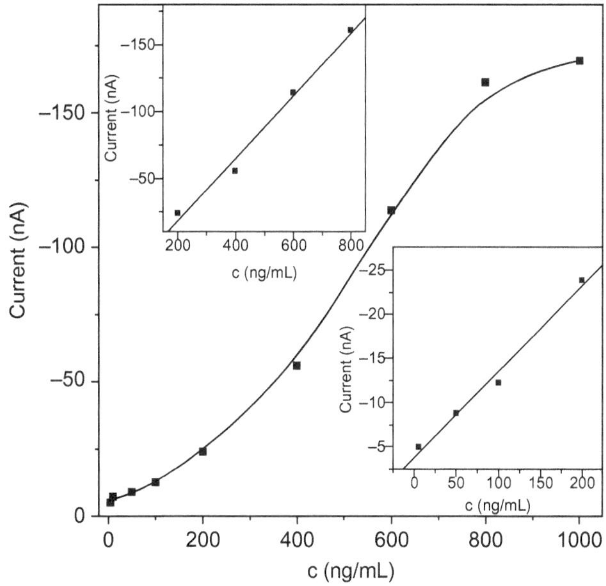

FIGURE 3.6

Response curves for MP in pH 7.0 PBS at MPDE-CdTe/Cys/Au_{nano}/MWCNT/GCE. Inset: the calibration curves for MP determination.

Source: From Ref. [29], with permission.

Scheme 3.2 illustrates the principle of an NP label-based electrochemical displacement immunoassay of TNT. This assay involves two immunoreactions: (1) anti-TNT antibodies linked to strepavidin-modified magnetic beads interact with TNT analog-poly[G]-silica NPs and form analog–antibody complexes on the surface of the magnetic beads; (2) TNT interacts with its specific antibody and replaces the analog, and it forms a TNT–antibody complex on the magnetic beads, releasing TNB-poly[G]-silica NPs into the test solution. After completing the displacement immunoassay, the solution containing the TNB-poly[G]-silica NPs is separated from magnetic beads with a magnet and introduced to the surface of the SPE and allowed to dry quickly. Then a phosphate buffer solution (PBS) solution containing $Ru(bpy)_3{}^{2+}$ is added, and a catalytic current resulting from guanine oxidation is measured [30–32]. This current is proportional to the amount of guanine on the surface of the SPE electrode, which in turn depends on the concentration of TNT in the sample. Therefore, this assay is quite suitable for quantitative analysis of TNT.

3.3.2 The analytical performance of TNT sensor

Figure 3.7 presents the typical Square-wave voltammograms (SWVs) of a TNT assay based on poly[G] functionalized silica NP labels in the presence and absence of TNT. In a control experiment (in the absence of TNT), a well-defined

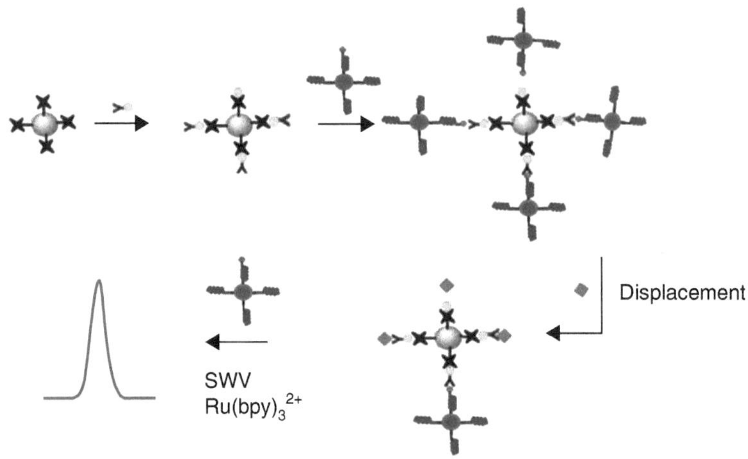

SCHEME 3.2

Schematic of the procedure and principle for competitive immunoassay of TNT on poly [G]-functionalized silica NP labels. Magnetic bead; TNT; biotin-conjugated anti-TNT antibody; Streptavidin; Silica NP; Poly (guanine); TNT analog (2,4,6-trinitrobenzene (TNB)).

Source: From Ref 3 with permission. (For interpretation of the references to color in this figure legend, the reader is referred to the web version of this book.)

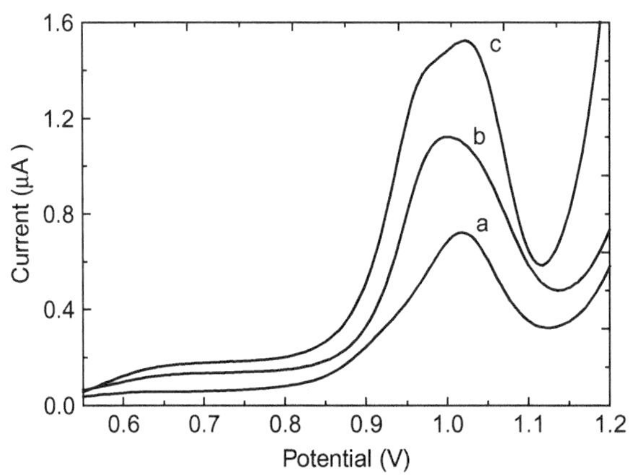

FIGURE 3.7

Square-wave voltammograms of the immunoassays obtained at SPE under different conditions: curve a, 0 ng/mL TNT, serving as a control; curve b, 1.0 ng/mL TNT; curve c, 10 ng/mL TNT. Electrochemical measurements were performed in 0.1 M PBS buffer containing 5.0 μM $Ru(bpy)_3Cl_2$.

Source: From Ref. [3], with permission.

peak (curve a) was observed on the bare SPE; the anodic current is almost identical to the one from the background current (the oxidation current of $Ru(bpy)_3^{2+}$) at the SPE. There is no catalytic current in the absence of TNT in the sample. However, a much higher peak current and a little negative shift of the peak was observed in the presence of TNT (curves b and c). The increase of the peak current and negative shift of the peak would be ascribed to catalytic oxidation of guanine, which came from TNT displacement immunoreactions during the assay. It can also be seen that the anodic current increased with the increase of TNT concentration from 1 to 10 ng/mL TNT (curves b and c, respectively). The results indicate that the assay is specific and suitable for quantification analysis.

The figure-of-merit (including dynamic range, detection limit, and reproducibility) of the assay based on a poly(guanine)-silica NP label was further examined with a wide range of TNT concentrations. Figure 3.8 shows the plot of background-subtracted peak current versus the logarithm of the concentration of TNT. As shown in Figure 3.8, a typical sigmodial curve is observed with different TNT concentrations (from 0.01 to 1.0 μg/mL). It can be seen that the current almost linearly increases with the increase of the logarithm of the TNT concentration from 0.5 to 50 ng/mL. The limit of detection (LOD) for this assay is ~0.1 ng/mL. This immunoassay method for TNT is quite sensitive and can be used to quantify a trace amount of TNT in water. Compared with other nanomaterial-based TNT sensors (e.g., gold/CNT-based), this method is more sensitive and can be combined with a flow-injection system for on-site TNT

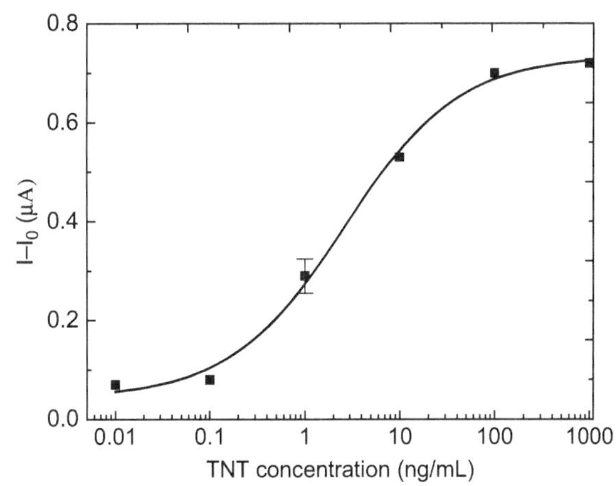

FIGURE 3.8

The plot of background-subtracted peak current versus logarithm of TNT concentration. The currents were obtained from square-wave voltammograms of the assay with different TNT concentrations (from 0.01 to 1000 ng/mL). The electrochemical measurements were carried out in 0.1 M PBS buffer containing 5.0 μM $Ru(bpy)_3Cl_2$.

Source: From Ref. [3], with permission.

detection. A series of six repetitive measurements of the 1.0 ng/mL TNT yielded reproducible SWV peaks with an RSD of 12%. The reproducibility of the method depends on the variation of displacement reaction conditions and the variations of electrodes. To improve the reproducibility of the method, it is crucial that the displacement reaction conditions (e.g., reaction temperature and time, the volume of magnetic beads) are carefully controlled. The results show that this electrochemical immunoassay can be used for sensitive quantitation of TNT.

3.4 Conclusions

We have discussed the enzyme/CNT-based amperometric biosensors and NPs-based immunosensors for sensitive detection of OP pesticides and explosives. The electrocatalytic activity of CNT leads to a greatly improved electrochemical detection of the enzymatically generated products with a low oxidation overvoltage, higher sensitivity, and stability. The simplicity and ease of operation of the enzyme biosensors shows great potential for on-site measurements.

Although this chapter only discusses the CNTs and silica NPs-based biosensors focusing on OPs and explosives, biosensors based on other nanomaterials, e.g., nanowires, quantum dots, and metal NPs for detection of environmental pollutants and toxins have been reported and documented in other review articles and book chapters for further reading [33–35]. Recently, graphene, emerging as a true two-dimensional material, has received increasing attention due to its unique physicochemical properties (high surface area, excellent conductivity, high mechanical strength, and ease of functionalization and mass production). Graphene has shown excellent performance in direct electrochemistry of enzyme, electrochemical detection of small biomolecules, electroanalysis (electrochemical sensors for bioanalysis and environmental analysis). As a competitor to CNTs, graphene has exhibited superior performance in these applications. Graphene-based electrochemical sensors and biosensors for environmental analysis and biomedical assays have been reported and summarized in recent review articles [36,37]. Overall, the nanomaterial-based biosensors open up a new avenue for simple, sensitive, on-site analysis of environmental samples.

Acknowledgments

This work is supported by the Department of Defense Strategic Environmental Research and Development Program and by Grant U01 NS058161-01 from the National Institute of Neurological Disorders and Stroke, the National Institutes of Health, and partially by the Centers for Disease Control/National Institute for Occupational Safety and Health grant R01 OH008173-01. The work was performed at the Environmental Molecular Sciences Laboratory, a national scientific user facility sponsored by the US Department of Energy (DOE) and located at Pacific Northwest National Laboratory, which is operated by Battelle for DOE under Contract DE-AC05-76RL01830.

References

[1] G.D. Liu, Y.H. Lin, Anal. Chem. 78 (2006) 835–843.

[2] Y.H. Lin, F. Lu, J. Wang, Electroanalysis 16 (2004) 145–149.

[3] J. Wang, G.D. Liu, H. Wu, Y.H. Lin, Anal. Chim. Acta. 610 (2008) 112–118.

[4] N.I. Sax, Dangerous Properties of Industrial Materials, second ed., Reinhold, New York, NY, 1963.

[5] D.L. Kaplan, A.M. Kaplan, Environ. Sci. Technol. 16 (1982) 566–571.

[6] W.D. Won, L.H. Disalvo, J. Ng, Appl. Environ. Microbiol. 31 (1976) 576–580.

[7] A. Preiss, M. Elend, S. Gerling, E. Berger-Preiss, K. Steinbach, Anal. Bioanal. Chem. 389 (2007) 1979–1988.

[8] J. You, M.J. Lydy, Int. J. Environ. Anal. Chem. 86 (2006) 381–389.

[9] J.S. Caygill, S.D. Collyer, J.L. Holmes, F. Davis, S.P. Higson, Analyst 138 (2013) 346–352.

[10] T.W. Chen, Z.H. Sheng, K. Wang, F.B. Wang, X.H. Xia, Chem.Asian J. 6 (2011) 1210–1216.

[11] D. Du, J. Wang, L.M. Wang, D.L. Lu, Y.H. Lin, Anal. Chem. 84 (2012) 1380–1385.

[12] X. Zhang, H. Wang, C.M. Yang, D. Du, Y.H. Lin, Biosens. Bioelectron. 41 (2013) 669–674.

[13] K.M. Mitchell, Anal. Chem. 76 (2004) 1098–1106.

[14] J. Wang, M. Pumera, M.P. Chatrathi, A. Escarpa, M. Musameh, G. Collins, et al., Anal. Chem. 74 (2002) 1187–1191.

[15] H. Matsuura, Y. Sato, T. Sawaguchi, F. Mizutani, Sens. Actuators B 91 (2003) 148–151.

[16] K.A. Joshi, M. Prouza, M. Kum, J. Wang, J. Tang, R. Haddon, et al., Anal. Chem. 78 (2006) 331–336.

[17] S.H. Chen, R. Yuan, Y.Q. Chai, L.Y. Zhang, N. Wang, X.L. Li, Biosen. Bioelectron. 22 (2007) 1268–1274.

[18] A.A. Ciucu, C. Negulescu, R.P. Baldwin, Biosen. Bioelectron. 18 (2003) 303–310.

[19] M. Musameh, J. Wang, A. Merkoci, Y.H. Lin, Electrochem. Commun. 4 (2002) 743–746.

[20] D.R.S. Jeykumari, S.S. Narayanan, Nanotechnol. 18 (2007)Art. No. 125501.

[21] J. Wang, G.D. Liu, M.H. Engelhard, Y.H. Lin, Anal. Chem. 78 (2006) 6974–6979.

[22] K.A. Joshi, J. Tang, R. Haddon, J. Wang, W. Chen, A. Mulchandani, Electroanalysis 17 (2005) 54–58.

[23] J. Wang, G.D. Liu, Y.H. Lin, Small 2 (2006) 1134–1138.

[24] D. Du, Z.X. Zou, Y. Shin, J. Wang, H. Wu, M.H. Engelhard, et al., Anal. Chem. 82 (2010) 2989–2995.

[25] D. Du, L.M. Wang, Y.Y. Shao, J. Wang, M.H. Engelhard, Y.H. Lin, Anal. Chem. 83 (2011) 746–752.

[26] S.B. Bird, T.D. Sutherland, C. Gresham, J. Oakeshott, C. Scott, M. Eddleston, Toxicology 247 (2008) 88–92.

[27] A. Mulchandani, P. Mulchandani, W. Chen, J. Wang, L. Chen, Anal. Chem. 71 (1999) 2246–2249.

[28] L. Viveros, S. Paliwal, D. McCrae, J. Wild, A.A. Simonian, Sens. Actuators B 115 (2006) 150–157.

[29] D. Du, W. Chen, W. Zhang, D. Liu, H. Li, Y.H. Lin, Biosens. Bioelectron. 25 (2010) 1370–1375.
[30] P.M. Armistead, H.H. Thorp, Anal. Chem. 72 (2000) 3764–3770.
[31] P.M. Armistead, H.H. Thorp, Anal. Chem. 73 (2001) 558–564.
[32] L.P. Zhou, J.F. Rusling, Anal. Chem. 73 (2001) 4780–4786.
[33] G. Liu, Y.H. Lin, Talanta 74 (2007) 308–317.
[34] J. Wang, G. Liu, Y. Lin, Nanotubes, nanowires, and nanocantilevers in biosensor development, in: Challa S.S.R, Kumar (Ed.), Nanomaterials for Biosensors, Nanotechnologies for Life Sciences, vol. 8, Wiley-VCH, Weinheim, Germany, 2007, pp. 56–100. Chapter 3.
[35] J. Wang, Small 1 (2005) 1036–1043.
[36] Y.Y. Shao, J. Wang, H. Wu, J. Liu, I.A. Aksay, Y.H. Lin, Electroanalysis 22 (2010) 1027–1036.
[37] D. Du, Y.Q. Yang, Y.H. Lin, MRS Bull. 12 (2012) 1290–1296.

CHAPTER

4 Dye Nanoparticle-Coated Test Strips for Detection of ppb-Level Ions in Water

Yukiko Takahashi

Department of Civil and Environmental Engineering, Nagaoka University of Technology, Nagaoka, Niigata, Japan

4.1 Introduction

Harmful inorganic substances have toxic and nondegradable properties that pose a particular hazard to bioorganisms and the ecosystems [1]. The World Health Organization (WHO) has recommended the water quality guidelines to be <50 ppb (μg/L) for toxic inorganic substances, including cadmium, lead, chromium(VI), mercury, selenium, and arsenic [2]. In line with this guideline, strict regulations for industrial effluents, drinking water, and natural waters have been adopted worldwide. High performance analytical instruments, such as atomic absorption spectroscopy (AAS), inductively coupled plasma optical emission spectrometry (ICP–OES), inductively coupled plasma mass spectrometry (ICP–MS), are extensively used as

Street, Sustich, Duncan and Savage. Nanotechnology Applications for Clean Water, 2nd Edition.
 DOI: http://dx.doi.org/10.1016/B978-1-4557-3116-9.00004-4

standard methods to determine trace levels of contaminant concentration. However, in addition to the costly initial/running expenses of instruments, specific technical skills are required for machine operation. Moreover, sample collection, transportation to the analytical laboratories, and quite-complicated sample pretreatment may take time. On-site analytical methods for harmful ions are desired for quick monitoring of industrial wastewater, evaluation of drinking water, urgent assessment of well water in times of disaster, as well as in environmental education in schools [3]. Despite the increasing demands for a simple test method of inorganic ions, the sensitivities of commercial test kits are insufficient to meet the criteria of water quality [4], limiting their usages to qualitative or semi-quantitative tests. Additionally, real water samples frequently contain interfering substances, and therefore high selectivity is required. Some of the optical sensors and ion-selective electrodes show quick response and high sensitivity, but are insufficient in selectivity for a target metal ion particularly in the presence of high levels of interfering ions [5]. Pretreatment of samples, including selective separation and enrichment of trace analyte, is recommended to attain high sensitivity and more reliable monitoring of water quality [6–8]. Solid phase extraction is the preferred technique for concentration of analytes. This technique combined with colorimetric determination has been proposed as a simple analytical method, where analyte ions form intensely colored complexes with the reagent fixed on the solid support [9,10].

4.2 Fundamental concept of dye nanoparticle-coated test strip

4.2.1 Structural features of dye nanoparticle-coated test strip

Recently, we reported a simple and versatile method for preparation of dye nanoparticle-coated test strips (DNTSs) for detection of inorganic ions [11]. A thin dye nanoparticle/nanofiber layer (400 nm–2 μm) located on the top surface of the support membrane provides a remarkably concentrated signaling surface composed of a 100% reagent layer. Therefore, almost all signals are available efficiently (Figure 4.1A), but in most conventional test papers, indicator dyes are distributed over the entire support (Figure 4.1B). For that reason, the signal obtained from the top surface must be less intense. It is noteworthy that the present membranes were prepared without any additives such as coating polymers or modifiers. Yet the reagent is not removed from the support by rubbing with a finger or by immersion into water. The cross-sectional SEM image indicates that the thickness of the dye layer is <1 μm (Figure 4.1C). Thus it provides a remarkably concentrated signaling surface composed of 100% pure indicator dye. In contrast, the surface concentration of dye is rather low in conventional test strips prepared by soaking of dye solution.

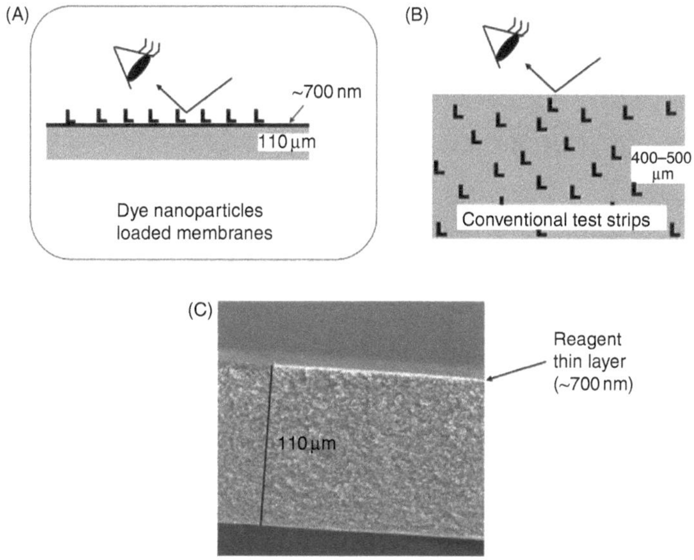

FIGURE 4.1

Conceptual illustrations of DNTSs (A) and conventional test strips (B), and a cross-sectional SEM image of DNTS (C).

Source: *Reproduced from Ref. [11]. Copyright WILEY-VCH Verlag GmbH & Co. KGaA. Reproduced with permission.*

4.2.2 Simple yet versatile fabrication methods of DNTSs

We designed two fabrication methods of DNTSs for the purpose of producing a large number of test strips with various organic indicator dyes. These methods, including two different preparation processes of nano-dispersion solution and simple filtration process of the dispersion with a fine membrane filter having microscopic pores, are shown in Figure 4.2.

In order to fabricate DNTSs with hydrophobic indicator dyes, a simple process referred to as the "reprecipitation method" [12,13] was proposed for the preparation of aqueous dye nano-dispersion (Figure 4.2, top). The typical preparative process involves injection of a water miscible organic solution of a dye compound into water under vigorous stirring. Dispersion of nano-sized dye compounds is formed immediately upon combining the aqueous solution. The growth time of the nano-dispersion solution was determined to be the time until the average size of dye nanoparticles become more than ca. 100 nm. After the growth period ended, the nano-dispersion is simply passed through with a fine membrane filter by suction. Then, by surface filtration mechanism, dye nanoparticles are uniformly and firmly attached to a membrane filter. The thicknesses of dye nanoparticle layers were experimentally observed to be between 400 and 700 nm, which

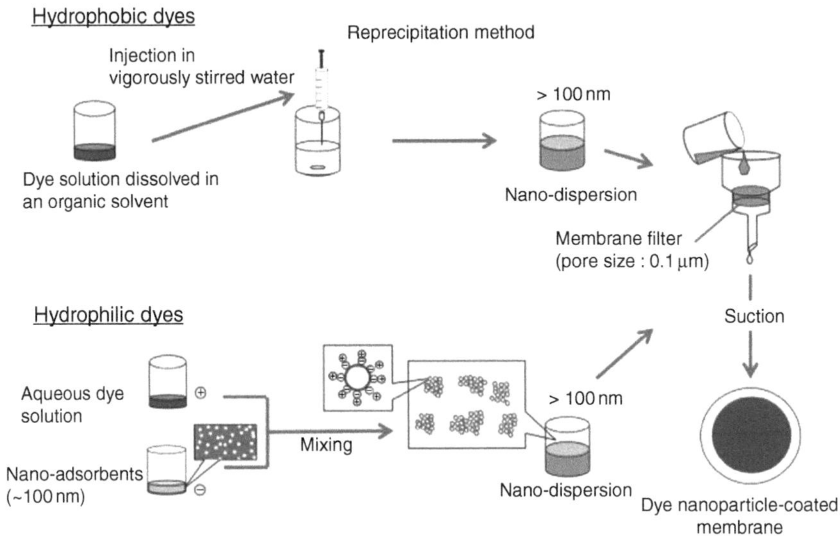

FIGURE 4.2

Two preparative methods of DNTSs for hydrophobic and hydrophilic dyes.

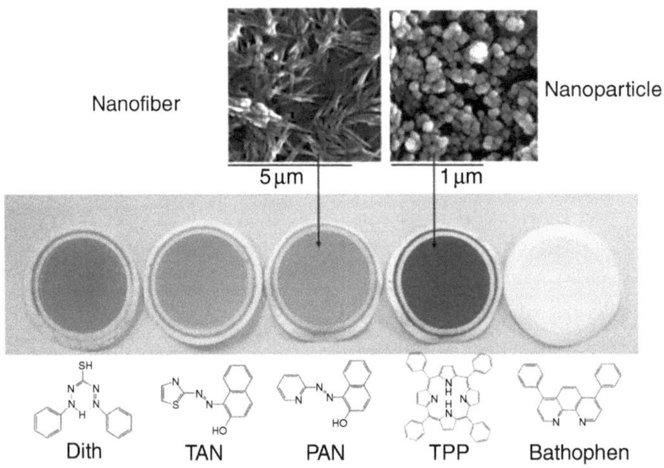

FIGURE 4.3

Photographs of dye nanoparticles/fibers loaded DNTSs and the SEM images.

Source: Reproduced from Ref. [11]. Copyright WILEY-VCH Verlag GmbH & Co. KGaA. Reproduced with permission.

depend largely on the amount of reagents loaded as well as on the molecular size (weight). The morphology of the nanoparticles varies from particle to fibrous depending on the molecular structure (Figure 4.3).

On the other hand, the "reprecipitation method" is inapplicable to hydrophilic charged dyes such as reagents having sulfone groups and pyridinium groups. For these hydrophilic indicator dyes, nanoscale ion exchangers are used as dye absorbents to produce dispersions of hydrophilic dye/nano absorbent nanocomposites. In principle, a charged dye adsorbs onto the surface of nano-adsorbent having the opposite charges (1–100 nm in diameter), followed by decrease in the surface charge of the adsorbent, and then aggregation occurs (Figure 4.2, bottom). A preparative procedure of hydrophilic dye/nano absorbent loaded DNTS is technically easy, only by mixing individual solutions under the appropriate pH.

For example, 10 mL of TMPyP/silica nanocomposite aqueous solution was prepared by mixing 100 μL of 2 mM TMPyP and 4×10^{-5} wt% of silica (ca. 10–14 nm in diameter) at pH 7.8 with 0.01 M Tris buffer solution [14]. Similarly, the growth time of nanocomposite in the aqueous solution was decided as the time until the average size of nanocomposites become more than ca. 100 nm. Then, by filtration in the similar way to that of hydrophobic dyes, roughly 100% of the nanocomposite was coated as a thin layer onto the surface of the membrane filter through surface filtration. Some metal oxides and modified latex nanoparticles, which are smaller than 100 nm in diameter and having sufficient positive or negative surface charges in aqueous solution, are promising candidates as nano-adsorbents for fabricating nanocomposite-coated DNTSs. The thicknesses of dye nanoparticle layers were experimentally observed to be ca. 1–2 μm, which depends on the amount and the size of the nano-adsorbent used.

A wide variety of analytical dye compounds can be converted as nanoparticles or nanocomposites in aqueous solution by these two methods and then DNTSs can be simply fabricated from them. The versatility of the preparative procedures of DNTSs enables a screening test of reagents against one target ion. Another outcome of the versatility is that DNTS is a novel platform technology to produce highly sensitive test strips for any target with common dye indicators.

4.2.3 Detection characteristics with DNTS

The DNTS is applicable not only to immersion test but also to filtration enrichment in which target ions are concentrated by passing a sample solution through it (Figure 4.4). Immersion test is just the same detection way as using any ordinary test papers, such as urine, pH, and ions, but it takes longer to complete sufficient color development with DNTS than those with other test papers. The major reason for this is that DNTSs detect a very low concentration of a target ion at ppb level. In addition, DNTSs are water permeable and insoluble in aqueous samples, and hence trace ions are enriched on them simply by filtration of the sample solution. Filtration enrichment is a unique approach provided by DNTS and based on the fact that the dye nanoparticle layer acts as a solid phase feasible to extract a target ion mainly by complex formation between a target ion and indicator dye. Generally, filtration enrichment is more sensitive than immersion test, and one of the examples performed by PAN (1-(2-pyridylazo)-2-naphthol) nanofiber DNTS is introduced in Section 4.4.1. Moreover,

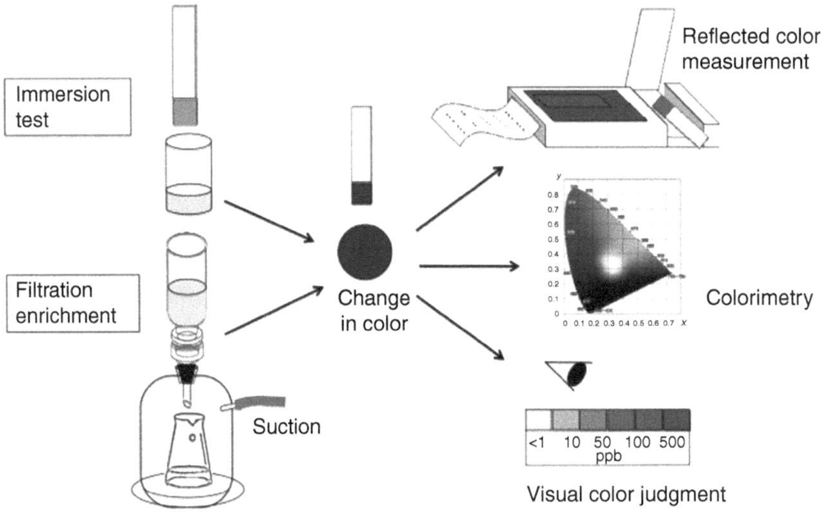

FIGURE 4.4

Two detection modes for sensing ions with DNTSs.

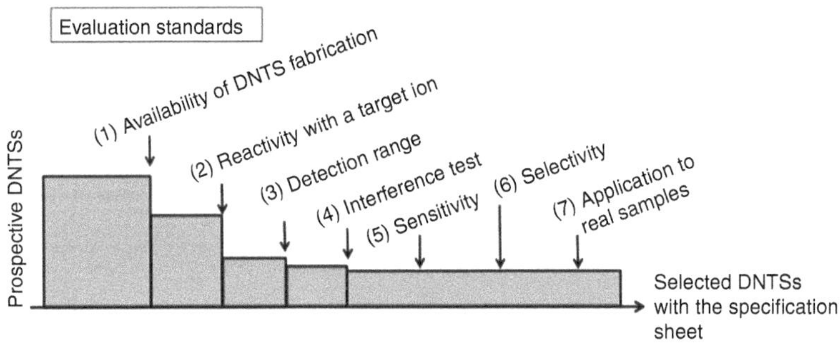

FIGURE 4.5

Seven evaluation standards for screening test of DNTSs against a particular ion.

filtration enrichment enables highly selective detection by optimizing the reaction conditions, such as solution pH, masking reagent, and flow rate, and the example carried out by dithizone nanofiber DNTS, described in Section 4.4.2.

4.3 The strategy to produce a suitable DNTS for a target ion

The versatility of the preparative procedures of DNTSs enables a screening test of a lot of reagents against a particular target. Figure 4.5 explains how to select a suitable DNTS among prospective DNTSs consisting of various candidate

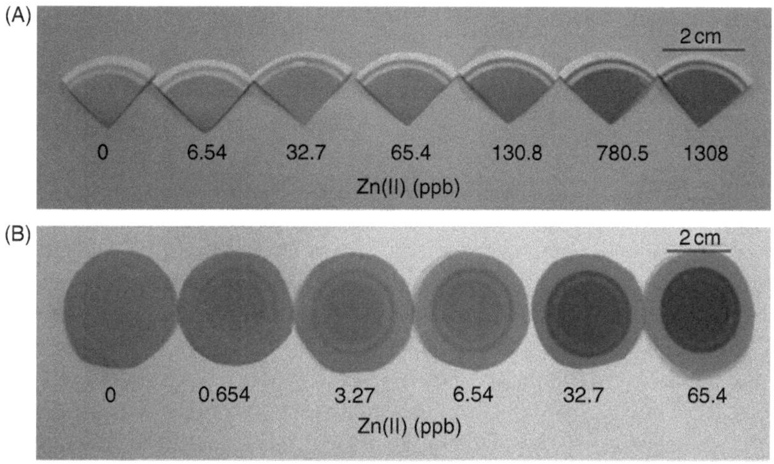

FIGURE 4.6

Detection of Zn^{II} by PAN nanofiber DNTS via immersion test and filtration enrichment. (A) Immersion test: A piece of test strip was dipped into 10 mL of aqueous solution containing Zn^{II} at pH 8.4 for 15 min. (B) Filtration enrichment: 100 mL of aqueous solution containing Zn^{II} at pH 8.4 was filtrated through the PAN membrane under ca. 6.9 mL/min flow rate.

Source: Reproduced from Ref. [11]. Copyright WILEY-VCH Verlag GmbH & Co. KGaA. Reproduced with permission.

reagents. Candidate dyes have been known and reported as sensitive indicators for a target ion in aqueous solution or organic solvent. For example, there were 15 candidates at the start of development of DNTSs for Cd(II) detection, finally 5-Br-PADAP nanoparticle DNTS and TMPyP/silica nanocomposite DNTS were selected through seven evaluation standards.

4.4 Detection of harmful ions in water with DNTSs

4.4.1 PAN nanofiber DNTS for Zn(II) detection

Membranes coated with PAN nanofibers were applied as test strips for the detection of Zn^{2+} in a test solution (pH 8.4) [11]. Figure 4.6A shows the color changes of PAN-coated strips in immersion test. Notably 65 ppb of Zn^{2+} was detected by a naked eye color test. Sub-ppb concentrations of Zn^{2+} were successfully detected by filtration enrichment of 100 mL of the sample solution (Figure 4.6B). This filtration enrichment procedure amplifies the signal intensity capable of eye detection of ppb-level metal ions. Practically, leaking of reagents out of the sample solution was observed to be negligible during dip test and sample filtration procedures. We also monitored the color change and the relative color intensity with reflectance absorption spectrometry. The peak at $\lambda_{max} = 550$ nm indicates that neutral $[Zn(PAN)_2]$ is formed on the membrane filter as the major species. The increasing peak intensity with increasing $Zn^{(II)}$ concentration enables quantitative determination of test samples by comparison with the calibration curve.

4.4.2 Dithizone nanofiber DNTS for Hg(II) detection

We demonstrated dithizone nanofiber thin film for highly sensitive and selective detection of Hg(II) [15,16]. Dithizone nanofiber was prepared by the reprecipitation method. Simply by filtration of the dispersion through the membrane filter dithizone nanofiber thin film of ca. 440 nm in thickness was deposited firmly and uniformly on one side of the membrane filter. Determination of Hg(II) at ppb level was achieved by filtration enrichment of a sample solution and successive colorimetric analysis. Consequently, Hg(II) was concentrated in the film as reddish brown complexes at pH 2.7. More than 90% of 10 ppb Hg(II) was retained in the film at the filtration rate of 1.3–9.3 mL/min. The presence of Na^+ (10,000 ppm), K^+ (5000 ppm), Ca(II) (5000 ppm), Al(III) (500 ppm), Cr(III) (200 ppm), Mn(II) (100 ppm), Fe(III) (100 ppm), Ni(II) (100 ppm), Cd(II) (10 ppm), and Pb(II) (100 ppm) in the acidic solution at pH 2.7 did not interfere with the detection of 10 ppb of Hg(II) at all (Table 4.1). The addition of the masking reagent, EDTA was very effective for Cu(II) and Zn(II) by forming water-soluble complex and passing through dithizone DNTS. Prefiltering, in

Table 4.1 Tolerance Limits on the Determination of 10 ppb of HgII[a]

Cations	Concentration (ppm)	Recovery (%)	Remarks
Na^+	10,000	99	[14]
K^+	5000	101	[14]
$Ca^{(II)}$	5000	99	[14]
$Al^{(III)}$	500	102	[15]
$Cr^{(III)}$	200	95	[15]
$Cr^{(VI)}$[b]	3	98	[15]
$Mn^{(II)}$	100	99	[15]
$Fe^{(III)}$	100	98	[14]
$Ni^{(II)}$	100	104	[15]
$Cd^{(II)}$	10	100	[14]
$Pb^{(II)}$	100	100	[14]
$Cu^{(II)}$[c]	6.4	102	[14]
$Zn^{(II)}$[c]	100	98	[14]
$Fe^{(II)}$[d]	70	102	[15]
Ag^+[e]	0.05	105	[15]

[a]0.01 mol/dm^3 glycine buffer (pH 2.2–2.9) was added.
[b]When 0.01 mol/dm^3 ascorbic acid was added, the allowance value in the presence of $Cr^{(VI)}$ was identical to that of $Cr^{(III)}$.
[c]2×10^{-4} mol/dm^3 EDTA was added.
[d]5 mM $KBrO_3$ was added.
[e]Pretreatment of a sample solution by filtration with a membrane filter (0.1 μm open pores) after the addition of 1×10^{-4} mol/dm^3 NaI.
Source: Reproduced from Ref. [15]. Copyright: The Japan Society for Analytical Chemistry. Reproduced with permission.

combination with the addition of iodine ion, also effectively masked Ag(I) interference. The interference from Cr(VI) and Fe(II) was also eliminated based on oxidation−reduction reaction. The present method succeeded in determining simulated wastewater, river water, and seawater spiked with 10 ppb Hg(II). Calibration curves for Hg(II) were obtained when the sample volume was 100 mL, and the 3σ detection limit was calculated to be 0.057 ppb ($n = 10$) by a thin layer chromatography (TLC) scanner at 500 nm.

Some other DNTSs were also effective in detecting targets at ppb level, for instance, TMPyP/silica nanocomposite DNTS for the determination of Cd(II) and Pb(II) ions by filtration enrichment [14], 5-Br-PADAP nanoparticle DNTS for Cd (II) by improved immersion test [17].

4.5 Conclusions and future perspectives

The attractive features of DNTS with a thin layer composed of dye nanoparticles or nanocomposites are its simplicity, extremely high sensitivity, and versatility. The versatility in particular means that DNTS is a novel platform to produce highly sensitive test strips for any target with common indicator dyes. In order to realize an "on-site" detection system by the present nanoparticle/fiber-based chemical sensors, a compact kit could be created, which would include a water sampler, pre-filtration membrane to remove insoluble particles, appropriate buffer, and masking reagents. A syringe type sampling system sandwiched with the membrane type color sensor is a primary candidate for simultaneous achievement of sampling, removal of interfering ions and signalization. In combination with a handy reflectometer or diffuse reflectance spectrometer, present nanostructured chemical sensors can provide a quantitative monitoring method of trace heavy metals (ppb level) that has not been achieved by conventional test kits.

Acknowledgments

This work was financially supported by the Environment Research and Technology Development Fund (B-1005) of the Ministry of the Environment, Japan and a grant (Kakenhi 25281038) from the Japan Society for the Promotion of Science (JSPS).

References

[1] W.W. Gorge (Ed.), Reviews of Environmental Contamination and Toxicology, Springer-Verlag, New York, 2000.

[2] World Health Organization, Guideline for Drinking Water Quality, fourth ed., WHO, Geneva, Switzerland, 2011.

[3] Y.A. Zolotov, V.M. Ivanov, V.G. Amelin, Test methods for extra-laboratory analysis, Trends Anal. Chem. 21 (2002) 302–319.

[4] M. Unger-Heumann, Strategy of analytical test kits, Fresenius J. Anal. Chem. 354 (1996) 803–806.

[5] I. Oehme, O.S. Wolfbeis, Optical sensors for determination of heavy metal ions, Mikrochim. Acta 126 (1997) 177–193.

[6] J. Minczewski, J. Chwastowska, R. Dybczynski, Separation and Preconcentration Methods in Inorganic Trace Analysis, Ellis Horwood Publ., Chichester, 1982, pp. 283–502.

[7] I. Liska, Fifty years of solid-phase extraction in water analysis-historical development and overview, J. Chromatogr. A 885 (2000) 3–16.

[8] E. Matoso, L.T. Kubota, S. Cadore, Use of silica gel chemically modified with zirconium phosphate for preconcentration and determination of lead and copper by flame atomic absorption spectrometry, Talanta 60 (2003) 1105–1111.

[9] J.S. Fritz, M. Macka, Solid-phase trapping of solutes for further chromatographic or electrophoretic analysis, J. Chromatogr. A 902 (2000) 137–166.

[10] D.B. Gazda, J.S. Fritz, M.D. Porter, Determination of nickel as the nickel glyoxime complex using colorimetric solid phase extraction, Anal. Chim. Acta 508 (2004) 53–59.

[11] Y. Takahashi, H. Kasai, H. Nakanishi, T.M. Suzuki, Test strips for heavy-metal ions fabricated from nanosized dye compounds, Angew. Chem. Int. Ed. 45 (2006) 913–916.

[12] H. Kasai, H.S. Nalwa, H. Oikawa, S. Okada, H. Matsuda, N. Minami, et al., A novel preparation method of organic microcrystals, Jpn. J. Appl. Phys. 31 (1992) L1132–L1134.

[13] H. Kasai, H.S. Nalwa, S. Okada, H. Oikawa, H. Nakanishi, Fabrication and spectroscopic characterization of organic nanocrystals, in: H.S. Nalwa (Ed.), Handbook of Nanostructured Materials and Nanotechnology, vol. 5, Academic Press, London, 2000.

[14] K.K. Latt, Y. Takahashi, Fabrication and characterization of a α,β,γ,δ-tetrakis(1-methylpyridinium-4-yl)porphine/silica nanocomposite thin-layer membrane for detection of ppb-level heavy metal ions, Anal. Chim. Acta 689 (2011) 103–109.

[15] Y. Takahashi, S. Danwittayakul, T.M. Suzuki, Dithizone nanofiber-coated membrane for filtration-enrichment and colorimetric detection of trace Hg(II) ion, Analyst 134 (2009) 1380–1385.

[16] Y. Takahashi, Elimination of interferences in the Mercury(II) determination with a ditizone nanofiber-coated test strip based on filtration-enrichment, Bunsekikagaku 61 (2012) 123–126.

[17] Y. Takahashi, S. Souma, Y. Wakui, Preparation of 2-(5-bromo-2-pyridylazo)-5-diethylaminophenol nanoparticle-coated test strip and its detection of trace Cd^{II} ion by immersion test, Bunsekikagaku 61 (2012) 229–234.

CHAPTER

5

Functional Nucleic Acid-Directed Assembly of Nanomaterials and Their Applications as Colorimetric and Fluorescent Sensors for Trace Contaminants in Water

Debapriya Mazumdar[1], Juewen Liu[2] and Yi Lu[3]

[1]ANDalyze, Inc. Chicago, IL, USA [2]University of Waterloo, Waterloo, ON, CA [3]Center of Advanced Materials for the Purification of Water with Systems, Department of Chemistry, University of Illinois at Urbana-Champaign, Urbana, IL, USA

Street, Sustich, Duncan and Savage. Nanotechnology Applications for Clean Water, 2nd Edition.
 DOI: http://dx.doi.org/10.1016/B978-1-4557-3116-9.00005-6

5.1 Detection of trace contaminants in water

Trace contaminants in our water resources are serious issues, as they come from both natural sources and human activities. The United States Environmental Protection Agency (US EPA) has set limits for about 90 contaminants in drinking water, which span an enormous range, from metal ions, radionuclides, volatile organics, synthetic organics, disinfectants, and their by-products, to viruses, bacteria, and other microbes. As highlighted in recent reviews, new classes of contaminants are added to the list every year [1]. These contaminants pose numerous health risks and their detection and quantification are essential to assess the risks, design removal strategies, and evaluate the effectiveness of the removal technology. The low quantities and high interference are the major obstacles associated with the detection and quantification.

To detect and quantify different contaminants, a number of analytical techniques are used. For example, inductively coupled plasma-mass spectrometry (ICP-MS) is generally used for the detection of inorganic metal species, liquid chromatography (LC)/MS is widely used for detection of organics, and matrix-assisted laser desorption/ionization (MALDI)-MS is increasingly being used for the detection of microorganisms [2]. These types of techniques are very sensitive and can detect a number of contaminants simultaneously; however they require expensive instrumentation and skilled operators, making on-site, real time detection difficult. In order to fulfill this demand, a number of materials that recognize a certain class of contaminants or a particular molecule have been developed into sensors that are often easy to use, inexpensive, and portable. In spite of this, there exists a need to obtain a general technology that incorporates both target recognition and signal generation for the rational design of sensors that can detect a wide range of contaminants. Recent development of functional nucleic acids and their application in directed assembly of nanomaterials such as gold nanoparticles and quantum dots have met the needs.

5.2 Functional nucleic acids for molecular recognition

Nucleic acids have recently emerged as an important platform for selective molecular recognition, one major requirement for sensors. Long considered as passive molecules for the storage of genetic information, RNA and DNA molecules with catalytic function similar to protein enzymes were discovered in the early 1980s and 1990s, respectively [3–5]. These enzymes are called ribozymes (catalytic RNA) and deoxyribozymes or DNAzymes (catalytic DNA). The nucleic acid enzymes usually require a metal ion co-factor to perform their catalytic function and can be tailored to be specific for a particular metal ion. In addition, nucleic acids that bind to a target molecule with high specificity and affinity (analogous to protein antibodies) have also been obtained, and these are called aptamers [6–9]. Nucleic acid enzymes and aptamers have also been fused to form a new class of allosteric enzymes called aptazymes [10,11]. Collectively, the nucleic acid enzymes, aptamers, and aptazymes are termed functional nucleic acids.

As a major component of sensors, nucleic acids possess many advantages. First, DNA/RNA targeting essentially any molecule of choice can be obtained through combinatorial selections [6–8,12], providing a unique opportunity to construct a general sensing platform for a broad range of analytes. This process is described in detail in the next section. Second, nucleic acids, particularly DNA, are very stable and can be denatured and renatured many times without losing their binding abilities, allowing a long shelf life. Third, nucleic acids have predictable base pairing interactions, which have been proven to be very useful for rational sensor design; such rational designs are difficult when making protein or organic molecule based sensors. Finally, DNA with a broad range of chemical modifications can be chemically synthesized with relatively low cost. These properties make DNA/RNA ideal candidates to create sensors. The examples discussed in this chapter mainly focus on DNA as the sensing molecule because DNA is much more stable than RNA and also less expensive, thus making it a more desirable candidate. It should however be noted that a large number of RNA aptamers and ribozymes are known and have been utilized to construct sensors. The stability of these nucleic acids can be further improved by chemical modifications and using nucleic acid analogs.

5.2.1 In vitro selection of functional nucleic acids that are selective for a broad range of target analytes

Although a number of naturally occurring ribozymes have been discovered in nature [3,4], DNAzymes and aptamers are obtained by a combinatorial biology technique called in vitro selection or systematic evolution of ligands by exponential enrichment (SELEX) [6–8,12]. This technique can be used to obtain nucleic acid sequences that recognize a target contaminant with sensitivity and specificity. Figure 5.1A is a schematic representation of the selection process. A large pool of nucleic acid sequences represented by the colored objects (approximately 10^{14}–10^{16} different sequences) is incubated with a target of interest in each round of selection. The "winner sequences," which bind to the target analyte (in the case of aptamer selection) or catalyze a reaction in the presence of the target (in the case of nucleic acid enzyme selection), are separated by various techniques such as gel

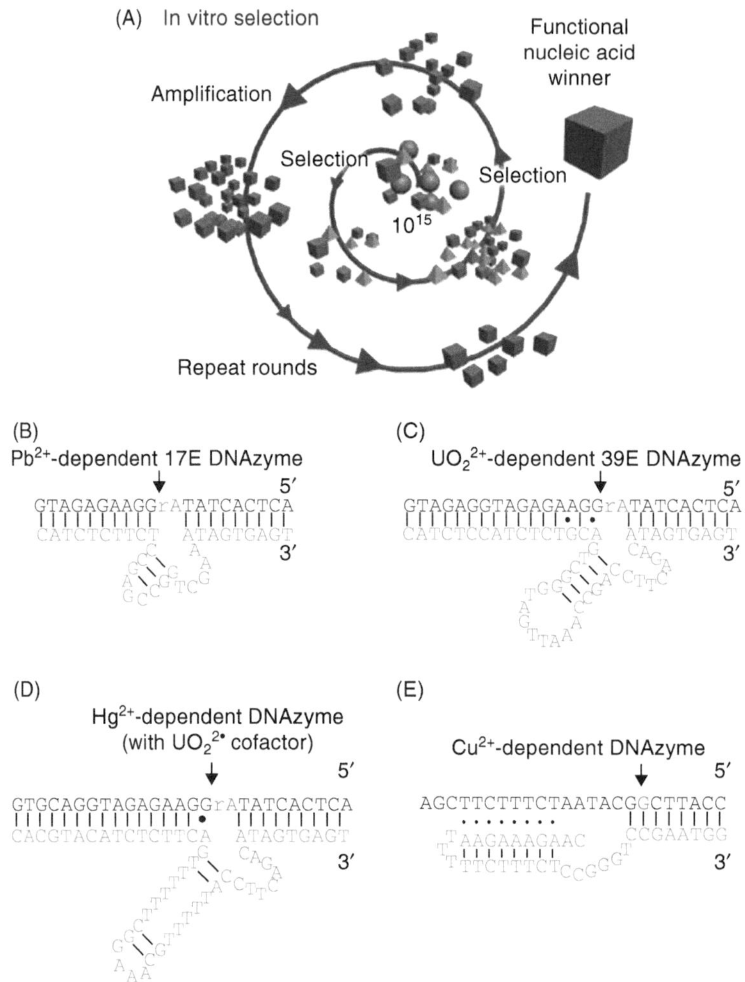

FIGURE 5.1

In vitro selection of functional nucleic acids. (A) Schematic depiction of general in vitro selection scheme for obtaining functional nucleic acid that interacts with a specific target (contaminant). (B-E) Predicted secondary structures of selected metal specific RNA cleaving DNAzymes obtained by in vitro selection. The black strands represent the substrate and the green strands are the enzyme. The cleavage site is depicted by the black arrow.

electrophoresis, column separation, and capillary electrophoresis. These "winners" are amplified using polymerase chain reactions (PCR) and used for the next round of selection. During each round of selection, the stringency can be increased by decreasing the interaction time between the target and the nucleic acid, or by decreasing the concentration of the target. Iterative rounds of selection are continued until the pool is sufficiently enriched with sequences of desired sensitivity and specificity (represented by blue cubes in Fig. 5.1A). This technique is particularly powerful as it provides a method for improving the specificity for a given target by incorporating rounds of negative selection, wherein the pool is incubated with potentially competing targets and the sequences that interact with these are removed from the pool [13]. Finally, the winner molecules that are isolated are identified by sequencing, and after some further biochemical and/or spectroscopic characterization they are used for different applications, particularly sensing [12,14–20], therapeutics [21–24], materials science, and nanotechnology [25–27].

5.2.2 Analytes or contaminants recognized selectively by functional nucleic acids

In vitro selection has been used to obtain a number of metal specific DNAzymes, such DNAzymes that are dependent on Pb^{2+} [28,29], Zn^{2+} [30], Co^{2+} [13], Cu^{2+} [31], UO_2^{2+} [32], Hg^{2+} [33], As^{5+} [33], some of which have been converted into fluorescent and colorimetric sensors as described in the following sections. A number of these metal ions, notably Pb^{2+}, Hg^{2+}, As^{5+}, are heavy metalions that are particularly toxic; UO_2^{2+} is a radionuclide. The maximum contamination level of these metal ions in drinking water is strictly regulated by the US EPA and few sensors can detect metal ions below those levels, and selectivity of those sensors should also be improved in order to be practically applicable. Therefore the utility of DNAzymes as toxic metal sensors is of great importance. The predicted secondary structures of a few DNAzymes are shown in Fig. 5.1B–E. The strands in green represent the enzyme and the strands in black are the nucleic acid substrate. All the DNAzymes are RNA cleaving enzymes that catalyze the cleavage of a single ribolinkage (represented by the arrow) embedded in the DNA substrate. Some of the fastest DNAzymes have a catalytic efficiency (k_{cat}/K_m) of $10^9\ M^{-1}\ min^{-1}$ [34], rivaling that of protein enzymes and thus they are ideal for fast sensing.

The list of aptamers obtained by in vitro selection is even longer, and more importantly the analytes recognized are far more diverse and include small molecules, antibiotics, proteins, nucleotides, and even viruses and bacteria cells and other microbes. As the nucleic acid equivalent of antibodies, this flexibility in the choice of targets for which aptamers can be obtained is a competitive advantage over antibodies for sensing applications [35,36]. Antibodies, on the other hand, cannot be raised against molecules too small to generate enough binding repertoires (e.g., metal ions not associated with any chelators), or compounds or proteins with poor immunogenic properties or with high toxicity. An online searchable aptamer database that contains detailed information of aptamer sequences for different analytes has been constructed by Ellington and coworkers [37]. Although a large majority of research in the field of aptamers is devoted to aptamers for therapeutic applications [21–24], their utility in nanotechnology and sensing has been widely explored [12,14–20].

Table 5.1 Partial List of Literature-Reported Functional Nucleic Acids that Target Water Contaminants or Water Contaminant Candidates

Contaminant Type	Examples and References
Metal ions	Pb^{2+} [28,29], Cu^{2+} [31], UO_2^{2+} [32], Hg^{2+} [33,38,39], As^{5+} [33], Zn^{2+} [30]
Radionuclides	UO_2^{2+} [32]
Toxins	Ricin [40], Abrin toxin [41], Microcystin [42]
Antibiotics	Vasopressin [43], Streptomycin [44], Tetracycline [45], Viomycin [46], Chloramphenicol [47]
Endocrine disrupting compounds and hormones	17β -estradiol [48], Thyroxine hormone [49,50]
Protein	HA1 proteins of H5N1 influenza virus [51]
Other small organic molecules	Cocaine [15], Cholic acid [52], (R)-thalidomide [53], Ethanolamine [54]
Cells and bacteria	Anthrax spores [55], Campylobacter jejuni [56]

Table 5.1 is a partial list of literature-reported functional nucleic acid targets that are considered contaminants in water or are being investigated as contaminants. The sensing of many pharmaceutical compounds, hormones, and receptors that are considered as emerging water contaminants benefit from the research into aptamers for therapeutic applications. Also, as can be seen in Table 5.1, aptamers are being developed for binding specifically bacterial cells, viral spores, and toxins that can be used as biological warfare agents.

It is quite clear that functional nucleic acids provide a unique recognition platform for a large range of different contaminants that are already known and are emerging every year.

5.3 Functional nucleic acid-directed assembly of nanomaterials for sensing contaminants

Since natural nucleic acids do not possess functional groups that can generate absorption in the visible region or fluorescence, external signaling labels need to be applied to convert them into sensors. To achieve this goal, many organic fluorophores and inorganic nanoparticles, which are summarized in the next few sections, have been employed.

5.3.1 Fluorescent sensors

Fluorescent sensors provide an opportunity for on-site and real time sensing as recent advances have led to development of hand-held fluorimeters that can be

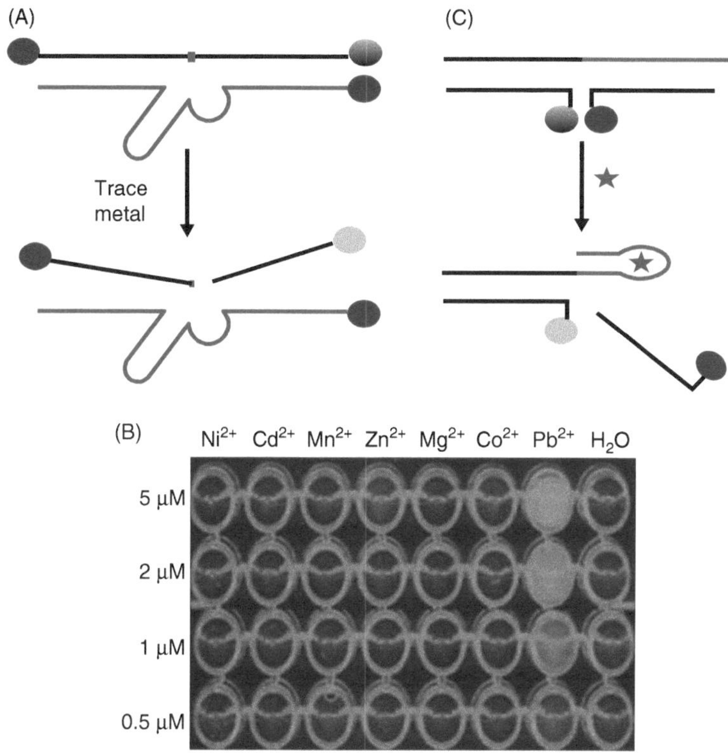

FIGURE 5.2

Fluorescent sensors using functional nucleic acid. (A) General design for a DNAzyme-based fluorescent sensor for metal ion detection. The black substrate strand is labeled with a fluorophore (green sphere) and quencher (brown sphere) and the green enzyme strand is labeled with a quencher (brown sphere). In the presence of specific metal ion, the enzyme catalyzed cleavage reaction causes the cleaved substrate fragments to be released and thus fluorescence is unmasked. (B) Fluorescent image of a microwell plate with the Pb^{2+}-dependent 17E DNAzyme incubated with different metal ions. (C) General design for aptamer-based fluorescent sensor. In the presence of target (red star), the aptamer sequence (green line) undergoes structure switching to wrap around the target, thereby disrupting the base pairing and releasing the quencher from this assembly.

used as a stand-alone device or connected to a laptop computer. Also, fluorescent sensors have the advantage of high sensitivity [57–60].

5.3.1.1 Sensing metal ions using DNAzyme based fluorescent sensors

Many metal specific DNAzymes have been successfully converted into fluorescent sensors using a catalytic beacon technology [61]. The general design is illustrated in Fig. 5.2A. The catalytic beacon is engineered to place a fluorophore (green sphere) on one substrate arm and a quencher (brown sphere) on the enzyme arm.

When the sensor is assembled, the fluorescence emission is quenched due to the proximity of the quencher to the fluorophore, brought about by DNA hybridization. A second quencher on the other arm of the substrate helps to reduce background fluorescence arising from non-hybridized substrate. In the presence of contaminant metal ion, the cleavage reaction causes the cleaved substrate fragment containing the fluorophore to be released into solution, thus enhancing the fluorescence strongly. The 17E DNAzyme was converted into a Pb^{2+} sensor with a detection limit of 10 nM [29], which is lower than the US EPA threshold for Pb^{2+} in water, set at 75 nM. Additionally, this sensor is over 80-fold more specific for Pb^{2+} over highly competing metal ions Co^{2+} and Zn^{2+} and over 1,000-fold more specific over other divalent metal ions including Mg^{2+} and Ca^{2+} that are found in water. Figure 5.2B is the fluorescent image of a microwell plate where the 17E DNAzyme is incubated with varying concentrations of Pb^{2+} and other competing metal ions. Concentration dependent enhanced fluorescence is only seen for Pb^{2+}. This technique has been tested on real-world water samples, such as from Lake Michigan, spiked with Pb^{2+}.

Following this, a number of other ions, UO_2^{2+} [32], Hg^{2+} [39], and Cu^{2+} [62], were converted into sensitive and selective fluorescent sensors. The performance of the UO_2^{2+} sensor is particularly impressive as the detection limit is reported to be 45 pM, which is not only lower than the EPA threshold (130 nM), but remarkably it is also lower than the detection limit for the widely used instrumental technique, ICP. Furthermore, the sensor has also been used to detect UO_2^{2+} extracted from soil of a nuclear waste site, as these sites are dangerous sources for contamination of groundwater and surface water with radionuclides.

A number of other designs for the position of the fluorophore–quencher have also been investigated and resulted in improvements in sensor designs [63]. In addition, new DNAzymes with fluorescently modified nucleotides were isolated by Li and coworkers [64,65] and utilized for metal sensing [66].

5.3.1.2 Sensing organic and biological molecules using aptamer-based fluorescent sensors

A general approach for converting aptamers into fluorescent sensors using the change in secondary structure of the aptamer upon target binding was reported by Li and coworkers (Fig. 5.2C). This method has been termed "structure switching signaling aptamers"[16]. A tripartite assembly is made between a fluorophore (green sphere) labeled DNA, a quencher (brown sphere) labeled DNA, and another linker DNA strand that contains the aptamer sequence and hybridizes to both the other strands, bringing the fluorophore and quencher into close proximity, which leads to fluorescence quenching. Introduction of the target causes the aptamer to wrap around the target causing the disruption of base pairing interactions between the quencher containing DNA and the linker, thereby leading to its dissociation and subsequent fluorescence enhancement. A number of alternates to this design have been reviewed [58] and aptamer based fluorescent sensors have been demonstrated for organic molecules, proteins, and cells.

5.3.2 Colorimetric sensors

Although fluorescent sensors are a very sensitive method for contaminant detection, they still require an instrument and the organic fluorophores can photo bleach relatively quickly. Colorimetric sensors eliminate the need for instruments and the following two sections discuss the use of metallic nanoparticles for the development of these sensors.

5.3.2.1 Sensing metal ions using DNAzyme/gold nanoparticle-based colorimetric sensors

Gold nanoparticles (AuNPs) make an attractive candidate for colorimetric labels as they have a very strong extinction coefficient ($10^8\ \Sigma\ cm^{-1}$ for 13 nm AuNPs, which is about three orders of magnitude higher than the best organic chromophores) and they display distance dependent optical properties. Dispersed nanoparticles are red in color, whereas aggregated nanoparticles are blue/purple in color. A quantitative analysis can be carried out by measuring the absorbance of the samples, for which portable colorimeters are also available. In 1996, Mirkin and coworkers utilized the DNA induced assembly of AuNPs to make a colorimetric sensor for nucleic acids [67]. Lu and coworkers expanded the scope of sensing to analytes beyond nucleic acids, by combining AuNPs with DNAzymes [68–71]. The sensing method is depicted using the Pb^{2+} dependent 17E DNAzyme as a representative example (Fig. 5.3A). AuNPs functionalized with short thiol modified DNA are assembled on the arms of the extended substrate, which is in turn hybridized to the enzyme. Since each AuNP is functionalized with many DNA strands, blue aggregates are formed. In the presence of Pb^{2+}, the enzyme catalyzed cleavage of the substrate will disassemble the aggregate producing red color. The color can be spotted on a thin layer chromatography (TLC) plate and one such representative test is shown in the inset of Fig. 5.3A. Red color of increasing intensity is seen with Pb^{2+}, whereas the sensors containing other metals are blue. The reaction is fast with color change occurring within 10 minutes under optimized conditions and the detection limit is approximately 100 nM.

A unique feature of this sensor is that the dynamic range of the sensor can be tuned by careful mutation of the DNA sequence, which is very useful for making sensor systems that can change colors at different threshold levels that match the maximum contaminant levels (MCLs) defined by EPA or CDC. The dynamic range of the Pb^{2+} sensor was shifted from 10 to 100 $\propto$M Pb^{2+} using this strategy [68].

5.3.2.2 Sensing organic and biological molecules using aptamer/gold nanoparticle-based colorimetric sensors

To detect contaminants beyond metal ions, aptamers have been used instead for directed assembly of nanomaterials such as gold nanoparticles [19]. Figure 5.3B is an illustration of the detection scheme using an adenosine (A) aptamer. Two types of oligonucleotide functionalized AuNPs (particles 1 and 2) are assembled on a linker DNA consisting of the adenosine aptamer. The addition of adenosine

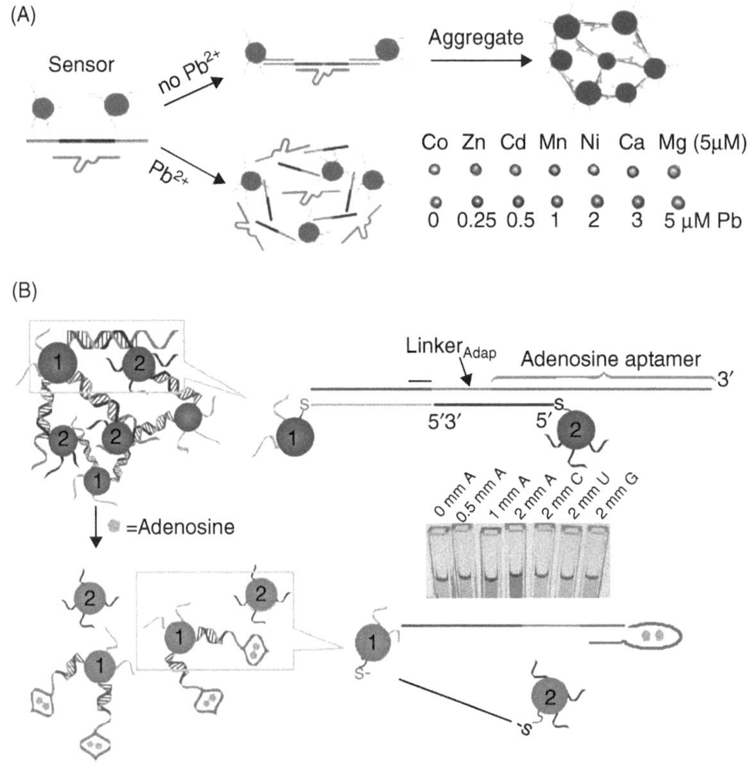

FIGURE 5.3

Colorimetric sensors using functional nucleic acid. (A) DNAzyme based colorimetric sensor for Pb^{2+} detection. In the absence of Pb^{2+}, the oligonucleotide functionalized gold nanoparticles (AuNPs) are assembled on the substrate to form blue aggregates. When Pb^{2+} is present, the substrate is cleaved and the aggregate is disassembled to yield red colored dispersed AuNPs. Inset: Picture of sensor incubated with increasing concentrations of Pb^{2+} and other metal ions. (B) Aptamer-based sensor for colorimetric detection of adenosine. Nanoparticles 1 and 2 are functionalized with two different DNA molecules through thiol-gold chemistry. The two kinds of AuNPs are linked by $Linker_{Adap}$ to form aggregates. In the presence of adenosine, the AuNPs disassemble to give dispersed red nanoparticles. Inset: Photograph of the samples with designated nucleoside added.

induces structure switching, leading to disassembly of the blue aggregate, producing a red color characteristic of dispersed AuNPs, which is not seen in the presence of control analytes (C), cytidine (C), uridine (U), and guanine (G) (inset of Fig. 5.3B). The generality of this method has been demonstrated by making a cocaine sensor in the same manner. Sensors that respond to multiple chemical stimuli have also been constructed by combining two aptamers in the same system

[72] such that the sensor responds either in the presence of both analytes just one of the analytes.

Recently, a new method has been reported for colorimetric sensing utilizing AuNPs, called the label free method [73]. Here the AuNPs do not need to be functionalized with thiol modified DNA, and thus there can be significant reduction in sensing costs as the DNA utilized will not require the chemical (thiol) modification required for attaching DNA to AuNPs. This method is based on the principle that single stranded DNA (ssDNA) can adsorb on AuNPs more effectively and therefore protect them from salt induced aggregation to a greater extent as compared to double stranded DNA or structured DNA, such as quadruplex DNA. Aptamer-based sensing has been demonstrated for some analytes by utilizing the change in the DNA secondary structure upon target binding [74–76].

5.4 Simultaneous multiplexed detection using quantum dots and gold nanoparticles

In order to probe complex chemical environments, it may be desirable to construct materials that are responsive to multiple chemicals in the same system, producing a unique signal for each chemical. Semiconductor quantum dots (QDs) are ideal as the emission wavelength of QDs can be tuned by varying their size, shape, and chemical composition, while keeping the excitation wavelength the same, thus making it possible to have QDs with different emission wavelengths to encode for different analytes. Multiplexed detection of four toxins was previously achieved using QDs linked to antibodies [77]. Liu et al. combined QDs and AuNPs with aptamer technology to demonstrate one-pot simultaneous detection of two analytes, adenosine, and cocaine (Fig. 5.4A) [78]. The detection scheme is based on disassembly of particles as described in Section 5.3.2. For adenosine detection, in addition to AuNPs 1 and 2, QD1 (emission at 525 nm), which is functionalized with the same DNA as AuNP 2, was also incorporated to make aggregates using the linker containing adenosine aptamer. In the assembled state, the emission of the QD was quenched because of energy transfer to nearby AuNPs [79–83]. The addition of adenosine disassembles the nanoparticles, giving increased emission intensity at 525 nm. Similarly, cocaine sensors were also prepared with QD2 (emission at 585 nm). When the sensors are mixed, different QDs can be excited at the same wavelength, as shown in Fig. 5.4B: two emission peaks at 525 and 585 nm were observed (solid line), corresponding to the adenosine and cocaine sensors, respectively. Addition of cytidine and uridine (dashed line) did not change the emission intensity of either peak. The addition of adenosine alone increased only the 525 nm peak (Fig. 5.4C); the addition of cocaine alone only increased the 585 nm peak (Fig. 5.4D); and the addition of both analytes resulted in enhancements in both peaks (Fig. 5.4E).

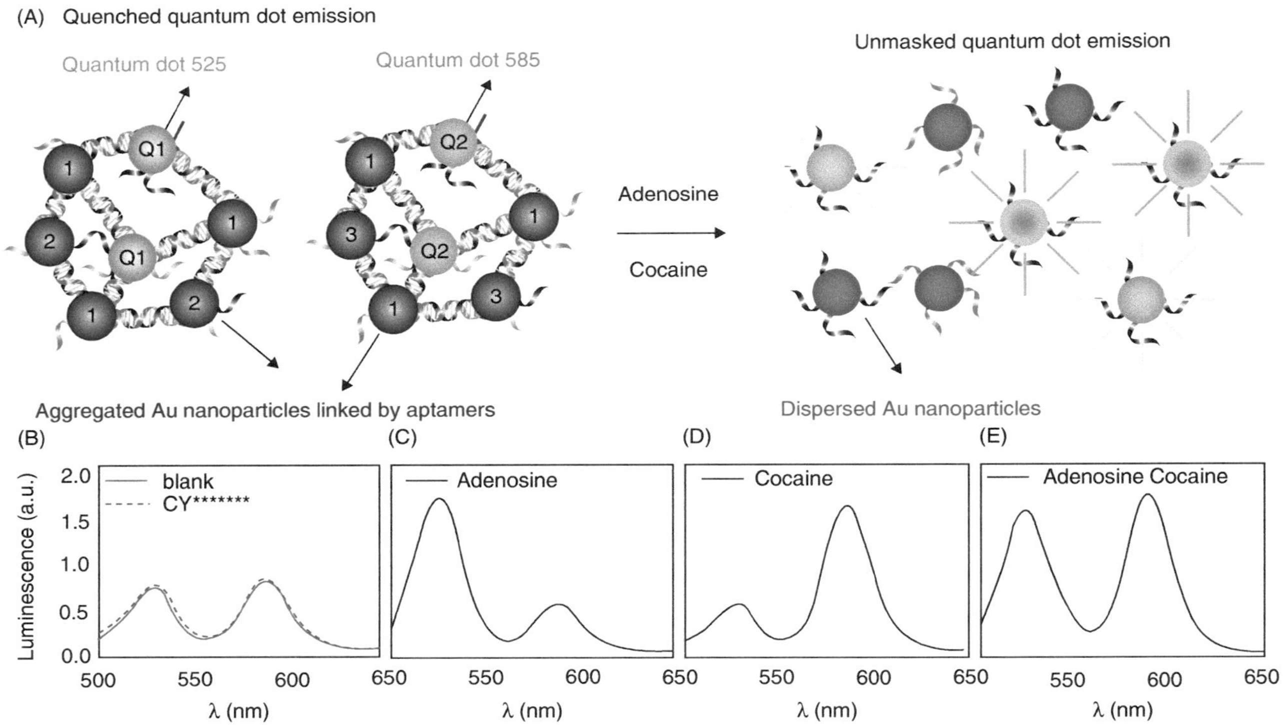

FIGURE 5.4

(A) Quantum dot (QD) encoded aptamer-linked nanoparticles for multiplex detection. QD emissions were initially quenched. The addition of target analytes allowed recovery of QD emission. (B–E) Luminescence spectra of mixed sensors in the presence of different analytes.

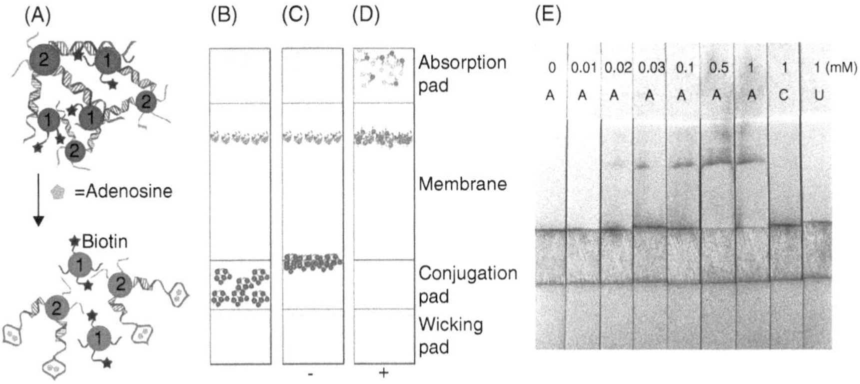

FIGURE 5.5

Aptamer/nanoparticle-based lateral flow device. (A) Adenosine induced disassembly of nanoparticle aggregates into red-colored dispersed nanoparticles. Biotin is denoted as a black star. Lateral flow devices loaded with the aggregates (on the conjugation pad) and streptavidin (on the membrane) before use (B), in a negative (C), or positive (D) test. (E) The lateral flow device is tested with varying concentrations of nucleosides.

5.5 Sensors on solid supports

5.5.1 Dipsticks

Although the colorimetric sensors provide a method of qualitative or semi-quantitative sensing without the requirement of an instrument, they still need precise transfer of solutions in small (usually microliter) quantities, which can be difficult for people without scientific training. In order to alleviate this problem, the aptamer–nanoparticle-based colorimetric tests can be converted into user-friendly "dipstick" tests using lateral flow devices that provide the reagents in a dry or nearly dry state immobilized on a pad. Several antibody-based dipstick tests are known, the home pregnancy test being one of the most common uses of this technology. The detection of DNA using lateral flow devices has also been demonstrated [84]. By utilizing the lateral flow devices for aptamer-based detection, the range of analytes that can be detected using this simple platform has been expanded [85]. A lateral flow device is constructed using four overlapping pads (wicking pad, conjugation pad, membrane, and absorption pad) placed on a plastic backing. The adenosine sensor consisted of the same components as described in Section 5.3.2, except in this case approximately 50 percent of DNA on particles 1 contained a biotin moiety, denoted by a black star (Fig. 5.5A). AuNP aggregates are dried on the conjugation pad of the devise and streptavidin is applied on the membrane (Fig. 5.5B). When the device is dipped into a solution without adenosine, the rehydrated aggregates migrate to the bottom of the membrane where they stop because of their large micrometer size (Fig. 5.5C). In the presence of adenosine, the dispersed nanoparticles [16,19] can migrate along the membrane and be captured by streptavidin to form a red line (Fig. 5.5D). Representative test results are depicted in Fig. 5.5E, where more intense red lines are seen with increasing concentrations of adenosine, but not with cytidine or uridine. These devices are also more sensitive than solution-based tests owing to the integration of binding, separation, and detection on a simple test-paper-like platform with no background interference.

5.5.2 Incorporation of sensors into devices

Whereas the functional nucleic acids described earlier are excellent for on-site and real-time applications, most sensing formats are for one-time use. In some applications, long-term and unattended monitoring is required. Furthermore, despite the fact that most sensors have quite high sensitivity and low detection limit, it is desirable to improve the detection limit even further. The immobilization of the fluorescent DNAzyme based sensors on gold surface and into micro- or nanofluidic devices has made it possible to achieve the goals mentioned earlier. For example, the detection limit for the fluorescent Pb^{2+} sensor was improved from 10 to 1 nM by reducing background noise, improving the signal to noise ratio. This was accomplished by rinsing away unhybridized DNA substrate from the immobilized DNAzyme sensor prior to sensing [86,87]. Furthermore, these sensors were also incorporated into micro- or nanofluidic devices, which require low sample volumes, low amounts of detection reagents, and which provide a possibility of facile regeneration and enhance the real-time long-term monitoring capability [88,89].

5.6 Other sensing schemes utilizing electrochemistry and magnetic resonance imaging

It is important to mention that apart from colorimetric and fluorescent sensors, other sensing methodologies are also being explored. Functional nucleic acids have been labeled with redox active groups, such that target binding produces an electrochemical signal (reviewed by Willner and Zayats [20]). For example, in one particular realization of this technology, an aptamer immobilized on an electrode is labeled with a redox active group such as ferrocene or methylene blue. Structural change due to target binding can bring this label closer to the electrode leading to greater electron transfer and thus an electrochemical signal [90–93]. The advantages of these sensors are that electronic devices can be miniaturized easily and these sensors can often be regenerated.

Recently, aptamers have also been combined with magnetic particles (superparamagnetic iron oxide nanoparticles or SPIO), which are contrast agents, for the purpose of constructing magnetic resonance imaging (MRI) sensors [94,95]. This method utilizes the change in the T2 relaxation time of the medium when the aptamer assembled SPIO goes from an aggregated (low T2, darker MR image) to a dispersed state (high T2, brighter MR image) in the presence of the target. Although MR imaging may not be feasible for on-site monitoring, this method has the potential to be used for three-dimensional imaging in environmental monitoring, such as mapping the removal of contaminants during water purification using columns. Additionally, since MR imaging depends on a nuclear signal, it is less likely to have interference from turbid matrices that can mask fluorescent or colorimetric signals.

5.7 Conclusions and future perspective

Since the development of in vitro selection in the 1990s to obtain functional nucleic acids with desired recognition properties, a number of sensing applications have emerged very rapidly. A variety of materials and devices have been successfully developed based on functional nucleic acids for the detection of trace contaminants in water. This chapter highlighted some of the major strengths of this novel approach of combining nucleic acids with nanotechnology: (1) Nucleic acids provide a general class of molecules that can be selected to recognize a variety of different contaminants. (2) In vitro selection can be utilized to tailor the sensitivity and specificity of the nucleic acid for the contaminant, so as to obtain better sensors. (3) Functional nucleic acids can be readily labeled with fluorophores and inorganic nanoparticles to obtain sensors with tunable dynamic ranges. (4) The sensors can be assembled into dipstick tests and devices for ease of use, longer shelf life, and regeneration.

In spite of the generality of nucleic acid sensors, there still exist challenges in selecting nucleic acids for certain kinds of analytes, such as anions like perchlorate or nitrate, which are negatively charged and thus are repelled by the negatively charged backbone of nucleic acids. It is important to explore newer selection strategies to overcome these and further expand the repertoire of analytes recognized.

Finally, further development in this field would require development of sensor arrays containing nucleic acids that recognize many contaminants for simultaneous monitoring purposes, similar to the microarray technology commonly used for nucleic acid detection.

Acknowledgments

The authors would like to thank Dr Daryl P. Wernette for assistance with the figures. This material is based upon work supported by the US National Science Foundation through the Science and Technology Center of Advanced Materials for the Purification of Water with Systems (WaterCAMPWS, CTS-0120978), the Strategic Environmental Research and Development Program, the U.S. Department of Energy (DE-FG02–08ER64568), and the the U.S. National Institute of Health (ES016865).

References

[1] S.D. Richardson, T.A. Ternes, Water analysis: emerging contaminants and current issues, Anal. Chem. 77 (2005) 3807–3838.

[2] S.D. Richardson, Water analysis: emerging contaminants and current issues, Anal. Chem. 79 (2007) 4295–4323.

[3] K. Kruger, P.J. Grabowski, A.J. Zaug, J. Sands, D.E. Gottschling, T.R. Cech, Self-splicing RNA: autoexcision and autocyclization of the ribosomal RNA intervening sequence of tetrahymena, Cell 31 (1982) 147–157.

[4] C. Guerrier-Takada, K. Gardiner, T. Marsh, N. Pace, S. Altman, The RNA moiety of ribonuclease P is the catalytic subunit of the enzyme, Cell 35 (1983) 849–857.

[5] R.R. Breaker, G.F. Joyce, A DNA enzyme that cleaves RNA, Chem. Biol. 1 (1994) 223–229.

[6] C. Tuerk, L. Gold, Systematic evolution of ligands by exponential enrichment: RNA ligands to bacteriophage T4 DNA polymerase, Science 249 (1990) 505–510.

[7] A.D. Ellington, J.W. Szostak, In vitro selection of RNA molecules that bind specific ligands, Nature 346 (1990) 818–822.

[8] D.S. Wilson, J.W. Szostak, In vitro selection of functional nucleic acids, Annu. Rev. Biochem. 68 (1999) 611–647.

[9] L. Gold, B. Polisky, O. Uhlenbeck, M. Yarus, Diversity of oligonucleotide functions, Annu. Rev. Biochem. 64 (1995) 763–797.

[10] R.R. Breaker, Engineered allosteric ribozymes as biosensor components, Curr. Opin. Biotechnol. 13 (2002) 31–39.

[11] J. Hesselberth, M.P. Robertson, S. Jhaveri, A.D. Ellington, In vitro selection of nucleic acids for diagnostic applications, Rev. Mol. Biotechnol. 74 (2000) 15–25.

[12] M. Famulok, Oligonucleotide aptamers that recognize small molecules, Curr. Opin. Struct. Biol. 9 (1999) 324–329.

[13] P.J. Bruesehoff, J. Li, A.J. Augustine, Y. Lu, Improving metal ion specificity during in vitro selection of catalytic DNA, Comb. Chem. High T. Scr. 5 (2002) 327–335.

[14] R.R. Breaker, DNA aptamers and DNA enzymes, Curr. Opin. Chem. Biol. 1 (1997) 26–31.

[15] M.N. Stojanovic, P. De Prada, D.W. Landry, Fluorescent sensors based on aptamer self-assembly, J. Am. Chem. Soc. 122 (2000) 11547–11548.

[16] R. Nutiu, Y. Li, Structure-switching signaling aptamers, J. Am. Chem. Soc. 125 (2003) 4771–4778.

[17] C.-C. Huang, Y.-F. Huang, Z. Cao, W. Tan, H.-T. Chang, Aptamer-modified gold nanoparticles for colorimetric determination of platelet-derived growth factors and their receptors, Anal. Chem. 77 (2005) 5735–5741.

[18] N.K. Navani, Y. Li, Nucleic acid aptamers and enzymes as sensors, Curr. Opin. Chem. Biol. 10 (2006) 272–281.

[19] J. Liu, Y. Lu, Fast colorimetric sensing of adenosine and cocaine based on a general sensor design involving aptamers and nanoparticles, Angew. Chem. Int. Ed. 45 (2006) 90–94.

[20] I. Willner, M. Zayats, Electronic aptamer-based sensors, Angew. Chem. Int. Ed. 46 (2007) 6408–6418.

[21] S.E. Osborne, I. Matsumura, A.D. Ellington, Aptamers as therapeutic and diagnostic reagents: problems and prospects, Curr. Opin. Chem. Biol. 1 (1997) 5–9.

[22] S.M. Nimjee, C.P. Rusconi, B.A. Sullenger., Aptamers: an emerging class of therapeutics, Annu. Rev. Med. 56 (2005) 555–583.

[23] J.F. Lee, G.M. Stovall, A.D. Ellington, Aptamer therapeutics advance, Curr. Opin. Chem. Biol. 10 (2006) 282–289.

[24] M. Famulok, J.S. Hartig, G. Mayer, Functional aptamers and aptazymes in biotechnology, diagnostics, and therapy, Chem. Rev. 107 (2007) 3715–3743.

[25] Y. Liu, C. Lin, H. Li, H. Yan, Aptamer-directed self-assembly of protein arrays on a DNA nanostructure, Angew. Chem. Int. Ed. 44 (2005) 4333–4338.

[26] Y. Lu, J. Liu, Functional DNA nanotechnology: emerging applications of DNAzymes and aptamers, Curr. Opin. Biotechnol. 17 (2006) 580–588.

[27] M. Famulok, G. Mayer, Chemical biology: aptamers in nanoland, Nature 439 (2006) 666–669.

[28] A.K. Brown, J. Li, C.M.B. Pavot, Y. Lu, A lead-dependent DNAzyme with a two-step mechanism, Biochemistry 42 (2003) 7152–7161.

[29] J. Li, Y. Lu, A highly sensitive and selective catalytic DNA biosensor for lead ions, J. Am. Chem. Soc. 122 (2000) 10466–10467.

[30] J. Li, W. Zheng, A.H. Kwon, Y. Lu, In vitro selection and characterization of a highly efficient Zn(II)-dependent RNA-cleaving deoxyribozyme, Nucleic Acids Res. 28 (2000) 481–488.

[31] N. Carmi, L.A. Shultz, R.R. Breaker, In vitro selection of self-cleaving DNAs, Chem. Biol. 3 (1996) 1039–1046.

[32] J. Liu, A.K. Brown, X. Meng, D.M. Cropek, J.D. Istok, D.B. Watson, et al., A catalytic beacon sensor for uranium with parts-per-trillion sensitivity and millionfold selectivity, Proc. Natl. Acad. Sci. USA 104 (2007) 2056–2061.

[33] R. Vannela, P. Adriaens, In vitro selection of Hg (II) and As (V)-dependent RNA-cleaving DNAzymes, Environ. Eng. Sci. 24 (2007) 73–84.

[34] S.W. Santoro, G.F. Joyce, A general purpose RNA-cleaving DNA enzyme, Proc. Natl. Acad. Sci. USA 94 (1997) 4262–4266.

[35] D.H.J. Bunka, P.G. Stockley, Aptamers come of age: at last, Nat. Rev. Micro. Vol. 4 (2006) 588–596.

[36] C.K. O'Sullivan, Aptasensors: the future of biosensing? Anal. Bioanal. Chem. 372 (2002) 44–48.

[37] J.F. Lee, J.R. Hesselberth, L.A. Meyers, A.D. Ellington, Aptamer database, Nucleic Acids Res. 32 (2004) D95–D100.

[38] Y. Miyake, H. Togashi, M. Tashiro, H. Yamaguchi, S. Oda, M. Kudo, et al., MercuryII-mediated formation of thymine-HgII-thymine base pairs in DNA duplexes, J. Am. Chem. Soc. 128 (2006) 2172.

[39] J. Liu, Y. Lu, Rational design of 'turn-on' allosteric DNAzyme catalytic beacons for aqueous mercury ions with ultrahigh sensitivity and selectivity, Angew. Chem. Int. Ed. 46 (2007) 7587–7590.

[40] J. Tang, J. Xie, N. Shao, Y. Yan, The DNA aptamers that specifically recognize ricin toxin are selected by two in vitro selection methods,", Electrophoresis Vol. 27 (2006) 1303–1311.

[41] J. Tang, T. Yu, L. Guo, J. Xie, N. Shao, Z. He, In vitro selection of DNA aptamer against abrin toxin and aptamer-based abrin direct detection, Biosens. Bioelectron. 22 (2007) 2456–2463.

[42] C. Nakamura, T. Kobayashi, M. Miyake, M. Shirai, J. Miyakea, Usage of a DNA aptamer as a ligand targeting microcystin, Mol. Cryst. Liq. Cryst. 371 (2001) 369–374.
[43] K.P. Williams, X.-H. Liu, T.N.M. Schumacher, H.Y. Lin, D.A. Ausiello, P.S. Kim, et al., Bioactive and nuclease-resistant L-DNA ligand of vasopressin, Proc. Natl. Acad. Sci. USA 94 (1997) 11285–11290.
[44] M. Bachler, R. Schroeder, U. von Ahsen, Strepto Tag: A novel method for the isolation of RNA-binding proteins, RNA 5 (1999) 1509–1516.
[45] C. Berens, A. Thain, R. Schroeder, A tetracycline-binding RNA aptamer, Bioorg. Med. Chem. 9 (2001) 2549–2556.
[46] M.G. Wallis, B. Streicher, H. Wank, U. von Ahsen, E. Clodi, S.T. Wallace, et al., In vitro selection of a viomycin-binding RNA pseudoknot, Chem. Biol. 4 (1997) 357–366.
[47] D.H. Burke, D.C. Hoffman, A. Brown, M. Hansen, A. Pardi, L. Gold, RNA aptamers to the peptidyl transferase inhibitor chloramphenicol, Chem. Biol. 4 (1997) 833–843.
[48] Y.S. Kim, H.S. Jung, T. Matsuura, H.Y. Lee, T. Kawai, M.B. Gu, Electrochemical detection of 17b-estradiol using DNA aptamer immobilized gold electrode chip, Biosens. Bioelectron. 22 (2007) 2525–2531.
[49] N. Kawazoe, Y. Ito, M. Shirakawa, Y. Imanishit, In vitro selection of a DNA aptamer binding to thyroxine, Bull. Chem. Soc. Jpn. 71 (1998) 1699–1703.
[50] D. Levesque, J.-D. Beaudoin, S. Roy, J.-P. Perreault, In vitro selection and characterization of RNA aptamers binding thyroxine hormone, Biochem. J. 403 (2007) 129–138.
[51] C. Cheng, J. Dong, L. Yao, A. Chen, R. Jia, L. Huan, et al., Potent inhibition of human influenza H5n1 virus by oligonucleotides derived by SELEX, Biochem. Biophys. Res. Comm 366 (2008) 670–674.
[52] T. Kato, T. Takemura, K. Yano, K. Ikebukuro, I. Karube, In vitro selection of DNA aptamers which bind to cholic acid, Biochim. Biophys. Acta 1493 (2000) 12–18.
[53] A. Shoji, M. Kuwahara, H. Ozaki, H. Sawai, Modified DNA aptamer that binds the (R)-isomer of a thalidomide derivative with high enantioselectivity, J. Am. Chem. Soc. 129 (2007) 1456–1464.
[54] D. Mann, C. Reinemann, R. Stoltenburg, B. Strehlitz, In vitro selection of DNA aptamers binding ethanolamine, Biochem. Biophys. Res. Comm. 338 (2005) 1928.
[55] J.G. Bruno, J.L. Kiel, In vitro selection of DNA aptamers to anthrax spores with electrochemiluminescence detection, Biosens. Bioelectron. 14 (1999) 457–464.
[56] S. McMasters, D.N. Stratis-Cullum, Evaluation of aptamers as molecular recognition elements for pathogens using capillary electrophoretic analysis. Proceedings of SPIE-The International Society for Optical Engineering, Vol. 6380, 63800B/1-63800B/8, 2006.
[57] R. Nutiu, Y. Li, Structure-switching signaling aptamers: transducing molecular recognition into fluorescence signaling, Chem. Eur. J. 10 (2004) 1868–1876.
[58] R. Nutiu, Y. Li, Aptamers with fluorescence-signaling properties, Methods 37 (2005) 16–25.
[59] Z. Cao, S.W. Suljak, W. Tan, Molecular beacon aptamers for protein monitoring in real-time and in homogeneous solutions, Curr. Proteomics 2 (2005) 31.
[60] E.J. Cho, M. Rajendran, A.D. Ellington, Aptamers as emerging probes for macromolecular imaging, Top. Fluoresc. Spectrosc. 10 (2005) 127–155.

[61] J. Liu, Y. Lu, Fluorescent DNAzyme biosensors for metal ions based on catalytic molecular beacons, Meth. Mol. Biol. 335 (2006) 275–288.
[62] J. Liu, Y. Lu, A DNAzyme catalytic beacon sensor for paramagnetic Cu2+ ions in aqueous solution with high sensitivity and selectivity, J. Am. Chem. Soc. 129 (2007) 9838–9839.
[63] W. Chiuman, Y. Li, Efficient signaling platforms built from a small catalytic DNA and doubly labeled fluorogenic substrates, Nucleic Acids Res. 35 (2007) 401–405.
[64] S.H.J. Mei, Z. Liu, J.D. Brennan, Y. Li, An efficient RNA-cleaving DNA enzyme that synchronizes catalysis with fluorescence signaling, J. Am. Chem. Soc. 125 (2003) 412–420.
[65] Z. Liu, S.H.J. Mei, J.D. Brennan, Y. Li, Assemblage of signaling DNA enzymes with intriguing metal-ion specificities and pH dependences, J. Am. Chem. Soc. 125 (2003) 7539–7545.
[66] Y. Shen, G. Mackey, N. Rupcich, D. Gloster, W. Chiuman, Y. Li, et al., Entrapment of fluorescence signaling DNA enzymes in sol-gel-derived materials for metal ion sensing, Anal. Chem. 79 (2007) 3494–3503.
[67] J.J. Storhoff, R. Elghanian, R.C. Mucic, C.A. Mirkin, R.L. Letsinger, One-pot colorimetric differentiation of polynucleotides with single base imperfections using gold nanoparticle probes, J. Am. Chem. Soc. 120 (1998) 1959–1964.
[68] J. Liu, Y. Lu, A colorimetric lead biosensor using DNAzyme-directed assembly of gold nanoparticles, J. Am. Chem. Soc. 125 (2003) 6642–6643.
[69] J. Liu, Y. Lu, Accelerated color change of gold nanoparticles assembled by DNAzymes for simple and fast colorimetric Pb2+ detection, J. Am. Chem. Soc. 126 (2004) 12298–12305.
[70] J. Liu, Y. Lu, Stimuli-responsive disassembly of nanoparticle aggregates for light-up colorimetric sensing, J. Am. Chem. Soc. 127 (2005) 12677–12683.
[71] J. Liu, Y. Lu, Design of asymmetric DNAzymes for dynamic control of nanoparticle aggregation states in response to chemical stimuli, Org. Biomol. Chem. 4 (2006) 3435–3441.
[72] J. Liu, Y. Lu, Smart nanomaterials responsive to multiple chemical stimuli with controllable cooperativity, Adv. Mater 18 (2006) 1667–1671.
[73] H. Li, L.J. Rothberg, Label-free colorimetric detection of specific sequences in genomic DNA amplified by the polymerase chain reaction, J. Am. Chem. Soc. 126 (2004) 10958–10961.
[74] L. Wang, X. Liu, X. Hu, S. Song, C. Fan, Unmodified gold nanoparticles as a colorimetric probe for potassium DNA aptamers, Chem. Comm. (2006) 3780–3782.
[75] W. Zhao, W. Chiuman, M.A. Brook, Y. Li, Simple and rapid colorimetric biosensors based on DNA aptamer and noncrosslinking gold nanoparticle aggregation, Chembiochem Vol. 8 (2007) 727–731.
[76] H. Wei, B. Li, J. Li, E. Wang, S. Dong, Simple and sensitive aptamer-based colorimetric sensing of protein using unmodified gold nanoparticle probes, Chem. Comm. 36 (2007) 3735–3737.
[77] E.R. Goldman, A.R. Clapp, G.P. Anderson, H.T. Uyeda, J.M. Mauro, I.L. Medintz, et al., Multiplexed toxin analysis using four colors of quantum dot fluororeagents, Anal. Chem. 76 (2004) 684–688.
[78] J. Liu, J.H. Lee, Y. Lu, Quantum dot encoding of aptamer-linked nanostructures for one pot simultaneous detection of multiple analytes, Anal. Chem. 79 (2007) 4120–4125.

[79] G. Mitchell, C.A. Mirkin, R.L. Letsinger, Programmed assembly of DNA functionalized quantum dots, J. Am. Chem. Soc. 121 (1999) 8122–8123.

[80] Z. Gueroui, A. Libchaber, Single-molecule measurements of gold-quenched quantum dots, Phys. Rev. Lett. 93 (166108) (2004) 1–4.

[81] R. Wargnier, A.V. Baranov, V.G. Maslov, V. Stsiapura, M. Artemyev, M. Pluot, et al., Energy transfer in aqueous solutions of oppositely charged CdSe/ZnS core/shell quantum dots and in quantum dot-nanogold assemblies, Nano. Lett. 4 (2004) 451–457.

[82] E. Oh, M.-Y. Hong, D. Lee, S.-H. Nam, H.C. Yoon, H.-S. Kim, Inhibition assay of biomolecules based on fluorescence resonance energy transfer (FRET) between quantum dots and gold nanoparticles, J. Am. Chem. Soc. 127 (2005) 3270–3271.

[83] L. Dyadyusha, H. Yin, S. Jaiswal, T. Brown, J.J. Baumberg, F.P. Booy, et al., Quenching of CdSe quantum dot emission, a new approach for biosensing, Chem. Comm. (2005) 3201–3203.

[84] K. Glynou, P.C. Ioannou, T.K. Christopoulos, V. Syriopoulou, Oligonucleotide-functionalized gold nanoparticles as probes in a dry-reagent strip biosensor for DNA analysis by hybridization, Anal. Chem. 75 (2003) 4155–4160.

[85] J. Liu, D. Mazumdar, Y. Lu, A simple and sensitive 'dipstick' test in serum based on lateral flow separation of aptamer-linked nanostructures, Angew. Chem. Int. Ed. 45 (2006) 7955–7959.

[86] C.B. Swearingen, D.P. Wernette, D.M. Cropek, Y. Lu, J.V. Sweedler, P.W. Bohn, Immobilization of a catalytic DNA molecular beacon on Au for Pb(II) detection, Anal. Chem. 77 (2005) 442–448.

[87] D.P. Wernette, C.B. Swearingen, D.M. Cropek, Y. Lu, J.V. Sweedler, P.W. Bohn, Incorporation of a DNAzyme into Au-coated nanocapillary array membranes with an internal standard for Pb(II) sensing, Analyst (2006).

[88] I.H. Chang, J.J. Tulock, J. Liu, W.-S. Kim, D.M. Cannon Jr., Y. Lu, et al., Miniaturized lead sensor based on lead-specific DNAzyme in a nanocapillary interconnected microfluidic device, Environ. Sci. Technol. 39 (2005) 3756–3761.

[89] K.A. Shaikh, K.S. Ryu, E.D. Goluch, J.-M. Nam, J. Liu, C.S. Thaxton, et al., A modular microfluidic architecture for integrated biochemical analysis, Proc. Natl. Acad. Sci. USA 102 (2005) 9745–9750.

[90] A.-E. Radi, J.L. Acero Sanchez, E. Baldrich, C.K. O'Sullivan, Reagentless, reusable, ultrasensitive electrochemical molecular beacon aptasensor, J. Am. Chem. Soc. 128 (2006) 117.

[91] J.L.A. Sanchez, E. Baldrich, A.E.-G. Radi, S. Dondapati, P.L. Sanchez, I. Katakis, et al., Electronic 'off-on' molecular switch for rapid detection of thrombin, Electroanalysis 18 (2006) 1957–1962.

[92] B.R. Baker, R.Y. Lai, M.S. Wood, E.H. Doctor, A.J. Heeger, K.W. Plaxco, An electronic, aptamer-based small-molecule sensor for the rapid, label-free detection of cocaine in adulterated samples and biological fluids, J. Am. Chem. Soc. 128 (2006) 3138.

[93] Y. Xiao, A.A. Rowe, K.W. Plaxco, Electrochemical detection of parts-per-billion lead via an electrode-bound DNAzyme assembly, J. Am. Chem. Soc. 129 (2007) 262.

[94] M.V. Yigit, D. Mazumdar, Y. Lu, MRI detection of thrombin with aptamer functionalized superparamagnetic iron oxide nanoparticles, Bioconjug. Chem. 19 (2008) 412–417.

[95] M.V. Yigit, D. Mazumdar, H.-K. Kim, J.H. Lee, B. Odintsov, Y. Lu, Smart 'turn-on' magnetic resonance contrast agents based on aptamer-functionalized superparamagnetic iron oxide nanoparticles, Chem. Bio. Chem. 8 (2007) 1675–1678.

PART

2

Separation Technologies

CHAPTER

Nanostructured Membranes for Water Purification

6

Omowunmi A. Sadik, Nian Du, Idris Yazgan and Veronica Okello

Department of Chemistry, Center for Advanced Sensors and Environmental Systems, State University of New York at Binghamton, Binghamton, NY, USA

6.1 Introduction

Nanoscale materials can be used to achieve wastewater purification, desalination, and inactivation of pathogenic bacteria and viruses, as well as for the adsorption and/or degradation of contaminants [1–16]. The development and sustainability of water treatment technologies have been mostly driven by three primary factors: the discovery of new rare contaminants, the promulgation of new water quality standards, and cost [17–23]. Most sources of water used for public drinking water supplies are not of suitable quality for consumption without some form of treatment [24–30].

The US EPA has ruled that all surface waters must be filtered and disinfected before consumption unless the purveyor can justify avoidance of filtration. Membrane-based water filtration (MWF) processes are currently among the most important and versatile technologies for conventional drinking water production, wastewater treatment, desalination, and water reuse. MWF processes offer unique advantages, such as continuous chemical-free operation, low-energy consumption, easy scale-up and hybridization with other processes, and highly automated process control. Commercially available membrane processes for water purification

Street, Sustich, Duncan and Savage. Nanotechnology Applications for Clean Water, 2nd Edition.
 DOI: http://dx.doi.org/10.1016/B978-1-4557-3116-9.00006-8

include electrodialysis, electrodeionization, reverse osmosis (RO), ultrafiltration (UF), and microfiltration. However, conventional membrane purification technologies are limited by membrane resistance to flow, imperfections, and fouling [31,32].

Moreover, due to the large pore sizes, conventional membrane filters cannot be used to filter submicron particles or engineered nanoparticles (ENPs). However, in their 1997 work on microporous polypropylene filters with surface charge modification, Kang and Shah demonstrated the use of filtration of ENPs with sizes between 60 and 200 nm [33]. These microporous filters did not work well in this case. The ENPs were captured on filter by adsorption based on surface charge. By using four layers of filters, the capture efficiency was boosted up to 95%. Research carried out by Liang et al. used carbonaceous nanofiber membranes for filtration of gold, silver, and silica ENPs [34]. These membranes were made of carbon nanofibers and had little interaction with the filtered ENPs. Although in the report, the pore size of membranes was not discussed, successful filtrations of ENPs with a wide range (5–150 nm) were demonstrated [34].

Recent work carried out by Krieg et al. showed the ability to trap particles with various sizes bigger than 5 nm using a 12-μm-thick nanoporous membrane [35]. By increasing the thickness to 45 μm, they discovered that the membrane could separate smaller particles (CdTe quantum dots (QDs) of 2–4 nm in size) and act as size-selective chromatography [35]. Several commercial membranes for UF may also be employed for the filtration of ENPs. Most materials that are used in UF are polymeric and naturally hydrophobic. Examples include polysulfone, polyethersulfone, polypropylene, or polyvinylidenefluoride. For example, the DOW™ UF double-walled hollow fiber membrane has uniform pore size of about 30 nm with high removal efficiency for bacteria and viruses. Another application recently reported was inorganic aluminum oxide membrane [36,37]. This membrane has a precise, nondeformable honeycomb pore structure with uniform pore size and an extraordinarily high pore density. Whatman Ltd/GE Healthcare has commercial membrane products named Anodisc™ available at three nominal pore sizes: 20, 100, and 200 nm. These aluminum oxide membranes can be used in micrometer and nanometer filtration [38].

Although these filters have been used for the isolation and separation of ENPs, their inconsistent range in pore size leads to inadequate filtration efficiency. Some extremely small ENPs could still penetrate the pores. Most importantly, these filters cannot be used for both separation and detection (Figure 6.1). Therefore, membrane filters with a broad range of pore sizes are needed for the isolation of diverse classes of ENPs, particularly with respect to size. Nanostructured materials that are specifically designed to purify difficult-to-treat feed water can effectively prevent constituents, which often pose problems to the conventional RO process. The ability to functionalize poly(amic acid) (PAA) membranes with natural polymers or biological building blocks to control the size, shape, and stability at the nanoscale level could potentially address the constraints of conventional MWF technologies.

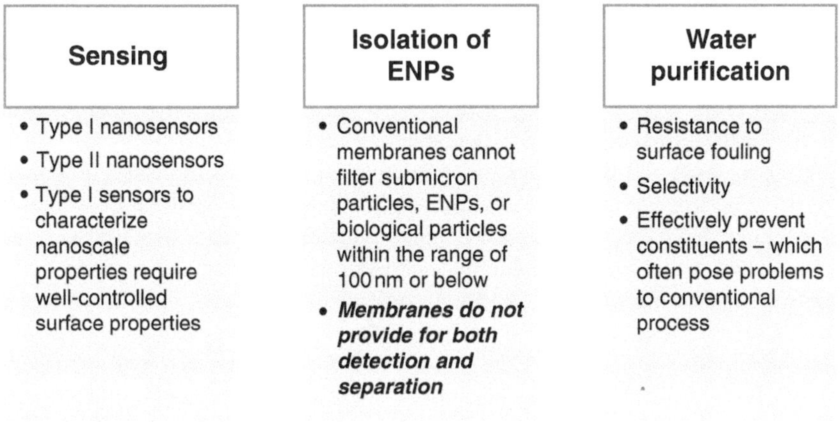

FIGURE 6.1

Research needs in the development of nanostructured materials for environmental applications.

We hereby present the use of nanostructured conducting phase-inverted poly(amic acid) (nPAA) membranes as novel filters for isolation and identification of ENPs especially in complex matrices (Figure 6.1).

6.2 Conducting PAA membranes

In general, nanosensors can be divided into two broad groups, namely nanotechnology-enabled sensors and nanoproperty-quantifiable sensors [39,40]. The former entails all sensors that are nano sized or have nanosized components in them, while the latter includes sensors that are used to measure nanoproperties. Since the dimensions of most biomolecules are on the same scale as nanomaterials, type I sensors have been greatly used to construct enzyme sensors, immunosensors, and genosensors, achieve direct wiring of enzymes and promote spectroelectrochemical reaction, impose barcodes for biomaterials and amplify signal of biorecognition events [41,42]. In addition, a number of research papers and reviews reporting on the use of nanomaterials for chemical and bioassays have been published [43]. The motivation behind this study was to incorporate novel materials that can serve as filters while providing intrinsic signals for quantitative analysis of the nanoscale materials itself. In that respect, there have been only a few sensors that are currently available to measure nanoscale properties especially in complex environmental matrixes. Type II sensors should be able to detect and accurately quantify nanomaterials in complex environmental samples. Recently, our laboratory showed that if beta-cyclodextrins (beta-CDs) are immobilized on an appropriate transducer such as a piezoelectric crystal, nanosized organics could be isolated and detected based on the ability of the CDs to complex the organics according to size exclusion mechanism [44]. The central idea is to utilize the ability of the CDs to complex a comparable sized fullerene (e.g., C_{60}), exclude it from the mixture of other fullerenes like C_{70}, hence measure the mass changes on a quartz crystal microbalance (QCM) according to the Sauebrey equation.

We have explored the feasibility of designing advanced conducting polymeric materials for sensing and remediation applications [46–68]. Examples include polyphenols, PAA nanomembranes, polyoxydianilines, and sequestered metal nanoparticles within electroconducting polypyrrole. However, the extreme hydrophobicity and insolubility of the polypyrrole matrix triggers protein adsorption and makes it less than ideal material for practical sensing applications. Alternative polymeric materials are needed that should possess a hydrophilic surface for low unspecific binding. Moreover, these materials should either possess the necessary functional groups needed for the attachment of biomolecules or can be easily functionalized. Hence we are focusing on creating advanced functional materials, which should be compatible with patterning of cells, neurons, and proteins as well as with modern electronics packaging.

PAA is a conducting electroactive polymer and its properties can be tuned by manipulating the delocalized π electron system for chemical and electrocatalytic applications. Its molecular structure and synthetic approach are shown in Figure 6.2. PAA provides a means of generating nanocomposites containing monodispersed metal particles while retaining its physical and chemical properties. It is versatile in both organic and inorganic solvents due to its chemical resistance. The carbonyl and amide functionalities in PAAs act as anchors that have

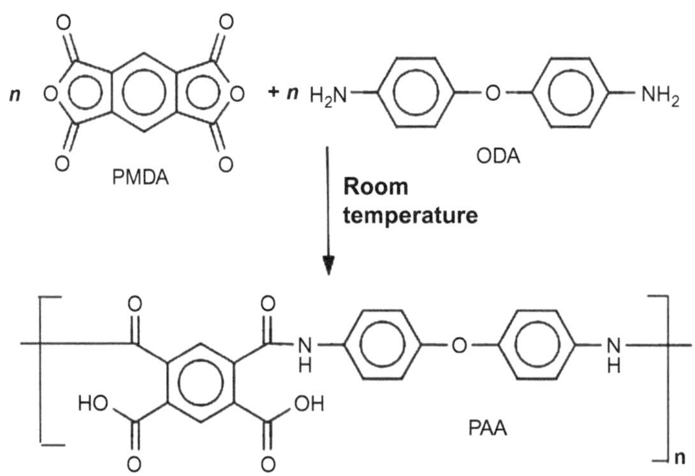

FIGURE 6.2

Synthesis and molecular structure of PAA. PMDA: Pyromelletic dianhydrate; ODA: 4,4'-oxydianiline.

been previously employed in the fabrication of flexible nanostructured PAA-silica (PSG) films [69,70].

Unlike other polymers such as self-assembled monolayers on mesoporous supports, dendrimers, carbon nanotubes, and metalloporphyrinogens, PAA has the following advantages: it is easily synthesized, soluble in both organic and inorganic solvents, able to retain the orientation of its functional groups, and displays both reduction and chelation properties due to the presence of both carboxylic and amide groups that can reduce and chelate metal ions. PAA has also the ability to generate nanocomposites containing monodispersed metal particles, i.e., in this case it acted as a stabilizer.

Figure 6.3 depicts how metal nanoparticle coordinates aryl ether rings using interconnecting PAA or ester rings perpendicular to the surface. The carbonyl and amide groups in PAA act as molecular anchors that bind the nanoparticle to the surface. Similarly, PAA can chelate Cr(III) using its amide and carbonyl functional groups forming an eight-membered ring making the metal ion less available in the aqueous phase.

The complexing power of PAA is significantly higher than that of the imide form, implying the ability of the carboxylic acid groups in PAA to exhibit polyfunctional behavior. Moreover, the possibility of creating monodispersed, nanoscale particles of noble metals can be used to create a high density of anchor groups for directed immobilization of Cr(VI). We reported the electrodeposited PAA films at low temperature (<100°C) with potentials for sensing biomolecules [71]. Compared to the imide form, the PAA have a significantly higher cation complexing property. This means that salts of noble metals having labile anions can react *in situ* to generate metal–polymeric blends. Thus, a single crosslink

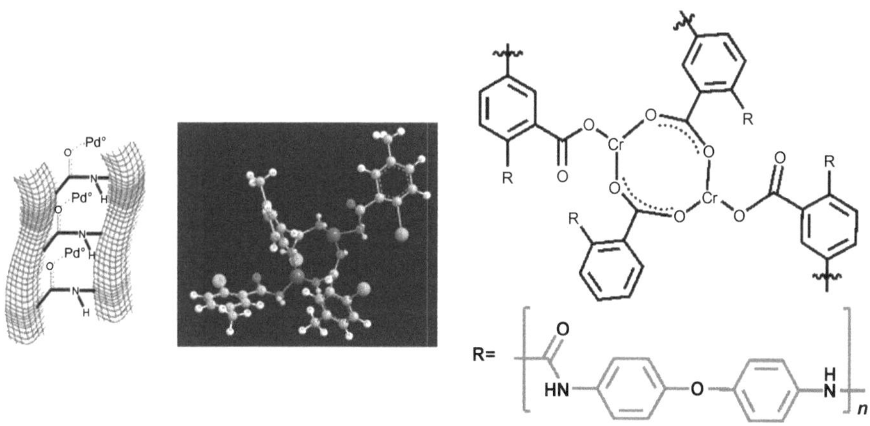

FIGURE 6.3

Model structures for the coordination of PAA with palladium nanoparticles (left) and chromium (III) following *in situ* catalytic reduction of chromium (VI) (right). Green, Cr(III); white, hydrogen; grey, carbon; blue, amide group; and red, carboxylic acid group.

between two molecules of PAA and noble metal composites can result in a stable, high molecular weight network. An average functionality in PAA can vary from 160 to 600 depending upon its molecular weight. Any reactive metal or complexing agents can crosslink the PAA to produce a high molecular weight crosslinked polymeric framework (Figure 6.2). In the following sections, we present the use of PAA membranes as filters for nanoparticles and for absolute removal of bacteria from water.

6.2.1 PAA membranes for nanofiltration of ENPs

We have shown that PAA nanomembranes can be used as nanofilters to isolate and remove silver nanoparticles and titanium dioxide particles in food supplements, aqueous environmental samples, and industrially relevant aerosol nanoparticles [72] with filtration efficiency of over 99%. We successfully synthesized a series of PAA and functionalized PAA membranes. We tested the PAA membranes as nanofilters to isolate and remove silver nanoparticles, QDs, and titanium dioxide particles in environmental samples. Filtration efficiency of over 99.98% was recorded for most contaminants. We observed high rates of particle removal using pressure-driven flow. Aqueous solutions of nanoparticles were filtered using the device illustrated in Figure 6.4A (left). PAA membranes with different porosities were used to target different nanoparticles and the performance of the PAA membranes (filtration efficiency) was compared with those of commercial Anodisc membranes (Table 6.1). The filtration of 20 nm QDs in aqueous solution was achieved using the membrane. The left cuvette contained orange QDs solution before filtration, and this solution turned colorless as shown on the right side after filtration. At the same time, the orange color (of the captured QDs) appeared on the light yellow PAA membrane.

In addition, Figure 6.4 shows the SEM data demonstrating that PAA membranes can be used to capture a broad range of ENPs including QDs, TiO_2, and silver nanoparticles. Following the capture of the nanoparticles, the simultaneous detection was achieved using differential pulse voltammetry. The metal particles were thoroughly oxidized to metal ions using the applied electrical potential. Results show that there was no electrochemical signal recorded either for the blank electrode or for the neat PAA membranes. However, upon coating the membranes with the nanoparticles, the redox peak was observed at ~ 100 mV and was linearly dependent on the concentration of the silver nanoparticles.

6.2.2 Application of PAA membranes for absolute disinfection of drinking water

We have integrated biological building blocks, such as glutaraldehyde and chitosan, with PAA to detect, disinfect, and completely remove pathogens in water. We described the synthesis of PAA and chitosan-modified PAA (PAA-CS) [73]. The resulting membrane was used for an absolute decontamination of bacteria in

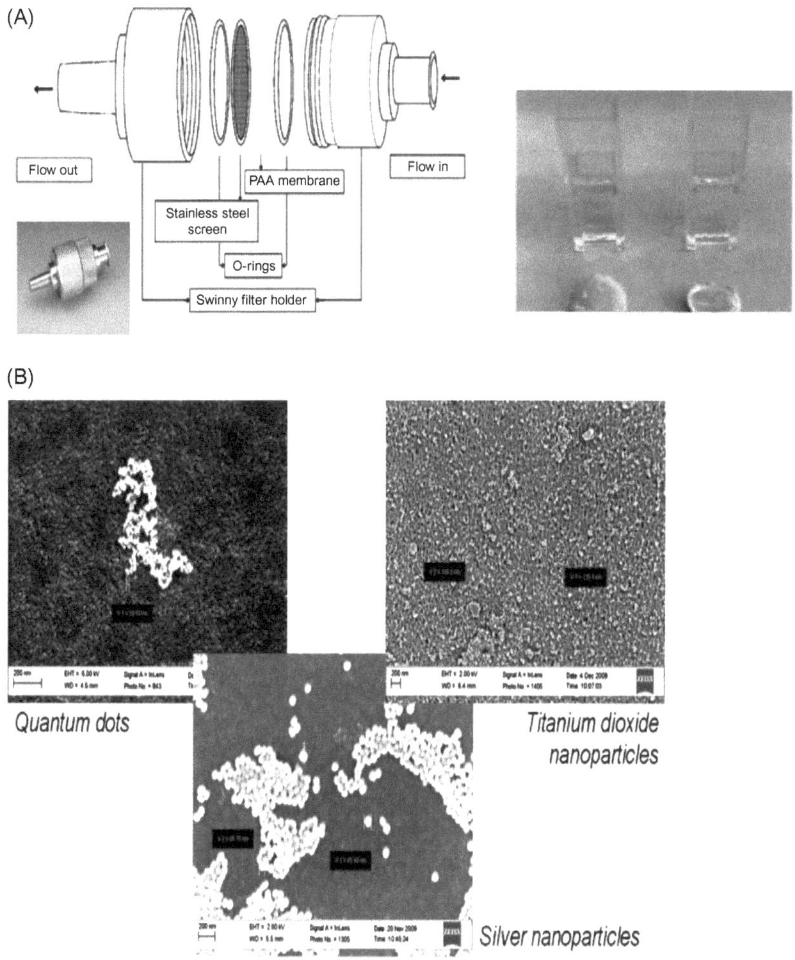

FIGURE 6.4

(A) Swinny filter holder and device setup. (Right) Images of 20 nm QDs in aqueous solutions before and after filtration using PAA membranes. (B) SEM micrographs of PAA membranes after being used to capture QDs (left), TiO_2 (right), and silver nanoparticles (bottom).

drinking water. The characterization data from Nuclear Magnetic Resonance (NMR), Fourier Transformation Infrared Spectroscopy (FT-IR), Scanning Electron Microscopy (SEM), and cyclic voltammetry confirmed the successful formation of electroactive, bifunctional, glutaraldehyde-linked PAA membranes. After the morphological, toxicological, and mechanical testing of the membranes using SEM and 3-(4,5-dimethylthiazol-2-yl)-2,5-diphenyltetrazolium bromide (MTT) assays, the results showed the successful formation of nontoxic, porous, free-standing, and

Table 6.1 Filtration Efficiency of Nanoporous PAA Membranes Versus Commercial Filters

	Average Percentage of Silver Nanoparticles Remaining in Filtrate (%)	Average Filtration Efficiency (%)	SD (%)
Whatman filter paper (FP)	98.76	1.24	1.33
Nylon filter membrane (NL)	55.48	44.52	3.15
PAA membrane (PAA)	1.50	98.50	0.10

mechanically strong PAA and PAA-CS membranes. The pore sizes were experimentally controlled from 4 to 35 nm, and the optimized membranes were tested against three of the most common drinking water contaminants, namely *Escherichia coli*, *Citrobacter freundii*, and *Staphylococcus epidermidis*. As a result, we were able to remove 100% of these microbial species through dead-end filtration and tangential flow filtration (Figure 6.5). The results were validated using plate counting and post-filtration SEM imaging.

6.3 Conclusions

We have developed a number of nanostructured materials as candidates for assessing the occurrence of water contamination. We have also reported the application of these materials for the design of nanostructured membrane filters and sensors. These membranes have shown good-to-excellent pathogen recovery efficiencies as well as a reasonable efficacy for sensing contaminants from water. Through this work, we envisage an advancement of innovative technologies for treating drinking water by simultaneously removing and inactivating classes of bacteria contaminants using low-cost, reusable, and sustainable biomembrane technologies. Nanotechnology applications for water treatment are not years away; they are already available and many more are likely to come on the market in the coming years. It is evident that nanotechnology research is being conducted in a broad spectrum of areas relevant to water treatment, including filters, catalysts, magnetic nanoparticles, and sensors. However, the maturity of research and development efforts is uneven across these areas with nanofiltration currently appearing most mature. Interest in the application of nanotechnology for water treatment appears to be driven by several factors including, but not limited to, reduced costs, improved ability to selectively remove contaminants, durability, and size of the device. While the current generation of nanofilters may be relatively simple, many researchers believe that future generations of nanotechnology-based water treatment devices will capitalize on the new properties of nanoscale materials. Advances through nanotechnology, therefore, may prove to be of significant interest to both developed and developing countries.

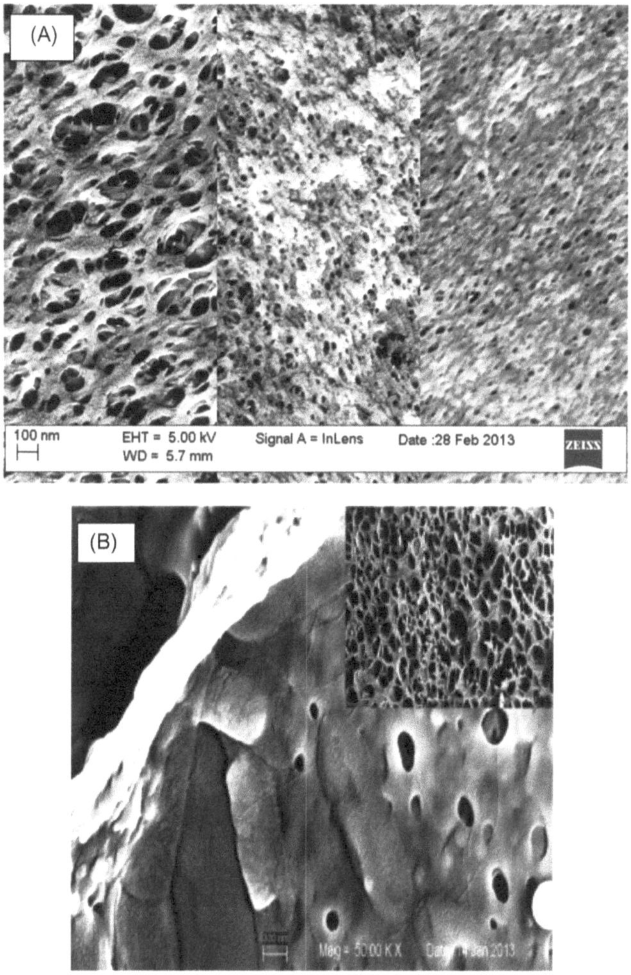

FIGURE 6.5

(A) SEM showing the effect of PAA concentration on pore size-increasing concentration (left to right) decreased pore size. (B) Nanostructured PAA membranes for capture and removal of pathogenic bacteria from drinking water.

Acknowledgments

The authors acknowledge the National Science Foundation CBET 1230189 for funding. N.D. also acknowledges the support of the IGERT fellowship of the National Science Foundation under Agreement # DGE-0654112, administered by the Nanobiotechnology Center at Cornell. US EPA and DOD are also acknowledged for funding. The Regional NMR facility (600 MHz instrument) at SUNY-Binghamton is supported by NSF (CHE-0922815).

References

[1] (a) M. Shannon, P. Bohn, M. Elimelech, J. Georgiadis, B. Marinas, A. Mayes, Science and technology for water purification in the coming decades, Nature 452 (2008) 301–310. (b) B. Karn, T. Kuiken, M. Otto, Nanotechnology and in situ remediation: a review of the benefits and potential risks, Environ. Health Perspect. 117 (2009) 1823–1831.

[2] (a) T.E. Cloete (Ed.), Nanotechnology in Water Treatment Applications, Caister Academic Press, Norfolk, UK, 2010. ISBN 978-1-904455-66-0. (b) M.S. Diallo, S. Christie, P. Swaminathan, J.H. Johnson, W.A. Godard III, Dendromer enhanced ultrafiltration. I. Recovery of Cu(II) from aqueous solutions using Gx-NH2 PAMAM dendromers with ethylene diamine core, Environ. Sci. Technol. 39 (2005) 1366–1377.

[3] M. Botes, T.E. Cloete, The potential of nanofibers and nanobiocides in water purification, Crit. Rev. Microbiol. 36 (1) (2010) 68–81.

[4] (a) T. Masciangioli, W. Zhang, Environmental technologies at the nanoscale, Environ. Sci. Technol. 37 (5) (2003) 73A–112A. (b) W.X. Zhang, J. Nanopart. Res. 5 (2003) 323–332. (c) M.R. Weisner, G.V. Lowry, P. Alvarez, D. Dionysios, P. Biswas, Environ. Sci. Technol. 40 (2006) 4336–4345.

[5] J. Theron, J. Walker, E. Cloete, Nanotechnology and water treatment: applications and emerging opportunities, Crit. Rev. Microbiol. 34 (1) (2008) 43–69.

[6] (a) M. Diallo, J. Duncan, N. Savage, A. Street, R. Sustich, Nanotechnology Applications For Clean Water, first ed., Elsevier, Maryland Heights, MO, 2008. (b) Netscribes, The US Market For Residential Water Treatment Equipment and Chemicals, ICON Group International, Inc, San Diego, CA, 2006. (c) P.M. Parker, The World Market For Machinery For Filtering And Purifying Water: 2007 Global Trade Perspective, ICON Group, Ltd, San Diego, CA, 2007.

[7] V.L. Colvin, The potential environmental impact of engineered nanomaterials, Nat. Biotechnol. 21 (10) (2003) 1166–1170.

[8] A. Seaton, K. Donaldson, Nanoscience, nanotoxicology, and the need to think small, The Lancet 365 (9463) (2005) 923–924.

[9] Z. Shi, J.T. Nurmi, P.G. Tratnyek, Effects of nano zero-valent iron (nZVI) on oxidation–reduction potential (ORP), Environ. Sci. Technol. 45 (2011) 1586–1592.

[10] J. Economy, J. Wang, C. Ba, Design of advanced membranes and substrates for water purification and desalination, in: M. Diallo, J. Duncan, N. Savage, A. Street, R. Sustich (Eds.), Nanotechnology Applications For Clean Water, first ed., Elsevier, Maryland Heights, MO, 2008. [Chapter 7].

[11] O. Bakajin, A. Noy, F. Fornasiero, C. Grigoropoulos, J. Holt, J. Bin In, et al., Nanofluidic carbon nanotube membranes: applications for water purification and desalination, in: M. Diallo, J. Duncan, N. Savage, A. Street, R. Sustich (Eds.), Nanotechnology Applications For Clean Water, first ed., Elsevier, Maryland Heights, MO, 2008. [Chapter 6].

[12] H Choi, R. Al-Abed, D. Dionysiou, Nanostructured titanium oxide film and membrane-based photocatalysis for water treatment, in: M. Diallo, J. Duncan, N. Savage, A. Street, R. Sustich (Eds.), Nanotechnology Applications For Clean Water, first ed., Elsevier, Maryland Heights, MO, 2008. [Chapter 3].

[13] E. Hoek, A. Ghosh, Nanotechnology-based membranes for water purification, in: M. Diallo, J. Duncan, N. Savage, A. Street, R. Sustich (Eds.), Nanotechnology Applications For Clean Water, first ed., Elsevier, Maryland Heights, MO, 2008. [Chapter 4].

[14] O. Renn, M.C. Roco, Nanotechnology and water purification: opportunities and challenges, J. Nanopart. Res. 8 (2006) 153–191.

[15] J.A. Hestekin, L.G. Bachas, D. Bhattacharyya, Poly (amino acid)-functionalized cellulose membranes: metal sorption mechanisms and results, Ind. Eng. Chem. Res. 40 (2001) 2668–2678.

[16] D.A. Butterfield, D. Bhattacharyya, S. Daunert, L. Bachas, Catalytic biofunctionalized membranes containing site-specific immobilized enzyme arrays: a review, J. Membr. Sci. 181 (2001) 29–37.

[17] W. Giger, et al., in: D.C. Washington (Ed.), Setting Priorities For Drinking Water Contaminants, Committee On Drinking Water Contaminants Of The US Academy Of Sciences, National Academy Press, 1999.

[18] W. Giger, Emerging chemical drinking water contaminants, in Identifying future drinking water contaminants, National Academy Press, Washington, DC, 1999, 112–119.

[19] United Nations Educational, Scientific, and Cultural Organization, (2003), World Water Assessment Program, Water for People, Water for Life—the United Nations World Water Development Report (Berghahn Books, Barcelona).

[20] R.B. Jackson, S.R. Carpenter, C.N. Dahm, D.M. McKnight, R.J. Naiman, S.L. Postel, et al. , Water in a changing world, Ecol. Appl. 11 (2001) 1027–1045.

[21] R.W. Macdonald, L.A. Barrie, T.F. Bidleman, M.L. Diamond, D.J. Gregor, R.G. Semkin, et al., Contaminants in the Canadian Arctic: 5 years of progress in understanding sources, occurrence and pathways, Sci. Total Environ. 254 (2000) 93–234.

[22] A guide to drinking water treatment technologies for households, Center For Disease Control, <http://www.cdc.gov/healthywater/drinking/travel/household_water_treatment.html/>.

[23] C.A. Impellitteri, C.L. Patterson, R.C.Haught, J.A. Goodrich, (2007), EPA/600/R-07/110, Small Drinking Water Systems: State Of The Industry And Treatment Technologies To Meet The Safe Drinking Water Act Requirements.

[24] S. Desai, D.A. Klanecky, Sustainability challenge: meeting the needs of the water-energy nexus, Chem. Eng. Prog. 4 (2011) 22–27.

[25] S.A. Deslauriers, M. Kanzaki, J.W. Bulkley, G.A. Keoleian, US Wastewater Treatment, Center for Sustainable Systems Fact Sheet, CSS04-14, University of Michigan, Ann Arbor, MI, 2004.

[26] European Commission Joint Research Centre, Institute for Environment and Sustainability 2010, International Life Cycle Data System (ILCD) Handbook, General guide for life cycle assessment – detailed guidance, first ed. European Commission Joint Research Centre, Institute for Environment and Sustainability: Luxembourg.

[27] J.B. Manley, P.T. Anastas, B.W. Cue, Frontiers in green chemistry: meeting the grand challenges for sustainability in R&D and manufacturing, J. Cleaner Prod. 16 (2008) 743–750.

[28] National Research Council (NRC), Enhancing The Effectiveness Of Sustainability Partnerships, National Academies Press, Washington, 2009. ISBN-13:978-0-309-12993-0.

[29] National Academy of Sciences, National Research Council, Safe Water From Every Tap: Improving Water Service To Small Communities, Committee On Small Water Supply Systems, National Academy Press, Washington, DC, 1997ISBN-10:0-309-05527-X.

[30] US Environmental Protection Agency (USEPA), Drinking Water Strategy. <http://water.epa.gov/lawsregs/rulesregs/sdwa/dwstrategy/index.cfm/>, 2011.

[31] M. Mulder, Basic Principles Of Membrane Technology, second ed., Kluwer Academic Publishers, Dordrecht, the Netherlands, 1996.

[32] J. Xi, D. Ho, B. Chu, T.D. Montemagno, Lessons learned from engineering biologically-active hybrid nano/micro devices, Adv. Mater. 15 (2005) 1233–1240.

[33] P.K. Kang, D.O. Shah, Filtration of nanoparticles with dimethyldioctadecylammonium bromide treated microporous polypropylene filters, Langmuir 13 (1997) 1820–1826.

[34] H.W. Liang, L. Wang, P-Y. Chen, H.T. Lin, L.F. Chen, D. He, et al., Carbonaceous nanofiber membranes for selective filtration and separation of nanoparticles, Adv. Mater. 22 (42) (2010) 4691–4695.

[35] E. Krieg, H. Weissman, E. Shirman, E. Shimoni, B. Rybtchinski, A recyclable supramolecular membrane for size-selective separation of nanoparticles, Nat. Nanotechnol. 6 (3) (2011) 141–146.

[36] A. Hormann, N. Teuscher, M. Pfannmöller, U. Rothe, A. Heilmann, Nanoporous aluminum oxide membranes for filtration and biofunctionalization, Small 3 (6) (2007) 1032–1040.

[37] M.G. Weinbauer, C. Beckmann, M.G. Hofle, Utility of green fluorescent nucleic acid dyes and aluminum oxide membrane filters for rapid epifluorescence enumeration of soil and sediment bacteria, Appl.Environ. Microbiol. 64 (12) (1998) 5000–5003.

[38] <http://www.whatman.com/products.aspx?PID = 193/> (accessed 18.02.13).

[39] O.A. Sadik, Detecting engineered nanomaterials, J. Environ. Monit. 9 (11) (2007) 1154.

[40] O.A. Sadik, S. Kikandi, C.Q. Wang, K. Varner, Sensors as tools for quantitation, nanotoxicity and nanomonitoring assessment of engineered nanomaterials, J. Environ. Monit. 11 (2009) 1782–1800.

[41] S. Andreescu, J. Njagi, C. Ispas, M.T. Ravalli, JEM spotlight: applications of advanced nanomaterials for environmental monitoring, J. Environ. Monit. (11) 27–40.

[42] D. Wei, MJA. Bailey, P. Andrew, T. Ryhanen, Electrochemical biosensors at the nanoscale, Lab on a Chip 9 (15) (2009) 2123–2131.

[43] C.W. Isaacson, M. Kleber, J.A. Field, Quantitative analysis of fullerene nanomaterials in environmental systems: a critical review, Environ. Sci. Technol. 43 (17) (2009) 6463–6474.

[44] S.N. Kikandi, V.A. Okello, Q. Wang, O.A. Sadik, K.E. Varner, S.A. Burns, Size-exclusive nanosensor for quantitative analysis of fullerene C-60, Environ. Sci. Technol. 45 (12) (2011) 5294–5300.

[45] M.A. Breimer, G. Yevgeny, S. Sy, O.A. Sadik, Nano Lett. 1 (2001) 305.

[46] D. Andreescu, A. Wanekaya, O.A. Sadik, J. Wang, Langmuir 21 (15) (2005) 6891–6899.

[47] D. Andreescu, O.A. Sadik, J. Electrochem. Soc. 152 (10) (2005) E299–E307.

[48] J. Karasinski, S. Andreescu, O.A. Sadik, Advanced electrochemical sensors for cell cancer monitoring, Methods 37 (1) (2005) 84–93.

[49] S. Andreescu, J. Karasinski, O.A. Sadik, Multiarray biosensors for toxicity monitoring and bacterial pathogens, in: G.K. Knopf, A.S. Bassi (Eds.), Smart Biosensors, CRC Press, Kentucky, 2007, pp. 521–538.

[50] O.A. Sadik, D. Witt, Novel monitoring strategies for environmental endocrine disruptors, Environ. Sci. Technol. – Feature Article 33 (17) (1999) A368.
[51] O.A. Sadik, J.M. Van Emon, Applications of electrochemical immunosensors for environmental monitoring, Biosens. Bioelectron. 11 (8) (1996) 1–11.
[52] S. Andreescu, O.A. Sadik, Trends & challenges in biochemical sensors for clinical and environmental monitoring, Pure. Appl. Chem. 76 (4) (2004) 861–878.
[53] O.A. Sadik, S. Andreescu, A. Wanekaya, Advanced environmental methodologies for agent detection, J. Environ. Monit. 6 (2004) 513–522.
[54] O.A. Sadik, W. Land, J. Wang, Targeting chemical and biological warfare agents at the molecular level, Electroanalysis 15 (4) (2003) 1149–1159.
[55] (a) S. Andreescu, O.A. Sadik, Correlation of analyte structures with biosensor responses using the detection of phenolic estrogens as a model example, Anal. Chem. 76 (3) (2004) 552–560. (b) S. Andreescu, O.A. Sadik, D. McGee, Autonomous multielectrode system for monitoring the interactions of isoflavonoids with lung cancer cells, Anal. Chem. 76 (2004) 2321–2330.
[56] S. Andreescu, O.A. Sadik, N. McGee, Effect of natural and synthetic estrogens on A549 lung cancer cells: correlation of chemical structures with cytotoxic effects, Chem. Res. Toxicol. 18 (3) (2005) 466–474.
[57] J. Karasinski, S. Andreescu, O.A. Sadik, B. Lavine, Multiarray sensors with pattern recognition for the detection, classification and differentiation of bacteria at species and subspecies levels, Anal. Chem. 77 (24) (2005) 7941–7949.
[58] J. Karasinski, S. Andreescu, L. White, Y. Zhang, O.A. Sadik, B.K. Lavine, et al., Detection and identification of bacteria using antibiotic susceptibility and a multiarray electrochemical sensor with pattern recognition, Biosens. Bioelectron. 22 (2007) 2646–2649.
[59] M. Breimer, G. Yevgheny, O.A. Sadik, Integrated fluorescence capillary DNA biosensor, Biosens. Bioelectron. 18 (2003) 1135–1147.
[60] F.S. Ligler, M. Breimer, J.P. Golden, D.A. Nivens, J.P. Dobson, O.A. Sadik, Integrating waveguide biosensor, Anal. Chem. 74 (2002) 713–719.
[61] A. Fatah, R. Arcilesi, T. Chekol, C. Latin, O.A. Sadik, A. Aluoch, Guide for the selection of biological agent detection equipment for emergency first responders, Guide, second ed., US Department of Homeland Security, Preparedness Directorate, Office of Grants and Training Systems Support Division, Washington, DC, 2007.
[62] I. Kowino, S. Mwilu, O.A. Sadik, Metal-enhanced biosensor for mismatch detection, Anal. Biochem. 369 (2007) 8–17.
[63] F. Yan, A. Erdem, B. Meric, K. Kerman, M. Ozsoz, O.A. Sadik, Electrochemical DNA biosensors for gene related microcystis species, Electrochem. Commun. 3 (2001) 224–228.
[64] O.A. Sadik, F. Yan, Novel biosensor for pathogenic toxins using fluorescent cyclic polypeptide conjugate, Chem. Commun. 13 (2004) 1136–1137.
[65] F. Yan, M. Ozsoz, O.A. Sadik, Electrochemical and conformational studies of microcystin-LR, Anal. Chim. Acta 409 (2000) 247–255.
[66] F. Yan, O.A. Sadik, Enzyme modulated cleavage of dsDNA for studying the interfacial biomolecular interactions, J. Am. Chem. Soc. 123 (2001) 11335–11340.
[67] O.A. Sadik, A. Wanekaya, L. Walker, M. Uematsu, L. Wong, M. Embrechts, Detection and classification of organophosphate nerve agent simulants using support vector machines with multiarray sensors, J. Chem. Inf. Model. 44 (2004) 499–507.

[68] W. Land, O.A. Sadik, D. Leibensperger, M. Breimer, Using support vector machines for development and testing intelligent sensors to combat terrorism, in: C.H. Dagli, A. L. Buczak, J. Ghosh, M.J. Embrechts, O. Ersoy, S.L. Kernel (Eds.), Using Support Vector Machines for Development and Testing Intelligent Sensors to Combat Terrorism, Artificial Neural Networks, Smart Engineering System Design, Neural Networks, Fuzzy Logic, Evolutionary Programming, Data Mining, and Complex Systems (ANNIE), vol. 12, [Peer-reviewed IEEE publication], 2002, pp. 811–816.

[69] N. Du, C. Wong, M. Feurstein, O. Sadik, C. Umbach, B. Sammakia, Flexible conducting polymers: effects of chemical composition on structural, electrochemical and mechanical properties, Langmuir 26 (2010) 14194–14202.

[70] D. Andreescu, A. Wanekaya, O.A. Sadik, J. Wang, Nanostructured polyamic membranes as electrode material, Langmuir 21 (15) (2005) 6891–6899.

[71] V.A. Okello, S. Gass, G. Pyrgiotakis, N. Du, K. Murdaugh, B. Deng, et al., Capture, isolation and detection of industrially-relevant engineered aerosol nanoparticles, Langmuir (2013).

[72] V. Okello, N. Du, B. Deng, O. Sadik, Environmental applications of poly (amic acid)-based nanomaterials, J. Environ. Monit. 13 (2011) 1236.

[73] I. Yazgan, N. Du, R. Congdon, V. Okello, O.A. Sadik, Absolute disinfection of drinking water using green nanostructured bio-membranes, J. Membr. Sci. (2014).

CHAPTER

7 Advances in Nanostructured Membranes for Water Desalination

Madhuleena Bhadra and Somenath Mitra

Department of Chemistry and Environmental Science, New Jersey Institute of Technology, University Heights, Newark, NJ, USA

7.1 Introduction

According to a report provided by the World Health Organization, 1.1 billion people lack access to adequate drinking water facilities. Furthermore, the anticipated 40–50% growth in the human population over the next 50 years, coupled with global industrialization and urbanization, will lead to freshwater shortage as a real threat. Since only 0.5% of the 1.4 billion cubic kilometers of water in the world is accessible as freshwater, the water scarcity will force us to use inferior quality and unconventional water sources such as seawater.

Street, Sustich, Duncan and Savage. Nanotechnology Applications for Clean Water, 2nd Edition.
 DOI: http://dx.doi.org/10.1016/B978-1-4557-3116-9.00007-X

Membrane separation technology can address and meet some of the challenges in water shortage. Since their inception in the early 1960s, membranes have revolutionized many separation processes [1]. Membranes can proficiently remove specific particles and molecules from liquids, and this process has been used to develop several water treatment technologies. In the areas of seawater desalination and saline water use, the two membrane-based techniques are reverse osmosis (RO) and membrane distillation (MD). The main barriers to expanding these technologies are high-energy requirements and infrastructure costs. The key to economical development of these technologies is the creation of novel membrane materials with high water permeability, superior selectivity, flux, fouling resistance, and stability.

In an effort to develop next-generation membranes with enhanced permeability, salt rejections, and low energy consumption, much effort has gone into both the synthesis of new membrane materials and new membrane architecture [2]. Of particular interest are the recent innovations in nanotechnology, where nanomaterials have been incorporated within membranes [2]. These could be successfully engineered for specific pore size, functionalities, physical, and chemical properties, thereby leading to the development of the next generation of desalination membranes. A variety of materials including carbon nanotubes (CNTs), zeolite, and nanoclay have been implemented in membrane structures for successful fabrication of nanostructured membranes. These nanomaterials have attracted significant attention due to their ability to display superior durability and separation characteristics. This chapter documents a review that intends to provide an in-depth insight into nanostructured membranes and their desalination applications by both RO and MD.

7.2 Desalination technologies

7.2.1 State of the art in RO

In RO, desalination is achieved by applying high pressure to feed water, forcing it through a semipermeable membrane that permits the flow of water molecules, while restricting the flow of salt ions. The RO process requires pressure higher than the osmotic pressure, which increases with salt concentration [3]. RO is a growing seawater desalination technology [4,5] because the energy cost is lower than thermal distillation processes such as multistage flash distillation. Currently, RO plants account for 41% of the world's desalination capacity and 80% of the number of operating plants.

RO processes were initially developed using cellulose acetate membranes and commercialized in the 1960; however, the flux was low and they were subject to biological degradation. Current state-of-the-art RO membranes are asymmetric polyamide (PA) and thin film composite (TFC) membranes fabricated using interfacial polymerization. These membranes consist of a hierarchical structure with a thin (100–1000 nm) PA selective layer fabricated on a porous polysulfone layer that offers mechanical support and minimizes pressure drop.

While RO systems have already been commercialized, improving flux, salt rejection, and fouling resistance in a cost-effective way is required to advance this technology to meet future needs. It is anticipated that higher permeability membranes may enable operation closer to the ideal osmotic pressure, thereby decreasing the energy cost; alternatively, such membranes may reduce capital cost by requiring less membrane area for a given desalination capacity. However, optimizing flux, salt rejection along with other properties such as mechanical stability and fouling resistance requires careful membrane design. This is a major challenge because improving one parameter tends to adversely affect the others [6]. Advances in nanotechnology have enabled unparalleled control over the fabrication of size-selective membranes with pore sizes in the sub-nanometer region which allow water molecules to pass through, while impeding the passage of ions that have a larger effective diameter due to their hydration shells. For example, the diameter of a hydrated sodium ion is ~7.6 Å; theoretically, if the pore diameter is smaller than that of a solvated ion but larger than a water molecule, the pore could act as a molecular sieve. It is anticipated that nanomaterials such as CNTs can serve as size exclusion membrane for small ionic species such as $Na+$ or $Cl-$. Additionally, simulation studies have also predicted that CNTs with diameter 0.34 nm with a negative charge will conduct $K+$ ions and exclude $Cl-$, whereas positively charged CNTs with diameter of 0.47 nm will conduct $Cl-$ and help in the possible exclusion of $K+$ ions [7,8]. In summary, while significant advances have been made in polymeric RO membrane technology, new approaches need to be exploited to develop the next-generation membranes. In the following sections, we review the progress in the fabrication and understanding of the transport characteristics of these nanomaterials to achieve improved desalination by RO.

7.2.2 State of the art in MD

MD is a thermally driven, membrane-based separation process which was introduced in late 1960s. In MD, hot aqueous solution is passed through the lumen of a porous hydrophobic hollow fiber [9]. While preventing the transport of the liquid phase, MD relies on the net flux of water vapor from the warm to the cool side of the membrane. Typically, MD is carried out at 60–90°C, which is notably lower than conventional distillation. Therefore, it has the potential to generate high-quality drinking water using only low-temperature heat sources such as waste heat from industrial processes and solar energy. In general, MD has received much attention as an alternative to thermal distillation and RO desalination of sea and brackish waters. Various types of MD have been known for the past several years: direct contact MD (DCMD), air gap MD (AGMD), sweeping gas MD (SGMD), and vacuum MD (VMD). These terms refer to the permeate side of the membrane. In all cases, the feed is in direct contact with the membrane. In DCMD, both sides of the membrane contact a liquid phase. The liquid on the permeate side is used as the condensing medium for the vapors. In AGMD the condensed permeate is not in direct contact with the membrane, in SGMD a sweep gas is used to remove the water vapor, and in VMD the permeate side has a vacuum.

A key component in MD is the membrane itself because it determines both flux and selectivity. As of now the throughput of MD processes is relatively low, since two major factors were impeding its development, namely membranes with adequate characteristics and unfavorable economics of the process compared to RO. Recent developments convey the coupling of MD with waste heat and renewable energy-driven systems such as geothermal and solar energy to provide greater efficiency. Some advantages of MD include the ability to operate at lower temperature, lower pressure, and the capacity to handle higher brine concentrations. In conclusion, the structural design and chemistry of membranes engineered for MD are critical to achieve high performance. In an effort to address such issues, recent times have witnessed the development of novel architectured nanostructured membranes which have enabled unprecedented control for membrane fabrication that is of great importance to enhance the desalination performances.

7.3 Nanostructured membranes

As previously mentioned, the two important membrane characteristics are flux and selectivity, which are typically controlled by physical and chemical characteristics, morphology, and presence/absence of pores. Typically, the assessment of permeability and selectivity has shown asymptotic limitations on the separation capability of conventional desalination membranes. Efforts to improve these have looked upon the development and incorporation of novel nanomaterials within the membrane structures in order to alter their structure and morphologies [2]. Recent advances have focused on incorporating nanomaterials such as nanozeolite, clay, and CNTs on the bulk membrane matrix or the surface, to enhance flux, selectivity, and fouling resistance for desalination applications.

7.3.1 Nanozeolite membranes

In 2001, molecular dynamics simulations showed that zeolite membranes previously applied for gas separations may be applicable for aqueous osmotic separations as well. Since then, nanozeolite membranes have been studied extensively for RO desalination [10–12]. Typically, the incorporation of zeolite can improve the separation performance due to the combined effect of molecular sieving, selective adsorption, or ion exchange. In this context, Li et al. [13] used mordenite framework inverted (MFI) type zeolites depicted in Figure 7.1, on a porous α-alumina support to develop 0.3-μm-thick membranes for RO desalination applications. Under an applied pressure of 2.07 MPa (20.7 bar) and with 0.1 M NaCl as feed water concentration, the modified membranes rejected 76.7% of Na^+ ions while permitting a water flux of 0.112 $kg/m^2/h$. The membrane was

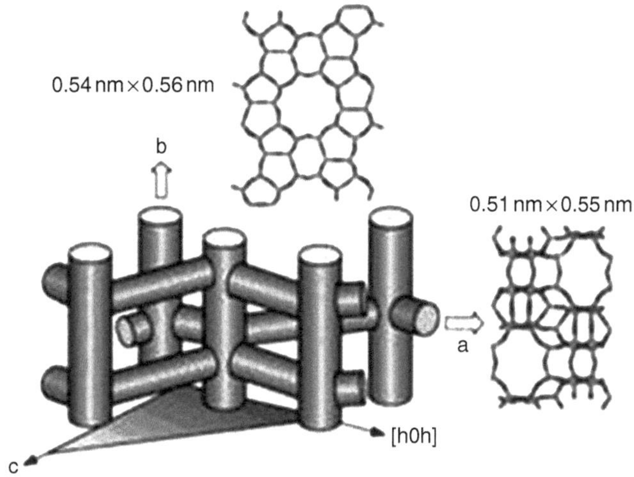

FIGURE 7.1

Structure of the MFI zeolite [13].

also tested with real RO feed water and the resulting permeate flux and rejection was comparatively lower. The cause for this lower rejection was recognized as ion transport across nanometer-sized interstitial defects, which were created during the membrane synthesis process. At this point, it is somewhat difficult to establish the role that zeolites might have in water transport and salt rejection. Additionally, if salt can transport around the zeolite crystal, perfect salt rejection cannot be achieved. Hence, further research is desirable to determine the specific transport mechanisms within the sub-nanometer zeolite pores and in the fabrication of such membranes.

7.3.2 Clay nanocomposite membranes

In addition to nanozeolites, nano clay offers interesting options for potential fillers for fabrication of nanostructured membranes. Prince et al. [14] carried out amalgamation of clay nanoparticles in polyvinylidene fluoride (PVDF) polymer solution, followed by electro spinning of the solution to produce a PVDF-clay nanocomposite membrane. Various concentrations (2, 4, and 8 wt%) of clay nanoparticles were incorporated into the nanocomposite membranes for DCMD applications. In their work, the authors observed that when the concentrations of the nanoparticles were increased, the hydrophobicity and melting point of the nanocomposite also increased. Highest water flux was observed for membranes containing 8 wt% clay. Further, the same group demonstrated that salt rejection of higher than 99.95% in DCMD process can be achieved by blending of clay in the nanocomposite nanomembranes when compared to control membrane of only 98.27%. This was due to prevention of pore wetting in clay composite membrane.

In conclusion, their study confirmed that one of the serious issues of membrane pore wetting in MD can be reduced by blending of nanoparticles and further surface properties of the membrane can also be altered by the incorporation of nanoparticles within the membranes.

7.3.3 CNT membranes

CNTs are an allotrope of carbon consisting of rolled up graphene sheets. There has been much interest in these materials because of their excellent adsorption, hydrophobicity, transport, separation characteristics, ease of functionalization, and fouling resistance. At present, few methods have been adopted to fabricate CNT membranes that could enable the development of a new generation of membranes whose permeation abilities, rejection properties, and robustness dramatically exceed those of conventional membranes for desalination applications. These fabrication approaches are documented in the following section.

7.3.3.1 CNT composite membrane

The first CNT nanocomposite membrane to reject NaCl was designed by Shawky et al. [15]. A polymer grafting process was used for the synthesis of MWCNT–PA composite membrane. In their study, the MWCNTs were initially dispersed in a solvent via ultrasonication prior to polymer grafting and membrane casting. Their method proposed that the addition of MWCNTs improved the rejection of both salt and organic matter relative to the 10% PA membrane base case. The nanocomposite membrane synthesized with 15 mg/g MWCNT in a 10% PA casting solution rejected NaCl and humic acid by factors of 3.17 and 1.67, respectively, relative to the unmodified PA membrane without MWCNTs, while membrane permeability decreased by 6.5%.

More recently, Dumee et al. [16] established CNT–buckypaper (BP) membranes by vacuum filtration of CNTs in a 99.8% pure propan-2-ol solution. Well-dispersed CNT solutions were obtained by repeated sonication at 15 min intervals. Following this procedure, they performed vacuum filtration with a 47-mm-diameter Millipore filtration unit with house line vacuum (95 kPa). The CNTs were filtered onto a poly (ethersulfone) (PES) 0.2 nm pore size Millipore membrane and then peeled off to form a self-supporting membrane for DCMD applications.

7.3.3.2 Aligned CNT membrane

Recent novel fabrication techniques that permit the assembly of vertically aligned (VA) CNTs in membrane platforms are enabling the testing of their transport performance. The first prototype for a VA–CNT membrane was introduced by Hinds's research group [17]. After being grown on an iron catalyst using the CVD process, multiwalled carbon nanotubes (MWNTs) were embedded in polymeric filler composed of polystyrene (PS). Hinds and coworkers performed a series of supplementary pressure-driven flow experiments with the MWNT/PS membrane. They found that water flow rates increased four- to fivefold over those of conventional fluid flow, which was estimated from the Hagen–Poiseuille equation. Originally, the Hinds's research group developed the CNT membrane as a chemically selective gatekeeper, which could separate different sized enzymes. Thus, they did not report ion selectivity, which is strongly related to desalination potential during the desalination process.

7.3.3.3 Interfacial polymerization synthesis of thin film nanocomposite membrane

Kim et al. [18] developed a new type of thin film nanocomposite (TFN) membrane, where MWNTs were entrapped during the formation by phase separation (Figure 7.2) of the support layer, while nanosilver (nAg) particles were incorporated as a thin film layer via interfacial polymerization. Microscopic characterization confirmed that MWNTs and nAg particles were distributed in the support layer and the thin film layer, respectively. MWNTs at a concentration of 5.0 wt% in the support layer and nAg particles at 10 wt% in the thin film layer enhanced the pure water permeability of the n-TFN membrane by 20–25%. Increase in pure water permeability and hydrophilicity of the n-TFN membrane were attributed to the diffusive effect of nanopores in the MWNTs.

7.3.3.4 Carbon nanotube immobilized membrane

Mitra and his group [19] have pioneered a novel architecture in producing CNT-based membranes for MD desalination. In their work, CNTs were immobilized within the pores of a commercially available membrane leading to the development of a unique membrane structure referred to as the carbon nanotube immobilized membrane (CNIM) (Figure 7.3). This was achieved by immobilizing CNT using dispersion in a polymer solution. Typically, the dispersion was injected into the lumen of a conventional hollow fiber under pressure. This served as the immobilization step, and the polymer served as the glue that typically held the

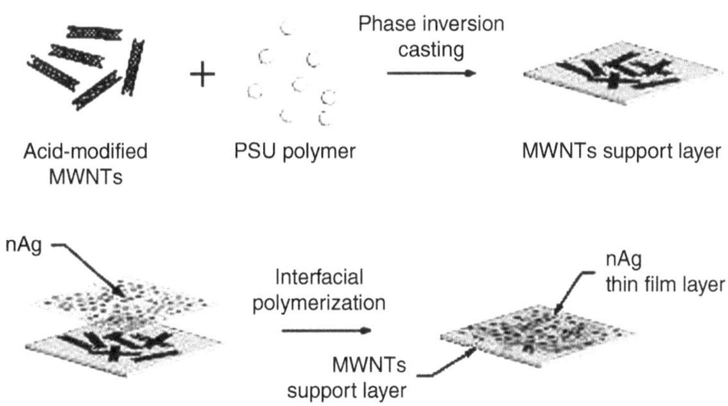

FIGURE 7.2

Schematic representation of n-TFN membrane synthesis [18].

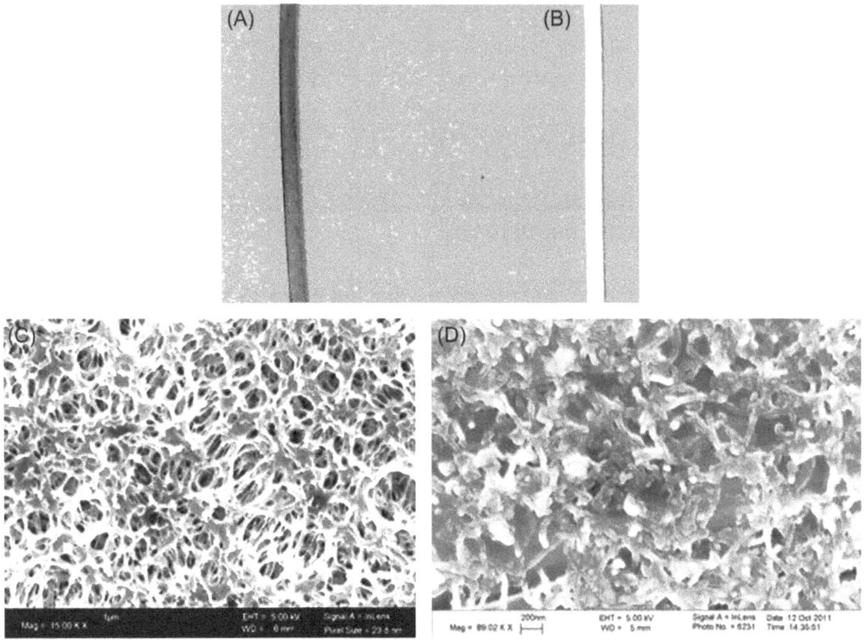

FIGURE 7.3

(A) Photograph of CNIM; (B) photograph of pure polypropylene [19]; (C) SEM image of unmodified polypropylene membrane; and (D) CNIM [19].

CNTs in place. Such membranes were robust, thermally stable, and possessed high selectivity. The goal here was to immobilize CNTs without covering its active surface with the polymer, or having a thick polymeric layer over it, so that their surface is free to interact directly with the water vapor. The membrane produced by this method has been used for other separations including extraction and pervaporation [20,21].

7.4 Application of nanostructured membranes

The nanostructured membranes are fairly innovative developments when it comes to desalination technologies. Some applications that show a great deal of promise are presented here. In the membrane desalination field, the largest application has been with the application of CNT membranes in RO and MD processes. This is attractive because the CNTs are exceedingly hydrophobic with high aspect ratios, smooth surface structure, excellent adsorption, and separation properties, which enhance permeability and selectivity. Additionally, the functionalization of CNT tip can introduce the required physicochemical characteristics into the membrane surface, which could lead to selectivity based upon physicochemical interaction of species with the functional group present over the CNT tip.

7.4.1 CNT membranes in RO

Recently developed novel fabrication techniques that permit the assembly of VA CNTs in membrane platforms are enabling the testing of their transport performance for RO applications. The size and uniformity of tubes which is required to achieve a desired salt rejection for water desalination using RO have been studied in depth by Corry et al. [22]. They studied the transport of water and ions through membranes formed from CNTs ranging in diameter from 6 to 11 Å using molecular dynamics simulations under hydrostatic pressure and equilibrium conditions. Ions face a large energy barrier and will not pass through the narrower CNTs studied ((5, 5) and (6, 6) "armchair" type tubes) but can pass through the wider (7, 7) and (8, 8) nanotubes. By measuring this conduction rate under a hydrostatic pressure difference, it was determined that membranes incorporating CNTs can, in principle, achieve a high degree of desalination far in excess of existing RO membranes.

Molecular dynamics simulations to examine water and ion transport showed that CNT functionalization [23] can lead to improved membrane performance. A range of differently charged polar functional groups were added to a 1.1-nm-diameter (8, 8) CNT that was previously found to be only moderately effective at rejecting salt ions. These CNTs were incorporated into membranes and simulations and conducted with a hydrostatic pressure difference to determine the ion rejection and flux of water passing through each as well as the energy barriers presented to ions and water molecules. It was observed that the performance of these membranes in the simulations is still many times more efficient than existing technologies and thus the inclusion of functionalized CNTs in desalination membranes may prove to be useful in achieving enhanced salt rejection and rapid water flow.

7.4.2 CNT membranes in MD

A novel desalination membrane process that also uses CNT-based membranes in MD is the CNIM [19]. It was postulated that the presence of CNTs serves as a sorbent to provide an additional pathway for solute transport. The CNTs favorably altered the water–membrane interactions to promote vapor permeability while preventing liquid penetration into the membrane pores due to their high hydrophobicity. A significant increase in the vapor permeation was observed. Figure 7.4A schematically depicts the transport and separation mechanism of the MD process with the presence of CNTs. Compared to conventionally used MD membranes, the CNIM can facilitate the MD process at relatively lower temperature. It also exhibits greater acceptability for high saline concentration up to the

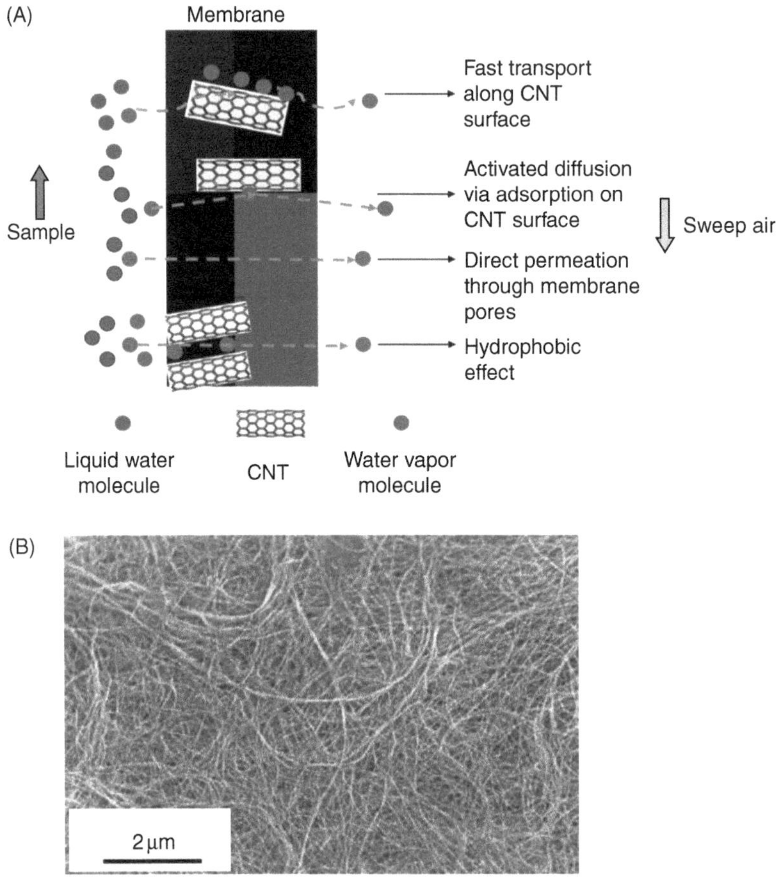

FIGURE 7.4

(A) Mechanisms of MD by CNIM made by immobilizing CNTs on polymeric membranes [19] and (B) SEM image of BP membrane surface made purely of CNTs [16].

equivalent of seawater. For a salt concentration of 34,000 mg/L and at 80°C, the nanotube incorporation led to a dramatic 1.85 times enhancement in flux.

A parallel development is the use of BP–CNTs as a membrane for desalination by MD. Dumee et al. [16] demonstrated that BPs made by self-supporting CNT–BP as observed from an SEM image (Figure 7.4B) can be used as potential high-performance membranes for desalination in a DCMD setup with excellent salt rejection (99%) and a flux rate of up to 12 $kg/m^2/h$ at a water vapor partial pressure difference of 22.7 kPa. Additionally, they observed that increasing the hydrophobicity of CNTs may slow down the rate of crack formation and may reduce surface pore wetting and thereby lead to increased water permeability. Hence, they carried out a stepwise CNT modification, where the CNTs, after treatment with ozone created hydroxyl and carboxyl functional groups, followed by reduction with $LiAlH_4$ (to transform carboxyl groups into hydroxyl functional groups) and subsequent reaction of hydroxyl groups with silage groups. Following this, the CNT–BP membranes fabricated with these modified CNTs showed increased hydrophobicity (contact angle of 140° compared to 125° for unmodified), thermal stability, and membrane limit entry pressure. The flux of the modified CNT–BP membranes nearly doubled when compared to unmodified CNT–BP. Additionally, the salt rejection of modified CNT–BP membranes by DCMD was higher (97%) than that of unmodified CNT–BP (95–97%). Overall, this study has suggested that life span, salt rejection and permeability of CNT–BP membranes can be enhanced by silane functionalization, which may be further improved by fluoro silane modification in the future.

7.5 Commercial efforts to date

Recently, a start-up company, NanoH_2O Inc. (El Segundo, CA), has commercialized the next-generation RO membrane [24]. Based on nanostructured materials and industry-proven polymer technology, NanoH_2O's Quantum Flux membranes appear to significantly improve desalination efficiency and productivity. Additionally, they proposed a method to protect against chlorine damage with a protective layer including reactive nitrogen which forms chloramines on the surface of the membranes that reduce chlorine penetration. The protective layer additionally provides substantial antifouling capabilities, whether used with a chlorinated or unclorinated feed stream, due to the antibacterial properties of chloramines. It is anticipated that more such membranes will be developed that will move forward the field of nanostructured membranes for seawater desalination.

On a similar note, recently, the excellent water transport properties of biological membranes have led to the innovative study of biomimetic nanostructured membranes incorporating aquaporins, which are proteins functioning as water-selective channels in biological cell membranes. Membranes incorporating bacterial aquaporin Z proteins have been reported [25] to show superior water transport efficiency and enhancements relative to conventional RO membranes. Aquaporins were incorporated into the walls of self-assembled polymer vesicles constituted of triblock copolymer, poly(2-methyl-2-oxazoline)-blockpoly (dimethylsiloxane)-block-poly(2-methyl-2-oxazoline). An initial permeability test was carried out on the aquaporin-triblock polymer vesicles by stopped-flow light-scattering experiments. The results show at least an order of magnitude improvement in permeability compared to commercially available TFC RO membranes. Although a salt separation test has yet to be reported, extremely high salt rejection is expected from aquaporins since their functional biological performance is to only allow the passage of water molecules. Hence, they represent an ideal opportunity for the future production of ultrapure water.

7.6 Future challenge of energy-efficient CNT membranes for desalination

There is a growing need to improve desalination technologies as the demand for clean water mounts, energy costs rise, and natural resources diminish. As current desalination technologies reach maturity, nanostructured membranes will play an increasingly important role in realizing next-generation desalination technologies. However, this offers many challenges since nanostructured membrane applications require the balancing of many parameters such as composite formation, nanomaterial functionalization, and adequate mechanical strength. However, the incorporation of nanomaterials with various functionalities that can positively alter both selectivity and permeability is probably one of the most promising approaches to the development of the next-generation desalination membranes. The nanomaterials offer major advantages by providing uniform pore size, high permeability, higher thermal stability, targeted functionalization, and they can be the nucleus of additional physical–chemical interactions such as enhancing the overall hydrophobicity, thus providing enhanced selectivity as well as flux. Considering all the above factors, much is to be done to move this technology forward to penetrate into the desalination market. For example, CNTs with a narrower distribution in pore sizes with an idyllic pore that might be less than 1 nm may be of great benefit. Regardless of all challenges, potential of nanostructured membranes is tremendous when it comes to generating higher flux than conventional membranes. Finally, there is always a question of economies of scale in real-world applications and process economics. In addition to this, the toxicity of the nanomaterials may also be an important consideration.

Acknowledgments

All authors acknowledge the support from the Electric Power Research Institute.

References

[1] M.C. Porter (Ed.), Handbook of Industrial Membrane Technology, Noyes Publications, Park Ridge, NJ, 1990.

[2] M. Bhadra, S. Mitra, Nanostructured membranes in analytical chemistry, Trends Anal. Chem. 45 (2013) 248–263.

[3] Kah Peng Lee, Tom C. Arnot, Davide Mattia, A review of reverse osmosis membrane materials for desalination—Development to date and future potential, J. Membr. Sci. 370 (2011) 1–22.

[4] A. Matin, Z. Khan, S.M.J. Zaidi, M.C. Boyce, Biofouling in reverse osmosis membranes for seawater desalination: phenomena and prevention, Desalination 281 (2011) 1–16.

[5] F. Macedonia, E. Curcio, E. Drioli, Integrated membrane systems for seawater desalination: energetic and exegetic analysis, economic evaluation, experimental study, Desalination 203 (2007) 260–276.

[6] L.F. Greenlee, D.F. Lawler, B.D. Freeman, B. Marrot, P. Moulin, Reverse osmosis desalination: water sources, technology, and today's challenges, Water research 43 (2009) 2317–2348.

[7] C. Peter, G. Hummer, Ion transport through membrane-spanning nanopores studied by molecular dynamics simulations and continuum electrostatics calculations, Biophys. J. 89 (2005) 2222–2234.

[8] J.H. Park, S.B. Sinnott, N.R. Aluru, Ion separation using a Y-junction carbon nanotube, Nanotechnology 17 (2006) 895–900.

[9] A. Alkhudhiri, N. Darwish, N. Hilal, Membrane distillation: a comprehensive review, Desalination 287 (2012) 2–18.

[10] B.H. Jeong, E.M.V. Hoek, Y. Yan, X. Huang, A. Subramani, G. Hurwitz, et al., Interfacial polymerization of thin film nanocomposites: a new concept for reverse osmosis membranes, J. Membr. Sci. 294 (2007) 1–7.

[11] M. Fathizadeh, A. Aroujalian, A. Raisi, Effect of added NaX nano-zeolite into polyamide as a top thin layer of membrane water flux and salt rejection in a reverse osmosis process, J. Membr. Sci. 375 (2011) 88–95.

[12] M.T.M. Pendergast, J.M. Nygaard, A.K. Ghosh, E.M.V. Hoek, Using nanocomposite materials technology to understand and control reverse osmosis membrane compaction, Desalination 261 (2010) 255–263.

[13] L. Li, J. Donga, T.M. Nenoffc, R. Lee, Desalination by reverse osmosis using MFI zeolite membranes, J. Membrn. Sci. 243 (2004) 401–404.

[14] J.A. Prince, G. Singh, D. Rana, T. Matsuura, V. Anbharasi, T.S. Shanmugasundaram, Preparation and characterization of highly hydrophobic poly vinylidene fluoride–clay nanocomposite nanofiber membranes (PVDF–clay NNMs) for desalination using direct contact membrane distillation, J. Membr. Sci. 397–398 (2012) 80–86.

[15] H.A. Shawky, S.R. Chae, S. Lin, M.R. Wiesner, Synthesis and characterization of a carbon nanotube/polymer nanocomposite membrane for water treatment, Desalination 272 (2011).

[16] L.F. Dumee, K. Sears, J. Schutz, N. Finn, C. Huynh, S. Hawkins, et al., Characterization and evaluation of carbon nanotube bucky paper membranes for direct contact membrane distillation, J. Membr. Sci. 351 (2010) 36–43.

[17] B.J. Hinds, N. Chopra, T. Rantell, R. Andrews, V. Gavalas, L.G. Bachas, Aligned multiwalled carbon nanotube membranes, Science 303 (2004) 62–65.

[18] E.S. Kim, G. Hwang, M. Gamal El-Din, Y. Liu, Development of nanosilver and multi-walled carbon nanotubes thin-film nanocomposite membrane for enhanced water treatment, J. Membr. Sci. 394 (2012) 37–48.

[19] K. Gethard, O.S. Khow, S. Mitra, Water desalination using carbon-nanotube enhanced membrane distillation, ACS Appl. Materials & Interfaces 3 (2011) 110.

[20] K. Gethard, O.S. Khow, S. Mitra, Carbon nanotube enhanced membrane distillation for simultaneous generation of pure water and concentrating pharmaceutical waste, Sep. Purif. Technolg. (2012) 239–245.

[21] M. Bhadra, S. Mitra, Carbon nanotube immobilized polar membranes for enhanced extraction of polar analytes, Analyst 137 (2012) 4464–4468.

[22] B. Corry, Designing carbon nanotube membranes for efficient water desalination, J. Phys. Chem. B. 112 (2008) 1427–1434.
[23] B. Corry, Water and ion transport through functionalised carbon nanotubes: implications for desalination technology, Energy. Environ. Sci. 4 (2011) 751–759.
[24] C.J. Kurth, J.A. Koehler, M. Zhou, B.A. Holmberg, R.L. Burk, Improved hybrid TFC RO membranes with nitrogen additives, US 2011005997, (2011).
[25] C.Y. Tang, Y. Zhao, R. Wang, C. Helix-Nielsen, A.G. Fane, Desalination by bio mimetic aquaporin membranes: review of status and prospects, Desalination 308 (2013) 34–40.

CHAPTER

Nanostructured Titanium Oxide Film- and Membrane-Based Photocatalysis for Water Treatment

8

Hyeok Choi[1], Abolfazl Zakersalehi[1], Souhail R. Al-Abed[2], Changseok Han[3,4] and Dionysios D. Dionysiou[3,4]

[1]Department of Civil Engineering, The University of Texas at Arlington, Arlington, TX [2]National Risk Management Research Laboratory, US Environmental Protection Agency, Cincinnati, OH [3]Environmental Engineering and Science Program, University of Cincinnati, Cincinnati, OH [4]Nireas – International Water Research Centre, University of Cyprus, Nicosia, Cyprus

8.1 TiO_2 photocatalysis and challenges

There have been tremendous efforts and achievements in the development of innovative and cost-effective modern treatment technologies to provide clean water. Among them, titanium oxide (TiO_2)-based advanced oxidation technologies (AOTs) and nanotechnologies (AONs), where reactive radical species readily attack organic chemicals in water, have received great attention for environmental remediation due to the environmentally benign properties of TiO_2 and the elimination of chemical additives [1,2]. When sufficient energy by photons from the ultraviolet (UV) spectrum of light is provided to TiO_2, electrons and holes are

Street, Sustich, Duncan and Savage. Nanotechnology Applications for Clean Water, 2nd Edition.
 DOI: http://dx.doi.org/10.1016/B978-1-4557-3116-9.00008-1

generated, and subsequently participate in redox reactions that form reactive species such as the hydroxyl radical, superoxide radical anion, and hydroperoxyl radical. Hydroxyl radicals have a high redox potential and are also nonselective in oxidation. They can oxidize almost all organic contaminants to their mineral products such as H_2O, CO_2, and other mineral species such as chloride ion and nitrate ion, a process known as mineralization [3]. In general, TiO_2 photocatalysis exhibits fast reaction kinetics for organic decomposition and thus requires short treatment times. However, for the widespread, full-scale application of this technology, several well-known concerns should be addressed, including the poor structural properties of TiO_2 material, its utilization as dispersed particles in suspension, its high energy requirement (i.e., UV radiation) for photoexcitation, and its nonselective attack behavior.

8.2 Sol–gel synthesis of porous TiO_2: surfactant self-assembling

Nanoscale control of the crystallographic, structural, and electronic properties of TiO_2 is of great interest since its photocatalytic activity is determined by the properties and availability of its surface that is irradiated by UV light. For example, the high specific surface area of TiO_2 nanoparticles is advantageous for better UV light utilization, concentration of photo-generated species and thus the rate of production of hydroxyl radicals, and reactant accessibility to the TiO_2 catalytic active sites. Among various synthesis methods of TiO_2, sol–gel technology, which involves the formation of solid inorganic materials from liquid molecular precursors, is popular for the fabrication of TiO_2 inorganic materials with engineered properties because of (i) room temperature wet-chemistry based synthesis, (ii) wide range selection of precursors and support materials, (iii) precise control of the properties of TiO_2 at the molecular level, and (iv) easy doping with other metals or nonmetals in TiO_2 matrix [4]. During the sol–gel synthesis of TiO_2, surfactant molecules as a pore directing agent in the TiO_2 inorganic matrix are introduced, as demonstrated in Figure 8.1 [5]. Under certain conditions, surfactants are known to self-assemble into various structures (micellar, hexagonal, lamellar) in a water-rich environment [6]. For example, when liquid phase TiO_2 precursors are added in the surfactant micellar aqueous solution, the TiO_2 precursors are hydrolyzed and condensed to form a solid TiO_2 inorganic network around the micelles, forming a surfactant organic/TiO_2 inorganic composite. During thermal treatment, the surfactant templates are pyrolyzed, leaving the TiO_2 inorganic matrix with a porous structure. In the case of water-poor conditions, the situation is reversed to form reverse micellar structures, then TiO_2 inorganic core/surfactant organic composites, and finally well-defined TiO_2 nanoparticles [7].

Based on this synthesis approach, in addition to the enhancement of its surface area, the structural properties of TiO_2, such as pore size and porosity, can be

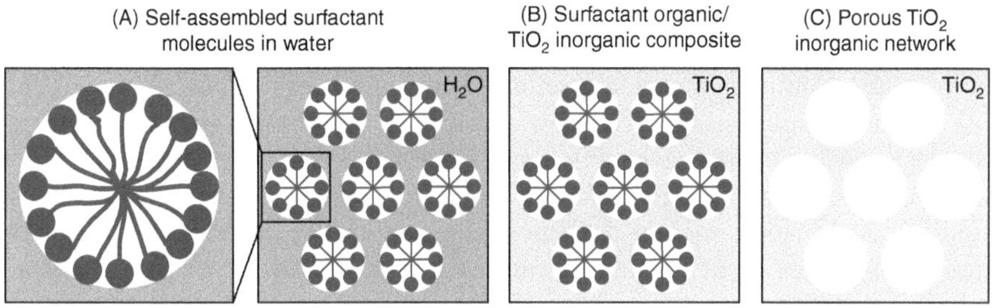

FIGURE 8.1

Schematic demonstrating a surfactant assembly-based sol–gel synthesis of mesoporous TiO_2: (A) micellar structure of surfactant molecules (hydrophilic surfactant head groups toward water molecules and hydrophobic tail groups inside) as a pore template in water-rich phase, (B) surfactant organic/TiO_2 inorganic composite after the liquid phase TiO_2 molecular precursors added are hydrolyzed to form solid inorganic TiO_2, and (C) mesoporous TiO_2 material after organic core is removed during heat treatment.

tailor-designed. Choi et al. [8–10] investigated the effect of surfactant type and concentration on the physicochemical properties of TiO_2 (i.e., physical structure, crystal phase, defect structure, impurities, bandgap energy) as well as its macroscopic phenomenological properties (i.e., homogeneity, morphology, light utilization, organic adsorption and degradation, hydrophilicity). The mesoporous TiO_2 with high surface area (~150 m^2/g) and porosity (~50%), uniform pore size distribution (2–8 nm), and small anatase crystal size (<9 nm) exhibits three to four times higher catalytic activity to degrade methylene blue dye than a conventional nonporous TiO_2 [9].

8.3 Immobilization of TiO_2 in the form of films and membranes

Recently, there have been great advances in the development of highly efficient TiO_2 films and membranes immobilized on support materials. TiO_2 particles in a system have been typically used for water treatment, thanks to their high catalytic surface area and absence of, or significantly much lower, mass transfer limitations. However, the nanosize white TiO_2 particles must be removed before the finished water is reused or discharged into the environment because of aesthetic reasons and most importantly because of possible toxicity of TiO_2 of nanosize dimensions [11,12]. Despite the very promising potential of immobilized type TiO_2 photocatalytic reactors in a variety of environmental applications, the low catalytic activity of immobilized TiO_2 due to limited activation of the TiO_2 film at the outer layer of the film, mass transfer limitations, and the partial loss and deactivation of TiO_2 have proven to be significant hurdles to the large-scale application of this technology for water treatment [13]. In order to overcome this, emphasis has been placed on the fabrication of TiO_2 films with the following properties: (i) highly porous and highly active, (ii) thin and transparent, (iii) uniform without detrimental defect structures, and (iv) mechanically stable [9,10,13]. For example, during sol–gel synthesis of TiO_2, (i) use of pore templates or particle size growth inhibitors as previously described, (ii) control of the fast hydrolysis reaction of reactive titanium alkoxide precursors, (iii) addition of chemical modifiers and binders such as diethanolamine, and/or (iv) peptization of agglomerated TiO_2 under acidic conditions, can result in uniformly dispersed TiO_2 nanocolloidal particles in a liquid phase [8–10,13,14]. Then the stable TiO_2 sol is coated onto various substrates (glass, stainless steel, alumina) and heat treated to form ultrathin (~100 nm) transparent mesoporous TiO_2 films.

More recently, attention has been also given to the fabrication of TiO_2 photocatalytic membranes, since their simultaneous physical separation and chemical decomposition functions toward water contaminants are interesting and promising for practical applications [10,15]. TiO_2 is immobilized on porous substrates, allowing water molecules to pass through the composite membrane. For this purpose, the TiO_2 coating procedure is repeated until the homogeneity of TiO_2 film layer on the support is achieved, which favorably increases organic selectivity but adversely decreases water permeability [16]. In order to minimize the hydraulic resistance through a membrane, an ideal membrane structure would exhibit gradual increase in pore size and porosity from the top to the bottom of the membrane, where the top portion typically determines its selectivity, unless many pinholes and defects are present there. Figure 8.2 shows such a hierarchical

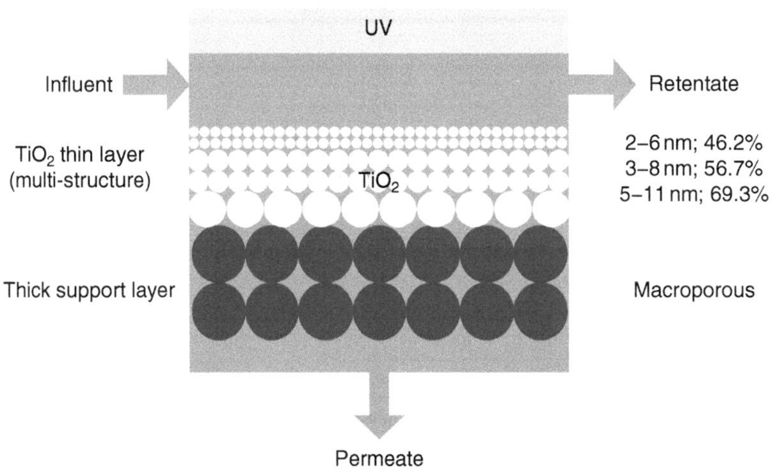

FIGURE 8.2

Schematic of TiO_2 multilayer structure showing hierarchical changes in pore size and porosity from the top to bottom of the composite membrane. Pore geometry is not necessarily similar to that illustrated here.

multilayer membrane, which was fabricated via a sol–gel method employing successive coatings of various TiO_2 sols with different surfactant properties (type and concentration) [17].

8.4 Activation of TiO_2 under visible light irradiation

The fact that irradiation of UV light with energy above the bandgap energy of TiO_2 is essential to photoexcite TiO_2 for the generation of the photo-generated species (electrons and holes) and subsequently the oxidizing hydroxyl radical species is a substantial limitation in utilizing solar light for TiO_2 activation since only 3–5% of the solar spectrum is in the UV range [1,2]. As a result, visible light activation of TiO_2 can facilitate and expedite the development of more sustainable processes for the remediation of contaminated water resources using solar light. Introduction of transition metals and nonmetal species, especially carbon (C), nitrogen (N), fluorine (F), and sulfur (S) to the TiO_2 lattice has proven to be effective for narrowing the bandgap of TiO_2 or introducing a localized midgap energy state into TiO_2 and thus utilizing visible light for TiO_2 excitation [18–24]. The use of titanium precursors combined with an N-containing ligand was proposed to increase N content in TiO_2 [25]. In another approach, N-containing surfactants like dodecylammonium chloride can be employed as a pore directing agent to control the structural properties of TiO_2 (as discussed in Section 8.2) and as a nitrogen dopant to enhance visible light response [26]. Instead of doping already synthesized TiO_2 with N, synthesis of TiO_2 and its N-doping is simultaneously processed and thus N atoms easily diffuse into tiny TiO_2 nuclei consisting of several unit cells. Recently, C and S have also been incorporated into the lattice of TiO_2 to extend the light absorption edge of TiO_2 toward the visible light region [21,22]. The C-containing pore directing agent, oleic acid, was used to provide C for the development of C-doped TiO_2 (C-TiO_2). For S-doped TiO_2 (S-TiO_2), sulfuric acid was employed as an *in situ* water forming reagent and a sulfur source. In both cases (C-TiO_2 and S-TiO_2), a reduced indirect bandgap energy of each catalyst was observed, which was 2.69 eV for C-TiO_2 and 2.94 eV for S-TiO_2, compared to that of conventional TiO_2 (3.2 eV for anatase phase). In particular, for the study of S-TiO_2, the calcination temperature was critical to maximize the S content in the lattice of TiO_2 for synthesis of highly visible light-active TiO_2. The S-TiO_2 calcined at 350°C showed the highest photocatalytic activity for water treatment [22]. Alternatively, two different nonmetal elements can be introduced to TiO_2 together in order to enhance visible light-induced photocatalytic activity of TiO_2. Two examples of codoped titania include N- and F-codoped TiO_2 (NF-TiO_2) and C- and N-codoped TiO_2 (CN-TiO_2), which were newly developed for the degradation of water contaminants [23,24]. For these studies, a C-based nonionic surfactant and a fluorosurfactant were employed as C and F precursors as well as pore directing reagents. As an N source, anhydrous ethylenediamine was introduced into the solution. In both cases, localized intra-gap states in TiO_2 were introduced, which resulted in enhanced photocatalytic activity for water treatment in the presence of visible light illumination. The ratio of fluorosurfactant in the solution was a critical parameter to provide larger Brunauer–Emmett–Teller (BET) surface area, high porosity, and high photocatalytic activity to NF-TiO_2 [23]. The calcination temperature significantly affected the synthesis of CN-TiO_2. The CN-TiO_2 calcined at 400°C showed the highest photocatalytic activity to decompose water contaminants under visible light illumination [24]. In addition, the use of pore directing reagents (i.e., surfactants) improved the physical and structural properties of nonmetal-doped TiO_2, including high porosity, larger BET surface area, and narrow pore size distribution. In the presence of visible light irradiation, all nonmetal-doped TiO_2 (C-TiO_2, N-TiO_2, S-TiO_2, NF-TiO_2, and CN-TiO_2) exhibited a high photocatalytic activity to decompose the cyanobacterial toxin microcystin-LR in contaminated water.

8.5 Selective decomposition of target contaminants

Since hydroxyl radicals generated at the TiO_2 surface non-selectively attack organic chemicals in water, the process cannot focus on the decomposition of target chemicals of concern. As a result, enhancing the preferential oxidation of target chemicals by preventing the nonspecific sorption and oxidation of coexisting non-target organic chemicals at the TiO_2 surface (e.g., NOM, organic interferences, proteins, and microorganisms) has been a challenging issue in TiO_2 photocatalysis applied for water treatment [27–29]. This becomes more critical, considering typical source water contains low concentrations of highly toxic substances (e.g., anthropogenic chemicals as target chemicals) together with high concentrations of less or nontoxic NOM as competing chemicals. Relative degradation rates between two contaminants can be altered by changing operational parameters such as temperature, solvent, and pH [30]. A variety of methods for synthesizing new TiO_2 materials and the ability to control their physical and chemical properties (e.g., particle size, crystal phase, and surface structure) can also improve selective oxidation [30]. These include preparation from different sources, doping with noble metals, over-coating with a specific material, encapsulation with a porous shell layer, impregnation onto inert organic/inorganic domains, and chemical imprinting [27–31]. For example, TiO_2 over-coated with a material that has specific affinity toward a certain contaminant can decompose the contaminant selectively. Most of the approaches basically utilize the heterogeneous nature of the photocatalytic process, i.e., sorption followed by oxidation. Controlling the mass transfer of contaminants is directly associated with their selective decomposition. However, studies which aim at achieving preferential photocatalytic degradation in a mixed stream are exceptionally rare so far. In many cases, studies represent an existing need to solve a specific problem. NOM oxidation suppression has not been a popular subject in TiO_2 photocatalysis probably because researchers have paid attention mainly to interpreting the mechanism of photocatalysis and enhancing TiO_2 catalyst reactivity.

8.6 Versatile environmental applications

In addition to the destruction of organic contaminants in water, TiO_2-based AONs can also be used to inactivate and kill pathogenic microorganisms in water [32]. Mesoporous TiO_2 materials exhibit significantly enhanced organic adsorption capacity and photocatalytic activity. TiO_2 photocatalytic membranes have an additional function—separation of organic molecules. Most importantly, TiO_2 membranes irradiated by UV show less flux decline and no significant fouling evolution over time. This is because of the photocatalytic activity of the TiO_2 skin layer, which results in the degradation and decomposition of organic foulants on the membrane surface [32]. In such membranes, the photocatalysis, disinfection, separation, and anti-biofouling functions are simultaneous. In addition, due to the synthesis of TiO_2 with nanosize dimensions, enhanced surface area, and controllability of other target-specific properties (e.g., point of zero charge, doping), the sol–gel methods are promising for the preparation of TiO_2 adsorbents for removing trace inorganic contaminants in water [33].

8.7 Suggestions and implications

The properties of TiO_2 catalyst for water treatment applications (i.e., destruction of organic contaminants, disinfection, adsorption of inorganic contaminants) can be improved when nanotechnological concepts and approaches are introduced for the synthesis of highly effective and tailor-designed TiO_2 materials. Sol–gel formulations based on surfactant micellar or reverse micellar solutions can be used as effective nanoreactors to tailor design the properties of TiO_2 films and membranes. Immobilization of TiO_2 nanomaterials can overcome the need of an additional process to separate and recover them after their water treatment applications. The inherent anti-biofouling properties of TiO_2 photocatalysts are promising for their application in the development of self-cleaning sensors and membranes. The multifunctional properties of TiO_2 films and membranes are of great interest in the field of water treatment and purification. N-TiO_2 has the potential for *in situ* remediation and disinfection of water contaminated with naturally occurring toxins, other organic harmful substances, and pathogenic microorganisms using solar light. The selectivity enhancement of the TiO_2 photocatalytic oxidation process toward target contaminants would be an ultimate goal in the area of photocatalysis to provide a sound solution to problematic treatment of water containing NOM. However, it should also be emphasized that the unique properties of TiO_2 of nanosize dimensions that may escape to the environment could also cause secondary environmental and health implications (i.e., nanotoxicity, adsorption carrier of organic and inorganic contaminants) [34]. Consequently, additional research on its environmental impacts and risks should be pursued in parallel with the development and environmental applications of new TiO_2 nanomaterials and reactors [35].

Acknowledgments

Dionysios D. Dionysiou acknowledges partial funding support for studies on TiO_2-based advanced oxidation nanotechnologies from the National Science Foundation through a CAREER award (BES 0448117) and the Cyprus Research Promotion Foundation through Desmi 2009–2010 which is cofunded by the Republic of Cyprus and the European Regional Development Fund of the EU under contract number NEA IPODOMI/STRATH/0308/09. Hyeok Choi appreciates the Faculty Research Enhancement Program Grant from the University of Texas at Arlington. Changseok Han is the recipient of the Graduate School Dean's Fellowship, which supports University of Cincinnati doctoral students in their final year of degree work. This chapter has not been subjected to internal policy review of the US Environmental Protection Agency. Therefore, the research results do not necessarily reflect the views of the agency or its policy.

References

[1] M.R. Hoffmann, S.T. Martin, W. Choi, D.W. Bahnemann, Environmental applications of semiconductor photocatalysis, Chem. Rev. 95 (1995) 69–96.

[2] J.M. Herrmann, Heterogeneous photocatalysis: state of the art and present applications, Top. Catal. 34 (2005) 49–65.

[3] A.L. Linsebigler, G.Q. Lu, J.T. Yates, Photocatalysis on TiO_2 surface: principles, mechanisms, and selected results, Chem. Rev. 95 (1995) 735–758.

[4] C.J. Brinker, G.W. Scherer, Sol–Gel Science: The Physics And Chemistry Of Sol–Gel Processing, first edition, Academic Press, San Diego, CA, 1990.

[5] D.M. Antonelli, J.Y. Ying, Synthesis of hexagonally packed mesoporous TiO_2 by a modified sol–gel method, Angewandte Chemie International Edition, England 34 (1995) 2014–2017.

[6] M.J. Rosen, Surfactants and Interfacial Phenomena, third edition, Wiley-Interscience, Hoboken, NJ, 2004.

[7] E. Stathatos, H. Choi, D.D. Dionysiou, A simple procedure of making room temperature mesoporous TiO_2 films with high purity and enhanced photocatalytic activity, J. Environ. Eng. Sci. 24 (2007) 13–20.

[8] H. Choi, E. Stathatos, D.D. Dionysiou, Effect of surfactant in a modified sol on the physicochemical properties and photocatalytic activity of crystalline TiO_2 nanoparticles, Top. Catal. 44 (2007) 513–521.

[9] H. Choi, E. Stathatos, D.D. Dionysiou, Synthesis of nanocrystalline photocatalytic TiO_2 thin films and particles using sol–gel method modified with nonionic surfactants, Thin Solid Films 510 (2006) 107–114.

[10] H. Choi, E. Stathatos, D.D. Dionysiou, Sol–gel preparation of mesoporous photocatalytic TiO_2 films and TiO_2/Al_2O_3 composite membranes for environmental applications, Appl. Catal., B 63 (2006) 60–67.

[11] T.C. Long, N. Saleh, R.D. Tilon, G.V. Lowry, B. Veronesi, Titanium dioxide (P25) produces reactive oxygen species in immortalized brain microglia (BV2): implications for nanoparticle neurotoxicity, Environ. Sci. Technol. 40 (2006) 4346–4352.

[12] L.K. Limbach, Y. Li, R.N. Grass, M.A. Hintermann, M. Muller, D. Gunther, et al., Oxide nanoparticle uptake in human lung fibroblasts: effects of particle size,

agglomeration, and diffusion at low concentrations, Environ. Sci. Technol. 39 (2005) 9370–9376.

[13] G. Balasubramanian, D.D. Dionysiou, M.T. Suidan, V. Subramanian, I. Baudin, J.M. Laıne, Titania powder modified sol–gel process for photocatalytic applications, J. Mater. Sci. 38 (2003) 823–831.

[14] C. Wang, Z.X. Deng, Y. Li, The synthesis of nanocrystalline anatase and rutile titania in mixed organic media, Inorg. Chem. 40 (2001) 5210–5214.

[15] M.A. Anderson, M.J. Gieselmann, Q. Xu, Titania and alumina ceramic membranes, J. Memb. Sci. 39 (1988) 243–258.

[16] L.Q. Wu, P. Huang, N. Xu, J. Shi, Effects of sol properties and calcination on the performance of titania tubular membranes, J. Memb. Sci. 173 (2000) 263–273.

[17] H. Choi, A.C. Sofranko, D.D. Dionysiou, Nanocrystalline TiO_2 photocatalytic membranes with a hierarchical mesoporous multilayer: synthesis, characterization, and multifunction, Adv. Funct. Mater. 16 (2006) 1067–1074.

[18] C. Wang, D.W. Bahnemann, J.K. Dohrmann, A novel preparation of iron-doped TiO_2 nanoparticles with enhanced photocatalytic activity, Chem. Commun. (2000) 1539–1540.

[19] R. Asahi, T. Morikawa, T. Ohwaki, K. Aoki, Y. Taga, Visible-light photocatalysis in nitrogen-doped titanium oxides, Science 293 (2001) 269–271.

[20] C. Burda, Y. Lou, X. Chen, A.C.S. Samia, J. Stout, J.L. Gole, Enhanced nitrogen doping in TiO_2 nanoparticles, Nano. Lett. 3 (2003) 1049–1051.

[21] G. Liu, C. Han, M. Pelaez, D. Zhu, S. Liao, V. Likodimos, et al., Synthesis, characterization and photocatalytic evaluation of visible light activated C-doped TiO_2 nanoparticles, Nanotechnology 23 (2012), pp. 294003 (10pp.).

[22] C. Han, M. Pelaez, V. Likodimos, A.G. Kontos, P. Falaras, K. O'Shea, et al., Innovative visible light-activated sulfur doped TiO_2 films for water treatment, Appl. Catal., B 107 (2011) 77–87.

[23] M. Pelaez, P. Falaras, V. Likodimos, A.G. Kontos, A.A. de la Cruz, K. O'Shea, et al., Synthesis, structural characterization and evaluation of sol–gel-based NF-TiO_2 films with visible light-photoactivation for the removal of microcystin-LR, Appl. Catal., B 99 (2010) 378–387.

[24] G. Liu, C. Han, M. Pelaez, D. Zhu, S. Liao, V. Likodimos, et al., Enhanced visible light photocatalytic activity of C–N-codoped TiO_2 films for the degradation of microcystin-LR, J. Mol. Catal. A: Chem. 372 (2013) 58–65.

[25] T. Sano, N. Negishi, K. Koike, K. Takeuchi, S. Matsuzawa, Preparation of a visible light-responsive photocatalyst from a complex of Ti^{4+} with a nitrogen-containing ligand, J. Mater. Chem. 14 (2004) 380–384.

[26] H. Choi, M.G. Antoniou, M. Pelaez, A.A. de la Cruz, J.A. Shoemaker, D.D. Dionysiou, Mesoporous nitrogen-doped TiO_2 for the photocatalytic destruction of the cyanobacterial toxin microcystin-LR under visible light, Environ. Sci. Technol. 41 (2007) 7530–7535.

[27] S. Ghosh-Mukerji, H. Haick, Y. Paz, Controlled mass transport as a means for obtaining selective photocatalysis, J. Photochem. Photobiol., A 160 (2003) 77–85.

[28] D. Sharabi, Y. Paz, Preferential photodegradation of contaminant by molecular imprinting on titanium dioxide, Appl. Catal., B 95 (2010) 169–178.

[29] Y. Paz, Composite titanium dioxide photocatalysts and the "adsorb & shuttle" approach: a review, Solid State Phenomena 162 (2010) 135–162.

[30] Y. Paz, Preferential photodegradation – why and how? C.R. Chim. 9 (2006) 774–787.
[31] R. Kato, N. Shimura, M. Ogawa, Controlled photocatalytic ability of titanium dioxide particle by coating with nanoporous silica, Chem. Lett. 37 (2008) 76–77.
[32] H. Choi, M.G. Antoniou, A.A. de la Cruz, E. Stathatos, D.D. Dionysiou, Photocatalytic TiO_2 films and membranes for the development of efficient wastewater treatment and reuse systems, Desalination 202 (2007) 199–206.
[33] D.E. Giammar, C.J. Maus, L. Xie, Effects of particle size and crystalline phase on lead adsorption to titanium dioxide nanoparticles, J. Environ. Eng. Sci. 24 (2007) 85–95.
[34] G. Jegadeesan, S.R. Al-Abed, V. Sundaram, H. Choi, K.G. Scheckel, D.D. Dionysiou, Spectroscopic investigation of arsenic adsorption and As(III) oxidation on amorphous TiO_2 nanoparticles, Water. Res. 44 (2010) 965–973.
[35] M.R. Wiesner, G.V. Lowry, P. Alvarez, D.D. Dionysiou, P. Biswas, Assessing the risks of manufactured nanomaterials, Environ. Sci. Technol. 40 (2006) 4336–4345.

CHAPTER

9 Nanotechnology-Based Membranes for Water Purification

Eric M.V. Hoek[1], Mary Theresa M. Pendergast[1] and Asim K. Ghosh[2]

[1]*Department of Civil and Environmental Engineering and California NanoSystems Institute, University of California, Los Angeles, CA, USA* [2]*Desalination Division, Bhabha Atomic Research Centre, Trombay, Mumbai, India*

9.1 Introduction

Almost 150 years ago, the physicist James Clerk Maxwell conceived his "sorting demon," a fictitious being that discriminates perfectly between molecules without the expenditure of work—the ideal membrane [1]. Since that time, various membrane processes have been developed, each with specific applications in mind. Membrane-based water purification processes are now among the most important and versatile technologies for conventional drinking water production, wastewater treatment, ultrapure water production, desalination, and water reuse. Commercially available membrane processes for water purification include electrodialysis, electro-deionization, reverse osmosis (RO), nanofiltration (NF), ultrafiltration, and microfiltration. Electrodialysis, electro-deionization, and NF are useful in demineralization, softening, and organic separations, but RO and ultrafiltration membranes

Street, Sustich, Duncan and Savage. Nanotechnology Applications for Clean Water, 2nd Edition.
 DOI: http://dx.doi.org/10.1016/B978-1-4557-3116-9.00009-3

are the workhorse technologies for desalination and water reuse. Other membrane-based processes such as forward osmosis, membrane distillation, and pervaporation are emerging, but have found limited application in practice.

In principle, intrinsic advantages of membrane processes include continuous, chemical-free operation, low-energy consumption, easy scale-up and hybridization with other processes, high process intensity (i.e., small land area per unit volume of water processed), and highly automated process control. General disadvantages of membrane processes are short membrane lifetime, limited chemical selectivity, concentration polarization, and membrane fouling [1]. Polarization and fouling of RO membranes require extensive physical and chemical feed water pretreatment (e.g., chlorination, in-line coagulation, flocculent aid addition, membrane filtration), low recovery operation, extensive chemical cleaning, and frequent operator intervention. RO processes further suffer from high intrinsic energy consumption, environmental issues associated with feed water intake and brine discharge, and the need for chemical conditioning of product water.

Nanotechnology promises to dramatically enhance many water purification technologies, such as adsorption, ion exchange, oxidation, reduction, filtration, membranes, and disinfection processes [2]; however, one key issue related to nanotechnology is the question of how to best apply it. Specifically, it is not clear how to interface nanoparticles with contaminants. At present, many expensive nanoparticles cannot be added to water like commodity chemicals and some nanoparticles could present new hazards to human health and the environment [3]. Thus, additional separation processes are required to recover nano-materials for risk avoidance and reuse. A promising approach is to immobilize nano-materials on or within a solid matrix, such as a membrane. The resulting membrane may exhibit improved separation performance, chemical, thermal, or mechanical stability, interfacial properties, or advanced functionality depending on the nano-material selected.

This chapter presents a review of six emerging nanotechnology-based membrane material concepts intended for use in water purification: (1) zeolite-coated ceramic membranes, (2) inorganic–organic thin film nanocomposite (TFN) membranes, (3) hybrid protein–polymer biomimetic membranes, (4) aligned carbon nanotube (CNT) membranes, (5) self-assembled block copolymer membranes, and (6) graphene-based membranes. We have selected these membrane nanotechnologies because each is touted as a possible low-energy replacement for conventional RO membranes. Our objective is to review the basic concepts underlying each technology, the respective materials and methods of fabrication, published separation performance, and current commercialization efforts.

9.2 Zeolite-coated ceramic membranes

Ceramic membrane development is currently seeking to form membranes with water permeability on the range of ultrafiltration with the solute selectivity of

NF or even RO membranes [4]. In 2001, molecular dynamics simulations showed that zeolite membranes—previously applied for gas separations—may be applicable for aqueous solution separations, in particular forward osmosis and RO [5]. Since then thin zeolite membranes have been studied for RO desalination of brackish water and a variety of wastewaters, showing promising rejection of salts and organics [6–11]. The basic ceramic membrane concepts and synthesis procedures are considered conventional; however, novel applications involve the use of nanoparticulate zeolites. Hydrophilic zeolite membranes applied for gas separations, such as pervaporation, are composed of a loose, thick zeolite film through which separation occurs; however, RO membranes require an ultrathin, dense layer and so pains must be taken to form nanoscale zeolite coating films for this technology to prove practical for water treatment applications.

Zeolites are naturally occurring aluminosilicate minerals with highly uniform subnanometer and nanometer scale crystalline structures. Typical zeolite membranes are amorphous silicate, aluminosilicate, or aluminophosphate crystalline structures formed via hydrothermal synthesis [9,10]. Other synthesis methods include *in situ* layer-by-layer crystallization and dry gel conversion in the presence of a template-water vapor [11]. Aluminosilicate crystals are intrinsically inert, imbuing these membranes with extreme thermal and chemical stability [12]. Zeolite crystals consist of a three-dimensional (3D) cross-linked $(Si/Al)O_4$ tetrahedral framework, which contains cavities that allow for the movement and containment of ions and water molecules [13]. The containment of molecules in a given zeolite framework is a function of temperature, water content, ion type, and the ratio of Si to Al atoms in the matrix [14]. Many natural zeolites can be produced synthetically, while additional structures, with no natural occurrence, have been synthesized and are characterized as zeolites based on their structures, such as zeolite-A produced by Linde Corporation [14].

Common zeolite materials employed in membranes include Mordenite Framework Inverted (MFI)-type, sodalite (SOD), and linde type A (LTA). Zeolite ZSM-5 (MFI)—the most commonly applied zeolite in membranes—contains straight channels in one direction and sinusoidal channels in the other direction [12]. Since these channels are not interconnected, orientation of the crystals with respect to the flow path is required. Hydroxyl SOD has also been applied in membrane materials [9]. SODs are not mineralogically defined as zeolites, but feldspathoids because in nature salt molecules are contained in their frameworks. The SOD cage, however, is quite common to zeolite structures and when crystalline networks are created with this cage structure, zeolitic properties are exhibited. The LTA structure is composed of SOD cages connected by truncated cubo-octahedron, forming an interconnected flow channel [14]. Zeolitic particles with interconnected cage structures like those in LTA offer simplified fabrication since alignment is unnecessary.

Pore size and framework density are the primary factors of concern when considering zeolites for water separations; pore size determines ion selectivity and framework density determines water permeability (Figure 9.1). Atoms other than Si and Al can be substituted into the cage structures of zeolites via ion exchange to imbue alternate charge and structural properties. Since the ability to act as a molecular sieve is due to the channel widths, changing the atoms in the framework, and thus the channel widths, will change the sieve properties [15]. Additionally, both the ion and water molecule mobility through a zeolite depend upon the relative density of the framework structure; open porous structures will facilitate less hindered transport [14]. This is indicated by the framework density, i.e., the number of Si or Al atoms per 1000 Å. The framework density (FD) values (normalized for ideal Si frameworks) are 18.4, 16.7, and 14.2 for MFI, SOD, and LTA, respectively [16], implying that LTA would be expected to have the largest water mobility.

The Si to Al ratio of a zeolite cage is the most important factor affecting chemical stability, hydrophilic properties, and occurrence of inter-crystalline defects. An increase in Si:Al implies a decrease in the overall surface charge on the framework [14]. The MFI-type zeolites are capable of a large range of Si:Al, from

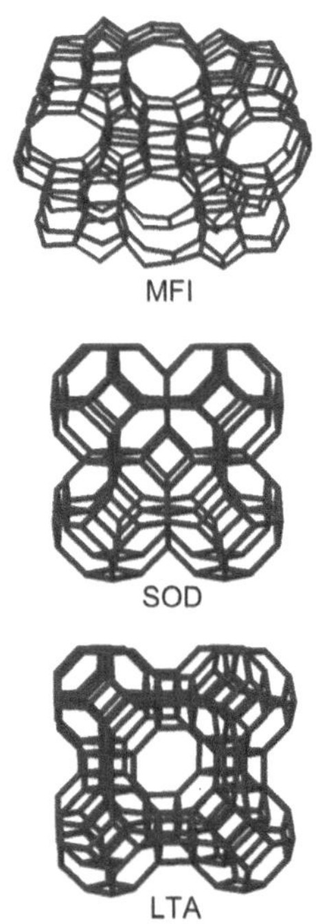

FIGURE 9.1

Framework structures of zeolites commonly used in membrane materials: MFI, SOD, and LTA. From the Structure Commission of the International Zeolite Association [16].

~30 in the ZSM-5 form to nearly pure Si for the isomorphous silicate-type MFI. Noack et al. [11] find that as the Si:Al decreases in MFI-type zeolites, water permeability and selectivity for water increase; however, defects simultaneously increase until a point where selectivity decreases. Additionally, the Si:Al allows for tuning of the surface properties and the resultant electrostatic double layer; such membranes could also be tuned for specific ion-selective applications [7].

By normalizing reported zeolite membrane permeabilities by their film thicknesses, comparisons can be drawn against conventional polymeric RO membranes on the basis of specific hydraulic resistance [17]. While most zeolite films (~3–50 μm thick) formed to date do not have overall permeabilities comparable to thin film composite (TFC) (~50–250 nm) RO membranes, the intrinsic permeabilities per thickness are comparable; in fact, membranes produced by Kazemimoghadam [9] appear to have intrinsic permeabilities three orders of magnitude higher than commercial seawater RO membranes. This implies that zeolite films could provide an economically viable alternative material for high-flux RO membranes with dramatically enhanced chemical, thermal, and mechanical stability. The challenge remains to improve control over crystal nucleation and growth to ensure defect-free ultrathin zeolite films, which may require abandoning or substantially modifying traditional hydrothermal synthesis methods [10]. No commercialization efforts have yet been seen.

9.3 Inorganic–organic TFN membranes

In general, nanocomposite materials are created by introducing nanoparticle materials (the dispersed "filler") into a macroscopic material (the continuous "matrix") [18]. The resulting nanocomposite material may exhibit drastically enhanced properties, such as mechanical properties (e.g., strength, modulus, and dimensional stability); chemical and thermal stabilities; permeability for gases, water, and hydrocarbons; electrical and thermal conductivities; surface properties, optical properties, or dielectric properties. For example, dispersing molecular sieve nanoparticles into polymers can produce mixed matrix membrane materials with improved gas mixture permselectivity [19]. In general, the nano-phase is dispersed into the matrix during processing and the mass fraction of nanoparticles introduced is very low (generally $<5\%$) due to the incredibly high surface area-to-volume ratio of nanoparticles.

Research and development of mixed matrix membranes is most extensive for gas separations, pervaporation, and fuel cell applications where the effort focuses on developing more efficient combinations of matrix and filler materials and into better controlling membrane formation. More recent research also focuses on incorporating nano-materials into microfiltration and ultrafiltration membranes to achieve a variety of goals, including enhanced permeability, decreased propensity for fouling, targeted pollutant degradations, and increased thermal and mechanical stabilities [20–26], while maintaining the ease of fabrication and low cost of their fully polymeric counterparts.

Nanoparticle additions have also been made to the selective layers of TFC membranes in order to take advantage of the properties of the nano-materials. The addition of nanoparticles into the polymerization process and through coatings and self-assembly atop membranes has brought about TFN membranes, which offer potential benefits of enhanced performance, reduced fouling, and other novel functionality [27–32]. As with TFC membranes, the two layers of TFN membranes can be independently tuned with nanoparticle additions to the support structure, selective thin film, or both.

In the earliest known study of TFN technology, Jeong et al. [27] describe the synthesis and characterization of zeolite–polyamide TFN membranes formed by interfacial polymerization. The general approach to TFN membrane formation is similar to that of traditional polyamide TFC membranes, but nanoparticles are dispersed in the initiator solution prior to interfacial polymerization, as depicted schematically in Figure 9.2. TFN membranes offer new degrees of freedom in designing NF and RO membranes because the nanoparticle and polymer phases can be independently tuned to impart a wide array of separation performance and novel functionality.

Jeong et al. [27] demonstrate increasing water permeability as a function of nanoparticle loading in MPD–TMC thin films with no observed decrease in rejection of salts and low molecular weight organic solutes. In addition, as nanoparticle loading increased, TFN membranes became increasingly more hydrophilic, negatively charged, and smooth than pure polyamide TFC counterparts

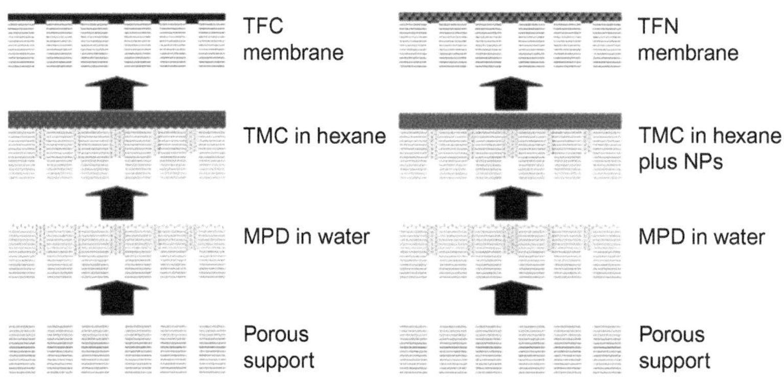

FIGURE 9.2

Schematic depiction of TFC and TFN membrane fabrication. MPD, *m*-phenylene diamine; TMC, trimesoyl chloride; TFC, thin film composite (pure polyamide film), and TFN, thin film nanocomposite (zeolite–polyamide nanocomposite film).

[27,28]. Lind et al. [30] evaluated the sizes of the nanoparticles used and found that larger nanoparticles produce membranes with highly favorable surface properties, while smaller nanoparticles produce increased permeability by causing larger characteristic pores. All TFN membranes are found to be less cross-linked than the TFC counterparts, suggesting a more open structure and implying that the nanoparticles used can be tailored to particular membrane applications to enhance membrane performance.

Additionally, resistance to physical compaction has been imparted to RO membranes by including nanoparticles both in the polysulfone support of TFC membranes and in the polyamide coating of TFN membranes [31,32]. In general, nanocomposite membranes have higher initial water permeation and less flux decline during compaction, implying lower overall operating energy is required. Permeability of the TFN membranes is intermediate between commercial seawater and brackish water RO membranes, while TFN membrane NaCl rejection is the highest. The seawater desalination performance of laboratory prepared TFC and TFN membranes was evaluated (Figure 9.3) along with commercially fabricated seawater RO membranes. Hand-cast TFC and TFN membranes rival commercial membranes (characterized in flat-sheet form by the authors and in spiral wound element by the manufacturers) simultaneously exhibit salt passages equivalent to commercial seawater RO membranes with one to three times higher permeability.

Li et al. [8] formed TFN membranes containing silver and titania nanoparticles with the goal of producing potentially antimicrobial and UV-active NF membranes. TFN NF membranes were prepared following the method of Jeong et al. [27].

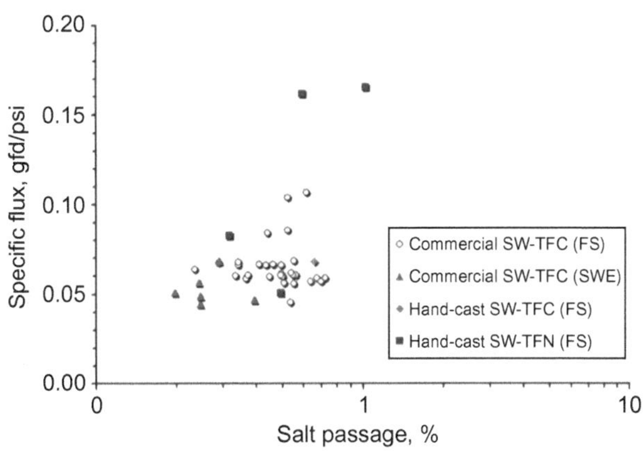

FIGURE 9.3

Specific flux and salt passage of commercial (SWRO) and hand-cast RO membranes. gfd, gallons per square foot per day; FS, flat-sheet membranes tested in a bench scale RO simulator, and SWE, manufacturer provided performance data derived from single element tests.

Silver nanocomposite membranes exhibited virtually identical water flux and salt rejection as pure polyamide equivalents. Most of the Ag^0 particles remained on the surface even after the performance test as confirmed by scanning electron microscopy (SEM), X-ray photoelectron microscopy (XPS), and atomic force microscopy (AFM). The authors describe "anti-biofouling tests of silver nanoparticle immobilized membranes," in which *Pseudomonas* (species unspecified) was deposited on as-synthesized silver–TFN membranes, and cultivated in the incubator at 37 °C and at 90 ± 5% humidity for 24 h. From visual inspection of SEM images, the authors speculate "colonies of *Pseudomonas* were formed in the membrane without Ag nanoparticles, whereas those in the Ag incorporated membrane were almost killed." However, no analyses are presented to quantify cell adhesion or to cell growth on the membranes. The same authors report that incorporation of titania nano-particles in MPD–TMC thin films increases water flux due to the enhanced hydrophilicity of the TFN membranes. Optimum membrane performance and good mechanical properties were achieved at a titania loading up to 5% (w/w). Above this critical concentration, poor mechanical properties and separation performance were reported.

A company was formed in California (see www.nanoh2o.com) to commercialize nanocomposite membranes for desalination and water purification. According to the company's web site, NanoH_2O's nanostructured membranes exhibit a wide array of advantageous membrane characteristics, including "improved permeability while maintaining requisite salt and contaminant rejection, both passive and active fouling resistance, as well as "tunable" membrane performance to address specific water chemistries. By leveraging existing membrane synthesis techniques, NanoH2O's advanced membrane technology requires few modifications to existing commercial manufacturing facilities—and fits within current desalination pressure vessels without alteration." Laboratory data presented above (Figure 9.3) suggests that nanocomposite RO membranes can be fabricated with similar rejection, but twice the permeability of commercially available seawater RO membranes. In addition, various nanoparticles can be selected to impart a wide array of antimicrobial and photo-catalytic capabilities, which may improve fouling resistance.

9.4 Hybrid protein–polymer biomimetic membranes

Biomimetic membranes are designed to imitate the highly selective transport of water and solutes across biological membranes. The general approach to forming biomimetic membranes is to incorporate active transport proteins (isolated from cell cultures) within a vesicular or planar lipid bilayer or a more stable synthetic analog, e.g., ABA triblock copolymer monolayers [33]. The protein may facilitate diffusion, cotransport, or counter-transport in the presence of concentration gradients, light energy, or chemical energy [34]. While single channel measurement in lipid bilayers has served as the basis for understanding membrane protein function, building functional devices incorporating such proteins requires high yield insertion and integration within a stable matrix [33,35]. Further stability of lipid or block copolymer membranes can be achieved by cross-linking the matrix via UV irradiation or by adding free-radical initiators such as azobisisobutyronitrile [1].

Perhaps the most promising approach for water purification is the development of aquaporin (AQP) protein–based biomimetic membranes. The interior of, for example, AQP-1 presents a 15 Å long, narrow channel through which water molecules line up and permeate by hopping along in single file [36–38]. Molecular dynamics simulations have been used to characterize osmotic and diffusion permeability coefficients for water through AQP-1 by applying chemical potential and hydrostatic pressure differences, respectively, across a single protein. Simulations agree well with previously published experimentally derived permeability coefficients and they reveal the osmotic permeability to be about an order of magnitude (~12–13 times) larger than the diffusion permeability. If this relationship is universal, AQP protein-based membranes might be better suited for forward osmosis separations (concentration-driven) rather than RO separations (pressure-driven). These and other simulations also suggest that water hopping rates in narrow CNTs are 7–14 times faster and thus potentially more promising for creating low-energy desalination membranes [39]. This is discussed in more detail below.

Water purification membranes built from AQP proteins embedded within lipid or copolymer films have been published in both the patent and open literature. For example, a patent was awarded in 2007 for methods of incorporating biological membrane proteins into a copolymer matrix to produce biomimetic membranes with a wide variety of functionalities [40]. In one form of the invention, a composite membrane incorporates two different proteins that cooperate to produce electricity from light. In another form, water transport proteins are embedded in a membrane to enable water purification. The patent claims, "The preferred form described has the form of a conventional filter disk. To fabricate such a disk, a 5 nm thick monolayer of synthetic triblock copolymer and protein is deposited on the surface of a 25 mm commercial ultrafiltration disk using a Langmuir–Blodgett trough. The monolayer on the disk is then cross-linked using UV light to the polymer to increase its durability. The device may be assayed by fitting it in a chamber that forces pressurized source water across the membrane." However, there is no guidance as to how one should select a synthetic triblock copolymer nor is there any data in support of the actual function of the embedded AQP.

Kumar et al. [41] characterize transport characteristics of AQP-Z containing symmetric poly-(2-methyloxazoline)–poly-(dimethylsiloxane)–poly-(2-methyloxazoline) ($PMOXA_{15}$–$PDMS_{110}$–$PMOXA_{15}$) triblock copolymer vesicles. AQP-Z is selected based on the ability for high water permeation and high selectivity. In addition, it is easy to purify and multiply using a recombinant *Escherichia coli* strain. The symmetric triblock copolymer with a high hydrophobic to hydrophilic block ratio was selected, reminiscent of a lipid-bilayer membrane. The authors report water permeability coefficients through AQP-containing vesicles that are almost 100 times higher than commercial RO membranes, whereas vesicles without AQPs exhibit permeabilities 10 times lower than RO membranes. The authors also claim high solute selectivity reporting, "The calculated reflection coefficients of salt, glycerol, and urea were greater than one, indicating higher relative rejection of these solutes (data not shown)." The concept of a "reflection coefficient" derives from nonequilibrium thermodynamic arguments, which give rise to values of the reflection coefficient ranging from 0 to 1 [42,43]. A value larger than one is theoretically impossible, which suggests that the authors' measurements or calculations may be improperly communicated in the chapter; regardless, this shows the functionality of AQPs in synthetic applications.

Transport across biological membranes is driven by osmotic pressure gradients, rather than mechanically applied pressure gradients as in industrial filtration processes. Kaufman et al. [44] demonstrate that lipid bilayers were formed atop NF membranes that can be operated under a mechanical driving force as RO membranes. NF membranes are chosen as the support because of their high permeability and low surface roughness that allow for minimal distortion of the lipid bilayer. AQP solutions (of protein PM28, the integral protein of a spinach leaf plasma membrane) are deposited onto commercially available NF membranes via vesicle fusion. Electrostatic interactions are tailored to optimize surface coverage with the lipid bilayer, and full, defect-free coverage is implied by the drastic decrease in permeability of the composite membrane.

Currently, scale-up is the primary concern for biomimetic membrane development. Some fabrication methods form free-standing lipid/polymer membranes and hydrogel-supported lipid layers [45–49]. Others, following the lead of Kaufman et al. [44], demonstrate the approaches to form AQP films atop robust conventional supports [50–52]. A recent approach has been shown by Zhao et al. [53] in which AQPs are first incorporated into proteoliposomes and the resulting structures are included in a cross-linked polyamide coating film (Figure 9.4). Similar

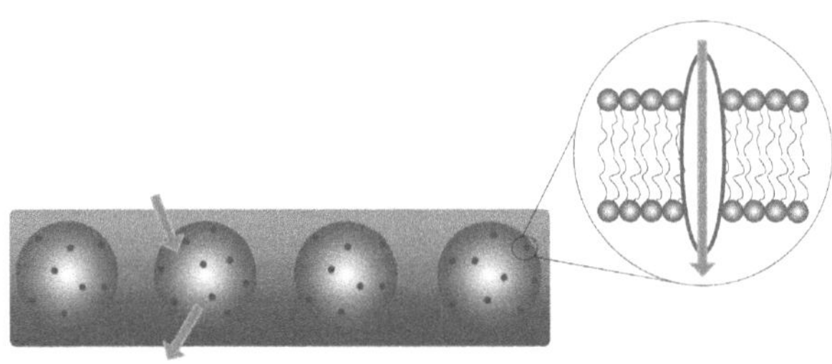

FIGURE 9.4

AQP-containing proteoliposomes are incorporated into the polyamide thin film during the interfacial polymerization process and act as preferential flow paths through the selective layer of the RO membrane. Image adapted from Zhao et al. [53].

to the zeolitic TFN membranes discussed above, here, the proteoliposomes form highly selective preferential flow paths for water passage through the selective layer. These membranes were shown to produce ~40% higher permeability than commercial brackish water RO membranes with similar or greater salt rejection. In addition, these membranes have been produced in 200 cm^2 sheets and hold promise for further scale-up, thanks to their utilization of the conventional interfacial polymerization method.

A company was formed in Denmark (see www.aquaporin.dk) to commercialize AQP membranes for water production via RO as well as forward osmosis and energy production by pressure retarded osmosis—as indicated by their patents [54,55]. Membranes are currently available in customizable sizes up to 20 cm × 25 cm and the company targets an automated roll-to-roll production by the end of 2013. The patent claims, "The invention primarily aims at developing an industrial water filtration membrane and device comprising AQPs incorporated into a membrane capable of purifying water with the highest purity, e.g., 100%. No techniques or filters known today can perform this task." Outside of this company's patent filings, there is little information available about how their membranes are made, tested, or fabricated in commercial-scale systems, but the recent work by Zhao et al. [53] indicates it may be possible. Further concerns about the long-term and chemical stability of these materials remain. Nonetheless, biomimetic membranes may be the closest real-world analog to Maxwell's sorting demon because of (a) their high selectivity and (b) the strong basis for further development of bio-inspired water purification membranes.

9.5 Aligned CNT membranes

It is not yet clear if robust protein-based membranes can be developed for practical water treatment applications—especially when subjected to hydraulic pressure, membrane fouling, and chemical cleaning over multiple years of service. Another approach toward bio-inspired membranes is the formation of aligned CNT membranes, reported to exhibit fast mass transport of both gases and water [56,57]. The biomimetic inspiration for aligned CNT membranes comes from studies that demonstrate single-walled CNT channels can be designed to mimic the rapid water transport through biological proteins in addition to gated/active transport [58–60]. Tremendous interest in aligned CNT membranes is evident from the plethora of studies that suggest dramatic fluxes of water through individual CNTs or aligned films of CNTs [61–70].

A molecular dynamics study suggests that membranes comprising subnanometer diameter CNTs can desalt water when used as RO membranes [71]. The narrow pores reject ions extremely well, but conduct water at 5–1000 times the rate of commercially available TFC RO membranes. These efficiencies may be achieved if aligned CNT membranes can be fabricated with <1 pore in 100 over the size of 10 Å in diameter. The primary causes of salt rejection and water transport in these studies is the narrow, smooth, nonpolar nature of the CNTs; hence, separation performance may not be specific to chirality number of concentric walls. An idealized structure is depicted in Figure 9.5, showing a highly regular and dense coverage of CNTs embedded in a stable polymeric matrix.

While the exceptional transport characteristics and gating potential of aligned CNT membranes are well recognized, the critical challenge for their application to water purification will be scaling up the synthesis of large membrane areas. Methods of fabricating aligned/oriented CNT membranes are well documented [72–78]. Few are likely to produce simple, fast, inexpensive, and scalable methods of preparing oriented CNT membranes. One approach uses an applied magnetic field to align nanotubes during polymer film casting as proposed by Smalley and co-workers [77,79]. Thick macroscopic membranes of magnetically aligned single-walled CNTs were produced via high-pressure filtration of aqueous surfactant-suspended single-walled nanotube in a magnetic field, resulting in membrane thicknesses of 10 μm and surface area of 125 cm^2. Field strengths of 7 and 25 T were used. Polarized Raman spectroscopy indicated good nanotube alignment (uniform anisotropy across their surface and throughout their thickness) at both intensities of magnetic field, but the authors did not evaluate filtration performance. Holt et al. [57] produce sub-2 nm diameter aligned double-walled CNT membranes through an automated microelectromechanical catalytic chemical vapor deposition (CVD) process. Pore densities are as high as 0.25×10^{12} pores per cm^2. Water flux through these CNT membranes is found to be at least three orders of magnitude higher than theoretical, Hagan–Poiseuille predictions; however, alignment via CVD is expensive, sensitive, and not yet applicable for large-scale fabrication.

A more practical approach, originally proposed by Deheer et al. [78] to form aligned CNT films for characterizing their optical and electronic properties, simply filters a suspension of CNTs onto a hydrophilic microfiltration membrane. Subsequently, Li et al. [74] used this approach to prepare a partially oriented

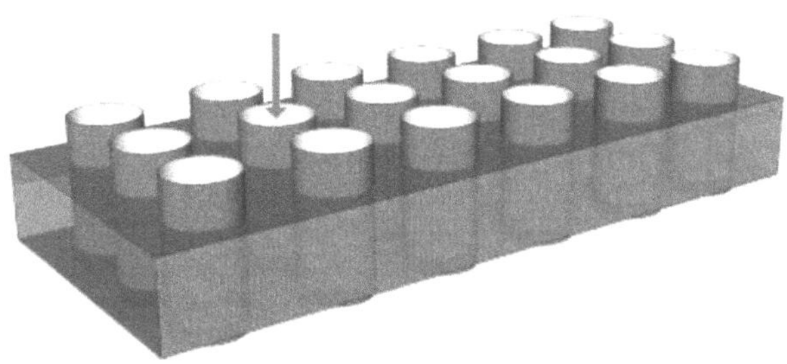

FIGURE 9.5

Conceptual image of an array of aligned CNTs embedded in a nonporous polymeric matrix for stability, providing fast mass transport flow paths for water and selectivity against salt ions.

super-hydrophobic films of CNTs, where CNTs were capped with platinum nanoparticles to facilitate their eventual use as proton-exchange fuel cell membranes. A cathode prepared from the oriented CNT film produced higher single-cell performance in lab-scale polymer electrolyte fuel cell than cathodes comprised of carbon black and disordered CNT films. Oriented CNT films also exhibit improved mass transport. A paper by Kim et al. [80] describes a similar filtration approach for fabricating aligned CNT polymer nanocomposite membranes for high-flux gas transport.

The lack of experimental desalination performance data for macroscopic aligned CNT membranes fabricated by scalable approaches sheds some doubt on their immediate viability for water purification. Moreover, unrelated research—describing a promising, practical approach to create water soluble CNTs—demonstrates that CNTs become highly carboxylated and sulfonated when exposed for only a few minutes to concentrated nitric and sulfuric acid mixtures (in the presence of heat) [81]. The energy barrier for salt partitioning into CNTs is based on a lack of polar functionality [71]; hence, CNT-based membranes may lose their favorable transport characteristics if exposed to acids. Additional practical considerations remain unclear, including: (1) the long-term stability of CNTs in water, (2) the possibility that solutes may partition into nanotubes and alter water or solute transport, and (3) the long-term fouling resistance and chemical tolerance of aligned CNTs membranes. Despite the lack of experimental desalination performance data, separation characteristics derived from molecular dynamics simulations and small-scale microfabricated films justify continued exploration of aligned CNT membranes.

9.6 Self-assembled block copolymer membranes

Another route to attaining highly porous and ordered structures is self-assembled nanoporous block copolymer membranes. Block copolymers are macromolecules composed of distinct block polymeric segments with the ability to self-assemble into highly ordered structures. By varying the conditions under which self-assembly occurs, various structures can be formed, including densely packed cylindrical pores ideal for water separation membranes, similar to that depicted in Figure 9.5. In the figure, the matrix material would be composed of one of the block segments and the aligned pores composed of a second distinct block segment. Self-assembly occurs when the characteristic differences between polymeric segments on single copolymer chains cause microphase separation to occur. For example, when a selective solvent is added to a solvent–nonsolvent system consisting of macromolecules of two distinct regions—one soluble, the other insoluble—a predictable arrangement will form based on the respective interactions of each polymer with the solvent [82].

Thanks to the tunability of each block, specific functionalities can be attained. Techniques for producing such membranes involve shear aligning (which typically produces thicker than desired films), controlled substrate–polymer interactions (which are effective, but difficult to control in large-scale production), and phase inversion (which is successful, but expensive as the block copolymer is used for both the support and the selective layers) [83]. In theory, aligned cylinders formed through nanostructuring of block copolymers could enable a fully polymeric analog to AQP or aligned CNT membranes, providing an opportunity to take advantage of nanopore performance, while maintaining ease and economy of large-scale polymeric membrane fabrication.

Successful fabrication of ultrafiltration and microfiltration self-assembled membranes has been seen in the literature [83–87]. Yang et al. [88] formed an NF membrane with an 80 nm thick selective layer of well-ordered 15 nm diameter cylindrical pores atop a (250 μm) conventional support, capable of filtering viruses. This approach holds benefits of a highly tunable top layer, with pores ranging from 10 to 40 nm, and the reliability of conventional supports. Using copolymers for the selective layer alone provides large cost savings and may pose an advantage for large-scale production [89]; however, the process is limited in scalability because of the difficulty of transferring films without damage to the porous structure. Other groups have removed the need for a transfer step by forming the self-assembled structures directly atop a conventional support or by forming block copolymer asymmetric membranes [84,90]. Kabuto et al. [91] have shown the ability to form asymmetric self-assembled membranes, from a commercially available copolymer, with a top layer of ordered pores as small as 3 nm. The challenge remains to control the pore size produced and attain nano-ranged pores capable of functioning as RO membranes. Additionally, control of the self-assembly and the ability to produce large, defect-free areas must be achieved in order to consider scale-up.

Membranes with aligned nanopores formed by self-assembly of block copolymers offer a significant promise as fully polymeric analogs to AQP and aligned CNT membranes. In principle, these structures could be fine-tuned for water filtration or RO applications if pore sizes can be accurately controlled and scale-up is made possible. They may also serve as ideal support membranes in high-flux, high-selectivity forward osmosis membranes for osmotic power production. Evidence of large-scale production of self-assembled membranes has not yet been realized.

9.7 Graphene-based membranes

Another carbon-based material that is currently being investigated for desalination membrane materials is graphene [92,93]. Graphene is a 2D sheet of hexagonally arranged sp^2-bonded carbon atoms, which is a single to several atomic layers thick, exhibits high mechanical strength, and is impermeable to molecules as small as helium [94,95]. Size-tunable nanopores can be formed in graphene through methods such as oxidation, electron beam irradiation, ion bombardment, doping, diblock copolymer templating, and chemical etching [96–102]. Graphene-based nanoporous films have been demonstrated to function as molecular sieves in applications such as DNA characterization [102]. In theory, regularly sized pores could be formed to create a novel RO membrane material (Figure 9.6).

A recent molecular dynamics study suggests that salt rejection is possible through graphene membranes with pore diameters up to 5.5 Å [93]. The results show potential permeability enhancements of two to three orders of magnitude over conventional seawater RO membranes. It is interesting to note, however, that there is almost no enhancement in permeability per unit thickness relative to conventional polyamide RO membranes, implying that grapheme is not intrinsically more permeable than commercial polymeric membranes. (This assumes a single atomic layer thickness of ~2 Å for graphene [93] and a thickness of ~200 nm for polyamide films.) This means that thin single sheets of graphene need to be produced in mechanically stable forms that can withhold applied pressures in order to realize these extreme flux enhancements. There is promise for large-scale production of graphene, since Bae et al. [103] have shown the ability to produce rectangular sheets as large as 30 in diagonal.

To date, no experimental measurements have been reported to confirm the theoretical transport results. This is an important next step in bringing graphene-based membranes to realization. To break into the membrane market, graphene sheets would need to be manufactured in large, defect-free areas. In addition, the materials will need to be shown to be resistant to fouling and scaling. Only then will it be possible to operate the membranes at higher than normal fluxes, leading to savings on both operating and capital expenses.

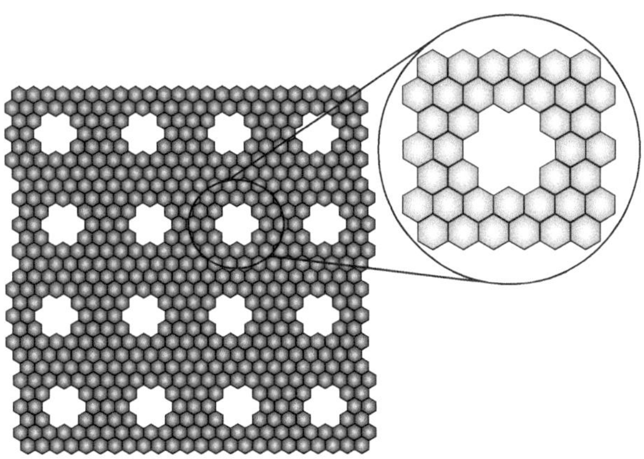

FIGURE 9.6

Conceptual image of a sheet of graphene with regular nanopores drilled into it.

9.8 Conclusions

In this chapter, we have discussed a range of nanotechnology-enabled materials touted as possible low-energy replacements for conventional RO membranes in desalination and water reuse applications. Each technology—zeolite-coated ceramic membranes, TFN membranes, protein–polymer biomimetic membranes, aligned CNT membranes, self-assembled block copolymer membranes, and graphene-based membranes—presents an opportunity for performance enhancements if realized; however, each also has unique limitations to overcome. Zeolite coatings promise high selectivity, with separation capabilities only realized with polymeric materials to date, as well as the chemical, thermal, and mechanical stabilities offered by conventional ceramic membranes; however, to date, practical fabrication has not been realized. The intrinsic permeabilities of these materials imply that they could provide economically viable alternatives for high-flux RO membranes with dramatically improved stability and cleanability; however, the challenge remains to improve control over crystal nucleation and growth to ensure defect-free ultrathin zeolite films, which may require abandoning or substantially modifying traditional hydrothermal synthesis methods. Mixed matrix membranes offer mechanically stable, compaction-resistant filtration membranes and support structures for high-pressure applications. TFNs seek to produce compaction-resistant membranes with silica, fouling-resistant membranes with nano-silver, self-cleaning photo-reactive membranes with titania nanoparticles, or highly permeable and selective membranes with molecular sieve zeolites. Data reveal that nanocomposite membranes exhibit up to three times higher permeability than current commercial polymeric RO membranes, with no change in salt rejection, and can additionally be fabricated with antimicrobial and photo-reactive functionality. Multiple authors have demonstrated macroscopic fabrication and practical desalination performance. The major advantage of these materials is the ability to fabricate them through standard practices and with current equipment since only an addition to the casting solutions is required. At this point, long-term stability is undocumented and needs to be explored, in particular for the surface functionalized versions that require exposed surface areas.

Biologically inspired membranes—AQPs, aligned CNTs, and block copolymers—seek to simultaneously improve selectivity and permeability. Each of these innovative materials concepts promises unique performance enhancements and each has unique hurdles to overcome before it is commercially viable. Biomimetic membranes can produce highly selective membranes potentially useful in both forward osmosis and RO applications. AQP-based lipid bilayer vesicles exhibit nearly 100 times higher water permeability than commercial RO membranes with near perfect salt rejection; however, macroscopic fabrication, practical desalination tests, and long-term stability are only beginning to be explored. CNT- and graphene-based membranes (theoretically) exhibit acceptable salt rejections with water permeabilities between 5 and 1000 times higher than commercial RO membranes. Self-assembled block copolymer membranes represent a fully polymeric approach to forming highly selective structures reminiscent of AQP- or CNT-based materials; however, controlling the formation and attaining pores small enough for RO separations remains a challenge. At present, macroscopic fabrication, practical desalination testing, and long-term stability are unproven for these carbon-based and self-assembled membranes; furthermore, all graphene-based membrane projections are based on computational studies alone. Commercial efforts are already under way for both nanocomposite and biomimetic membranes. Currently, there is no evidence of zeolite ceramic, CNT, self-assembled, or graphene membrane commercialization.

References

[1] M. Mulder, Basic Principles of Membrane Technology, Kluwer Academic Publishers, London, 1996.

[2] N. Savage, M.S. Diallo, Nanomaterials and water purification: opportunities and challenges, J. Nanopart. Res. 7 (2005) 331–342.

[3] O. Renn, M.C. Roco, Nanotechnology and the need for risk governance, J. Nanopart. Res. 8 (2006) 153–191.

[4] E.M.V. Hoek and A. Jawor, Nano-filtration separations, in Dekker Encyclopedia of Nanoscience and Nanotechnology, 2002.

[5] J. Lin, S. Murad, A computer simulation study of the separation of aqueous solutions using thin zeolite membranes, Mol. Phys. 99 (2001) 1175–1181.

[6] L. Li, J. Dong, T. Nenoff, R. Lee, Desalination by reverse osmosis using MFI zeolite membranes, J. Membrane Sci. 243 (2004) 401–404.

[7] M. Duke, J. O'Brien-Abraham, N. Milne, B. Zhu, J. Lin, J. Diniz da Costa, Seawater desalination performance of MFI type membranes made by secondary growth, Separ. Purif. Technol. 68 (2009) 343–350.

[8] L. Li, N. Liu, B. McPherson, R. Lee, Enhanced water permeation of reverse osmosis through MFI-type zeolite membranes with high aluminum contents, Ind. Eng. Chem. Res. 46 (2007) 1584–1589.

[9] M. Kazemimoghadam, New nanopore zeolite membranes for water treatment, Desalination 251 (2010) 176–180.

[10] I. Kumakiri, T. Yamaguchi, S. Nakao, Application of a zeolite A membrane to reverse osmosis process, J. Chem. Eng. Jpn. 33 (2000) 333–336.

[11] M. Noack, P. Kölsch, V. Seefeld, P. Toussaint, G. Georgi, J. Caro, Influence of the Si/Al-ratio on the permeation properties of MFI-membranes, Micropor. Mesopor. Mat. 79 (2005) 329–337.

[12] N. Liu, L. Li, B. McPherson, R. Lee, Removal of organics from produced water by reverse osmosis using MFI-type zeolite membranes, J. Membrane Sci. 325 (2008) 357–361.

[13] R.F. Lobo, Handbook of Zeolite Science and Technology, Marcel Dekker, Inc., New York, NY, 2003.

[14] A. Dyer, An Introduction to Zeolite Molecular Sieves, John Wiley & Sons Ltd., Chichester, UK, 1988.

[15] Y.Y. Li, T. Chung, S. Kulprathipanja, Novel Ag^+-zeolite/polymer mixed matrix membranes with a high CO_2/CH_4 selectivity, AIChE J 53 (2007) 610–616.

[16] C. Baerlocher and L. B. McCusker, *Database of Zeolite Structures*, <http://www.iza-structure.org/databases/>.

[17] M.M. Pendergast, E.M.V. Hoek, A review of water treatment membrane nanotechnologies, Energy Environ. Sci. 4 (2011) 1946–1971.

[18] S. Komarneni, Nanocomposites, J. Mater. Chem. 2 (1992) 1219–1230.

[19] T.T. Moore, R. Mahajan, D.Q. Vu, W.J. Koros, Hybrid membrane materials comprising organic polymers with rigid dispersed phases, AIChE J 50 (2004) 311–321.

[20] J.S. Taurozzi, H. Arul, V.Z. Bosak, A.F. Burban, T.C. Voice, M.L. Bruening, et al., Effect of filler incorporation route on the properties of polysulfone-silver nanocomposite membranes of different porosities, J. Membrane Sci. 325 (2008) 58–68.

[21] L. Yan, Y.S. Li, C.B. Xiang, Preparation of poly(vinylidene fluoride)(PVDF) ultrafiltration membrane modified by nano-sized alumina (Al2O3) and its antifouling research, Polymer 46 (2005) 7701–7706.

[22] Z.F. Fan, Z. Wang, N. Sun, J.X. Wang, S.C. Wang, Performance improvement of polysulfone ultrafiltration membrane by blending with polyaniline nanofibers, J. Membrane Sci. 320 (2008) 363–371.

[23] T.H. Bae, T.M. Tak, Effect of TiO2 nanoparticles on fouling mitigation of ultrafiltration membranes for activated sludge filtration, J. Membrane Sci. 249 (2005) 1–8.

[24] L. Wu, M. Shamsuzzoha, S.M.C. Ritchie, Preparation of cellulose acetate supported zero-valent iron nanoparticles for the dechlorination of trichloroethylene in water, J. Nanopart. Res. 7 (2005) 469–476.

[25] B. Zhu, T. Lim, Catalytic reduction of chlorobenzenes with Pd/Fe nanoparticles: reactive sites, catalyst stability, particle aging, and regeneration, Environ. Sci. Technol. 41 (2007) 7523–7529.

[26] K. Ebert, D. Fritsch, J. Koll, C. Tjahjawiguna, Influence of inorganic fillers on the compaction beaviour of porous polymer based membranes, J. Membrane Sci. 233 (2004) 71–78.

[27] B.H. Jeong, E.M.V. Hoek, Y.S. Yan, A. Subramani, X.F. Huang, G. Hurwitz, et al., Interfacial polymerization of thin film nanocomposites: a new concept for reverse osmosis membranes, J. Membrane Sci. 294 (2007) 1–7.

[28] A.K. Ghosh, B.H. Jeong, X.F. Huang, E.M.V. Hoek, Impacts of reaction and curing conditions on polyamide composite reverse osmosis membrane properties, J. Membrane Sci. 311 (2008) 34–45.

[29] S. Lee, H. Kim, R. Patel, S. Im, J. Kim, B. Min, Silver nanoparticles immobilized on thin film composite polyamide membrane: characterization, nanofiltration, antifouling properties, Polym. Advan. Technol. 18 (2007) 562–568.

[30] M. Lind, A. Ghosh, A. Jawor, X. Huang, W. Hou, Y. Yang, et al., Influence of zeolite crystal size on zeolite-polyamide thin film nanocomposite membranes, Langmuir 25 (2009) 10139–10145.

[31] M.M. Pendergast, J.M. Nygaard, A.K. Ghosh, E.M.V. Hoek, Using nanocomposite materials technology to understand and control reverse osmosis membrane compaction, Desalination 261 (2010) 255–263.

[32] M.M. Pendergast, A.K. Ghosh, E.M.V. Hoek, Separation performance and interfacial properties of nanocomposite reverse osmosis membranes, Desalination 308 (2013) 180–185.

[33] D. Ho, B. Chu, J.J. Schmidt, E.K. Brooks, C.D. Montemagno, Hybrid protein-polymer biomimetic membranes, IEEE T. Nanotechnol. 3 (2004) 256–263.

[34] J.Z. Xi, D. Ho, B. Chu, C.D. Montemagno, Lessons learned from engineering biologically active hybrid nano/micro devices, Advan. Funct. Mater. 15 (2005) 1233–1240.

[35] D. Wong, T.J. Jeon, J. Schmidt, Single molecule measurements of channel proteins incorporated into biomimetic polymer membranes, Nanotechnology 17 (2006) 3710–3717.

[36] F.Q. Zhu, E. Tajkhorshid, K. Schulten, Theory and simulation of water permeation in aquaporin-1, Biophys. J. 86 (2004) 50–57.

[37] B.L. de Groot, H. Grubmuller, Water permeation across biological membranes: mechanism and dynamics of Aquaporin-1 and GlpF, Science 294 (2001) 2353–2357.

[38] K. Murata, K. Mitsuoka, T. Hirai, T. Walz, P. Agre, J.B. Heymann, et al., Structural determinants of water permeation through Aquaporin-1, Nature 407 (2000) 599–605.

[39] F.Q. Zhu, K. Schulten, Water and proton conduction through carbon nanotubes as models for biological channels, Biophys. J 85 (2003) 236–244.

[40] Biomimetic membranes, US Patent No. 7208089, 2007, C.D. Montemagno, J.J. Schmidt, S.P. Tozzi.

[41] M. Kumar, M. Grzelakowski, J. Zilles, M. Clark, W. Meier, Highly permeable polymeric membranes based on the incorporation of the functional water channel protein Aquaporin Z, P. Natl. Acad. Sci. USA 104 (2007) 20719–20724.

[42] O. Kedem, A. Katchalsky, Thermodynamic analysis of the permeability of biological membranes to non-electrolytes, Biochimica Et Biophysica Acta 27 (1958) 229–246.

[43] K.S. Spiegler, O. Kedem, Thermodynamics of hyperfiltration (reverse osmosis) - Criteria for efficient membranes, Desalination 1 (1966) 311–326.

[44] Y. Kaufman, A. Berman, V. Freger, Supported lipid bilayer membranes for water purification by reverse osmosis, Langmuir 26 (2010) 7388–7395.

[45] J.S. Hansen, M. Perry, J. Vogel, T. Vissing, C.R. Hansen, O. Geschke, et al., Development of an automation technique for the establishment of functional lipid bilayer arrays, J. Micromech.Microeng. 19 (2009) 025014.

[46] J. Vogel, M. Perry, J.S. Hansen, P. Bolinger, C. Nielsen, O. Geschke, A support structure for biomimetic applications, J. Micromech.Microeng. 19 (2009) 025026.

[47] A. González-Pérez, K.B. Stibius, T. Vissing, C.H. Nielsen, O.G. Mouritsen, Biomimetic triblock copolymer membrane arrays: a stable template for functional membrane proteins, Langmuir 25 (2009) 10447–10450.

[48] M. Roerdink Lander, S. Ibragimova, C. Rein, J. Vogel, K. Stibius, O. Geschke, et al., Biomimetic membrane arrays on cast hydrogel supports, Langmuir 27 (2011) 7002–7007.

[49] S. Ibragimova, K. Stibius, P. Szewczykowski, M. Perry, H. Bohr, C. Hélix-Nielsen, Hydrogels for in situ encapsulation of biomimetic membrane arrays, Polym. Advan. Technol. 23 (2012) 182–189.

[50] P.S. Zhong, T.-S. Chung, K. Jeyaseelan, A. Armugam, Aquaporin-embedded biomimetic membranes for nanofiltration, J. Membrane Sci. 407–408 (2012) 27–33.

[51] G. Sun, H. Zhou, Y. Li, K. Jeyaseelan, A. Armugam, T.-S. Chung, A novel method of AquaporinZ incorporation via binary-lipid Langmuir monolayers, Colloid Surface B 89 (2012) 283–288.

[52] H. Wang, T.S. Chung, Y.W. Tong, K. Jeyaseelan, A. Armugam, Z. Chen, et al., Highly permeable and selective pore-spanning biomimetic membrane embedded with Aquaporin Z, Small 8 (2012) 1185–1190.
[53] Y. Zhao, C. Qiu, X. Li, A. Vararattanavech, W. Shen, J. Torres, et al., Synthesis of robust and high-performance aquaporin-based biomimetic membranes by interfacial polymerization-membrane preparation and RO performance characterization, J. Membrane Sci. 423–424 (2012) 422–428.
[54] Membrane for filtering water, Application No.: PA 2005 00740 (US), PCT/DK2006/000278 (International), 2005, P. H. Jensen, D. Keller and C. H. Nielsen.
[55] Biomimetic water membrane comprising aquaporins used in the production of salinity power, US Patent No. 8,123,948 B2, 2012, P. H. Jensen.
[56] D.S. Sholl, J.K. Johnson, Making high-flux membranes with carbon nanotubes, Science 312 (2006) 1003–1004.
[57] J.K. Holt, H.G. Park, Y.M. Wang, M. Stadermann, A.B. Artyukhin, C.P. Grigoropoulos, et al., Fast mass transport through sub-2-nanometer carbon nanotubes, Science 312 (2006) 1034–1037.
[58] V.J. van Hijkoop, A.J. Dammers, K. Malek, M.O. Coppens, Water diffusion through a membrane protein channel: a first passage time approach, J. Chem. Phys. 127 (2007) 085101.
[59] P. Nednoor, N. Chopra, V. Gavalas, L.G. Bachas, B.J. Hinds, Reversible biochemical switching of ionic transport through aligned carbon nanotube membranes, Chem. Mater. 17 (2005) 3595–3599.
[60] P. Nednoor, V.G. Gavalas, N. Chopra, B.J. Hinds, L.G. Bachas, Carbon nanotube based biomimetic membranes: mimicking protein channels regulated by phosphorylation, J. Mater. Chem. 17 (2007) 1755–1757.
[61] I.C. Yeh, G. Hummer, Nucleic acid transport through carbon nanotube membranes, P. Natl. Acad. Sci. USA 101 (2004) 12177–12182.
[62] Z.K. Wang, L.J. Ci, L. Chen, S. Nayak, P.M. Ajayan, N. Koratkar, Polarity-dependent electrochemically controlled transport of water through carbon nanotube membranes, Nano Lett. 7 (2007) 697–702.
[63] S.R. Majumder, N. Choudhury, S.K. Ghosh, Enhanced flow in smooth single-file channel, J. Chem. Phys. 127 (2007) 054706.
[64] M. Majumder, N. Chopra, R. Andrews, B.J. Hinds, Nanoscale hydrodynamics - enhanced flow in carbon nanotubes, Nature 438 (2005) 44.
[65] M. Majumder, N. Chopra, B.J. Hinds, Effect of tip functionalization on transport through vertically oriented carbon nanotube membranes, J. Am. Chem. Soc. 127 (2005) 9062–9070.
[66] H.M. Liu, S. Murad, C.J. Jameson, Ion permeation dynamics in carbon nanotubes, J. Chem. Phys. 125 (2006) 084713.
[67] J.Y. Li, Z.X. Yang, H.P. Fang, R.H. Zhou, X.W. Tang, Effect of the carbon-nanotube length on water permeability, Chinese Phys. Lett. 24 (2007) 2710–2713.
[68] J.Y. Li, X.J. Gong, H.J. Lu, D. Li, H.P. Fang, R.H. Zhou, Electrostatic gating of a nanometer water channel, P. Natl. Acad. Sci. USA 104 (2007) 3687–3692.
[69] S. Joseph, R.J. Mashl, E. Jakobsson, N.R. Aluru, Electrolytic transport in modified carbon nanotubes, Nano Lett. 3 (2003) 1399–1403.
[70] A. Burykin, A. Warshel, Membranes assembled from narrow carbon nanotubes block proton transport and can form effective nano-filtration devices, J. Comp. Theor. Nanos. 3 (2006) 237–242.

[71] B. Corry, Designing carbon nanotube membranes for efficient water desalination, J. Phys. Chem. B 112 (2008) 1427–1434.
[72] W.L. Mi, Y.S. Lin, Y.D. Li, Vertically aligned carbon nanotube membranes on macroporous alumina supports, J. Membrane Sci. 304 (2007) 1–7.
[73] C. Basheer, A.A. Ainedhary, B.S.M. Rao, S. Valliyaveettil, H.K. Lee, Development and application of porous membrane-protected carbon nanotube micro-solid-phase extraction combined with gas chromatography/mass spectrometry, Anal. Chem. 78 (2006) 2853–2858.
[74] W.Z. Li, X. Wang, Z.W. Chen, M. Waje, Y.S. Yan, Carbon nanotube film by filtration as cathode catalyst support for proton-exchange membrane fuel cell, Langmuir 21 (2005) 9386–9389.
[75] J.K. Holt, A. Noy, T. Huser, D. Eaglesham, O. Bakajin, Fabrication of a carbon nanotube-embedded silicon nitride membrane for studies of nanometer-scale mass transport, Nano Lett. 4 (2004) 2245–2250.
[76] B.J. Hinds, N. Chopra, T. Rantell, R. Andrews, V. Gavalas, L.G. Bachas, Aligned multiwalled carbon nanotube membranes, Science 303 (2004) 62–65.
[77] D.A. Walters, M.J. Casavant, X.C. Qin, C.B. Huffman, P.J. Boul, L.M. Ericson, et al., In-plane-aligned membranes of carbon nanotubes, Chem. Phys. Lett. 338 (2001) 14–20.
[78] W.A. Deheer, W.S. Bacsa, A. Chatelain, T. Gerfin, R. Humphrey-Baker, L. Forro, et al., Aligned carbon nanotube films - production and optical and electronic-properties, Science 268 (1995) 845–847.
[79] M.J. Casavant, D.A. Walters, J.J. Schmidt, R.E. Smalley, Neat macroscopic membranes of aligned carbon nanotubes, J. Appl. Phys. 93 (2003) 2153–2156.
[80] S. Kim, J.R. Jinschek, H. Chen, D.S. Sholl, E. Marand, Scalable fabrication of carbon nanotube/polymer nanocomposite membranes for high flux gas transport, Nano Lett. 7 (2007) 2806–2811.
[81] Y.B. Wang, Z. Iqbal, S. Mitra, Rapidly functionalized, water-dispersed carbon nanotubes at high concentration, J. Am. Chem. Soc. 128 (2006) 95–99.
[82] T. Smart, H. Lomas, M. Massignani, M.V. Flores-Merino, L.R. Perez, G. Battaglia, Block copolymer nanostructures, Nano Today 3 (2008) 38–46.
[83] W. Phillip, B. O'Neill, M. Rodwogin, M. Hillmyer, E. Cussler, Self-assembled block copolymer thin films as water filtration membranes, ACS Appl. Mater. Interf 2 (2010) 847–853.
[84] K.V. Peinemann, V. Abetz, P.F.W. Simon, Asymmetric superstructure formed in a block copolymer via phase separation, Nat. Mater. 6 (2007) 992–996.
[85] Isoporous Membrane and Method of Production Thereof, International Patent No. PCT/EP2007/006759, 2008, K.-V. Peinemann, V. Abetz, P. F. W. Simon and G. Johannsen.
[86] X. Qiu, H. Yu, M. Karunakaran, N. Pradeep, S.P. Nunes, K.-V. Peinemann, Selective separation of similarly sized proteins with tunable nanoporous block copolymer membranes, ACS Nano 7 (2012) 768–776.
[87] W.A. Phillip, R. Mika Dorin, J. Werner, E.M.V. Hoek, U. Wiesner, M. Elimelech, Tuning structure and properties of graded triblock terpolymer-based mesoporous and hybrid films, Nano Lett. 11 (2011) 2892–2900.
[88] S.Y. Yang, I. Ryu, H.Y. Kim, J.K. Kim, S.K. Jang, T.P. Russell, Nanoporous membranes with ultrahigh selectivity and flux for the filtration of viruses, Advan. Mater. 18 (2006) 709–712.

[89] M.J. Fasolka, A.M. Mayes, Block copolymer thin films: physics and applications, Annu. Revi. Mater. Res 31 (2001) 323–355.
[90] D. Fierro, K. Buhr, C. Abetz, A. Boschetti-de-Fierro, V. Abetz, New insights into the control of self-assembly of block copolymer membranes, Aust. J. Chem. 62 (2009) 885–890.
[91] T. Kabuto, Y. Hashimoto, O. Karthaus, Thermally stable and solvent resistant mesoporous honeycomb films from a crosslinkable polymer, Adv. Funct. Mater 17 (2007) 3569–3573.
[92] T. Humplik, J. Lee, S.C. O'Hern, B.A. Fellman, M.A. Baig, S.F. Hassan, et al., Nanostructured materials for water desalination, Nanotechnology (2011) 22.
[93] D. Cohen-Tanugi, J.C. Grossman, Water desalination across nanoporous graphene, Nano Lett. 12 (2012) 3602–3608.
[94] A.K. Geim, Graphene: status and prospects, Science 324 (2009) 1530–1534.
[95] C. Lee, X. Wei, J.W. Kysar, J. Hone, Measurement of the elastic properties and intrinsic strength of monolayer graphene, Science 321 (2008) 385–388.
[96] J. Zhang, H.L. Zou, Q. Qing, Y.L. Yang, Q.W. Li, Z.F. Liu, et al., Effect of chemical oxidation on the structure of single-walled carbon nanotubes, J. Phys. Chem. B 107 (2003) 3712–3718.
[97] A. Hashimoto, K. Suenaga, A. Gloter, K. Urita, S. Iijima, Direct evidence for atomic defects in graphene layers, Nature 430 (2004) 870–873.
[98] D.C. Bell, M.C. Lemme, L.A. Stern, J.R. Williams, C.M. Marcus, Precision cutting and patterning of graphene with helium ions, Nanotechnology (2009) 20.
[99] D. Wei, Y. Liu, Y. Wang, H. Zhang, L. Huang, G. Yu, Synthesis of N-doped graphene by chemical vapor deposition and its electrical properties, Nano Lett. 9 (2009) 1752–1758.
[100] M. Bieri, M. Treier, J. Cai, K. Ait-Mansour, P. Ruffieux, O. Groening, et al., Porous graphenes: two-dimensional polymer synthesis with atomic precision, Chem. Commun. (2009) 6919–6921.
[101] M. Kim, N.S. Safron, E. Han, M.S. Arnold, P. Gopalan, Fabrication and characterization of large-area, semiconducting nanoperforated graphene materials, Nano Lett. 10 (2010) 1125–1131.
[102] S. Garaj, W. Hubbard, A. Reina, J. Kong, D. Branton, J.A. Golovchenko, Graphene as a subnanometre trans-electrode membrane, Nature 467 (2010) 190–193.
[103] S. Bae, H. Kim, Y. Lee, X. Xu, J.-S. Park, Y. Zheng, et al., Roll-to-roll production of 30-inch graphene films for transparent electrodes, Nat. Nanotechnol. 5 (2010) 574–578.

CHAPTER

Multifunctional Nanomaterial-Enabled Membranes for Water Treatment

10

Volodymyr V. Tarabara

Department of Civil and Environmental Engineering, Michigan State University, East Lansing, Michigan, USA

10.1 Introduction

Development of new materials is an area of water quality engineering where nanotechnology is expected to make a critical impact. The impact is likely to be especially significant for "high-tech" treatment technologies such as the membrane-based separation. The newly available nanomaterials and nanotechnology tools can enable the synthesis of novel polymers and ceramics, as well as polymer–inorganic composites to manufacture higher performance membranes with increased permeability, selectivity, and resistance to fouling. The expanding

Street, Sustich, Duncan and Savage. Nanotechnology Applications for Clean Water, 2nd Edition.
 DOI: http://dx.doi.org/10.1016/B978-1-4557-3116-9.00010-X

choice of functional nanoparticles opens up another intriguing opportunity—the bottom-up design of nanoparticle-enabled multifunctional membranes—more complex devices capable of performing multiple tasks. Combining unique functionalities afforded by nanoparticles with other functions in one hybrid process can enhance the overall treatment efficiency and remove excessive redundancy. A smaller environmental footprint is another increasingly important advantage of combining several functions in one hybrid membrane-based process.

Whereas a conventional membrane is a device with static properties that is designed for one function—that of separation—a multifunctional membrane couples separation with one or more additional functions. The additional functions can either be static or be triggered by an external agent or field inducing desirable changes in membrane properties. (This is different in the case of a conventional membrane, wherein changes are typically caused by such phenomena as membrane fouling, ageing, and chemical or mechanical damage; the effects of such phenomena are detrimental to the membrane performance and result in a decrease in rejection or permeate flux, or both.) Examples of additional functions include sorption [1–5], catalysis [6,7], controlled release [8–10], molecular recognition [11], or sensing [13]. An important example of membranes with a dynamic function is stimuli-responsive (or adaptive) membranes that combine sensing and valve functionalities. Adaptive membranes reversibly adjust their permeability or rejection [12] in response to external stimuli such as pH [13–18], temperature [19–25], redox environment [26,27], or the presence of a specific dissolved or suspended species in the permeating solution [28–30].

A distinction can be drawn between the case when additional functions are external to the membrane and the case when the functions are "built-in." Examples of hybrid processes involving membranes with an external additional function include membrane bioreactors [31], sorption/ultra- and microfiltration systems [32–37], and photocatalytic membrane reactors where photocatalysis is coupled with membrane distillation [38,39] or ultrafiltration [40,41]. In contrast, built-in additional functions are integrated into the membrane as inherent features of its structural design; such membranes are true multifunctional devices with the highest potential for synergistic coupling between functions and between functions and the membrane structure. The following discussion will focus on membranes with built-in additional functions.

10.2 Nanostructured membranes with functional nanoparticles

Polymers and ceramics, the two main classes of membrane materials, have distinct advantages as platforms for the design of multifunctional membranes. Due to their high chemical and thermal stability, ceramic membranes can support additional functions and processes in highly oxidizing environments and under conditions of extreme pH and temperatures. Polymers, on the other hand, offer great design flexibility and are generally less expensive. Polymeric multifunctional membranes can be prepared by modifying materials traditionally used in membrane manufacture. Surface functionalization of existing membranes and the use of block copolymers and hybrid polymers as membrane materials are examples of such methods [42]. Another approach is that of developing new hybrid membrane materials by incorporating (metal oxide, metal, carbon, or polymeric) functional nanoparticles into the polymeric matrix.

10.2.1 Overview of recent progress in the development of multifunctional membranes

Several groups have investigated the coupling of photocatalysis and membrane separation by employing titanium oxide (TiO_2) particles in conjunction with dense polymeric (e.g., polyamide [43,44], porous sulfonated polyethersulfone [45,46], poly(phthalazinone ether sulfone ketone) [47], and polysulfone [40]) membranes. In these studies TiO_2 nanoparticles were either self-assembled onto polymeric membrane surface [43–46,48,49] or incorporated into the membranes during membrane casting [40,45,47]. In all cases, an improvement in the antifouling properties of resulting nanostructured membranes was reported.

Polymer–inorganic porous composite membranes incorporating bimetallic (Fe/Ni [50,51], Fe/Pd [52]) and zero-valent iron [53] nanoparticles have been developed and applied to the reductive degradation of halogenated organic solvents. Bhattacharyya et al. reported on the formation of Fe/Ni nanoparticles in porous cellulose acetate membranes by in situ reduction during membrane casting, wherein the reducing agent was a component of the non-solvent bath [50]. The same group synthesized nanosized Fe/Pd bimetallic particles in the pores of poly(vinylidene fluoride) microfiltration membranes by using a different, two-step approach that involved (1) chemical reduction of ferrous ions bound to the surface of membrane pores followed by (2) partial reduction of Pd^{2+} with Fe nanoparticles [52]. Ritchie et al. incorporated Fe/Ni nanoparticles into cellulose acetate membranes by transferring nanoparticles from aqueous suspension into the organic solvent, producing an organosol, and then adding the sol to the membrane casting mixture prior to phase inversion [51]. In a separate study by the same group, zero-valent iron nanoparticles were incorporated into porous cellulose acetate membranes by dispersing nanoparticles in water–oil emulsion, mixing with cellulose acetate–acetone solution, and then casting the membrane using the phase inversion process [53].

Nanoparticles can also be introduced as components of polyelectrolyte multilayer films (PMFs). These PMFs are constructed on the membrane surface using layer-by-layer adsorption of oppositely charged electrolytes [54–56]. Catalytic and other functional nanoparticles can be incorporated into PMFs by co-deposition [57] or by in situ synthesis within the film [58]. The PMFs can be prepared to cover only the upstream surface of the membrane or the entire internal surface of the porous matrix [59,60]. Importantly, when supported by a porous membrane, PMFs are known to have water permeabilities and ion rejections typical for nanofiltration membranes. The possibility to control the composition of the PMF in terms of its polymeric constituents and nanoparticle fillers and to regulate PMF's separation properties presents unique opportunities for the design of nanoparticle-enabled membranes.

Metal–polymer nanocomposite membranes can be prepared by the in situ reduction of unbound ionic precursors in the process of phase inversion. Silver–polymer membranes, for example, were synthesized by reducing ionic silver chemically [8,9], by ultraviolet irradiation [61], or by heat [62]. Yu et al. used jet–wet spinning technique to prepare silver-loaded cellulose acetate and polyacrylonitrile hollow fiber UF membranes. Nanoparticles were produced by the reduction of Ag^{+} by dimethylformamide, a component of the spinning solution. Such silver–polymer composites can function as controlled release devices wherein silver is released from the porous matrix. Under flow-through conditions, the inhibitory effect of silver on the biofilm growth on the membrane surface was observed [8,9]. Silver–polymer nanocomposites with embedded [63] or self-assembled [64] silver nanoparticles also find applications as flow-through sensors.

10.2.2 Porous polymer nanocomposite membranes: structural aspects

The studies referenced earlier can be grouped into four categories based on whether particles and membranes were preformed or were synthesized in-situ during the formation of the nanocomposite. Introduction of pre-formed or in-situ synthesized nanoparticles on the pore surface of existing membranes has the advantages of improved reproducibility of nanocomposite preparation, the possibility to reuse the membrane matrix or regenerate nanoparticles, and better accessibility of immobilized nanopartiles to the reactants in the permeate flow. By contrast, when preformed or generated in-situ nanoparticles are incorporated into the forming membrane matrix, the structure of the resulting membrane is generally a function of the type, the load, and the method of incorporation of nanoparticles. The appeal of the latter approach is in the additional possibilities for the design of nanocomposite structure and coupling between the flow and nanoparticle reactivity. An interesting example is the study where magnetite nanoparticles were embedded in polysulfone matrix to produce a stimulus-responsive UF membrane with flux and rejection dependent on the intensity of the external magnetic field [65].

The advances in the development of dense polymer nanocomposites (e.g. [66,67]) and, more specifically, mixed-matrix gas separation membranes [68,69] could provide guidance for the development of dense nanocomposite membranes with salt-rejecting capability for water softening and desalination. In a recent study, zeolite–polyamide nanocomposite membranes were prepared by interfacial polymerization of thin film nanocomposites and competitive fluxes and rejections [70]. Less is known about the processes governing formation of porous nanocomposite membranes. Most studies published so far (reviewed in [79]) involved the incorporation of passive fillers into polymeric matrices and indicated that the presence of nanoparticles may affect both the porosity (pore size and density) of the membrane skin and the macrovoid morphology of the asymmetric support layer. Each aspect is important: the former determines permeability, retention, and selectivity of membranes whereas the latter is known to control their compaction behavior [71,72]. Incorporation of increasing loadings of silica nanoparticles into polysulfone membranes was reported to result in higher porosity and interconnectedness of macrovoid walls [73], increased density and size of surface pores, and thicker skin layer [74]. For clay−polysulfone composite membranes, skin layer thickness was found to decrease with increasing clay loadings [75]. Macrovoid suppression and increased porosity and thickness of the membrane's skin layer for higher filler concentrations was also reported for the hybrid TiO_2−polysulfone UF membranes [76]. Similarly, incorporation of activated carbon was found to diminish macrovoid formation in composite polysulfone membranes; the asymmetry and mean pore size of membranes increased with an increase in the size of carbon filler particles [77]. By contrast, when polysulfone was embedded with carbon particles in the form of (acid-treated) multiwall carbon nanotubes, the average porosity of the composites membrane was found to peak at the nanotube loading of 1.5 percent, which also corresponded to the highest permeate flux. Interestingly, the only pronounced change in membrane macroscopic morphology upon nanotube incorporation was an increase in surface roughness [78]. The results of the studies published so far strongly imply that the microstructural changes induced by nanoparticle incorporation can potentially be controlled to prepare membranes with desirable properties.

10.2.3 Example: effect of filler incorporation route on the structure and biocidal properties of polysulfone-silver nanocomposite membranes of different porosities

Silver-polysulfone composite membranes of three types differing in skin porosity and macrovoid structure were prepared to explore the coupled effects of host matrix porosity and nanoparticle growth conditions on the properties of resulting nanocomposites [79]. Silver nanoparticles were either synthesized ex situ and then added to the casting solution as organosol or produced in the casting solution via in situ reduction of ionic silver by the polymer solvent. Polysulfone was chosen as the matrix material due to its wide use in the preparation of UF membranes. Nanoscale silver was chosen for two reasons. First, there are established protocols for the preparation and characterization of silver sols in various dispersion media, including organic solvents that can reduce silver and are used in membrane preparation. Second, the presence and availability of silver can be detected by the extent of its biocidal effect, which provides a convenient framework for the evaluation of the accessibility of silver in polymer matrices and

Table 10.1 Components of Membrane Casting Mixture (% Mass)

Component (%)	Membrane					
	PSf I	PSf II	PSf III	PSf/Ag I	PSf/Ag II	PSf/Ag III
PSf	19.97	10.08	10.12	19.20	9.88	9.92
DMAC	45.04	70.78	60.87	43.32	69.38	59.66
DMF	18.97	19.14	19.22	18.24	18.76	18.84
PEG	16.02	—	—	15.40	—	—
2-propanol	—	—	9.79	—	—	9.59
$AgNO_3$	—	—	—	3.84	1.98	1.98

DMAC: N,N-Dimethylacetamide; DMF: Dimethylformamide; PSf: Polysulfone.

serves as additional motivation in view of potential use of silver-filled nanocomposites as biofouling-resistant surfaces.

All membranes were prepared by the wet phase inversion process wherein cast films of polysulfone (PSf) dissolved in a mixture of N,N-dimethylacetamide (DMAC) and dimethylformamide (DMF) were immersed in a non-solvent with respect to polysulfone. DMAC served as the primary polymer solvent whereas DMF acted as the reducing agent for silver. Membranes of three distinct types (Types I, II, and III) were obtained by varying the composition of the casting mixture and the composition and temperature of the non-solvent in the immersion bath (Table 10.1). Either polyethylene glycol (PEG) or 2-propanol was added to the casting mixture as a pore forming agent (porogen). Either water or propanol was used as the non-solvent medium.

Two silver incorporation approaches were adapted. In the first approach, Ag nanoparticles were synthesized ex situ and were added to the casting solution as Ag-DMF organosol. The organosol was prepared by adding $AgNO_3$ to DMF (reducing agent) and heating the solution under intense stirring conditions [80]. The second approach involved an in situ reduction of ionic Ag^+ by DMF in the membrane casting mixture. In this case, $AgNO_3$ was first dissolved in DMF at room temperature to minimize Ag reduction.

The $AgNO_3$/DMF solution was added to the casting mixture, which was then heated under intense stirring to initiate the reduction of silver ions to Ag^0, with the concomitant formation of silver nanoparticles. Reduction was allowed to proceed for 1 minute prior to casting the membrane.

The impact of silver incorporation on membrane properties was pronounced only for nanocomposites with lower porosity of the host matrix (Fig. 10.1). For such membranes, the incorporation of nanoscale silver caused macrovoid broadening, an increase in surface pore size and density, and a significant decrease in the hydraulic resistance accompanied by only a minor decrease in rejection. The distribution of silver within membranes depended on the extent of silver reduction

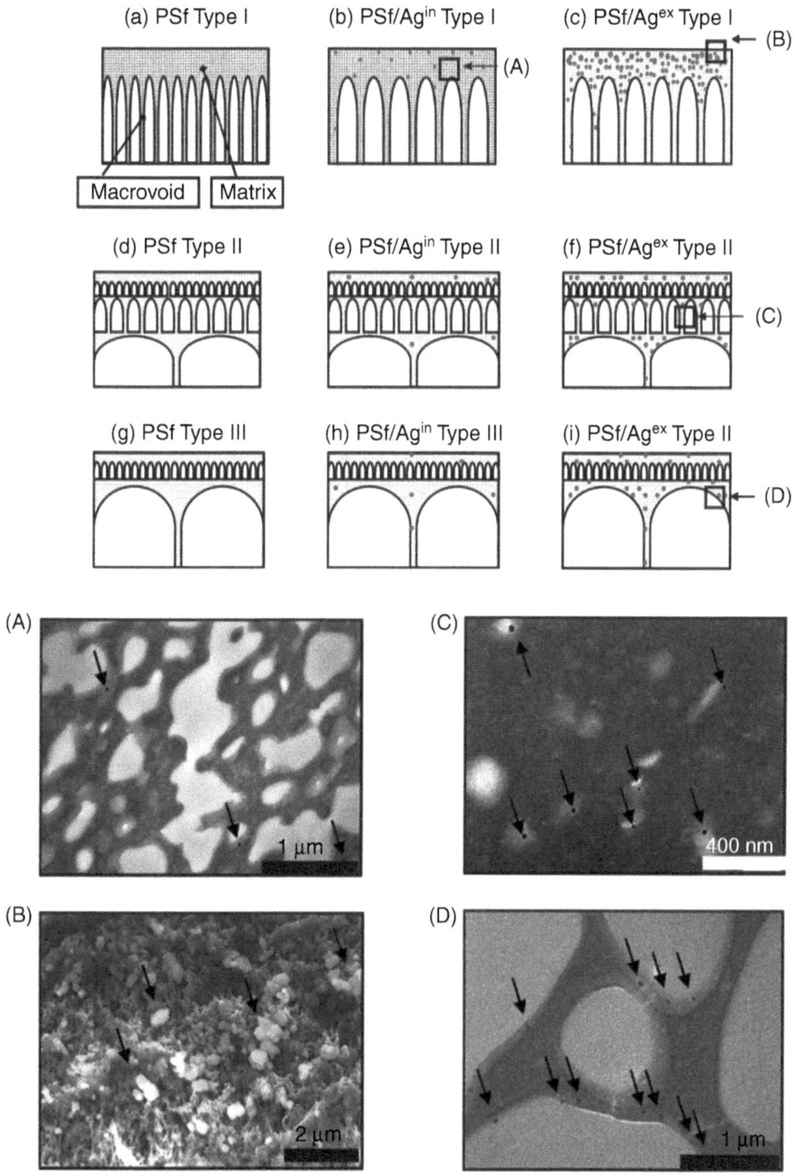

FIGURE 10.1

Distribution of silver nanoparticles in the silver–polysulfone nanocomposites prepared using different methods of silver incorporation. Red dots correspond to silver nanoparticles. Insets (A), (B), and (C) show cross-section transmission electron microscope images. Arrows point to nanoparticles embedded within the polysulfone matrix. Inset (D) shows a scanning electron microscope image of the top cross-section layer, with arrows pointing to observed larger silver crystals.

prior to membrane casting. In all cases, nanoparticles seemed to be preferentially concentrated along the internal pore surface (Fig. 10.1, insets A, B, and C), with ex situ types showing apparent higher coverage densities. The findings indicate that the mere presence of silver nanoparticles in the casting mixture, regardless of their size, is enough to induce morphological changes in the morphology of the polymeric matrix for sufficiently dense membranes. The observed morphologies can be interpreted as resulting from three processes that simultaneously take place in the casting mixture: (1) nanoparticle formation via the reduction of ionic silver by DMF, and (2) nanoparticle re-dissolution into the DMAC, a component of the casting mixture that cannot reduce silver but can dissolve it, and (3) demixing and polymer precipitation.

Biofouling tests were conducted to determine the biocidal effectiveness of the prepared nanocomposites. A thick biofilm could be observed on surfaces of all PSf membranes whereas on the nanocomposite surfaces the bacterial growth was inhibited (Fig. 10.2). There was an evident decrease in bacterial growth on nanocomposites of both PSf/Ag^{in} and PSf/Ag^{ex} of all porosities with respect to their silver-free counterparts. It was also observed that for the very porous membranes, simulated by Type II membranes operated with the more porous side facing the feed, the biocidal efficacy was insensitive to the method of silver incorporation. Because the skin, now facing the permeate side, provided an absolute barrier to the bacteria, the number of inoculated bacteria retained by the membrane was the same as that in all other biofouling assays performed in this work, enabling a straightforward comparison with the Type II membranes inoculated in the conventional way. In summary, it was demonstrated that the reduction of ionic silver by polymer solvents during the phase inversion process can be employed to synthesize bioactive silver–polymer nanocomposites of a range of porosities. The morphology of the resulting nanocomposite is a result of the dynamic interplay

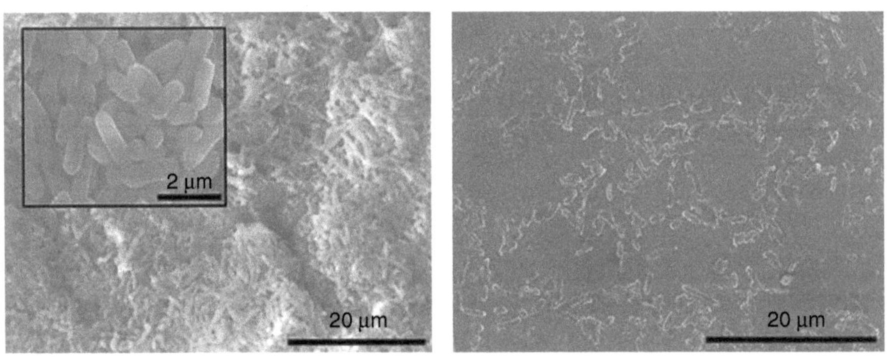

FIGURE 10.2

Scanning electron microscope micrographs illustrating the inhibition biofilm growth on the surface of PSf/Ag^{in} nanocomposite (right) compared to the biofilm developed on the surface of the silver-free polysulfone membrane (left) under the same conditions.

between the polymer coagulation and nanoparticle formation processes. The built-in antibacterial capacity due to the gradual release of ionic silver by the prepared nanocomposites can be effective in reducing intrapore biofouling in porous membranes of a wide porosity range. Such nanocomposites could in principle be used as materials for macroporous membrane spacers to inhibit the biofilm growth on downstream membrane surfaces.

10.2.4 Example: Self-cleaning membrane for ozonation–ultrafiltration hybrid process

Ozone is known to preferentially oxidize electron-rich moieties and carbon–carbon double bonds of organic molecules. For example, ozone exhibits high reactivity with natural organic matter (NOM), complex macromolecules with a high percentage of aromaticity [83]. Accordingly, ozone can be an attractive approach to the mitigation of membrane fouling by NOM—a long-standing problem that decreases the efficiency of membrane-based water treatment processes. However, the application of ozone in conjunction with membrane processes has been limited by the low resistance of most polymeric membranes to ozone. A hybrid process that combines ozonation with ceramic membrane UF (Fig. 10.3) was proposed as an approach to overcome this limitation [84,85] by benefiting from the chemical resistance of ceramic membranes that is significantly higher than that of their polymeric counterparts [84,86]. It has been reported that in the hybrid system ozonation was effective in mitigating fouling due to the oxidation of NOM by ozone and/or hydroxyl radicals [81–84,86]. The effect of oxygen sparging on the permeate flux was shown to be limited (Fig. 10.4), which

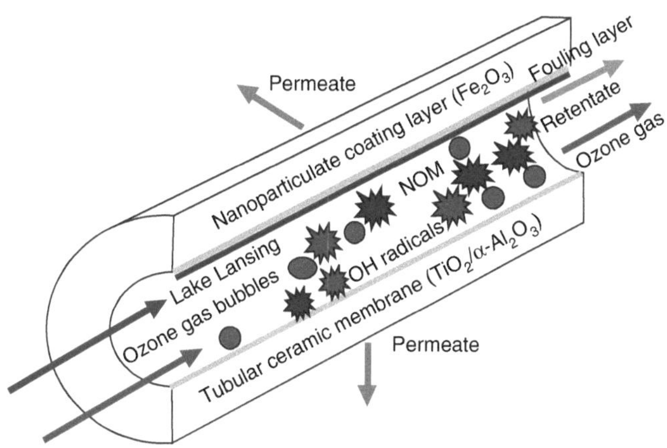

FIGURE 10.3

Schematic of the ceramic catalytic membrane reactor for the hybrid ozonation–ultrafiltration process [81].

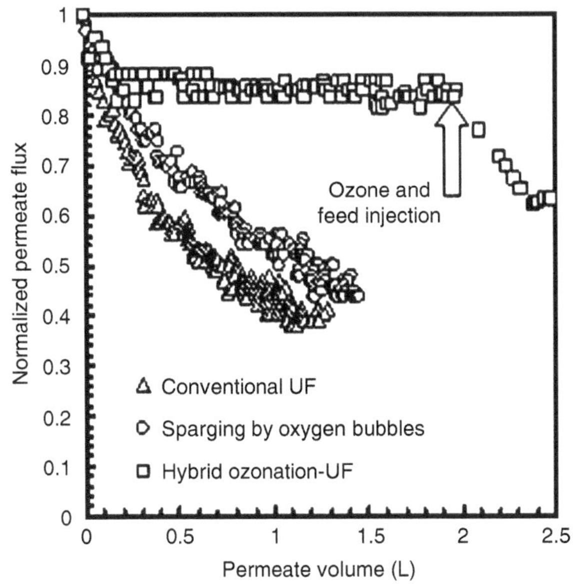

FIGURE 10.4

Permeate flux decline under conditions of oxygen sparging or ozonation [82]. Cross-flow velocity: 0.88 m/s; transmembrane pressure: 0.68 bar; gas phase ozone concentration: 9.5 g/m^3; gas flow rate: 0.2 L/min.

indicated that the improved performance seen in the hybrid system was indeed the results of the oxidation of NOM by ozone or secondary oxidants.

By introducing nanoparticles such as Fe_2O_3 and MnO_2 at the membrane surface, the efficiency of the hybrid process can be significantly enhanced due to both the catalytic effect of the nanoparticles and more targeted oxidation of the NOM portion that is concentrated at or near the membrane surface contributing to membrane fouling. In the most recent studies on this topic [81,82], it was demonstrated that the fouling behavior is strongly dependent upon ozone concentration (Fig. 10.5) and hydrodynamic conditions (Fig. 10.6). Higher permeate flux was observed at higher cross-flow velocities and ozone concentrations, and at lower transmembrane pressures.

The efficiency of the hybrid process appears to be controlled by two factors: (1) dissolved ozone availability, which is a function of ozone demand of the feed water and the ozone absorption rate, and (2) accessibility of ozone to the NOM and/or catalytic membrane surface. In turn, the accessibility of the membrane surface is controlled by the electrostatic interactions between NOM and the ceramic membrane and is influenced by the feed water pH and the concentration of calcium ions [82]. A unique aspect of this hybrid process is that due to the effect of catalytic ozonation at the membrane surface, the surface remains relatively foulant-free; therefore, in the absence of the fouling layer, foulant-membrane interactions remain important for extended periods of membrane

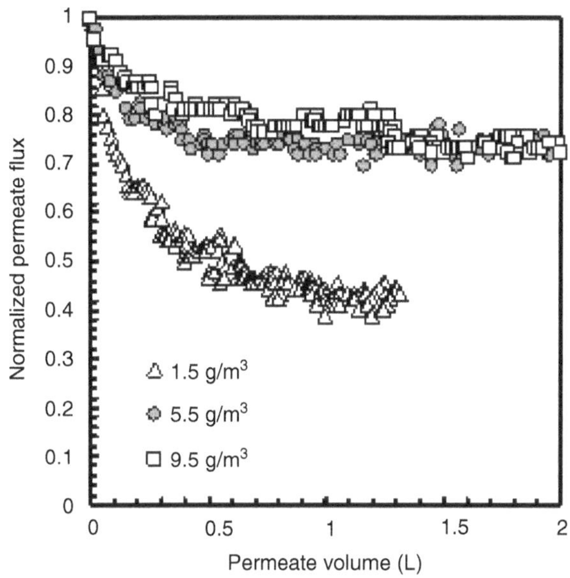

FIGURE 10.5

Effect of ozone dosage on permeate flux at cross-flow velocity of 0.47 m/s and transmembrane pressure of 0.68 bar (gas flow rate is 0.2 L/min).

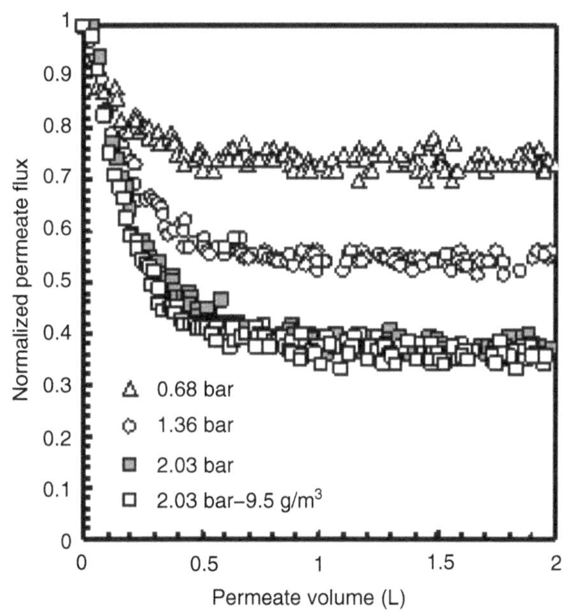

FIGURE 10.6

Effect of transmembrane pressure on permeate flux at an ozone concentration of 5.5 g/m^3 and a cross-flow velocity of 0.47 m/s.

operation. This, in turn, increases the relevance of membrane surface engineering for longer-term membrane operation.

10.3 Potential future research directions

Improvements in our knowledge of materials science aspects of nanocomposite formation and of fundamental processes governing transport through reactive nanostructures are likely to guide the development of nanoparticle-enabled multifunctional membranes. Better understanding of the mechanisms of reactions at the nanostructured surface and within the confines of membrane nanopores is needed to describe the reactive transport in membranes. The progress in membrane materials could be expected in directions of increased structural and functional complexity (e.g., evolving, adaptive, or programmable structures), leveraging structure–function coupling (e.g., nanoparticle-directed healing of structural defects, or, conversely, removal of compromised or even sacrificial portions of the membrane matrix) and process intensification (e.g., use of membrane asymmetry to establish and separate in space a hierarchical sequence of membrane functions). Of immediate interest is elucidating the roles of key variables that are unique for nanocomposite membranes: (1) nanoparticle dispersion and distribution within host matrices such as polymers, multilayer polyelectrolyte films, grafted polymeric films, and so on, (2) nanoparticle–matrix adhesion, (3) changes in the structure (porosity, connectivity) and properties (permeability, hydrophilicity, tensile strength, compressibility) of the host matrix, especially if nanoparticles were introduced into the forming matrix, (4) changes in the nanoparticle properties, especially if nanoparticles were synthesized in preformed or forming porous matrix.

Acknowledgments

This material is based upon work supported by the National Science Foundation under grant nos. 0530174 and 0506828.

References

[1] S.M.C. Ritchie, L.G. Bachas, T. Olin, S.K. Sikdar, D. Bhattacharyya, Surface modification of silica and cellulosic-based microfiltration membranes with functional polyamino acids for heavy metal sorption, Langmuir 15 (1999) 6346–6357.

[2] S.M.C. Ritchie, K.E. Kissick, L.G. Bachas, S.K. Sikdar, C. Parikh, D. Bhattacharyya, Polycysteine and other polyamino acid functionalized microfiltration membranes for heavy metal capture, Environ. Sci. Technol. 35 (2001) 3252–3258.

[3] J.A. Hestekin, L.G. Bachas, D. Bhattacharyya, Poly(amino acid)-functionalized cellulosic membranes: metal sorption mechanisms and results, Ind. Eng. Chem. Res. 40 (2001) 2668–2678.
[4] V. Smuleac, D.A. Butterfield, S.K. Sikdar, R.S. Varma, D. Bhattacharyya, Polythiol-functionalized alumina membranes for mercury capture, J. Membr. Sci. 251 (2005) 169.
[5] M.S. Diallo, S. Christie, P. Swaminathan, J.H. Johnson Jr., W.A. Goddard III, Dendrimer enhanced ultrafiltration. 1. recovery of Cu(II) from aqueous solutions using Gx-NH2 PAMAM dendrimers with ethylene diamine core, Environ. Sci. Technol. 39 (2005) 13661377.
[6] D.A. Butterfield, D. Bhattacharyya, S. Daunert, L. Bachas, Catalytic biofunctional membranes containing site-specifically immobilized enzyme arrays: a review, J. Membr. Sci. 181 (2001) 29–37.
[7] J.G. Sanchez Marcano, T.T. Tsotsis, Catalytic Membrane Reactors and Membrane Reactors, Wiley-VCH, Weinheim, 2002.
[8] D.G. Yu, M.Y. Teng, W.L. Chou, M.C. Yang, Characterization and inhibitory effect of antibacterial PAN-based hollow fiber loaded with silver nitrate, J. Membr. Sci. 225 (2003) 115.
[9] W.L. Chou, D.G. Yu, M.C. Yang, The preparation and characterization of silver-loading cellulose acetate hollow fiber membrane for water treatment, Polym. Adv. Technol. 16 (2005) 600–607.
[10] R. Bhattacharya, T.N. Phaniraj, D. Shailaja, Polysulfone and polyvinyl pyrrolidone blend membranes with reverse phase morphology as controlled release systems: experimental and theoretical studies, J. Memb. Sci. 227 (2003) 23.
[11] S.A. Piletsky, T.L. Panasyuk, E.V. Piletskaya, I.A. Nicholls, M. Ulbricht, Receptor and transport properties of imprinted polymer membranes—a review, J. Membr. Sci. 157 (1999) 263.
[12] A. Kumar, A. Srivastava, I.Y. Galaev, B. Mattiasson, Smart polymers: physical forms and bioengineering applications, Prog. Polym. Sci. 32 (2007) 1205.
[13] B. Adhikari, S. Majumdar, Polymers in sensor applications, Prog. Polym. Sci. 29 (2004) 699.
[14] T. Peng, Y.L. Cheng, pH-responsive permeability of PE-g-PMAA membranes, J. Appl. Polym. Sci. 76 (2000) 778–786.
[15] L. Ying, P. Wang, E.T. Kang, K.G. Neoh, Synthesis and characterization of poly (acrylic acid)-graft-poly(vinylidene fluoride) copolymers and pH-sensitive membranes, Macromol. 35 (2002) 673–679.
[16] Y. Wang, Z. Liu, B. Han, Z. Dong, J. Wang, D. Sun, et al., pH sensitive polypropylene porous membrane prepared by grafting acrylic acid in supercritical carbon dioxide, Polym. 45 (2004) 855.
[17] J.F. Hester, S.C. Olugebefola, A.M. Mayes, Preparation of pH-responsive polymer membranes by self-organization, J. Membr. Sci. 208 (2002) 375.
[18] H. Iwata, I. Hirata, Y. Ikada, Atomic force microscopic analysis of a porous membrane with pH-sensitive molecular valves, Macromol. 31 (1998) 3671–3678.
[19] H. Iwata, M. Oodate, Y. Uyama, H. Amemiya, Y. Ikada, Preparation of temperature-sensitive membranes by graft polymerization onto a porous membrane, J. Membr. Sci. 55 (1991) 119–130.
[20] L. Liang, X. Feng, L. Peurrung, V. Viswanathan, Temperature-sensitive membranes prepared by UV photopolymerization of N-isopropylacrylamide on a surface of porous hydrophilic polypropylene membranes, J. Membr. Sci. 162 (1999) 235.

[21] N.I. Shtanko, V. Ya Kabanov, P. Yu Apel, M. Yoshida, A.I. Vilenskii, Preparation of permeability-controlled track membranes on the basis of 'smart' polymers, J. Membr. Sci. 179 (2000) 155.

[22] W. Lequieu, N.I. Shtanko, F.E. Du Prez, Track etched membranes with thermo-adjustable porosity and separation properties by surface immobilization of poly (N-vinylcaprolactam), J. Membr. Sci. 256 (2005) 64.

[23] S. Petrov, T. Ivanova, D. Christova, S. Ivanova, Modification of polyacrylonitrile membranes with temperature sensitive poly(vinylalcohol-co-vinylacetal), J. Membr. Sci. 261 (2005) 1.

[24] W.Y. Wang, L. Chen, X. Yu, Preparation of temperature sensitive poly(vinylidene fluoride) hollow fiber membranes grafted with N-isopropylacrylamide by a novel approach, J. Appl. Polym. Sci. 101 (2006) 833–837.

[25] L.Y. Chu, T. Niitsuma, T. Yamaguchi, S. Nakao, Thermoresponsive transport through porous membranes with grafted PNIPAM gates, AICHE J. 49 (2003) 896–909.

[26] X. Liu, K.G. Neoh, E.T. Kang, Redox-sensitive microporous membranes prepared from poly(vinylidene fluoride) grafted with viologen-containing polymer side chains, Macromol. 36 (2003) 8361–8367.

[27] Y. Ito, S. Nishi, Y.S. Park, Y. Imanishi, Oxidoreduction-sensitive control of water permeation through a polymer brushes-grafted porous membrane, Macromol. 30 (1997) 5856–5859.

[28] S.A. Piletsky, T.L. Panasyuk, E.V. Piletskaya, A.V. El'skaya, R. Levi, I. Karube, et al., Imprinted membranes for sensor technology: opposite behavior of covalently and noncovalently imprinted membranes, Macromol. 31 (1998) 2137–2140.

[29] T. Yamaguchi, T. Ito, T. Sato, T. Shinbo, S. Nakao, Development of a fast response molecular recognition ion gating membrane, J. Am. Chem. Soc. 121 (1999) 4078–4079.

[30] N. Minoura, K. Idei, A. Rachkov, Y.W. Choi, M. Ogiso, K. Matsuda, Preparation of azobenzene-containing polymer membranes that function in photoregulated molecular recognition, Macromol. 37 (2004).

[31] W. Yang, N. Cicek, J. Ilg, State-of-the-art of membrane bioreactors: worldwide research and commercial applications in North America., J. Membr. Sci. 270 (2006) 201.

[32] C. Anselme, I. Baudin, and M.R. Chevalier. Drinking water production by ultrafiltration and PAC adsorption: first year of operation of a large capacity plant. In American Water Works Association Membrane Technology Conference Long Beach, CA, 1999.

[33] C. Campos, B.J. Mariñas, V.L. Snoeyink, I. Baudin, J.M. Laîné, PAC-membrane filtration process I: model development, J. Environ. Eng. 126 (2000) 97–103.

[34] C. Campos, B.J. Mariñas, V.L. Snoeyink, I. Baudin, J.M. Laîné, PAC-membrane filtration process II: model application, J. Environ. Eng. 126 (2000) 104–111.

[35] M.M. Benjamin, M. Zhang, C. Li, and Y. Chang, eds., Combining Adsorbents With Membranes for Water Treatment, Amer, Water Works Assn. 2002.

[36] M.M. Clark, W.Y. Ahn, X. Li, N. Sternisha, R.L. Riley, Formation of polysulfone colloids for adsorption of natural organic foulants, Langmuir 21 (2005) 7207–7213.

[37] J. Kim, Z. Cai, M.M. Benjamin, Effects of adsorbents on membrane fouling by natural organic matter, J. Membr. Sci. 310 (2008) 356.

[38] S. Mozia, M. Tomaszewska, A.W. Morawski, Photocatalytic membrane reactor (PMR) coupling photocatalysis and membrane distillation—effectiveness of removal of three azo dyes from water, Catal. Today 129 (2007) 3.

[39] K. Azrague, P. Aimar, F. Benoit-Marquie, M.T. Maurette, A new combination of a membrane and a photocatalytic reactor for the depollution of turbid water, Appl. Catal. B: Environ. 72 (2007) 197.

[40] R. Molinari, L. Palmisano, E. Drioli, M. Schiavello, Studies on various reactor configurations for coupling photocatalysis and membrane processes in water purification, J. Membr. Sci. 206 (2002) 399.

[41] R. Molinari, M. Borgese, E. Drioli, L. Palmisano, M. Schiavello, Hybrid processes coupling photocatalysis and membranes for degradation of organic pollutants in water, Catal. Today 75 (2002) 77.

[42] M. Ulbricht, Advanced functional polymer membranes, Polym. 47 (2006) 2217.

[43] S.H. Kim, S.Y. Kwak, B.H. Sohn, T.H. Park, Design of TiO2 nanoparticle self-assembled aromatic polyamide thin-film-composite (TFC) membrane as an approach to solve biofouling problem, J. Memb. Sci. 211 (2003) 157.

[44] S.Y. Kwak, S.H. Kim, S.S. Kim, Hybrid organic/inorganic reverse osmosis (ro) membrane for bactericidal anti-fouling. 1. Preparation and characterization of TiO_2 nanoparticle self-assembled aromatic polyamide thin-film-composite (TFC) membrane, Environ. Sci. Technol. 35 (2001) 2388–2394.

[45] T.H. Bae, T.M. Tak, Effect of TiO_2 nanoparticles on fouling mitigation of ultrafiltration membranes for activated sludge filtration, J. Membr. Sci. 249 (2005) 1.

[46] T.H. Bae, I.C. Kim, T.M. Tak, Preparation and characterization of fouling-resistant TiO2 self-assembled nanocomposite membranes, J. Membr. Sci. 275 (2006) 1.

[47] J.B. Li, J.W. Zhu, M.S. Zheng, Morphologies and properties of poly(phthalazinone ether sulfone ketone) matrix ultrafiltration membranes with entrapped TiO2 nanoparticles, J. Appl. Polym. Sci. 103 (2006) 3623–3629.

[48] T.H. Bae, T.M. Tak, Preparation of TiO2 self-assembled polymeric nanocomposite membranes and examination of their fouling mitigation effects in a membrane bioreactor system, J. Membr. Sci. 266 (2005) 1.

[49] M.L. Luo, J.Q. Zhao, W. Tang, C.S. Pu, Hydrophilic modification of poly(ether sulfone) ultrafiltration membrane surface by self-assembly of TiO_2 nanoparticles, Appl. Surf. Sci. 249 (2005) 76.

[50] D.E. Meyer, K. Wood, L.G. Bachas, D. Bhattacharyya, Degradation of chlorinated organics by membrane-immobilized nanosized metals, Environ. Program 23 (2004) 232–242.

[51] L. Wu, S.M.C. Ritchie, Removal of trichloroethylene from water by cellulose acetate supported bimetallic Ni/Fe nanoparticles, Chemosphere 63 (2006) 285.

[52] J. Xu, D. Bhattacharyya, Fe/Pd nanoparticle immobilization in microfiltration membrane pores: synthesis, characterization, and application in the dechlorination of polychlorinated biphenyls, Ind. Eng. Chem. Res. 46 (2007) 2348–2359.

[53] L.F. Wu, M. Shamsuzzoha, S.M.C. Ritchie, Preparation of cellulose acetate supported zero-valent iron nanoparticles for the dechlorination of trichloroethylene in water, J. Nanopart. Res. 7 (2005) 469–476.

[54] G. Decher, Fuzzy nanoassemblies: toward layered polymeric multicomposites, Sci. 277 (1997) 1232–1237.

[55] D. Lee, R.E. Cohen, M.F. Rubner, Antibacterial properties of Ag nanoparticle loaded multilayers and formation of magnetically directed antibacterial microparticles, Langmuir 21 (2005) 9651–9659.
[56] Z. Li, H. Huang, T. Shang, F. Yang, W. Zheng, C. Wang, et al., Facile synthesis of single-crystal and controllable sized silver nanoparticles on the surfaces of polyacrylonitrile nanofibres, Nanotechnol. 17 (2006) 917–920.
[57] A.A. Mamedov, A. Belov, M. Giersig, N.N. Mamedova, N.A. Kotov, Nanorainbows: graded semiconductor films from quantum dots, J. Am. Chem. Soc. 123 (2001) 7738–7739.
[58] J. Dai, M.L. Bruening, Catalytic nanoparticles formed by reduction of metal ions in multilayered polyelectrolyte films, Nano Lett. 2 (2002) 497–501.
[59] J. Xu, A. Dozier, D. Bhattacharyya, Synthesis of nanoscale bimetallic particles in polyelectrolyte membrane matrix for reductive transformation of halogenated organic compounds, J. Nanopart. Res. 7 (2005) 449–467.
[60] D.M. Dotzauer, J. Dai, L. Sun, M.L. Bruening, Catalytic membranes prepared using layer-by-layer adsorption of polyelectrolyte/metal nanoparticle films in porous supports, Nano Lett. (2006) 2268–2272.
[61] W.K. Son, J.H. Youk, T.S. Lee, W.H. Park, Preparation of antimicrobial ultrafine cellulose acetate fibers with silver nanoparticles, Macromol. Rapid Commun. 25 (2004) 1632–1637.
[62] K.H. Hong, J.L. Park, I.H. Sul, J.H. Youk, T.J. Kang, Preparation of antimicrobial poly(vinyl alcohol) nanofibers containing silver nanoparticles, J. Polym. Sci, Part B: Polym. Phys. 34 (2006) 2468–2474.
[63] Y. Kurokawa, Y. Imai, Surface-enhanced Raman scattering (SERS) using polymer (cellulose acetate and Nafion) membranes impregnated with fine silver particles, J. Membr. Sci. 55 (1991) 227–233.
[64] J. Taurozzi, V.V. Tarabara, Silver nanoparticle arrays on track etch membrane support as flowthrough optical sensors for water quality control, Environ. Eng. Sci. 24 (2007) 122–137.
[65] P. Jian, H. Yahui, W. Yang, L. Linlin, Preparation of polysulfone-Fe_3O_4 composite ultrafiltration membrane and its behavior in magnetic field, J. Membr. Sci. 284 (2006) 9–16.
[66] J.N. Coleman, U. Khan, Y.K. Gun'ko, Mechanical reinforcement of polymers using carbon nanotubes, Adv. Mater. 18 (2006) 689–706.
[67] A. Heilmann, Polymer Films with Embedded Metal Nanoparticles, Springer, 2002.
[68] R.H.B. Bouma, A. Checchetti, G. Chidichimo, E. Drioli, Permeation through a heterogeneous membrane: the effect of the dispersed phase, J. Membr. Sci. 128 (1997) 141–149.
[69] C.M. Zimmerman, A. Singh, W.J. Koros, Tailoring mixed matrix composite membranes for gas separations, J. Memb. Sci. 137 (1997) 145–154.
[70] B.H. Jeong, E.M.V. Hoek, Y. Yan, A. Subramani, X. Huang, G. Hurwitz, et al., Interfacial polymerization of thin film nanocomposites: a new concept for reverse osmosis membranes, J. Memb. Sci. 294 (2007) 1.
[71] G. Jonsson, Methods for determining the selectivity of reverse osmosis membranes, Desalin. 24 (1977) 19–37.
[72] K.M. Persson, V. Gekas, G. Tragardh, Study of membrane compaction and its influence on ultrafiltration water permeability, J. Memb. Sci. 100 (1995) 155.

[73] P. Aerts, E. Van Hoof, R. Leysen, I.F.J. Vankelecom, P.A. Jacobs, Polysulfone-aerosil composite membranes: Part 1. The influence of the addition of aerosil on the formation process and membrane morphology, J. Memb. Sci. 176 (2000) 63.

[74] P. Aerts, I. Genne, S. Kuypers, R. Leysen, I.F.J. Vankelecom, P.A. Jacobs, Polysulfone-aerosil composite membranes: Part 2. The influence of the addition of aerosil on the skin characteristics and membrane properties, J. Memb. Sci. 178 (2000) 1.

[75] O. Monticelli, A. Bottino, I. Scandale, G. Capannelli, S. Russo, Preparation and properties of polysulfone-clay composite membranes, J. Appl. Polym. Sci. 103 (2007) 3637–3644.

[76] Y. Yang, H. Zhang, P. Wang, Q. Zheng, J. Li, The influence of nano-sized TiO_2 fillers on the morphologies and properties of PSF UF membrane, J. Memb. Sci. 288 (2007) 231.

[77] L. Ballinas, C. Torras, V. Fierro, R. Garcia-Valls, Factors influencing activated carbon-polymeric composite membrane structure and performance, J. Phys. Chem. Solids 65 (2004) 633–637.

[78] J.H. Choi, J. Jegal, W.N. Kim, Fabrication and characterization of multi-walled carbon nanotubes/polymer blend membranes, J. Memb. Sci. 284 (2006) 406–415.

[79] J.S. Taurozzi, H. Arul, V.Z. Bosak, A.F. Burban, T.C. Voice, M.L. Bruening, et al., Effect of filler incorporation route on the properties of polysulfone-silver nanocomposite membranes of different porosities. J. Membr. Sci. 325 (2008) 58–68.

[80] I. Pastoriza-Santos, L.M. Liz-Marzan, Formation and stabilization of silver nanoparticles through reduction by N,N-Dimethylformamide, Langmuir 15 (1999) 948–951.

[81] J. Kim, S.H.R. Davies, M.J. Baumann, S.J. Masten, V.V. Tarabara, Effects of Solution Chemistry on NOM Fouling in a Hybrid Ozonation–Ultrafiltration Water Treatment System, AEESP meeting, Blacksburg, VA, 2007.

[82] J. Kim, S.H.R. Davies, M.J. Baumann, V.V. Tarabara, S.J. Masten, Effect of ozone dosage and hydrodynamic conditions on the permeate flux in a hybrid ozonation-ceramic ultrafiltration system treating natural waters, J. Membr. Sci. 311 (2007) 165–172.

[83] U. von Gunten, Ozonation of drinking water: Part I. Oxidation kinetics and product formation, Water. Res. 37 (2003) 1443–1467.

[84] B. Schlichter, V. Mavrov, H. Chmiel, Study of a hybrid process combining ozonation and microfiltration/ultrafiltration for drinking water production from surface water, Desalin. 68 (2004) 307–317.

[85] B.S. Karnik, S.H.R. Davies, K.C. Chen, D.R. Jaglowski, M.J. Baumann, S.J. Masten, Effect of ozonation on the permeate flux of nanocrystalline ceramic membranes, Water. Res. 39 (2005) 728–734.

[86] B.S. Karnik, S.H.R. Davies, M.J. Baumann, S.J. Masten, Fabrication of catalytic membranes for the treatment of drinking water using combined ozonation and ultrafiltration, Environ. Sci. Technol. Vol. 39 (19) (2005) 7656–7661.

CHAPTER

11 Nanofluidic Carbon Nanotube Membranes: Applications for Water Purification and Desalination

Olgica Bakajin[3], Aleksandr Noy[1], Francesco Fornasiero[1], Costas P. Grigoropoulos[2], Jason K. Holt[1], Jung Bin In[1,2], Sangil Kim[1] and Hyung Gyu Park[1]

[1]Molecular Biophysics and Functional Nanostructures Group, Chemistry, Materials, Earth, and Life Sciences Directorate, Lawrence Livermore National Laboratory, Livermore, CA, USA [2]Department of Mechanical Engineering, University of California at Berkeley, Berkeley, CA, USA [3]Chief Executive Officer at Porifera Inc., San Francisco, CA, USA

Street, Sustich, Duncan and Savage. Nanotechnology Applications for Clean Water, 2nd Edition.
 DOI: http://dx.doi.org/10.1016/B978-1-4557-3116-9.00011-1

11.1 Introduction: carbon nanotube membrane technology for water purification

The availability of safe, clean, and inexpensive water has emerged as an issue that defines global problems in the twenty-first century. Water shortages are some of the root causes of societal disruptions such as epidemics, environmental disasters, tribal and ethnic conflicts, growth shortfalls, and even countrywide political destabilization. Membrane-based filtration is the current leading energy-efficient technology for cleanup and desalination of brackish water, recycled water, and seawater. Membrane-based filtration offers other advantages, as filtration through the tight membrane pores can also remove dangerous impurities, such as As, as well as toxic large organic compounds. Factors that limit the efficiency of the membrane purification technologies include the membrane resistance to the flow, membrane fouling, and membrane imperfections that lead to incomplete rejection or to a drop in the membrane rejection properties over time. The latest technological developments and high-efficiency energy recovery systems in particular have pushed the current efficiency of reverse osmosis (RO) membranes to a very impressive 4 kWh/m^3 [1]; however, this number is still well above the theoretical minimum energy cost of 0.97 kWh/m^3 for 50 percent recovery [2,3]. To move further, we need to develop transformative membrane technologies that utilize fundamentally new transport and filtration mechanisms for drastic gains in transport efficiency.

Carbon nanotube (CNT) membranes are promising candidates for one such solution primarily because of their transport characteristics. The inner cavity of a CNT forms a natural pore with very small diameter that can in some instances be smaller than 1 nm. Moreover, smooth hydrophobic surfaces of the nanotubes lead to nearly frictionless flow of water through them, enabling transport rates that are orders of magnitude higher than transport in conventional pores. Finally, the structure of CNTs permit targeted specific modifications of the pore entrance without destroying the unique properties of the inner nanotube surface. The combination of these three factors could enable a new generation of membranes whose transport efficiency, rejection properties, and lifetimes drastically exceed those of the current membranes.

This chapter presents a brief overview of the basic physical processes that govern the structure and transport of water inside CNT pores, basic properties that make nanotube pore technologies attractive for water purification and desalination, the fabrication approaches for producing CNT membranes, and the experimental observations of water transport and ion exclusion properties in CNT membranes.

11.2 Basic structure and properties of carbon nanotubes

By now the CNT has firmly established itself as the iconic molecule of nanoscience [4]. A CNT is simply a nanometer-sized rolled-up atomically smooth graphene sheet that forms a perfect seamless cylinder (Fig. 11.1) capped at the

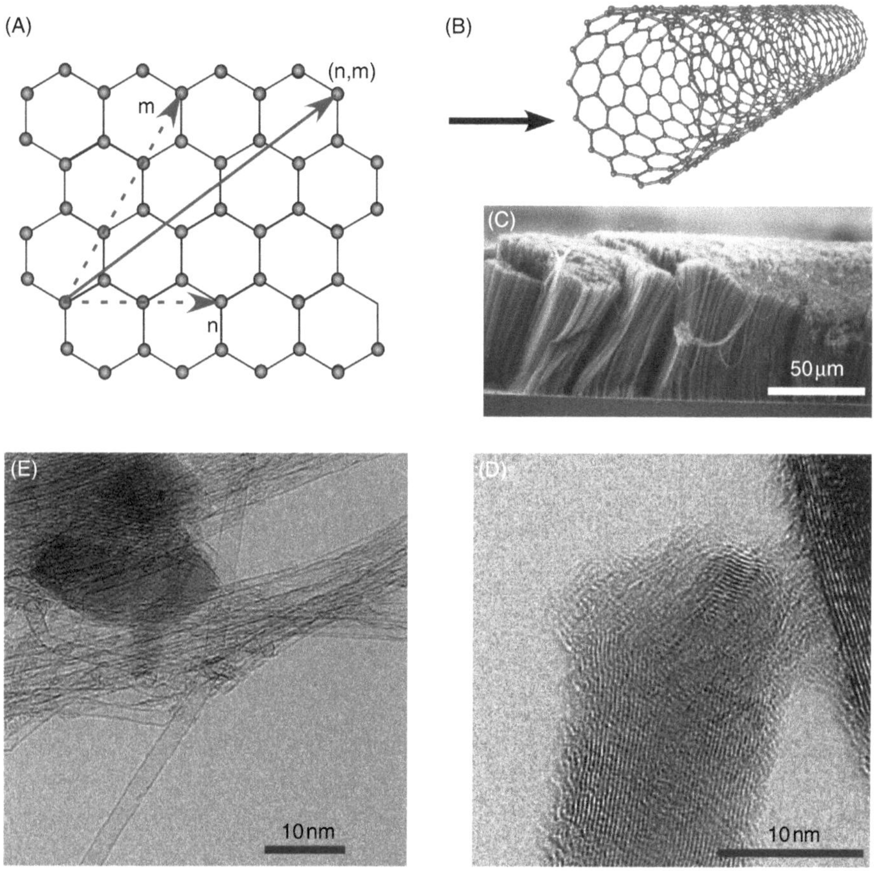

FIGURE 11.1

Structure and morphology of carbon nanotubes (CNTs). (A) Schematic representation of a graphene sheet and a CNT roll-up vector. The roll-up vector is perpendicular to the axis of the CNT. (B) A 3-D model of a single-wall CNT. (C) A scanning electron microscope (SEM) image of a vertically aligned array of multi-wall CNT grown on a silicon substrate. (SEM images: M. Stadermann, O. Bakajin, A. Noy, LLNL.) (D,E) Transmission electron microscope (TEM) images of single-wall (e) and multi-wall (d) CNTs. (TEM images: J. Plitzko, A. Noy, LLNL.).

ends by fullerene caps. It is common to characterize the structure of the nanotube by its rolled-up vector (n,m), called chirality or helicity, which defines the position of the matched carbon rings during the roll-up of the graphene sheet [5]. A CNT can have one (as in case of a single-walled CNT), or several concentric graphitic shells (as in case of multi-walled nanotubes) and it can reach up to several millimeters in length, yet retain a diameter of only a few nanometers.

Several methods of CNT production currently exist. In the laboratory environment, catalytic chemical vapor deposition (CVD) is preferred over other methods, such as arc discharge and laser ablation, because it produces higher quality CNTs. CVD reactors can produce individual isolated nanotubes, as well as densely packed vertically aligned arrays (Fig. 11.1C). Unfortunately, the ultimate goal of the CNT synthesis producing a uniform population of nanotubes with a given chirality still remains elusive. Several studies indicated that the size of the catalyst particle during the growth stage determines the size of the CNT to less than 10 percent [6]; yet efforts to control the size of the CNTs with greater precision have been largely unsuccessful. Thus, synthesizing a vertically aligned CNT array with a narrow distribution of sizes still remains a difficult endeavor requiring considerable process development and optimization efforts [6–8]. The task of describing the details of CNT synthesis goes well beyond the scope of this chapter; therefore we refer the readers to a number of reviews on this subject [9–11].

11.3 Water transport in carbon nanotube pores: an MD simulation view

11.3.1 Water inside carbon nanotubes

The task of observing and understanding fluid and gas flows in CNT pores raises a set of unique fundamental questions [12]. First, it is surprising that hydrophilic liquids, especially water, enter and fill very narrow and hydrophobic CNTs. If the water does enter the CNTs, what influence does extreme confinement have on the water structure and properties? It is important to evaluate how these changes in structure influence the rate, efficiency, and selectivity of the transport of liquids and gases through the CNTs. As it is often the case, molecular dynamics (MD) simulations have provided some of the first answers to these questions. G. Hummer and colleagues used MD simulations to observe (Fig. 11.2A and B) filling of the (6,6) nanotube (0.81 nm in diameter and 1.34 nm in length) with water molecules [13]. Surprisingly, they found that water filled the empty cavity of the CNT within a few tens of picoseconds and the filled state continued over the entire simulation time (66 ns). More importantly, the water molecules confined in such a small space formed a single-file configuration that is unseen in the bulk water. Several experimental studies also provide some evidence of water filling of CNTs [14–17]. Further analysis of the simulation results of Hummer and colleague showed that water molecules inside and outside the nanotube were in thermodynamic equilibrium. This observation illustrates one of the more important and counterintuitive phenomena associated with nanofluidic systems: nanoscale confinement leads to the narrowing of the interaction energy distribution, which then causes the lowering of the chemical potential [13]. In other words, confining the liquid inside a nanotube channel actually lowers its free energy! Further simulations by the same group showed that the filling equilibrium

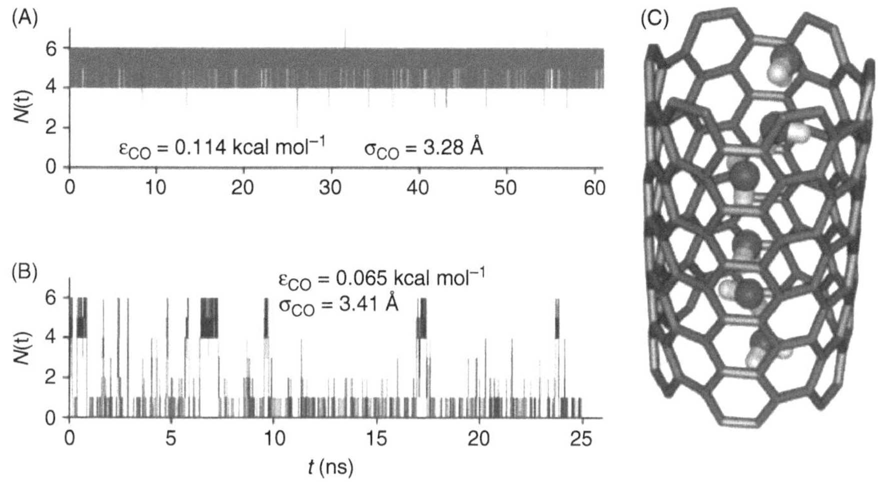

FIGURE 11.2

MD simulations of water transport in carbon nanotubes (CNTs). (A,B). Number *N* of water molecules inside an 8.1 Å diameter nanotube as a function of time for sp^2 carbon parameters (A) and reduced carbon–water attractions (B). (C) Structure of the hydrogen-bonded water chain inside the CNT.

Reproduced from [13] with permission © 2001 Nature Publishing Group.

was very sensitive to the water–nanotube interactions parameters: a 40 percent reduction in the depth of the carbon–water interaction potential resulted in the emptying of the CNT cavity, or a 25 percent reduction resulted in the fluctuation between filled and empty states (bi-stable states) [13,18]. This sharp transition between the two states has been observed for other hydrophobic nanopores as well [19–22]. MD simulations have also studied the dependence of the CNT hydration on other properties of CNTs such as the nanotube wall flexibility [23], charge [24,25], chirality [25,26], length [18], and diameter [24,27,28].

11.3.2 Carbon nanotubes as biological channel analogs

MD simulations show that a defining feature of the water structure in CNTs is the formation of the hydrogen-bonded "water wires" (Fig. 11.2C) oriented along the nanotube axis [12,13,29]. Such one-dimensional hydrogen bonded structures are highly reminiscent of the water wires observed in the biological channels specializing in water transport, such as aquaporins [30]. In fact, the similarity between aquaporin channels and CNT channels goes further. Similar to the hydrophobic interior of CNTs, the inner cavity of the aquaporins is lined with hydrophobic residues that facilitate the formation of the one-dimensional hydrogen-bonded water chains [30,31]. Weak interactions of water molecules with the hydrophobic walls combine with the smooth nature of the nanotube walls to enable nearly frictionless transport of water in nanotubes channels. Kalra and Hummer et al. showed that water moves very rapidly through a nanotube channel under osmotic pressure [29]. They observed that friction at the channel walls in that system was so low that water transport was no longer governed by the Hagen–Poisselle flow, but instead depended mainly on the events at the nanotube entrance and exit [29]. As a result, calculated rates of water transport approached 5.8 water molecules/ns/nanotube. Interestingly, these rates are comparable to the water transport rates achieved in aquaporins [32]. Other MD simulations also observed fast water transport through CNTs [33–35].

11.4 Fabrication of carbon nanotube membranes

Verification of these seemingly exotic predictions of fast transport through CNTs that emerged from the MD simulations required fabrication of a robust test platform: a CNT membrane. Such membranes typically consist of an aligned array of CNTs encapsulated by a filler (matrix) material, with the nanotube ends opened at the top and bottom. Although there are likely many ways to produce a structure of this type (a notable early result by Martin and coworkers was based on fabrication of amorphous CNTs within porous alumina membrane template [36]), the approach that has proven most fruitful to date involves growing an aligned array of CNTs, followed by infiltration of a matrix material in the gaps between the CNTs (Fig. 11.3A). Extremely high aspect ratio of the gaps between the nanotubes in the array (of order 1,000 length/diameter or larger) presents a great fabrication challenge for this approach. Fortunately, researchers have developed successful strategies to overcome this challenge.

11.4.1 Polymeric/CNT membranes

Hinds' group at the University of Kentucky has pioneered a membrane fabrication strategy based on polymer encapsulation of CNT arrays [37]. They infiltrated multi-wall CNT arrays with polystyrene solution that after evaporation produced a high-density multi-wall CNT membrane of ca. 7 nm pore size (Fig. 11.3D). As the process occurs in the liquid phase, care was necessary to ensure that the CNTs do not bundle together upon solvent evaporation.

11.4.2 Silicon nitride CNT membranes

Our group at the Lawrence Livermore National Laboratory (LLNL) developed a process for encapsulation of a vertically aligned array of CNTs with low-stress silicon nitride by a low-pressure chemical vapor deposition process [38,39]. This is a method widely used for a host of microfabrication processes and it produces an extremely conformal coating around CNTs (Fig. 11.3C). The membrane produced is robust and is capable of withstanding pressure gradients in excess of

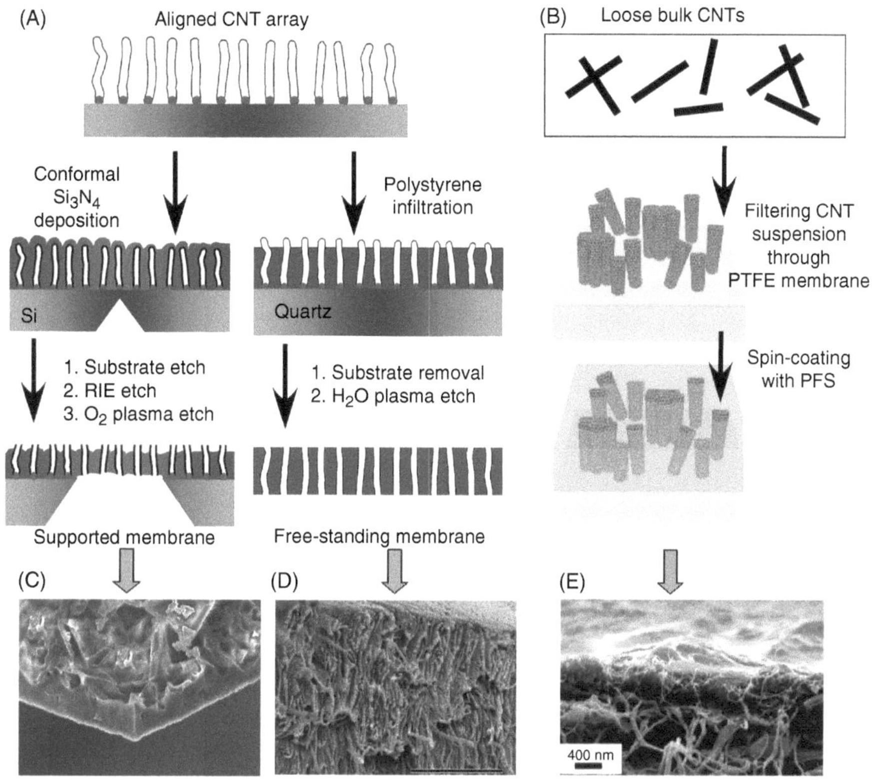

FIGURE 11.3

Fabrication of carbon nanotube (CNT) membranes. (A,B) Process flow diagrams for fabrication of CNT membranes using nanotube array encapsulation with Si_3N_4 or polymers (A), and filtration-assisted alignment (B). (C–E) Scanning electron microscope images of (C) Si_3N_4-encapsulated membrane, from [39]; (D) polystyrene encapsulated membrane, from [37]; and (E) filtration-assisted assembly membrane, from [40]. © 2006, 2004 American Association for the Advancement of Science and 2001 American Chemical Society, respectively.

1 atmosphere. Subsequent to encapsulation, the membrane undergoes a series of etching steps to selectively remove excess silicon nitride from the tips of the CNTs, followed by oxygen plasma to uncap the CNTs. Transmission electron micrographs of thinned-down sections of our double-wall CNT (DWCNT) membranes (Fig. 11.4) suggest that they consist of pores less than 2 nm diameter, consistent with diameters of as-grown nanotubes, and no nano- or microvoids apparent in the structure. We have demonstrated fabrication of membranes with two different CNT pore diameters: double-wall at 1.1 nm $<D<$ 2 nm and multiwall at approximately 6.5 nm.

11.4.3 CNT polymer network fabrication

A considerably different approach to producing an aligned CNT–polymer composite membrane was recently described by E. Marand and coworkers (Fig. 11.4B) [40]. Amine-functionalized CNTs were dispersed in tetrahydrofuran and subsequently filtered through a hydrophobic (0.2 μm) PTFE poly(tetrafluoroethylene) filter, leading to alignment within the membrane pores (Fig. 11.3E). Spin coating with a dilute polymer solution (polysulfone) produced a mechanically stable thin film structure with the CNT tips protruding from the top of the membrane. Membranes produced with this method exhibited enhancements in gas transport rates and non-Knudsen selectivities for binary gas mixtures. This approach has the advantage of being potentially more scalable and economical than direct growth CVD of CNTs on a substrate, although at the current stage of development the nanotube densities (and thus the available pore density) are much smaller than for the membranes produced by CNT array encapsulation.

11.5 Experimental observations of water transport in double-wall and multi-wall carbon nanotube membranes

We also observed high rates of water transport through the double-wall sub-2-nm CNT membranes using pressure-driven flow [39] . Similarly high rates were also observed by Majumder et al. for multi-walled nanotube membranes with larger pore diameters [41]. As previously discussed, the single largest uncertainty in quantifying the flux through individual pores lies in the determination of the active pore density (i.e., those nanotubes that are open and span the membrane). Majumder et al. estimated the active pore densities by quantifying diffusion of small molecules through the CNTs. They measured enhancements of four to five orders of magnitude compared to Hagen–Poiseulle formalism. As described in the previous section, we estimated the upper bounds to the pore densities so that our measurements represent lower boundary estimates. The transport rates that we measured reveal a flow enhancement that is at least two to three orders of magnitude faster than no-slip, hydrodynamic flow calculated using Hagen–Poiseuille equation (Fig. 11.5). The calculated slip length for sub-2-nm CNTs is as large as hundreds of nanometers, which is almost three orders of magnitude larger than the pore size and is almost on the order of the overall nanotube length. In contrast, the polycarbonate membrane with a pore size of 15 nm reveals a much smaller slip length of just 5 nm! This comparison suggests that slip flow formalism may not be applicable to water flow through CNTs, possibly due to length scale confinement [41,42] or to partial wetting between water and the CNT surface [43]. Interestingly, the measured water flux compares well with that predicted by the MD simulations [29]. The simulations predict a flux of 12 water molecules/nm^2 (of nanotube cross-sectional area)/ns; our measured flux,

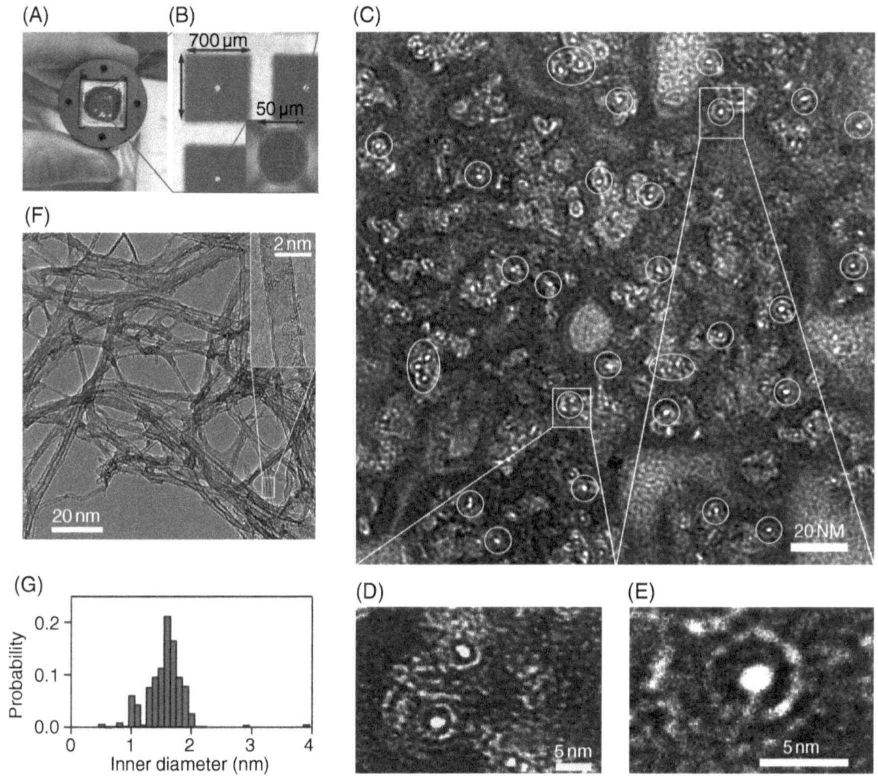

FIGURE 11.4

Sub-2-nm carbon nanotube (CNT) membranes. (A) A photograph of a CNT membrane chip in the sample holder. (B) Optical micrographs of the regions of the chip that contain CNT membrane windows. (C–E) High-resolution transmission electron microscope (TEM) images of the thinned cross-sections of the membrane showing sub-2-nm pores. (F,G) HR-TEM characterization of the CNT size: (F) A TEM image of the dispersed carbon nanotubes from the vertically aligned array used for membrane fabrication. (G) A histogram of measured inner diameters of carbon nanotubes.

From [39], © 2006 American Association for the Advancement of Science.

extrapolated to the simulation pressure drop, corresponds to 10–40 water molecules/nm^2/ns [39]. Moreover, the measured absolute flow rates of at least 0.9 water molecule/nanotube is similar to the rate of 3.9 molecule/pore measured for aquaporins. The comparison to the aquaporins in not straightforward since the diameters of our CNTs are twice that of aquaporins and the CNTs are considerably longer, to name just a few differences. Therefore, we cannot yet imply that the same mechanism is responsible for transport in our CNTs and aquaporins. Nevertheless, our experiments demonstrate that the water transport through CNTs starts to approach the efficiency of biological channels.

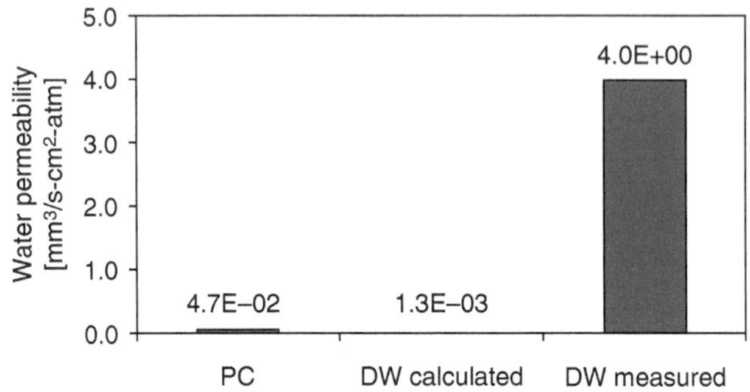

FIGURE 11.5

Water transport in sub-2-nm carbon nanotube (CNT) membranes. Comparison of the water flux predicted for a polycarbonate membrane (left), and a double-wall CNT membrane (center) with the flux measured for the double-wall CNT membrane (right).

11.6 Nanofiltration properties of carbon nanotube membranes

11.6.1 Size exclusion experiments in the 1–10 nm size range

Nearly frictionless graphitic walls of CNT composite membranes offer the unique combination of extremely fast flow and very small pore size, which potentially give them a tremendous advantage over traditional membrane materials for energy-efficient, low-cost ultrafiltration and nanofiltration applications. Several experimental studies have used concentration gradient or pressure driven flow to determine the size exclusion properties of CNT composite membranes. Diffusion studies by Hinds' group showed that 10 nm gold nanoparticles were completely excluded by a multiwall CNT (MWCNT)-membrane with pore inner diameter of ca. 7 nm, whereas small dyes (0.5–2 nm) diffuse with low hindrance [37,41]. Our aligned DWCNT array membranes completely excluded 2 nm and 5 nm gold nanoparticles in pressure-driven filtration experiments as expected from the transmission electron microscope (TEM) measurements of the pore diameter (Fig. 11.4F and G). Note that these membrane still exhibited extremely fast water permeation [39].

11.6.2 Ion exclusion in carbon nanotube membranes

With the diameters in the nanometer regime and high water permeabilities, CNTs are a promising platform for ion removal from water, as required for desalination and demineralization. MD simulations [44,45] predict that if the nanotube is uncharged, size-based exclusion of small ionic species such as Na^+, K^+, or Cl^- requires CNT diameters of about 0.4 nm. These diameters are comparable to the hydrated ion size. At this scale, the ion is forced to lose part of its hydration shell to enter the CNT, implying a very high energy barrier to cross the membrane (approximately 120 kJ/mole). For slightly larger pore sizes (>1 nm), this free energy penalty decays almost to zero (approximately 5 kJ/mole), allowing small ion free access. Molecular simulations [45] have also predicted that CNTs with a diameter of 0.34 nm, decorated with negative charges along the walls, will conduct K^+ ions while excluding Cl^-, whereas positively charged CNTs with a diameter of 0.47 nm diameter will exclude K^+ ions while conducting Cl^-. Experimental verification of MD simulation prediction has not been achieved yet as CNT membranes with such small pore openings have not been successfully fabricated to date. Joseph et al. showed by simulations that the presence of charged groups on the open CNT tips [46] induces preferential ion transport for CNT with a diameter of 2.2 nm. These results in particular suggest that dedicated functionalization of small-diameter CNT membranes (such as the membranes demonstrated by Holt et al. [39]) may enable the control of ionic flow or even the exclusion of very small ions, a particularly exciting prospect for water purification and desalination.

Recently our team has performed the first evaluation of ion exclusion in CNT membranes [47]. That study found that CNT membranes excluded a significant fraction of the ions in the feed solution (in some cases more than 90 percent). The exclusion characteristics of the membrane were also a strong function of the solution ionic strength with lower ionic strength exhibiting higher exclusion ratios. The last property provides a strong clue that electrostatic interactions on the CNT mouth play an important role. Further investigation showed that the rejection properties of the membrane obey the predictions of the Donnan equilibrium model. The exclusion properties observed for the CNT membrane are similar to the exclusion properties of the nanofiltration membranes with pores in the same size range. Notably CNT membranes are much thicker than the active layers of current nanofiltration membranes, yet they exhibit much higher water flux!

11.7 Altering transport selectivity by membrane functionalization

An important avenue for controlling transport through nanotube channels involves using chemical modifications of the nanotube to alter the channel permeability or exclusion characteristics. To maintain efficient flow through the nanotube these treatments need to preserve the fundamental smoothness and hydrophobicity of the nanotube walls; therefore, we will concentrate on the modification strategies that target only the entrance and the exit of the nanotube. In fact, most membrane fabrication strategies facilitate this approach by using an oxidation step to remove the fullerene cap from the nanotube. These procedures typically produce carboxylic groups at the mouth of the nanotube [48], which not only render the pore mouth negatively charged, but also serve as the target for a variety of chemical modification approaches [49,50].

Most of the recent progress in this area has been associated with the work of Hinds and coworkers who used carbodiimide chemistry to attach a variety of organic and biological molecules to the mouth of CNTs. Interestingly, they observed that, for the polymer matrix based aligned nanotube membranes, the idealized picture of the oxidation step producing only a ring of carboxylic acid groups on the nanotube end and leaving the rest of the nanotube intact misrepresents the reality. Experiments on decorating the nanotube surface with gold nanoparticles showed that the oxidation step produced reactive groups in the nanotube regions that were up to 700 nm away from the tip of the nanotube [50,51], although after only 50 nm of separation from the tip the functional group density was already significantly reduced. Researchers argue [50] that these apparent large penetration depths are consistent with the observation that the manufacturing process produces exposed CNT tips above the polystyrene matrix.

Majumder et al. used carbodiimide chemistry to attach aliphatic chains, charged dye molecules, and polypeptides to the polystyrene-based CNTs membranes [50]. These modifications have a measurable effect on the flux of the two large organic cations [methyl viologen (Mv^{2+}), and $Ru(bpy)_3{}^{2+}$] used for these experiments; for example, attaching a C_{40} alkane chain reduced the flux of Mv^{2+} by six times. Interestingly, the researchers did not observe a clear trend for the effect of the modification on the flux of the test species: for example, attaching a bulky charged organic dye to the mouth of the CNT membrane actually increased the ion flux, presumably due to the interactions of the dye molecules with the oppositely charged ions [50]. Also the relationship between the size of the modifier group and the effect on the charged species flux was complex; the authors of the report speculated that longer hydrophobic aliphatic chains prefer to orient along CNT walls and thus have a reduced effect on the overall flux.

11.8 Is energy-efficient desalination and water purification with carbon nanotube membranes possible and practical?

The definitive answer to this question will emerge only from continued research and development on the CNT membrane prototype. Yet, several conclusions can be reached even today. The most promising property of CNT membranes for water purification applications is their extremely high permeability. This property should translate into more water per unit of applied pressure, more efficient, smaller purification units and ultimately into lower purification or desalinations costs. Rich possibilities for chemical functionalization, coupled with the rather unique ability to manipulate only the chemistry at the nanotube mouth open up the possibility of producing membranes tailored for specific applications (e.g., RO desalination or impurity purification) while maintaining the basic membrane structure and high permeability.

However, a true assessment of the potential impact of CNT membranes on water purification (and specifically on water desalination) applications requires a more comprehensive comparison of the membrane characteristics with the general requirements of the membrane purification process. At least in the case of RO desalination, the process efficiency comes from three main sources: capital costs, energy costs, and operation costs (which include costs for pretreatment, posttreatment, and membrane cleaning and regeneration). It is instructive to evaluate the potential of the CNT membrane technology against these three areas. We must first note that the CNT technology is still in its infancy; therefore, most attempts at quantitative evaluation will face large uncertainties associated with predicting the future technological milestones, or the fact that some of the major membrane characteristics (e.g., fouling properties) have not been sufficiently evaluated. Another large source of uncertainty is the lack of availability and cost estimates for a manufacturing process that allows scale-up of membrane fabrication. However, we still can reach some qualitative conclusions based even on the limited set of data that is available now. The high flux of CNT membranes provides a clear advantage for both the energy costs and the capital costs, as the same amount of product water could be obtained with smaller driving pressures and less membrane area. However some of the other important advantages of CNT membrane technology could come from the factors contributing to the third cost factor: the operation costs. The uniform pore size of CNT membranes could simplify or even eliminate the requirements for complicated multistage pretreatment efforts. The membrane pore surface is also rather chemically inert, which could increase the membrane lifetime against the harsh agents used for pretreating water before RO or other purification steps. Unlike most polymeric membrane surfaces, the CNT membrane surface is hydrophilic; therefore, it could offer an increased resistance to fouling, as well as easier cleanup by rinsing or backwashing. These factors could all contribute to an increased membrane lifespan and ultimately to operation cost savings.

If we consider these factors, it becomes clear that the real impact of CNT membrane technology may lie in its potential to improve all of the major areas that contribute to the costs of water purification processes. Clearly, much work needs to be done before these promises translate into field applications. Researchers need to develop approaches for fabricating CNTs with an even narrower distribution of the pore sizes, ideally with pores that are less than 1 nm. Targeted chemical modification of the pore entrances should improve dramatically the rejection characteristics of the membrane. Further studies are necessary to quantify the membrane fouling resistance and useful lifespan. Finally, development of large-scale, low-cost manufacturing processes is imperative to ensure that CNT membrane technology can achieve significant penetration into the water purification market. These are all challenging tasks, yet the potential of CNT membrane technology is high enough for us to have no doubts that it will find its place in the arsenal of the water purification techniques available for mankind.

Acknowledgments

This chapter has been partially adapted from an invited review written by this team for the Nano Today magazine in 2007. C.G. and O.B. acknowledge support from NSF NER 0608964; A.N., C.G., and O.B. acknowledge support from NSF NIRT CBET-0709090. All authors (except C.G.) acknowledge internal developmental funding support from LLNL. Parts of this work were performed under the auspices of the US Department of Energy by Lawrence Livermore National Laboratory under Contract DE-AC52-07NA27344.

References

[1] S.A. Avlonitis, K. Kouroumbas, N. Vlachakis, Energy consumption and membrane replacement cost for seawater RO desalination plants, Desalination 157 (1–3) (2003) 151–158.

[2] K.S. Spiegler, Y.M. El-Sayed, A Desalination Primer: Introductory Book for Students and Newcomers to Desalination, Balaban Desalination Publications, 1994.

[3] M. Shannon, P. Bohn, M. Elimelech, J. Georgiadis, B. Marinas, A. Mayes, Science and technology for water purification in the coming decades, Nature 452 (2008) 301–310.

[4] M. Terrones, Science and technology of the twenty-first century: synthesis, properties and applications of carbon nanotubes [Review], Annu. Rev. Mater. Res. 33 (2003) 419–501.

[5] M.S. Dresselhaus, R. Saito, Physical Properties of Carbon Nanotubes, Imperial College Press, 1998.

[6] T. Yamada, T. Namai, K. Hata, D.N. Futaba, K. Mizuno, J. Fan, et al., Size-selective growth of double-walled carbon nanotube forests from engineered iron catalysts, Nat. Nanotechnol. 1 (2) (2006) 131.

[7] D.N. Futaba, K. Hata, T. Yamada, T. Hiraoka, Y. Hayamizu, Y. Kakudate, et al., Shape-engineerable and highly densely packed single-walled carbon nanotubes and their application as super-capacitor electrodes, Nat. Mater. 5 (12) (2006) 987–994.

[8] C.L. Cheung, A Kurtz, H. Park, C.M. Lieber, Diameter-controlled synthesis of carbon nanotubes, J. Phys. Chem. B 106 (10) (2002) 2429–2433.

[9] A. Loiseau, X. Blase, J-C. Charlier, P. Gadelle, C. Journet, C. Laurent, et al., Synthesis methods and growth mechanics, Lect. Notes Phys. 677 (2006) 49–130.

[10] N.R. Franklin, Q. Wang, T.W. Tombler, A. Javey, M. Shim, H. Dai, Integration of suspended carbon nanotube arrays into electronic devices and electromechanical systems, Appl. Phys. Lett. 81 (5) (2002) 913–915.

[11] A.A. Puretzky, D.B. Geohegan, S. Jesse, I.N. Ivanov, G. Eres, In situ measurements and modeling of carbon nanotube array growth kinetics during chemical vapor deposition, Appl. Phys. A-Mater. Sci. Process. 81 (2) (2005) 223–240.

[12] G. Hummer, Water, proton, and ion transport: from nanotubes to proteins, Mol. Phys. 105 (2) (2007) 201–207.

[13] G. Hummer, J.C. Rasaiah, J.P. Noworyta, Water conduction through the hydrophobic channel of a carbon nanotube, Nature 414 (6860) (2001) 188–190.

[14] A.I. Kolesnikov, J.M. Zanotti, C.K. Loong, P. Thiyagarajan, A.P. Moravsky, R.O. Loutfy, et al., Anomalously soft dynamics of water in a nanotube: a revelation of nanoscale confinement, Phys. Rev. Lett. 93 (3) (2004) 35503.
[15] N. Naguib, H. Ye, Y. Gogotsi, A.G. Yazicioglu, C.M. Megaridis, M. Yoshimura, Observation of water confined in nanometer channels of closed carbon nanotubes, Nano Lett. 4 (11) (2004) 2237–2243.
[16] Y. Maniwa, K. Matsuda, H. Kyakuno, S. Ogasawara, T. Hibi, H. Kadowaki, et al., Water-filled single-wall carbon nanotubes as molecular nanovalves, Nat. Mater. 6 (2) (2007) 135–141.
[17] E. Mamontov, C.J. Burnham, S.H. Chen, A.P. Moravsky, C.K. Loong, N.R. de Souza, et al., J. Chem. Phys. 124 (2006) 194703.
[18] A. Waghe, J.C. Rasaiah, G. Hummer, Filling and emptying kinetics of carbon nanotubes in water, J. Chem. Phys. 117 (23) (2002) 10789.
[19] O. Beckstein, P.C. Biggin, M.S.P. Sansom, A hydrophobic gating mechanism for nanopores, J. Phys. Chem. B 105 (51) (2001) 12902–12905.
[20] O. Beckstein, M.S.P. Sansom, Liquid–vapor oscillations of water in hydrophobic nanopores, PNAS 100 (12) (2003) 7063–7068.
[21] R. Allen, S. Melchionna, J.P. Hansen, Intermittent permeation of cylindrical nanopores by water, Phys. Rev. Lett. 89 (17) (2002) 175502.
[22] R. Allen, J-P. Hansen, S. Melchionna, Molecular dynamics investigation of water permeation through nanopores, J. Chem. Phys. 119 (7) (2003) 3905.
[23] S. Andreev, D. Reichman, G. Hummer, Effect of flexibility on hydrophobic behavior of nanotube water channels, J. Chem. Phys. 123 (19) (2005) 194502.
[24] J. Wang, Y. Zhu, J. Zhou, X.-H. Lu, Diameter and helicity effects on static properties of water molecules confined in carbon nanotubes, Phys. Chem. Chem. Phys. 6 (4) (2004) 829–835.
[25] C.Y. Won, S. Joseph, N.R. Aluru, Effect of quantum partial charges on the structure and dynamics of water in single-walled carbon nanotubes, J. Chem. Phys. 125 (11) (2006) 114701–114709.
[26] L-L. Huang, Q. Shao, L-H. Lu, X-H. Lu, L-Z. Zhang, J. Wang, et al., Helicity and temperature effects on static properties of water molecules confined in modified carbon nanotubes, Phys. Chem. Chem. Phys. 8 (33) (2006) 3836–3844.
[27] R.J. Mashl, S. Joseph, N.R. Aluru, E. Jakobsson, Anomalously immobilized water: a new water phase induced by confinement in nanotubes, Nano Lett. 3 (5) (2003) 589–592.
[28] C.Y. Won, N.R. Aluru, Water permeation through a subnanometer boron nitride nanotube, J. Am. Chem. Soc. 129 (10) (2007) 2748–2749.
[29] A. Kalra, S. Garde, G. Hummer, Osmotic water transport through carbon nanotube membranes, Proc. Natl. Acad. Sci. 100 (18) (2003) 10175–10180.
[30] H. Sui, B-G. Han, J.K. Lee, P. Walian, B.K. Jap, Nature 414 (6866) (2001) 872–878.
[31] K. Murata, K. Mitsuoka, T. Hirai, T. Walz, P. Agre, J.B. Heymann, et al., Structural determinants of water permeation through aquaporin-1, Nature 407 (6804) (2000) 599–605.
[32] P. Agre, M.J. Borgnia, M. Yasui, J.D. Neely, J. Carbrey, D. Kozono, et al., Discovery of the aquaporins and their impact on basic and clinical physiology, Aquaporins (2001) 1–38.

[33] E.M. Kotsalis, J.H. Walther, P. Koumoutsakos, Multiphase water flow inside carbon nanotubes, Int. J. Multiphase Flow 39 (2004) 995–1010.
[34] I. Hanasaki, A. Nakatani, Flow structure of water in carbon nanotubes: Poiseuille type or plug-like? J. Chem. Phys. 124 (14) (2006) 144708.
[35] A. Striolo, The mechanism of water diffusion in narrow carbon nanotubes, Nanoletters 6 (4) (2006) 633–639.
[36] G. Che, B.B. Lakshmi, C.R. Martin, E.R. Fisher, R.S. Ruoff, Chemical vapor deposition based synthesis of carbon nanotubes and nanofibers using a template method, Chem. Mater. 10 (1) (1998) 260–267.
[37] B.J. Hinds, N. Chopra, T. Rantell, R. Andrews, V. Gavalas, L.G. Bachas, Aligned multiwalled carbon nanotube membranes, Science 303 (5654) (2004) 62–65.
[38] J. Holt, A. Noy, T. Huser, D. Eaglesham, O. Bakajin, Fabrication of a carbon nanotube-embedded silicon nitride membrane for studies of nanometer-scale mass transport, Nano Lett. 4 (11) (2004) 2245–2250.
[39] J.K. Holt, H.G. Park, Y. Wang, M. Stadermann, A.B. Artyukhin, C.P. Grigoropoulos, et al., Fast mass transport through sub-2-nanometer carbon nanotubes, Science 312 (5776) (2006) 1034–1037.
[40] S. Kim, J.R. Jinschek, H. Chen, D.S. Sholl, E. Marand, Scalable fabrication of carbon nanotube/polymer nanocomposite membranes for high flux gas transport, Nano Lett. 7 (9) (2007) 2806–2811.
[41] M. Majumder, N. Chopra, R. Andrews, B.J. Hinds, Nanoscale hydrodynamics: enhanced flow in carbon nanotubes, Nature 438 (7064) (2005) 44.
[42] C. Cottin-Bizonne, S. Jurine, J. Baudry, J. Crassous, F. Restagno, E. Charlaix, Nanorheology: an investigation of the boundary condition at hydrophobic and hydrophilic interfaces, Eur. Phys. J. E-Soft Matter. 9 (1) (2002) 47–53.
[43] V.S.J. Craig, C. Neto, D.R.M. Williams, Shear-dependent boundary slip in an aqueous Newtonian liquid, Phys. Rev. Lett. 87 (5) (2001) 54504.
[44] C. Peter, G. Hummer, Ion transport through membrane-spanning nanopores studied by molecular dynamics simulations and continuum electrostatics calculations, Biophys. J. 89 (4) (2005) 2222–2234.
[45] J.H. Park, S.B. Sinnott, N.R. Aluru, Ion separation using a Y-junction carbon nanotube, Nanotechnology 17 (3) (2006) 895–900.
[46] S. Joseph, R.J. Mashl, E. Jakobsson, N.R. Aluru, Electrolytic transport in modified carbon nanotubes, Nano Lett. 3 (10) (2003) 1399–1403.
[47] F. Fornasiero, H-G. Park, J.K. Holt, M. Stadermann, C. Grigoropoulos, A. Noy, et al., Ion exclusion by sub 2-nm carbon nanotube pores, PNAS (2008), Available from: http://dx.doi:10.1073/pnas.0710437105. (Published online before print June 6, 2008.)
[48] H. Hiura, T.W. Ebbesen, K. Tanigaki, Opening and purification of carbon nanotubes in high yields, Adv. Mater. 7 (3) (1995) 275–276.
[49] S.S. Wong, E. Joselevich, A.T. Woolley, C.L. Cheung, C.M. Lieber, Covalently functionalized nanotubes as nanometrel-sized probes in chemistry and biology, Nature 394 (6688) (1998) 52–55.
[50] M. Majumder, N. Chopra, B.J. Hinds, Effect of tip functionalization on transport through vertically oriented carbon nanotube membranes, J. Am. Chem. Soc. 127 (25) (2005) 9062–9070.
[51] N. Chopra, M. Majumder, B. Hinds, Bifunctional carbon nanotubes by sidewall protection, Adv. Funct. Mater. 15 (5) (2005) 858–864.

CHAPTER

Design of Advanced Membranes and Substrates for Water Purification and Desalination

12

James Economy, Jinwen Wang and Chaoyi Ba

Center of Advanced Materials for the Purification of Water with Systems, Department of Materials Science and Engineering, University of Illinois at Urbana-Champaign, Urbana, IL, USA

12.1 Overview

Since the introduction of Loeb–Sourirajan membranes in the early 1960s, membrane processes have received widespread applications due to their low energy demand. The Loeb–Sourirajan membranes are formed by the immersion of a casting solution into water. These membranes, sometimes known as phase separation membranes, are asymmetric porous membranes [1]. By using these porous membranes as mechanical supports, John Cadotte deposited a thin selective layer to make a thin film composite (TFC) membrane [2]. Most pressure-driven membranes are in these two categories, that is, phase separation membranes and TFC membranes. According to pore diameters, there are four categories of pressure-driven membranes—these include microfiltration (MF), ultrafiltration (UF),

Street, Sustich, Duncan and Savage. Nanotechnology Applications for Clean Water, 2nd Edition.
 DOI: http://dx.doi.org/10.1016/B978-1-4557-3116-9.00012-3

nanofiltration (NF), and reverse osmosis (RO) membranes. Nanofiltration, defined by many "as a process between Ultrafiltration (UF) and Reverse Osmosis (RO)," is a relatively recent technology, largely developed over the past decade. Typically, NF membranes have sodium chloride rejections between 20 and 80 percent and molecular weight cutoffs for dissolved organic solutes of 200–1000 Dalton. Their low-pressure operation (4–14 bars) provides increased energy savings with significantly lower installation and operating costs. A number of polymer materials have been investigated for NF applications [3,4]. Cellulose acetate (CA) membrane and polyamide TFC membrane have been the most successful:

1. Cellulose acetate membranes. Cellulose is a linear, rod-like material that renders membranes mechanically robust. Cellulose acetate was the first high-performance NF/RO membrane material discovered. Today, CA membranes still maintain a small fraction of the market because they are easy to make, mechanically tough, and resistant to degradation by chlorine up to 1 ppm. However, CA membranes slowly hydrolyze over time and are stable only in the pH range of 4–6. Salt rejection decreases as temperature increases and therefore CA membranes are generally not used above 35°C. In addition, they are easily attacked by bacteria.
2. Polyamide TFC membranes. A more successful, commercially available NF/RO membrane is the TFC aromatic polyamide membrane. This type of membrane has higher salt rejections and fluxes than CA membranes. On the other hand, aromatic polyamides have several disadvantages, including:
   a. Low resistance to fouling. Membrane fouling (scale, silt, biofouling, organic fouling, etc.) is the main cause of permeate flux decline and loss of water quality. Fouling can decrease total throughput more than 30 percent, and further increase capital costs by 15 percent resulting from membrane cleaning.
   b. Limited oxidant tolerance due to the existence of secondary amides and electron-rich aromatic rings [5]. Selectivity is rapidly and permanently lost once exposed to feed water containing more than a few ppb levels of chlorine or hypochlorite disinfectants. It creates major problems in designing for effective pretreatment of chlorine to destroy microorganisms.
   c. Chemical and thermal instabilities. The substrate layer of TFC membranes, usually polysulfone, is attacked by many solvents, which limits deposition of a permselective, ultrathin membrane from solvent systems [such as Dimethylformamide (DMF), and N-methyl-pyrrolidone (NMP)]. Most TFC membranes can only be used below 30–50°C because of limitations of the polysulfone. On the other hand, hot wastewater from the food, chemical, and petroleum processing industries is discharged directly and the energy lost is estimated at one to two percent of the total energy consumption in United States annually. Developing chemically stable substrates, which operate at higher temperatures, will result in substantial energy savings. Therefore, the development of more thermally and chemically resistant selective top layers and substrates is critically required for water purification.

In this chapter we describe the development of four new systems that address these problems: (i) Novel method to make a continuous micro-mesopore membrane with tailored surface chemistry for use in nanofiltration; (ii) Deposition of polyelectrolyte complex films under pressure and from organic solvents; (iii) Solvent resistant hydrolyzed polyacrylonitrile (PAN) membranes; (iv) Polyimides (PIs) membranes for nanofiltration.

12.2 Novel method to make a continuous micro-mesopore membrane with tailored surface chemistry for use in nanofiltration

We have demonstrated a new method to conveniently make NF membranes with tailored surface chemistry from UF membranes. Because of the large pore size in the UF membrane it is easier to modify the pore surface chemistry. After the introduction of functional groups, meso-macropores in UF membranes were shrunk into micro-mesopores by surface tension forces at the vapor–liquid interface within the capillary pores during heat treatment. On the other hand, with use of specific reagents such as Lewis acids, which can act as both attracting centers and fillings, micro-mesopores will be kept open and lead to denser concentrations of functional groups on the wall. Due to both interactions, NF membranes with controlled highly dense pore surface functional groups were prepared directly from UF membranes during heat treatment. Cationic and anionic exchange NF PAN membranes were easily prepared from UF membranes by the aforementioned method in the presence of $ZnCl_2$. By the reaction between nitrile groups and NaOH or 3-(dimethylamino)propyl amine, negative or positive functional groups were introduced into the meso-macropores of PAN UF membranes. These meso-macropores were then reduced into the range of micro-mesopores by taking advantage of surface tension forces within the capillary pores during heat treatment in the presence of $ZnCl_2$. A typical procedure of this method is illustrated in Fig. 12.1 [6,7].

This method could be extended to other polymer systems such as polystyrene or poly (styrene-co-acrylonitrile). In these systems, UF membranes could be precipitated from non-solvents besides water by the phase separation technique. These non-solvents can cause functional groups of the polymers to migrate onto the meso-macro continuous pore surface as in the case of water and nitrile groups of PAN. Also, the large pore size in the UF membrane allows easier modification of the pore surface chemistry. With the help of templates that interact with the functional groups on the meso-macro pore surface, as in the case of $ZnCl_2$ with nitrile groups, continuous micropores with highly dense pore surface functional groups could be formed during the heating and evaporation of the solvent of the template, for example, water and $ZnCl_2$, once the surface tension force exceeds the modulus of the polymers.

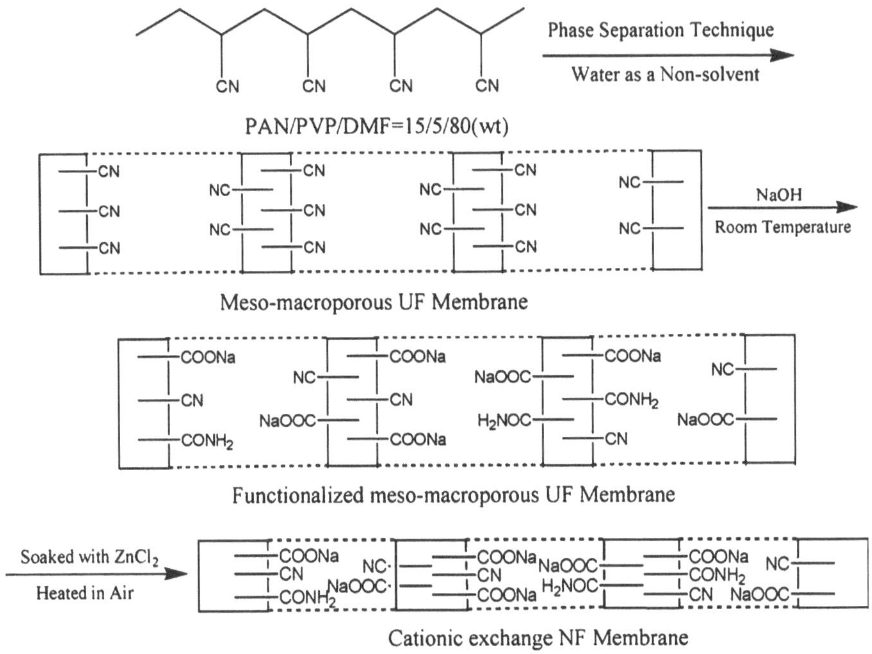

FIGURE 12.1

Preparation of cationic exchange polyacrylonitrile nanofiltration membranes. DMF: Dimethylformamide; NF: Nanofiltration; PAN: polyacrylonitrile; PVP: Polyvinylpyrrolidone; UF: Ultrafiltration.

12.3 Deposition of polyelectrolyte complex films under pressure and from organic solvents

The alternating physisorption of oppositely charged polyelectroytes on porous supports is a relatively new technique that provides a simple way to create ultrathin polyelectrolyte multilayers (PEM) [8,9]. There are many advantages of PEM films for NF membranes [10]. This "layer-by-layer" or "electrostatic self-assembly" (ESA) method affords control over thickness, charge density, and composition of the selective skin layer in NF membranes. Moreover, a wide range of polyelectrolytes is available to form PEM films. Therefore, flux, selectivity, and possibly fouling rates of NF membranes could be tailored by judicious selection of constituent polyelectrolytes. Despite the versatility of PEM films, studies for NF application thus far primarily focused on poly (styrene sulfonate)/poly(allylamine hydrochloride) (PSS/PAH) [11], PSS/Chitosan [12], PSS/poly(diallyldimethyl ammonium chloride) (PDADMAC) [12], hyaluronic acid (HA)/Chitosan [12], and poly (vinyl amine)/poly (vinyl sulfate) [13]. Bruening et al. [11,12] reported that the PSS/PDADMAC films showed limited salt rejection of up to 40 percent for a

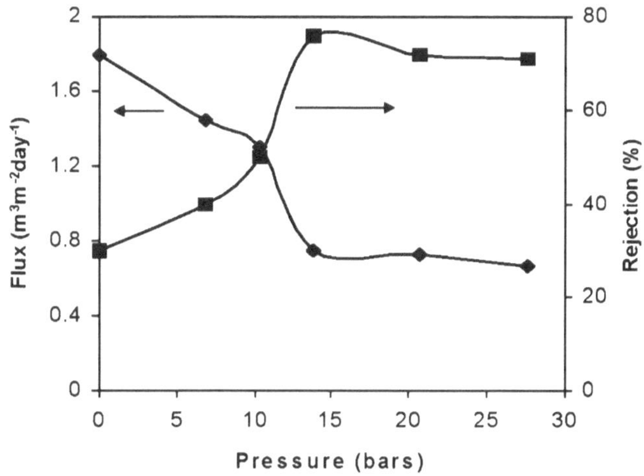

FIGURE 12.2

Performance of polyelectrolyte multilayers prepared at different pressures.

feed concentration of 0.01 M NaCl. The PEM membranes using Chitosan as polycation showed even lower solute rejections probably because Chitosan has lower charge density than PDADMAC and thus fewer ionic cross-links could be formed in the films. Although Tieke et al. [13] achieved 84 percent rejection for 0.01 M NaCl, the flux was relatively low (approximately 0.013 $m^3\ m^{-2}\ day^{-1}$ at 5 bars) since he deposited 60-bilayer poly (vinyl amine)/poly (vinyl sulfate) films on PAN/poly(ethylene terephthalate (PET) supporting membranes. In addition, all these PEM films were prepared in water. The formation of PEM films in organic solvents for NF application has received much less attention.

We have demonstrated the preparation of PEM films having a high salt rejection (up to 80 percent) with only several bilayers (below five bilayers) of sulfonated poly (ether ether ketone) (sPEEK) and branched polyethylenimine (PEI). It was found that the salt rejection was increased two or three times by pressing water through the membrane during the film deposition as shown in Fig. 12.2. One possible explanation of this effect might be the survival of the fittest. When the film was formed, there were some weaker regions in every single layer. These regions could not withstand the pressure applied. Whereas in the next deposition cycle, stronger layers might form on these damaged regions. Therefore, membranes suitable for the corresponding pressure survived. The final membrane is more compact than those prepared without pressure. We also deposited PEM consisting of sPEEK and PEI from organic solvents, for example, methanol. The salt rejection can be up to 80 percent for 2 g/L NaCl solution. This PEM is expected to be more stable in aqueous environments since the sPEEK we used is insoluble in water [14].

12.4 Solvent resistant hydrolyzed polyacrylonitrile membranes

Although polymeric membranes are primarily used with aqueous systems, numerous separation technologies for solutions polluted by organic solvents could be optimized by the application of solvent resistant polymeric membranes. The rapidly expanding interest in the separation of low molecular weight substances from nonaqueous solutions by membrane technologies also demands solvent-stable membranes. It should be noted that the substrate layer of TFC membranes, usually polysulfone, is attacked by many solvents. This limits deposition of a permselective, ultrathin membrane from solvent systems (such as DMF, NMP) and limits the development of new RO/NF membranes.

Several approaches have been reported to develop PAN membranes with improved chemical stability not only to common solvents but also to polar aprotic solvents such as DMF, N,N-dimethylacetamide (DMAc), and NMP. These approaches used systems such as PAN/metal ethoxides [15], PAN/sodium hydroxide and heating [16,17], poly(acrylonitrile–co-vinylacetate)/hydroxylamine/sodium carbonate/cyanuric chloride [18], and poly(acrylonitrile–co-glycidylmethacrylate) or poly(acrylonitrile-co-vinylbenzylchloride)/multifunctional amines [19–21] to form cross-linked networks. However, we have found that PAN membranes become resistant to all solvents just after treatment with NaOH. By this simple method we can develop NF/UF membranes or substrates with excellent solvent resistance for TFC RO/NF membranes. The degree of swelling of hydrolyzed PAN membranes in DMF is shown in Fig. 12.3, where (A) is the degree of swelling of membranes with the polyester support. In general, the swelling ratio in length decreases with the hydrolysis. As expected, the degree of swelling in length is very small, even zero, due to the limitation from the polyester support. In the direction of thickness, the membrane swells and the swell ratio is almost stable between 21 and 33 hours of hydrolysis. Interestedly, after about 40 hours of hydrolysis, there are negative swelling ratios, that is, membranes shrink. Figure 12.3(B) shows the degree of swelling of membranes without the polyester support. Again, the swelling ratio in length decreases with the hydrolysis. In the direction of thickness, the swell ratio is close to a constant for membranes hydrolyzed between 24 and 33 hours. After 40 hours of hydrolysis, membranes shrink in both directions. The phenomena discussed earlier indicate that the cross-linking degree increases as the hydrolysis progresses [22].

12.5 Polyimides membranes for nanofiltration

Currently, it is still a challenge to develop chemically and thermally stable polymeric membranes that can withstand harsh environments such as pH extremes, Cl_2, organic solvents, and very low or high temperatures, and so on, while maintaining the separation ability of the commercial TFC membranes [23].

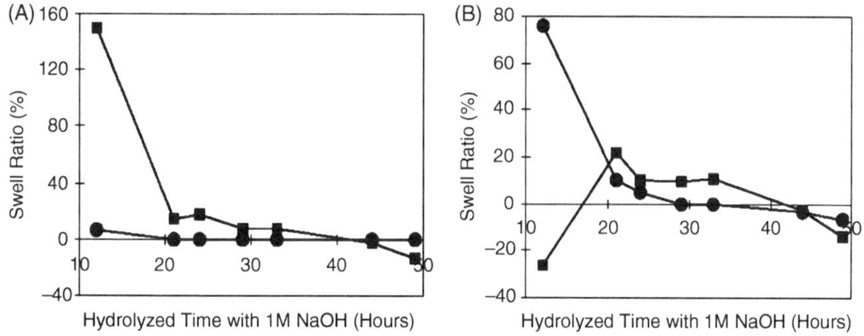

FIGURE 12.3

Effects of hydrolysis time on the length swell (●) and thickness swell (■) of polyacrylonitrile ultrafiltration membranes in DMF for 72 hours with (A) and without (B) nonwoven polyester support.

Some high-performance polymers such as PIs, PEI, polybenzimidazole (PBI), and so on have been considered for developing a new generation of membranes for water purification and desalination. For example, PBI has been fabricated into hollow-fiber nanofiltration membranes for separation of multivalent ions, chromate, and cephalexin [24–26].

Due to their unique properties, PIs have been developed into membrane for separation of gases, vapors, and liquids [27]. Polyimide membranes exhibit excellent resistance to almost all chemical agents. Their thermal resistance allows separations to be performed for a long time at elevated temperatures. Specifically, poly(pyromellitic dianhydride-co-4,4′-oxydianiline) (PMDA/ODA PI) has a very high glass transition temperature of approximately 400°C and excellent resistance to most nonoxidizing acids at room temperature and almost all organic solvents [28]. However, because the PMDA/ODA PI is both insoluble and infusible, any products, including membranes, can only be processed from its precursor, polyamic acid (PAA), which requires an additional step to transform the PAA into PI by either thermal or chemical treatments. So far, studies have focused on its fabrication into asymmetric membranes for separation of gases and organics [29–33]. A dense symmetric membrane has also been prepared from PAA solution via solvent evaporation followed by chemical imidization. The fully imidized membrane showed very high salt rejection but low water flux [34].

Neutralization of the acid groups prior to imidization would improve the hydrolytic stability and physical properties of the PAA. The neutralizers can be tertiary amines [35] or metal ions [36,37]. However, the PAA amine salt polymers are soluble in water and thus may not be suitable for developing asymmetric membranes by phase inversion method. On the other hand, the metal ions, such as Zn^{2+}, Co^{2+}, Ni^{2+}, Cu^{2+}, can strongly coordinate with the carboxylic groups, which would increase the PAA's mechanical properties and the solution viscosity. In our research, zinc chloride was added to the PAA/NMP solution to prepare a

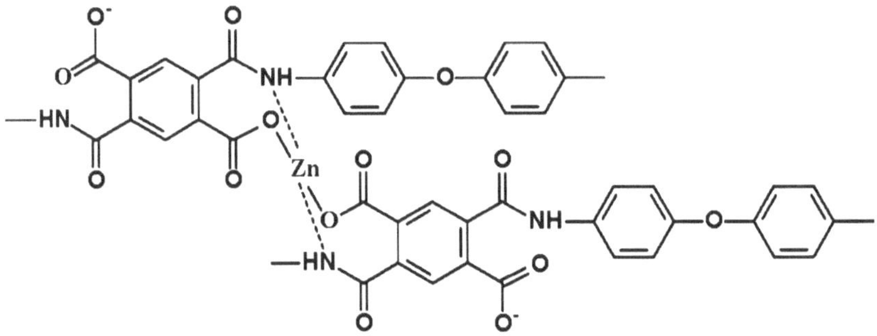

FIGURE 12.4

Interaction between Zn^{2+} and polyamic acid.

casting solution for developing a PI membrane. It has been shown that by adding the $ZnCl_2$, viscosity of the casting solution increased, probably because of the complex formation between PAA and Zn^{2+} as shown in Fig. 12.4.

The increase of viscosity of the casting solution had a strong influence on membrane morphology and permeability. Our preliminary data showed that with $ZnCl_2$ additive, macrovoids in the membranes were depressed and water permeability was improved. The resultant membranes developed under optimized conditions can be used for a substrate to prepare thermally stable composite membranes for NF/RO application [38].

A unique feature of PIs is their chemical reaction with amines, which results in the opening of some of the imide functions to form ortho-diamide functions. This reaction can be used to introduce functional groups onto the PI membrane surface by choosing suitable amines. In particular, by reacting with multifunctional amines (di- or higher functional), PIs can be cross-linked so that the PI membranes would show improved separation efficiency and solvent resistance. For example, W. Albrecht et al. studied the surface functionalization of poly(ether imide) membranes by di- and multivalent amines. It was shown that when using high molecular weight PEI as modifying agent, the poly(ether imide) membranes became insoluble even in the polar aprotic solvents, such as DMAc, and high contents of amine groups were detected [39]. These amine groups made the membrane more hydrophilic and positively charged as shown by contact angle and streaming potential studies [39,40].

Lenzing P84 is a soluble co-polyimide of 3, 3′, 4, 4′-benzophenone tetracarboxylic dianhydride with 80 percent toluenediisocynate and 20 percent methylphenylenediisocyanate (BTDA-TDI/MDI) with a Tg of 315°C. Chemical modification of the P84 membranes has been performed by treatment with diamines for use either for pervaporation dehydration of isopropanol [41] or for nanofiltration in polar aprotic solvents, such as DMF, NMP [42]. The diamine-modified P84 membranes were good for separation of organics because of the improved organic resistance. For electrolyte removal, however, membrane charge would play an important role because of the Donnan exclusion effect. For example, the commercial membranes are generally negatively charged [43] and multivalent anions, like sulfate and phosphate, are rejected more effectively [44].

In our work [45], by modifying the P84 membranes using PEI, positively charged nanofiltration membranes were obtained, which showed 50.9 ± 5.1 percent rejection to 2.0 g/L NaCl solution with a flux of $1.2 \pm 0.1\ m^3\ m^{-2}\ day^{-1}$ when tested at 13.8 bar and room temperature. After modification, the membranes were cross-linked by the PEI and showed good resistance to organic solvents. Development of the PEI-modified P84 membranes not only provides a solvent resistant and positively charged NF membrane for removal of heavy metal ions (>98 percent rejection of Zn^{2+}, Cu^{2+}, or Fe^{3+}) but also affords the design of a completely new type of anti-fouling membrane consisting of an erasable polyelectrolyte protective coating. This area appears to be of great commercial value.

12.6 Conclusions

The functionalities on the pore wall surface are very important to facilitate the flow of water and inhibit transport of salt for NF applications. We developed several ways to design and prepare a microporous polymeric membrane with a high density of functional groups.

One approach we considered involved the use of surface tension forces within the capillary pores during heat treatment in the presence of Lewis acids (e.g., $ZnCl_2$). Asymmetric PAN NF membranes with tailored surface chemistry were conveniently prepared from PAN UF membranes by this new technique. In addition, the solvent resistance of hydrolyzed PAN membranes was studied.

The other approach we developed involved the use of pressure during the alternating physisorption of oppositely charged polyelectrolytes on porous supports. For example, new PEMs as selective skins in composite membranes for NF were prepared by alternating layer-by-layer deposition of sulfonated sPEEK and PEI. During this electrostatic self-assembly deposition, it was demonstrated that the use of pressure could increase the salt rejection of the PEMs by one to two times.

Finally, PMDA/ODA PI membrane was developed from its precursor, PAA, using $ZnCl_2$ as additive. It was found that $ZnCl_2$ could improve membrane's hydrolytic stability, mechanical strength, and permeability. In addition, NF membranes were developed by chemical modification of P84 copolyimide using PEI. This membrane showed good desalination performance and solvent stability.

References

[1] S. Loeb, S. Sourirajan, in: R.F. Gould (Ed.), Saline Water Conversion II, Advances in Chemistry, Series Number 38, American Chemical Society, Washington, D.C., 1963, pp. 117–132.

[2] L.T. Rozelle, J.E. Cadotte, K.E. Cobian, C.V.J. Kopp, in: S. Sourirajan (Ed.), Reverse Osmosis and Synthetic Membranes, National Research Council Canada, 1977, pp. 249–262.

[3] R.W. Baker (Ed.), Membrane Technology and Applications, John Wiley & Sons, 2004.

[4] A.I. Schäfer, A.G. Fane, T.D. Waite (Eds.), Nanofiltration—Principles and Applications, Elsevier, 2005.

[5] J. Glater, S. Hong, M. Elimelech, The search for a chlorine-resistant reverse osmosis membrane, Desalination 95 (1994) 325–345.

[6] J. Wang, Z. Yue, J. Economy, J. Membr. Sci. 308 (2008) 191–197.

[7] J. Wang, Z. Yue, J.S. Ince, J. Economy, J. Membr. Sci. 286 (2006) 333–341.

[8] P. Bertrand, A. Jonas, A. Laschewsky, R. Legras, Macromol. Rapid Commun. 21 (2000) 319–348.

[9] S.K. Tripathy, J. Kumar, H.S. Nalwa (Eds.), Handbook of Polyelectrolytes and Their Applications, American Scientific Publishers, 2002.

[10] A.S. Michaels, H.J. Bixler, R.W. Hausslein, S.M. Fleming, Research and Development Progress Report for Office of Saline Water, U.S. Department of the Interior, 1965, p. 149.

[11] B.W. Stanton, J.J. Harris, M.D. Miller, M.L. Bruening, Langmuir 19 (2003) 7038–7042.

[12] M.D. Miller, M.L. Bruening, Langmuir 20 (2004) 11545–11551.

[13] W. Jin, A. Toutianoush, B. Tieke, Langmuir 19 (2003) 2550–2553.

[14] J. Wang, Z. Yue, J. Economy, Preprints of Extended Abstracts Presented at the ACS National Meeting, ACS, Division of Environmental Chemistry, vol. 47(1), 2007, pp. 1178–1182.

[15] C. Linder, M. Nemas, M. Perry, R. Ketraro, European Patent 0392982, 1990.

[16] H.-G. Hicke, I. Lehman, G. Malsch, M. Ulbricht, M. Becker, J. Membrane Sci. 198 (2002) 187–196.

[17] C. Linder, M. Nemas, M. Perry, R. Katraro, U.S. Patent 5039421, 1991.

[18] C. Linder, M. Perry, M. Nemas, R. Ketraro, U.S. Patent 5032282, 1991.

[19] C. Linder, G. Aviv, M. Perry, R. Ketraro, European Patent 0061610, 1982.

[20] H.G. Hicke, I. Lehann, M. Becker, M. Ulbricht, G. Malsch, D. Paul, U.S. Patent 6159370, 2000.

[21] A. Glaue, G. Malsch, R. Swoboda, T. Weigel, DE Application 10 (138) (2001) 318.

[22] J. Wang, Z. Yue, J. Economy, Solvent resistant polyacrylonitrile membranes, Preprints of Extended Abstracts Presented at the ACS National Meeting, American Chemical Society, Division of Environmental Chemistry, vol. 48(1), 2008, pp. 645–649.

[23] R.J. Petersen, J. Membr. Sci. 83 (1993) 81–150.

[24] K.Y. Wang, Y. Xiao, T.-S. Chung, Chem. Eng. Sci. 61 (2006) 5807–5817.

[25] K.Y. Wang, T.-S. Chung, J. Membr. Sci. 281 (2006) 307–315.

[26] K.Y. Wang, T.-S. Chung, AIChE J. 52 (2006) 1363–1377.

[27] H. Ohya, V.V. Kudryavtsev, S.I. Semenova, Polyimide Membranes: Applications, Fabrications, and Properties, Gordon and Breach, Tokyo, 1996.

[28] T. Takekoshi, in: M.K. Ghosh, K.L. Mittal (Eds.), Polyimides—Fundamentals and Applications, Marcel Dekker, New York, 1996.

[29] U. Razdan, S.V. Joshi, V.J. Shah, Curr. Sci. 85 (2003) 761.

[30] S.I. Semenova, H. Ohya, T. Higashijima, Y. Negshi, J. Membr. Sci. 67 (1992) 29–37.
[31] H. Yanagishita, D. Kitamoto, K. Haraya, T. Nakane, T. Okadab, H. Matsuda, et al., J. Membr. Sci. 188 (2001) 165–172.
[32] H. Yanagishita, J. Arai, T. Sandoh, H. Negishi, D. Kitamoto, T. Ikegami, et al., J. Membr. Sci. 232 (2004) 93–98.
[33] S. Niyogi, B. Adhikari, Eur. Polym. J. 38 (2002) 1237–1243.
[34] A. Walch, H. Lukas, A. Klimmek, W. Pusch, J. Polym. Sci. Polym. Letters (Ed.), 12 (1974) 697–710.
[35] Y. Ding, B. Bikson, J.K. Nelson, Macromolecules 35 (2002) 905–911.
[36] B.A. Zhubanov, I.A. Arkhipova, I.D. Shalabaeva, Acta Polymerica 39 (1988) 443–445.
[37] X. Yu, B.P. Grady, R.S. Reiner, S.L. Cooper, J. Appl. Polym. Sci. 47 (1993) 1673–1683.
[38] C. Ba, J. Economy, Preparation of polyimide for use as substrate in a thermally stable composite membrane, Preprints of Extended Abstracts Presented at the ACS National Meeting, American Chemical Society, Division of Environmental Chemistry, vol. 47(1), 2007, pp. 1167–1170.
[39] W. Albrecht, B. Seifert, T. Weigel, M. Schossig, A. Hollander, T. Groth, et al., Macromol. Chem. Phys. 204 (2003) 510–521.
[40] C. Trimpert, G. Boese, W. Albrecht, K. Richau, T. Weigel, A. Lendlein, et al., Macromol. Biosci. 6 (2006) 274–284.
[41] X. Qiao, T.-S. Chung, AIChE J. 52 (2006) 3462–3472.
[42] Y.H. See Toh, F.W. Lim, A.G. Livingston, J. Membr. Sci. 301 (2007) 3–10.
[43] A. Bhattacharya, P. Ghosh, Rev. Chem. Eng. 20 (2004) 111–173.
[44] J.M.M. Peeters, J.P. Boom, M.H.V. Mulder, H. Strathmann, J. Membr. Sci. 145 (1998) 199–209.
[45] C. Ba, J. Economy, Preparation of nanofiltration membranes by chemical modification of P84 polyimide membranes using polyethylenimine, Preprints of extended abstracts presented at the ACS national meeting, American Chemical Society, Division of Environmental Chemistry, vol. 48(1), 2008, pp. 619–622.

CHAPTER

Customization and Multistage Nanofiltration Applications for Potable Water, Treatment, and Reuse

13

Curtis D. Roth[1], Saik Choon Poh[2] and Diem X. Vuong[1]

[1]*DXV Water Technologies, LLC, Tustin, CA, USA*

[2]*CH2MHILL, Los Angeles, CA, USA*

13.1 Potable water

13.1.1 Nanofiltration membranes as a water treatment solution

Nanofiltration membranes are defined as having a pore size in the order of nanometers (nm) (1×10^{-9} m). As a comparison, the atomic radius of a sodium ion and a chlorine ion is about 0.97 nm (0.97×10^{-9} m) and 1.8 nm (1.8×10^{-9} m), respectively. This demonstrates that nanofiltration membranes are near the range to remove rather small ions.

However, the term nanofiltration is really a misnomer. As the nanofiltration membranes are charged, the removal mechanism is not purely filtration as with ultrafiltration membranes, but also osmotic. This makes them a true hybrid, bridging ultrafiltration and reverse osmosis membranes in the range of membrane treatment options.

In general, the primary factors that affect the performance of the membranes include the membrane material (charge of the membrane), concentration

Street, Sustich, Duncan and Savage. Nanotechnology Applications for Clean Water, 2nd Edition.
 DOI: http://dx.doi.org/10.1016/B978-1-4557-3116-9.00013-5

polarization at the membrane face (buildup of concentration at the membrane face), and fouling of the membrane to name a few. As such, pore size alone does not predict the removal of constituents. Adding more complication to the problem, every manufacturer's membranes are slightly different, meaning there is no simple method for predicting removals. Pilot testing of the nanofiltration membranes is imperative when designing a system to target certain constituents.

Membrane manufacturers do discuss ranges for constituent removal based on atomic mass. Most indicate that nanofiltration membranes remove compounds/ions with a molecular weight greater than 300–400 g/mol. It should be noted that this number represents the size for complete (or nearly so) removal. For partial removals, the range extends down to less than 100 g/mol. This is evidenced by a figure published by Koch Membranes on their website that shows the range of nanofiltration membranes from 100 to 20,000 g/mol. In addition, testing completed by the Long Beach Water Department (Long Beach, California) confirms the smaller molecular weight as they have shown significant removal (up to 90 percent) of aqueous salts, which range in molecular weight from about 60 to 500 g/mol. With this range of removals, nanofiltration membranes provide engineers and scientists with a unique opportunity to produce customized water products.

Currently, there is a wide range of nanofiltration membranes available in the market, providing numerous options for targeting mass removals of specific or groups of constituents. Although current reverse osmosis technology can remove a larger mass of the same constituents as compared to nanofiltration membranes, this may not be an ideal solution. Reverse osmosis membranes remove not only the constituents of concern, but also dissolved minerals and hardness to such an extent that the product water is aggressive toward metal and concrete conduits. Treatment systems must then add some of these minerals and hardness back. The result is "wasted" effort to overshoot the water quality objective and then bring it back to a level that can be used.

With new breakthroughs in membrane technology, constituent specific membranes are a possibility. Membrane companies have developed nanofiltration membranes capable of 95 percent salt removal, increasing from the 20 percent that the early nanofiltration membranes had. In addition, research in the use of embedded "nanoparticles" in reverse osmosis membranes has shown promise in decreasing the transmembrane pressure, thus increasing the flux. This type of technology may provide similar improvements to nanofiltration flux and removal efficiencies.

13.1.2 Nanofiltration of freshwater sources

Current treatment of freshwater sources has focused on removal of particles, microbes, and viruses. As much of our potable water sources double as our drainage systems (stormwater, agricultural, and wastewater effluents), more concern is being raised about pollutants such as pesticides, heavy metals, endocrine disrupters, and pharmaceuticals. Conventional treatment methods do not remove these types of constituents at appreciable levels. Membrane technologies can remove all of these to very low levels. As discussed earlier, reverse osmosis membranes can remove these, but will generally overshoot the goals and require significant chemical stabilization downstream. Nanofiltration membranes can achieve the same results with a potentially significant increase in process efficiency. Table 13.1 shows some of the constituents of concern for water systems and their molecular weights.

Membrane manufacturers produce a variety of nanofiltration membranes that target different molecular weights. As an example, Dow Filmtec offers nanofiltration membranes to target approximately 90, 200, and 270 g/mol. This provides design engineers with many options for single pass applications. However, when one includes multiple pass applications, the options are endless.

Consider the following scenario as an example: A constituent of concern (Compound A) needs to be removed to meet drinking water standards. Based on testing of the source water, 85 percent (by mass) of Compound A needs to be removed to meet the requirements. Pilot testing is conducted resulting in the following single pass removals: reverse osmosis membrane, 99.9 percent; nanofiltration membrane no. 1, 75 percent; nanofiltration membrane no. 2, 61 percent; nanofiltration membrane no. 3, 40 percent. Based on this type of result the options include single pass reverse osmosis or the two pass combination of nanofiltration membranes as shown in Fig. 13.1.

Table 13.1 Drinking Water Constituents of Concern

Compounds	Approximate Molecular Weight (g/mol)
NaCl	58
$CaCO_3$ (representative for hardness)	100
Disinfection by-product Contaminants (trihalomethanes, haloacetic acids, bromates, chlorite)	50–150
Volatile organic compounds (VOC)	60–250+
Endocrine disrupters	90–400+
Arsenic*	75
Pharmaceuticals	150–1,000+
Copper*	64
Heavy metals (lead, mercury, etc.)	150–250+
Pesticides	175–500+
Organic molecules	30–1,000+
Microorganisms	*Approximate size (nm)*
Cryptosporidium	3,000+
Coliforms	50–200
Viruses	10–100

**Actual molecular weight will vary depending on actual compound formed.*

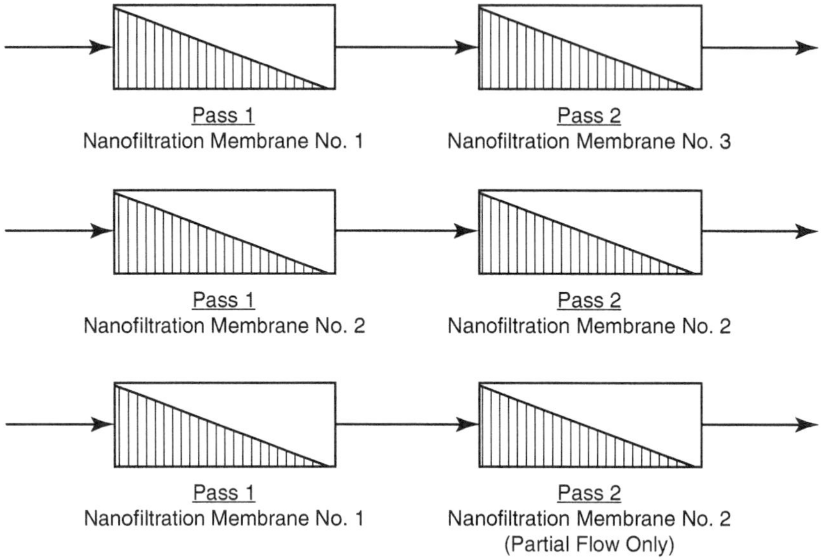

FIGURE 13.1

Two pass combinations of nanofiltration membranes.

As can be seen, there are a large number of combinations that can be made to provide the required results. Pilot testing of these combinations would allow for the determination of the most consistent, energy efficient, and easily maintainable configuration, thus providing the best system for the application.

13.1.3 Nanofiltration for seawater desalination

Of all membrane applications for purifying water, seawater desalination is the most complex. In recent years, many communities have constructed seawater desalination facilities using conventional seawater reverse osmosis membrane processes. The pressures required for desalination using these conventional reverse osmosis membranes are extreme—55 to 83 bars. In addition, the raw seawater and brine are highly corrosive requiring extensive use of austenitic and duplex stainless steels. The result is an energy consumptive, maintenance intensive, and costly process.

One community, however, decided to take a different proactive approach toward supplying their future water needs. The Long Beach Water Department in Long Beach, California, began a testing program in order to identify potential methods and processes to simplify and reduce the cost of seawater desalination. As a result of this testing, they developed a novel dual-stage (or pass) process using nanofiltration membranes.

With the support of their board of directors, the Long Beach Water Department staff built and tested a pilot scale plant with a capacity of 34 m^3/day. The results of this testing proved the concept and showed significant potential.

As significant funds for research are generally not available in the municipal setting, Long Beach Water Department entered into a partnership with the United States Bureau of Reclamation and the Los Angeles Department of Water and Power to continue the research through the construction and operation of a 1,136 m^3/day prototype (side by side with a reverse osmosis system) testing facility. Having been completed in 2007, the preliminary results at the facility have shown some potential advantages to using a dual-stage nanofiltration process over a traditional reverse osmosis process. These include the following:

1. Energy savings from lower pressure requirements. The pressure required for a traditional reverse osmosis desalination process typically ranges from 55 to 83 bars, whereas the dual-stage nanofiltration process ranges from 38 bars in the first stage to 17 bars in the second [1]. The result is an energy saving of approximately 10 to 20 percent.
2. The dual-stage nanofiltration system has two barriers, compared with only one for the traditional reverse osmosis process. This provides system redundancy to increase the consistency of water quality. The second stage is used as a polishing step to fine-tune water quality objectives. This system flexibility has shown the potential for being able to adjust to the wide variations in feed water salinity.
3. Traditional, single-stage reverse osmosis desalination systems typically do not address the World Health Organization (WHO) limit of 0.5 mg/L for boron, a naturally occurring mineral found in seawater at concentrations between 4 and 6 mg/L [1]. A dual-stage nanofiltration allows for pH adjustment between the two passes, which forms borate ions ($H_2BO_3^-$) that are physically large enough for the nanofiltration membranes to reject via a physical barrier method. The same method can be applied to the single-stage reverse osmosis, but would require a significantly more chemical addition as it occurs upstream of the process.

More information on the results and progress of the Long Beach Water Department testing can be obtained from their website (www.lbwater.org), from the American Water Works Association Research Foundation, or the United States Bureau of Reclamation.

13.2 Water treatment and reuse

13.2.1 Nanofiltration for wastewater treatment and reuse

Wastewater treatment for discharge or reuse has a different set of issues than does drinking water treatment. Historically, wastewater treatment has focused on removal of solids, organics, and microorganisms. In the near future, regulations may no longer be limited to just these types of contaminants. As an example, the United States Environmental Protection Agency is now implementing Total Maximum Daily Loads (TMDL) for water bodies within the United States. The TMDL studies have resulted in more limits for compounds such as ammonia, nitrates, and nitrites. In the future it is possible that further study may lead to limits for contaminants such as heavy metals, pharmaceuticals, and endocrine disrupters, just to name a few. Table 13.2 lists the sizes of some of the contaminants of concern for wastewater. Although there is some improvement that can be achieved through source control prior to waste discharge, it is likely that much of these compounds will need to be removed at the wastewater treatment plant. These types of limits may not be restricted to water that will be reused, but also for water that is discharged.

As with drinking water, reverse osmosis is an option for meeting these increasingly stringent requirements. However, there are several problems with this approach. First, it is likely that reverse osmosis will overshoot the requirements for significant stabilization, not just for protection of pipes and channels, but also for protection of habitat. Second, it is very costly, both in terms of capital and operation. Producing such extremely high quality water for discharge can be considered by some to be inefficient. However, use of this type of water for replenishment of potable sources, although technically feasible, has proven to be difficult from a public perception standpoint.

Table 13.2 Wastewater Contaminants of Concern

Compounds	Approximate Molecular Weight (g/mol)
$CaCO_3$ (representative for hardness)	100
Disinfection by-product Contaminants (trihalomethanes, haloacetic acids, bromates, chlorite)	50–150
Volatile organic compounds (VOC)	60–250+
Ammonia, nitrates, nitrites	36–62
Heavy metals (lead, mercury, etc.)	150–250+
Endocrine disrupters	90–400+
Pharmaceuticals	150–1,000+
Organic molecules	30–1,000+
Microorganisms	*Approximate size (nm)*
Cryptosporidium	3,000+
Coliforms	50–200
Viruses	10–100
Proteins / polysaccharides	2–10
Enzymes	2–5

Ultrafiltration membranes provide another alternative and are being used at many wastewater treatment plants to meet the tertiary filtration requirements. Ultrafiltration membranes do not, however, help much with any but the very large contaminants as they do not have any charge and thus osmotic removal capabilities.

As with water treatment, the broad range of nanofiltration membranes available, all with different characteristics, can provide a much better set of solutions to meet the more stringent discharge requirements. In single or multiple passes, nanofiltration membranes can be configured to target certain contaminants more efficiently than reverse osmosis. In addition, this type of configuration can prove to be the ultimate in flexibility to provide designer waters for reuse customers. Consider the following example: A wastewater treatment plant has installed a "loose" nanofiltration membrane (target for constituents at 270 g/mol) to meet effluent and reuse requirements. This level of water quality is sufficient for all of their reuse customers, except for two that need enhanced hardness removal to prevent scaling. As a solution, the wastewater treatment plant could install a second pass nanofiltration membrane unit for the reuse portion of the effluent stream or could install small versions of these units at the point of delivery.

In general terms, this flexibility will result in cost savings and improved efficiency. In the future, this increased efficiency will allow more municipalities to provide cleaner effluents and reuse larger quantities for a palatable cost.

Reference

[1] American Water Works Association, A Novel Approach to Seawater Desalination Using Dual Stage Nanofiltration, AWWA Research Foundation, Denver, 2006.

CHAPTER

Commercialization of Nanotechnology for Removal of Heavy Metals in Drinking Water

14

Lisa Farmen

Crystal Clear Technologies Inc., Portland, OR, USA

14.1 Issues that need to be addressed

Hurricane Katrina more than proved that the municipal water treatment infrastructure in the United States can be taken out of service, in this case by a natural disaster, that it will directly impact the ability to provide clean drinking water to the American people, and that the disruption in the tap water supply may extend far more than just a few days. It also demonstrated the need for developing robust technology that can remove biological and chemical contaminants from water without generating a wastestream and purify a contaminated raw water source into Environmental Protection Agency (EPA) certified drinking water.

The United States is predominantly reliant on municipal water to provide clean water for the American people, and water treatment plants are reliant on utility power. The impact of losing both municipal water supply and power is a critical national need (CNN). In the developing world there is a global water crisis—it is estimated that by 2015, two-thirds of the world's population will be without a source of clean water and one-third will not have a reliable source of water at all.

Street, Sustich, Duncan and Savage. Nanotechnology Applications for Clean Water, 2nd Edition.
 DOI: http://dx.doi.org/10.1016/B978-1-4557-3116-9.00014-7

Both in the United States and in other countries, there is a significant need to be able to passively purify a raw water source into clean drinking water and to have a direct impact on the global water crisis; this technology must be at a price affordable to the majority of the world's population. These are the driving forces to innovative and affordable new nanotechnology for water treatment.

In January 2006, after a 5-year advance notice, the US EPA lowered the maximum recommended level for arsenic in drinking water from 50 ppb to 10 ppb; California set a "Public Health Goal" for arsenic at 4 ppt; New Jersey lowered its limit to 5 ppb. Of note is that EPA's goal for both arsenic and lead in drinking water is "zero," meaning non-detection or less than 0.2 ppb. On August 17, 2006, the US Geological Survey published results of water tests from wells in the United States and Puerto Rico: 11 percent exceeded the EPA limit for arsenic; uranium and mercury also exceeded the standard for a smaller percentage (see Fig. 14.1).

Crystal Clear Technologies, Inc. (CCT) received a Phase I Small Business Innovation Research grant (SBIR) from the National Science Foundation (NSF) in 2005 to develop nanotechnology for the effective removal of both biological and chemical contaminants in water. This was a Security Technologies solicitation to answer a question posed by the NSF: "Where would the American Public go to obtain clean drinking water if the municipal water infrastructure in the US was taken out of service?" Since the majority of bottled water in the United States is sourced from municipal tap water, this showed a clear CNN for development of new technology for water purification that can purify contaminated raw water (i.e., rivers) into certified clean drinking water.

A Phase II SBIR grant from the NSF to commercialize the technology was won in 2006. This grant required that the proof of technology and intellectual property be

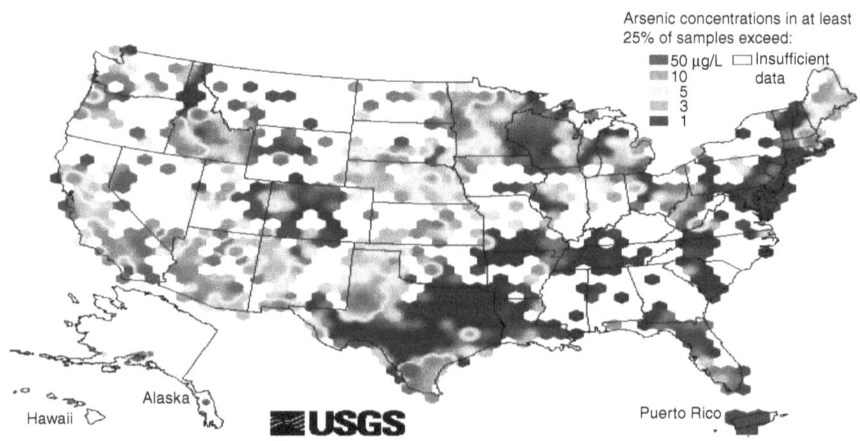

FIGURE 14.1

Map of arsenic contaminated well and raw water in the United States [4].

in place, along with a commercialization plan and manufacturing and distribution partnerships. A third Phase II(R) SBIR grant was won in 2007 to build the computational chemistry model with a NSF Engineering Research Center at Kansas University.

14.2 General approaches

Since the EPA lowered the Maximum Contaminant Levels (MCL) limit, arsenic contamination in raw water has generated a tremendous amount of attention with regard to enhancing the ability to remove it from water in municipal systems, but also to removing it from large volumes of water affordably, while minimizing the potential waste material generated from current adsorptive technology.

The University of Oregon (UO) has been working on arsenic chelation chemistries for binding ionized arsenic in water. Their extensive work on chelation and ligand chemistries allowed the creation of the partnership with CCT to extend that basic research with CCT's NSF Phase I grant into a commercially viable technology that met the need for removing arsenic and other metal contaminants.

The UO, in conjunction with CCT funded through the NSF SBIR grant, has developed nanocoating technology using a supramolecular design strategy for forming arsenic-based assemblies, which relies on self-assembly of As(III) with thiol ligands, and hence, the reversibility and lability of arsenic–sulfur bonds [As_2L_3]. The design strategy incorporates the unusual, but predictable trigonal-pyramid coordination geometry of As(III) featuring a stereochemically active lone pair when coordinated by sulfur based ligands [1].

The selection of As(III) as a design component is because of its unusual coordination geometry and a general lack of specific chelators for a highly toxic ion that, when hydrated, is recognized as a known carcinogen. Arsenic is a naturally occurring contaminant abundant in the earth's crust and is commonly present in raw water in two ionized forms, As(III) and As(V) [1].

Surface functionalization of high surface area substrates with monofunctional ligands bonds to the surface area of the substrate at a current density loading of 3–4 ligands/nm^2. CCT's technology roadmap identifies high surface area substrates, such as titania, alumina, and boehmite, which have surface areas ranging from 180 m^2/g up to 300 m^2/g and a synthetic substrate over 600 m^2/g, to the current ligand density loading of 3–4 ligands/nm^2 up to 6–8 ligands/nm^2, at the same time we are reducing the cost per kilogram.

Surface functionalization of various native substrates with a monofunctional ligand is determined by thermo-gravimetric analysis (TGA). A sample TGA given in Fig. 14.2 shows ligand loading of the monofunctional ligand onto an alumina substrate under STP conditions over a 30-minute and 2-hour time period.

The UO has been exploring the coordination chemistry of the metalloid arsenic for use in the formation of discrete supramolecular assemblies. Their recent

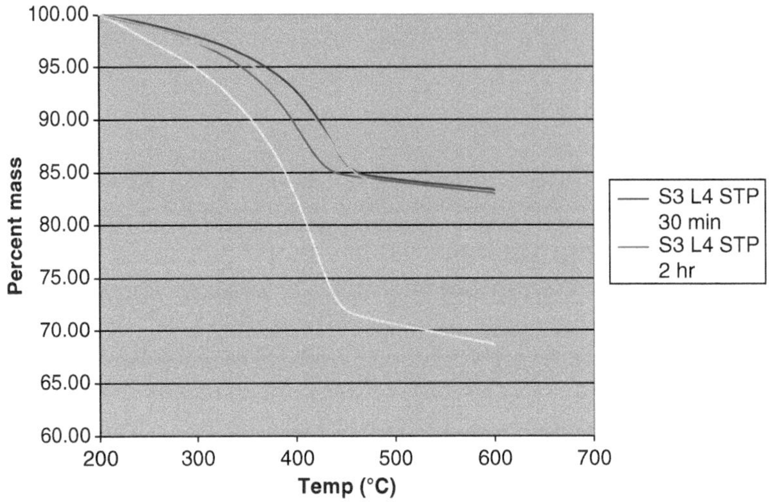

FIGURE 14.2

Monofunctional ligand loading on an alumina substrate under standard temperature and pressure (STP) conditions. The dark curve denotes the mass loss under a TGA analysis of an alumina substrate once the monofunctional ligand pyrolizes from the substrate after 30 minutes. The light colored curve is the same TGA analysis after 2 hours.

work has demonstrated that As(III) in combination with appropriately designed bridging thiolate ligands can self-assemble into a discrete dinuclear As_2L_3.Thus the use of As(III) provides access to a structure type mostly unavailable with the transition metals. Herein we expand the use of the reversibility and lability of As-S bonds in the preparation of supramolecular structures. The preparation of isomeric dinuclear arsine macrocycles from a dithiol and an As(III) source is described in Fig. 14.3 [1].

The NMX™ technology developed in conjunction with the UO uses inexpensive nanocoatings, ligands, which bind to the surface of low-cost substrates, such as minerals (alumina, boehmite, titania, zeolite), transforming these materials into high-capacity, target-specific adsorptive media that sequester high concentrations of metals. Adsorbed metals, such as arsenic, uranium, mercury, cadmium, selenium, and lead bind to the NMX™ substrate covalently and do not leach. They are, however, available for subsequent recycling in some instances. There is no wastestream generated or water wasted with the CCT technology. This aspect of the technology makes CCT's approach to low-cost water purification environmentally attractive.

CCT's filter media can be optimized for different contaminants and applications. The "self-assembling" nanocoating media can be "recharged" in situ, significantly enhancing its capacity, lifetime, and cost effectiveness over conventional media. Initial testing indicates that CCT's technology is so effective in removing

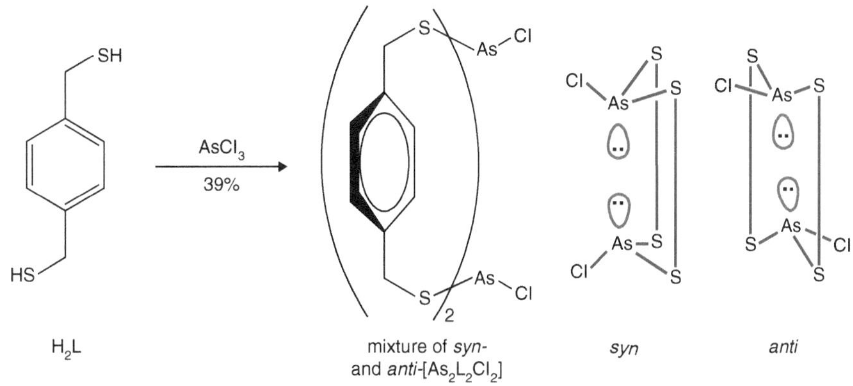

FIGURE 14.3

Synthesis of [As_2L_3] assemblies.

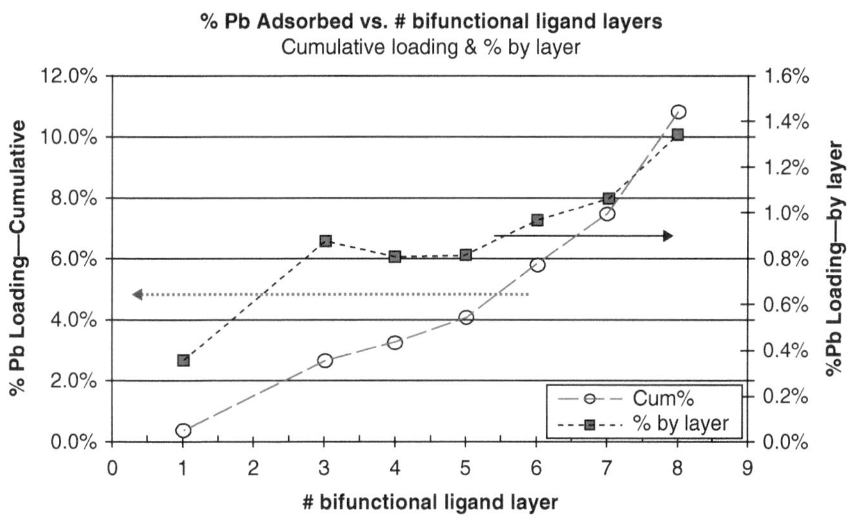

FIGURE 14.4

Cumulative lead layering on a titanium oxide (TiO_2) substrate.

inorganic contaminants, such as heavy metals, that spent filter media, literally, can be "mined" to recover metals, eliminating a source of hazardous waste and lowering the overall treatment cost. Figure 14.4 shows the cumulative lead loading, by layer, of NMX™ filter media recharged with a bifunctional ligand and additional metal layers bonded to the same filter media.

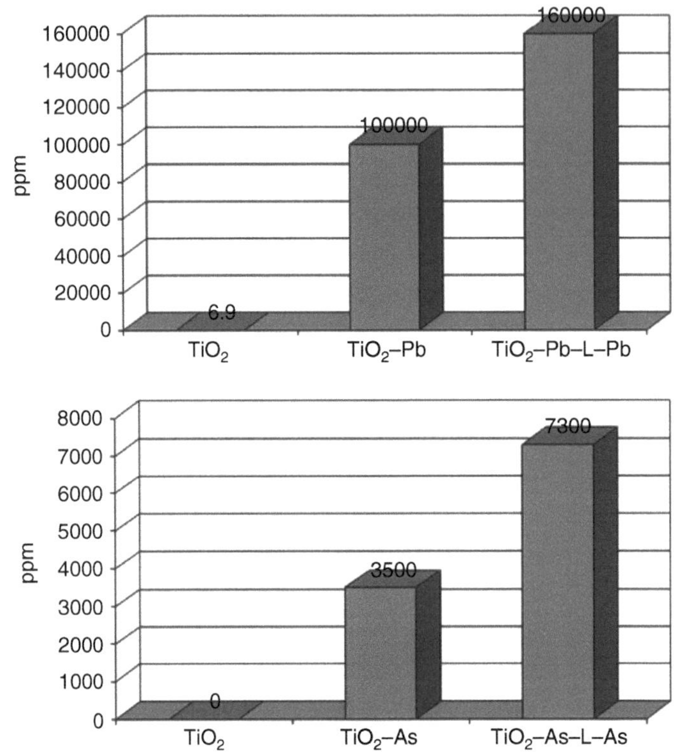

FIGURE 14.5

Layered lead and arsenic on titanium oxide (TiO_2) substrates [3].

14.3 Specific technology used by CCT and results

Our first-generation technology uses a bifunctional ligand, which has two receptor sites for target species, capable of attaching a target contaminate at either end; the molecular structure prevents both sites from adhering to the same substrate. In effect, we have a self-assembling "double stick" molecule with active receptor sites on both ends with the contaminant covalently bonded in between both ligand layers.

Figure 14.5 shows two layers of metal bonded to a titania substrate, separated by a bifunctional ligand layer. Lead layered 10 percent on the first layer, with 6 percent lead loading on the second, for a cumulative loading of 16 percent. Arsenic (III,IV) loaded 0.35 percent on the first layer, with a cumulative loading for two layers of 0.73 percent arsenic. Selenium, not graphed, layered a second layer of 3.2 percent over an initial layer of lead.

Recent work has demonstrated that As(III), in combination with appropriately designed bridging thiolate ligands (L) can self-assemble into a discrete dinuclear As_2L_3 structure, supported by a secondary supramolecular interaction between the

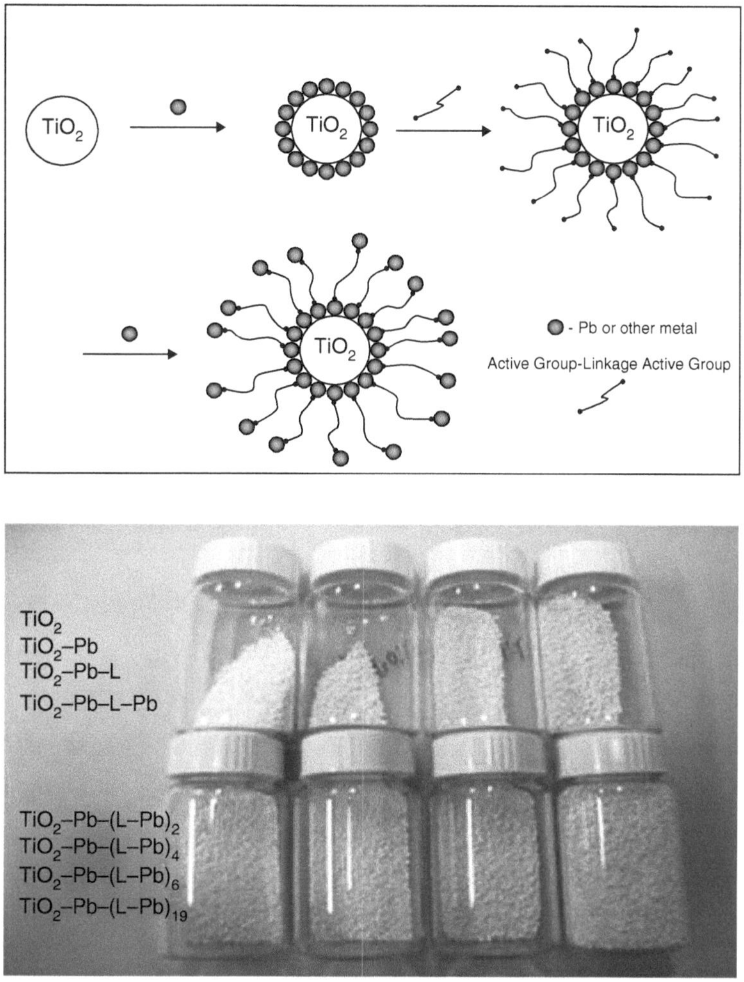

FIGURE 14.6

Layering of lead (19 layers) on a titanium oxide (TiO_2) substrate.

lone pair of As(III) and the aromatic ring of (L). These secondary bonding interactions (SBIs) can take many forms, as shown in the assembly of a naphthalene-imide with As(III), where weak arsenic-oxygen interactions support a highly symmetric complex in solution. The SBIs similarly work to confer selectivity for a target analyte [2].

The graphic in Fig. 14.6 shows the layering of the titanium oxide (TiO_2) substrate with lead on both layers, separated by a bifunctional ligand layer. We have successfully bonded 19 layers of lead on a titania substrate as shown with sample TiO_2–Pb–(L–Pb)$_{19}$. Figure 14.6 shows the layering graphically and the samples with 19 layers of lead.

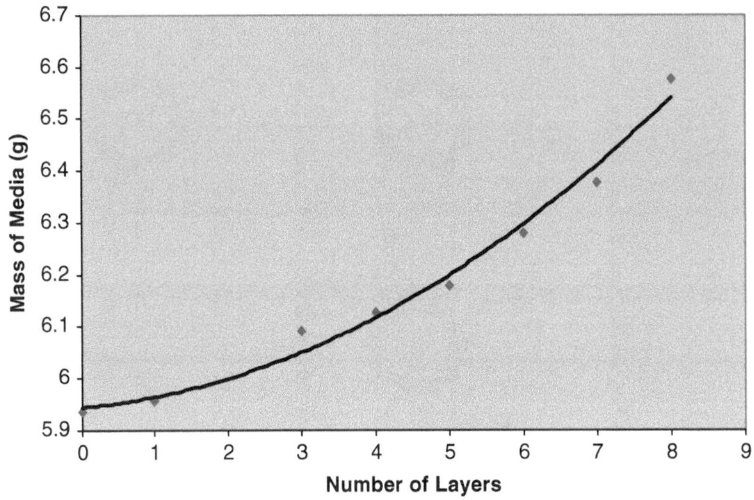

FIGURE 14.7

Plot of the mass of a multiple layer material as a function of the number of layers constructed. A generally exponential increase in mass is observed. An error in sampling prevented acquisition of mass data for the second layer.

The metal-ligand bonding complexes are strong enough that the adsorbed NMX™ media passes EPA's Toxicity Characteristic Leachate Procedure (TCLP) test, which allows for safe disposal of spent filter media if the metal is not reclaimed from the media.

The general strategy for construction of multiple layers is flexible in that the base layer can have a variable composition. For example, metal oxide adsorbents can be used directly to bind metal contaminants that can then be subsequently regenerated. Alternatively, these oxide substrates can be first modified with organic ligands that are suitable for binding metals. Similarly, these materials can also be regenerated once a layer of metal has been adsorbed on the surface.

In the initial studies, M1, Metsorb® TiO_2 granules were used as the substrate to directly bind metal from solution. In the experimental design, the granules are placed in a teabag and soaked in alternating solutions: to adsorb lead, the solution is 1 M $Pb(NO_3)_2$; to regenerate with nanocoating ligand, an α, α'-dimercapto-*p*-xylene solution is used. The teabag was steeped in the appropriate solution for 60 minutes and then dried under vacuum for 12 hours. The mass of the media was obtained after each round of soaking.

As can be seen in Fig. 14.7, there is a generally exponential increase in mass upon loading multiple layers of ligand and lead. It is posited that this behavior is due to the approximately spherical morphology of the substrate granules. Surface area increases as the square of the radius; thus although each added layer leads to an increase in the radius, it also leads to a proportionally larger increase in surface area.

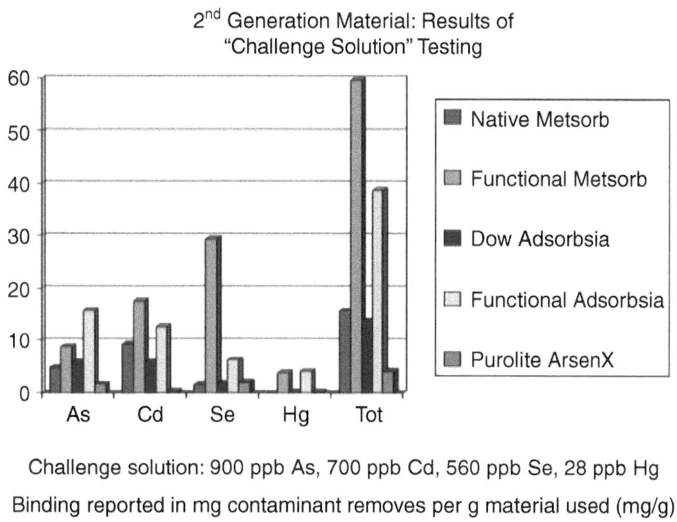

FIGURE 14.8

Surface functionalization of commercially available substrates with CCT's monofunctional ligand chemistry.

This increase in surface area provides more physical space on which the material can adsorb contaminant. Initial Time-of-Flight (TOF) – Secondary Ion Mass Spectroscopy (SIMS) results (not shown) indicate that there is indeed a layered structure formed upon the successive addition of ligand and lead to the TiO_2 substrate.

Noting an obvious gradation in color change upon adsorption of the lead to the substrate, an experiment was performed in which a teabag full of material was quickly subjected to alternating solutions of lead and ligand. As can be seen in Fig. 14.6, the native TiO_2 was white in color; however, as layers of ligand and lead were added and Pb–S bonds were formed, the color of the material took on a yellow appearance. The intensity of yellow color increased as more layers were assembled, perhaps coming to a maximum in between 10 and 20 layers.

Second-generation materials involve the functionalization of mineral substrates with monofunctional ligands in order to bind specific metal contaminants. The advantage to this approach is that existing, commercially available substrates can be modified, resulting in new materials that remove metal contaminants more efficiently that the native material. The metal removal capacities for the second-generation nanocoating of commercially available substrates can be seen in the bar chart in Fig. 14.8.

By changing the functional group of the ligand, the substrate can be "tuned" to bind specific contaminants. Thiols have shown preference for Pb or As, whereas hydroxyl groups have a preference for U. This allows a mixed waste containing both As and U to be separated.

CCT is examining eight commercially available substrates and ten ligands; eight of the ligands are "monofunctional," being applied to a meso-porous substrate initially; two are bifunctional, being applied after the initial monofunctional ligand and first metal adsorption layer. Teraphthalic Acid, 2L2 is both a monofunctional and bifunctional ligand based on the substrate and metal adsorbed. The mono-ligands all have a carboxyl group, -COOH, on one side and typically a thiol, -SH, group on another. Currently we project that a carboxyl group is best for binding to an alumina surface but that a different functional group will be better for attaching to a titania surface. Optimistically, the computational model being developed with the University of Kansas computational chemistry department will identify additional favorable candidates for titania attachment. Two key issues in ligand selection, beyond the obvious attachment issue, are cost and toxicity. We continue to be extremely sensitive to both these issues in choosing ligand candidates.

Currently we use a TGA scan to determine the ligand loading achieved. Ligand loading as a weight percentage of substrate ranges from less than 3 percent to about 40 percent. A 10 percent loading number corresponds to 100 grams of ligand per kilogram of substrate. At a 75 percent metal adsorption loading efficiency for L6, a nano-ligand based filter would adsorb about 0.7 moles of metal per kilogram of substrate at full adsorption; that could range anywhere from about 53 gram for arsenic to about 146 gram for lead. The specific isotherm and starting concentrations of metals would determine the exact loading. Recently we have become aware of "interference" effects that are quite common with ion exchange resins; we were not certain whether or not a ligand-based adsorption would have the same issue.

A TGA scan is the weight of the sample versus temperature as the temperature is increased at a preset rate. Water vaporizes first; then the ligand begins to pyrolyze and subsequently vaporize as the carbon atoms oxidize and the gaseous elements are released. Since each sample contains a varying amount of water and has a different initial weight, all of the data is normalized to the weight at 200°C; this eliminates the surface water and sample size variations. Since we see variation in the substrate itself versus temperature, the ligand + substrate data is normalized to the substrate only data versus temperature as well. The substrate only variation is thought to be caused by "sub-surface" water; as the substrate is mesoporous and the water is deeply trapped, additional energy is required to release it.

The loading of ligand is measured by two methods—mass of the final product and TGA analysis. In the reaction for the As sequestering material, 0.200 g of boehmite was reacted with an excess of ligand to yield 0.426 g of product. The mass gain of 0.226 g corresponds to an approximately 53 percent loading of ligand. Although the mass of the final product provides an estimate of the ligand gained, TGA is a more accurate method for this determination. The TGA profile for this material is shown in Fig. 14.9. An initial mass loss due to water occurs over the temperature range of 25–200°C. The next loss in mass is due to thermal decomposition of the attached ligand, which occurs from 200°C to 400°C. This loss corresponds to a mass percentage of approximately 30 percent, which is the

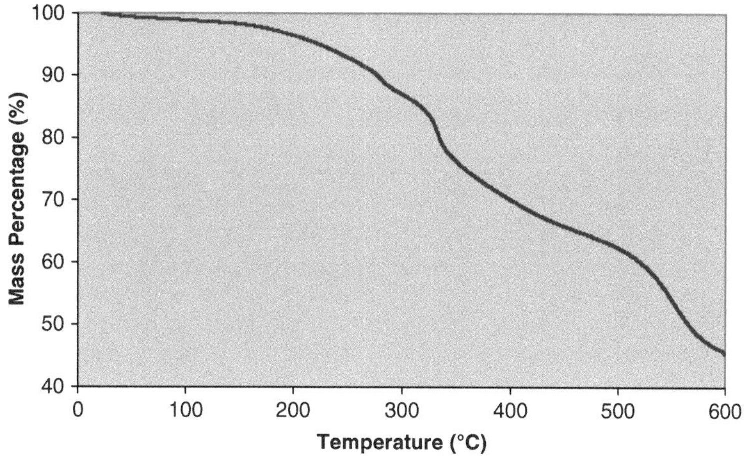

FIGURE 14.9

Thermo-gravimetric analysis (TGA) profile for the As-selective material. The mass loss of approximately 30 percent over the temperature range of 200–400°C is attributable to the ligand that was loaded onto the substrate.

estimate of ligand loaded onto the substrate. The final mass loss is due to the condensation of water from the framework of the boehmite substrate as it is degraded and converted to γ-alumina.

The points plotted for the ligand data are: $[L^*(T_i)/S^*(T_i)]$ where $L^* = L(T_i)/L(T_{200})$ and $S^* = S(T_i)/S(T_{200})$, where $L(T_i)$ is the ligand plus substrate data point at temperature T_i and $S(T_i)$ is the substrate only data at temperature T_i. $L^*(T_i)$ and $S^*(T_i)$ data points are normalized to their respective weights at $T_i = 200°C$. By definition, $S^*(T_{200})$ and $L^*(T_{200})$ equal 1.0 or 100 percent.

Infrared (IR) spectroscopy was also used to characterize these materials. In Fig. 14.10, the transmission IR spectra for the boehmite substrate, the thiol ligand, and the resulting material are shown. Note that the spectrum for the material is a rough composite of the spectra of the two precursors.

14.3.1 Synthesis and characterization of materials

CCT's technology roadmap shows that with different ligands and selective groups for binding specific contaminants, one can separate a mixed waste containing both arsenic and uranium onto separate functionalized substrates.

The ligands designed for use in these materials have different donor atoms (oxygen or sulfur) that are better suited to attract uranium and arsenic, respectively. To generate these materials, the ligands of choice were individually anchored to a boehmite substrate by refluxing the two precursors in water at 100°C for 16 hours. Boehmite, S1, is a hydrous aluminum oxide material

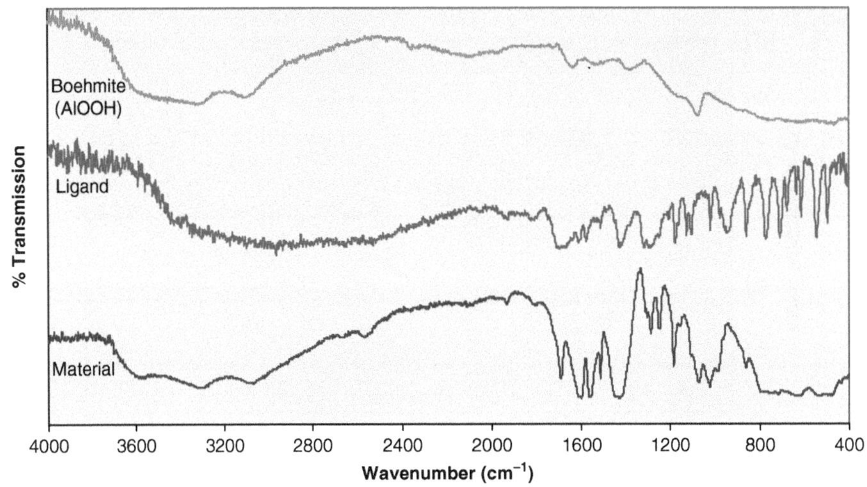

FIGURE 14.10

Infrared spectra of the thiol ligand based material for As removal. The spectrum of the material is a superimposition of the two precursor spectra.

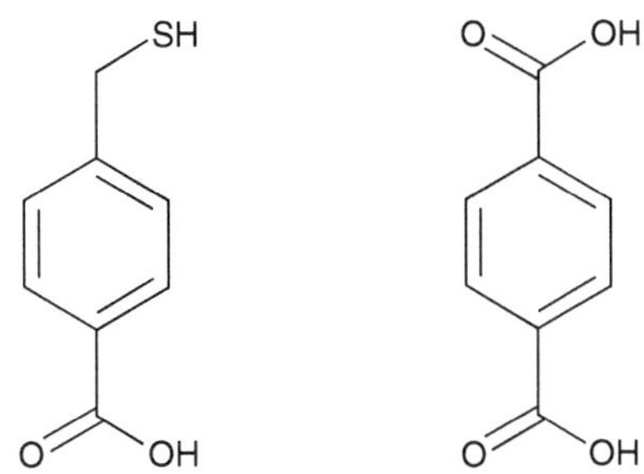

FIGURE 14.11

Ligands L1 (*p*-mercaptomethyl benzoic acid) and 2L2 (terephthalic acid). Each is attached to the substrate through a carboxylic acid functional group; the other group is free to attract metal in solution.

represented by the formula Al(O)OH. The two ligands, *p*-(mercaptomethyl)benzoic acid, L1, and terephthalic acid, 2L2, are shown in Fig. 14.11. Each ligand molecule has at least one carboxylic acid functional group that is reacted in order to attach the ligand to the hydrous surface of the mineral (boehmite) substrate. Additionally, the ligands have either a thiol (sulfur) functionality to attract arsenic or another carboxylic acid group to attract uranium. In the loading reaction, the thiol (or the second carboxylic acid) is unaffected and is therefore free to bind a contaminant.

Table 14.1 Arsenic and Uranium Binding Constants Observed for the Two Different Materials at the Concentrations Specified for the Separation Experiment

	Binding on As-selective Media (mg/g)	Binding on U-selective Media (mg/g)
100 ppb As	6.02	1.26
50 ppb U	0.115	1.85

To demonstrate the ability of selective ligands to separate a mixed waste, experiments were run on solutions of arsenic and uranium in water, with the goal being to separate 100 ppb arsenic and 50 ppb uranium. In Table 14.1, the individual binding constants for each metal on both of the metal-specific materials are provided. Figures 14.12–14.14 show the results of these experiments.

Since the largest binding was observed for 100 ppb As using the material designed for its removal, the experiment was designed to first treat the mixed waste stream with the As-selective media. In the initial experiment, 1 L of waste was used, supplying 0.1 mg arsenic in total, and requiring 17 mg of media to completely remove the arsenic. This mass of media would also remove approximately 0.002 mg of the 0.05 mg total uranium supplied in the 1 L of solution. Note that this is only 4 percent of the total uranium available, allowing for separation of the two elements at a level better than the 75 percent/25 percent goal. Once the solution has been treated with the As-selective media, the solids will be filtered off and the remaining solution will then be treated with the U-selective media for removal of the second metal. Approximately 27 mg of this material would be required to completely remove the uranium at a concentration of 50 ppb.

14.3.2 Metal binding tests

In a typical experiment for metal binding capacity, a specified mass of adsorbent material was suspended in a solution of the desired contaminant for 2 hours by mechanical stirring, followed by vacuum filtration of the media from the solution. The concentration of the contaminant in solution was determined in the "before" and "after" condition by means of Inductively Coupled Plasma (ICP) – Mass Spectrometry (MS) analysis. From these measurements, adsorption isotherms could be generated and future experiments designed. The As-specific material was tested versus a 50/50 v/v mixture of As(III)/As(V) contaminant over the concentration range of 10–500 ppb As. Individual stock solutions of As(III) as sodium arsenite ($NaAsO_2$) and As(V) as sodium arsenate heptahydrate ($Na_2HAsO_4 \cdot 7H_2O$) were diluted to prepare the desired solutions for these tests. As can be seen in Fig. 14.12A, the adsorption increased as the initial concentration of the metal increased, leveling off near 7 mg As/g media.

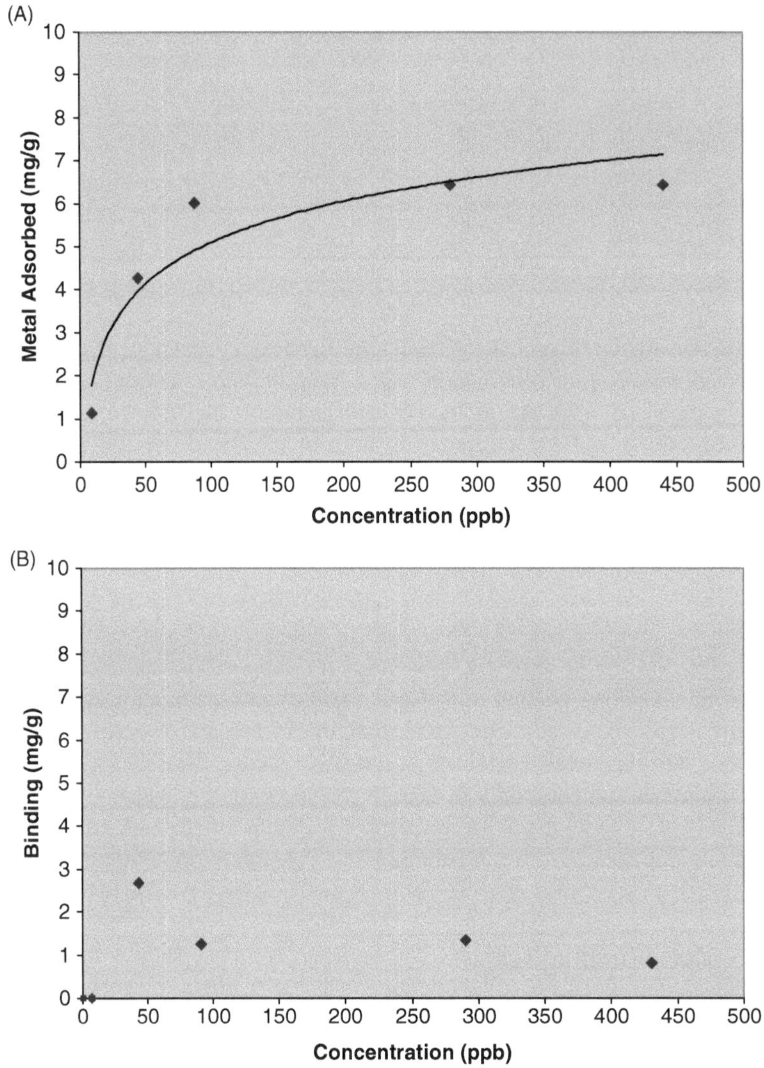

FIGURE 14.12

Adsorption isotherms for arsenic removal using the As-specific material (A) and the U-specific material (B). The adsorption of As at 100 ppb is roughly six times greater for the As-selective media.

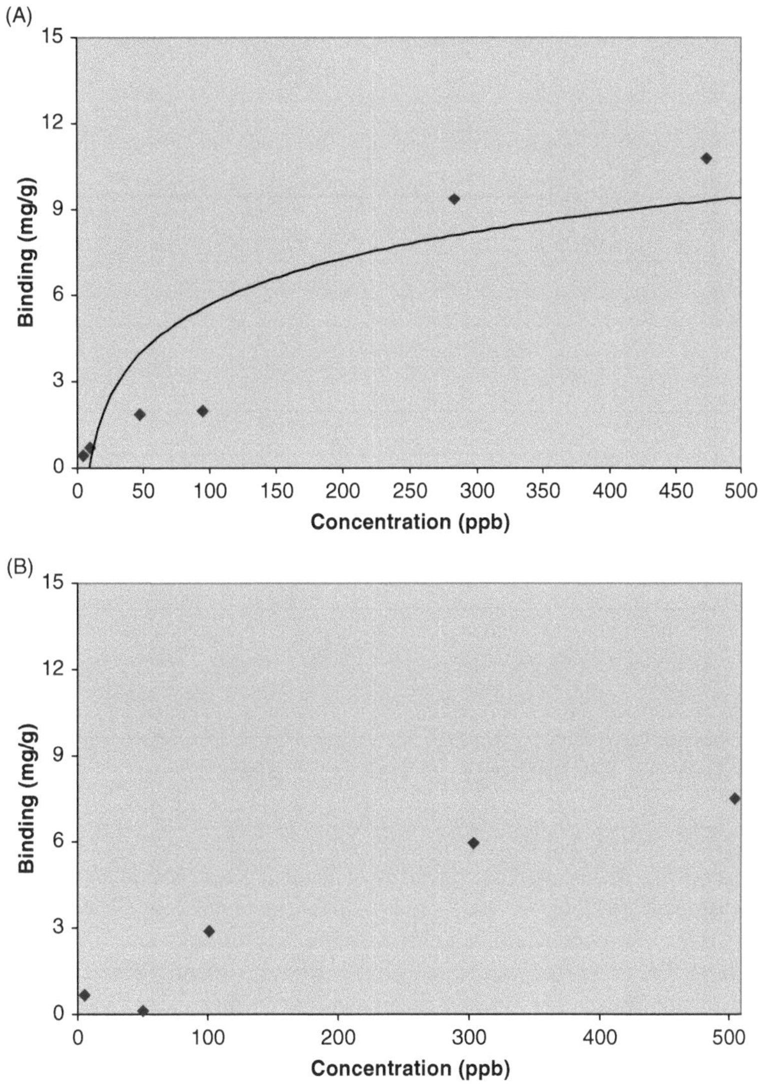

FIGURE 14.13

Adsorption isotherms for uranium removal using U-specific material (A) and the As-specific material (B). The adsorption of U at 50 ppb is roughly 10 times greater for the U-selective media.

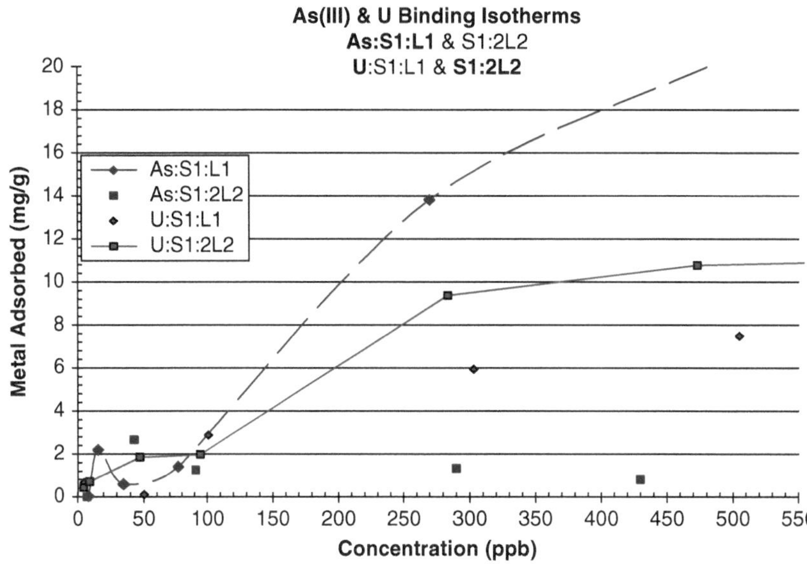

FIGURE 14.14

Binding isotherms for the two ligands used in the As and U separation experiments.

14.4 Moving technology to the next phase

This is the real challenge. Developing new nanocoating technology using ligands to bind heavy metals and delivering on proof of concept and technology is without question the first step. This proof of concept is what has to be delivered to attract customers for them to verify and validate the technology. These validation samples have to be conducted under "real world" conditions.

We have functionalized various substrates with our second-generation technology and demonstrated that we can greatly improve other commercially available adsorptive media on the market today for removing heavy metals from water. These validation samples are intended to demonstrate that CCT's monofunctional ligands can functionalize other companies' materials and improve them at an affordable cost. Bonding the monofunctional ligand nanocoatings on a synthetic, high surface area substrate can be commercialized as a competitive adsorptive media.

Our bifunctional ligand nanocoatings are capable of adding additional layers of metal to an already spent adsorptive media, allowing it to be recharged and reused. This technology will allow wastewater, laden with heavy metals, to be mined for their metal value and for reclaiming the wastewater.

CCT's technology road map includes additional ligands, functional groups, and nanocoatings for removing perchlorate, nitrates, fluorides, and endocrine disruptors from personal care products, from raw water.

As demand increases for the finite supply of water, being able to remove contaminants from raw water passively, without generating a wastestream or wasting any water, while meeting the price point that emerging economies can afford, will be the driving force behind further development of nanotechnology for water purification.

References

[1] W.J. Vickaryous, R. Herges, D. Johnson, Arsenic-π interactions stabilize a self-assembled As_2L_3 supramolecular complex, Angew. Chem. (International Edition), 43 (43) (2004) 5831–5833.

[2] Arsenic facts: <http://www.oregon.gov/DHS/ph/cdsummary/2007/ohd5606.pdf>.

[3] Lead facts: <http://www/wwsp.edu/geo/courses/geog100/Lead-InEnv.htm>.

[4] Maps: <http://www/water/usgs.gov/nawqa/trace.a>.

CHAPTER

15 Water Treatment by Dendrimer-Enhanced Filtration: Principles and Applications

Mamadou S. Diallo[1,2]

[1]Materials and Process Simulation Center, Division of Chemistry and Chemical Engineering, California Institute of Technology, Pasadena, CA, USA
[2]Department of Civil Engineering, Howard University, Washington, DC, USA

15.1 Introduction

The availability of clean water has emerged as one of the most serious problems facing the global economy in the twenty-first century. Water treatment systems typically involve a series of coupled processes, each designed to remove one or more different substances in the source water, with the particular treatment process being based on the molecular size and properties of the target contaminants. Membrane processes such as microfiltration (MF), ultrafiltration (UF), nanofiltration (NF), and reverse osmosis (RO) are emerging as key components of water treatment, reuse, and desalination systems throughout the world [1–3]. RO is very effective at retaining dissolved inorganic and small organic molecules. NF can effectively remove hardness (e.g., Ca(II)) and natural organic matter. However, high pressures (100–1,000 psi) are required to operate both RO and NF membranes. Conversely, UF and MF membranes require lower pressure

Street, Sustich, Duncan and Savage. Nanotechnology Applications for Clean Water, 2nd Edition.
 DOI: http://dx.doi.org/10.1016/B978-1-4557-3116-9.00015-9

(5–60 psi) but unfortunately cannot retain dissolved ions and organic solutes. Advances in nanochemistry such as the invention of dendritic nanopolymers are providing unprecedented opportunities to develop enhanced UF and MF processes for recovering dissolved ions from aqueous solutions. Dendritic nanopolymers are highly branched 3D globular nanoparticles with controlled composition and architecture and sizes in the range of 1–100 nm [4]. They consist of three components: a core, interior branch cells, and terminal branch cells.

Dendritic nanopolymers include hyperbranched polymers, dendrigraft polymers, dendronized linear polymers, dendrimers, and many other supramolecular assemblies such as core-shell tecto(dendrimers) and dendrimer-like star polymers (Fig. 15.1) [4]. They exhibit a number of critical physicochemical properties that make them attractive as separation and reaction media for water purification. Dendritic nanopolymers can encapsulate a broad range of solutes in water including cations (e.g., copper, silver, gold, and uranium), anions (e.g., perchlorate, nitrate, and phosphate), and organic compounds (e.g., pharmaceuticals and pesticides) [4–6]. Dendritic nanopolymers can serve as nanoscale reactors and catalysts [4]. They can also bind and deactivate bacteria and viruses [4]. Their globular shape and large size makes them easier to filter than linear polymers [5,6]. Diallo and coworkers are exploiting these unique properties of dendritic nanopolymers to

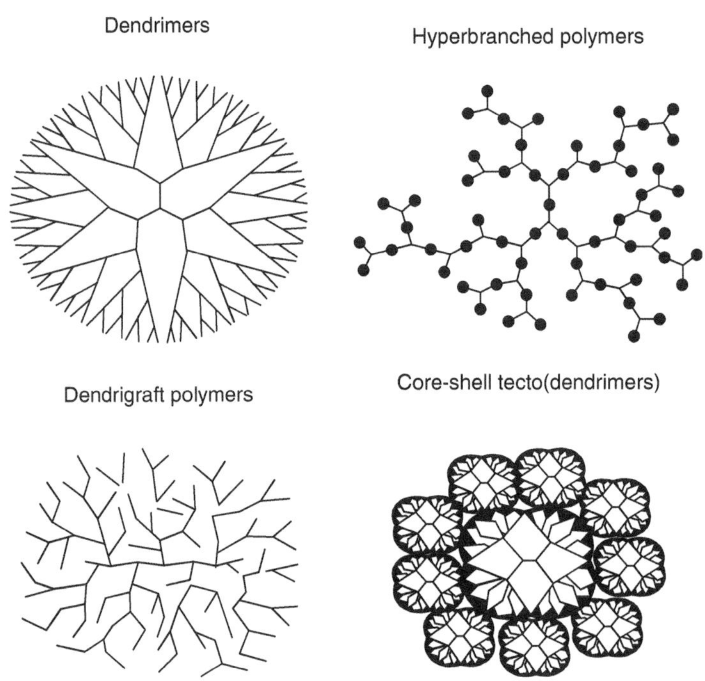

FIGURE 15.1

Selected classes of dendritic nanopolymers.

develop enhanced UF and MF processes for recovering dissolved cations and anions from aqueous solutions [5,6]. This chapter gives an overview of the principles and applications of dendrimer-enhanced filtration (DEF) [5,6]. The use of DEF in the treatment of industrial wastewater contaminated by heavy metals (e.g., copper) and radionuclides (e.g., uranium) and the remediation of groundwater contaminated by anions (e.g., perchlorate) are highlighted.

15.2 Dendrimers as recyclable ligands for cations

Chelating agents are widely used in a variety of environmental and industrial separation processes. These include (i) selective extractants in hydrometallurgy, (ii) metal ion binding functionalities for ion exchange resins, and (iii) high-capacity polymeric ligands for water treatment [7]. The complexation of metal ions is an acid–base reaction that depends on several parameters including (i) metal ion size and acidity, (ii) ligand molecular architecture and basicity, and (iii) solution physicochemical conditions [7]. Although macrocyles and their "open chain" analogues (unidentate and polydentate ligands) have been shown to form stable complexes with a variety of metal ions [7], their limited binding capacity (i.e., 1:1 complexes in most cases) is a major impediment to their utilization as high-capacity chelating agents for industrial and environmental separations. Their relatively low molecular weights also preclude their effective recovery from industrial wastewater streams by low cost membrane-based techniques (e.g., UF). The invention of dendrimers is providing unprecedented opportunities to develop high–capacity, recyclable chelating agents with high molar mass and well-defined molecular composition, size, and shape.

Poly(amidoamine) (PAMAM) dendrimers provide good model systems for probing the aqueous coordination chemistry of cations with dendritic nanopolymers. These dendrimers were the first dendrimer family to be synthesized, characterized, and commercialized. PAMAM dendrimers (Fig. 15.2) possess amide, tertiary and primary amine groups arranged in regular "branched upon branched" patterns, which are displayed in geometrically progressive numbers as a function of generation level. This high density of nitrogen ligands in concert with the possibility of attaching various functional groups such as amines, carboxyl, and so on to PAMAM dendrimers make them particularly attractive as high-capacity chelating agents for cations including transition metals, lanthanides, and actinides [8,9].

Diallo et al. [8,9] have carried out an extensive study of Cu(II) and U(VI) binding to PAMAM dendrimers of different generations and terminal groups. Figure 15.3A and B shows the effects of metal ion dendrimer loading and solution pH on the extent of binding (EOB; i.e., number of moles of bound metal ions per mole of dendrimer] and fractional binding (FB) of Cu(II) in aqueous solutions of a G4-NH_2 EDA core PAMAM dendrimer. The tertiary amine groups of this dendrimer have a pKa of 6.30–6.85 [9]. Conversely, the pKa of its primary amine groups is 9.0–10.2 [9]. At pH 9, the EOB of Cu(II) increases linearly with metal

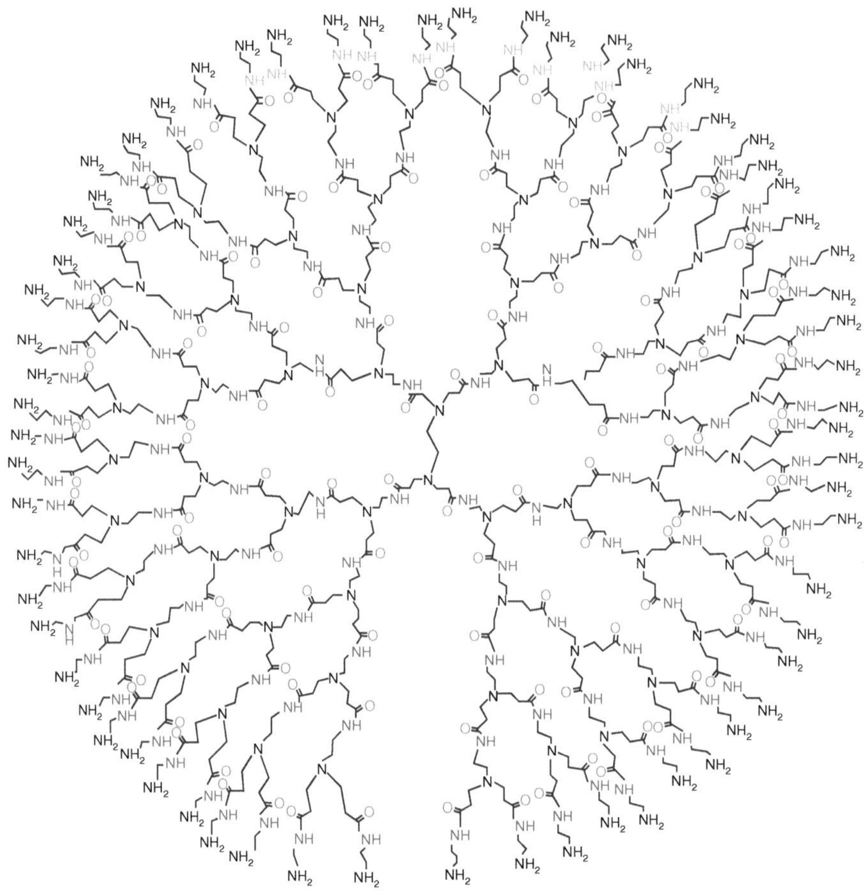

FIGURE 15.2

Structure of G4-NH_2 poly(amidoamine) dendrimer.

ion dendrimer loading within the range of tested metal ion dendrimer loadings. In all cases, 100 percent of the Cu(II) ions are bound to the dendrimers. This behavior is attributed to the low extent of protonation of the dendrimer amine groups. When these groups become fully protonated at pH 5.0, no binding of Cu(II) is observed (Fig. 15.3A). A more complex metal ion uptake behavior is observed at pH 7.0. In this case, the EOB of Cu(II) in aqueous solutions of the G4-NH_2 PAMAM dendrimer go through a series of two distinct binding steps as metal ion dendrimer loading increases (Fig. 15.3A). A more detailed discussion of Cu(II) coordination with PAMAM dendrimers is given elsewhere [9].

Figure 15.4A and B highlights the binding of U(VI) to G4-NH_2 PAMAM dendrimer in deionized water and NaCl solutions [8]. At pH 7.0 and 9.0, the G4-NH_2 PAMAM dendrimer can bind up to 220 U(VI) ions without reaching saturation.

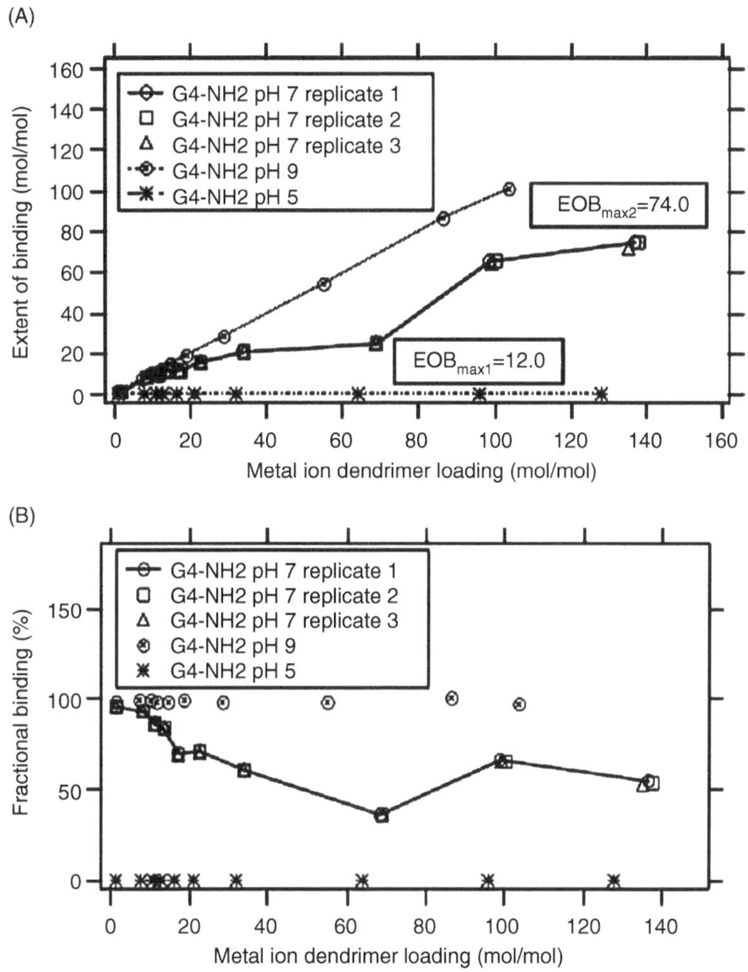

FIGURE 15.3

(A) Extent of binding of Cu(II) in aqueous solutions of G4-NH_2 poly(amidoamine) (PAMAM) dendrimer at room temperature [9]. (B) Fractional binding of Cu(II) in aqueous solutions of G4-NH_2 PAMAM dendrimer at room temperature [9].

The uranyl FB is greater than 92 percent in all cases. At pH 3.0, Fig. 15.4A also shows significant binding of U(VI) to the G4-NH_2 PAMAM dendrimer (with FB approximately 76–87 percent and EOB up to 180) even though its tertiary and primary amine groups are fully protonated in this case. Note that no binding of Cu(II) by the dendrimer was observed at pH 5.0 (Fig. 15.3A). This strongly suggests that uranyl complexation by the G4-NH_2 PAMAM dendrimer at pH 3.0 and 5.0 involves the deprotonation of its amine groups followed by coordination with

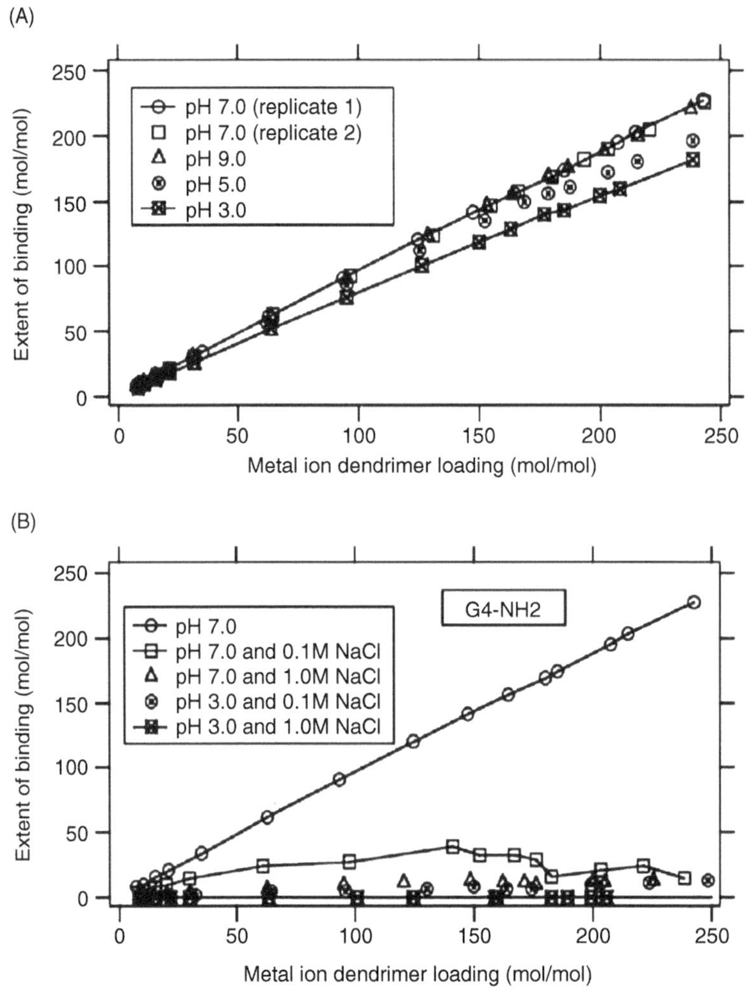

FIGURE 15.4

(A) Extent of binding of U(VI) in aqueous solutions of G4-NH_2 poly(amidoamine) (PAMAM) dendrimer at room temperature [8]. (B) Effect of NaCl on the extent of binding of U(VI) in aqueous solutions of G4-NH_2 PAMAM dendrimer at room temperature [8].

the UO_2^{2+} metal ion. Diallo et al. [8] were able to suppress the uptake of U(VI) by the G4-NH_2 PAMAM in aqueous solutions containing at least 0.1 M (5.8 g/L) of sodium chloride at pH 3.0 (Fig. 15.4B). The overall results of the metal binding experiments strongly suggest that dendritic nanopolymers such as PAMAM can serve as high-capacity, selective, and recyclable chelating ligands for transition metal ions (e.g., Cu(II)) and actinides (e.g., U(VI)) [8,9].

15.3 Dendrimers as recyclable ligands for anions

Anions have emerged as major water contaminants throughout the world. In the United States, the discharge of anions such as perchlorate (ClO_4^-) and nitrate (NO_3^-) into publicly owned treatment works, surface water, groundwater, and coastal water systems is having a major impact on water quality. Although significant research efforts have been devoted to the design and synthesis of selective chelating agents for cation separations [7], anion separations have comparatively received limited attention [10,11]. The design of selective ligands for anions is a challenging undertaking. Unlike cations, anions have filled orbitals and thus cannot covalently bind to ligands [10,11]. Anions have a variety of geometries (e.g., trigonal for NO_3^- and tetrahedral for ClO_4^-) and are sensitive to solution pH in many cases [3,4]. Thus, shape-selective and pH-responsive receptors may be needed to effectively target anions. The charge-to-radius ratios of anions are also lower than those of cations. Thus, anion binding to ligands through electrostatic interactions tends to be weaker than cation binding. Anion binding and selectivity also depend on (i) anion hydrophobicity and (ii) solvent polarity [3,4]. As a first step toward the development of high-capacity, selective, and recyclable dendritic ligands for anions such as perchlorate, Diallo et al. [12] tested the hypothesis that dendrimers with hydrophobic cavities and positively charged internal groups should selectively bind ClO_4^- over more hydrophilic anions such as Cl^-, NO_3^-, SO_4^{2-}, and HCO_3^-. They measured the uptake of ClO_4^- by the fifth generation (G5-NH_2) poly(propyleneimine) (PPI) dendrimer with a diamobutane core and terminal NH_2 groups (Fig. 15.5) in deonized water and model electrolyte solutions as a function of (i) anion–dendrimer loading, (ii) solution pH, (iii) background electrolyte concentration, and (iv) reaction time [12].

Figure 15.6A and B shows the effects of anion–dendrimer loading and solution pH on the EOB and FB of ClO_4^- to a G5-NH_2 PPI dendrimer in aqueous solutions at room temperature and reaction time of 1 hour. The pKa of the dendrimer tertiary and primary amine groups are, respectively, equal to 6.10 and 9.75 [12]. At pH 4.0, 99 percent of the tertiary amine groups of the G5-NH_2 PAMAM are protonated [12]. As shown in Fig. 15.6B, 98 percent of ClO_4^- are bound to the dendrimer in this case when the anion–dendrimer loading is approximately 0.31. The corresponding EOB is approximately 9.0. The FB of ClO_4^- in aqueous solutions of the G5-NH_2 PPI dendrimer is approximately 53 percent at pH 7.0 with an anion–dendrimer loading of 0.31. In this case, only 11 percent of the tertiary amine groups of the dendrimer are protonated, whereas 99 percent of its primary amine groups remain protonated. Note that the maximum EOB of ClO_4^- (approximately 2.5 at anion–dendrimer loading of 32.0) decreases by a factor of 4 at pH 7.0 compared to that at pH 4.0. The maximum EOB of ClO_4^- (approximately 2.20) is also smaller at pH 9.0 and anion–dendrimer loading of 32.0 (Fig. 15.6A) even though approximately

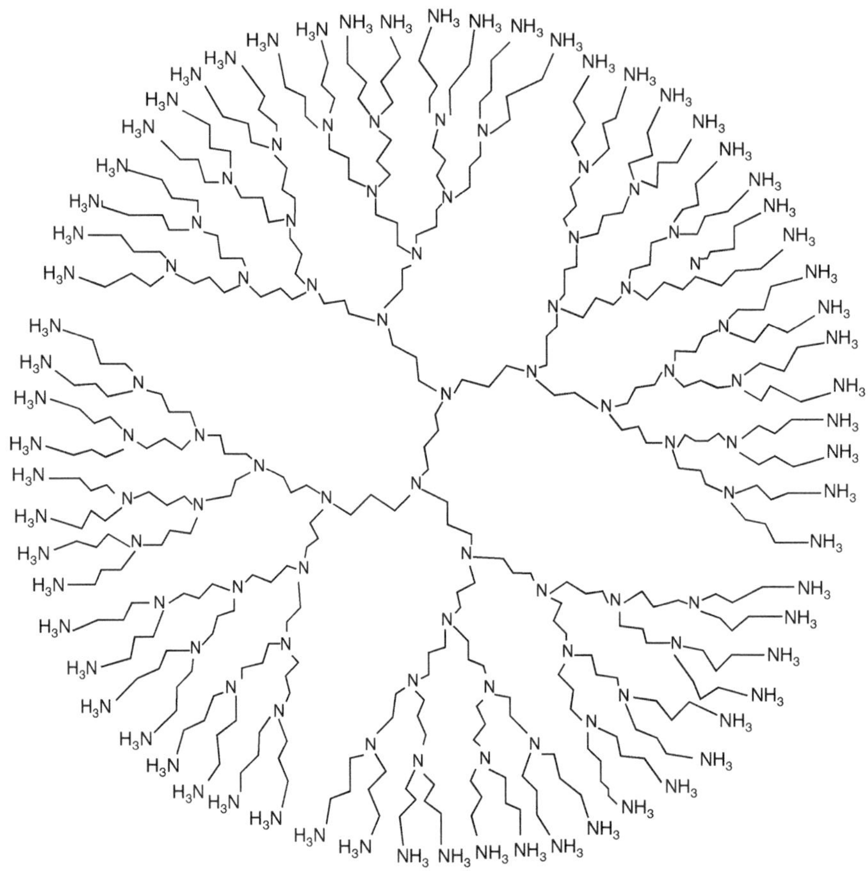

FIGURE 15.5

Structure of G5-NH_2 poly(propyleneimine) dendrimer [8].

81 percent of the dendrimer NH_2 groups remain protonated. This suggests that electrostatic interactions between ClO_4^- and protonated NH_2 groups of the dendrimer do not have a significant effect on perchlorate uptake. Figure 15.6A also shows some binding of ClO_4^- (with a maximum EOB of approximately 1.29 at anion–dendrimer loading of 32.0) at pH 11.0 when both the tertiary and primary amines of the G5-NH_2 PPI dendrimer are unprotonated. A more detailed discussion of the mechanisms of perchlorate binding to the G-NH_2 PPI dendrimer is given elsewhere [12]. The overall results of the anion binding experiments suggest that dendritic macromolecules such as the G5-PPI NH_2 PPI dendrimer provide ideal building blocks for the development of high capacity, selective, and recyclable ligands for anions such as ClO_4^-.

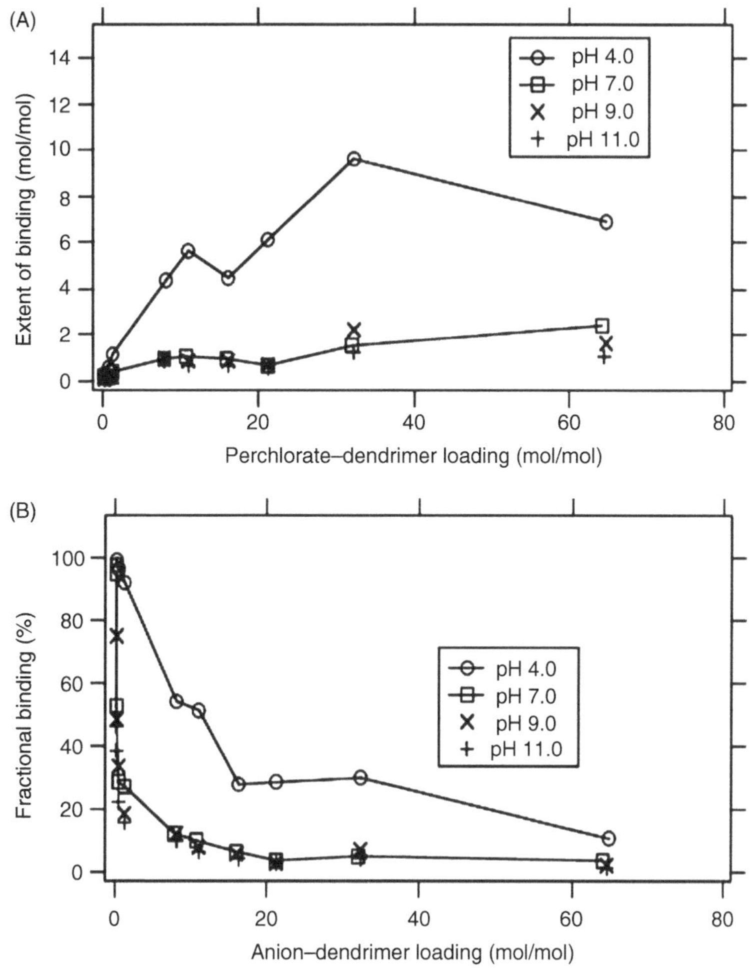

FIGURE 15.6

(A) Extent of binding of ClO_4^- in aqueous solutions of G5-NH_2 PPI dendrimer at room temperature. (B) Fractional binding of ClO_4^- in aqueous solutions of G5-NH_2 PPI dendrimer at room temperature.

Data for (A) and (B) taken from [12].

15.4 Dendrimer-enhanced filtration: overview and applications

As stated in the introduction, the invention of dendritic nanopolymers is providing unprecedented opportunities to develop enhanced UF and MF processes for recovering dissolved ions from aqueous solutions. The DEF process developed by Diallo [5,6] is structured around three unit operations (Fig. 15.7): (i) treatment

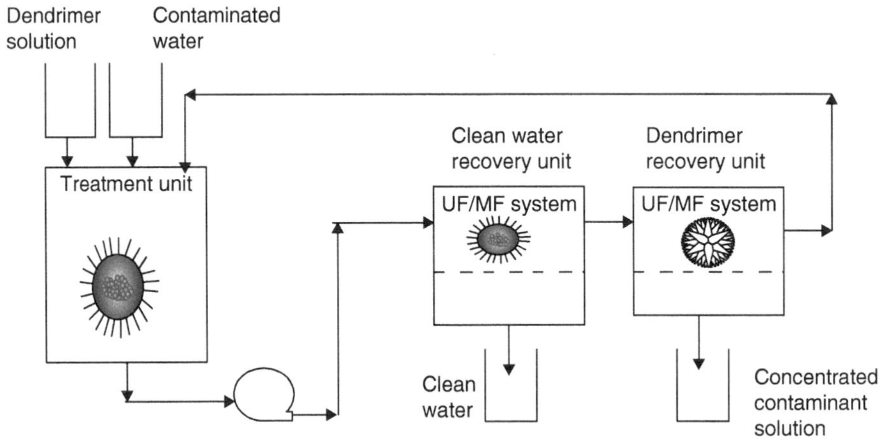

FIGURE 15.7

Water treatment by dendrimer enhanced filtration.

Adapted from [5]. UF: Ultrafiltration; MF: Microfiltration.

unit; (ii) clean water recovery unit; and (iii) dendrimer recovery unit. In the treatment unit, contaminated water is mixed with a solution of functionalized dendritic nanopolymers to carry out the specific reactions of interest (e.g., cation and anion binding). Following completion of the reaction, the complexes of nanopolymers + bound contaminants are sent to the clean water recovery unit where they are filtered using UF or MF to recover the clean water. The resulting concentrated solution of nanopolymers + target substance is subsequently sent to the recovery unit. This system consists of an UF or MF unit in which the bound target substance is separated from the nanopolymers by, for example, changing the acidity (i.e., pH) of the solution. Finally, the recovered concentrated solution of contaminants is collected for disposal or subsequent processing whereas the nanopolymers are recycled [5,6]. The key novel feature of the proposed DEF process is the combination of dendritic polymers with multiple chemical functionalities with the well-established technology of UF and MF. This allows for the development of a new generation of water treatment processes that are flexible, reconfigurable, and scalable [5]. The flexibility of DEF is illustrated by its modular design approach. DEF systems can be designed to be "hardware invariant" and thus reconfigurable in most cases by simply changing the "dendrimer formulation" and process conditions for the targeted contaminants [5]. Because DEF is a membrane-based process, it is scalable and could be used to develop small and mobile water treatment systems as well as large and fixed treatment systems.

As a proof-of-concept study, Diallo et al. [6] have combined bench scale measurements of metal ion binding to Gx-NH_2 PAMAM dendrimers with dead-end UF experiments to assess the feasibility of using DEF to recover Cu(II) from aqueous solutions. On a mass basis, the Cu(II) binding capacities of the Gx-NH_2 PAMAM

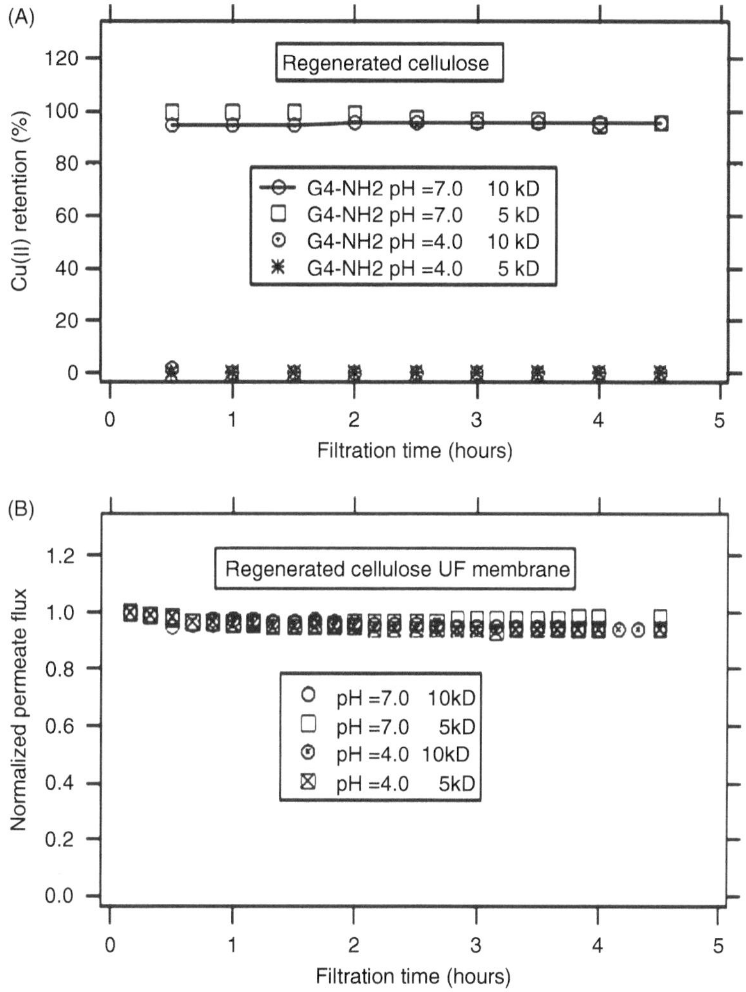

FIGURE 15.8

(A) Retention of Cu(II) complexes with G4-NH_2 poly(amidoamine) (PAMAM) dendrimer by regenerated cellulose ultrafiltration (UF) membranes [6]. (B) Normalized permeate flux of aqueous solutions of Cu(II) + G4-NH_2 PAMAM dendrimer through regenerated cellulose membranes [6].

dendrimers are much larger and more sensitive to solution pH than those of linear polymers with amine groups [6]. Separation of the dendrimer–Cu(II) complexes from solutions can be achieved simply by UF (Fig. 15.8A). The metal ion laden dendrimers can then be regenerated by decreasing the solution pH to 4.0–5.0 [6], thus enabling the recovery of the bound Cu(II) and recycling of the dendrimer. Dendritic nanopolymers such the Gx-NH_2 PAMAM dendrimers have also much less tendency to pass through the pores of UF membranes than do linear polymers of similar molar mass because of their much smaller polydispersity and globular shape [6]. As shown in Fig. 15.8B, the Gx-NH_2 EDA core PAMAM have also a very low tendency to foul the commercially available regenerated cellulose (RC) membranes. Dendritic nanopolymers have also much smaller intrinsic viscosities than linear polymers with the same molar mass because of their globular shape [6]. Thus, comparatively smaller operating pressure, energy consumption, and loss of ligands by shear-induced mechanical breakdown could be achieved with dendritic polymers in cross-flow UF systems typically used industrial water treatment [2]. These unique properties of the Gx-NH_2 EDA core PAMAM dendrimers make DEF (Fig. 15.7) an attractive process for recovering metal ions such as Cu(II) from contaminated water. Other applications of DEF including the recovery of ClO_4^- and U (VI) from aqueous solutions are discussed elsewhere [8,12].

15.5 Summary and outlook

Dendritic nanopolymers are among the most chemically and structurally diverse classes of nanomaterials available to date. These "soft" nanoparticles, with sizes in the range of 1–100 nm, can serve as high-capacity, recyclable ligands for cations and anions. The DEF process exploits these unique properties of dendritic nanopolymers to develop a new generation of low-pressure filtration processes for treating water contaminated by toxic metal ions (e.g., Cu(II) and U(VI)) and oxyanions (e.g., ClO_4^-). A start-up company (Aqua Nano Technologies) has been set up in California to commercialize the DEF technology. Initial target applications include:

1. Recovery of perchlorate from contaminated groundwater.
2. Recovery of uranium from in situ mining leach solutions and contaminated groundwater.
3. Recovery of metal ions (e.g., copper, silver, nickel, and zinc) from industrial wastewater.

Acknowledgments

I thank the US National Science Foundation (NSF Grants CTS-0086727, CTS-0329436, and NIRT CBET-0506951) and the US Environmental Protection Agency (NCER STAR Grant R829626) for funding my research on the use of dendritic nanopolymers as functional materials for water purification. Partial funding for this research was also provided by the Department of Energy (Cooperative Agreement EW15254), the W.M. Keck Foundation, and the National Water Research Institute (Research Project Agreement NO 05-TT-004).

References

[1] US Bureau of Reclamation and Sandia National Laboratories, Desalination and Water Purification Technology Roadmap: A Report of the Executive Committee, Water Purification Research and Development Program Report No. 95, US Department of Interior, Bureau of Reclamation, January 2003.

[2] L.J. Zeman, A.L. Zydney, Microfiltration and Ultrafiltration: Principles and Applications, Marcell Dekker, New York, 1996.

[3] N. Savage, M.S. Diallo, Nanomaterials and water purification, J. Nanopart. Res. 7 (4–5) (2005) 331–342.

[4] J.M.J. Frechet, D.A. Tomalia (Eds.), Dendrimers and other Dendritic Polymers, John Wiley & Sons, New York, 2001.

[5] M.S. Diallo,Water treatment by dendrimer enhanced filtration, US Patent Application, US 1006/0021938 A1, 2006.

[6] M.S. Diallo, S. Christie, P. Swaminathan, J.H. Johnson Jr., W.A. Goddard III, Dendrimer enhanced ultrafiltration. 1. Recovery of Cu(II) from aqueous solutions using Gx-NH2 PAMAM dendrimers with ethylene diamine core, Environ. Sci. Technol. 39 (5) (2005) 1366–1377.

[7] A.E. Martell, R.D. Hancock, Metal Complexes in Aqueous Solutions, Plenum Press, New York, 1996.

[8] M.S. Diallo, A. Wondwossen, J.H. Johnson Jr., W.A. Goddard III, Dendritic chelating agents. 2. U(VI) binding to poly(amidoamine) and poly(propyleneimine) dendrimers in aqueous solutions, Environ. Sci. Technol. 42 (2008) 1572–1579.

[9] M.S. Diallo, S. Christie, P. Swaminathan, L. Balogh, X. Shi, W. Um, et al., Dendritic chelating agents. 1. Cu(II) binding to ethylene diamine core poly(amidoamine) dendrimers in aqueous solutions, Langmuir 20 (7) (2004) 2640–2651.

[10] P.D. Beer, P.A. Gale, Anion recognition and sensing: the state of the art and future perspectives, Angew. Chem. Int. Ed. Engl. 40 (2001) 486–516.

[11] K. Gloe, H. Stephan, M. Grotjahn, Where is the anion extraction going? Chem. Eng. Technol. 26 (2003) 1107.

[12] M.S. Diallo, K. Falconer, J.H. Johnson Jr., W.A. Goddard III, Dendritic anion hosts: perchlorate binding to G5-NH2 poly(propyleneimine) dendrimer in aqueous solutions, Environ. Sci. Technol. 41 (18) (2007) 6521–6527.

CHAPTER

16 Detection and Extraction of Pesticides from Drinking Water Using Nanotechnologies

T. Pradeep[1], Anshup[2] and M.S. Bootharaju[3]

[1]DST Unit of Nanoscience and Thematic Unit of Excellence, Department of Chemistry, Indian Institute of Technology Madras, Chennai, India [2]InnoNano Research Private Limited, An IIT Madras-Incubated Company Focusing on Technologies for Clean Drinking Water, Chennai, India [3]Department of Chemistry, Indian Institute of Technology Madras, Chennai, India

Street, Sustich, Duncan and Savage. Nanotechnology Applications for Clean Water, 2nd Edition.
 DOI: http://dx.doi.org/10.1016/B978-1-4557-3116-9.00016-0

16.1 Introduction

With the rapid increase in industrialization and associated environmental degradation, water bodies are threatened with alarming levels of contamination. The type of contaminants varies from industrial solvents (e.g., chlorophenol) to electronic waste (e.g., heavy metals). In the context of the developing world, especially India, greater risks are due to insecticides and pesticides. Unlike the immediate effects of industrialization faced by the cities and neighboring towns, the impact of agricultural chemicals is faced by population living in villages. A review of commonly used pesticides, their structure, and the health effects is presented in Table 16.1. It is quite evident that many such pesticides contain highly toxic recalcitrant groups and hence are extremely difficult to break through normal synthetic routes of degradation.

One such industrialization program from which India witnessed unintended environmental degradation is the Green revolution. This massive program, while making India self-reliant in its food requirements, over many decades also resulted in long-term environmental degradation. The rush to achieve drastic increases in agricultural productivity by the incessant use of pesticides, over prolonged duration, started showing its effects. Another classic example of an environmentally unfriendly use of chemicals is dichlorodiphenyltrichloroethane (DDT). Across many developing countries, the incessant use of DDT to control malaria, whilst saving a large number of lives, is leading to a greater trouble as it has entered into the food chain.

The global utilization of pesticides amounts to 5.05 billion metric tons (year: 2001) valued at $31.76 billion [2]. The per hectare consumption of agrochemicals in India is 570 g compared to 2500 g in the United States, 3000 g in Europe, and 12,000 g in Japan. Though consumption in India compared with other developed economies is generally low, there are many areas in India where the use of pesticides has been excessively high. Many of the organic pesticides are highly stable in the environment and highly insoluble in water (e.g., the solubility of endosulfan is 2 ppm). However, the dispersions used are a much higher concentration, leading to saturation solubility. The other most common pesticides such as chlorpyrifos (CP), malathion, and DDT are similar; applied concentrations are very large and solubility is fairly low. As a result, pesticide levels detected in the surface waters of India vary from ppt to ppm levels. While there are no nationwide details available on pesticide contamination in drinking water, regional surveys conducted across the country definitely point toward high pesticide contamination in drinking water [3–5]. In villages, where endosulfan was aerially sprayed for 25 years, it caused hitherto unknown effects, including genetic deformations [6]. It is also imperative to realize that with the rising population in many developing countries, the use of pesticides and other agrochemicals is projected to increase proportionately to ensure food security for the people. The problem outlined above will intensify as the water resources are limited, and

Table 16.1 Information about Organic Pesticides Commonly Used in India (Representative Group)

Pesticide	Type	Toxicity	Application	Health Effects
DDT 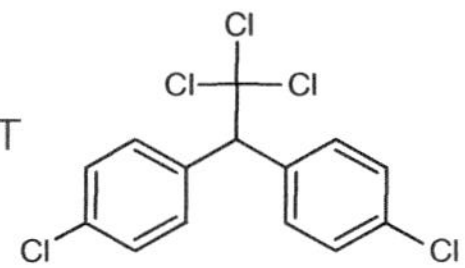	OC	MH	Very effective against insects, domestic insects, and mosquitoes	Chronic liver damage, endocrine and reproductive disorders, immunosuppression, cytogenic effects.
Endosulfan 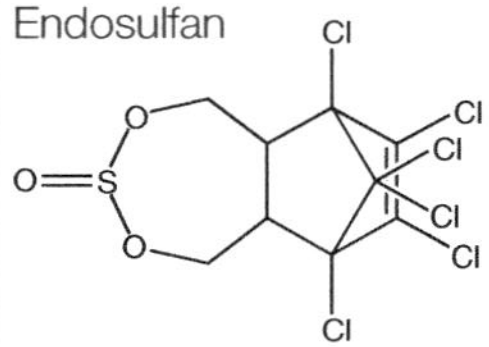	OC	MH	Broad spectrum control against insect pests	Affects kidney, fetus, reproductive system, and liver. Mutagenic.
Malathion 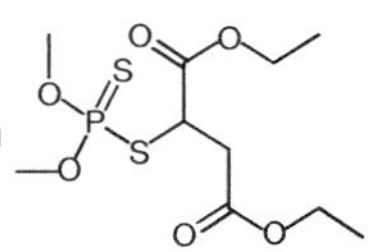	OP	SH	Controls aphids, thrips, red spider mites, and leafhoppers	Carcinogen. Mutagenic. Neurological disorders, allergy, immunodepressant.
Chlorpyrifos 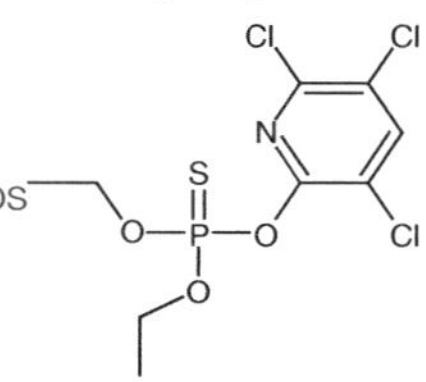	OP	MH	Controls domestic pests	Neurobehavioral disorders, e.g., persistent headaches, blurred vision, memory, concentration.

(*Continued*)

Table 16.1 (Continued)

Pesticide	Type	Toxicity	Application	Health Effects
Carbaryl 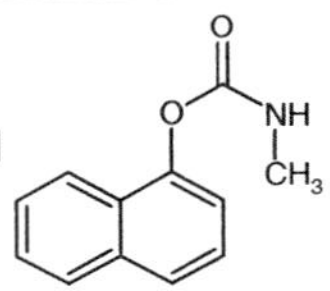	C	MH	Controls pests of fruits, vegetables, and cotton	Mutagenic. Affects kidney, nervous system. Produces *N*-nitrocarbaryl, a carcinogen.
Carbofuran 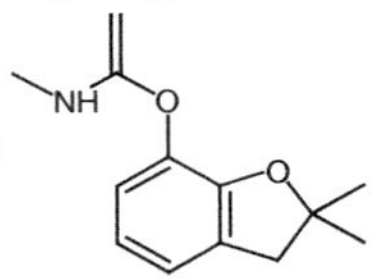	C	HH	Broad spectrum control against insects and mites	Cholinesterase inhibitor affecting nervous system function.
Simazine 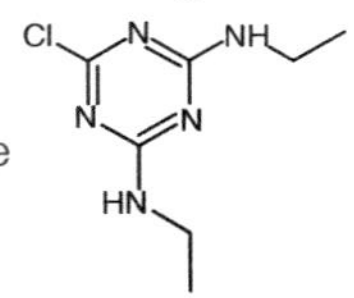	TD	–	Controls weeds in maize, sugarcane, citrus, coffee, and tea	Cancer of testes.

Abbreviations: OC, organochlorine; OP, organophosphorus; C, carbamate; TD, triazine derivative; HH, highly hazardous; MH, moderately hazardous; SH, slightly hazardous (definitions based on WHO classification [1]).

available water will become increasingly contaminated from additional anthropogenic sources. Without quality drinking water, all developmental programs become only proposals, and practical improvement in human development becomes a distant dream. It is in this context that newer technologies for cleaner and affordable water must be investigated.

16.2 The need for nanomaterials and nanotechnology

In the course of our evolving understanding of the effect of water quality on health, quality standards for drinking water are being revised threefold: allowed limits for well-known contaminants are being revised (e.g., lead), the definition of contaminant is becoming specific (e.g., maximum concentration of organic residues vs. pesticides) and newer contaminants (e.g., hexahydro-1,3,5-trinitro-1,3,5-triazine (RDX)) are being brought under regulation. In this course of evolution, it is likely that the contaminant concentrations will reach molecular limits; i.e., the level at which they pose risks could become a few molecules/ions/atoms per glass of drinking water consumed. As a result, the technology to remove these should also reach molecular limits. There must be highly efficient molecular capturing agents which would grab them during single encounter events. The other equally important requirements of such a solution are minimum electricity, economical and environmentally benign. Importantly, the chemistry should be effective even at parts per billion concentrations. In most chemical processes we see around us, the concentrations involved are sufficiently large so that the encounters between the reacting partners are numerous. On the contrary, at a dilution of 10^9, the encounters also get reduced by such orders. All of these pose numerous requirements, difficult to meet entirely.

Where do we go for such a technology? Nature suggests that such a technology exists right around us; this is nanotechnology. Think of watermelons, storing high-quality water in a highly efficient fashion. Considering the feed water, the quality of water saved in tender coconuts is phenomenal. This occurs after a series of filtration and transportation events, in which everything unwanted is rejected. The problem in using this technology is that we know very little about how to imitate this in the large scale. However, we have already seen that understanding of the transport of water across membranes resulted in reverse osmosis. Understanding the molecular transport in cells and role of molecules and molecular materials will help us in filtering water in the most energy efficient fashion. Thus nanotechnology will offer the most energy efficient and clean filtration mechanism in the future. In the interim period, nanomaterial-based approaches may be tried, especially because they offer new possibilities [7,8]. This chapter looks at some of those possibilities in the limited context of organochlorine pesticides.

16.3 Earlier efforts for pesticide removal

Chemistry based on bulk materials has primarily utilized the properties of adsorption, photocatalysis, membrane separation, or biodegradation. Many excellent articles and books have appeared in the past describing the chemistry of each method in detail. A brief overview of such methodologies is presented here. It is important to realize that each of these approaches target many contaminants present in water. Thus, the pesticide removal efficiency decreases significantly in the presence of competing species.

16.3.1 Surface adsorption

The most commercially utilized material for surface adsorption is activated carbon, representing all the carbonaceous materials derived from things such as charcoal, husk, wood, and coal. In all of its forms, it presents a relatively high surface area (>500 m^2/g) due to the presence of macro-, meso-, and microporous structure. Availability of this high surface area and the inherent tendency of carbonaceous materials to absorb many organic residues lead to the excellent adsorption characteristics of activated carbon (Figure 16.1). Surface activation through thermal heating or chemical activation through the use of acids/bases can further increase its absorption properties.

The adsorption of pesticides on activated carbon is driven through two major factors: the nature of the porous structure and the presence of competing organic species in water. The nature of the porous structure can be modified to a large extent to suit the molecule of interest, e.g., the size of pesticide molecules ranges anywhere between 0.5 and 2 nm, whereas the size of natural organic matter vary between 1 and 100 nm.

While activated carbon is one of the most versatile materials for water purification, the use of it suffers from the inherent disadvantage of the necessity to

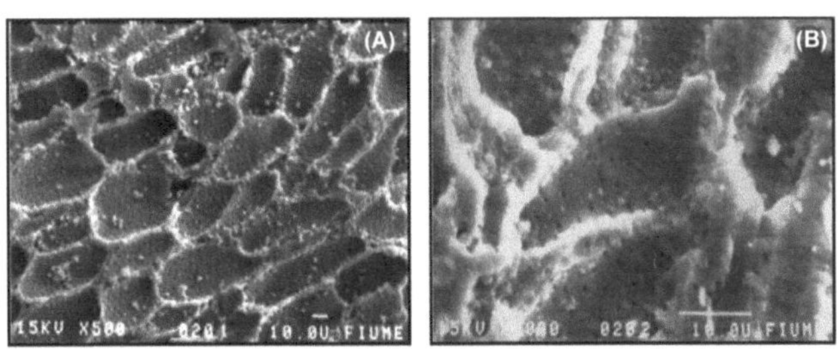

FIGURE 16.1

Scanning electron microscopy (SEM) micrographs of (A) the rough areas of the surface from a granular activated carbon and (B) linear accumulation pattern of the contaminants (phenolphthalein) on the edges of the rough surface.

Reprinted with permission from Ref. [9]. Copyright (2002) with permission from Elsevier.

reprocess the adsorbent and to dispose of residuals in an environmentally responsible manner.

16.3.2 Biological degradation

The inspiration for biological degradation originates from Nature. Nature has been doing it for many decades during which pesticide residues never appeared in the drinking water, and biological degradation by far the most efficient route of environmental cleanup.

The mechanism followed by Nature is the biodegradation of pesticides using microorganisms such as bacteria and fungi [10]. It is suggested that the metabolism of pesticides usually involves a three-phase process as described in Table 16.2. The metabolism of pesticides is dependent on a multitude of factors: environmental conditions (temperature, moisture, pH, etc.), nature of pesticide (hydrophilicity, functional groups, etc.), and metabolism of microorganisms.

16.3.3 Membrane filtration

With the advent of smaller pore size membranes in the early 1970s and their comprehensive use for filtration of contaminated water from different sources (wastewater, domestic water, seawater, etc.), attempts were made to utilize the low pore size and charge properties of membranes for filtering pesticides from water. In principle, membrane-based filtration of pesticides is based on three parameters:

1. *Size distribution of the pores of the membrane and contaminant species:* The retention capacity of a membrane for a contaminant molecule is measured through the fraction of pores on the membrane that are smaller than the molecule.
2. *Surface charge of membrane along with dipole moment of the contaminant species:* Molecules having a dipole can easily orient to have favorable charge interaction between the membrane and the molecule leading to its permeation. Correspondingly, the retention capacity of a membrane is seriously hampered in the case of the presence of a competing polar contaminant in water.
3. *Surface adsorption capacity of the membrane:* The adsorption characteristics of organic matter on membrane surfaces are governed by a variety of factors: organic matter concentration, its nature and mass distribution, calcium ion concentration, and physical and chemical properties of the membranes. Usually the membrane surfaces are negatively charged, which implies co-ions are repelled while counterions are attracted to it, leading to the formation of an electrical double layer at a membrane surface.

Overall, in spite of its ability to remove a broad spectrum of organic and inorganic contaminants, the membrane filtration technique is usually affected by permeate flux deterioration caused by blockages in the membrane transport pores, originating from adsorption of contaminants on membrane surfaces (Figure 16.2).

Table 16.2 Summary of the Three Phases of Pesticide Metabolism [11]

Characteristics	Initial Properties	Phase I	Phase II	Phase III
Reactions	Parent compound	Oxidation, hydrolysis, reduction	Intra- or extracellular conjugation to xylose, methyl, or acetyl groups	Secondary conjugation or incorporation into biopolymers
Solubility	Lipophilic	Amphophilic	Hydrophilic	Hydrophilic or insoluble
Phytotoxicity	Toxic	Modified or less toxic	Greatly reduced or nontoxic	Nontoxic
Mobility	Selective	Modified or reduced	Limited or immobile	Immobile
Bioavailability	Readily absorbed in gastrointestinal tract of animals	Readily absorbed in gastrointestinal tract of animals	Less absorption	Limited absorption or unavailable

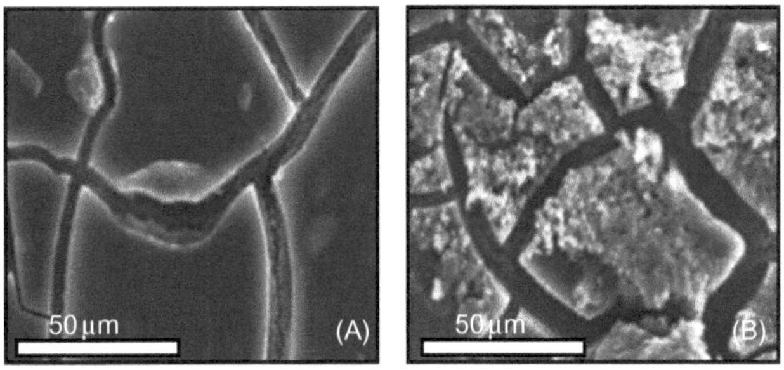

FIGURE 16.2

SEM micrographs of NF-270 membrane surfaces (A) clean and (B) treated with 10 mg/L humic acids.

Reprinted with permission from Ref. [12]. Copyright (2006) with permission from Elsevier.

16.4 Nanomaterials-based chemistry: recent approaches

16.4.1 Homogeneous versus heterogeneous chemistry

Contaminants such as pesticides are removed through homogeneous or heterogeneous chemistry. In the homogeneous process, the molecules of relevance are degraded by the nanoparticles dispersed in the solution phase. This methodology is attractive because it utilizes all the available surface area offered by nanoparticles. However, a major issue of concern is the possible presence of nanoparticles, and thereby the contaminants in the purified water. This is important as the dispersed nanoparticles cannot be easily separated. Heterogeneous chemistry utilizes supported nanoparticles. Here, highly dispersed nanoparticles on supports such as oxides, polymers, and fibers are used and water is passed through such media. The binding between the particle and the support is strong enough to avoid leaching of the particles into water. Additionally, with a decrease in the particle size, metal particles gradually become unstable without protecting agents, due to increasing surface energy. Consequently, immobilization of metal nanoparticles on a suitable high-surface-area solid helps in avoiding agglomeration, even though it leads to changes in the properties and behavior of the isolated particles [13].

As mentioned, there are numerous methods to attach nanoparticles on supports. Some of these strategies are discussed below.

16.4.1.1 Impregnation

This is one of the earliest methods used for particle immobilization on supports. In a generalist representation of the method, the aqueous solution of the metal precursor salt (e.g., $HAuCl_4$, $AgNO_3$, H_2PtCl_4) is added to the cleaned supports (e.g., Al_2O_3, SiO_2, TiO_2). Surface cleaning helps in the removal of trapped air and surface impurities. Subsequently, the solution is subjected to the reducing medium (e.g., use of reducing agents such as $NaBH_4$, H_2, sodium citrate, or thermal decomposition). Later the solution is subjected to high-temperature heating for surface activation. This method suffers from an inherent disadvantage as it results in the formation of large and thus catalytically inactive, particles. This is attributed to the presence of inorganic anions in the solid leading to sintering during thermal activation [14]. One of the major advantages of this method is that it leads to high surface loading of nanoparticles (Figure 16.3A).

16.4.1.2 Ion exchange

In this method, the cations present on the surface or porous structure of the support are replaced by metal cations followed by calcination at high temperature and reduction. This procedure leads to the formation of extremely small metal particles. This method suffers from a disadvantage of low cation exchange efficiency leading to a limited number of cationic sites for nanoparticle formation.

The method is illustrated through a representative example: In a typical procedure to synthesize Na–Y zeolite, aluminum powder and sodium chloride were dissolved in an appropriate amount of tetramethylammonium hydroxide (TMAOH) solution, followed by the addition of tetraethyl orthosilicate. The resulting solution was left for ageing which results in a gel and was heated at 100°C for a few days. The solid obtained after heating is centrifuged and washed with distilled water (composition: SiO_2–Al_2O_3–$(TMA)_2O$–Na_2O–H_2O). To load the metal nanoparticles on the zeolite support, the functionalized zeolite was dispersed in metal nanoparticle solution and it was centrifuged and washed (Figure 16.3B) [15].

16.4.1.3 Coprecipitation

The salts of first-row metals of the Transition Series in Groups 4–12 and a few other metals such as aluminum and magnesium form hydroxides or hydrated

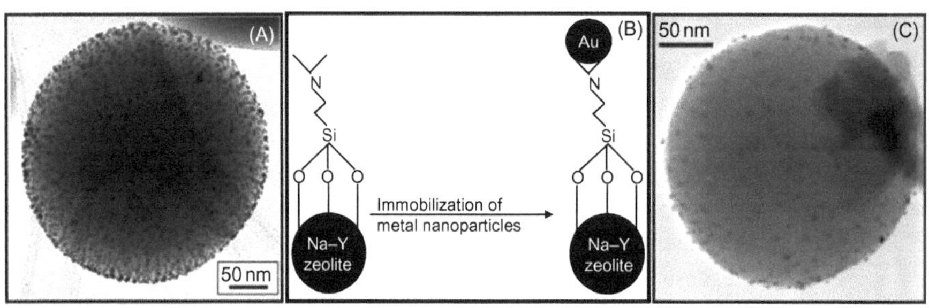

FIGURE 16.3

(A) TEM image of gold nanoparticles deposited on silica using incipient wetness impregnation method. (B) Direct assembly of gold nanoparticles on amine-functionalized zeolite particle. (C) TEM image of a single platinum nanoparticle deposited on carbon support through low-temperature chemical vapor deposition (CVD) in a fluidized bed reactor.

(A) Reproduced with permission from Ref. [14]. Copyright 2007 by Springer-Verlag. (B) Reprinted from Ref. [15]. Copyright (2004) Wiley-VCH. (C) Reproduced from Ref. [16] with permission of the Royal Society of Chemistry.

oxides in basic medium. Coprecipitation of two hydroxides has often been used to prepare supported base metal catalysts, usually with alumina acting as support.

The method is illustrated through a representative example: An aqueous solution of metal salt is mixed with the nitrate form of the support and mild basic solution of ammonium carbonate is added for pH control. The resultant coprecipitate is dried under vacuum, followed by calcination at high temperature.

16.4.1.4 Deposition–precipitation

In this method, the precursor for the active species is brought out of solution in the presence of a suspension of the support (e.g., by raising the pH in order to precipitate the salt as a hydroxide), the surface of the support acts as a nucleating agent, and the method, if properly performed, leads to the greater part of the active precursor being attached to the support. The secret lies in preventing the precipitation away from the support surface by avoiding local high concentrations of hydroxide. The chemistry as it applies to the active component therefore closely resembles the one followed in coprecipitation. Deposition–precipitation has the advantage over coprecipitation in that the entire active component remains on the surface of the support and none is buried within it. In addition, a narrower particle size distribution is achieved.

The method is illustrated through a representative example: The nitrate salts of metal are mixed with oxide-based supports. The pH of the solution is adjusted using mild basic reagents. The resulting product is centrifuged and washed to remove excess anions present in the solution, followed by calcination at high temperature.

16.4.1.5 Vapor phase deposition and grafting

In this method, a stream of volatile metal compound is flowed through a porous support, assisted by an inert gas. The chemical interaction between metal salt with the porous support leads to the formation of a precursor. The method is illustrated through a representative example (Figure 16.3C): Carbon spheres, which are used for deposition of metal nanoparticles, are produced through iron salt catalyzed high-temperature reaction between methane and hydrogen gas. The as-produced material was washed with acetone so as to remove heavy hydrocarbon contamination. The nitric acid-based oxidative treatment is used for increasing the concentration of grafting sites on the surface, decreasing the surface area and increasing the microporous volume without affecting the structure. The resulting carbon support is then mixed with organometallic compounds (of desired metal nanoparticle) and carried through a fluidized bed CVD reactor using inert carrier gas.

16.4.2 Variety of nanosystems

There are numerous variety of nanomaterials used for environmental detoxification. A broad category of these materials are nanometals, nanooxides, and nanoclays. There are also more traditional categories of materials such as nanomembranes, nanopores, and zeolites. These latter categories of materials remove a given species by providing a physical barrier and such materials are not discussed here.

16.4.2.1 Nanometals

The most important materials studied in this regard are zero-valent iron (ZVI), which is known to degrade a diverse variety of toxic molecules. Due to the low cost and easy production, ZVI is an industrially feasible remediation methodology. The attractiveness of nanometals arises from the dramatic increase in its specific surface area (owing to small particle size) and increase in the surface energies which helps in the generation of electrons and its consequent transfer to organic species (explained in detail in a later section).

The other category of nanomaterials is noble metals, which we shall discuss in detail in a later section. Similarly, several other metals such as copper, zinc, and tin have also been successfully tested for decomposition of chlorinated hydrocarbons.

The mechanism of reactivity for ZVI is similar to the mechanism of corrosion (i.e., oxidation of iron). The mechanism involves the generation of electrons which in turn reduces the organic species through dechlorination (Figure 16.4):

$$Fe^0 + RCl + H^+ \rightarrow Fe^{2+} + RH + Cl^-$$

The use of ZVI for commercial-scale pesticide removal is detrimentally impacted by the following challenges:

- Reaction by-products originating during the decomposition of chlorinated hydrocarbons. Nano-sized metallic iron has been reported to show low

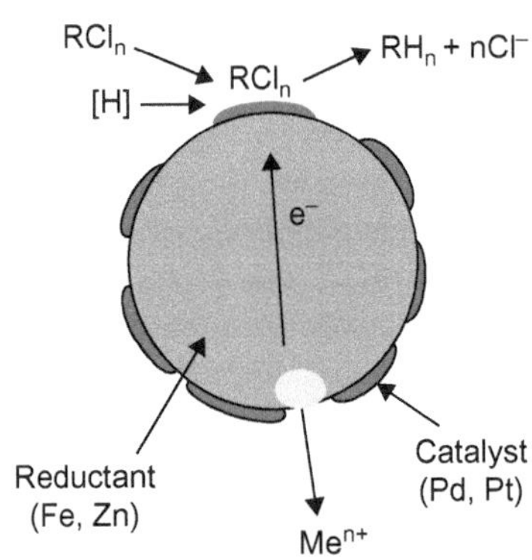

FIGURE 16.4

Schematic of a nanoscale bimetallic particle for the treatment of chlorinated solvents. Within a bimetallic complex, one metal (Fe, Zn) serves primarily as electron donor, while the other (Pd, Pt) works as a catalyst.

Reprinted with permission from Ref. [17]. Copyright (1998) with permission from Elsevier.

reactivity toward lightly chlorinated hydrocarbons. It may also be possible that some of the reaction by-products are highly toxic in nature or have high solubility in water (e.g., 1,1-dichloro-2,2-bis (p-chlorophenyl) ethane, dechlorinated derivative of DDT, has 50 times more solubility in water compared to DDT).

- Continuous decrease in the reactivity of metallic iron due to the presence of hydroxyl radicals in the solution, leading to the formation of iron hydroxides on the nanometal surfaces. This phenomenon has been reported to occur faster at high pH. To avoid this difficulty, the surface of nano-sized ZVI has been coated with other well-known metal catalysts such as palladium. The surface protection of ZVI with palladium also leads to enhancement in catalytic properties due to the action of palladium.

16.4.2.2 Nanooxides

The common oxide materials of industrial relevance are nano-TiO_2 and nano-ZnO, both of which are semiconducting oxides. These have been used to decompose a broad range of pesticides, organic dyes, and industrial solvents by photocatalysis. TiO_2, in its bulk and nano-forms, is one of the most researched materials for photocatalytic activity because of its stability under harsh conditions, low solubility in water, possibility of investigation using both fixed bed and suspension forms and commercial availability. Significant research attempts have been made to understand the factors affecting the photocatalytic activity of TiO_2. The topic is separately discussed as a chapter in this book.

16.4.2.3 Nanoclays

Clay is one of the naturally occurring materials which has been utilized by Nature for maintaining flow of groundwater (due to relatively high impermeability to water), removing toxic species present in the water (due to presence of surface charge on the clay structure), and imparting rigidity to natural structures (due to natural plasticity behavior). Clays are alumino silicates with a planar silicate structure. There are three main categories of clay: kaolinite, montmorillonite–smectite, and illite, among which the first two are the most widely studied.

The kaolinite group includes the two forms of minerals: dioctahedral minerals (dioctahedral sites occupied by Al^{3+}, e.g., kaolinite and dickite) and trioctahedral minerals (dioctahedral sites occupied by Mg^{2+} and Fe^{2+}, e.g., antigorite and chamosite). Usually the structure contains silicate sheets (Si_2O_5) bonded to aluminum oxide/hydroxide layers ($Al_2(OH)_4$) called gibbsite layers. The primary structural unit of this group is a layer composed of one octahedral sheet condensed with one tetrahedral sheet.

The smectite group refers to a family of nonmetallic clays primarily composed of hydrated sodium calcium aluminum silicate. Montmorillonite belongs to the family of 2:1 smectite minerals having two tetrahedral sheets with the unshared vertex of each sheet pointing toward each other and forming each side of the octahedral sheet. Depending on the composition of the tetrahedral and octahedral sheets, the layer will have no charge, or will have a net negative charge (balanced by interlayer cations). A permanent negative charge is located on the clay layers because of isomorphic substitution of Al^{3+} atoms in the octahedral layer by Mg^{2+} and Fe^{2+} and of Si^{4+} atoms in the tetrahedral layer by Al^{3+}. Montmorillonite being dioctahedral clay contains one vacant site in every three octahedral positions. The exchangeable cations can easily be replaced by positively charged species. Additionally, the outer silicate surface of a montmorillonite cell contains Si atoms bound to hydroxyls (Si–OH), which help in the adsorption of the organic species (Figure 16.5A).

There are many approaches for increasing the adsorption of a specific molecule, e.g., to increase the adsorption of organic species, the inorganic cations are replaced by quaternary ammonium cationic surfactants (containing long chain alkyl group to make it more organophilic). Similarly, depending on the nature of charge on clay layer, metal cations (e.g., lead, arsenic, and cadmium) can be picked from the water through intercalation (Figure 16.5B) [19].

16.4.2.4 Dendrimers

Dendrimers represent a novel category of polymeric molecules, with multiple functionalities. They can be utilized in twin ways: as a stabilizer to metal nanoparticles and as a contaminant-specific functionalized group on a nanoparticle surface. The two most commonly used dendrimers are polyamidoamine (PAMAM) dendrimers and polypropylene imine dendrimers.

The terminal functional group of dendrimers can easily be utilized for engineering the interaction between molecules of interest with dendrimers, e.g., long aliphatic chains can be activated on dendrimer surfaces so as to render the dendrimers lipophilic, without affecting the binding properties of organic species with dendritic macromolecules. Similarly, the dielectric gradient between the core and the surface (owing to the presence of different reacting species) can be tuned for the encapsulation of the incoming guest molecule. Similarly, the target molecules can be engulfed in the cavity of the dendrimer by providing appropriate molecular functionality within the cavity. These kinds of approaches have not been developed for pesticide molecules, but interesting possibilities have been shown using metal ions and nanoparticles (Figure 16.6).

16.4.2.5 Carbon-based materials

Carbon is one of the most abundant elements in the earth's crust. Carbon-based nanomaterials including graphene, graphene oxide (GO), reduced graphene oxide (RGO), and carbon nanotubes (CNTs) were explored extensively in the recent past. Graphene is a two-dimensional single sheet of sp^2 carbon atoms with an unusually large surface area (theoretically, 2630 m^2/g which is greater than CNTs and graphite whose surface areas are of the order of 1315 and ~10 m^2/g, respectively), electrical conductivity, etc. Partially functionalized graphene with carboxylic, epoxy, and ether groups (GO and RGO) which can be dispersed in water are utilized for water purification. Unusual adsorption capacities of GO and RGO

(A)

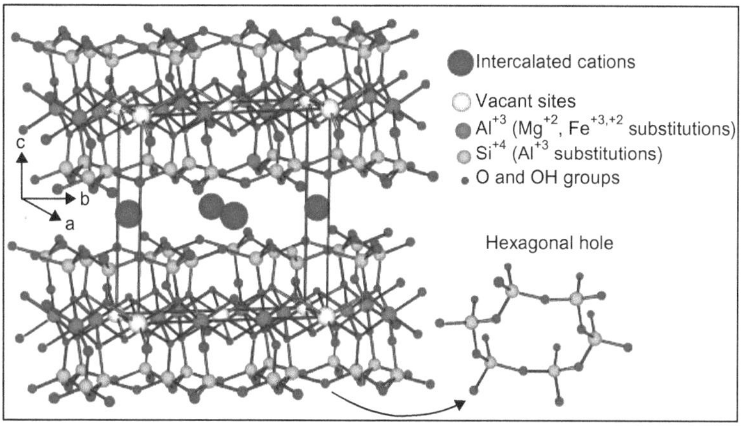

(B)

FIGURE 16.5

(A) Schematic representation of structure of montmorillonite showing octahedral and tetrahedral sheets, vacant sites, and intercalated cations. (B) SEM picture of a natural clay sediment. Inset: Lead map made by characteristic X-ray photons.

(A) Reprinted from Ref. [18]. Used with permission from IOP publishing.

(B) Reprinted with permission from Ref. [19]. Copyright (2003) with permission from Elsevier.

FIGURE 16.6

Interaction studies of dendrimer with metal ions and their subsequent reduction to prepare encapsulated nanoparticles.

Adapted from Ref. [20]. Copyright (2003) American Chemical Society.

for water-assisted adsorption of pesticides including chlorpyrifos, endosulfan, and malathion are reported [21]. Graphene-coated silica [22] and river sand [23] have been shown to remove diverse organophosphorous pesticides from water. Interaction of a strong π-bonding network in graphene and electron donating functional groups in pesticide molecule is the likely reason for pesticide adsorption on graphene surface. Directly synthesized graphene–chitosan composite [24] has been utilized for electrochemical detection and solid phase extraction of methyl parathion down to a level of 0.8 ng/mL.

16.5 Pesticide removal from drinking water: a case study

16.5.1 Noble metal nanoparticle-based mineralization of pesticides

One of the areas of interest of our research group is the interaction of noble metal nanoparticles with organochlorine and organophosphorus pesticides. Specifically, the nanoparticles used for study are of gold and silver, whereas the pesticides are endosulfan, malathion, and chlorpyrifos. This study is a consequence of our findings on the possibility of degrading different halocarbons using noble metal nanoparticles [25].

It was discovered that mineralization of halocarbons happens in two steps: adsorption of halocarbons on the nanoparticle surface and its consequent mineralization to metal halides and amorphous carbon [25]. A few of the important revelations from this study are:

- *Environmentally benign nature of the reaction products and zero production of by-products:* One of the key issues to resolve in tackling the pesticide contamination problem in its entirety is the elimination of reaction products. The products from such a catalytic decomposition are metal halides and amorphous carbon, both of which are environmentally benign. Using advanced mass spectroscopic techniques, it was established that there are no reaction by-products left in the solution and mineralization of pesticides is complete for the molecules investigated.

- *Demonstration of size selective reactivity for nanoparticles:* The experiments were conducted with bulk form of noble metals and it was concluded that this property is demonstrated only in the case of nanoparticles. The absence of such a property being exhibited by the bulk form of noble metals is also explained from thermochemical calculations (explained later).
- *Capability to target a broad range of organochlorine and organophosphorus pesticides:* The reactivity was studied with a broad range of halocarbons (e.g., benzyl chloride, chloroform, and bromoform), organochlorine pesticides (e.g., endosulfan), and organophosphorus pesticides (e.g., chlorpyrifos, malathion). Using sensitive instrumentation, it was established that all such organics undergo complete mineralization.

Based on the comprehensive study conducted at various reaction conditions, it is postulated that the catalytic decomposition of halocarbons is initiated through the transfer of electrons from metal nanoparticles to the solvent, which in turn causes the mineralization of halocarbons. The scheme of reactions is suggested to be the following (not balanced by stoichiometry as it is difficult to predict the number of metal atoms in a nanoparticle) [25]:

$$Ag\ (nano) \rightarrow Ag^{+}(nano) + e^{-}$$

$$(CH_3)_2CHOH \rightarrow (CH_3)_2CO + 2H^{+} + 2e^{-}$$

$$Ag + CCl_4 \rightarrow AgCl + C$$

In this specific case, presented above, $(CH_3)_2CHOH$ was used as a solvent for the bulk mineralization of CCl_4, studied in a $(CH_3)_2CHOH/H_2O$ solvent mixture. From a thermodynamic standpoint, the reaction of halocarbons with silver is not favored due to small positive reaction enthalpy (standard enthalpy values for CCl_4 and AgCl are −128.2 and −127.0 kJ/mol, respectively). It is suggested that the nano-form of the noble metal particles along with the presence of energetic surface atoms helps in overcoming the thermochemical and entropic barriers.

It is easy to comprehend that the decrease in particle size leads to drastic changes in the rate kinetics of the reaction owing to an increase in the availability of reaction sites on the metal nanoparticle surface. The other variable component of the reaction kinetics is the activation energy which is known to be affected by many parameters such as nature of reactants, crystallinity of the reactive surfaces, ease of charge transfer, surface impurities such as dopants. The reaction outlined above can result in the removal of the metal nanoparticle as ions. As a result, it has been used for the creation of nanoshells starting from core–shell nanoparticles [26]. Such nanoshells have been used for biological applications, namely for the delivery of drug molecules into bacteria [27]. These mineralization reactions can be used to probe the porosity of the core–shell structures [28].

16.5.1.1 The mechanism of nanoparticle reactivity

The dramatic reactivity change at the nanoscale is largely attributed to the perfection achieved in the ordering of atoms (or molecules) [29] and change in surface energy. The bottom-up approach of material synthesis helps to minimize the imperfection and defects present on the surface of the particle. The high degree of atomic order in crystalline nanoparticles helps to control the physical and chemical behaviors and consequently creates extremely novel properties [30]. At the same time, it must be remembered that the surface defects play a dominant role in the reactivity of many systems, especially oxide nanoparticles.

16.5.1.2 Nanoparticle activity based on energy gap

The underlying building block of any material, be it nanoparticle or bulk material, is atoms and molecules, each of which have discrete energy levels or orbitals. However, owing to the complex nature of interactions between multiple atoms and molecules present in the material, a variety of energy bands are formed. In the case of metals, transfer of valence electrons to the conduction band occurs, leading to many novel properties. For metal nanoparticles, delocalized electrons can be excited even by visible light (in some cases, also by near infrared) and it is reflected in the form of surface plasmon resonance which is characterized through distinct visible color for many nanoparticle solutions (Figure 16.8). The spacing, δ, between the adjacent energy levels in a band is given by the approximate relationship $\delta \approx E_F/n$, where E_F is the Fermi level energy and n is the number of atoms in the particle. As the particle size decreases, at a critical point, the energy band spacing becomes greater than the thermal energy, kT, where the atoms begin to behave as individual species, and the particle may lose its metallic properties. A spacing δ of 2.5×10^{-2} eV exceeds the thermal energy at room temperature when n is less than 400 (i.e., 2 nm in diameter) [31].

16.5.1.3 Mode of decomposition for organic species

For a large variety of organic molecules, the principal mode of decomposition is electron transfer, from the metal particles. As the oxidation of metals is generally favored and as these processes are facilitated in aqueous medium, there is availability of electrons at metal particle surfaces. Oxidation of nanoparticles is even more facile, in comparison to bulk. Remember that oxidation is the removal of an electron from the whole nanoparticle and not from an atom. This oxidation can also be tuned by varying the particle dimension, by providing appropriate functionalization and by attaching the particle on suitable substrates. The oxidized species can be one of the many different oxides or hydroxides, which may have reduced solubility in water.

The electron removed from the particle surface can be made available for appropriate reduction chemistry. For molecules such as halocarbons, the electron is transferred to the halogen making the C–X bond weak, resulting in the formation of the halide ion. The organic species can participate in various other reactions, depending on the medium or other reagents present. The halide ion formed can also produce the metal halide and it again is dissolved in the medium or precipitated depending on the solubility product.

16.5.1.4 New insights on surface chemistry of nanoparticles/clusters

Mechanistic details of the interaction of nanomaterials with pesticide molecules have recently been derived. CP undergoes hydrolysis in the presence of silver nanoparticles to give 3,5,6-trichloro-2-pyridinol (TCP) which is relatively less toxic compared to chlorpyrifos and diethyl thiophosphate (DETP) (Figure 16.7). Degradation of CP proceeds through the formation of AgNP − S surface complex, which is confirmed by Raman spectroscopy. In this complex, the P − O bond cleaves to yield a stable aromatic species, TCP [32]. The rate of degradation of CP increases with increase in temperature and pH. Mineralization of halocarbons, an important category of volatile organic compounds (e.g., carbon tetrachloride, benzylchloride, chloroform), is also understood to some extent. Monolayer-protected noble metal quantum clusters (especially clusters of silver) which are smaller in size compared to nanoparticles show efficient reactivity with halocarbons than nanoparticles [33]. It is found that the solvent plays an important role in degradation. Electrons released during the oxidation of 2-propanol in presence of metal clusters were abstracted by chlorocarbons to give Cl^- ions. The monolayer protecting the cluster core was replaced by Cl^- leading to destabilization. As a consequence of this, more stable AgCl crystals were obtained.

16.5.1.5 Discharge of metal ions in water

The presence of metal ions in the solution in water is an important concern, especially in the case of toxic chemicals. However, for Fe, Ag, TiO_2, ZnO, and CeO_2 derived materials, the solubility of the species concerned is low, and therefore ionic concentration is not beyond the permissible limits. This is very important in

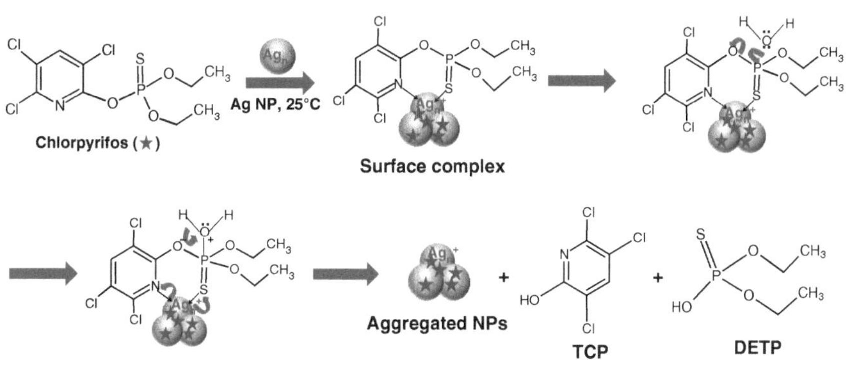

FIGURE 16.7

Schematic representation of degradation of CP on Ag NPs surface.

Adapted from Ref. [32]. Copyright (2012) American Chemical Society.

designing water quality applications. As water in general is contaminated with several ionic species, the nature of the metal ion can be different from case to case. The solubility and stability of ions is strongly dependent on pH. Therefore, the latter is an important parameter in deciding the type of the ion. The important issue of concern is the coverage of the surface with the oxides or hydroxides form inhibiting the electron transfer, thereby reducing the surface activity. Highly dispersed particles of smaller dimensions make it possible to use all the available metals for the chemistry concerned.

16.5.1.6 Effect of atomic arrangement on nanoparticle reactivity

While the bulk form of gold is expected to be face-centered cubic, the Au_{32} cluster is predicted to be an empty icosahedron. Control over synthesis mechanisms has enabled the fine-tuning of the properties of the material through changes in the lattice structure. Nanoparticles of different shapes have differences in the exposed surfaces. This also leads to differences in atomic distribution across the nanoparticle surface, which in turn affects the electron transfer rate kinetics between metal nanoparticles and corresponding organic species. Accordingly, the nanoparticles have been reported to have higher catalytic activity when they are present in the tetrahedral structure versus cubic or spherical structure [34]. This is attributed to enhancement in chemical reactivity at the sharp edges and corners and can easily be correlated with the number of atoms found at the respective places (Figure 16.8B). The 4.5-nm-sized tetrahedral nanoparticle is composed entirely of (111) facets with sharp edges and corners, which comprise ~28% of the total atoms and ~35% of the surface atoms. The 7.1-nm-sized cubic nanoparticle is composed entirely of (100) facets with a smaller fraction of atoms on their edges and corners, which comprise ~0.5% of total atoms and ~4% of surface atoms.

It is well known that high-index planes generally exhibit much higher catalytic activity than the most common stable planes, owing to high density of catalytic sites. On the contrary, due to high surface energy for high-index planes, crystals usually grow perpendicular to high-index planes. The novel nanoparticle synthesis mechanisms have however solved this problem. This has contributed significantly to improvement in high-efficiency catalysis [35].

As the particle size decreases and it reaches low nanometer dimensions, the atoms at the surface of the particle start to show enhanced vibrational motion normal to the surface, which creates enhanced surface mobility for the atoms. This is clearly evident through the changes in the physical properties of the material (e.g., the 2-nm gold particles melt at about 300 K whereas the bulk melting point is 1337 K).

16.5.2 Detection of ultralow pesticide contamination in water

While research efforts led to thorough investigations on multiple methods for degradation of such compounds, it also became clear that focused scientific research is required for easy-to-use ultralow concentration pesticide detection protocols. Many new scientific tools developed over the past 25 years have played a major

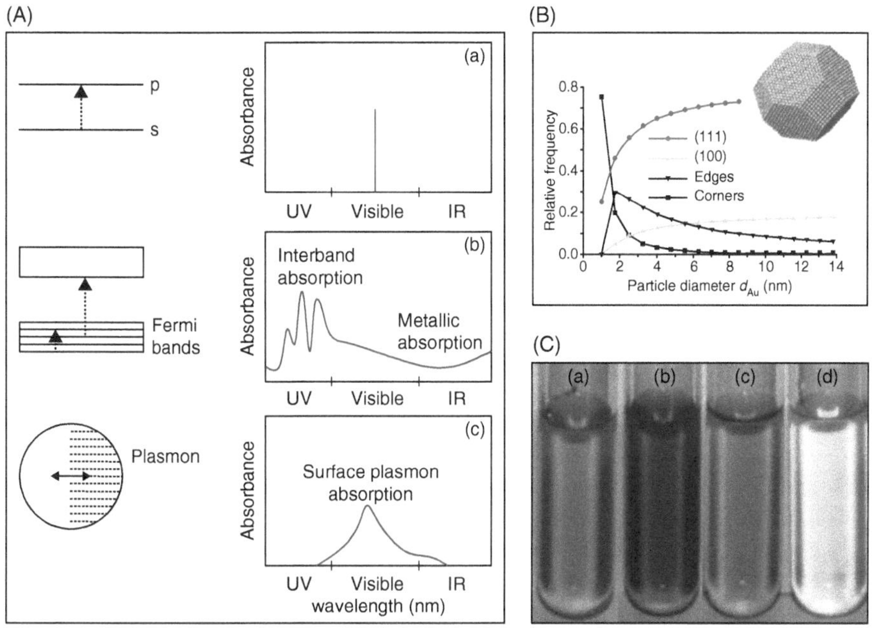

FIGURE 16.8

(A) Schematic representation of band transitions and corresponding UV–visible–NIR spectrum for (a) atom, (b) metal, and (c) clusters. (B) Graphical representation of percentage number of atoms found on various facets of a nanoparticle, calculated using a theoretical modeling scheme for cubo-octahedron model (shown in the inset) [36]. (C) Changes in the color of the nanoparticle solution on addition of organochlorine molecule. (a) Citrate stabilized gold nanoparticle solution, (b) gold nanoparticle solution incubated with endosulfan, (c) citrate stabilized silver nanoparticle solution, and (d) silver nanoparticle solution incubated with carbon tetrachloride.

(B) Reprinted with permission from Ref. [36]. Copyright (2002) with permission from Elsevier. (C) reproduced from Refs. [25,39]. Reproduced by permission of the Royal Society of Chemistry.

role in extremely low concentration detection of chemical species [37]. To facilitate widespread monitoring for pesticides in different sources of water, detection protocols need extensive simplification. Two approaches toward low concentration detection are described here.

16.5.2.1 Pesticide interaction with biomolecules

The approach to biomolecule-based pesticide detection is as follows: A biological recognition element (enzyme, antibody, receptor, or microorganism) is interfaced to a chemical sensor (i.e., analytical device to measure optical, amperometric, potentiometric properties) that together reversibly respond in a concentration-dependent manner to a chemical species.

In one of the methods utilizing pesticide interaction with biomolecules [38], acetyl cholinesterase (an essential enzyme for neurological functions) is studied along with organochlorine pesticides. It has been found that the activity state of the enzyme is impaired in the presence of pesticides. The phosphorylation reaction between the pesticide and AChE happens through interaction of the pesticide with the OH bond on the serine of AChE, leading to inhibition of the reaction with acetylcholine chloride (ACh). The pesticide as an inhibitor can be determined by means of measuring the kinetic performance of the initial velocity of the reaction catalyzed by this enzyme before and after an incubation step with the pesticide.

$$Acetylcholine + H_2O \xrightarrow{AChE} Acetic\ Acid + Chloine$$

16.5.2.2 Pesticide interaction with metal nanoparticles

It is a well-known phenomenon that metal nanoparticles undergo chemical state transformation during mineralization of pesticides, an example of which is the conversion of metal to metal salt (i.e., $Ag\ (0) + Cl^- \rightarrow AgCl + e^-$). It is also a known fact that chemistry at the nano-regime is extremely sensitive even at ultra-low concentrations.

In one of the methods to demonstrate the applicability of metal nanoparticles [39], endosulfan (an organochlorine pesticide) solution was mixed with a citrate stabilized gold nanoparticle solution. It was found that while there were significant changes in the UV-visible absorption characteristics of the solution, a more defining change was in the color of the solution. Intensity of color change is nearly quantitative and therefore offers a colorimetric method for detection (Figure 16.8C). It is also vital to understand that a broad range of pesticides can be targeted for colorimetric detection, at very low concentrations [25,39].

16.5.2.3 New approaches for pesticide detection

New approaches for low concentration detection of pesticides are an important aspect of water research. A number of techniques utilizing nanomaterial surfaces are being studied for sensing pesticide molecules; these techniques utilize changes in physical and chemical properties of nanomaterials (e.g., optical absorption, photoluminescence, surface enhancement in Raman signals) upon interaction with pesticide molecules. The fluorescence turn-on mechanism induced by simple ligand exchange has been developed to detect a range of the organophosphorothioate pesticides in nanomolar levels [40]. Here, green emission of CdTe quantum dots (QDs) was quenched by coordination of dithizone molecule with QDs owing to fluorescence resonance energy transfer (FRET). After the introduction of organophosphorothioate molecule, hydrolyzate of organophosphorothioate replaces the existing ligand which collapses the FRET resulting in the green emission of QDs. The emission intensity was directly proportional to the pesticide quantity. The detection limit of CP was as low as 0.1 nM. This method was tested to detect CP residues in apple down to 5.5 ppb level which is below US EPA standard (10 ppb). In another example, surface plasmon resonance for metal nanoparticles is utilized for pesticide detection. Metal nanoparticles (especially Ag and Au) are well known to enhance the Raman signals of analyte molecules at their hot spots due to electromagnetic field and chemical enhancement mechanisms. The shell thickness-dependent Raman enhancement of thiocarbamate and organophosphorous compounds by Au/Ag core–shell nanostructures has been demonstrated [41]. The core–shell nanoparticles are better candidates compared to mono metallic (Ag/Au) nanoparticles for surface-enhanced Raman spectroscopy (SERS). With the increase of silver thickness to 7 nm, there was a wide range and strong surface plasmon resonance in 320–560 nm range. This results in the strong enhancement in Raman signals of pesticides enabling the detection of pesticides in nanogram levels. Biosensors and electrochemistry-based sensors are also available.

16.5.3 Technology to product: a snapshot view

The ultimate objective of a scientific discovery is to create a value for society—the utilizable form of the technology in a form of a product. A scientific discovery is a fusion of novel materials, their characterization, understanding the process of interaction and postulation. Product development is a blend of technology optimization, engineering the prototype, cost-effectiveness, customer satisfaction, and strict adherence to quality control norms. The authors' work in the area of product development for pesticide removal from drinking water is utilized here for providing an overview of the development phase and associated challenges. The underlying technology utilized for this is the application of noble metal nanoparticles for the mineralization of pesticides.

The first phase of the product development program was to extend the applicability of metal nanoparticles for other pesticides of relevance to Indian drinking water. This step demonstrated the capability of metal nanoparticles for complete mineralization of organochlorine, organophosphate, and organosulfur pesticides (Figure 16.9A and B). The second phase of this program was to adapt the technology to the needs of the domestic customers, i.e., online, point-of-use drinking water purifier. This necessitated the study of nanoparticles incorporated on suitable supports, i.e., activated alumina, magnesia, or carbon (Figure 16.9C). It is important to understand the need for a suitable support, i.e., nonreactivity between support and nanomaterials, availability of large nanomaterial surface for pesticide adsorption, extremely low solubility of support in drinking water, and nontoxicity. This phase also involved the development of a pretreatment mechanism for contaminated water using activated carbon since the excess free chloride ions (and chlorine) adsorb on nanoparticle surfaces, rendering those inactive for the pesticide mineralization reaction. The third phase of this program was to develop a commercial-scale production plant for producing large quantities of noble metal nanoparticles coated on activated alumina (Figure 16.9D). The large-scale production of nanoparticles is usually challenged by issues such as precise control over

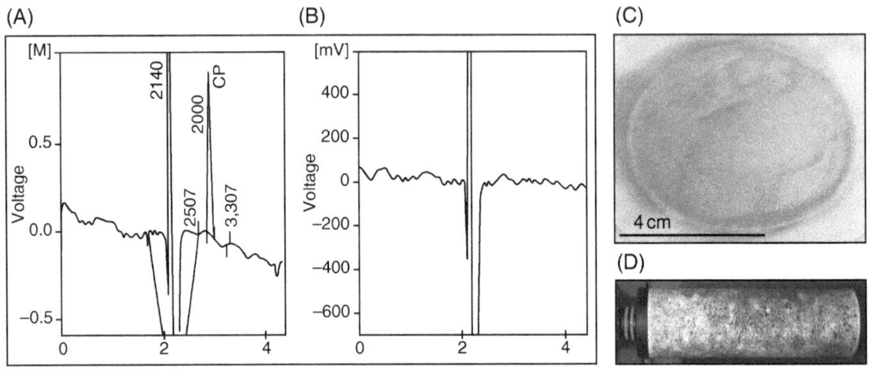

FIGURE 16.9

(A) Gas chromatogram of chlorpyrifos solution (50 ppb in water) extracted with hexane. The peak at 2.933 is that of chlorpyrifos (labeled CP) and that at 2.14 is that of the solvent. (B) Complete disappearance of the peak due to chlorpyrifos after passing through the nanomaterial column. (C) Silver nanoparticles coated on activated alumina and (D) a cartridge prepared from nanosilver supported on alumina, for use in an online domestic water purifier for pesticide removal.

A and B reproduced with permission from Ref. [45]. Copyright: American Scientific Publishers.

the processes, commercial viability of the process, availability of sophisticated instrumentation for production, and workforce development for continuous production. This program utilized wet chemistry-based synthesis for the production of noble metal nanoparticles, as it offers a larger degree of control over the production processes. One such innovation involved cost-effective and controlled immobilization of metal nanoparticles on alumina support [42–44]. The fourth phase of this program was to develop a suitable mechanism for environmentally responsible disposal of residuals and recovery of precious metals. This entailed study of cost-effective separation of the alumina support, the metal chloride nanoparticle and organic compounds bound on the nanomaterial surface.

16.6 Future directions

In the last 20 years, we have developed a thorough understanding of the chemistry behind nanomaterials, their synthesis, characterization, and new applications. This has been made possible due to the advent of new scientific instrumentation to study nanomaterials. This capability has also been enhanced greatly due to the availability of researchers above a critical mass, who have worked on the properties of nanoscale materials. Going forward, research in nanotechnology will proceed twofold: While on the one side, attempts are being made to develop novel materials with drastically different chemical and physical properties; on the other side, nano research is constantly being driven to develop new applications for societal problems, for many of which science has not been able to create a cost-effective, versatile, and simple solution. Availability of pure water, air, and food for all are the biggest of such problems before us. It is imperative that while the first avenue will require nanoscience to continuously evolve as an interdisciplinary science, the second avenue will require nanoscience to synergistically collaborate with engineering to create new values to society. In the near term, based on the current understanding of the chemistry behind pesticide mineralization, the following five objectives will need to be achieved:

1. extension of pesticide mineralization study to other pesticides of relevance;
2. development of novel materials for mineralization of pesticides, especially inorganic pesticides;
3. development of nano-based technology for pesticide removal into a commercial community-scale purifier and industrial purifier;
4. extension of low-cost mechanisms for low concentration detection of pesticides and similar chemical agents;
5. extension of water purification work using nanomaterials to a broader objective of how to protect the environment from such contaminants.

Looking at the landscape of human beings' interaction with organic compounds, it is likely that certain transformational technologies are also bound to be developed using nanomaterials. One such example is the development of a nanomaterial-based artificial renal dialyzer for use in low-cost, efficient, and swift filtration of renal discharge. The primary toxic organic compound filtered by the human body is also a carbamate pesticide (urea). If developed, this may play a pivotal role in saving the huge loss of human life due to kidney failure in many developing countries. Another such example is the development of a nanomaterial-based anti-chemical warfare agent kit. Among the group of toxic compounds used as chemical warfare agents, many of those contain C–Cl bonds, which can easily be mineralized using nanomaterials.

It is very important to realize that it has taken over 150 years for pesticides to appear in drinking water, simply because Mother Nature has been continuously filtering it out from groundwater resources. Looking at the pace of development in pesticide removal using man-made technologies, it is quite encouraging to see that nano-based technology has the capacity to offer solutions for many such problems. The first product for removing pesticides using nanomaterial reached commercialization in a period of 4 years, starting from basic research. During this period, many hurdles on the synthesis, formulation, and design of the products have been overcome. This is indeed a testimony to the capabilities of nanotechnology.

16.7 Summary

As we progress through rapid phases of industrialization and economic upliftment, it is only natural to expect increasing ecological contamination, even though equilibrium with Nature is the most desired form. Science and technology will continue to innovate and discover novel materials with drastically different properties, against some of which we may not have any natural defense. The discovery of pesticides and other similar synthetic organic compounds is a representative example. Hence, it becomes clear that without disturbing the path of scientific discovery for novel materials and their commercialization, we need to focus our attention on developing preventive technologies for inhibiting their detrimental impact on the environment.

While traditional synthetic protocols have helped in controlling the spread of toxic species, the finest method has been practiced by Mother Nature for ages. Although we have just started to understand and implement the chemistry of biological degradation, it is certain that Nature's path will remain by far the most desired method for detoxification of the environment. The underlying theme of nanotechnology has been immensely influenced by Nature's methodology: complete mineralization of pesticides to nontoxic forms, environmentally benign nature of reaction end products, and large conversion efficiency with minimum catalysts. Thus one thing becomes clear to us. To do things—whether chemical synthesis, energy transfer, or water filtration in the most efficient fashion—we need to go by Nature's way; the path of nanotechnology has to be toward achieving that.

A multitude of factors have been attributed to chemical reactivity enhancement of nanoparticles: small particle size leading to increase in specific surface area, increase in the surface energies for easier delocalization of electrons and perfection in atomic organization. Besides increased reactivity, availability of different forms of nanomaterials (e.g., nanooxides, nanometals, dendrimers, nanoclay) and enhanced knowledge about nanoparticle incorporation on support structures has opened new avenues for science to thoroughly investigate the potential of nanotechnology for large-scale environmental detoxification.

Where do we go from here? The discovery of novel nanomaterials with radically different properties, study of nanomaterials for existing societal problems, technology transfer of nano research to commercial applications, and penetration of nanotechnology in human life are all happening. In addition, science has to clarify an important and simple question related to human existence: At any point in future, can widespread use of nanomaterials also raise unacceptable risks to human life? [46] Currently science is clueless to this question even though there are more reasons to believe in the environment-friendliness of nanomaterials. The same properties of nanomaterials (e.g., increased surface area, chemical reactivity) which result in their unique ability to degrade toxic species present in the environment may also render them toxic to humans. While it is suggested that the small particle size of the nanomaterial will provide them ready access to the cytoplasm and thus interfere with bioreactions, it is also important to understand that cellular gateways are also specific to the nature of the incoming species. Nevertheless, the issue is also compounded by the sudden increase in the variety of novel nanomaterials (with a variety of functional groups) and unavailability of scientific tools for thorough study of their impact on the human body. Thus the issue of nanotoxicity is being approached from two pathways: interaction of nanoparticles for disease detection/drug delivery and animal toxicity studies of nanomaterials. The issue of nanotoxicity is of strategic interest as it immensely impacts the future of nanotechnology by influencing the government, policy makers, regulators, society, and science.

References

[1] The WHO recommended classification of pesticides by hazard and guidelines to classification [Internet]. [cited 2007 November 08]. Available from: <http://www.who.int/ipcs/publications/pesticides_hazard_rev_3.pdf>.

[2] Pesticides Industry Sales and Usage: 2000 and 2001 Market Estimates [Internet]. [cited 2007 November 14]. Available from: <http://www.epa.gov/oppbead1/pestsales/01pestsales/market_estimates2001.pdf>.

[3] G. Shukla, A. Kumar, M. Bhanti, P.E. Joseph, A. Taneja, Organochlorine pesticide contamination of ground water in the city of Hyderabad, Environ. Int. 32 (2006) 244–247.

[4] H.B. Mathur, H.C. Agarwal, S. Johnson, N. Saikia, Analysis of pesticide residues in blood samples from villages of Punjab, Down to Earth 14 (2) (2005) 27–38.

[5] H.B. Mathur, S. Johnson, A. Kumar, Analysis of pesticide residues in soft drinks, Down to Earth 15 (6) (2006) 28–36.

[6] Centre for Science and Environment report on pesticide contamination in Kasargod [Internet]. [cited 2007 November 17]; Available from: <www.cseindia.org/html/lab/lab_pesticide_analysis_result_kasargod.htm>.

[7] T. Pradeep, Anshup, Noble metal nanoparticles for water purification: a critical review, Thin Solid Films 517 (2009) 6441–6478.

[8] M.U. Sankar, S. Aigal, A. Chaudhary, Anshup, S.M. Maliyekkal, A.A. Kumar, et al., Biopolymer reinforced synthetic granular nanocomposites for affordable point-of-use water purification, Proc. Natl. Acad. Sci. 110 (2013) 8459–8464.

[9] B. Tansel, P.S.E.M. Nagarajan, study of phenolphthalein adsorption on granular activated Carbon, Adv. Environ. Res. 8 (2004) 411–415.

[10] J.A. Bumpus, M. Tien, D. Wright, S.D. Aust, Oxidation of persistent environmental pollutants by a white rot fungus, Science (1985) 1434–1436.

[11] J.C. Hall, R.E. Hoagland, R.M. Zablotowicz, Pesticide Biotransformation in Plants and Micro-organisms: Similarities and Divergences, first ed., Oxford University Press, UK, 2001.

[12] K.V. Plakas, A.J. Karabelas, T. Wintgens, T. Melin, A study of selected herbicides retention by nanofiltration membranes—The role of organic fouling, J. Membr. Sci. 284 (2006) 291–300.

[13] G.C. Bond, in: G. Ertl, H Knozinger, J. Weitkamp (Eds.), Handbook of Heterogeneous Catalysis, VCH, Weinheim, 1997, Vol. 2, Section 3, pp. 752–770.

[14] H. Hofmeister, P.T. Miclea, M. Steen, W. Morke, H. Drevs, Structural characteristics of oxide nanosphere supported metal nanoparticles, Top. Catal. 46 (2007) 11–21.

[15] S. Phadtare, V.P. Vinod, K. Mukhopadhyay, A. Kumar, M. Rao, R.V. Chaudhari, et al., Immobilization and biocatalytic activity of fungal protease on gold nanoparticle-loaded zeolite microspheres, Biotechnol. Bioeng. 85 (2004) 629–637.

[16] P. Serp, R. Feurer, Y. Kihn, P. Kalck, J.L. Faria, J.L. Figueiredo, Novel carbon supported material: highly dispersed platinum particles on carbon nanospheres, J. Mater. Chem. 11 (2001) 1980–1981.

[17] W.X. Zhang, C.B. Wang, H.L. Lien, Treatment of chlorinated organic contaminants with nanoscale bimetallic particles, Catal. Today 40 (1998) 387–395.

[18] A. Bakandritsos, A. Simopoulos, D. Petridis, Iron changes in natural and Fe(III) loaded montmorillonite during carbon nanotubes growth, Nanotechnology 17 (2006) 1112–1117.

[19] N.M. Nagy, J. Konya, M. Beszeda, I. Beszeda, E. Kalman, Z.s. Keresztes, et al., Physical and chemical formations of lead contaminants in clay and sediment, J. Colloid. Interface. Sci. 263 (2003) 13–22.

[20] R.W.J. Scott, H. Ye, R.R. Henriquez, R.M. Crooks, Synthesis, characterization, and stability of dendrimer-encapsulated palladium nanoparticles, Chem. Mater. 15 (2003) 3873–3878.

[21] S.M. Maliyekkal, T.S. Sreeprasad, D. Krishnan, S. Kouser, A.K. Mishra, U.V. Waghmare, et al., Graphene: a reusable substrate for unprecedented adsorption of pesticides, Small 9 (2012) 273–283.

[22] X. Liu, H. Zhang, Y. Ma, X. Wu, L. Meng, Y. Guo, et al., Graphene-coated silica as a highly efficient sorbent for residual organophosphorus pesticides in water, J. Mater. Chem. A. 1 (2013) 1875–1884.

[23] S.S. Gupta, T.S. Sreeprasad, S.M. Maliyekkal, S.K. Das, T. Pradeep, Graphene from sugar and its application in water purification, ACS Appl. Mater. Interfaces 4 (2012) 4156–4163.

[24] S. Yang, S. Luo, C. Liu, W. Wei, Direct synthesis of graphene–chitosan composite and its application as an enzymeless methyl parathion sensor, Colloid Surf. B. 96 (2012) 75–79.

[25] A.S. Nair, T. Pradeep, Halocarbon mineralization and catalytic destruction by metal nanoparticles, Curr. Sci. 84 (2003) 1560–1564.

[26] M.J. Rosemary, A.S. Nair, PG. Reddy, I. MacLaren, S. Baskaran, T. Pradeep, Fluorescent nanobubbles, Trans. Mater. Res. Soc. Jpn. 29 (2004) 33–36.

[27] M.J. Rosemary, I. MacLaren, T. Pradeep, Investigations of the antibacterial properties of ciprofloxacin@SiO_2, Langmuir 22 (2006) 10125–10129.

[28] A.S. Nair, V. Suryanarayanan, R.T. Tom, T. Pradeep, Porosity of core shell nanoparticles, J. Mater. Chem. 14 (2004) 2661–2666.

[29] J.M. Buriak, Chemistry with nanoscale perfection, Science 304 (2004) 692–693.

[30] T.S. Sreeprasad, A.K. Samal, T. Pradeep, Body or tip controlled reactivity of gold nanorods and their conversion to particles through other anisotropic structures, Langmuir 23 (2007) 9463–9471.

[31] G.C. Bond, D.T. Thompson, Catalysis by gold, Catal. Rev.-Sci. Eng. 41 (1999) 319–388.

[32] M.S. Bootharaju, T. Pradeep, Understanding the degradation pathway of the pesticide, chlorpyrifos by noble metal nanoparticles, Langmuir 28 (2012) 2671–2679.
[33] M.S. Bootharaju, G.K. Deepesh, T. Udayabhaskararao, T. Pradeep, Atomically precise silver clusters for efficient chlorocarbon degradation, J. Mater. Chem. A. 1 (2013) 611–620.
[34] R. Narayanan, M.A. El-Sayed, Shape-dependent catalytic activity of platinum nanoparticles in colloidal solution, Nano Lett. 4 (2007) 1343–1348.
[35] N. Tian, Z.Y. Zhou, S.G. Sun, Y. Ding, Z.L. Wang, Synthesis of tetrahexahedral platinum nanocrystals with high-index facets and high electro-oxidation activity, Science 316 (2007) 732–735.
[36] S. Schimpf, M. Lucas, C. Mohr, U. Rodemerck, A. Bruckner, J. Radnik, et al., Supported gold nanoparticles: in-depth catalyst characterization and application in hydrogenation and oxidation reactions, Catal. Today 72 (2002) 63–78.
[37] J. Slobodnik, B.L.M. van Baar, U.A.T. Brinkman, Column liquid chromatography-mass spectrometry: selected techniques in environmental applications for polar pesticides and related compounds, J. Chromatogr. A 703 (1995) 81–121.
[38] S. Andreescu, J.L. Marty, Twenty years research in cholinesterase biosensors: From basic research to practical applications, Biomol. Eng. 2 (2006) 1–15.
[39] A.S. Nair, R.T. Tom, T. Pradeep, Detection and extraction of endosulfan by metal nanoparticles, J. Environ. Monit. 5 (2003) 363–365.
[40] K. Zhang, Q. Mei, G. Guan, B. Liu, S. Wang, Z. Zhang, Ligand replacement-induced fluorescence switch of quantum dots for ultrasensitive detection of organophosphorothioate pesticides, Anal. Chem. 82 (2010) 9579–9586.
[41] B. Liu, G. Han, Z. Zhang, R. Liu, C. Jiang, S. Wang, et al., Shell thickness-dependent raman enhancement for rapid identification and detection of pesticide residues at fruit peels, Anal. Chem. 84 (2012) 255–261.
[42] T. Pradeep, A.S. Nair with IIT Madras, A method of preparing purified water from water containing pesticides (chlorpyrifos and malathion), Indian patent (2007) 6 (7), June 2, 2006.
[43] Extraction of malathion and chlorpyrifos from drinking water by nanoparticles, US 7,968,493 A method for decontaminating water containing pesticides, EP 17,15,947 with IIT Madras.
[44] A. Sreekumaran Nair, T. Pradeep, A method to produce supported noble metal nanoparticles in commercial quantities for drinking water purification, 1879/CHE/2007 dated December 10, 2007 with IIT Madras and Eureka Forbes Limited.
[45] A.S. Nair, T. Pradeep, Extraction of chlorpyrifos and malathion from water by metal nanoparticles, J. Nanosci. Nanotechnol. 7 (2007) 1871–1877.
[46] A.D. Maynard, RJ. Aitken, T. Butz, V. Colvin, K. Donaldson, G. Oberdorster, et al., Safe handling of nanotechnology, Nature 444 (2006) 267–269.

Further Reading

[1] R. Carson, Silent Spring. 40th Anniversary, Mariner Books, USA, 1962.
[2] G.T. Miller, Sustaining the Earth, seventh ed., Thompson Advantage Books, USA, 2004.

[3] T. Colborn, D. Dumanoski, J.P. Myers, Our Stolen Future, first ed., Penguin Group, USA, 1997.

[4] The Nobel Prize in Physiology or Medicine 1948 [Internet]. [cited 2007 November 18]. Available from: <http://nobelprize.org/nobel_prizes/medicine/laureates/1948/>.

[5] 2002 Pesticide use maps [Internet]. [cited 2007 December 15]. Available from: <http://water.usgs.gov/nawqa/pnsp/usage/maps/compound_listing.php?year = 02>.

CHAPTER

Nanomaterials-Enhanced Electrically Switched Ion Exchange Process for Water Treatment

17

Yuehe Lin[1], Daiwon Choi[2], Jun Wang[2] and Jagan Bontha[2]

[1]Chemistry Department, King Abdulaziz University, Jeddah, Saudi Arabia; School of Mechanical and Materials Engineering, Washington State University, Pullman, WA, USA

[2]Pacific Northwest National Laboratory, Richland, WA, USA

17.1 Introduction

Electrically switched ion exchange (ESIX) technology combines ion exchange and electrochemistry to provide a selective, reversible method for the removal of target species from wastewater. In this technique, an electroactive ion exchange layer is deposited on a conducting substrate, and ion uptake and elution are controlled directly by modulation of the potential of the layer. ESIX offers the advantages of no secondary waste generation. We have improved upon the ESIX process by modifying the conducting substrate with carbon nanotubes (CNTs) prior to the deposition of the electroactive ion exchanger. The nanomaterial-enhanced electroactive ion exchange technology will remove cesium-137, chromate, and perchlorate rapidly from wastewater. The high porosity and high surface area of the electroactive ion

Street, Sustich, Duncan and Savage. Nanotechnology Applications for Clean Water, 2nd Edition.
 DOI: http://dx.doi.org/10.1016/B978-1-4557-3116-9.00017-2

exchange nanocomposites results in high loading capacity and minimizes interferences for nontarget species. Since the ion adsorption/desorption is controlled electrically without generating a secondary waste, this electrically active ion exchange process is a novel technology that will greatly reduce operating costs.

17.2 Principle of the electrically switched ion exchange technology

The ESIX technology, being developed at the Pacific Northwest National Laboratory (PNNL), provides a more economical remediation alternative [1–5]. It is a novel technique that combines both the principles of electrochemistry and ion exchange for the removal of toxic ions from waste effluents. This technology utilizes the redox reaction of electrically conductive material to regulate the uptake or elution of various ions to/from and into aqueous solution to maintain charge neutrality. By modulating the potential and time of the ESIX material, selective ion exchange of various cations or anions can be achieved. The main advantage of ESIX over conventional systems in wastewater treatment applications is in selectivity and reversibility for ion separation that lowers costs and minimizes secondary waste generation typically associated with conventional regenerated ion exchange processes.

For example, cation separation is done by applying a cathodic potential to the film sufficient to induce electrochemical reduction of an electroactive species, X, which forces a cation from the waste solution into the film following Equation 17.1. Cation unloading is done by applying an anodic potential to the electrodes; reoxidation of the electroactive film instigates cation out of the film and into the external elution solution following reversed direction of Equation 17.1 [1–4]. A similar mechanism applies to anion removal following Equation 17.2. Thus, the use of regenerant chemicals and the multiple steps following regeneration, each resulting in a huge amount of secondary waste that needs to be disposed of, is eliminated. Electrically switched ion exchange enables the reuse of the ion exchange medium up to fifteen hundred cycles as opposed to conventional ion exchange where inorganic ion exchangers are more often used only once and discarded.

$$e^- + X + M^+ \overset{V}{\leftrightarrow} |X^- M^+| \tag{17.1}$$

$$X + M^- \overset{V}{\leftrightarrow} |X^+ M^-| + e^- \tag{17.2}$$

As in conventional ion exchange, selectivity of X for the metal ion of interest is important whenever the waste solution contains more than one cation. If X has a greater selectivity for the cation of interest M_1^+ than a second cation M_2^+, the film may first be "activated" by reduction in the presence of a solution of M_2^+. Introducing the waste solution results in uptake of M_1^+ by ion exchange for M_2^+ where M_2^+ is displaced into the waste solution Equation 17.3.

$$M_2^+ X^- + M_1^+ \overset{V}{\leftrightarrow} |M_1^+ X_-| + M_3^+ \tag{17.3}$$

Competition for binding sites will occur, and loading will be driven by thermodynamics. Therefore, to successfully remove the target cation, films must bind that ion preferentially, which depends on the electroactive property of X.

17.3 Nanomaterials-enhanced electrically switched ion exchange for removal of radioactive cesium-137

Significant amounts of cesium-137, an important fission by-product, are present in radioactive liquid waste generated during the reprocessing of nuclear fuel by the Purex process. Removing the long-lived ($T_{1/2} = 32$ years) radioisotope of cesium from the radioactive waste is a challenging project and several techniques, such as ion exchange, sorption, solvent extraction, and precipitation processes, have been investigated. One such process is ESIX using electroactive nickel hexacyanoferrates (NiHCF) developed at PNNL.

Transition metal hexacyanoferrates (MHCFs) are an important class of inorganic polynuclear mixed-valence compounds because of their interesting properties including electrocatalysis, electrochromicity, ion exchange selectivity, sensing, and magnetism. Therefore, preparation and application of Prussian blue (PB) and its analogous forms such as samarium hexacyanoferrate, NiHCF, and cobalt hexacyanoferrate have been extensively studied. Also, thin films of PB and related analogous complex salts are also promising candidates for separating membranes capable of ion sieving and electrochemical-controllable ion exchange because of their zeolitic structure. Hexacyanoferrates are well-known ion exchangers with high selectivity over cesium in concentrated sodium solutions but hexacyanoferrates used in conventional packed column ion exchange are difficult to elute [1–5]. Therefore, columns are normally discarded once used. However, when hexacyanoferrates are applied as electroactive films, depending on the film deposition formulation and substrate, highly reversible uptake and elution of cesium can be achieved from fifteen hundred to over four thousand cycles [2,4]. When a cathodic potential is applied to the NiHCF film, Fe^{3+} (ferricyanide) is reduced to the Fe^{2+} state (ferrocyanide); thus, a cation must be intercalated into the film to maintain its charge neutrality. In practice, the reduction step is usually conducted in a sodium solution as shown in Equation 17.4. The film is then exposed to the waste solution containing cesium, which loads into the film via ion exchange for sodium following Equation 17.5. When an anodic potential is applied, a cesium cation is released from the film as shown in Equation 17.6. Therefore, NiHCF has been coated on the electrode as shown in Fig. 17.1 for ESIX of Cs [2].

$$CsNiFe^{III}(CN)_6 + Na^+ + Na^+ + e^- \rightarrow CsNaNiFe^{II}(CN)_6 \quad (17.4)$$

$$CsNaNiFe^{II}(CN)_6 + Cs^+ \rightarrow Cs_2NiFe^{II}(CN)_6 + Na^+ \quad (17.5)$$

$$Cs_2NiFe^{II}(CN)_6 \rightarrow CsNiFe^{III}(CN)_6 + Cs^+e^- \quad (17.6)$$

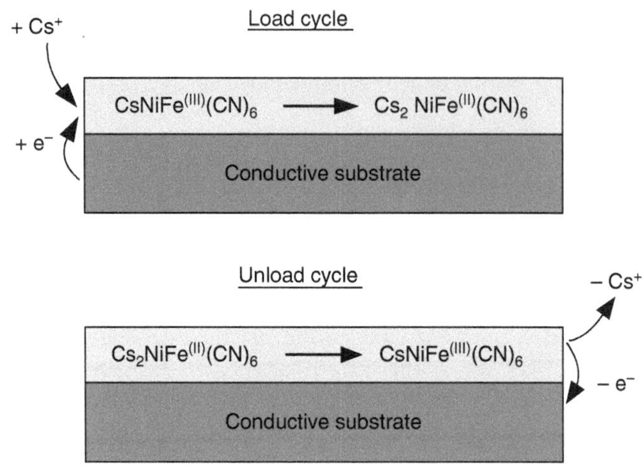

FIGURE 17.1

Electrically switched ion exchange (ESIX) mechanism of nickel hexacyanoferrate thin film electrode for Cs^+ loading and unloading.

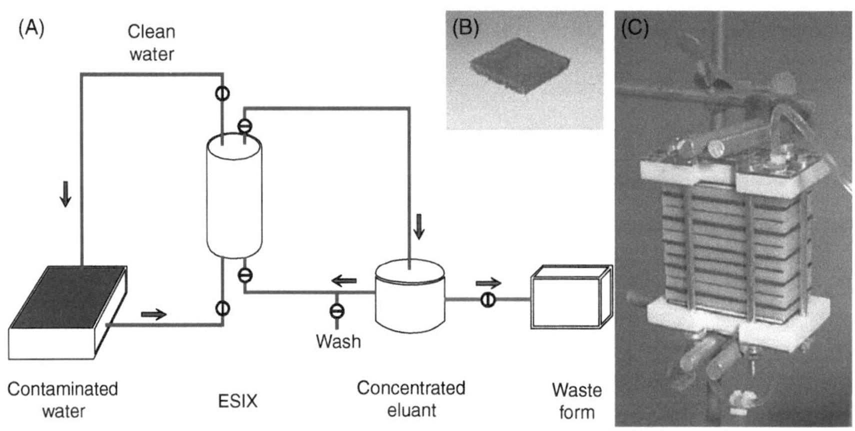

FIGURE 17.2

Schematic illustration of (A) a bench-scale electrically switched ion exchange (ESIX) system based on (B) nickel hexacyanoferrate coated nickel foam electrodes, and (C) five-electrode cell stacks used as the ESIX system.

Figure 17.2 shows bench-scale flow cell testing of ESIX systems developed at PNNL using NiHCF films deposited on high surface area nickel foam electrodes with a nominal surface area per volume of 40 cm^2/cm^3 (60 ppi) [4]. The flow tests using NiHCF material were performed to evaluate the feasibility of the approach as an alternative to conventional ion exchange processes.

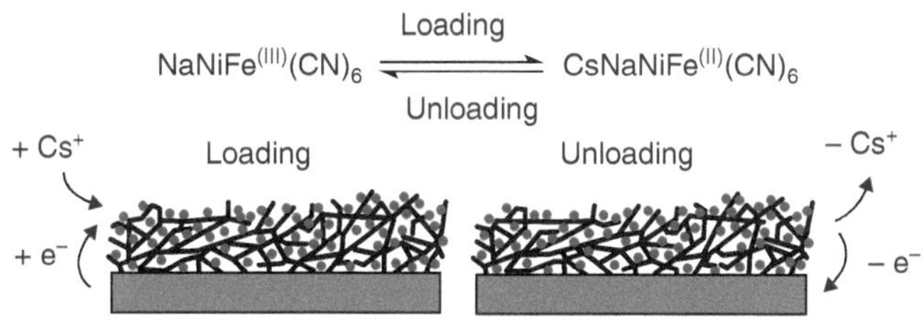

FIGURE 17.3

Schematic illustration for electrically switched ion exchange based on carbon nanotubes, polyanline, and nickel hexacyanoferrate nanocomposite.

Data from [5] with permission.

Bench-scale ESIX flow system studies showed no change in capacity or performance of the ESIX films at a flow rate up to 113 bed volumes/h (BV/h). A comparison of results for a stacked five-electrode cell versus a single-electrode cell showed enhanced breakthrough performance. In the stacked configuration, breakthrough began at approximately 120 BV for a feed containing 0.2 ppm cesium at a flow rate of 13 BV/h. A case study for the KE Basin (a spent nuclear fuel storage basin) on the Hanford Site demonstrated that wastewater could be processed continuously with minimal waste generation, reduced disposal costs, and lower capital expenditures.

Novel nanocomposite materials comprised of high surface area of CNTs, chemically stable polyaniline (PANI), and ion exchange properties of NiHCF have been synthesized through electrodeposition [5,6]. Figure 17.3 illustrates the concept of electrically switched ion exchange based on CNT/PANI/NiHCF nanocomposites. High surface area of the porous CNT film leads to the high loading capacity for the NiHCF nanoparticles, which in turn leads to the high ion exchange capacity of the NiHCF/CNT nanocomposite film. The presence of the PANI polymer film further improves the stability of the nanocomposite film, which undergoes more than five hundred cycles retaining 92 percent of its initial capacity.

Recently, NiHCF nanotubes have been fabricated by an electrokinetic method based on the distinct surface properties of porous anodic alumina [7]. By this method, nanotubes formed rapidly with the morphologies replicating the nanopores in the template. The electrochemical measurements show that the NiHCF nanotubes exist only in the form of $K_2Ni[Fe(CN)_6]$ with excellently stable cesium-selective ion exchange ability due to single composition and unique nanostructure. NiHCF nanotubes modified electrodes retains 95.3 percent of its initial value after five hundred potential cycles. Even after fifteen hundred and three thousand cycles, the NiHCF nanotubes still retain 92.2 percent and 82.9 percent, respectively, of their ion exchange capacity.

17.4 Nanomaterials-enhanced electrically switched ion exchange for removal of chromate and perchlorate

Wide use of heavy metals and its related compounds by industries has resulted in the pollution of the environment. These inorganic toxins are of considerable concern since they are nonbiodegradable, highly toxic with possible carcinogenic effect. Among the heavy metals, chromium is one of the most widely used metals in many of the industrial processes such as tanning, electroplating, making printed circuit boards, painting, and steel fabrication. Most of the chromium is discharged into aqueous waste as Cr(III) and Cr(VI). Cr(VI), which is the more toxic of the two, is present as dichromate ($Cr_2O_7^{2-}$), chromate (CrO_4^{2-}), or hydrogen chromate ($HCrO_4^{-}$). The current maximum contamination level (MCL) for total chromium in drinking water in the United States is regulated by the United States Environmental Protection Agency (EPA) at 100 μg/L.

Perchlorate ion also poses significant health concern since it can block the uptake of iodine in the thyroid gland and thereby affect the production of thyroid hormones. Perchlorate anion is a critical component in combat and training munitions. In addition, perchlorate salts are extensively used as chemical reagents in the production of leather, rubber, fabrics, paints, and aluminum. As a result, perchlorate contamination is now recognized as a widespread concern affecting many water utilities. Recently, EPA, based on a recommendation by the National Research Council (NRC), set the safe dose for perchlorate at 0.70 μg/kg of body weight per day. However, due to its solubility and nonreactivity, perchlorate is a very stable substance in aquatic systems and is, therefore, difficult to remove.

Conventional chromium and perchlorate removal methods include adsorption, chemical precipitation, biological degradation, ion exchange, and electrochemical method. However, these technologies have technical limitations, generate substantial secondary wastes, and are costly. Therefore, innovative, cost-effective, and green technology for the treatment of perchlorate needs to be developed. Electrochemical ion exchange treatment is a promising technique that offers several advantages over other techniques for remediation of metals from contaminated effluents or wastewaters.

For chromate removal, arc-assisted carbon (AC), activated carbon (RC), and carbon aerogel have been used for ion exchange electrode [8,9]. Hexavalent chromium, [Cr(VI)] may exist in the aqueous phase in different anionic forms, such as $Cr_2O_7^{2-}$, CrO_4^{2-}, or $HCrO_4^{-}$, with total chromate concentrations and pH dictating which particular chromate species is predominant. Equation 17.7 shows that equilibrium relationship between the different chromium anions:

$$2CrO_4^{2-} + 2H \leftrightarrow (2HCrO_4^{-}) \leftrightarrow Cr_2O_7^{2-} + H_2O \tag{17.7}$$

Carbon obtained by the arc-assisted evaporation of graphite rods (AC) exhibited a high selectivity toward anions of hexavalent chromium from aqueous solutions in the pH range of 3–9 [8]. There is almost no removal of trivalent chromium or other metal cations such as Pb^{+} and Zn^{+} at low pH values (i.e., <9) since trivalent chromium is present as a cation in solution whereas hexavalent chromium exists as an anion. This is consistent with the hypothesis that the surface of AC carries positive charges that facilitate the removal of the anion of the hexavalent chromium but prevent the removal of metal cations. At high pH, the positive surface charges are neutralized by the presence of OH^{1}; hence, the metal cation removal efficiency is increased by an ion exchange adsorption mechanism.

Commercial carbon (RC), on the other hand, has a negatively charged surface due to the presence of oxygen functional groups of "oxo" (C_xO_y) type produced during its the activation at a very high temperature. Therefore, metal cation removal is facilitated. An increase in the pH of the solution contributes to an increase of the surface negative charge density, resulting in an even greater removal efficiency of lead, zinc, and trivalent chromium metal cations.

Recently, carbon aerogel electrodes were applied for chromium removal from wastewater since carbon aerogel possess unique thermal, mechanical, and electrical properties, which are directly related to their unusual nanostructure composed of interconnected particles with microscopic interstitial pores [9]. Carbon aerogel is an ideal electrode material because of its low electrical resistivity (≤40 mΩcm), high specific surface area (400–1100 m^2/g), and controllable pore size distribution (50 nm). The effect of pH (2–7), concentration 2–8 mg/l, and charge 0.3–1.3 A h was investigated where the chromium ion removal was significantly increased at reduced pH and high charge conditions. The metal concentration in the wastewater can be reduced by 98.5 percent under high charge (0.8A h) and acidic conditions (pH = 2). Also, the elution solution can be used repeatedly, thereby minimizing secondary wastes at the largest degree and reducing costs greatly.

For perchlorate removal, electrically conductive polypyrrole (PPy) polymer as an electroactive ion exchange layer has been deposited onto a conducting substrate [10,11]. Figure 17.4 shows the schematic illustration of PPy preparation and the electrically controlled anion exchange for the separation of the perchlorate ion

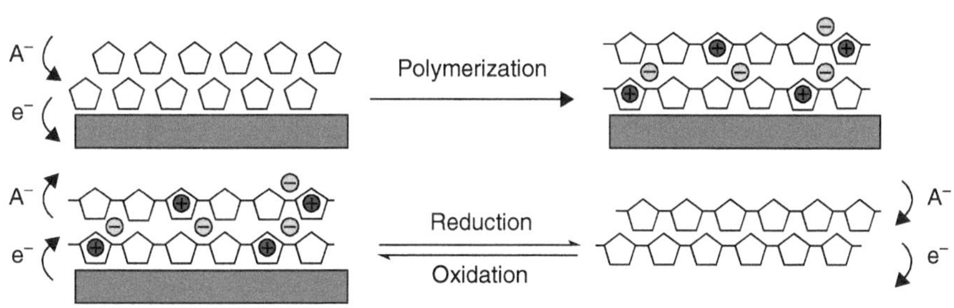

FIGURE 17.4

Schematic illustration for the polymerization of electrically conductive polypyrrole (PPy) and anion intake and elution with the oxidation and reduction of PPy film. Three or four pyrrole units involving one positive charge are considered in oxidized PPy.

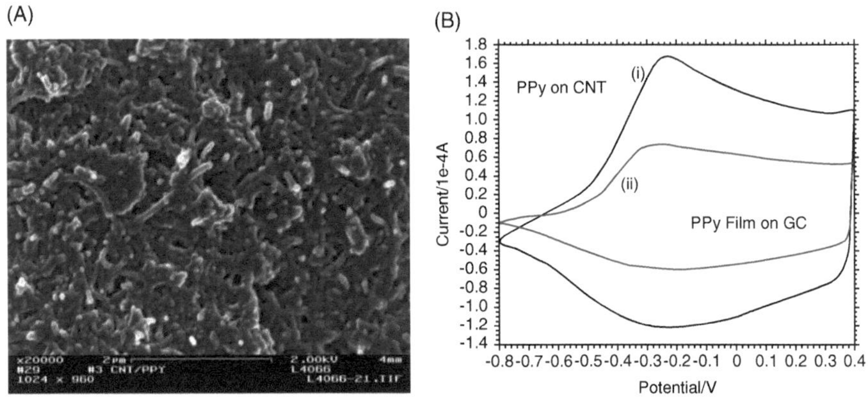

FIGURE 17.5

(A) Scanning electron microscope image of polypyrrole (PPy) film on carbon nanotube (CNT)/ glassy carbon (GC) electrode surface prepared by controlling the potential at 0.7 V, 100 seconds, and (B) a comparison of the tenth cycle behavior of (i) CNT–PPy and (ii) GC/PPy.

Data from [11] with permission.

from wastewaters using PPy. The pyrrole monomer polymerizes into PPy during oxidation in supporting electrolyte with anions. After electrode has been coated with PPy, anions can be reversibly ion exchanged by applying anodic and cathodic potentials. Therefore, the ClO_4^- ion uptake occurs during electrochemical oxidation of the electroactive species by applying an anodic potential on the film in the solution containing ClO_4^-, which forces the ClO_4^- from the waste solution into the film. Elution occurs when the potential is switched to cathode, forcing the ClO_4^- out of the film into the elution solution. Therefore, the eluant can be repeatedly used to significantly reduce the quantity of secondary waste generated. Although the ion exchange property of PPy has been realized, its capacity is limited by only one positive charge per three or four pyrrole units. Besides, it is difficult for the dopant anions to diffuse in and out of the polymer due to the poor mass transfer properties of the PPy films. At PNNL, in order to improve the mass transfer properties of the PPy, and thereby the ion exchange capacity, PPy was deposited on to high surface area CNT substrates. Carbon nanotubes are one of the novel nanostructure forms of carbon with very high surface area and good conductivity that offers an idea matrix for depositing PPy film. According to a survey by Cientifica, the world's leading nanotechnology information company, CNT costs are expected to decrease by a factor 10–100 in the next 5 years making CNT-composite a more practical solution for large-scale waste processing.

Figure 17.5 shows the microstructure and cyclic voltammogram of CNT/PPy nanocomposite electrode. Also shown in Fig. 17.5B is cyclic voltammogram of PPy deposited on a glassy carbon (GC) electrode. It can be seen that significantly higher electrochemical current density was delivered by creating higher active area using CNT substrate. The CNT/PPy electrode showed higher perchlorate ion

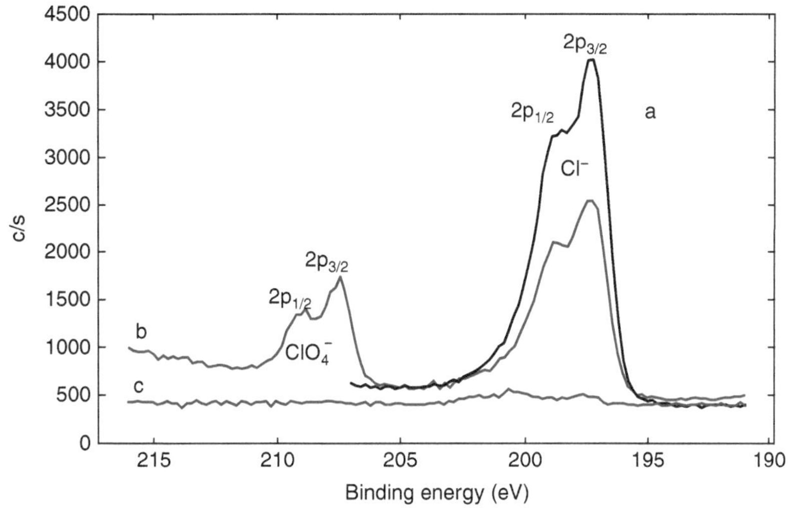

FIGURE 17.6

High-resolution X-ray photoelectron spectroscopy Cl spectrum of polypyrrole (PPy) film prepared (a) in 0.2 M NaCl, (b) after control of the electrode at 0.4 V for 300 seconds in a solution containing 0.02 M $NaClO_4$ and 0.2 M NaCl, and (c) after a cathode potential at −0.8 V was applied on the film for 300 seconds in a solution of 0.2 M NaCl.

Data from [10] with permission.

selectivity over chlorine ion and X-ray photoelectron spectroscopy analysis (Fig. 17.6) indicated reversible elution of absorbed ions, making PPy a promising candidate material for perchlorate removal [10,11]. When ion-exchange capacity and reversibility of pure PPy and CNT/PPy nanocomposite thin films were compared using electrochemical quartz microbalance (EQCM) in NaCl, $NaClO_4$, $NaNO_3$, Na_2SO_4, and $NaHCO_3$ electrolyte solution, the results indicated that the ion exchange kinetics of PPy improved by more than ten times when CNT substrate was used. In addition, the ion exchange reached saturation within 100 seconds for CNT/PPy. Finally, the anion loading capacity (mole percent) of CNT/PPy was in order of $Cl^- > ClO_4^- > NO_3^- > SO_4^{2-} > CO_3^{2-}$ whereas selectivity of anions was in order of $ClO_4^- > NO_3^- > SO_4^{2-} > Cl^- > CO_3^{2-}$.

17.5 Conclusions

Electrically switched ion exchange technology has recently been applied for removing cesium, chromate, and perchlorate ions from wastewater. The efficiency of these ESIX systems can be significantly improved through nanotechnology by providing better electrochemical stability and a larger contact area with wastewater. Therefore, the combination of novel electroactive ion exchange material with nanotechnology will lead to the development of more efficient and economic wastewater treatment systems based on ESIX technology.

Acknowledgments

This work is supported by the US Department of Defense Strategic Environmental Research and Development Program (Project No ER-1433). The research described in this chapter was performed at the Environmental Molecular Science Laboratory, a national scientific facility sponsored by DOE's Office of Biological and Environmental Research and located at Pacific Northwest National Laboratory (PNNL). PNNL is operated by Battelle for DOE under Contract DE-AC05-76L01830. The authors would like to acknowledge Dr. Mike A. Lilga for helpful discussion.

References

[1] M.A. Lilga, R.J. Orth, J.P.H. Sukamto, S.M. Haight, D.T. Schwartz, Metal ion separations using electrically switched ion exchange, Sep. Purif. Technol. 11 (1997) 147–158.

[2] M.A. Lilga, R.J. Orth, J.P.H. Sukamto, S.D. Rassat, J.D. Genders, R. Gopal, Cesium separation using electrically switched ion exchange, Sep. Purif. Technol. 24 (2001) 451–466.

[3] S.D. Rassat, J.H. Sukamto, R.J. Orth, M.A. Lilga, R.T. Hallen, Development of an electrically switched ion exchange process for selective ion separations, Sep. Purif. Technol. 15 (1999) 207–222.

[4] M.A. Lilga, R.J. Orth, and J.P.H. Sukamto. Electrically switched cesium ion exchange FY98 Final Report, PNNL-12002, Pacific Northwest National Laboratory. 1998.

[5] Y. Lin, X. Cui, Novel hybrid materials with high stability for electrically switched ion exchange: carbon nanotubes/polyaniline/nickel hexacyanoferrate nanocomposites, Chem. Commun. 17 (2005) 2226–2228.

[6] Y. Lin, X. Cui, Electrosynthesis, characterization, and application of novel hybrid materials based on carbon nanotube-polyaniline-nickel hexacyanoferrate nanocomposites, J. Mater. Chem. 16 (2006) 585–592.

[7] W. Chen, X.H. Xia, Highly stable nickel hexacyanoferrate nanotubes for electrically switched ion exchange, Adv. Funct. Mater. 17 (2007) 2943–2948.

[8] S.B. Lalvani, T. Wiltowski, A. Hubner, A. Weston, N. Mandich, Removal of hexavalent chromium and metal cations by a selective and novel carbon adsorbent, Carbon 36 (7–8) (1998) 1219–1226.

[9] P. Rana, N. Mohan, C. Rajagopal, Electrochemical removal of chromium from wastewater by using carbon aerogel electrodes, Water Res. 38 (2004) 2811–2820.

[10] Y. Lin, X. Cui, J. Bontha, Electrically controlled anion exchange based on polypyrrole and carbon nanotubes nanocomposite for perchlorate removal, Environ. Sci. Technol. 40 (2006) 4004–4009.

[11] X. Cui, M.H. Engelhard, Y. Lin, Preparation, characterization and anion exchange properties of polypyrrole/carbon nanotube nanocomposites, J. Nanosci. Nanotechnol. 6 (2006) 547–553.

PART

3

Transformation Technologies

CHAPTER

Nanometallic Particles for Oligodynamic Microbial Disinfection

18

Gordon Nameni[1] and James Economy[2]

[1]Accelerate DBA, Milwaukee, WI, USA [2]Center of Advanced Materials for Purification of Water with Systems, Department of Materials Science and Engineering, University of Illinois at Urbana-Champaign, Urbana, IL, USA

18.1 Introduction

Recent published data from the World Health Organization (WHO) and from other epidemiological researchers suggest that the health consequences associated with the microbiological contamination of water supplies remain a pressing global issue facing our society in the new millennium [1–3]. Worldwide, approximately 1.8 million die per year from waterborne infections [4]. Despite the growth of the water utilities industry and consequent use of chemical disinfection with new methods such as ozonation or ultraviolet (UV) radiation, the incidence of health-related microbiological contamination suggests that more research is necessary to further address the problem.

The explosive growth in nanotechnology research has opened the doors to new strategies using nanometallic particles for oligodynamic disinfection.

Street, Sustich, Duncan and Savage. Nanotechnology Applications for Clean Water, 2nd Edition.
 DOI: http://dx.doi.org/10.1016/B978-1-4557-3116-9.00018-4

The excellent microbicidal properties of the oligodynamic nanoparticles qualify their use as viable alternatives for water disinfection.

The following text first explores an economic analysis of the current chemical-based disinfection infrastructure with the associated global health consequences as a driver for the development of alternative disinfection systems based on oligodynamic particles. A review of the most noteworthy oligodynamic nanoparticle systems, including their synthesis and performance, is presented to provide an insight into their use in alternative disinfection systems that can alleviate the infrastructure burden from chemical-based disinfection and yet maintain or improve the elimination of waterborne disease-causing microbes.

18.2 Economic impact of modern disinfection systems

A review of the water disinfection infrastructure in many developed nations reveals the indelible link between economic vitality and municipal-based disinfection systems. The worldwide market for water equipment and chemicals in 2007 is estimated at $76.3 billion [5]. The more developed regions, North America and Europe, account for the lion's share of expenditures, owning 49 percent as shown in Fig. 18.1.

Within the United States, the drinking water infrastructure market is predicted to grow to $138 billion by 2015 [6]. Whereas other parts of the world are experiencing similar investment and growth in their water procurement related industries, Africa, the region that consistently ranks high in incidence of infections related to waterborne contamination, only imports 3.4 percent of the world's supply of machinery for filtering and purifying water as can be seen from Table 18.1.

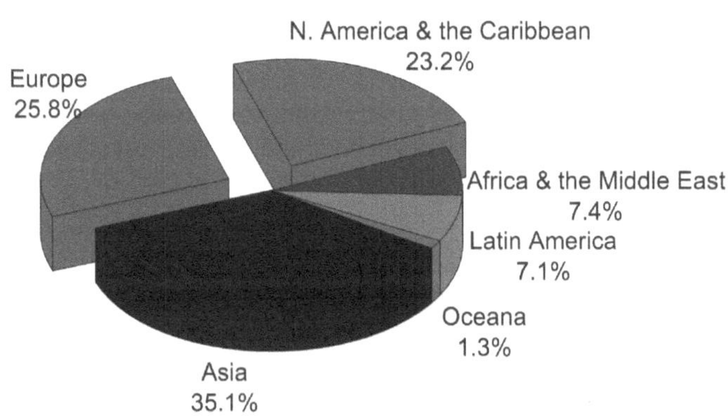

FIGURE 18.1

Worldwide market potential for water equipment and chemicals in 2007 [5].

Table 18.1 World Market for Imported Machinery for Filtering and Purifying in the Year 2007 [7]

Region	Value ($M)	% of World
Europe	1,402.42	41.0
Asia	768.66	22.5
North America and Caribbean	744.88	21.7
Latin America	180.10	5.3
The Middle East	119.15	3.5
Africa	115.36	3.4
Oceana	90.38	2.6
Total	3,420.95	100.0

Based on the market value provided in Table 18.1 and recent statistics from the United Nations' world population database, the African market results in $0.13 per person compared to $1.92 per person for the European market for filtering and purifying water machinery [8]. The low figure for Africa (less than 14x difference with Europe) strongly suggests that new research efforts toward increased effectiveness in disinfection technology at par costs would have a sizable impact for this region in the world.

18.3 Health impact of water disinfection shortfalls

The impact from microbiological contamination of waterborne supplies remains a global problem despite the growing world market for water infrastructure [5]. Globally, 80 percent of all sickness and disease results directly or indirectly from poor water supply and 19 percent of all deaths are due to waterborne infections. In developing countries, 30,000 people die per day from poor water supply and sanitation [9]. This corresponds to 10.9 million per year.

Although the incident rates are more prominent in developing countries, health-related problems from the microbiological contamination of water supplies in the developed nations still occur quite periodically. In the United States, an estimated 19.5 million illnesses per year occur from waterborne infections [4]. Furthermore, groundwater, which makes up 37.5 percent of the public water supply, was associated with seven times more waterborne disease-causing outbreaks than surface water [10]. Out of the documented outbreaks in the 2003–2004 survey period, 51 percent of contaminated water supplies were out of the jurisdiction of a water utility [2]. These numbers suggest that research into alternative disinfection systems that stand apart from municipal water treatment could reduce the incidence of water-related infections in the United States and in developing countries. Equally, such systems could positively impact developing regions in the world with similar problems due to waterborne infections.

18.4 Modern disinfection systems

Both groundwater and surface water sources are equally susceptible to microbiological contamination from bacteria, protozoans, and viruses [10]. Despite advances in disinfection technology, previously undiscovered pathogens and known pathogens moving into new areas continue to pose a great threat to human health, globally. Even developed nations with more advanced water disinfection infrastructure continue to report disease outbreaks from waterborne infections [4]. Thus, next generation disinfection technologies not only have to eliminate the emerging pathogens, but also remain cost effective for large-scale adoption.

A number of disinfection technologies, including both POU and point-of-entry (POE), have been deployed to address the issue with waterborne microbiological contamination. Suitability of disinfection technology is dependent on the application (POU or POE). Point-of-entry applications are primarily deployed to disinfect water that does not come from a municipal supply, whereas POU applications are typically used in a setting against water that may have already been chemically treated. Disinfection technologies that find use in POE applications include chemical-based disinfection agents (chlorine, chlorine dioxide, potassium permanganate, hydrogen peroxide, ozone) and UV radiation. Ultraviolet radiation is also used in POU applications along with filtration [6]. Table 18.2 summarizes the advantages and disadvantages of recent technologies that have been qualified as potential alternatives to the standard chlorine chemical disinfection, which is toxic and produces carcinogenic and mutagenic disinfection by-products.

It should be noted that the cost for chemical disinfection of municipal water in the United States is approximately \$0.40 m^{-3} [6].

18.5 Nanometallic particles in alternative disinfection systems

In total, qualified alternative disinfectants listed in Table 18.2 have been widely successful. In the past few decades, researchers have displayed the efficacy of metal ions as bactericides in water disinfection [13]. Recent research in the synthetic fabrication, however, has produced nanometallic particles with microbicidal action that are not only as effective, but also offer other advantages in synthetic production, dosing control, and cost. Nanometallic particles are expected to play an important role in the future of water purification because of their high reactivity resulting from their large surface-to-volume ratios [13].

This new class of nanometallic particles produces antimicrobial action referred to as oligodynamic disinfection for their ability to inactivate microorganisms at low concentrations. Various oligodynamic metals exhibit microbicidal, bactericidal, and viricidal properties; however, reducing the size of the metals to the nanoscale produces tremendous advantages in disinfection capacity due to the

Table 18.2 Summary of Advantages and Disadvantages of Previously Qualified Alternative Disinfectants [6,11,12]

Method of Disinfection	Advantages	Disadvantages
Chlorine dioxide	Stronger than chlorine with shorter contact time Wide microbicidal action (algae, fungi, yeast, virus, and mold spores) Less formation of chlorinated organics Good taste, odor, and color control Strong oxidant	Expensive Production of chlorine, chlorite, and chlorate (hazardous compounds) in manufacturing process
Ozone	Faster and stronger than chlorine Wide microbicidal action (algae, fungi, yeast, virus, and mold spores) Effective against Cryptosporidium, Giardia, and Legionella Good taste, odor, and color control Strong oxidant (can oxidize trihalomethane precursors)	No residual effect Unknown organic reaction products
Ultraviolet	Wide microbicidal action (algae, fungi, yeast, virus, and mold spores) No disinfection byproducts (DBPs)	No residual effect Longer exposure time for effectiveness

greater surface area, contact efficiency, and often better elution properties. These qualities enable these materials to be considered as viable alternative disinfectants. A summary of reported nanometallic particles for disinfection of water in order of strength include, silver (Ag), copper (Cu), zinc (Zn), titanium (Ti), and cobalt (Co). New combinatorial oligodynamic materials consisting of these nanometallic particles have been deployed among a number of substrates for their use in water disinfection [14–16]. Such materials as Ag deposited on titanium oxide, and Ag-coated iron oxide had displayed faster kinetics and greater efficiency in eliminating bacteria. The statistical significance of the synergistic effects is not clear but researchers do report marked improvement in bactericidal effectiveness among these systems.

18.5.1 Silver nanoparticles

The authors will focus on the silver-based systems since silver is the most widely studied oligodynamic material due to its wide range in microbicidal effectiveness, low toxicity, and ease of incorporation on various substrates in a host of dynamic disinfection applications. The main known negative health effect from silver is argyria, which is an irreversible darkening of the skin and mucous membrane resulting from overexposure to ionic silver (Ag(I), Ag^+) [17].

18.5.2 Synthesis

Typically, the silver nanoparticles that are deployed onto a variety of substrates are derived from silver salts (silver nitrate ($AgNO_3$), silver chloride (AgCl), silver bromide (AgBr), and silver iodide (AgI)). The excellent solubility of $AgNO_3$ has enabled its use across a variety of disinfection applications [18]. The concentration of $AgNO_3$ ranges from 0.0001 to 0.7 mol L^{-1} based on the target bactericidal strength [19]. In many of the cases, a variety of solvents are used to dissolve the salts; however, water appears to be the solvent of choice for most systems using $AgNO_3$. A variety of substrates have been reported to support silver after immersion under various conditions (varying time, mechanical agitation, and temperature) in the silver salt solution [19,20]. Silver deployed on granulated activated carbon, activated carbon fibers (ACF), polyurethane, zeolites, and ceramics in POE and POU applications displayed effective inactivation of pathogens in water [18,20–22]. A variety of chemical agents are used to aid in the reduction of ionic silver (Ag^+) to metallic silver nanoparticles (Ag^0) but most frequently a thermal reduction process greater than 70°C is employed [21,23].

Researchers report a strong dependency between the Ag content on the substrate and particle size with the concentration of the salt, immersion time, and reduction temperature [19,24]. Though not much difference has been reported in the inactivation efficiency between the systems containing ionic silver and metallic silver, there is a marked difference in the mechanism of inactivation between the two species of silver [25]. The particle size of the silver supported by ACF ranged as low as 70 nm from 0.01 mol L^{-1} concentration to several microns at 0.7 mol L^{-1} (Fig. 18.2). The silver supported on fiberglass displayed a similar trend. Silver particles immersed in 0.5 mol L^{-1} $AgNO_3$ displayed a particle size range of 20–300 nm (Fig. 18.3). The greater the silver content, the lower the surface area of the supporting substrate.

18.5.3 Utility

The systems supported with nanometallic silver particles are effective in reducing the presence of target microorganisms in a wide variety of water disinfection applications. The minimum inhibitory concentration (MIC) varied depending on the substrate the silver was deployed from and the form of the silver (Table 18.3).

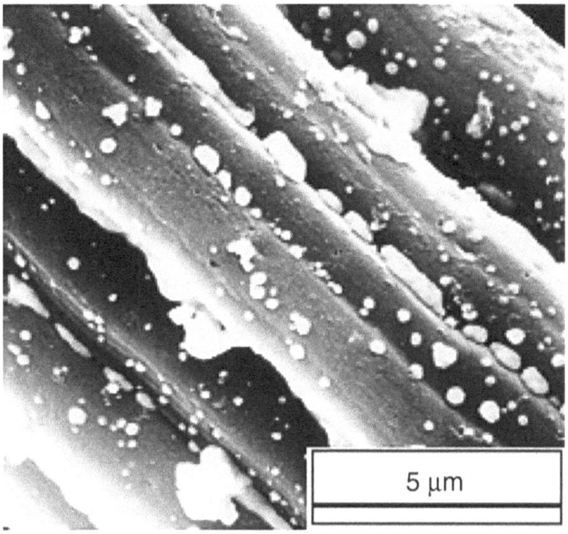

FIGURE 18.2

SEM image of activated carbon supporting silver. Sample was immersed in 0.4 mol L^{-1} $AgNO_3$ for 15 h and reduced at 300°C for 1.5 h.

Image modified from Wang et al. [19].

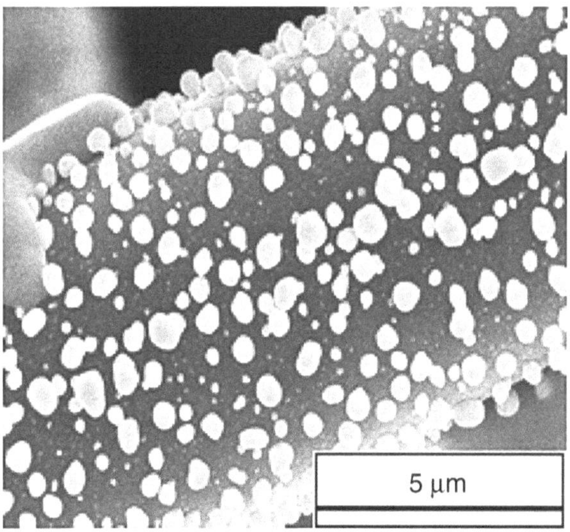

FIGURE 18.3

SEM image of fiberglass supporting silver. Sample was immersed in 0.5 mol L^{-1} $AgNO_3$ for 12 h and reduced at 300°C for 0.5 h.

Image modified from Wang et al. [19].

Table 18.3 Minimum Inhibitory Concentration (MIC) for *E. coli* as a Function of Silver Form Following a 24 h Incubation Period at 25°C [17]

Form of Silver	MIC, Total Silver (mg Ag/L)
Ag(I), Ag^+	$0.10 \geq MIC \geq 0.05$
Organic Ag	$0.10 \geq MIC \geq 0.05$
Ag_2O	$0.14 \geq MIC \geq 0.07$
AgCl(s)	$0.40 \geq MIC \geq 0.20$
Ag^0	MIC* >82

**This value represents the total silver measured via atomic absorption spectrophotometry following digestion. Since the source does not indicate the size of the nanoparticles, it is presumed that more efficient bacterial inactivation can be achieved with silver nanoparticles that are deployed on a high-surface substrate with improved contact efficiency.*

The mechanisms for the bactericidal activity of silver have been well reported in the literature [20,26,27]. They include:

1. Reaction with thiol groups.
2. Reaction with amino acids and proteins.
3. Binding to critical enzyme functional groups.
4. Inhibition of the cellular respiratory chain.
5. Inhibition of cellular phosphate uptake.
6. Binding/densification of DNA.

Though many forms of silver have found use in disinfection applications, which include swimming pools and hospital hot water systems, silver nanoparticles find the most usage in POU applications including activated-carbon-based and ceramic water purification filters. POU filters composed of granular activated carbon impregnated with silver have received ample attention in the past decade owing to their high surface area and pore size distribution that allow silver to be easily entrapped in the pores and later desorbed [19,28]. Carbon-based substrates lower the impact of the silver nanoparticles. With a loading of 0.05 wt percent Ag impregnation in an ACF with extremely high surface area (1200 m^2/g), the fastest time achieved for complete bacterial elimination was 30 minutes [28]. The silver impregnated carbon-based filters displayed only bacteriostatic performance since they were not able to completely eliminate microbial regrowth in POU devices. Such performance related issues can be addressed by deploying silver nanoparticles on inorganic-based substrates and by using combinations of oligodynamic nanoparticles [24,28].

As can be seen in Fig. 18.4, silver nanoparticles impregnated onto an inorganic fiberglass substrate (nonwoven Craneglas® 230, Crane Nonwovens, Inc., 7 percent PVA binder, diameter 6.5 μm) produce far superior bactericidal activity over carbon-based systems with even greater silver loading. Such materials can be implemented in POU or POE applications that provide an alternative means

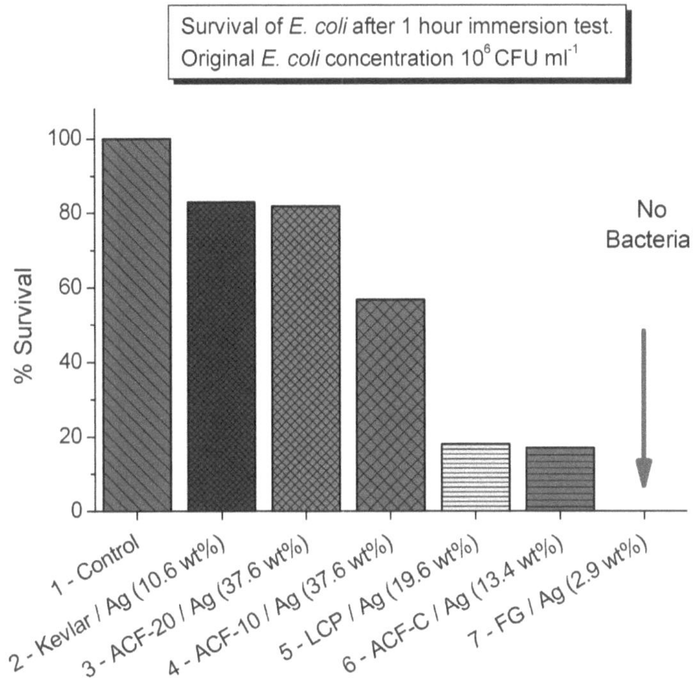

FIGURE 18.4

Comparison of bactericidal effect of various substrates impregnated with silver nanoparticles.

for water purification without the massive infrastructure burden that the chemical disinfectants require.

In light of the promise that oligodynamic nanometallic particles offer an alternative disinfection systems, there are certain drawbacks that must be taken into consideration. The silver desorption rate is a crucially important parameter that must be thoroughly characterized before such systems are deployed in the field. Silver attrition on the POU devices is also an issue as both variables affect the mean silver content available for microbicidal action [17,19]. The US Environmental Protection Agency (EPA) mean contamination level (MCL) for silver in water is 100 ppb; hence, sufficient attention must be paid [29]. Furthermore, the native constituent of water also has an affect on the disinfection ability of the silver nanoparticles [17]. Due to silver's wide microbicidal range, a number of nonpathogenic bacteria, which provide useful environmental functions, as well as pathogenic bacteria are indiscriminately inactivated [30]. Lastly, certain microbes have been shown to build resistance to silver [31].

Though these are significant points of contention with silver nanoparticles, they can be addressed with specific research. For example, to address the attrition issue, the localized binding between the silver nanoparticles and the substrate can be improved by optimizing synthetic conditions and substrate chemistry.

The issue of increased silver resistance among microorganisms can be addressed with the deployment of combinatorial oligodynamic nanometallic particles that have displayed enhanced synergistic disinfection activity. For example, fiberglass impregnated with silver and copper nanoparticles displayed enhanced disinfection efficiency against gram negative *E. coli* (Fig. 18.5).

It has also been demonstrated that fiberglass impregnated with silver nanoparticles and iron oxide displays an increased bactericidal effectiveness while maintaining robust viricidal properties (Fig. 18.6) [32].

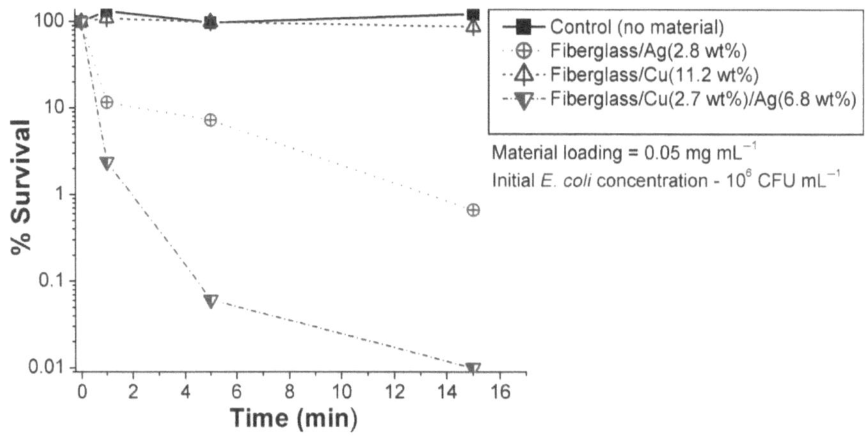

FIGURE 18.5

Bactericidal activity of fiberglass impregnated with silver and copper nanoparticles.

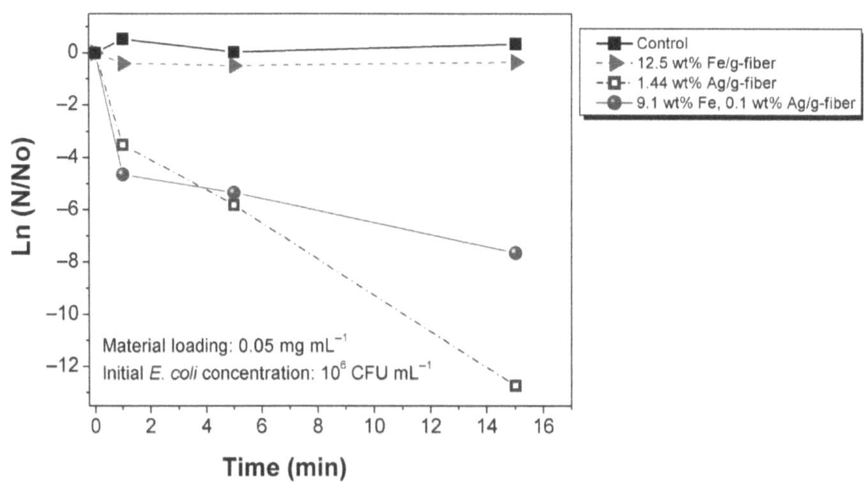

FIGURE 18.6

Bactericidal activity of fiberglass impregnated with iron oxide and silver nanoparticles.

18.6 Conclusions

The heavy expenditures in the water infrastructure industry and the ongoing global health-related consequences of microbiological contamination of water supplies clearly indicate that as a global society there needs to be a change in how water disinfection is addressed. Much of the world still does not have the industrial infrastructure to support chemical-based disinfection of water. Hence, alternative disinfection systems that enable developed countries, developing nations, and transitional regions to address their disinfection needs with a less intensive infrastructure and in a more cost effective manner must be investigated.

The emergence of oligodynamic nanometallic particles in water disinfection offer a tremendous opportunity for the development of disinfection systems that are extremely effective and tunable for small or large communities without the infrastructure burden of modern chemical-based disinfection systems. In particular, the excellent disinfection properties of the nanometallic particles, particularly when they are used in combination with each other [16,28], warrant the consideration of this new class of materials for alternative disinfection systems in developing nations where their oligodynamic properties may be the only barrier against pathogenic strains of bacteria and other disease-causing microbes.

Researchers have cited the potential ecotoxity and increase in microbial resistance to oligodynamic metals as an argument against their use [31,33]. Specifically regarding silver, the WHO has published data that indicate the no observed adverse effect level (NOEL) for silver is 10 g [34]. The WHO has reported that at the current recommended limit for silver in water (100 ppb), it would take 70 years to reach half of the NOEL level for silver in humans.

As the planet's water resources shrink in the future, there are invariably two outcomes that must occur for there to be continued safe consumption: (i) water becomes more expensive and; (ii) more cost-effective methods of disinfection are invented. The development of oligodynamic nanometallic particles enables the progression toward the second option.

References

[1] World Health Organization, World Health Report, World Health Organization, Geneva, 2005.

[2] J.L. Liang et al., Surveillance for waterborne disease and outbreaks associated with drinking water and water not intended for drinking—United States, 2003–2004 MMWR, SS-12, 55, (2004) pp. 31–68.

[3] B.G. Blackburn et al., Surveillance for waterborne disease and outbreaks associated with drinking water—United States, 2001–2002 MMWR, SS-8, 53, (2004) pp. 23–45.

[4] K.A. Reynolds, Water quality monitoring: lessons from the developing world, Water Conditioning Purif. 39 (12) (2007) 66–68.

[5] P.M. Parker, The 2007–2012 World Outlook for Water Equipment and Chemicals, ICON Group International, Inc., San Diego, 2006.

[6] Netscribes, The US Market for Residential Water Treatment Products, s.l, Don Montuori Publishing, 2005.

[7] P.M. Parker, The World Market for Machinery for Filtering and Purifying Water: A 2007 Global Trade Perspective, 2007, ICON Group, Ltd., San Diego, 2007.

[8] Population Division of the Department of Economic and Social Affairs of the United Nations Secretariat, World Population Prospects: The 2006 Revision and World Urbanization Prospects: The 2005 Revision, United Nations Population Division, New York, 2008.

[9] World Health Organization, World Health Report, World Health Organization, Geneva, 2002.

[10] K. Thomas, The Residual Water Treatment Market, Kalorama Information, LLC, New York, 1999.

[11] A.C. Anderson, R.S. Reimers, P. DeKernion, A brief review of the current status of alternatives to chlorine disinfection of water, Am. J. Public Health 92 (11) (1982) 1290–1293.

[12] E.P.A., Waste Water Technology Fact Sheet: Chlorine Disinfection, Washington D.C., United States Environmental Protection Agency, 1999.

[13] P. Pradeep, T. Jain, Potential of silver nanoparticle-coated polyurethane foam as an antibacterial water filter, Biotech. Bioeng. 90 (1) (2005) 59–63.

[14] K.D. Kim, et al., Formation and characterization of Ag-deposited TiO2 nanoparticles by chemical reduction method, Scr. Materiala 54 (2006) 143–146.

[15] J. Keleher, B. Bashant, N. Heldt, L. Johnson, Y. Li, Photo-catalytic preparation of silver-coated TiO_2 particles for antibacterial applications, World J. Microbiol. Biotechnol. 18 (2002) 133–139.

[16] J. Kim, M. Cho, B. Oh, S. Choi, J. Yoon, Control of bacterial growth in water using synthesized inorganic disinfectant, Chemosphere 55 (2004) 775–780.

[17] M.A. Butkus, L. Edling, M.P. Labare, The efficacy of silver as a bactericidal agent: advantages, limitations and considerations for future use, J. Water Supply: Res. Technol.—AQUA 52 (6) (2003) 407–416.

[18] H. Remy, Treatise on Inorganic Chemistry, New York, s.n., 1956.

[19] Y.L. Wang, et al., Preparation and characterization of antibacterial viscose-based activated carbon fiber supporting silver, Carbon 36 (1) (1998) 1567–1571.

[20] I. Yoshihiro, Bactericidal activity of Ag zeolite mediated by reactive species under aerated conditions, J. Inorg. Biochem. 92 (2002) 37–42.

[21] V. Kumar, et al., Highly efficient Ag/C catalyst prepared by electrodeposition method in controlling microorganisms in water, J. Mol. Cat. A: Chem. 223 (2004) 313–319.

[22] L. Logenberger, G. Mills, Formation of metallic particles in aqueous solutions by reactions of metal complexes with polymers, J. Phys. Chem. 99 (1995) 475–478.

[23] R.J. Chung, et al., Antimcrobial hydroxyapatite particles synthesized by a sol–gel route, J. Sol–Gel Sci. Technol. 33 (2005) 229–239.

[24] G. Nangmenyi, E. Mintz, W. Xao, J. Economy, Bactericidal acitivty of Ag nanoparticles coated on a fiberglass substrate, J. Water Health (2007) submitted.

[25] J. Marones, J. Elechiguerra, A. Camacho, K. Holt, J. Kouri, J. Ramirez, et al., The bactericidal effect of Ag nanoparticles, Nanotechnology 16 (2005) 2346–2353.

[26] M. Yamanaka, K. Hara, J. Kudo, Bactericidal action of a silver ion solution on Escherichia coli, studied by energy-filtering transmission electron microscopy and proteomic analysis, App. Environ. Microbiol. 71 (11) (2005) 7589–7593.

[27] Q.L. Feng, G. Chen, F. Cui, T. Kim, J. Kim, A mechanistic study of the antibacterial effect of silver ions on Escherichia coli and Staphylococcus aureus, J. Biomed. Mater. Res. 52 (2000) 662–668.

[28] H. Le Pape, et al., Evaluation of the anti-microbial properties of an activated carbon fibre supporting silver using a dynamic method, Carbon 40 (2002) 2947–2954.

[29] Office of Water, National Primary Drinking Water Standards, s.l., United States Environmental Protection Agency, 2003. <http://www.epa.gov/safewater>.

[30] N. Simonetti, G. Simonetti, F. Bougnol, M. Scalzo, Electrochemical Ag+ for preservative use, App. Environ. Microbiol. 58 (12) (1992) 3834–3836.

[31] S. Silver, Bacterial silver resistance: molecular biology and uses and misuses of silver compounds, FEMS Microbiol. Rev. 27 (2003) 341–353.

[32] G. Nangmenyi, X. Liand J. Economy, Antibacterial Activity of Ag Fe2O3 Impregnated Fiberglass Systems, New Orleans, American Chemical Society Press, 2008.

[33] K. Pelkonanen, H. Heinonen-Tanskei, O. Hanninen, Accumulation of silver into cerebelum and musculus soleus in mice, Toxicology 186 (2003) 151–157.

[34] World Health Organization, Guidelines for Drinking Water Quality, Vol. 1, WHO Press, Geneva, 2006.

CHAPTER

Nanostructured Visible-Light Photocatalysts for Water Purification 19

Qi Li[1], Pinggui Wu[2] and Jian Ku Shang[1]

[1]*Center of Advanced Materials for Purification of Water with Systems, Department of Materials Science and Engineering, University of Illinois at Urbana-Champaign, Urbana, IL, USA* [2]*Superior Graphite Company, Chicago, IL, USA*

19.1 Visible-light photocatalysis with titanium oxides

Since the discovery of photoelectrochemical splitting of water on *n*-titanium oxide (n-TiO_2) electrodes by Fujishima and Honda [1] in 1972, there has been much interest in semiconductor-based materials as photocatalysts for both solar energy conversion [2–11] and environmental applications [12–15]. The basic principles of semiconductor-based photocatalysis are schematically illustrated in Fig. 19.1 [16]. The photocatalytic process is initiated by the absorption of photons with energy equal to or greater than the semiconductor band gap, which is shown in the enlarged section of Fig. 19.1. After light irradiation, electrons (e^-) are excited to jump from the valence band (VB) to the conduction band (CB), leaving holes (h^+) in the valence band. The fate of the separated e^- and h^+ pair can follow several pathways. The electron and hole can recombine either on the semiconductor particle surface (path A) or in the bulk of the semiconductor particle (path B) in a

Street, Sustich, Duncan and Savage. Nanotechnology Applications for Clean Water, 2nd Edition.
 DOI: http://dx.doi.org/10.1016/B978-1-4557-3116-9.00019-6

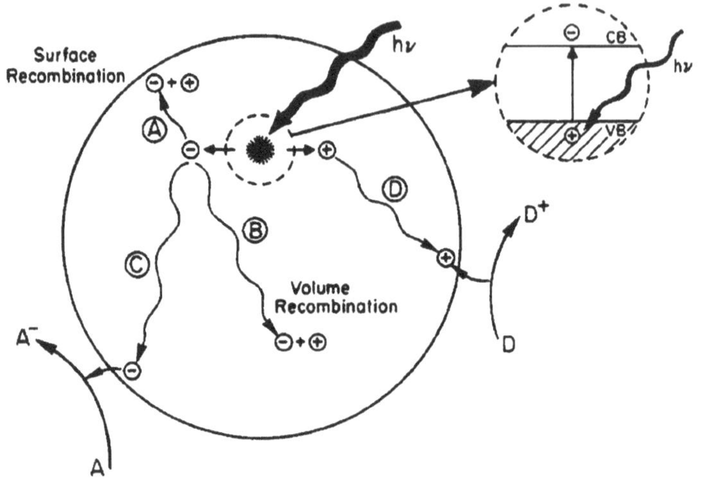

FIGURE 19.1

Schematic illustration of photoexcitation in a semiconductor photocatalyst followed by deexcitation events. CB: Conduction band; VB: Valence band.

Adapted from [16].

few nanoseconds with the release of heat. Under this situation, no free radicals are produced to continue the secondary reactions and there will be no photocatalytic activity. If the electron and hole migrate to the semiconductor surface without recombination, the semiconductor can donate the surface electron to reduce the electron acceptor species (A) (path C), whereas an electron from the electron donor (D) species combines with the surface hole that oxidizes the electron donor species (path D). Due to the limited lifetime of the electron and hole pair, the electron acceptor and donor species should be adsorbed or be close to the semiconductor particles.

Among various semiconductor-based photocatalysts, TiO_2 is the leading candidate for industrial use because of its high chemical stability, good photoactivity, relatively low cost, and nontoxicity [17]. However, the photocatalytic capability of TiO_2 is limited to only ultraviolet (UV) light (wavelength, $l, < 400$ nm), which provides sufficient energy to excite electrons across the relatively wide band gap of 3.2 eV in the TiO_2 anatase crystalline phase. Since UV accounts for about 3–4 percent of the solar spectrum, it is of great interest to develop photocatalysts that can absorb visible light so that a major part of the solar spectrum could be used for photocatalytic reactions. Previous approaches to extend the absorption band-edge of TiO_2 from UV to visible-light region have included doping transition metals into TiO_2 [18–21], or reducing TiO_2 to TiO_x ($x < 2$) [22–23]. However, for lack of reproducibility and chemical stability, these approaches have not found widespread industrial applications.

Recently, Asahi et al. [24] reported that nitrogen doping of n-TiO_2 can extend the optical absorbance of TiO_2 into the visible-light region and bring about photodegradation of methylene blue and gaseous acetaldehyde in the visible region of $l < 500$ nm. Figure 19.2A shows the optical absorption spectra of their undoped and nitrogen-doped TiO_2 (TiON) thin films, which shows clearly that $TiO_{2-x}N_x$ films noticeably absorb light at less than 500 nm, whereas the TiO_2 films just absorb UV light. Photodecomposition of gaseous acetaldehyde was used in their study to evaluate the photocatalytic activity of their $TiO_{2-x}N_x$ powder sample, in comparison with TiO_2 powder. Figure 19.2B shows that the photocatalytic activity of the $TiO_{2-x}N_x$ sample is superior to that of the TiO_2 sample in the visible range of irradiation, whereas both samples yield similar UV activity. This observation is in accordance with their optical absorbance demonstrated in Fig. 19.2A.

Since the report of TiON by Asahi et al., anion-doped TiO_2 has attracted great interest. Both experiments and theoretical calculations demonstrated that anionic nonmetal dopants, such as nitrogen [24–36], carbon [37–44], sulfur [45–49], or fluorine [50], can extend the photocatalytic activity of TiO_2 into the visible-light region with good stability, photocatalytic efficiency, and ease of the doping process [51]. In the early work of Asahi et al. [24], it was believed that the N 2p and the O 2p states were mixed in TiON, leading to a narrower band-gap energy of TiO_2 and subsequent visible-light absorption. However, this proposed mechanism

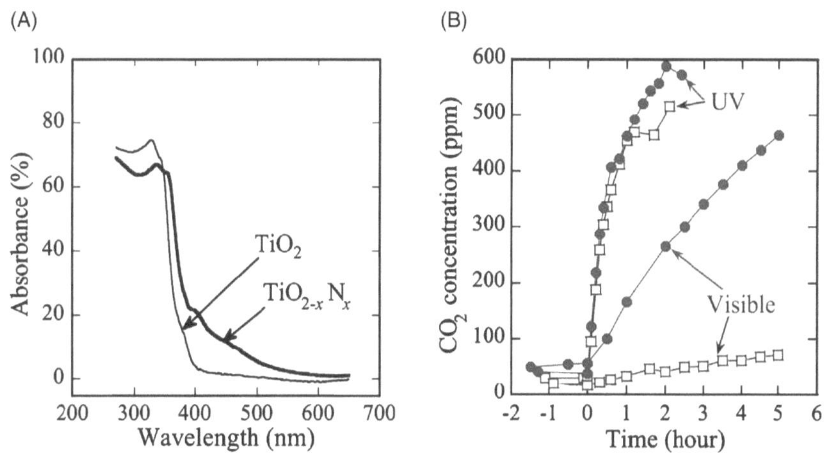

FIGURE 19.2

(A) Optical absorption spectra of undoped and nitrogen-doped titanium oxide (TiO_2) thin films. (B) CO_2 evolution as a function of irradiation time (light on at zero) during the photodegradation of acetaldehyde gas (with an initial concentration of 485 ppm) under UV irradiation (BL with a peak at 351 nm and the light power of 5.4 mW/cm^2) and visible irradiation (fluorescent light cut by the optical high-path filter [SC42, Fuji Photo Film], with a peak intensity at 436 nm and a light power of 0.9 mW/cm^2]. (Solid circles: TiO_2 powder; Open square: nitrogen-doped TiO_2 powder.)

Adapted from [24].

was challenged by Irie et al. [25]. They measured the quantum yield values during the photodecomposition of gaseous 2-propanol (IPA) under the same absorbed photon number of either visible or UV light, and found the quantum yield values from irradiating with visible light were always lower than with UV light. Thus, they suggested that the improved visible-light absorption was due to the isolated N 2p states above the valence-band maximum of TiO_2, which is consistent with several other recent experimental studies [26,51–52]. In a recent theoretical study by Lee et al. [53], their first-principles density-functional calculations demonstrated that with a small amount of nitrogen dopant, N 2p states lay in the band gap of TiO_2 and the mixing of N with O 2p states was too weak to produce a significant band-gap narrowing.

Although anion-doping, especially nitrogen-doping, was demonstrated as an effective approach to extend the photocatalytic activity of TiO_2 into the visible-light region, the photocatalytic efficiency of anion-doped TiO_2 reported in the early studies was relatively low under visible-light illumination. A number of major challenges must be overcome to make visible-light-activated photocatalysts suitable for industrial applications. First, both doping process and concentration control varied over a wide range and therefore systematic studies were needed to provide a reliable production route of photocatalysts with high quality. Second, although anion-doping introduces visible-light absorption capability to TiO_2 by creating isolated anion dopant states within the band gap of TiO_2, it also creates the serious problem of massive charge carrier recombination, thus largely limiting their photoactivity. In order to improve photoactivity, efforts should be made to reduce the charge carrier recombination. Third, the photocatalyst must be made into engineering forms suitable for broad environmental applications such as water purification.

19.2 Sol–gel fabrication of nitrogen-doped titanium oxide nanoparticle photocatalysts

Since the early work of Asahi et al. [24] on TiON, it has been extensively studied to extend its photocatalytic activity into the visible-light region [24–36]. In these previous works, TiON was fabricated into both powder and thin film. Compared with TiON thin films, TiON powder photocatalysts offer the advantages of high surface area, low cost, and suitability for large-scale production. Among various synthesis methods for preparing TiON powders, sol–gel based processes [28–29,32] seem to have the most potential. Fully developed, these processes can provide a reliable production route of TiON powders at relatively low cost through careful control of initial precursor composition and ratio.

Recently, we reported a systematic study of the precursor ratio effect on the structure, composition, and optical properties of sol–gel derived TiON nanoparticles [54]. In that study, tetramethylammonium hydroxide (TMA) and titanium tetraisopropoxide (TTIP) were used as precursors to synthesize TiON nanoparticle photocatalyst. The preparation of TiON precursor was done at room temperature in a sol–gel process as follows. First, TMA was dissolved in Ethyl alcohol (EtOH) at a mol ratio of 1:10. The solution was stirred magnetically for 5 minutes, and then TTIP was added into the solution with various TMA/TTIP mol ratios at 1:3, 1:5, and 1:10, respectively. For each TMA/TTIP mol ratio, the mixture was loosely covered and stirred continuously until a homogenous gel was formed. The hydrolysis of the precursor was initiated by exposure to moisture in air. The gel was aged in air for several days to allow further hydrolysis and drying. Then, the xerogel was crushed into fine powders and calcined at various temperatures in air for 3 hours to obtain the nanoparticle photocatalysts. For comparison, TiO_2 precursor was prepared by the same sol–gel process as mentioned earlier, but without the addition of TMA. Calcination of the TiO_2 xerogel was conducted at 400°C in air for 3 hours.

X-ray diffraction (XRD) patterns of obtained powders are shown in Fig. 19.3. After being calcinated in air at 400°C for 3 hours, TiO_2 powders are well crystallized into the anatase-type crystal structure. At the initial TMA/TTIP mol ratio of 1:10, the TiON powder also shows the anatase-type structure, but the XRD peaks are broad and the peak heights are weaker than those of TiO_2 powders, indicating only partial crystallization. For TiON powders made at higher initial TMA/TTIP mol ratios, no discernable reflection peaks could be identified, implying little or no crystallization. It is clear that the introduction of N into the TiO_2 structure (the addition of TMA) disrupts the crystallization of these sol–gel powders. For TiON nanoparticle powders calcinated in air at higher temperatures, an anatase-type

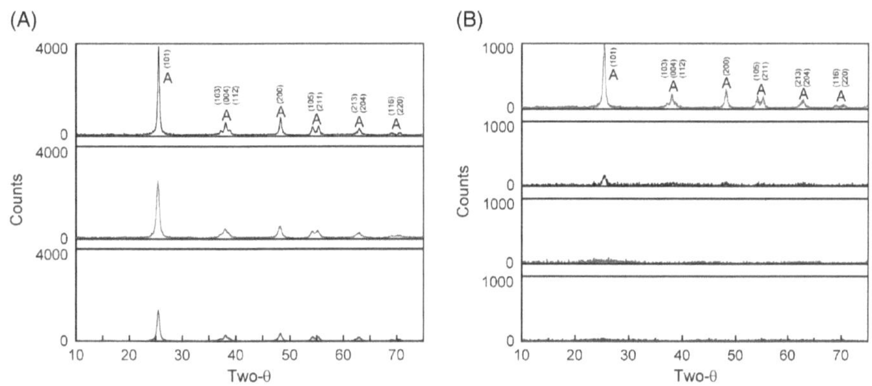

FIGURE 19.3

X-ray diffraction patterns of nitrogen-doped titanium oxide (TiON)/titanium oxide powders [TiO_2: TiON with various initial tetramethylammonium hydroxide/titanium tetraisopropoxide ratios at 1:3, 1:5, and 1:10] obtained by calcinating xerogels in air for 3 hours at (A) 400°C, and (B) 500°C, respectively.

Adapted from [54].

structure was obtained for all powders within the temperature range investigated (from 430 to 500°C). As an example, Fig. 19.3B presents the XRD patterns of TiON powders calcinated in air at 500°C. With the increase of the calcination temperature, the XRD peak intensity increases, indicating improved crystallization. For each calcination temperature, powders made at lower initial TMA/TTIP ratios have stronger peak intensity and sharper peaks, which confirm that the addition of TMA has the effect of disrupting the crystallization of sol–gel TiON nanoparticle powders. As a result, a higher calcination temperature would be needed to achieve complete crystallization. Figure 19.4 shows scanning electron microscope (SEM) and transmission electron microscope (TEM) images of a highly crystallized TiON sample, which consist of nanosized particles with non-uniform shapes.

The composition of the powder was examined by X-ray photoelectron spectroscopy (XPS). Figure 19.5A summarizes the N/Ti atomic ratio of these powders. It is clear that with the increase of calcination temperature, the N/Ti atomic ratio decreases for all these powders, whereas a higher initial TMA/TTIP ratio in the TiON precursors results in a faster decline of N/Ti ratio. This is further illustrated in Fig. 19.5B, where the residual N is lower at higher initial TMA/TTIP ratios. Therefore, a high initial TMA/TTIP ratio in the TiON precursor does not necessarily result in a high N/Ti atomic ratio in the TiON powders because the rate of N loss is faster in the calcination process. To obtain a higher N/Ti atomic ratio,

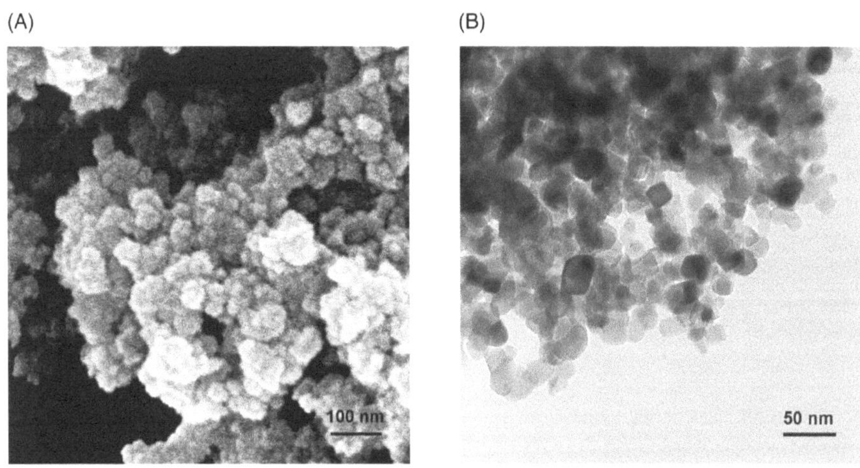

FIGURE 19.4

(A) Scanning electron microscope (SEM) and (B) transmission electron microscope (TEM) images of nitrogen-doped titanium oxide powders with initial tetramethylammonium hydroxide/titanium tetraisopropoxide ratios at 1:5 and obtained by calcinating xerogels in air for 3 hours at 500°C.

Adapted from [54].

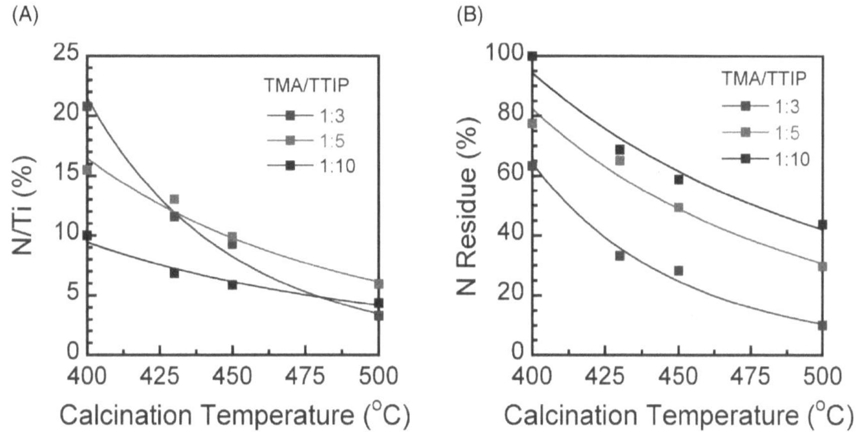

FIGURE 19.5

(A) N/Ti atomic ratio, and (B) the residue percentage of N content within the total tetramethylammonium hydroxide (TMA) addition in obtained nitrogen-doped titanium oxide (TiON) powders with various initial TMA/titanium tetraisopropoxide ratios at 1:3, 1:5, and 1:10, respectively. (The lines merely guide the eye.)

Adapted from [54].

the initial TMA/TTIP ratio should be controlled carefully in combination with calcination temperature.

The optical absorption of TiON nanoparticle powders was characterized by the diffuse reflectance measurements. The optical absorbance is approximated from the reflectance data by the Kubelka–Munk function [55], as given by Equation 19.1:

$$F(R) = \frac{(1-R)^2}{2R} \tag{19.1}$$

where R is the diffuse reflectance. Figure 19.6A shows the light absorbance of TiON powders obtained by calcinating xerogels in air at 500°C for 3 hours, compared with a commercial Degussa P25 powder. P25 shows the characteristic spectrum with the fundamental absorbance stopping edge at approximately 400 nm. TiON powders, however, show a clear shift into the visible-light range (> 400 nm). With the increase of the N content, more visible-light absorbance is observed. Figure 19.6B shows the Tauc Plot [55] ($(F(R)^*hv)^n$ vs. hv) constructed from Fig. 19.6A to determine the semiconductor band gap. The band gap of the Degussa P25 powder is approximately 3.20 eV, whereas TiON powders show a smaller band gap, and thus are well suited for the visible-light activation in photocatalysis. With the increase of the N content (caused by the various initial TMA/TTIP ratios), the band gap of TiON powder decreases steadily from approximately 3.05 to 2.90 eV, in agreement with the band gap narrowing in TiON films [56].

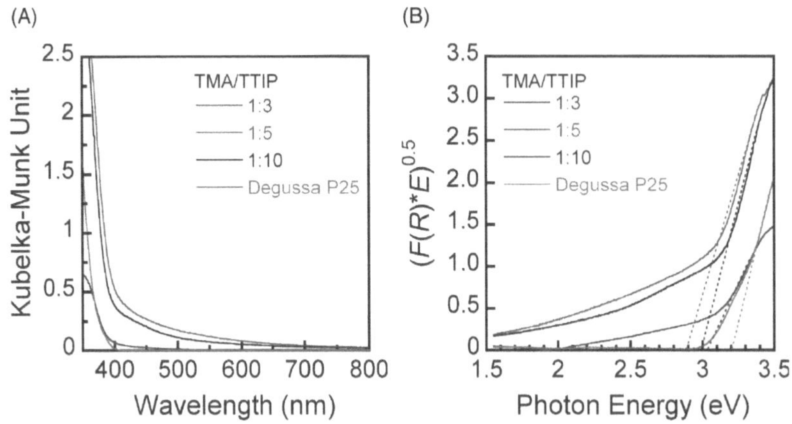

FIGURE 19.6

(A) Optical absorbance (in terms of Kubelka–Munk equivalent absorbance units) of nitrogen-doped titanium oxide (TiON) powders obtained by calcinating xerogels in air at 500°C for 3 hours with various initial tetramethylammonium hydroxide (TMA)/titanium tetraisopropoxide (TTIP) ratios at 1:3, 1:5, and 1:10, compared with the optical absorbance of Degussa P25 powder. (B) Tauc Plot constructed from (A). Band-gap values are determined from the extrapolation of the linear Tauc Region line to the photon energy abscissa.

Adapted from [54].

From such a systematic study, TMA as a nitrogen source is found to retard the crystallization of sol–gel TiON powders so that a higher calcination temperature is required for the crystallization of TiON powders, especially at high initial TMA/TTIP ratios. The increase of the calcination temperature promotes the crystallization level, but reduces N concentration in the TiON powders. The rate of N loss depends on the initial TMA/TTIP ratio, with a higher TMA/TTIP ratio resulting in a faster N loss. These sol–gel TiON nanoparticle photocatalysts have shown visible-light absorbance. For a strong visible-light absorbance, the initial TMA/TTIP ratio and calcination temperature must be controlled together to achieve a high N/Ti atomic ratio and a complete crystallization of TiON nanoparticles. Under our experimental conditions, the best initial TMA/TTIP ratio was determined to be 1:5 at a calcination temperature of 500°C.

19.3 Metal-ion-modified nitrogen-doped titanium oxide photocatalysts

Although nitrogen-doping has been demonstrated to provide visible-light photocatalytic activity to TiO_2 by introducing isolated anion dopant states within the band gap of TiO_2, the intraband states also create the serious problem of massive charge carrier recombination, thus largely limiting their photoactivity. It is generally believed that metal ion dopants may influence the light absorption and photoreactivity of TiO_2 most significantly by acting as electron (or hole) traps [18–21,57]. The trapping of charge carriers can decrease the e^-/h^+ pair recombination rate and subsequently increase the lifetime of charge carriers, which is beneficial to improving the photoreactivity of TiO_2. Thus, metal ions were added to TiON to reduce the charge carrier recombination and the photoactivity of TiON [58].

Palladium was added to TiON to form palladium-modified N-doped TiO_2 (TiON/PdO) nanoparticle photocatalysts. The concentration of palladium was controlled by the precursor ratio in a sol–gel process. The sol–gel process uses the optimized processing parameters for TiON synthesis [54], only adding palladium acetylacetonate (using CH_2Cl_2 as the solvent) as the palladium source. Figure 19.7A shows a typical XRD pattern of TiON/PdO powders. Major XRD

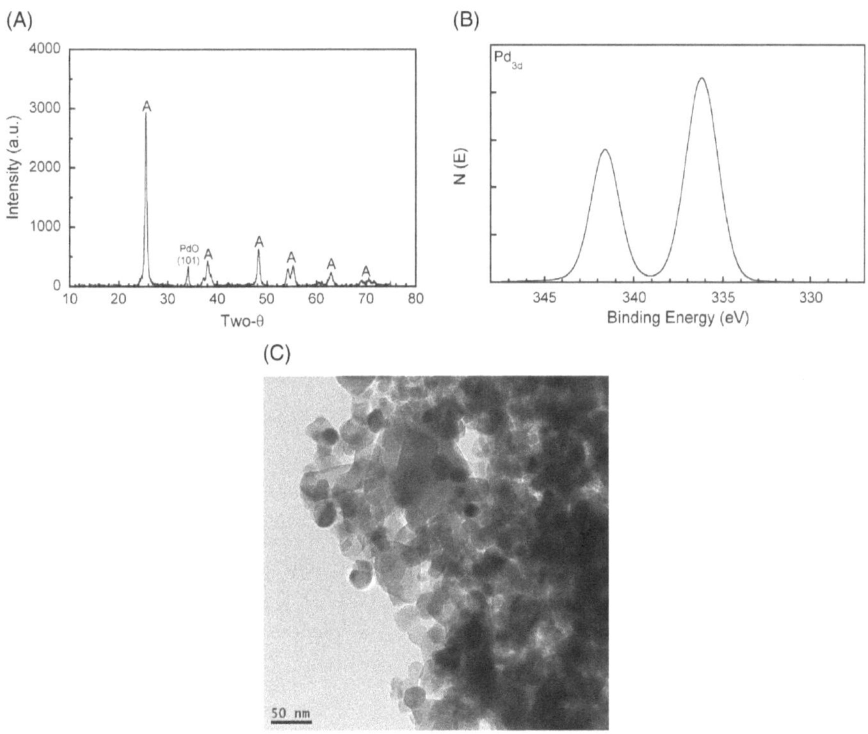

FIGURE 19.7

(A) Typical X-ray diffraction patterns of nitrogen-doped titanium oxide (TiON)/palladium-modified (PdO) powders. (B) High-resolution X-ray photoelectron spectroscopy scan over Pd 3d peaks of TiON/PdO nanoparticle powders. (C) TEM image of TiON/PdO nanoparticle powders.

Adapted from [58].

peaks belong to anatase with no rutile phase observed. Besides the anatase-type structure, a peak assigned to PdO (101) can be clearly identified, suggesting that palladium additive exists as PdO and is not incorporated into the anatase structure. High-resolution XPS scan over Pd 3d (Fig. 19.7B) peaks shows that the binding energy of Pd $3d_{5/2}$ is approximately 336.20 eV, which can be attributed to PdO species. This observation is in accordance with our XRD experiment result. The Brunauer-Emmett-Teller (BET) surface specific areas of these powders are approximately 50 m^2/g, corresponding to an average particle diameter approximately 30 nm. Figure 19.7C shows the TEM image of a TiON/PdO nanoparticle sample. The sample was composed of nanosized particles with nonuniform shapes. TEM revealed a particle size very similar to those obtained from the BET measurements.

The optical properties of these TiON/PdO nanoparticle powders were investigated by measuring the diffuse reflectance spectra. From the reflectance data, optical absorbance can be approximated by the Kubelka–Munk function [55]. Figure 19.8A shows the light absorbance of TiON/PdO nanoparticle powders, compared with TiON nanoparticle powders with similar nitrogen dopant concentration (N/Ti atomic ratio at approximately 0.06). TiON powders show a clear shift into the visible-light range ($>$400 nm) as expected, which can be attributed to the nitrogen doping. TiON/PdO powders show much higher visible-light

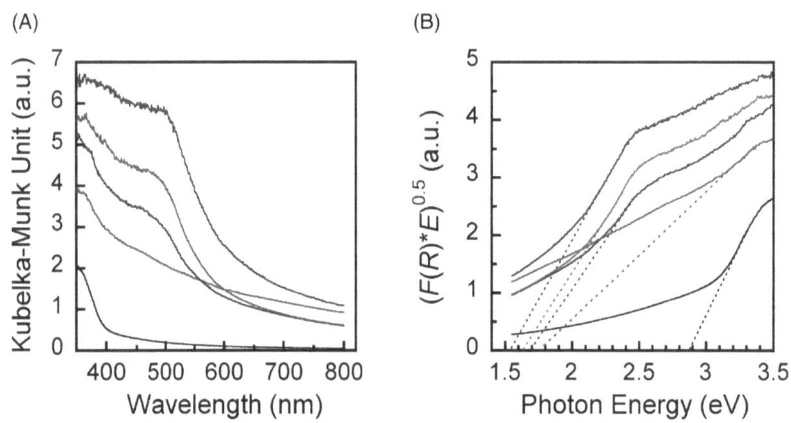

FIGURE 19.8

(A) Optical absorbance (in terms of Kubelka–Munk equivalent absorbance units) of nitrogen-doped titanium oxide (TiON)/palladium-modified (PdO) powders, compared with the optical absorbance of TiON powder. (B) Tauc Plot constructed from (A). Band-gap values are determined from the extrapolation of the linear Tauc Region line to the photon energy abscissa.

Adapted from [58].

absorption than TiON powders. The visible-light absorbance increased with the increase of the Pd ion concentration, which suggests that Pd additive promotes visible-light absorption in TiON nanoparticle photocatalysts. Figure 19.8B shows the Tauc Plot [55] constructed from Fig. 19.8A to determine the semiconductor band gap. TiON has a band gap at approximately 2.87 eV, which is consistent with its visible-light absorbance ability. TiON/PdO powders show a much smaller band gap, which explains their better visible-light absorbance ability. With the increase of the Pd content from 0.1 to 2.0 percent, the band gap of TiON/PdO powders decreases from approximately 1.76 to 1.54 eV.

To understand the mechanism of metal ion modification in this novel material system, TiON/PdO thin film samples with similar structure, composition, and light absorption properties to TiON/PdO nanoparticle were synthesized by ion-beam-assisted deposition (IBAD) [59]. The film samples were used in the mechanism study because they are easier to be rid of surface contamination and to be analyzed by various material characterization tools than nanoparticle powders. In the TiON/PdO nanoparticle photocatalyst, palladium additions exist as PdO species and are not incorporated into the anatase structure. In situ XPS study revealed changes in the valence states of the palladium species when illuminated. In the in situ experiment, the sample was first placed in the dark for 3 hours right before the XPS high-resolution scan over Pd 3d peak in the dark. Then, XPS scans were performed with a visible-light illumination on the sample simultaneously after 0.5 hour, 1 hour, and 2 hours. Differences in the Pd 3d peak shape and position were observed between scans in the dark and under visible-light illumination. For comparison purpose, the same in situ XPS study was also conducted on the TiO_2/PdO sample. Figure 19.9 shows the comparison of XPS

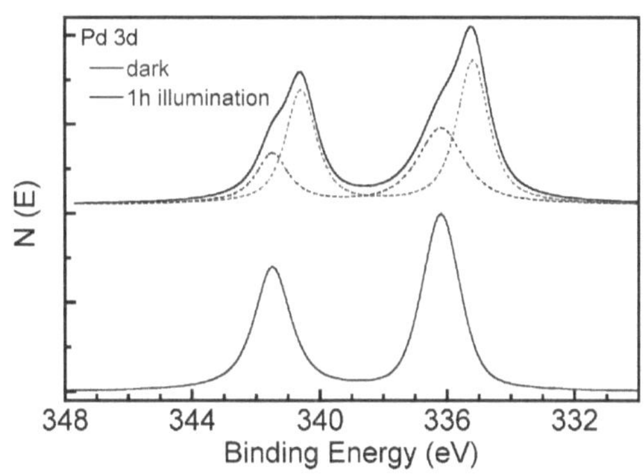

FIGURE 19.9

In situ X-ray photoelectron spectroscopy high-resolution scan over Pd 3d peaks on nitrogen-doped titanium oxide (TiON)/TiON thin film in dark and after 1-hour visible-light illumination. Note that the one dashed curve fits PdO 3d peaks, whereas the other dashed curve fits Pd^{2+} 3d peaks after 1-h visible-light illumination.

Adapted from [59].

high-resolution scans over Pd 3d peaks in the dark and after 1-hour visible-light illumination, respectively. In the dark, the binding energy of Pd $3d_{5/2}$ is approximately 336.20 eV, indicating that Pd dopant exists as PdO. Under visible-light illumination, however, Pd peaks are broadened and the binding energy of Pd $3d_{5/2}$ shifts to approximately 335.30 eV. The broad Pd $3d_{5/2}$ is best fit as a combination of Pd^{2+} $3d_{5/2}$ (peak at 336.2 eV) and Pd^0 $3d_{5/2}$ (peak at 335.2 eV), which clearly demonstrates that a portion of PdO semiconductor nanoparticles in TiON/PdO thin film is reduced to metallic Pd nanoparticles under visible-light illumination. The change in the valence state of Pd is consistent with the notion that metal-ion dopants in TiO_2 can act as electron traps to alter the electron–hole pair recombination rate [18–21,57]. The metallic nanoparticles can also create the surface plasma resonance originated from collective oscillation of free electrons to enhance the visible-light absorption of TiON/PdO as observed in Fig. 19.8. In contrast, TiO_2/PdO sample shows no change in Pd 3d peak shape and position under the same visible-light illumination condition because TiO_2 thin film absorbs mainly UV light. Thus, no significant interaction should be expected. The binding energy of Pd $3d_{5/2}$ in TiO_2/PdO sample is approximately 336.20 eV, indicating that Pd dopant exists as PdO in the dark or under visible-light illumination.

The photocatalytic activities of TiON/PdO nanoparticle photocatalysts were characterized by photocatalytic degradation of humic acid (HA) under visible-light illumination, with the TiON nanoparticle photocatalyst as a comparison basis. Photocatalytic degradation of HA was conducted by exposing the HA solution to various photocatalysts under visible light ($l >$ 400 nm) for varying time intervals. After the centrifugation to recover photocatalysts, the light absorption of the clear solution was measured and the remaining percentage of HA in the solution was calculated by the ratio between the light absorptions of photocatalyst-treated and untreated HA solutions. Figure 19.10A shows representative light absorption spectra of HA solutions with different photocatalyst treatment times. As the illumination time increased, light absorption decreased steadily for HA solutions treated by TiON/PdO photocatalysts at their l_{max} (approximately 280 nm), indicating more and more HA was degraded under the photocatalyst treatment. Figure 19.10B summarizes the residual percentage of HA as a function of visible-light illumination time for TiON/PdO or TiON photocatalysts in a semi-logarithmic plot format. A first-order exponential decay of HA was observed, which can be fitted into Equation 19.2:

$$RP = ae^{-bt} \tag{19.2}$$

where RP is the residual percentage of HA, t is the visible-light illumination time, a and b are the first-order exponential fitting constants. Constant b, the decay rate, can be used as a parameter to compare the photodegradation efficiency of different photocatalysts. An enhanced photodegradation efficiency was observed within a certain range of palladium additive concentration. With the addition of palladium ion, the decay rate constant b increases from 0.012 for TiON to 0.018

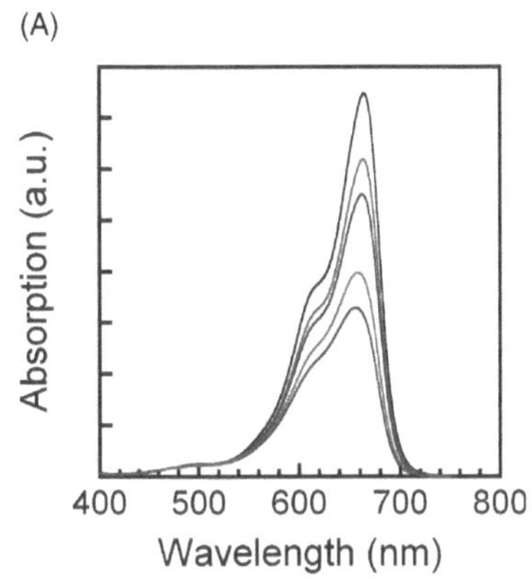

FIGURE 19.10

(A) Typical absorption spectra of humic acid (HA) solution during photodegradation under visible-light illumination. (Note that the black line stands for 0 hours, red line for 2 hours, blue line for 5 hours, and green line for 10 hours.) (B) HA residue percentage versus visible-light (> 400 nm) illumination time with nitrogen-doped titanium oxide (TiON), TiON/0.1 percent PdO, TiON/0.5 percent PdO, TiON/1.0 percent PdO, and TiON/2.0 percent PdO, respectively. (Note that the lines are fitted into the first-order exponential formula.)

Adapted from [58].

for TiON/PdO containing 0.5 percent PdO, representing an approximately 50 percent increase. When the palladium dopant concentration was further increased, an inverse effect was observed that the decay rate constant *b* decreased to 0.009 for TiON/PdO with 2.0 percent PdO, an approximately 25 percent decrease when compared to the TiON with no PdO additives.

It is widely reported that transition-metal-ion additives play a complex role in affecting the photocatalytic activity of TiO_2 [18–21,57]. Besides acting as a charge trapping center, they can also serve as charge recombination center and affect the charge release and migration. In the work by Choi et al. [19], the ability of a metal-ion additive to function as an effective trap in TiO_2 was dependent on its concentration, as found here for the metal-ion-modified TiON. Thus, although stronger visible-light absorbance may be achieved by increasing the Pd additive concentration, there is no simple relationship between visible-light absorbance and photocatalytic efficiency. Other metal-ion-modifications of TiON, such as Ag^+, Nd^{3+}, Cu^{2+}, Y^{3+}, Ce^{4+}, and Fe^{3+}, had also been explored. They demonstrated similar enhancements on the photoactivity of TiON under visible-light illumination, whereas different performances were observed between different species. To achieve high photocatalytic efficiency in this transition-metal-ion-modified TiON photocatalysts, a careful optimization of the transition-metal-ion-additive concentration is needed.

19.4 Nanostructured nitrogen-doped titanium-oxide-based photocatalysts

In the previous sections, we had discussed the synthesis of TiON-based photocatalysts as nanoparticles or thin films made up of nanosized grains. By the combination of the sol–gel process with a template technique, nanostructured TiON-based photocatalysts have been made into various forms to satisfy the requirement from different applications. For example, it is desirable to immobilize photocatalysts onto large porous templates for water treatment purposes because the immobilization removes the need of recycling photocatalysts in the powder form from the aqueous environment, reducing the cost and yet increasing the treatment efficiency.

Figure 19.11 shows SEM images of TiON/PdO photocatalytic fibers created on active carbon glass fiber (ACGF) templates [60]. At the low magnification (Fig. 19.11A), a nonwoven fiber network is observed, which reflects the ACGF template structure. On the individual glass fiber, a thin layer of TiON/PdO photocatalyst was immobilized. High-magnification SEM image of the TiON/PdO fiber surface (Fig. 19.11B) shows that the TiON/PdO photocatalytic layer has a mesoporous structure, and the average particle size is several nanometers. Thus, the TiON/PdO fiber network has a combination of macroporous (from ACGF template matrix) and mesoporous (from coated photocatalytic layer) structures. This dual porous structure is beneficial for achieving relatively high disinfection efficiency because it provides access to the entire photocatalyst contact area for contaminants present in water.

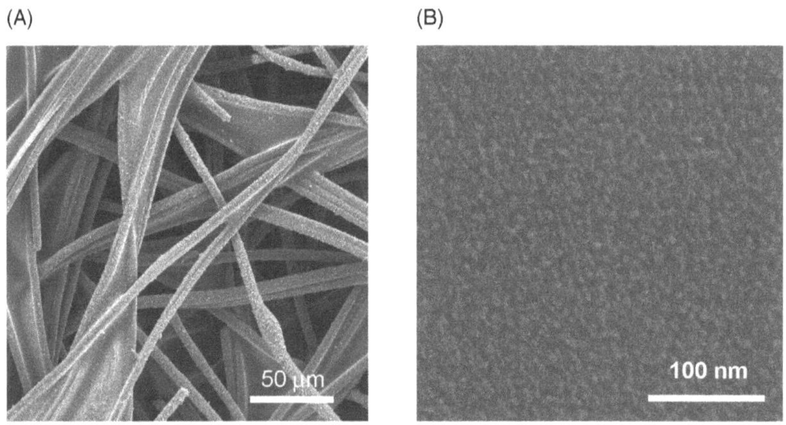

FIGURE 19.11

Low-magnification (A) and high-magnification (B) scanning electron microscope images of nitrogen-doped nitrogen-doped titanium oxide (TiON)/palladium-modified (PdO) photocatalytic fiber, respectively.

Adapted from [60].

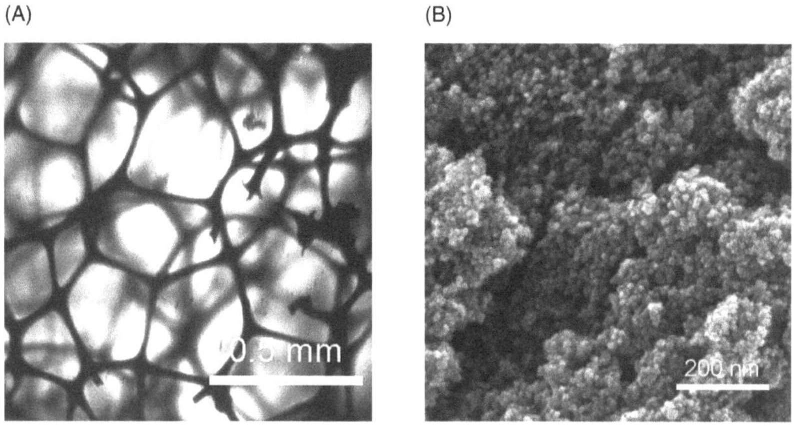

FIGURE 19.12

Optical (A) and scanning electron microscope (B) images of nitrogen-doped titanium oxide (TiON)/palladium-modified (PdO) photocatalytic foam, respectively.

Adapted from [61].

Similar dual porous structure may also be formed in a monolithic TiON foam grown on polymer templates [61]. As shown in Fig. 19.12, the foam contains an open network of large pores (hundreds of micrometers in diameter, Fig. 19.12A), with heavily populated mesopores in the walls (Fig. 19.12B). The mesoporous structure improves the contact efficiency between contaminants present in water and photocatalysts, whereas the macropores reduce the pressure drop across the foam in a dynamic reactor.

19.5 Environmental properties of nitrogen-doped titanium-oxide-based photocatalysts

A major potential application of semiconductor-based photocatalysts is in the control of environmental contaminants in water or air [12–15]. The basis for photocatalytic control is the production of highly reactive oxidants, such as OH radicals, for oxidization of organic pollutants, disinfection of microorganisms, and degradation of hazardous disinfection by-products (DBPs) and disinfection by-product precursors (DBPPs). In these previous studies, TiO_2 was used as the photocatalyst, but its photocatalytic capability is limited to activation by UV light.

With the newly developed TiON-based photocatalysts, photocatalytic degradation and disinfection can be implemented with visible light. The replacement of UV by visible light offers potential for low-cost environmental measures, especially for water treatment, where UV access is rather limited. In Section 19.3, the removal of organic contaminations had been demonstrated by the photodegradation of HA by TiON/PdO under visible-light illumination. In this section, examples will be given of pathogen disinfection using visible-light photocatalysis.

Figure 19.13A shows the survival ratio of *Escherichia coli* (*E. coli*) bacteria versus visible-light irradiation time following different photocatalytic treatments under visible-light illumination [62]. *E. coli* is a model gram-negative bacterium, which is widely used as a bacterial indicator. TiON/PdO shows a fast disinfection efficiency against *E. coli*. After 30-minute visible-light illumination, the survival ratio of *E. coli* drops to approximately 10^{-8}. Figure 19.13B shows the survival ratio of *Pseudomonas aeruginosa* bacteria versus visible-light irradiation time with different photocatalyst treatments under visible-light illumination [62]. This gram-negative bacterium—one of the most formidable biofilm-forming organisms—is ubiquitous in the natural environment. Fast disinfection efficiency was also achieved by the TiON/PdO photocatalyst.

Figure 19.14 shows the survival ratio of *Bacillus subtilis* spores versus visible-light irradiation time with different photocatalyst treatments under visible-light illumination [63]. Spores are among the most resistant living structures as they can survive harsh treatments that destroy most other bacteria, including the bacteria that form the spores, and pose a great potential threat to public health because they can lead to a high mortality rate and transmit rapidly in aerosol form. Although TiON does not have sufficient power to inactivate spores, TiON/PdO offers fast disinfection performance under visible-light illumination. After several hour treatments, over 99 percent inactivation can be achieved. Following the

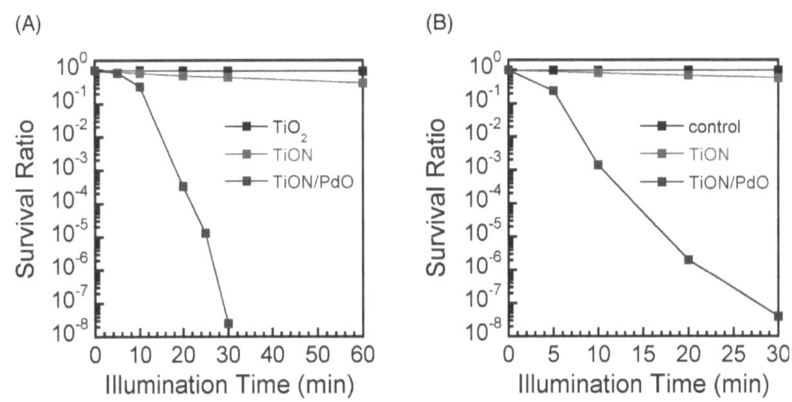

FIGURE 19.13

(A) Survival ratio of *E. coli* cells versus visible-light irradiation time on different photocatalysts titanium oxide (TiO2), nitrogen-doped TiO2 (TiON), and TiON/palladium-modified (PdO) photocatalysts. The cell suspension had an initial concentration ca.107 cfu/mL. (B) Survival ratio of *P. aeruginosa* cells versus visible-light irradiation time on different photocatalysts TiON and TiON/PdO.

Adapted from [62].

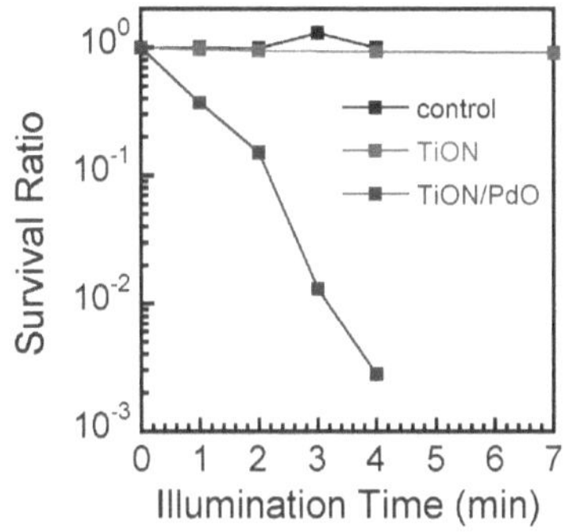

FIGURE 19.14

Survival ratio of *B. subtilis* spores versus visible-light irradiation time on different treatments: control without photocatalyst (■),nitrogen-doped nitrogen-doped titanium oxide (TiON) (■), and TiON/palladium-modified photocatalyst (■).

Adapted from [63].

visible-light photocatalytic treatment with TiON/PdO, both the spore coat and core suffer from morphological changes, which indicate that the spores are indeed killed by the photocatalytic oxidation on TiON/PdO photocatalyst under visible-light illumination.

19.6 Conclusions and future directions

Nanostructured TiON-based semiconductor photocatalysts have been developed with various compositions and in various forms, which extend the absorption band-edge of TiO_2 into the visible-light region with improved stability, photocatalytic efficiency, and ease of the doping process. Metal-ion modification has strong effects on the optical and photocatalytic properties of TiON photocatalysts, and greatly enhances photoactivity of TiON under visible-light illumination if the metal-ion-additive concentration is optimized. By combining sol–gel process with templating, nanostructured photocatlysts have been synthesized into fibrous and porous forms to meet the specific requirements for water and air purification. With TiON-based photocatalysts, visible-light-induced photocatalytic degradation and disinfection have been successfully achieved. These nanostructured photocatalysts have shown great promise in a wide range of optical and photochemical solutions, especially for environmental controls.

Although visible-light photocatalysts have shown great promise in degrading organics and microbials in water, a number of great challenges still remain and require attention in the future work. First of all, the mechanism of the interaction between microorganisms with photocatalysts needs to be examined in detail to gain a clear understanding of the disinfection effect and provide guidelines for future water treatment design based on photocatalysis. Second, much of the work so far has been focused on TiO_2 and very few other material systems have been investigated. There is a strong need to explore other material systems with improved photocatalytic performance under visible-light illumination. Third, there are early indications that chemical modifications of the photocatalysts may enhance photocatalytic activities selectively. If confirmed, it would be possible to tailor the photocatalyst to specific environmental applications. Fourth, new and improved methods are needed to synthesize photocatalysts into various forms at low costs to meet the needs of different environmental applications. For example, various forms of semiconductor-based photocatalysts (particularly nanotubes, nanowires, and nanosized three-dimensional structures) may be explored as photoelectrocatalytic sensors for rapid monitoring and control of environmental pollutants. Finally, broader environmental applications should be explored. For example, the photocatalytic disinfection of various microorganisms, such as fungi or viruses, should be investigated, which is very important to the human health, environmental protection, and agriculture area.

References

[1] A. Fujishima, K. Honda, Electrochemical photolysis of water at a semiconductor electrode, Nature 238 (1972) 37–38.

[2] J.R. Bolton, Solar photoproduction of hydrogen: a review, Solar Energy 57 (1996) 37–50.

[3] S.U.M. Khan, J. Akikusa, Photoelectrochemical splitting of water at nanocrystalline n-Fe_2O_3 thin-film electrodes, J. Phys. Chem. B. 103 (1999) 7184–7189.

[4] J. Akikusa, S.U.M. Khan, Stability and photoresponse of nanocrystalline n-TiO_2 and n-TiO_2/Mn_2O_3 thin film electrodes during water splitting reactions, J. Electrochem. Soc. 145 (1998) 89–93.

[5] S.U.M. Khan, J. Akikusa, Photoelectrolysis of water to hydrogen in p-SiC/Pt and p-SiC/n-TiO_2 cells, Int. J. Hydrogen. Energy. 27 (2002) 863–870.

[6] O. Khaselev, J.A. Turner, A monolithic photovoltaic–photoelectrochemical device for hydrogen production via water splitting, Science 280 (1998) 425–427.

[7] S. Licht, B. Wang, S. Mukerji, T. Soga, M. Umeno, H. Tributsch, Efficient solar water splitting, exemplified by RuO_2-catalyzed AlGaAs/Si photoelectrolysis, J. Phys. Chem. B. 104 (2000) 8920–8924.

[8] B. O'Regan, M. Gratzel, A low-cost, high-efficiency solar cell based on dye-sensitized colloidal TiO_2 films, Nature 353 (1991) 737–740.

[9] G.R. Comte, M. Gratzel, A contribution to the optical design of dye-sensitized nano-crystalline solar cells, Solar Energy Mater. Solar Cells 58 (1999) 321–336.

[10] G. Sauve, M.E. Cass, S.J. Doig, I. Lauermann, K. Pomykal, N.S. Lewis, High quantum yield sensitization of nanocrystalline titanium dioxide photoelectrodes with cis-dicyanobis(4,4'-dicarboxy-2,2'-bipyridine)osmium(II) or tris(4,4'-dicarboxy-2,2'-bipyridine)osmium(II) complexes, J. Phys. Chem. B. 104 (2000) 3488–3491.

[11] M.R. Hoffman, S.T. Martin, W. Choi, D.W. Bahnemann, Environmental applications of semiconductor photocatalysis, Chem. Rev. 95 (1995) 69–96.

[12] N.S. Frank, A.J. Bard, Heterogeneous photocatalytic oxidation of cyanide and sulfite in aqueous solutions at semiconductor powders, J. Phys. Chem. 81 (1977) 1484–1488.
[13] A. Hagfeldt, M. Graetzel, Light-induced redox reactions in nanocrystalline systems, Chem. Rev. 95 (1995) 49–68.
[14] H. Einaga, S. Futamura, T. Ibusuki, Photocatalytic decomposition of benzene over TiO_2 in a humidified airstream, Phys. Chem. Chem. Phys. 1 (1999) 4903–4908.
[15] A. Fujishima, T.N. Rao, D.A. Tryk, Titanium dioxide photocatalysis, J. Photochem. Photobiol. C 1 (2000) 1–21.
[16] A.L. Linsebigler, G. Lu, J.T. Yates Jr, Photocatalysis on TiO_2 surfaces: principles, mechanisms, and selected results, Chem. Rev. 95 (1995) 735–758.
[17] K. Hashimoto, H. Irie, A. Fujishima, TiO_2 photocatalysis: a historical overview and future prospects, Jpn. J. Appl. Phys. 44 (2005) 8269–8285.
[18] A.K. Ghosh, H.P. Maruska, Photoelectrolysis of water in sunlight with sensitized semiconductor electrodes, J. Electrochem. Soc. 124 (1977) 1516–1522.
[19] W. Choi, A. Termin, M.R. Hoffmann, The role of metal ion dopants in quantum-sized TiO_2: correlation between photoreactivity and charge carrier recombination dynamics, J. Phys. Chem. 98 (1994) 13669–13679.
[20] M. Anpo, Photocatalysis on titanium oxide catalysts: approaches in achieving highly efficient reactions and realizing the use of visible light, Catal. Surv. Jpn. 1 (1997) 169–179.
[21] V. Subramanian, E. Wolf, P.V. Kamat, Semiconductor–metal composite nanostructures: to what extent do metal nanoparticles improve the photocatalytic activity of TiO_2 films? J. Phys. Chem. B. 105 (2001) 11439–11446.
[22] R.G. Breckenridge, W.R. Hosler, Electrical properties of titanium dioxide semiconductors, Phys. Rev. 91 (1953) 793–802.
[23] D.C. Cronemeyer, Infrared absorption of reduced rutile TiO_2 single crystals, Phys. Rev. 113 (1959) 1222–1226.
[24] R. Asahi, T. Morikawa, T. Ohwaki, K. Aoki, Y. Taga, Visible-light photocatalysis in nitrogen-doped titanium oxides, Science 293 (2001) 269–271.
[25] H. Irie, Y. Watanabe, K. Hashimoto, Nitrogen-concentration dependence on photocatalytic activity of $TiO_{2-x}N_x$ powders, J. Phys. Chem. B. 107 (2003) 5483–5486.
[26] T. Lindgren, J.M. Mwabora, E. Avendaño, J. Jonsson, C.G. Granqvist, S.E. Lindquist, Photoelectrochemical and optical properties of nitrogen doped titanium dioxide films prepared by reactive DC magnetron sputtering, J. Phys. Chem. B. 107 (2003) 5709–5716.
[27] Irie, Watanabe, and Hashimoto, Nitrogen-concentration dependence. pp. 5483–5486.
[28] S. Sakthivel, H. Kisch, Photocatalytic and photoelectrochemical properties of nitrogen-doped titanium dioxide, ChemPhysChem. 4 (2003) 487–490.
[29] C. Burda, Y. Lou, X. Chen, A.C.S. Samia, J. Stout, J.L. Gole, Enhanced nitrogen doping in TiO_2 nanoparticles, Nano. Lett. 3 (2003) 1049–1051.
[30] S. Yin, Q. Zhang, F. Saito, T. Sato, Preparation of visible light-activated titania photocatalyst by mechanochemical method, Chem. Lett. 32 (2003) 358–359.
[31] O. Diwald, T.L. Thompson, T. Zubkov, E.G. Goralski, S.D. Walck, J.T. Yates, Photochemical activity of nitrogen-doped rutile $TiO_2(110)$ in visible light, J. Phys. Chem. B. 108 (2004) 6004–6008.
[32] S.W. Yang, L. Gao, New method to prepare nitrogen-doped titanium dioxide and its photocatalytic activities irradiated by visible light, J. Am. Ceramic Soc. 87 (2004) 1803–1805.

[33] C.D. Valentin, G. Pacchioni, A. Selloni, Origin of the different photoactivity of N-doped anatase and rutile TiO_2, Phys. Review B. 70 (2004) 85–116.
[34] C. Lettmann, K. Hildenbrand, H. Kisch, W. Macyk, W.F. Maier, Visible light photodegradation of 4-chlorophenol with a coke-containing titanium dioxide photocatalyst, Applied Catalysis B. 32 (2001) 215–227.
[35] S. Livraghi, A. Votta, P. Cristina, E. Giamello, The nature of paramagnetic species in nitrogen doped TiO_2 active in visible light photocatalysis, Chem. Commun. (2005) 498–500.
[36] S. Livraghi, M.C. Paganini, E. Giamello, A. Selloni, C.D. Valentin, G. Pacchioni, Origin of photoactivity of nitrogen-doped titanium dioxide under visible light, J. Am. Chem. Soc. Vol. 128 (2006) 15666–15671.
[37] S.U.M. Khan, M. Al-Shahry, W.B. Ingler, Efficient photochemical water splitting by a chemically modified n-TiO_2, Science 297 (2002) 2243–2245.
[38] S. Sakthivel, H. Kisch, Daylight photocatalysis by carbon-modified titanium dioxide, Angewa. Chemie Int. Ed. 42 (2003) 4908–4911.
[39] H. Irie, Y. Watanabe, K. Hashimoto, Carbon-doped anatase TiO_2 powders as a visible-light sensitive photocatalyst, Chem. Lett. 32 (2003) 772–773.
[40] K. Noworyta, J. Augustynski, Spectral photoresponses of carbon-doped TiO_2 film electrodes, Electrochem. Solid-State Let. 7 (2004) E31–E33.
[41] H. Wang, J.P. Lewis, Effects of dopant states on photoactivity in carbon-doped TiO_2, J. Phys. Condens. Matter 17 (2005) L209–L213.
[42] W. Wang, P. Serb, P. Kalck, J.L. Faria, Visible light photodegradation of phenol on MWNT-TiO_2 composite catalysts prepared by a modified sol–gel method, J. Mol. Cat. A: Chem. 235 (2005) 194–199.
[43] L. Lin, W. Lin, Y.X. Zhu, B.Y. Zhao, Y.C. Xie, Y. He, et al., Uniform carbon-covered titania and its photocatalytic property, J. Mol. Catal. A: Chem. 236 (2005) 46–53.
[44] M.E. Rincon, M.E. Trujillo-Camacho, A.K. Cuentas-Gallegos, Sol–gel titanium oxides sensitized by nanometric carbon blacks: comparison with the optoelectronic and photocatalytic properties of physical mixtures, Catal. Today 107–108 (2005) 606–611.
[45] T. Umebayashi, T. Yamaki, H. Itoh, K. Asai, Band gap narrowing of titanium dioxide by sulfur doping, Appl. Phys. Lett. 81 (2002) 454–456.
[46] T. Umebayashi, T. Yamaki, S. Tanaka, K. Asai, Visible light-induced degradation of methylene blue on S-doped TiO_2, Chem. Lett. 32 (2003) 330–331.
[47] T. Umebayashi, T. Yamaki, S. Yamamoto, A. Miyashita, S Tanaka, T. Sumita, et al., Sulfur-doping of rutile-titanium dioxide by ion implantation: photocurrent spectroscopy and first-principles band calculation studies, J. Appl. Phys. 93 (2003) 5156–5160.
[48] T. Ohno, T. Mitsui, M. Matsumura, Photocatalytic activity of S-doped TiO_2 photocatalyst under visible light, Chem. Lett. 32 (2003) 364–365.
[49] T. Yamamoto, Y. Yamashita, I. Tanaka, F. Matsubara, A. Muramatsu, Electronic states of sulfur doped TiO_2 by first principles calculations, Mater. Trans. 45 (2004) 1987–19890.
[50] J.C. Yu, J.G. Yu, W.K. Ho, Z.T. Jiang, L.Z. Zhang, Effects of F^- doping on the photocatalytic activity and microstructures of nanocrystalline TiO_2 powders, Chem. Mater. 14 (2002) 3808–3816.
[51] G.R. Torres, T. Lindgren, J. Lu, C.G. Granqvist, S.E. Lindquist, Photoelectrochemical study of nitrogen-doped titanium dioxide for water oxidation, J. Phys. Chem. B. 108 (2004) 5995–6003.

[52] R. Nakamura, T. Tanaka, Y. Nakato, Mechanism for visible light responses in anodic photocurrents at N-doped TiO_2 film electrodes, J. Phys. Chem. B. 108 (2004) 10617–10620.

[53] J.Y. Lee, J. Park, J.H. Cho, Electronic properties of N- and C-doped TiO_2, Appl. Phys. Lett. 87 (2005) 011904.

[54] Q. Li, R. Xie, E.A. Mintz, J.K. Shang, Effect of precursor ratio on synthesis and optical absorption of TiON photocatalytic nanoparticles, J. Am. Ceram. Soc. 90 (2007) 1045–1050.

[55] J. Tauc, R. Grigorovici, A. Vancu, Optical properties and electronic structures of amorphous germanium, Phys. Status Solidi 15 (1966) 627–637.

[56] P.G. Wu, C.H. Ma, J.K. Shang, Effects of nitrogen doping on optical properties of TiO_2 thin films, Appl. Phys. A. 81 (2005) 1411–1417.

[57] S.I. Shah, W. Li, C.P. Huang, O. Jung, C. Ni, Study of Nd^{3+}, Pd^{2+}, Pt^{4+}, and Fe^{3+} dopant effect on photoreactivity of TiO_2 nanoparticles, Proc. Natl. Acad. Sci. U.S.A. 99 (2002) 6482–6486.

[58] Q. Li, R. Xie, E.A. Mintz, J.K. Shang, Enhanced visible-light photocatalytic degradation of humic acid by palladium oxide-sensitized nitrogen-doped titanium oxide, J. Am. Ceram. Soc. 90 (2007) 3863–3868.

[59] Q. Li, W. Liang, J.K. Shang, Enhanced visible-light absorption from PdO nanoparticles in nitrogen-doped titanium oxide thin films, Appl. Phys. Lett. 90 (2007) 63–109.

[60] Q. Li, M.A. Page, B.J. Marinãs, J.K. Shang, Treatment of coliphage MS2 with palladium-modified nitrogen-doped titanium oxide photocatalyst illuminated by visible light, Environ. Sci. Technol. 42 (2008) 6148–6153.

[61] P. Wu, PhD dissertation, University of Illinois at Urbana–Champaign, May 2007.

[62] P. Wu, R. Xie, J.A. Imlay, J.K. Shang, Visible-light-induced photocatalytic inactivation of bacteria by composite photocatalysts of palladium oxide and nitrogen-doped titanium oxide, Appl. Catal. B: Environ. (2008) submitted.

[63] P. Wu, R. Xie, J.K. Shang, Enhanced visible-light photocatalytic disinfection of bacterial spores by palladium-modified nitrogen-doped titanium oxide, J. Am. Ceram. Soc. 91 (2008) 2957–2962.

CHAPTER

20 Nanotechnology-Enabled Water Disinfection and Microbial Control: Merits and Limitations

Shaily Mahendra[1], Qilin Li[2], Delina Y. Lyon[3], Lena Brunet[4] and Pedro J.J. Alvarez[2]

[1]Department of Civil and Environmental Engineering, University of California at Los Angeles, Los Angeles, CA, USA [2]Department of Civil and Environmental Engineering, Rice University, Houston, TX, USA [3]Shell Oil Company Houston, TX, USA [4]Institut Francais du Petrole, Lyon, France

20.1 Introduction

Waterborne infectious diseases continue to be the leading cause of death in many developing nations. According to the recent WHO report [1], approximately four billion cases of diarrhea are reported each year, causing 1.8 million deaths, 90 percent of which are among children under the age of five. About 10 percent of the developing world population is infected by intestinal worms; six million people are blind from trachoma with five hundred million of the population at risk from this disease. In addition, recent outbreaks of infectious diseases have occurred in industrialized nations, following floods and other extreme events [2,3]. Water

Street, Sustich, Duncan and Savage. Nanotechnology Applications for Clean Water, 2nd Edition.
 DOI: http://dx.doi.org/10.1016/B978-1-4557-3116-9.00020-2

disinfection serves not only to protect the health of consumers but also to lower the cost of maintaining any water-driven system. Undesirable biofilm formation in water distribution and storage systems and in industrial cooling facilities greatly increases the cost of maintaining these systems and hinders efficient water conservation. Such microbial growth can also have detrimental economic consequences associated with infrastructure corrosion, secondary contaminants, taste and odor problems, and increased friction and energy loss during water flow. The importance of water disinfection and biofilm control cannot be overstated.

Current disinfection and microbial control approaches often rely on chemical oxidants such as free chlorine, chloramines, and ozone, which are quite effective at killing bacteria and inactivating viruses in water treatment plants. However, they are ineffective against cyst-forming protozoa such as *Giardia* and *Cryptosporidium*. In addition, these chemical disinfection methods often produce harmful disinfection by-products, such as carcinogenic trihalomethanes when free chlorine reacts with natural organic matter (NOM), N-nitrosodimethylamine (NDMA) when monochloramine or chlorine react with ammonia and organic amines, and bromate when ozone reacts with bromide [4,5]. Therefore, there is a need to consider innovative approaches that enhance the reliability and robustness of disinfection while avoiding unintended adverse health effects. It is in this context that antimicrobial nanomaterials may play an important future role.

The demand for nanotechnology-enabled microbial control may also increase due to a greater need for decentralized or point-of-use water treatment and reuse systems. Water distribution systems in many cities are approaching or exceeding their economic life and cannot support rapid metropolitan expansion. The high energy consumption and health risks associated with microbial contamination during water distribution through centralized systems call for technology and infrastructure reform. One solution is to develop novel decentralized water treatment technologies to alleviate dependence on major infrastructure and exploit local alternative water sources and recycling opportunities. Future high-performance, small-scale, and point-of-use systems incorporating antimicrobial nanomaterials may increase the robustness of water distribution networks for neighborhoods and buildings not connected to a central network, and for emergency response following catastrophic events.

This chapter reviews the applicability of several nanomaterials for water disinfection and biofouling control, and highlights future research needed to enhance the efficacy and sustainability of nanotechnology-enabled microbial control.

20.2 Current and potential applications

Nanomaterials with similar or higher disinfection efficacies than conventional water treatment would make excellent alternative treatments or could be used in conjunction with existing technologies such as ultraviolet (UV) disinfection. They could also be used for biofouling control for water filtration membranes and other

Table 20.1 Applications of Nanomaterials Utilizing Antimicrobial Properties

Nanomaterial	Antimicrobial Mechanism	Current Applications	Potential Future Applications
nAg	Release of Ag^+ ions, disruption of cell membrane and electron transport	Potable water filters, clothing, medical devices, coatings, washing machines, refrigerators, food storage	Surface coatings, membranes
TiO_2	Production of ROS, cell membrane and cell wall damage	Air purifiers, water purifiers	Solar and UV disinfection of water and wastewater, reactive membranes, hollow fibers, biofouling-resistant surfaces
CNT	Physically compromise cell envelope	None	Biofouling-resistant membranes, carbon hollow fibers, packing in fixed bed columns

CNT: Carbon nanotubes; ROS: Reactive oxygen species; UV: Ultraviolet.

surfaces in water treatment reactors and distribution pipelines. The antimicrobial mechanisms as well as current and potential applications of several nanomaterials in microbial control are summarized in Table 20.1. Their merits and limitations relevant to water disinfection are discussed later.

20.2.1 Nanosilver

Silver is the most commonly used nanomaterial for microbial control. Several antimicrobial mechanisms of nanosilver (nAg) have been postulated, such as adhesion to cell surface altering the membrane properties, penetration inside bacteria resulting in DNA damage, and the release of antimicrobial Ag^+ ions [6,7]. Currently, there are over one hundred consumer products that contain nAg as an antimicrobial agent including nutrition supplements, food storage containers, kitchenware, refrigerators, textiles, laundry additives, washing machines, paints, faucets, sanitizers, contact lens solutions, catheters, and wound dressings [8–10]. Several home water purification systems utilizing nAg are available on the market, for example, Aquapure®, Kinetico®, and QSI-Nano®. These systems can remove 99.99 percent of pathogenic bacteria, viruses, protozoa, and cysts. The leaching of nAg particles or Ag^+ ions in polished water is undetectable. Future applications of nAg include coating of pipes in distributions systems to prevent regrowth of pathogens, and incorporation into membranes for large-scale water filtration [11,12].

20.2.2 Titanium oxide

The antibacterial activity of titanium oxide (TiO_2) is related to ROS production, especially peroxide and hydroxyl radicals under UV-A (320–400 nm) irradiation via both oxidative and reductive pathways [13]. However, bactericidal activity of TiO_2 (330 nm average aggregate size) has also been observed in the dark, indicating that other mechanisms may be involved [14]. Commercial water purification systems based on TiO_2 photocatalysis already exist (e.g., Purifics®). Studies on the photocatalytic disinfection efficiency of TiO_2 are relatively few, but have demonstrated the potential benefits of using TiO_2 for drinking water disinfection (Table 20.1). The most promising property of TiO_2-based disinfection is probably its photoactivation by sunlight. Complete inactivation of fecal coliforms was achieved in 15 minutes at an initial bacterial concentration of 3000 cfu/100 mL in a study using water stored in a plastic container that was coated inside with TiO_2 and exposed to sunlight [15]. Disinfection systems such as this will be especially useful in developing countries where infrastructure and electricity for water treatment are not available. However, TiO_2-based solar disinfection is in general a very slow process due to the small fraction of UV-A in solar radiation. Success in research on metal or nitrogen doping to improve visible light absorbance of TiO_2 is critical to the application of TiO_2 solar disinfection. Recently, it was demonstrated that doping TiO_2 with silver greatly improved UV-A photocatalytic bacterial inactivation by TiO_2 (30 nm) [16]. In another study, 1 percent nAg in P-25 TiO_2 (w/w) reduced the reaction time required for complete removal of 10^7cells/mL *E. coli* from 65 to 16 minutes [17]. Silver is believed to enhance photoactivity by facilitating electron-hole separation and/or providing more surface area for adsorption. Therefore, silver doping is expected to enhance disinfection by TiO_2 (e.g., 17 nm) under all wavelengths of UV as well as solar radiation [18].

20.2.3 Fullerenes

Fullerenes are not currently used in water disinfection, but certain types of fullerenes have potential applicability. Hydroxylated C_{60} or fullerol, which is relatively nontoxic [19], exhibits photochemical activity that could be exploited for disinfection or degradation [20]. However, it is neither an inexpensive nor readily available alternative to TiO_2, which is a more established and stronger photo-oxidant. Compared to TiO_2, fullerenes produce significantly lower amounts of hydroxyl radicals, which are the strongest oxy-radicals. Another obstacle to the use of fullerol in water treatment is the difficulty in immobilizing, separating, and recycling fullerol nanoparticles. No method currently exists to easily and cost-efficiently remove these small, light nanomaterials. However, grafting functionalized fullerenes to a surface could be considered. C_{60} encapsulated in polyvinyl pyrrolidone (PVP) exhibits antibacterial activity [21] and photoactivity, albeit with undetermined toxicity to humans. The functional groups on its organic cage might facilitate its anchorage to a surface without losing its antibacterial properties, a property desirable in disinfection applications involving fixed beds, membranes, or surfaces.

Carbon nanotubes (CNTs) represent another class of fullerenes that have been reported to exhibit antimicrobial properties. Knowing that a physical contact might be needed to kill bacteria [22,23], CNTs can be exploited in several ways for disinfection applications. First, single-walled nanotubes (SWNT) could be coated and immobilized on filters [22]. Additionally, multi-walled CNTs could be made into hollow fibers [24]. These nanotube filters were able to remove microbial contaminants such as *E. coli* and poliovirus, and they also offer several advantages including increased mechanical strength, heat resistance, and easy cleaning. Bundles of nonaligned single- or multi-walled nanotubes contained within a filter, could also be considered as packing in a fixed-bed filter (Table 20.1).

20.2.4 Combining current technologies with nanotechnology

The most economic and, therefore, most likely use of nanotechnology for water treatment would involve incorporating it into existing treatment strategies. The use of nanomaterials could enhance the performance of chlorination, advanced oxidation processes, and membrane filtration systems, especially in large centralized water treatment systems. In contrast, smaller point-of-use systems in the near future could be based entirely on nanotechnology.

While chlorination and ozonation are effective for the removal of bacteria and viruses, they are ineffective against cyst-forming protozoa such as *Giardia* and *Cryptosporidium*. UV-C can kill these organisms, but UV alone is relatively ineffective against viruses unless the contact time and energy output are significantly increased. This suggests an opportunity to exploit the photosensitivity of nanomaterials, such as some fullerenes [20] and TiO_2 that produce ROS to enhance UV disinfection. Large-scale UV reactors internally coated with TiO_2 have already been shown to enhance water disinfection rate [25]. Additionally, TiO_2 (e.g., P-25) can degrade a wide range of organic contaminants including natural organic matter, a major membrane foulant [26].

The increasing application of membranes for drinking water and wastewater treatment [27] promises another attractive application of nanomaterials in water treatment. In spite of the advantages membrane systems offer, the inherent problem of organic fouling and biofouling poses the biggest obstacle to their broader application. Nanomaterials can be incorporated into membranes to enhance their mechanical strength and anti-fouling capacities. Polymeric and ceramic membranes containing TiO_2 (8–10 nm) were found to be highly efficient in destroying a number of organic contaminants and pathogenic microorganisms in the presence of UV-A irradiation; these membranes are hence less vulnerable to organic and biological fouling [28,29]. Similarly, nanocomposite membranes incorporating other functional (e.g., catalytic, photocatalytic, and antimicrobial) nanoparticles into water treatment membranes can be developed. The nanoparticles immobilized on membranes will reduce membrane fouling by degrading organic and biological foulants, as well as remove contaminants that are not rejected by membranes. For photocatalysts, an outside-in submerged microfiltration (MF) or ultrafiltration (UF) membrane reactor configuration can be utilized to allow introduction of light using submerged sources such as optical fibers.

20.3 Outlook on the role of nanotechnology in microbial control: limitations and research needs

The relatively high potency of some antimicrobial nanoparticles and their increasing availability and affordability make them attractive for water disinfection applications. Some antimicrobial nanomaterials offer potential advantages over traditional chemical disinfectants that are prone to generating harmful disinfection by-products and experience short-lived reactivity. Nanomaterial disinfectants do not provide a disinfection residual in distribution systems; consequently, they might not be suitable alone in municipal water treatment facilities. Most likely, nanoparticles will be used to enhance water disinfection applications, such as UV inactivation of viruses, solar disinfection of bacteria, and membrane filtration processes that are more resistant to biofouling. Although potential economic and logistic limitations currently preclude the widespread application of such nanoparticles, a potential paradigm shift toward decentralized water treatment and reuse systems will likely stimulate research activity in nanotechnology-enabled microbial control. The future research discussed here is likely to overcome many of the current technical limitations and help discern viable applications for nanotechnology to enhance disinfection and sustainable water management.

Significant limitations exist for the use of nanomaterials for microbial control. For nanomaterials in suspension, design obstacles include loss of antimicrobial activity in the presence of high NOM and salt concentrations, nanoparticle aggregation, and loss of nanoparticles from the system if not retained or recycled. For example, although the buckminsterfullerene water suspension (THF/nC_{60}) is noted for its strong antibacterial activity [30], it is an unlikely water treatment candidate not only due to its potential human cytotoxicity [19], but also because of the loss of antimicrobial effect due to interactions with NOM that decrease bioavailability or the presence of salts that promote coagulation and precipitation [31]. These limitations are likely to be shared by other nanoparticle suspensions. Furthermore, good dispersion of nanoparticles in water is required for full utilization of the reactive surfaces, and an efficient separation process is required downstream to retain the nanoparticles. One approach that might enhance retention and recycling of suspended nanoparticles is to mount them onto magnetic platforms such as the magnetite nanoparticles that were recently used to remove arsenic from potable water [32]. Since magnetite nanoparticles can be separated from water by a relatively low magnetic field, they could be used as a platform to develop multifunctional nanocomposite materials (Fig. 20.1) that are subject to magnetic separation. This

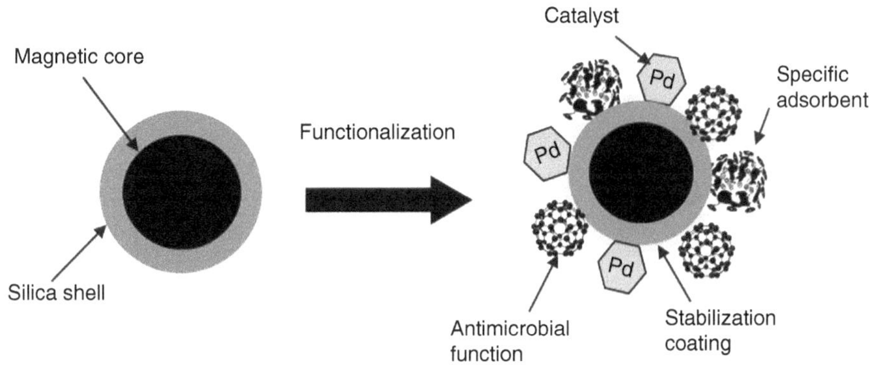

FIGURE 20.1

Multifunctional magnetic nanoparticles for antimicrobial applications.

would enable both chemical disinfection and photocatalytic destruction of waterborne pathogens while ensuring retention of the nanomaterials.

Immobilization of nanomaterials on reactor surfaces or water filtration membranes eliminates the need for separation, but the efficiency of disinfection may be compromised by the lower effective nanomaterial dose, reduced access to light source, and sometimes loss of reactive surface area. When coated on surfaces in contact with potable water to prevent microbial attachment and biofilm formation, antimicrobial nanoparticle coatings are likely to rapidly lose their effectiveness due to adsorption of extracellular polymeric material and occlusion by precipitating debris. In addition, nanoparticles can escape the reactor and enter drinking water if not properly immobilized, which is not desirable for both economic and human or environmental health concerns. Research on methods to anchor nanoparticles to reactor surfaces or to the selective layer of filtration membranes, and to separate and retain suspended nanoparticles, will be of paramount importance to decrease costs associated with premature loss and potential environmental impacts.

From an economic perspective, research needs to be conducted on the scalability and competitiveness of using antimicrobial nanoparticles for disinfection and microbial control, especially in comparison to established methods such as chlorination, ozonation, and UV treatment. Such economic analysis should consider relevant externalities associated with the potential environmental impacts in the event that the nanoparticles escape disinfection reactors. These concerns and the design difficulties make it premature to recommend the broad application of nanomaterials as disinfectants in water treatment. Nevertheless, it is likely that future research will overcome many of the current technical limitations and help discern viable applications for nanotechnology to enhance disinfection and sustainable water management.

References

[1] WHO, Water, Sanitation and Hygiene Links to Health: Facts and Figures, 2004, cited February 3, 2008, available from <http://www.who.int/water−sanitation−health/en/factsfigures04.pdf>.

[2] CDC, Infectious disease and dermatologic conditions in evacuees and rescue workers after Hurricane Katrina—multiple states, August−September, 2005, J. Am. Med. Assoc. 294 (2005) 2158−2216.

[3] S.E. Hrudey, E.J. Hrudey, Safe Drinking Water: Lessons from Recent Outbreaks in Affluent Nations, International Water Association, London, UK, 2004.

[4] E. Lefebvre, P. Racaud, T. Parpaillon, A. Deguin, Results of bromide and bromate monitoring at several water treatment plants, Ozone Sci. Eng. 17 (1995) 311−327.

[5] U. von Gunten, J. Hoigne, Bromate formation during ozonation of bromide-containing waters: interaction of ozone and hydroxyl radical reactions, Environ. Sci. Technol. 28 (7) (1994) 1234−1242.

[6] A. Gupta, S. Silver, Silver as a biocide: will resistance become a problem? Nat. Biotechnol. 16 (1998) 888.

[7] J.R. Morones, J.L. Elechiguerra, A. Camacho, K. Holt, J.B. Kouri, J.T. Ramirez, et al., The bactericidal effect of silver nanoparticles, Nanotechnol. 16 (10) (2005) 2346−2353.

[8] S.A. Blaser, M. Scheringer, M. MacLeod, K. Hungerbuhler, Estimation of cumulative aquatic exposure and risk due to silver: contribution of nano-functionalized plastics and textiles, Sci. Total Environ. 390 (2008) 396−409.

[9] X.C. Chen, H.J. Schluesenera, Nanosilver: a nanoproduct in medical application, Toxicol. Lett. 176 (1) (2008) 1−12.

[10] A.D. Maynard, Nanotechnologies: overview and issues, Nanotechnology: Toxicological Issues and Environmental Safety, Washington, D.C., Project on Emerging Nanotechnologies, Woodrow Wilson International Center for Scholars (2007) 1−14.

[11] M. Bosetti, A. Masse, E. Tobin, M. Cannas, Silver coated materials for external fixation devices: in vitro biocompatibility and genotoxicity, Biomaterials 23 (3) (2002) 887−892.

[12] W.L. Chou, D.G. Yu, M.C. Yang, The preparation and characterization of silver-loading cellulose acetate hollow fiber membrane for water treatment, Polym. for Adv. Technol. 16 (8) (2005) 600−607.

[13] Y. Kikuchi, K. Sunada, T. Iyoda, K. Hashimoto, A. Fujishima, Photocatalytic bactericidal effect of TiO_2 thin films: dynamic view of the active oxygen species responsible for the effect, J. Photochem. Photobiol. A 106 (1997) 51−56.

[14] L.K. Adams, D.Y. Lyon, P.J.J. Alvarez, Comparative ecotoxicity of nanoscale TiO_2, SiO_2, and ZnO water suspensions, Water. Res. 40 (19) (2006) 3527−3532.

[15] S. Gelover, L.A. Gomez, K. Reyes, M.T. Leal, A practical demonstration of water disinfection using TiO_2 films and sunlight, Water. Res. 40 (2006) 3274−3280.

[16] K. Page, R.G. Palgrave, I.P. Parkin, M. Wilson, S.L.P. Savin, A.V. Chadwick, Titania and silver-titania composite films on glass-potent antimicrobial coatings, J. Mater. Chem. 17 (1) (2007) 95−104.

[17] M.P. Reddy, A. Venugopal, M. Subrahmanyam, Hydroxyapatite-supported $Ag-TiO_2$ as escherichia coli disinfection photocatalyst, Water. Res. 41 (2007) 379−386.

[18] H.M. Sung-Suh, J.R. Choi, H.J. Hah, S.M. Koo, Y.C. Bae, Comparison of Ag deposition effects on the photocatalytic activity of nanoparticulate TiO_2 under visible and UV light irradiation, J. Photochem. Photobiol. A 163 (1–2) (2004) 37–44.

[19] C.M. Sayes, J.D. Fortner, W. Guo, D. Lyon, A.M. Boyd, K.C. Ausman, et al., The differential cytotoxicity of water-soluble fullerenes, Nano. Lett. 4 (10) (2004) 1881–1887.

[20] E.M. Hotze, J. Labille, P. Alvarez, M.R. Wiesner, Mechanisms of photochemistry and reactive oxygen production by fullerene suspensions in water, Environ. Sci. Technol. 42 (2008) 4175–4180.

[21] D.Y. Lyon, L.K. Adams, J.C. Falkner, P.J.J. Alvarez, Antibacterial activity of fullerene water suspensions: Effects of preparation method and particle size, Environ. Sci. Technol. 40 (14) (2006) 4360–4366.

[22] S. Kang, M. Pinault, L.D. Pfefferle, M. Elimelech, Single-walled carbon nanotubes exhibit strong antimicrobial activity, Langmuir 23 (2007) 8670–8673.

[23] R.J. Narayan, C.J. Berry, R.L. Brigmon, Structural and biological properties of carbon nanotube composite films, Materials Science and Engineering B 123 (2005) 123–129.

[24] A. Srivastava, O.N. Srivastava, S. Talapatra, R. Vajtai, P.M. Ajayan, Carbon nanotube filters, Nat. Mater. 3 (2004) 610–614.

[25] K. Sunada, Y. Kikuchi, K. Hashimoto, A. Fujishima, Bactericidal and detoxification effects of TiO_2 thin film photocatalysts, Environ. Sci. Technol. 32 (5) (1998) 726–728.

[26] X.H. Huang, M. Leal, Q.L. Li, Degradation of natural organic matter by TiO_2 photocatalytic oxidation and its effect on fouling of low-pressure membranes, Water. Res. 42 (2008) 1142–1150.

[27] M. Marcucci, I. Ciabatti, A. Matteucci, G. Vernaglione, Membrane technologies applied to textile wastewater treatment, Ann. NY Acad. Sci. 984 (2003) 53–64.

[28] H. Choi, E. Stathatos, D. Dionysiou, Photocatalytic TiO_2 films and membranes for the development of efficient wastewater treatment and reuse systems, Desalination 202 (2007) 199–206.

[29] S.H. Kim, S.Y. Kwak, B.H. Sohn, T.H. Park, Design of TiO_2 nanoparticle self-assembled aromatic polyamide thin-film-composite (TFC) membrane as an approach to solve biofouling problem, J. Memb. Sci. 211 (2003) 157–165.

[30] D.Y. Lyon, D.A. Brown, P.J.J. Alvarez, Implications and potential applications of bactericidal fullerene water suspensions: effect of nC(60) concentration, exposure conditions and shelf life, Water. Sci. Technol. 57 (2008) 1533–1538.

[31] D. Li, D.Y. Lyon, Q. Li, P.J.J. Alvarez, Effect of soil sorption and aquatic natural organic matter on the antibacterial activity of a fullerene water suspension, Environmental Toxicology and Chemistry 27 (2008) 1888–1894.

[32] C.T. Yavuz, J.T. Mayo, W.W. Yu, A. Prakash, J.C. Falkner, S. Yean, et al., Low-field magnetic separation of monodisperse Fe_3O_4 nanocrystals, Science 314 (5801) (2006) 964–967.

CHAPTER

21 Possible Applications of Fullerene Nanomaterials in Water Treatment and Reuse

So-Ryong Chae, Ernest M. Hotze and Mark R. Wiesner

Department of Civil and Environmental Engineering, School of Engineering, Duke University, Durham, NC, USA

21.1 Introduction

Nano-engineered materials are likely to find numerous applications that will improve environmental technologies and help protect public health, including industrial separations, potable water treatment, chemical synthesis, energy generation and transmission, ground water remediation, and air quality control, to name a few.

Fullerenes are a class of molecules composed entirely of carbon. The first of these molecules, Buckminsterfullerene, was discovered in 1985 and contains 60 carbons in the form of a hollow spherical cage consisting of 12 pentagonal and 20 hexagonal faces [1]. Other spherical fullerenes or "buckyballs" have since been synthesized, the smallest containing 20 carbons. Nonspherical fullerenes have also been synthesized including cylinders (carbon nanotubes-CNTs), lobed structures, bowls, and dendrimers, to name a few. Further variations on fullerenes include the addition of an almost infinite variety of functionalities ranging from

Street, Sustich, Duncan and Savage. Nanotechnology Applications for Clean Water, 2nd Edition.
 DOI: http://dx.doi.org/10.1016/B978-1-4557-3116-9.00021-4

simple hydroxylation to the grafting of deoxyribonucleic acid (DNA) molecules. Driven by immediate applications and the utility of undifferentiated material for subsequent modification, there has been a significant commercial emphasis placed on the production of buckyballs and CNTs.

Fullerene-based nanomaterials are emerging in a variety of potential applications, including cosmetics, energy production [2], semiconductors [3], and medical treatments [4]. Estimates of the size of the current nanotechnology market range from 30 to 45 billion dollars [5]. In environmental engineering, fullerenes have been proposed as a basis for developing new technologies for nanomaterial-enabled oxidation and disinfection, improved membrane processes, adsorbents, and biofilm-resistant surfaces [6].

This chapter details recent progress toward the development of these proposed applications. We examine the development of fullerene composite materials using CNTs to strengthen membranes and modify membrane surface chemistry. We also explore the use of fullerene nanomaterials to generate reactive oxygen species (ROS) as the basis for a range of new technologies including in situ generation of oxidants to destroy trace organic compounds, new strategies for disinfection, the inhibition of biofilm development, and reduced biofouling. The use of fullerenes in conjunction with ultraviolet (UV) irradiation is considered as an advanced disinfection process (ADP) for viral inactivation.

21.2 Chemistry of fullerene nanomaterials

Fullerenes may exist in a number of geometries such as nanotubes and spherical cages (C_{60}), both of which have been tested for use in environmental application research (Fig. 21.1). The latter form has an extremely low solubility in water [7]

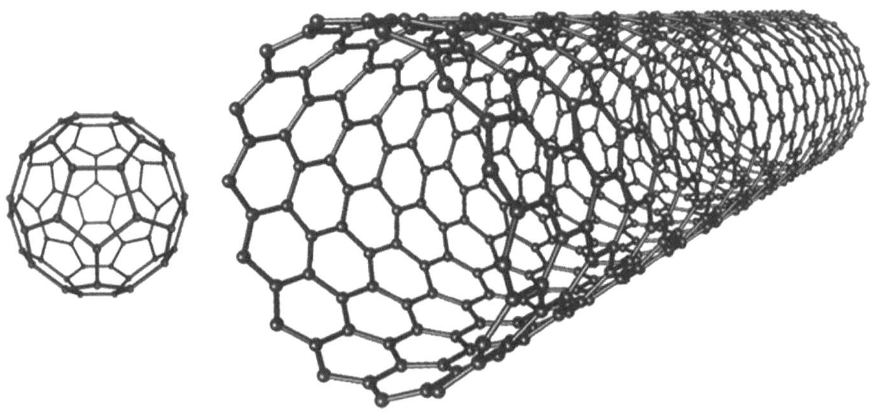

FIGURE 21.1

Simulated structures of spherical (C_{60}) (left) and tubular (single-walled carbon nanotube, SWCNT) fullerenes (right).

and therefore they must be modified on the surface [8], clustered (nC_{60}) [9], or mixed with a surfactant or stabilization agent [10] for significant concentrations to be reached in water. As a variation on fullerenes, CNTs are limited by similar solubility constraints and also need to be functionalized or coated to increase their affinity for water [11]. This is a primary consideration when adapting fullerene nanomaterials for environmental applications.

Two varieties of CNTs are multi-walled carbon nanotubes (MWCNT) and single-walled carbon nanotubes (SWCNT). SWCNT can be visualized as graphite sheets that have been rolled up and seamlessly attached via carbon bonds. The nature of this rolling (chirality) determines whether these materials are metallic (conducting electricity) or are semiconductors [12]. MWCNT are made up of two or more such tubes wrapped around each other, similar to the layers of an onion. Of the two, SWCNT are more attractive for use in applications due to their purity and uniformity where tensile strength (stronger than steel), electrical conductivity (comparable to copper), and thermal conductivity are all improved over MWCNT. However, functionalizing the surface of the SWCNT to improve its affinity for water (or for some other reason) may entail a sacrifice in the strength or conductivity of the SWCNT.

While the surface chemistries of nanotubes and C_{60} are quite similar in the aqueous environment their photochemistries are very different. In particular, C_{60} is a photoactive molecule. Photoactive materials may be classified as photocatalysts such as titanium dioxide [13] or photosensitizers such as the dye Rose Bengal (RB) [14] and the molecule C_{60}. Photocatalytic behavior is governed by the band gap that can be described as the energy difference between the valence band (Fermi level of the highest energy electrons) and the conduction band (lowest energy unoccupied molecular orbitals). Light striking a photocatalyst will promote an electron across this band gap creating holes in the molecular orbitals it leaves behind. Both the promoted electrons and the holes can react in solution to form ROS. In contrast, electrons in photosensitizers are excited within their molecular orbitals. Sensitized electrons can behave according to two mechanisms: type I electron transfer involving a donor molecule and type II primarily involving energy transfer to a ground state oxygen. Both of these pathways will potentially produce ROS in solution. The net effect of both these processes is to convert light energy into oxidizing chemical energy. This chemical energy can then potentially be used to oxidize contaminants or disinfect microorganisms. In the case of pure C_{60} the photosensitization process happens efficiently due to its ability to perform intersystem crossing with close to zero loss of energy [15]. However, in aqueous systems C_{60} will form clusters. Cluster formation will tend to decrease the lifetime of the triplet state and therefore the formation of ROS [16]. The addition of functional groups and coatings that reduce clustering may increase ROS production [16–18] although the addition of functional groups to the C_{60} cage tends to reduce the quantum yield of the triplet state that is most responsible for ROS formation. A diagram of possible C_{60} photosensitization pathways is given in Fig. 21.2.

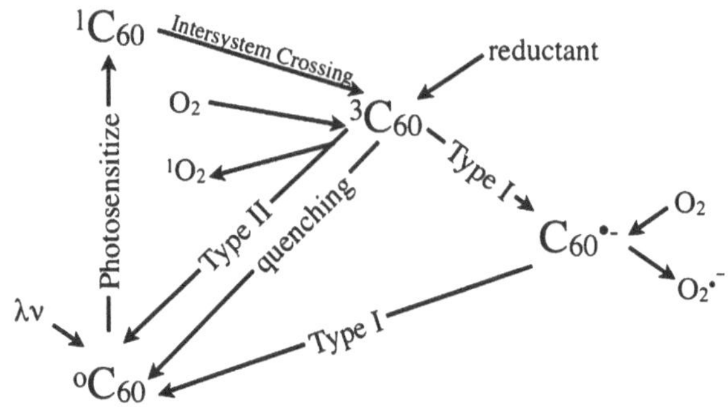

FIGURE 21.2

Potential photosensitization pathways of C_{60}. Intersystem crossing occurs very efficiently due to the unique geometry of the molecule. Then C_{60} is added to aqueous systems where clustering, functionalization, and coatings will reduce the efficiency of some of these processes.

21.3 Applications of fullerene nanomaterials

21.3.1 Membrane fabrication using fullerene nanomaterials

Membrane technologies are playing an increasingly important role as processes for removing salts, particles, organic matter, or gases from water. The performance of membranes is intimately linked to the materials they are made from. The composition of the membrane will determine important properties such as rejection (selectivity), propensity to foul, mechanical strength, and reactivity.

Fullerene nanomaterials have unique properties of strength, ability to tailor size, flexibility in modifying functionality, and electron affinity that have created much excitement around their potential for new membrane-based technologies. The small and controllable diameter of fullerene nanotubes suggests that membranes made from these materials in a fashion where fluid flows through the center of the CNT might be highly selective. The small diameter of the CNT also implies a high resistance to flow through a membrane composed of such nanometer-sized pores. Surprisingly, molecular modeling indicates that water should be able to flow much faster through hydrophobic CNTs due to the formation of ordered hydrogen bonds [19]. The hydrophobic surface in the interior of a defect-free CNT appears to allow for a nearly frictionless flow [19]. Visualization of water within CNTs confirms the lack of interaction between water molecules and the interior surface of CNTs [20] and experiments using membranes composed of aligned CNTs have confirmed that flow through the CNTs is orders of magnitude greater than that predicted by Poiseuille flow through tubes composed of conventional materials [21,22].

There are also promising applications for fullerene–polymer composites in pressure-driven membranes. The strength of CNTs, coupled with reported antibacterial properties, suggest that fullerene–polymer composites may find use in

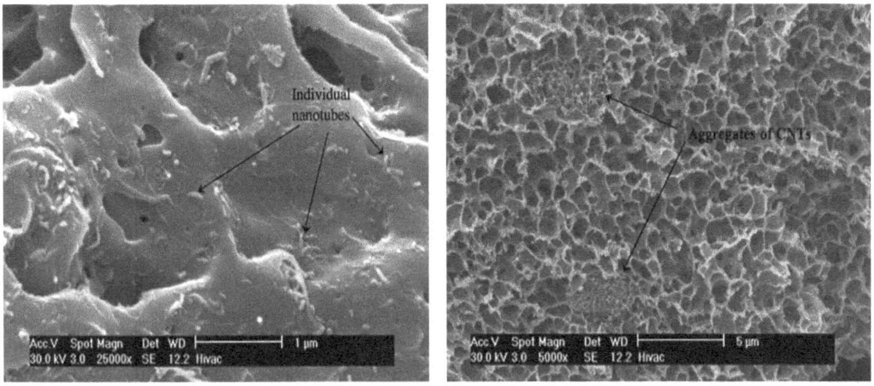

FIGURE 21.3

Surface morphology of a membrane containing semi-dispersed carbon nanotubes (CNTs) by scanning electron microscopy.

Adapted from Brunet et al. [24].

creating membranes that resist breakage or inhibit biofouling. The incorporation of C_{60} into polymeric membranes has been observed to effect membrane structure and rejection [23].

We have explored the use of MWCNTs in the development of advanced composite membranes for applications in water treatment. MWCNT (4% w/w) were incorporated as a composite into polysulfone ultrafiltration (UF) membranes, prepared according to the wet-phase inversion method. The dispersion of the nanotubes and the morphology of the membranes were observed by scanning electron microscopy (Fig. 21.3, adapted from Brunet et al. [24]). The membranes were characterized for surface roughness, contact angle, permeability, and mechanical properties. A partial de-aggregation of the nanotubes leads to individual nanotubes within the polymer as well as bundles nested in the pores. After addition of MWCNTs, the asymmetric structure of the membrane and the permeability were not disturbed, neither was the hydrophobicity, but the roughness increased. However, the tensile strength of the composite membrane was not improved in this case, suggesting the need to match CNT functionalization to the polymer to more evenly distribute the nanotubes throughout the composite.

Future convergence between nanochemistry and membrane science will likely yield a generation of active membrane systems. Nanomaterials might also be incorporated into membranes to impart properties that are activated by an electrical or chemical signal. Living organisms are the ultimate nanotechnology. The ability of cell membranes to selectively transport materials, often against concentration gradient, and to avoid fouling is impressive. As the field of nanochemistry advances, engineered biomimetic systems based on selective transport or rejuvenating layers of self-organizing materials may be developed for performing critical separations in energy and environmental applications.

21.3.2 Oxidation of organic compounds

The ROS producing properties [11,15,25] of fullerenes might be harnessed to generate oxidizing species to enhance destruction of organic compounds in water [26]. We are exploring the oxidation of probe organic compounds by ROS generated from a suspension of hydroxylated fullerene (fullerol, $C_{60}(OH)_{22-24}$)) prepared using a sonication method that rapidly produces stable suspensions of fullerol aggregates [27]. The efficiency of contaminant destruction under UV irradiation can be compared with that of a known photosensitizer that produces singlet oxygen, RB [28]. Results to date show proof of concept for this approach, although the degree of compound destruction is modest and compound-specific. Specificity of the reaction may be exploited to achieve destruction of one compound while avoiding undesirable oxidation by-products that arise from reactions with other materials. Figure 21.4 illustrates the relative destruction of 2-chlorophenol (2-CP) by equal concentrations of fullerol and RB when irradiated with UV light. For the conditions applied, approximately 17 percent of the 2-CP was destroyed after 30 minutes of irradiation. RB produced a slightly greater degree of photo-induced degradation (28 percent). For further study, it is important to understand oxidation kinetics of various organic compounds at various temperatures and pHs. Finally, the formation of ROS from fullerol suggests the possibility of engineering fullerol-sensitized systems to destroy organic compounds or possible disinfection in water and wastewater.

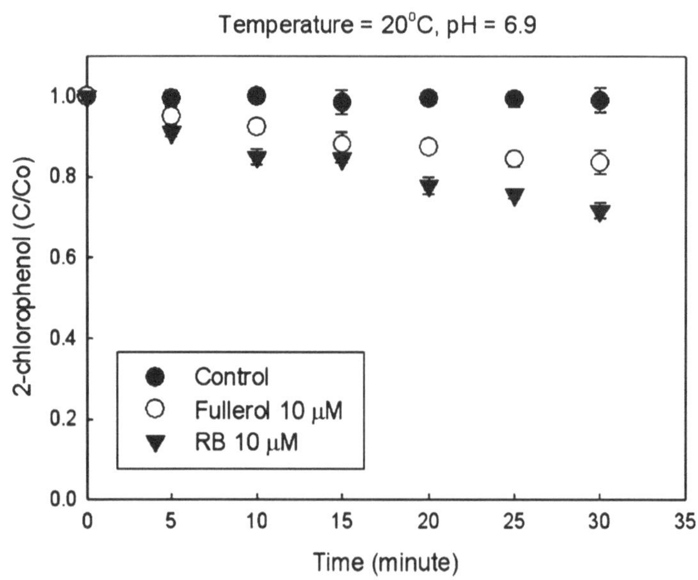

FIGURE 21.4

Oxidation of 2-chlorophenol using photosensitized fullerol. RB: Rose Bengal dye.

21.3.3 Bacterial and viral inactivation

Fullerenes have also been considered for their possible antimicrobial properties. The medical literature details the ability of C_{60} and C_{70} fullerenes to cleave DNA and inactivate viruses, bacteria, and kill tumor cells [29,30] suggesting that they might be used for disinfection [28] or to produce surfaces resistant to microbial growth. We consider here the example of inactivation of waterborne bacterial viruses [27] and the development of anti-fouling agents for membranes used in water and wastewater treatment where biofouling is known to be a critical limitation [31]. As an illustration of the latter case, we have modified ceramic membranes by depositing a layer of C_{60} and then observed impact on bacterial attachment and metabolic activity. Bacterial activity was monitored in terms of the total number of bacteria present (DAPI) to those that were metabolically active (CTC). As shown in Fig. 21.5, the total number of bacterial colonies of *E. coli* K12 and the fraction of viable bacteria as measured by the CTC to DAPI ratio (CTC/DAPI) decreased rapidly as the amount of C_{60} on the membranes increased. Thus, both the affinity for bacterial attachment to the membranes and bacterial viability decreased with increasing amounts of C_{60} present on the membranes. It is possible that extracellular polymeric substances (EPS) and soluble microbial product (SMP) will hinder contact between bacteria and C_{60} and therefore limit long-term efficiency of C_{60} in anti-biofouling activity. Such "fouling of anti-foulants" remains an active area of research.

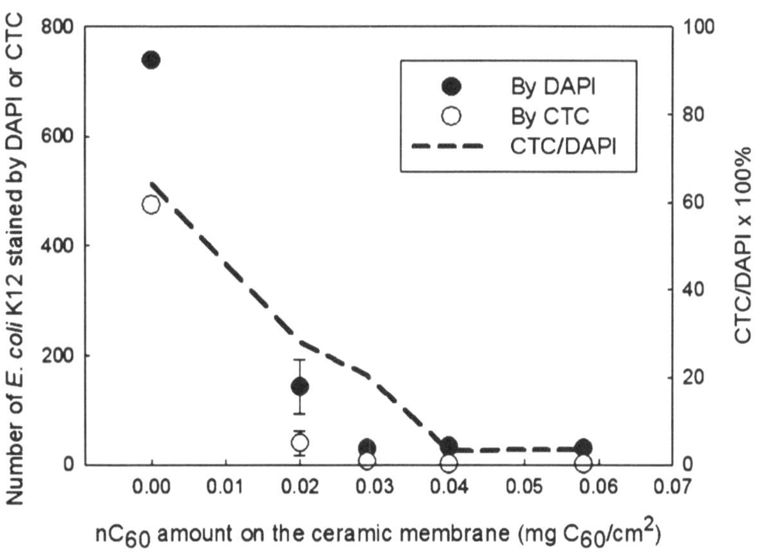

FIGURE 21.5

Effects of the amount of nC_{60} on bacterial attachment and growth. CTC: Metabolically active bacteria; DAPI: Total number of bacteria present.

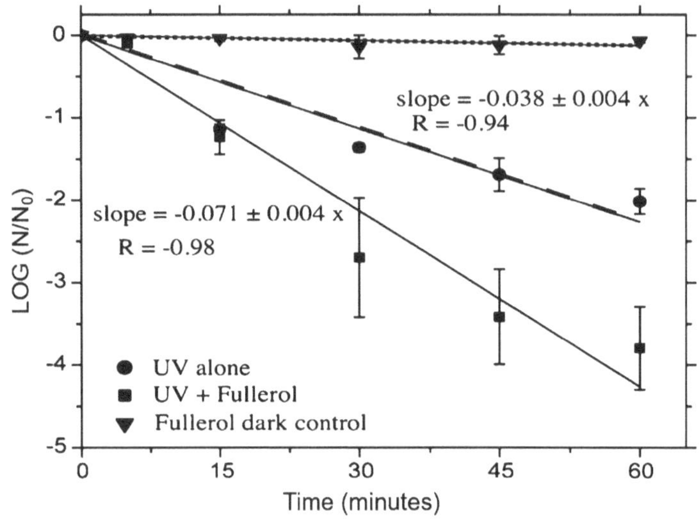

FIGURE 21.6

MS2 inactivation rate in the presence of 1 μM photoactivated fullerol.

Adapted from Badireddy et al. [27].

The impact of fullerol on virus was studied using the MS2 bacteriophage, a virus that is used extensively for studies of water pollution and control [32–34], and it is similar in morphology to hepatitis A virus and poliovirus. A simple dilution and enumeration technique allowed for viruses to be counted before and after exposure to fullerol and low pressure UV light (approximately 365 nm). As seen in Fig. 21.6, the addition of as little as 1 μM fullerol into the suspension nearly doubled the log inactivation rate of the MS2 phage in UV light alone.

21.4 Summary

Fullerenes are versatile, new materials with properties that suggest great potential for improving water treatment technologies. However, our ability to manipulate and fully exploit the properties of these materials remains limited, requiring further innovations to develop efficient formats for immobilizing, recovering fullerenes. More revolutionary changes in water treatment using these materials are likely to be based on the ability to conceive of and build devices that actively function at the scale of the material itself. Such devices might combine functions such as energy harvesting and treatment, detection or control.

Acknowledgements

This study was supported in part by ONR grant 05032901, NSF grant BES-0508207, and partial funding by a grant from the Korea Research Foundation (MOEHRD) (KRF-2006-D00125).

References

[1] H.W. Kroto, J.R. Heath, S.C. Obrien, R.F. Curl, R.E. Smalley, C_{60}: buckminsterfullerene, Nature 318 (6042) (1985) 162–163.

[2] P.V. Kamat, M. Haria, S. Hotchandani, C_{60} cluster as an electron shuttle in a Ru(II)-polypyridyl sensitizer-based photochemical solar cell, J. Phys. Chem. B 108 (17) (2004) 5166–5170.

[3] N. Saran, K. Parikh, D.S. Suh, E. Muñoz, H. Kolla, S.K. Manohar, Fabrication and characterization of thin films of single-walled carbon nanotube bundles on flexible plastic substrates, J. Am. Chem. Soc. 126 (14) (2004) 4462–4463.

[4] T. Da Ros, M. Prato, Medicinal chemistry with fullerenes and fullerene derivatives, Chem. Commun. 21 (8) (1999) 663–669.

[5] S. Mize, Nanotechnology Opportunity Report, London, CMP-Cientifica, 2002.

[6] J.-Y. Bottero, J. Rose, M.R. Wiesner, Nanotechnologies: tools for sustainability in a new wave of water treatment processes, Integr. Environ. Assess. Manage. 2 (4) (2006) 391–395.

[7] R.S. Ruoff, D.S. Tse, R. Malhotra, D.C. Lorents, Solubility of C_{60} in a variety of solvents, J. Phys. Chem. 97 (13) (1993) 3379–3383.

[8] C. Cusan, T. Da Ros, G. Spalluto, S. Foley, J.M. Janto, P. Seta, et al., A new multi-charged C_{60} derivative: Synthesis and biological properties, Eur. J. Org. Chem. (17) (2002) 2928–2934.

[9] J.D. Fortner, D.Y. Lyon, C.M. Sayes, A.M. Boyd, J.C. Falkner, E.M. Hotze, et al., C_{60} in water: Nanocrystal formation and microbial response, Environ. Sci. Technol. 39 (11) (2005) 4307–4316.

[10] Y.N. Yamakoshi, T. Yagami, K. Fukuhara, S. Sueyoshi, N. Miyata, Solubilization of fullerenes into water with polyvinylpyrrolidone applicable to biological tests, J. Chem. Soc., Chem. Commun. 21 (4) (1994) 517–518.

[11] V.C. Moore, M.S. Strano, E.H. Haroz, R.H. Hauge, R.E. Smalley, J. Schmidt, et al., Individually suspended single-walled carbon nanotubes in various surfactants, Nano Lett. 3 (10) (2003) 1379–1382.

[12] A. Hirsch, M. Brettreich, Fullerenes: Chemistry and Reactions, Wiley-VCH, Weinheim, 2005.

[13] A. Wold, Photocatalytic properties of TiO_2, Chem. Mater. 5 (3) (1993) 280–283.

[14] D.C. Neckers, Rose Bengal, J. Photochem. Photobiol. A: Chem. 47 (1) (1989) 1–29.

[15] J.W. Arbogast, A.P. Darmanyan, C.S. Foote, Y. Rubin, F.N. Diederich, M.M. Alvarez, et al., Photophysical properties of C_{60}, J. Phys. Chem. 95 (1) (1991) 11–12.

[16] T. Anderson, K. Nilsson, M. Sundahl, G. Westman, O. Wennerstrom, C_{60} Embedded in γ-cyclodextrin: a water-soluble fullerene, J. Chem. Soc., Chem. Commun. (1992) 604–606.

[17] D.M. Guldi, K. Asmus, Photophysical properties of mono- and multiply-functionalized fullerene derivatives, J Phys Chem A 101 (1997) 1472–1481.

[18] E.M. Hotze, J. Labille, P. Alvarez, M.R. Wiesner, Mechanisms of photochemistry and reactive oxygen production by fullerene suspensions in water, Environ. Sci. Technol. 42 (11) (2008) 4175–4180.

[19] G. Hummer, J.C. Rasaih, J.P. Noworyta, Water conduction through the hydrophobic channel of a carbon nanotube, Nature 414 (2001) 188–190.

[20] N. Naguib, H.H. Ye, Y. Gogotsi, A.G. Yazicioglu, C.M. Megaridis, M. Yoshimura, Observation of water confined in nanometer channels of closed carbon nanotubes, Nano Lett. 4 (11) (2004) 2237–2243.

[21] J.K. Holt, H.G. Park, Y. Wang, M. Stadermann, A.B. Artyukhin, C.P. Grigoropoulos, et al., Fast mass transport through sub-2-nanometer carbon nanotubes, Science 312 (5776) (2006) 1034–1037.

[22] M. Majumder, N. Chopra, R. Andrews, B.J. Hinds, Nanoscale hydrodynamics: enhanced flow in carbon nanotubes, Nature 438 (7064) (2005) 44.

[23] G. Polotskaya, Y. Biryulin, V. Rozanov, Asymmetric membranes based on fullerene-containing polyphenylene oxide, Fullerenes Nanotubes Carbon Nanostruct. 12 (1–2) (2004) 371–376.

[24] L. Brunet, D.Y. Lyon, K. Zodrow, J.-C. Rouch, B. Caussat, P. Serp, et al., Properties of membranes containing semi-dispersed carbon nanotubes, Environm. Eng. Sci. 25 (4) (2008) 565–575.

[25] B. Vileno, P.R. Marcoux, M. Lekka, A. Sienkiewicz, T. Feher, L. Forro, Spectroscopic and photophysical properties of a highly derivatized C_{60} fullerol, Adv. Funct. Mater. 16 (1) (2006) 120–128.

[26] J.Y. Bottero, J. Rose, M.R. Wiesner, Nanotechnologies: Tools for sustainability in a new wave of water treatment processes, Integr. Environ. Assess. Manage. 2 (4) (2006) 391–395.

[27] A.R. Badireddy, E.M. Hotze, S. Chellam, P. Alvarez, M.R. Wiesner, Inactivation of bacteriophages via photosensitization of fullerol nanoparticles, Environ. Sci. Technol. 41 (18) (2007) 6627–6632.

[28] K.D. Pickering, M.R. Wiesner, Fullerol-sensitized production of reactive oxygen species in aqueous solution, Environ. Sci. Technol. 39 (5) (2005) 1359–1365.

[29] N. Tsao, T.Y. Luh, C.K. Chou, J.J. Wu, Y.S. Lin, K.Y. Lei, Inhibition of group A streptococcus infection by carboxyfullerene, Antimicrob. Agents Chemother. 45 (6) (2001) 1788–1793.

[30] Y. Yamakoshi, N. Umezawa, A. Ryu, K. Arakane, N. Miyata, Y. Goda, et al., Active oxygen species generated from photoexcited fullerene (C_{60}) as potential medicines: $O_2^{-\oplus}$ versus O_2^1, J. Am. Chem. Soc. 125 (42) (2003) 12803–12809.

[31] J. Mallevialle, P. Odendaal, M.R. Wiesner, Membrane Processes in Water Treatment, McGraw-Hill, New York, N.Y, 1996.

[32] L.E. Batch, C.R. Schulz, K.G. Linden, Evaluating water qualify effects on UV disinfection of MS2 coliphage, J. Am. Water Works Assoc. 96 (7) (2004) 75–87.

[33] T. Kohn, K.L. Nelson, Sunlight-mediated inactivation of MS2 coliphage via exogenous singlet oxygen produced by sensitizers in natural waters, Environ. Sci. Technol. 41 (1) (2007) 192–197.

[34] B.X. Mi, B.J. Marinas, J. Curl, S. Sethi, G. Crozes, D. Hugaboom, Microbial passage in low pressure membrane elements with compromised integrity, Environ. Sci. Technol. 39 (11) (2005) 4270–4279.

CHAPTER

22 Heterogeneous Catalytic Reduction for Water Purification: Nanoscale Effects on Catalytic Activity, Selectivity, and Sustainability

Timothy J. Strathmann[1], Charles J. Werth[1] and John R. Shapley[1,2]

[1]*Department of Civil and Environmental Engineering and Center of Advanced Materials for the Purification of Water with Systems, University of Illinois at Urbana–Champaign, Urbana, IL, USA* [2]*Department of Chemistry and Center of Advanced Materials for the Purification of Water with Systems, University of Illinois at Urbana–Champaign, Urbana, IL, USA*

22.1 Introduction

Advances in analytical chemistry have led in recent years to an increased awareness that surface and groundwater supplies are being affected by a myriad of contaminants originating from widely diverse sources, including improper waste disposal (e.g., halogenated solvents, perchlorate), direct releases (e.g., pesticides, animal waste), and incomplete wastewater treatment (e.g., pharmaceuticals, endocrine disrupting compounds) [1–3]. The freshwater scarcity in many regions of the United States is forcing utilities to face the challenge of providing safe drinking water from lower quality source waters, including those with known contamination problems and those receiving significant inputs of wastewater effluent [4].

Street, Sustich, Duncan and Savage. Nanotechnology Applications for Clean Water, 2nd Edition.
 DOI: http://dx.doi.org/10.1016/B978-1-4557-3116-9.00022-6

Meeting this ever increasing challenge with conventional treatment technologies alone is likely to be difficult and costly, and new treatment technologies that are effective, economical, and sustainable must be developed.

Reductive catalysis is a developing water treatment technology that employs heterogeneous metal catalysts (e.g., Pd, Ni) to convert dihydrogen gas to adsorbed atomic hydrogen H(ads), a powerful reducing agent that reacts readily with oxidized functional groups. Heterogeneous catalytic treatment processes have several potential advantages over other treatment technologies, including (i) high selectivity for a given target, (ii) fast rates, (iii) use of less hazardous chemical additives, (iv) lower net energy inputs, and (v) low production of residuals or concentrates that need further treatment. Recent research developments have demonstrated relatively efficient metal particle catalysts for reductive destruction of a variety of water contaminants, such as halocarbons [5–7], nitrate [8,9], perchlorate [10], and N-nitrosamines [11,12]. However, a lack of information about the long-term sustainability of these catalytic treatment processes is a major obstacle to their widespread adoption for remediation applications. Recent advances in nanotechnology and the concomitant improving molecular-scale understanding of interfaces and interfacial processes are driving current research into the characterization of new heterogeneous catalysts together with their mechanisms of deactivation and the pathways for regeneration. Two examples of heterogeneous reductive catalytic systems examined at the nanoscale level in studies emanating from the authors' laboratories will be illustrated in this chapter.

22.2 Catalytic hydrodehalogenation: iodinated X-ray contrast media

A large body of literature has been reported on the kinetics, mechanistic features, and field applications of catalytic hydrodehalogenation processes for water purification [13–23]. Figure 22.1 depicts the general hydrodehalogenation process in

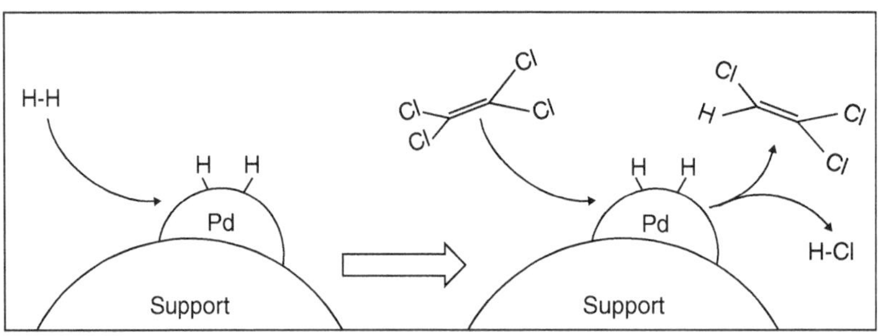

FIGURE 22.1

Simplified pictorial depiction of a catalytic hydrodehalogenation reaction on supported palladium nanoparticle catalyst surface.

which halide release is accompanied by a formal reduction of the associated carbon and C-H bond formation. Here, we highlight results from a recent application of catalytic hydrodehalogenation to treat iodinated X-ray contrast media (ICM) (see Fig. 22.2) [24]. Readers are encouraged to consult the wider body of literature for a more in-depth understanding of hydrodehalogenation processes.

Iodinated X-ray contrast media are radiopaque agents used in brain and body imaging diagnostic procedures. Diatrizoate and iopromide are representative ionic and neutral ICM, respectively. Recent studies indicate that the 2,4,6-triiodinated ring structure of ICM is recalcitrant to treatment by conventional wastewater and drinking water treatment processes, and these compounds are highly persistent in receiving water bodies [25]. Although advanced oxidation processes (AOPs) have been shown to be capable of partially mineralizing ICM [26], treatment of trace micropollutants such as ICM in complex matrices with AOPs is inefficient because of the non-selectivity of the active OH radical oxidizing agent. Reductive catalytic treatment technologies are more selective, enabling targeted treatment of chemicals containing reactive moieties such as C-halogen bonds.

A recent study demonstrated that hydrodehalogenation of ICM in water with hydrogen occurs readily over supported Pd and porous Ni catalysts [24]. Diatrizoate and iopromide undergo complete dehalogenation, leading to release of three stoichiometric equivalents of iodide. Liquid chromatography-mass spectrometry confirmed that loss of the parent ICM and iodide release are accompanied by sequential formation and decay of two organic intermediates followed by formation of a stable final product, each of which differs in mass by 126 (atomic weight corresponding to $-I^- + H^+$; see Fig. 22.3).

Kinetic studies showed that ICM degradation in hydrogen-saturated aqueous batch reactors ($P(H_2) = 1$ atm) can be described by a first-order rate law, and the kinetic trends were consistent with a single-site Langmuir–Hinshelwood–Hougen–Watson model that assumes competitive adsorption of the two reactive species [27]. Pseudo-first-order rate constants k(cat) for diatrizoate and iopromide reactions with a 5 percent Pd/alumina catalyst (normalized by the mass of palladium) were very similar (k(cat) = 3.1 and 3.6 L/s g-Pd, respectively). A 1 percent

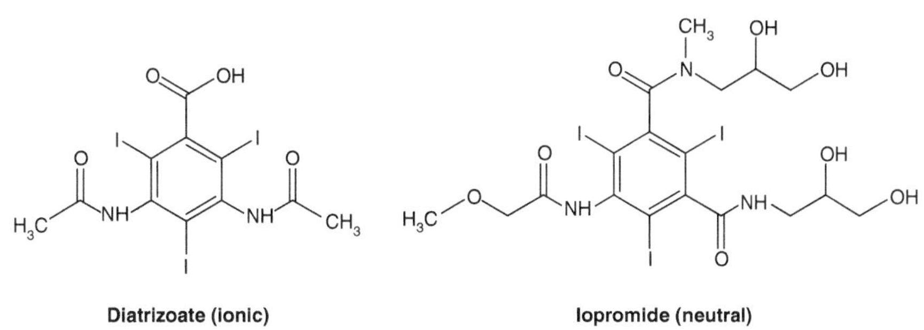

FIGURE 22.2

Molecular structures for two representative iodine contrast media (ICM).

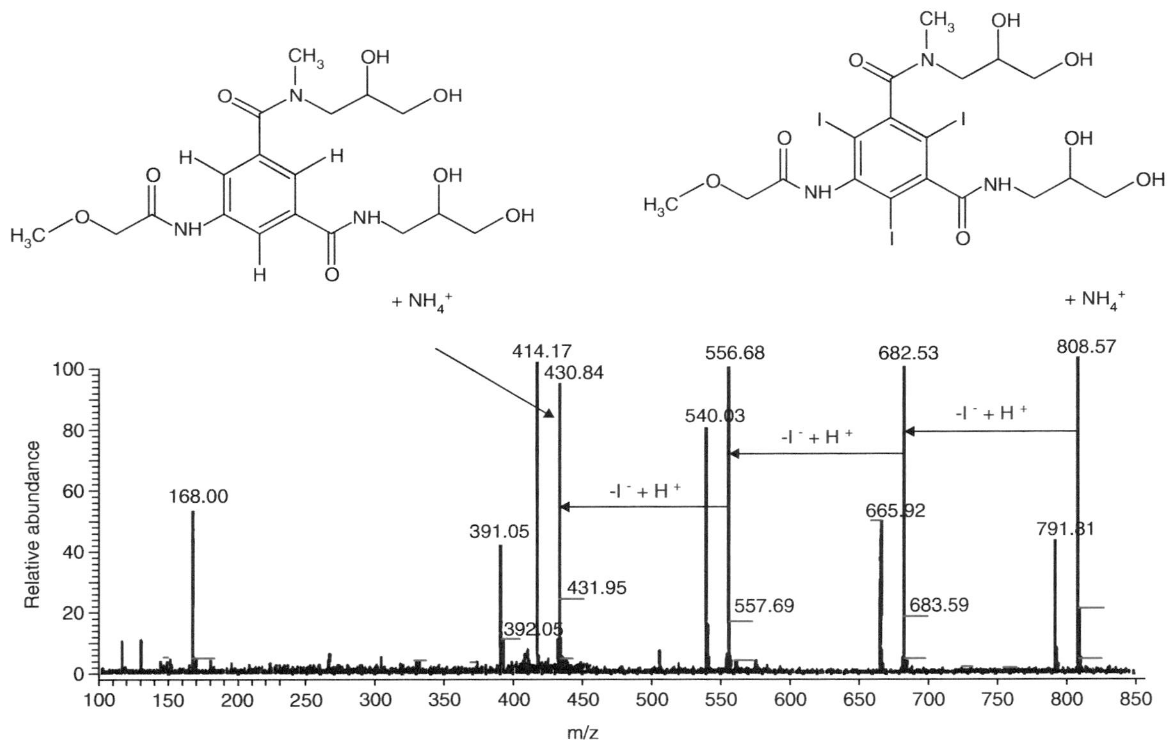

FIGURE 22.3

Mass spectrum (m/z) of partially reacted sample from Pd-on-alumina catalyst system containing a mixture of iopromide and dehalogenated products. Each sequential deiodination process is shown by arrows indicating replacement of I^- in the structure by H^+.

Pd/alumina catalyst was nearly four times as active on a palladium-mass basis (k(cat) = 13.4 L/s g-Pd), possibly due to greater Pd dispersion on the support material. A porous Ni catalyst (i.e., Raney Ni) also catalyzed the hydrodehalogenation of diatrizoate, but it was less active than the Pd-based catalyst on a metal-mass basis: k(cat) = 0.0216 L/s g-Ni. Although less active than Pd, Ni is a financially attractive option for practical treatment processes because of its much lower cost.

The influence of environmental factors and nontarget water constituents on Pd-catalyzed ICM hydrodehalogenation was also investigated. Varying pH (3–11), ionic strength ($< 0.001-0.05$ M), and the presence of many common inorganic constituents (10 mM Mg^{2+}, Ca^{2+}, Cl^-, Br^-, SO_4^{2-}, NO_3^-, and HCO_3^-) had little effect on diatrizoate reduction. The presence of elevated concentrations of iodide ion, natural organic matter (NOM), and (bi)sulfide ion inhibited the ICM hydrodehalogenation reaction. Iodide appeared to partially inhibit the catalytic process by competitive adsorption with the target reactant, and catalyst activity was rapidly restored by rinsing iodide-exposed catalysts with deionized water. The activity of NOM-deactivated catalysts could also be fully restored by rinsing the catalysts in alkaline solution (pH 11 for 2 hours), again indicating that the NOM is reversibly adsorbed to the catalyst surface. The activity of (bi)sulfide-exposed catalysts was only partially restored by rinsing with a sodium hypochlorite solution (28 mM NaOCl, pH 7.8, 2 hours). (Bi)sulfide-promoted inhibition of Pd catalyst activity is well documented and is attributed to formation of strong PdS chemisorption complexes on catalyst surfaces [17,28]. In addition to demonstrating an effective treatment process for ICM, results obtained from this study indicate that reductive catalysis may be effective for treating a wide range of wastewater-derived halogenated contaminants, including regulated and emerging disinfection by-products (such as iodoacetic acids [29]) as well as ubiquitous personal care products (e.g., triclosan).

22.3 Selective catalytic nitrate reduction

Nitrate (NO_3^-) is the world's most ubiquitous groundwater contaminant, and a serious threat to drinking water supplies [30], as it is reduced to toxic nitrite in the human body. Maximum contaminant levels (MCLs) in the United States are 10 mg/L as N for NO_3^- (44 mg/L as NO_3^-) and 1.0 mg/L as N for NO_2^- (3.1 mg/L as NO_2^-) [31]. Ion exchange technology is the primary method used to capture nitrate in drinking water plants [32]. The technology is effective and reliable, but it produces a concentrated nitrate brine that must be further treated. Reduction of nitrate in bioreactors has also been demonstrated. However, there is resistance to this approach because of concerns regarding pathogen growth, greater turbidity, and higher organic loadings. These factors all increase chlorine demand that can lead to the production of chlorinated organic disinfection by-products, which are frequently carcinogenic.

An alternative and promising approach to remove nitrate from water is selective catalytic reduction in the presence of hydrogen as electron donor. This process requires a second, oxygen atom accepting metal to augment the hydrogen activating capability of a typical catalytic metal such as palladium, and it has been demonstrated with Pd–Cu, Pd–In, and Pd–Sn catalysts loaded onto a variety of supports, including alumina and activated carbon [8,9,33–35]. End products of NO_3^- reduction are NH_3 and dinitrogen (N_2); observed intermediates are nitrite (NO_2^-) and nitrous oxide (N_2O) [35–37]. Balanced equations for NO_3^- and NO_2^- reduction are given in Equations 22.1–22.3. Note the higher hydrogen demand involved in forming the undesirable by-product ammonia relative to forming the desired product dinitrogen. Thus, the availability of hydrogen to the catalyst during operation may have a significant impact on the overall selectivity of the reduction process.

$$NO_3^- + 1.0H_2 \rightarrow NO_2^- + H_2O \quad (22.1)$$

$$NO_2^- + 1.5H_2 \rightarrow 0.5\ N_2^- + H_2O + OH^- \quad (22.2)$$

$$NO_2^- + 3.0H_2 \rightarrow NH_3 + H_2O + OH^- \quad (22.3)$$

Catalytic nitrate reduction in pure water with hydrogen is relatively fast, with half-lives of minutes in batch reactions [38]. Catalytic nitrate reduction in natural waters can be considerably slower. For example, elevated biocarbonate (HCO_3^-), elevated chloride (Cl^-), reduced sulfur species (e.g., sulfide, S^{2-}), and humic acid, all decrease catalytic NO_3^- reduction rates, and increased production of ammonia is observed as well. The most serious fouling (i.e., fouling at the lowest concentrations) occurs for reduced sulfur species, which are known poisons for Pd-based catalysts [6,17,39–41]. There are also a number of ions in natural waters that have been found not to measurably affect catalytic NO_3^- reduction at concentrations typical of drinking water applications; these are SO_4^{2-}, Cl^-, and CO_3^{2-} [38,42,43].

A recent study with a 5 percent Pd–1 percent In on alumina catalyst focused specifically on the question of catalyst fouling with sulfide in order to better understand why reduced sulfur species inhibit NO_3^- reduction rates and increase NH_3 production [28]. Results from X-ray photoelectron spectroscopy (XPS) indicated that (bi)sulfide binds strongly to both Pd and In sites (Fig. 22.4). On a fresh catalyst, Pd is present in elemental form (BE 335.2 eV) and In is mainly present as an oxide (In_2O_3, BE 444.6 eV). After sulfide fouling, Pd is mainly present as PdS (BE 336.9 eV), whereas In is predominantly present as In_2S_3 (BE 445.2 eV).

Pseudo first-order rates constants for NO_3^- reduction decrease smoothly with greater sulfide loadings for both Pd and In. These results indicate that nitrate reduction rates decrease with sulfide fouling because Pd and In sites bound to sulfur are not active. Ammonia production also increases with catalyst exposure to higher sulfide concentrations. A possible explanation is that with increased sulfide

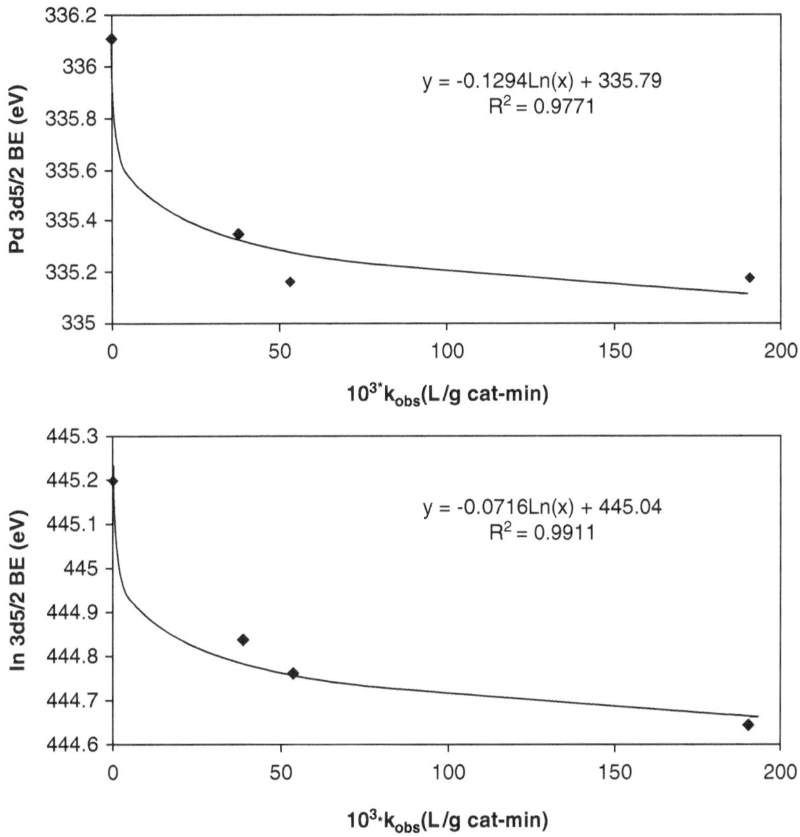

FIGURE 22.4

Correlations between Pd 3d(5/2) (upper) and In 3d(5/2) (lower) XPS binding energies in sulfide-exposed catalysts and the corresponding rate constant for nitrate reduction.

fouling, active Pd and In sites are located further apart on the alumina support and nitrogen–nitrogen pairing (which is necessary for N_2O and N_2 formation) becomes less likely.

A number of oxidants have been evaluated as regenerants for Pd-based catalysts fouled with sulfide [17,28]. Heating the dry catalyst in excess air and wet catalyst regeneration using dissolved oxygen, hydrogen peroxide, or sodium hypochlorite are effective for Pd-only catalysts. Hypochlorite was also able to partially restore the nitrate reduction capability for a sulfide-fouled, alumina-supported Pd–Cu catalyst; however, with each regeneration cycle, copper was lost to dissolution, and less nitrate reduction capacity was restored [28]. Heating alumina supported Pd–In catalysts in air removed much of the bound sulfide and restored NO_3^- reduction rates; however, this procedure may not be very practical due to the energy demands associated with drying and heating. Wet catalyst regeneration using dissolved oxygen or hydrogen peroxide proved ineffective for the Pd–In catalysts. However, regeneration using hypochlorite was able to restore the NO_3^- reduction rate of the Pd–In catalyst nearly to the fresh catalyst value. Concomitant XPS measurements on the regenerated catalysts showed complete return to metallic Pd and nearly complete recovery of the In_2O_3 state of the fresh catalyst [28].

22.4 Conclusions and prospects

The studies discussed in this chapter indicate that reductive catalysis is an emerging technology that shows great promise for treating a variety of recalcitrant micropollutants present in water sources, including halocarbons and nitrate as well as nitrosamines and perchlorate. With appropriate additional research, such processes may well become viable treatment technologies in the future.

One of the major obstacles to widespread adoption of heterogeneous catalytic treatment processes is uncertainty about their sustainability in practical treatment applications. Surface fouling and deactivation processes caused by interactions with deleterious nontarget constituents and reaction by-products is a significant concern. Thus, further research is needed to identify the mechanisms and environmental factors responsible for catalyst fouling and deactivation. This information can then be used to design effective strategies for improving catalytic process sustainability, either through preventative or regenerative means. The development of practical treatment processes would also benefit from further research on contaminant treatment and field applications in more complex matrices that are representative of real world water sources (e.g., groundwater or highly treated wastewater effluent).

Another perceived obstacle to the widespread use of reductive catalytic treatment processes is the high cost of precious metals commonly employed in the catalysts (e.g., Pd, Pt). Thus, further research is needed to identify and assess the behavior of non-precious metal substitutes (e.g., Ni). Although safety concerns may place limits on the practical application of traditional porous Ni catalysts, new non-pyrophoric Ni catalysts that may become economically attractive alternatives are currently being developed [44,45]. For example, a simple comparison between the cost-normalized rate constants for reduction of NDMA, a toxic nitrosamine compound, by Ni and Pd catalysts [11,12] illustrates the potential cost savings of the former. By using market prices for Ni (5 c//g) and Pd ($11.70/g) on May 11, 2007, the cost-normalized rate constants for NDMA reduction using porous Ni and Pd–Cu bimetal catalysts were determined to be 1500 and 5.7 L $\$^{-1}$ h^{-1}, respectively. Of course, the long-term costs of a catalytic process depend not just on the initial cost of the catalyst but also on its amortized cost over a number of possible regeneration and/or replacement cycles. Thus, catalyst improvement and sustainability depend critically on developing a detailed understanding of the deactivation and regeneration processes that can occur at the nanoscale level with the catalytic material.

References

[1] D.W. Kolpin, E.T. Furlong, M.T. Meyer, E.M. Thurman, S.D. Zaugg, L.B. Barber, et al., Pharmaceuticals, hormones, and other organic wastewater contaminants in US streams, 1999–2000: a national reconnaissance, Environ. Sci. Technol. 36 (2002) 1202–1211.

[2] T. Ternes, Analytical methods for the determination of pharmaceuticals in aqueous environmental samples, Trends Anal. Chem. 20 (2001) 419–434.

[3] S.D. Richardson, Disinfection by-products and other emerging contaminants in drinking water, Trends Anal. Chem. 22 (2003) 666–684.

[4] T. Ternes, A. Joss, H. Siegrist, Scrutinizing pharmaceuticals and personal care products in wastewater treatment, Environ. Sci. Technol. (2004) 38393A–38399AA.

[5] S. Kovenklioglu, Z. Cao, D. Shah, R.J. Farrauto, Direct catalytic hydrodechlorination of toxic organics in wastewater, AIChe J. 38 (1992) 1003–1012.

[6] C.G. Schreier, M. Reinhard, Catalytic hydrodehalogenation of chlorinated ethylenes using palladium and hydrogen for the treatment of contaminated water, Chemosphere 31 (1995) 3475–3487.

[7] C. Schüth, S. Disser, F. Schüth, M. Reinhard, Tailoring catalysts for hydrodechlorinating chlorinated hydrocarbon contaminants in groundwater, Appl. Catal. B Environ. 28 (2000) 147–152.

[8] S. Hörold, K.-D. Vorlop, T. Tacke, M. Sell, Development of catalysts for selective nitrate and nitrite removal from drinking water, Catal. Today 17 (1993) 21–30.

[9] A. Pintar, J. Batista, J. Levec, T. Kajiuchi, Kinetics of the catalytic liquid-phase hydrogenation of aqueous nitrate solutions, Appl. Catal. B Environ. 11 (1996) 81–98.

[10] K.D. Hurley, J.R. Shapley, Efficient heterogeneous catalytic reduction of perchlorate in water, Environ. Sci. Technol. 41 (2007) 2044–2049.

[11] M.G. Davie, M. Reinhard, J.R. Shapley, Metal-catalyzed reduction of N-nitrosodimethylamine with hydrogen in water, Environ. Sci. Technol. 40 (2006) 7329–7335.

[12] A.J. Frierdich, J.R. Shapley, T.J. Strathmann, Rapid reduction of N-nitrosamine disinfection byproducts in water with hydrogen and porous nickel catalysts, Environ. Sci. Technol. 42 (2008) 262–269.

[13] F. Alonso, I.P. Beletskaya, M. Yus, Metal-mediated reductive hydrodehalogenation of organic halides, Chem. Rev. 102 (2002) 4009–4091.

[14] Y. Ambroise, C. Mioskowski, G. Djega-Mariadassou, B. Rousseau, Consequences of affinity in heterogeneous catalytic reactions: Highly chemoselective hydrogenolysis of iodoarenes, J. Org. Chem. 65 (2000) 7183–7186.

[15] F.-D. Kopinke, K. Mackenzie, R. Koehler, A. Georgi, Alternative sources of hydrogen for hydrodechlorination of chlorinated organic compounds in water on Pd catalysts, Appl. Catal. A Gen. 271 (2004) 119–128.

[16] G.V. Lowry, M. Reinhard, Hydrodehalogenation of 1- to 3-carbon halogenated organic compounds in water using a palladium catalyst and hydrogen gas, Environ. Sci. Technol. 33 (1999) 1905–1909.
[17] G.V. Lowry, M. Reinhard, Pd-catalyzed TCE dechlorination in groundwater: Solute effects, biological control, and oxidative catalyst regeneration, Environ. Sci. Technol. 34 (2000) 3217–3223.
[18] K. Mackenzie, H. Frenzel, F.-D. Kopinke, Hydrodehalogenation of halogenated hydrocarbons in water with Pd catalysts: reaction rates and surface competition, Appl. Catal. B Environ. 63 (2006) 161–167.
[19] W.W. McNab, R. Ruiz, M. Reinhard, In-situ destruction of chlorinated hydrocarbons in groundwater using catalytic reductive dehalogenation in a reactive well: testing and operational experiences, Environ. Sci. Technol. 34 (2000) 149–153.
[20] M.O. Nutt, J.B. Hughes, M.S. Wong, Designing Pd-on-Au bimetallic nanoparticle catalysts for trichloroethene hydrodechlorination, Environ. Sci. Technol. 39 (2005) 1346–1353.
[21] C. Schüth, M. Reinhard, Hydrodechlorination and hydrogenation of aromatic compounds over palladium on alumina in hydrogen-saturated water, Appl. Catal. B Environ. 18 (1998) 215–221.
[22] F.J. Urbano, J.M. Marinas, Hydrogenolysis of organohalogen compounds over palladium supported catalysts, J. Mol. Catal. A. 173 (2001) 329–345.
[23] G. Yuan, M.A. Keane, Liquid phase catalytic hydrodehalogenation of 2,4-dichlorophenol over supported palladium: an evaluation of transport limitations, Chem. Eng. Sci. 58 (2003) 257–267.
[24] L.E. Knitt, J.R. Shapley, T.J. Strathmann, Rapid metal-catalyzed hydrodehalogenation of iodinated X-ray contrast media, Environ. Sci. Technol. 42 (2008) 577–583.
[25] T.A. Ternes, R. Hirsch, Occurrence and behavior of X-ray contrast media in sewage facilities and the aquatic environment, Environ. Sci. Technol. 34 (2000) 2741–2748.
[26] M.M. Huber, S. Canonica, G.-Y. Park, U. Von Gunten, Oxidation of pharmaceuticals during ozonation and advanced oxidation processes, Environ. Sci. Technol. 37 (2003) 1016–1024.
[27] M.A. Vannice, Kinetics of Catalytic Reactions, Springer-Kluwer, New York, 2005.
[28] B.P. Chaplin, J.R. Shapley, C.J. Werth, Regeneration of sulfur fouled bimetallic Pd-based catalysts, Environ. Sci. Technol. 41 (2007) 5491–5497.
[29] M.J. Plewa, E.D. Wagner, S.D. Richardson, A.D. Thruston, Y.-T. Woo, A.B. McKague, Chemical and biological characterization of newly discovered iodoacid drinking water disinfection byproducts, Environ. Sci. Technol. 38 (2004) 4713–4722.
[30] United States Environmental Protection Agency, National Water Quality Inventory, EPA 816-R-00-013, Washington, D.C., USEPA Office of Water, 2000.
[31] United States Environmental Protection Agency, National Primary Drinking Water Regulations: Contaminant Specific Fact Sheets, Inorganic Chemicals, Technical Version, EPA 811-F-95-002a-T, Washington D.C., USEPA Office of Water, 1995.
[32] A. Kapoor, T. Viraraghavan, Nitrate removal from drinking water-review, J. Environ. Eng. 123 (1997) 371–380.
[33] A. Pintar, J. Batista, I. Muševič, Palladium-copper and palladium-tin catalysts in the liquid phase nitrate hydrogenation in a batch-recycle reactor, Appl. Catal. B Environ. 52 (2004) 49–60.

[34] R. Gavagnin, L. Biasetto, F. Pinna, G. Struckul, Nitrate removal in drinking waters: The effect of tin oxides in the catalytic hydrogenation of nitrate by Pd/SnO_2 catalysts, Appl. Catal. B Environ. 38 (2002) 91–99.

[35] Y. Yoshinaga, T. Akita, I. Mikami, T. Okuhara, Hydrogenation of nitrate in water to nitrogen over Pd–Cu supported on active carbon, J. Catal. 207 (2002) 37–45.

[36] U. Prüsse, M. Hähnlein, J. Daum, K.-D. Vorlop, Improving the catalytic nitrate reduction, Catal. Today 55 (2000) 79–90.

[37] J. Daum, K.-D. Vorlop, Kinetic investigation of the catalytic nitrate reduction: construction of the test reactor system, Chem. Eng. Technol. 22 (1999) 199–202.

[38] B.P. Chaplin, E. Roundy, K.A. Guy, J.R. Shapley, C.J. Werth, The effects of natural water ions and humic acid on nitrate reduction using an alumina supported Pd-Cu catalyst, Environ. Sci. Technol. 40 (2006) 3075–3081.

[39] C. Bartholomew, Mechanisms of catalyst deactivation, Appl. Catal. A Gen. 212 (2001) 17–60.

[40] P.A. Gravil, H. Toulhoat, Hydrogen, sulphur and chlorine coadsorption on Pd(111): A theoretical study of poisoning and promotion, Surf. Sci. 430 (1999) 176–191.

[41] J. Oudar, Sulfur adsorption and poisoning of metallic catalysts, Catal. Rev.-Sci. Eng. 22 (1980) 171.

[42] L. Lemaignen, C. Tong, V. Begon, R. Burch, D. Chadwick, Catalytic denitrification of water with palladium-based catalysts supported on activated carbons, Catal. Today 75 (2002) 43–48.

[43] A. Pintar, J. Batista, J. Levec, Integrated ion exchange/catalytic process for efficient removal of nitrates from drinking water, Chem. Eng. Sci. 56 (2001) 1551–1559.

[44] I. Mikami, Y. Yoshinaga, T. Okuhara, Rapid removal of nitrate in water by hydrogenation to ammonia with Zr-modified porous Ni catalysts, Appl. Catal. B Environ. 49 (2004) 173–179.

[45] J. Geng, D.A. Jefferson, B.F.G. Johnson, The unusual nanostructure of nickel–boron catalyst, Chem. Commun. (2007) 969–971.

CHAPTER

23 Enhanced Dechlorination of Trichloroethylene by Membrane-Supported Iron and Bimetallic Nanoparticles

S.M.C. Ritchie

Department of Chemical and Biological Engineering, University of Alabama, Tuscaloosa, AL, USA

23.1 Introduction

The application of zero-valent metal nanoparticles to water treatment is a field of considerable interest. These materials are very effective reducing agents for the dechlorination of organics [1,2]. However, their application is complicated due to their rapid degradation due to oxidation in air [3]. In addition, they are not amenable to more traditional treatment technologies such as packed beds [4]. Deposition of the nanoparticles on a support material may promote increased implementation. Some supports that have been explored in the literature include

Street, Sustich, Duncan and Savage. Nanotechnology Applications for Clean Water, 2nd Edition.
 DOI: http://dx.doi.org/10.1016/B978-1-4557-3116-9.00023-8

solid and porous materials and membranes [5,6]. Solid supports limit pressure drop in a packed bed, but have limited surface area. The surface area can be improved by using a porous support; however, this introduces mass transfer resistance and reduces efficiency. Membranes improve surface area and avoid mass transfer limitations because they are very thin (200 μm) and contaminated water can permeate the membrane by convection under a pressure gradient [7,8].

There are several advantages to supporting nanoparticles in a membrane. First, the nanoparticles may be synthesized in controlled conditions in large batches [7,8]. Reactants are fed to a reactor vessel in solution, and mass transfer is aided by mixing. This also makes it easier to control the particle composition, particularly when synthesizing bimetallic nanoparticles. Post-coating of iron nanoparticles keeps the second metal at the nanoparticle surface where it can catalyze the reduction reaction. Second, the membrane serves as an absorbent for the contaminant. Most chlorinated organics have very limited solubility in water; in contrast, their solubility in rubbery polymers is very high [9]. The choice of polymer will affect the extent of absorption. Supporting nanoparticles in a polymeric membrane provides a synergy between rapid contaminant sorption and reaction at the nanoparticle surface. It also permits control of the morphology of the support, as this can change quite drastically with the method of polymer phase inversion. Denser polymer films may provide an effective diffusion barrier for oxygen, thereby inhibiting oxidation of the nanometal surface. The purpose of this chapter is to describe how membrane-supported nanoparticles are formed, as well as to examine some unique aspects of these materials and their potential advantages for water treatment.

23.2 Nanoparticle formation

There are numerous methods for synthesis of zero-valent nanoparticles. The most common methods for synthesizing iron nanoparticles are in solution or in a water-in-oil microemulsion [10–12]. In both cases, a solution of sodium borohydride is added to an iron solution, and nanoparticles are formed as zero-valent iron nanoparticles nucleate and grow. Iron salts may also be dissolved in a polymer phase [13]. Nanoparticles form when the material is soaked in a reducing solution. Another metal can be added by posttreatment, where zero-valent iron acts as a reducing agent for the second metal [8]. The second metal may also be present with the iron salt and the metals are simultaneously reduced. Once formed, it is important to use the materials or store them in a solvent to avoid oxidation by atmospheric oxygen or hydrolysis by water [3].

23.2.1 Solution and emulsion techniques

The simplest procedure for the formation of zero-valent iron nanoparticles involves the addition of a solution of sodium borohydride to a solution of iron chloride [7]. The addition is performed with continuous mixing to ensure uniform distribution of reducing agent. The reducing agent is also added in excess to prevent oxidation of the nanoparticles prior to application. Nanoparticles formed in this way tend to have a broad distribution of sizes. This is due to variations in mixing, distribution of reactants, and particle agglomeration. The transmission electron micrograph (TEM) shown in Fig. 23.1 is an excellent example of the broad size distribution obtained using the solution technique. Notice that the particle size range is very broad, and that the nanoparticles are up to 100 nm in diameter. The last point is of particular concern because the benefits of using nanoscale iron decrease exponentially as the particle size increases [7].

Emulsions retain most of the simplicity of solution techniques, but improve control of particle size [8]. A microemulsion is formed in two stages. The oil phase is a three-component mixture of solvent, co-solvent, and surfactant. An aqueous metal solution is added to the oil phase and mixed to form an emulsion. The co-solvent should be water-soluble (typically an alcohol), and is used to swell the micelles. These are actually reverse micelles as the hydrophilic ends of the surfactant face the center of the micelle and the lipophilic ends face the exterior of the micelle and the solvent phase [14]. The reverse micelle size is controlled by the surfactant concentration and the amount of co-solvent. An aqueous solution of sodium borohydride is next added to the microemulsion, resulting in iron reduction, nucleation, and nanoparticle formation. The nanoparticle size tends to be much more uniform than solution-phase techniques because each reverse micelle contains a limited amount of metal, and the surfactant minimizes aggregation and

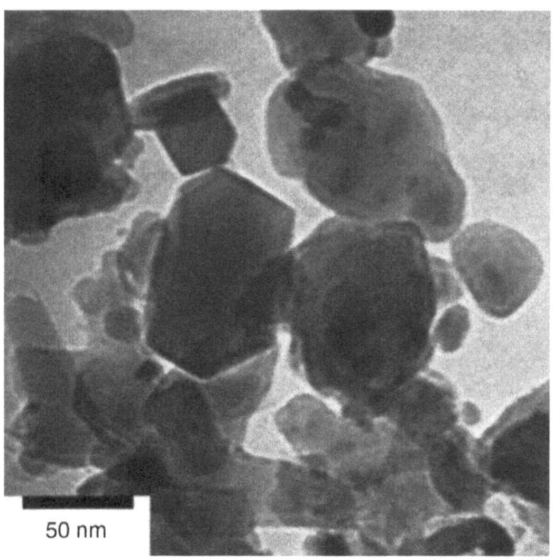

FIGURE 23.1

Transmission electron micrograph bright field image of unsupported iron nanoparticles [7].

growth of the nanoparticles. A TEM of a typical batch of iron nanoparticles synthesized using this technique is shown in Fig. 23.2. Note that the particle size is much more uniform than for solution techniques. More importantly, the particles are truly nanoscale, with an approximate size of 7–11 nm.

23.2.2 In situ formation of nanoparticles

Another technique for the synthesis of supported nanoparticles involves their formation in a solid polymer [15,16]. One of the problems with nanoparticles is their tendency to aggregate prior to immobilization in a support. Surfactants on the surface of the nanoparticle can inhibit aggregation, but this mechanism fails when nanoparticles are dried during deposition on a support. In contrast, organic salts are easily dispersed in a polymer solution. Once the solid-phase polymer film is formed by phase inversion, it can be post-treated with a reducing agent, such as soaking in an aqueous solution containing sodium borohydride. Metal ions are reduced in place to form nanoparticles. The metal concentration and low temperature operation limit large aggregate formation in the thin film. Nanoparticles synthesized by this method are not the focus of this chapter.

23.2.3 Addition of secondary metals

The activity of iron nanoparticles is greatly enhanced by the addition of a second metal [8,10,17]. Transition metals and noble metals, such as nickel and palladium, catalyze the oxidation of iron and the transfer of electrons to adsorbed chlorinated organics. This is a surface reaction, and therefore the second metal tends to be

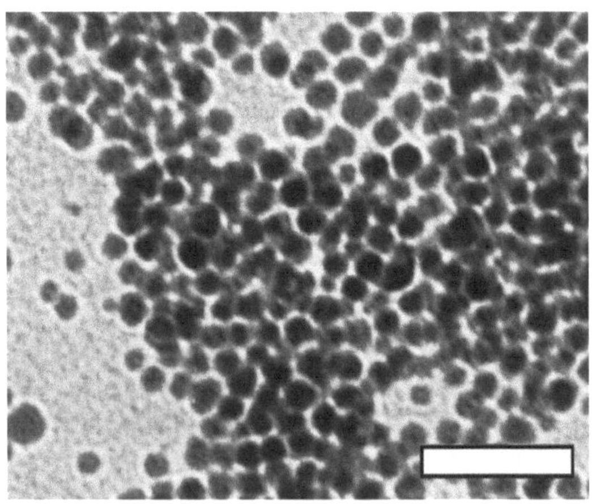

FIGURE 23.2

Transmission electron micrograph bright field image of unsupported iron nanoparticles that form in a water-in-oil microemulsion. The bar represents 100 nm [8].

most effective when it is present at the nanoparticle surface. Therefore, although the second metal can be reduced simultaneously with iron, it is generally more effective when it is added by post-coating iron nanoparticles [8]. Post-coating is achieved by dispersion of iron nanoparticles in an aqueous solution containing the second metal. Zero-valent iron serves as a reducing agent for the second metal, which deposits on the nanoparticle surface. Subsequent treatment with sodium borohydride reduces any oxidized iron.

Simultaneous reduction of a second metal can be applied to solution and emulsion techniques. The second metal is added as a salt to the aqueous iron solution, and the synthesis proceeds in a similar fashion as previously stated for iron nanoparticles. The nanoparticle size will be dictated by the synthesis technique, and therefore is similar to sizes for single metal nanoparticles. TEM images of nickel–iron nanoparticles synthesized by solution and emulsion techniques are shown in Fig. 23.3. The nanoparticles formed by microemulsion are approximately the same size as the iron nanoparticles shown in Fig. 23.2.

23.2.4 Preserving zero-valence

Iron is a reducing agent when used for water treatment. Therefore, the iron is oxidized when nanoparticles are used for dechlorination. Nanoparticles are susceptible to oxidation and hydrolysis because of their high surface area. When using unsupported nanoparticles, it is crucial to use the nanoparticles as soon as they are synthesized to maximize the use of metal for dechlorination. When supporting nanoparticles, it is important to protect the nanoparticles from these side reactions as well. Solvent exchange is a simple technique to preserve the zero-valence of iron and bimetallic nanoparticles. First, the nanoparticles are separated using a magnetic field. In the laboratory, this is achieved by aggregation on a magnetic

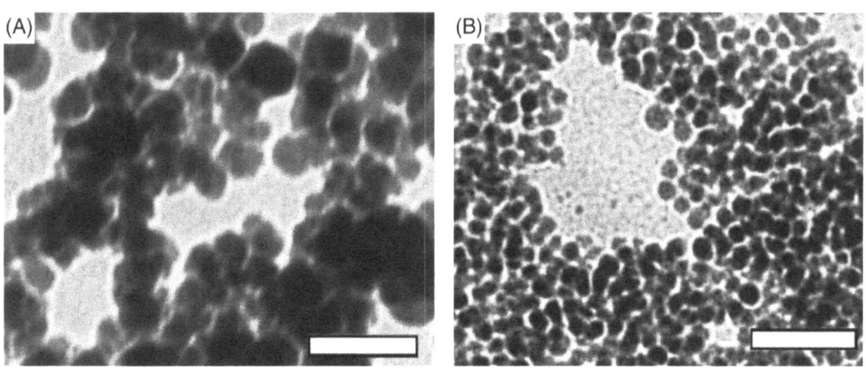

FIGURE 23.3

Transmission electron micrograph bright field images of unsupported nickel–iron nanoparticles formed in (A) solution, and in (B) water-in-oil microemulsion. The bar represents 100 nm [8].

stir bar. The supernatant is then decanted to reduce the amount of residual water in the reactor. Not all of the supernatant is removed to avoid exposure of the nanoparticles to air. Methanol is added to the reactor, the system is mixed, and a magnetic field is used to separate the nanoparticles. The decant procedure is repeated two more times, resulting in a nanoparticle–methanol slurry. The nanoparticles are stable for months in this format, and are also suitable for dispersion in a polymer solution to make thin films. Once in the thin film, the polymer acts as a diffusion barrier for oxygen.

23.3 Polymers

Although numerous polymers could be used to support nanoparticles, a good starting point is to examine polymers used in the membrane industry. These are generally soluble in common solvents such as acetone, dimethyl formamide (DMF), or *n*-methylpyrrolidone (NMP). However, it is important that each solvent is soluble in water. This is critical for the phase inversion step of thin film formation. A thin film of the polymer–solvent solution is placed in water. The polymer has no solubility in water. Therefore, as the solvent is extracted by water, the polymer concentration in the thin film increases until it reaches saturation and precipitates. The rate of solvent extraction will affect the properties of the thin film, including the porosity and the pore size. If the solvent is allowed to evaporate from the surface prior to immersion in water, the thin film will have a dense surface layer with negligible porosity [18].

Two polymers widely used in the membrane industry include cellulose acetate (CA) and polyethersulfone (see Table 23.1). Cellulose acetate is one of the most

Table 23.1 Chemical Structures of Membrane Polymers

Polymer	Structure
Cellulose triacetate (CTA)	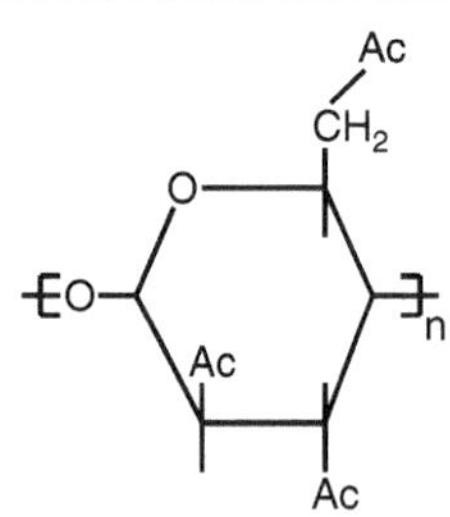
Poly(ether sulfone) (PES)	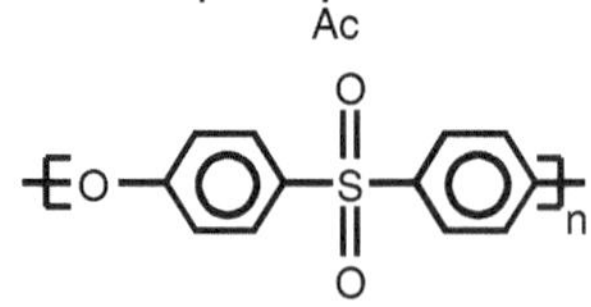

Table 23.2 Solubility Parameters for Membrane Polymers [24,25]

Polymer	δ ($MPa^{1/2}$)
PES	22.9
CTA	25–28

widely used polymers in the membrane industry. It is relatively inexpensive, widely available, and soluble in acetone. The porosity of CA thin film is easily manipulated based on polymer concentration, water temperature, and evaporation time prior to phase inversion [18]. Thin films composed of CA are packaged into a variety of membrane modules, including pleated depth filters, spiral modules, and hollow fibers.

Polyethersulfone (PES) is also widely used in the membrane industry. It is more expensive than CA; however, it has better thermal and chemical resistance. Polyethersulfone is soluble in DMF, and its porosity in a thin film can be manipulated in a similar fashion to CA. Thin films composed of PES are highly pliable, and therefore can be packaged in a variety of membrane modules. Due to the aromatic backbone structure of PES, it tends to be more hydrophobic than CA. The solubility parameter of each polymer is shown in Table 23.2, and provides a quantitative measure of hydrophobicity. The solubility parameter is lower for more hydrophobic polymers. For water treatment applications, a decrease in the solubility parameter would decrease membrane permeability [18]. However, organic contaminant absorption would increase. This will be examined in more detail in a later section of this chapter.

23.4 Composite material

Membrane-supported nanoparticles are a composite material based on the distribution of nanoparticles in a thin film of polymer. The polymer may be porous or dense, and ideally the nanoparticles are uniformly distributed throughout the material. The composite is formed by phase inversion of a dispersion of nanoparticles in a polymer solution. The metal content of a typical composite membrane is 6.8 wt%. Nanoparticles are distributed throughout the thin film, and may be exposed in the membrane pores or completely imbedded in the polymer. When the thin film is porous, we have observed some oxidation of the nanoparticles during dry storage in air. Dense films showed negligible oxidation.

Membrane-supported nanoparticles are formed through a series of steps. First, a concentrated nanoparticle suspension is added to a polymer solution. The nanoparticle suspension is in methanol, and therefore should be concentrated to minimize the amount of methanol being added to the polymer solution. This is necessary because methanol is a non-solvent for the polymer, and will cause the polymer to precipitate if too much is present. Care should be taken when mixing the suspension and the polymer solution to avoid entrainment of air. Rotation of the mixing vessel is generally sufficient to get good mixing. However, if air bubbles are entrained, the mixture can be stored overnight and gas bubbles will rise to the surface and exit the mixture. A small vacuum can also be pulled on the mixture. This causes expansion of entrained bubbles and faster separation. Only a small vacuum should be applied to avoid solvent evaporation.

The next step of composite formation is casting of the polymer film. On a laboratory scale, this is most easily accomplished using a handheld "doctor blade." This device is a gate that is mounted on two runners. The height of the gate is set using two micrometers mounted at each end of the gate. The suspension is poured onto a clean glass plate, and the doctor blade is dragged over the pool of suspension. A thin film of uniform thickness is then formed on the plate. The film is then allowed to evaporate slightly to enhance the surface density prior to immersion of the glass plate into a non-solvent bath. For membrane-supported nanoparticles, it is better to use ethanol for the non-solvent. This prevents hydrolysis of the nanoparticle surface during phase inversion. In addition, once the solid thin film forms, storage of the membrane in methanol preserves the zero-valence of the nanoparticles.

Differences in the polymer film will affect water treatment. For example, the density of the film will affect the transport of contaminants to the nanoparticle surface. The TEM in Fig. 23.4 shows the cross-section of a porous membrane. Nanoparticles are located in the pores of the material. Because the material is

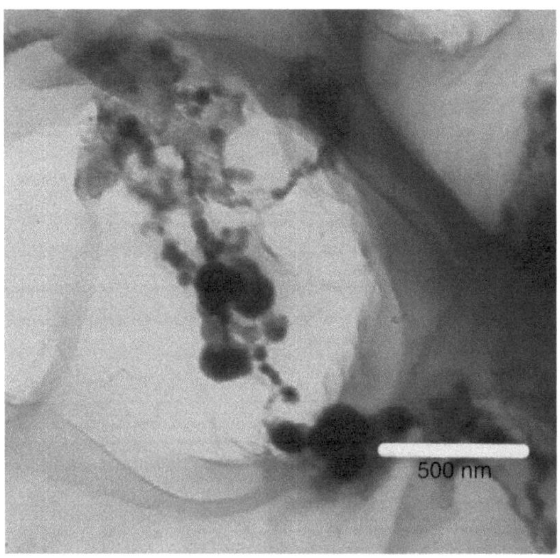

FIGURE 23.4

Transmission electron micrograph bright field image of membrane-supported iron nanoparticles.

porous, contaminant transport to the metal surface would occur as contaminated water permeated the membrane under a pressure gradient. However, the nanoparticles would also be exposed to water and dissolved oxygen. If the film is dense, however, the contaminant would be transported by diffusion to the nanoparticle surface under a concentration gradient. Water dissolved in the polymer would also be transported by diffusion. Oxygen exposure should be minimal due to low concentration in the aqueous phase, and therefore negligible driving force for mass transfer. The longer times associated with transport by diffusion would be mitigated by producing thinner films. In all cases, there are two mechanisms for separation of the contaminant: reaction and absorption. These mechanisms will be examined in more detail in the next section.

23.5 Water treatment

Zero-valent iron nanoparticles have been used extensively for water treatment [1,2,10]. One common treatment technique is achieved by the injection of nanoparticles into groundwater. As nanoparticles migrate through the subsurface, chlorinated organics adsorb on the surface of the nanoparticles and are reduced by zero-valent iron. The iron is oxidized during the reaction and is released to the environment. Unused iron is also oxidized and released to the environment. When the nanoparticles are composed solely of iron, this is an acceptable trade-off. An intractable and toxic contaminant is destroyed in exchange for a slightly higher level of iron ions. Iron is relatively inexpensive, and procedures for synthesis of iron nanoparticles are fairly simple.

Of course, this process does have some disadvantages. In spite of the very active nature of zero-valent iron, it is not sufficiently active to rapidly destroy all chlorinated organics. For example, polychlorinated biphenyls (PCBs) in sediments have been partially dechlorinated using high concentration slurries of iron nanoparticles (20 wt%) in 30 percent methanol/water mixture [19]. However, besides the very high requirement of nanoparticles, the half-life of the PCBs varied from 2.4 months to 77 years. Even if there is some dechlorination, the by-products of reduction may be more toxic than the parent compound. By contrast, palladized iron nanoparticles can provide 90 percent dechlorination of 2,2'-dichlorobiphenyl to biphenyl in less than 2 hours [16]. The addition of a second metal can increase the activity of the nanoparticles [20], but this adds significant expense (palladium) or introduces a new contaminant to the environment (e.g., nickel). In the latter case, depending on the metal the new contaminant is much worse since nature is much better suited to degrade organics.

If nanoparticles are to be applied in a more traditional engineering sense with ex situ treatment, they must be made amenable to large scale processing. Hence, characteristics such as shelf-life, column operation in a packed bed, and regeneration of the material become more important. Methods for synthesis of composite materials must be scalable and safe. For example, large scale storage and handling of reducing agents at remote sites is inherently unsafe. The following sections describe dechlorination results using membrane-supported nanoparticles produced using scalable processes. The nanoparticles are produced in scalable batch processes, and the composite materials may be synthesized using conventional membrane production processes [8]. All membranes are amenable to packaging in conventional membrane modules. The use of ex situ treatment methods (not injection), also permits better control of nanoparticle and secondary metal fate and transport to the environment.

23.5.1 Metal particle composition

The composition of zero-valent metal nanoparticles has a direct effect on nanoparticle activity. For degradation of trichloroethylene (TCE), nickel has a positive effect [21]. Unfortunately, there is also evidence that nickel is released to solution. The use of palladium has a similar positive effect at much lower content (wt%). However, there is a cost associated with the improved activity. In this section, representative results for nanoparticles formed in microemulsion and supported in CA will be presented.

23.5.1.1 Supported iron nanoparticles

Unsupported iron nanoparticles are effective for the dechlorination of TCE. Therefore, it is important that any support material does not significantly detract from the activity of the nanoparticles. Table 23.3 shows data comparing reaction rate constants for unsupported and supported iron nanoparticles, as well as results for iron microparticles. Clearly, nanoparticles are more active than microparticles on a mass basis. However, of more interest is the comparison between unsupported and supported nanoparticles. In that case, the mass normalized reaction rate constant was 36 percent of that for unsupported nanoparticles. Given the larger average size of the supported nanoparticles and the lower dose, the support did not impose significant mass transfer resistance.

23.5.1.2 Supported nickel–iron nanoparticles

The primary reason for adding a second metal to iron nanoparticles is to increase activity and hence the rate of reaction. The data shown in Table 23.3 represent reaction times of 48 hours or longer to achieve complete dechlorination. In

Table 23.3 Effect of Support and Particle Size on Degradation of Trichloroethylene by Zero-Valent Iron Particles

Iron Loading (mg)	Particle Size (nm)	Mass Normalized Reaction Rate Constant k_m ($l \cdot h^{-1}\ g^{-1}$)
64 (supported)	10–80	0.024
200 [12]	10	0.067
10,000 [22]	150,000	0.000088

groundwater applications, particularly for injected nanoparticles, long reaction times are acceptable due to the relative slow migration of groundwater through the porous subsurface. However, for ex situ treatment, long reaction times correspond to very large reactors and major capital expense. In addition, slow kinetics can result in higher concentrations of intermediates. For example, vinyl chloride is a potential by-product of TCE degradation by zero-valent iron, and vinyl chloride is more toxic than TCE [26].

Nickel is a popular choice for addition to iron nanoparticles because it is a good catalyst for hydrogenation in the form of Raney nickel. The standard reduction potential of nickel is −0.25 V. This is lower than the standard reduction potential for iron (− 0.44 V), and therefore zero-valent iron will reduce divalent nickel to the zero-valent form during nickel–iron nanoparticle synthesis. However, if the surface of the iron particle is completely coated with nickel, the nanoparticle activity will be of the same order as iron. This is shown by the data in Fig. 23.5.

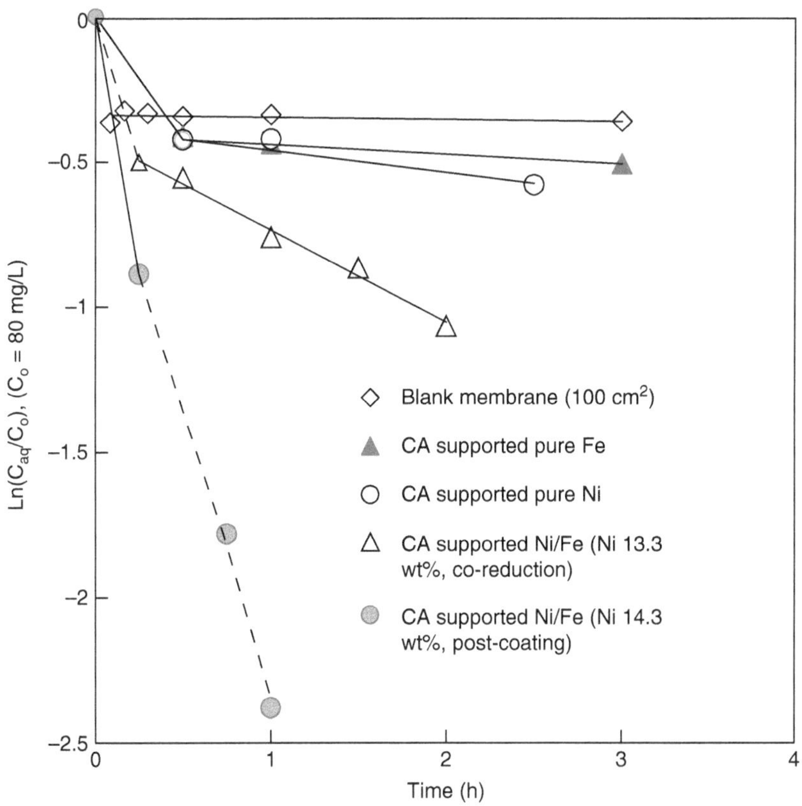

FIGURE 23.5

Representative results for dechlorination of trichloroethylene with membrane-supported nickel–iron nanoparticles [8].

There is essentially no difference between membrane-supported iron or nickel nanoparticles. However, when there is only partial coverage of the nanoparticle surface, nickel catalyzes oxidation of the iron and electron transfer to adsorbed TCE. Consequently, the reaction rate increases by more than an order of magnitude compared to single metal nanoparticles.

The amount of nickel at the surface will affect the reaction rate constant [8,21]. Fig. 23.5 contains data for two sets of nickel–iron nanoparticles. In one case, the nickel was co-reduced with iron during nanoparticle synthesis. Due to the lower redox potential of nickel, it should be more concentrated toward the center of the particle. Therefore, the nickel concentration at the surface would be much lower than for post-coated nanoparticles. The data in Fig. 23.5 show that concentration of nickel at the nanoparticle surface is advantageous to distribution of nickel throughout the particle.

23.5.1.3 Supported palladium–iron nanoparticles

One of the major problems with adding a second metal to improve the activity of iron nanoparticles is oxidation and loss of the second metal. In batch studies, oxidized iron and nickel are ubiquitous in the aqueous phase for unsupported and supported nanoparticles. Some representative data for solution metal concentrations for batch dechlorination of TCE are shown in Fig. 23.6. Iron is oxidized during the reaction, and therefore its solution concentration ($Fe2+$/$Fe3+$) increases over time. Pure nickel nanoparticles behave in a similar fashion. Bimetallic nanoparticles will lose iron and nickel to solution. The nickel concentration, however, is constant over time and is dependent on the dose. This may be due to some reduction of dissolved nickel at the nanoparticle surface by zero-valent iron. It is clear, however, that the use of nickel–iron nanoparticles will always result in a loss of this toxic metal to the environment, and therefore alternative strategies to improve the reaction rate are needed.

Although some researchers have made an effort to recapture oxidized metal by incorporating polyelectrolytes into the support [15], most researchers have sought to overcome this problem by switching to a more noble metal such as palladium. The benefits to using palladium include better activity, significantly lower content of second metal, and negligible loss of toxic metal to the environment [23]. The expense is significantly higher than nickel; however, if the material is maintained in a controlled process, recovery of the metal would help to minimize operating costs. The last point is a major disadvantage to injection of palladium–iron nanoparticles.

Representative results for TCE dechlorination by membrane-supported palladium–iron nanoparticles are shown in Fig. 23.7. The rate of dechlorination was highly dependent on the metal dose. This has been attributed to limited active sites in the membrane. The palladium content of the post-coated iron nanoparticles was only 1.9 wt%. Comparable behavior for nickel–iron nanoparticles required nearly 10-times the amount of nickel. Given the reported results of Bhattacharyya for the degradation of PCBs using palladium–iron nanoparticles

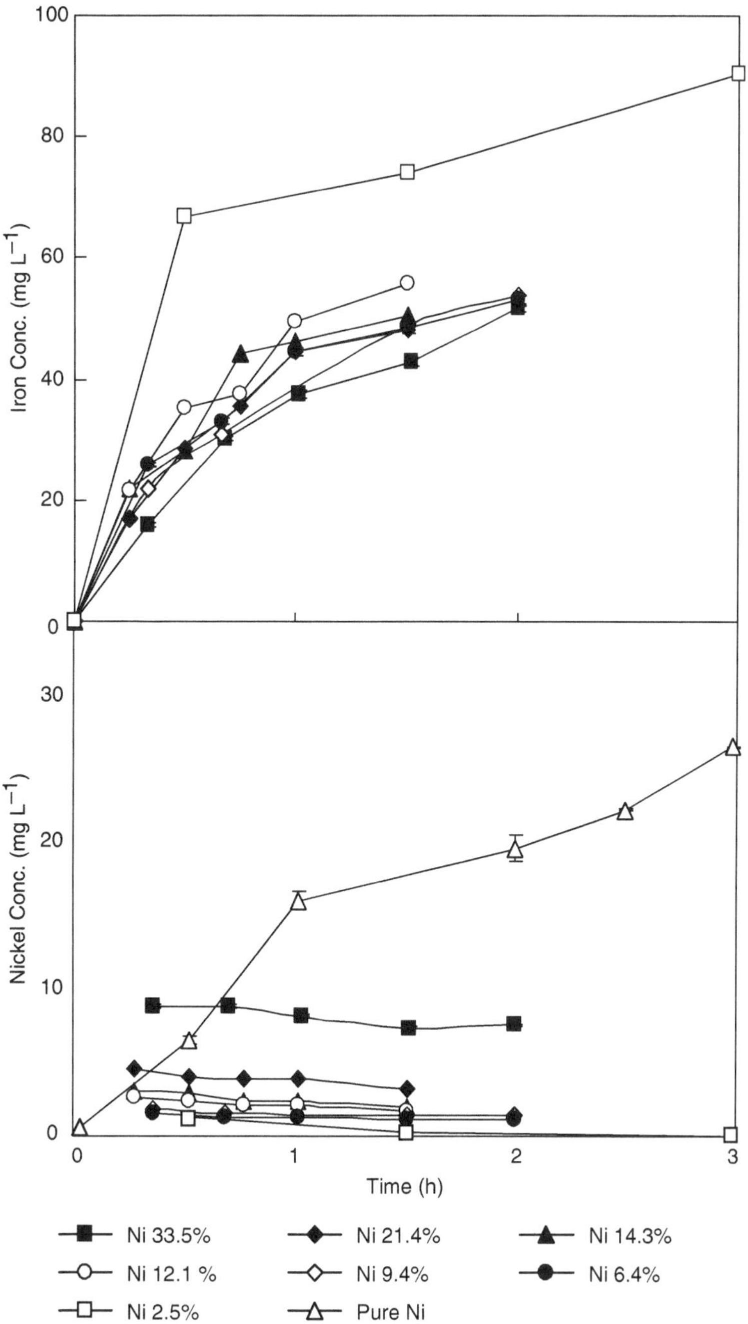

FIGURE 23.6

Solution metal concentrations for batch reduction of trichloroethylene using membrane-supported nanoparticles [8].

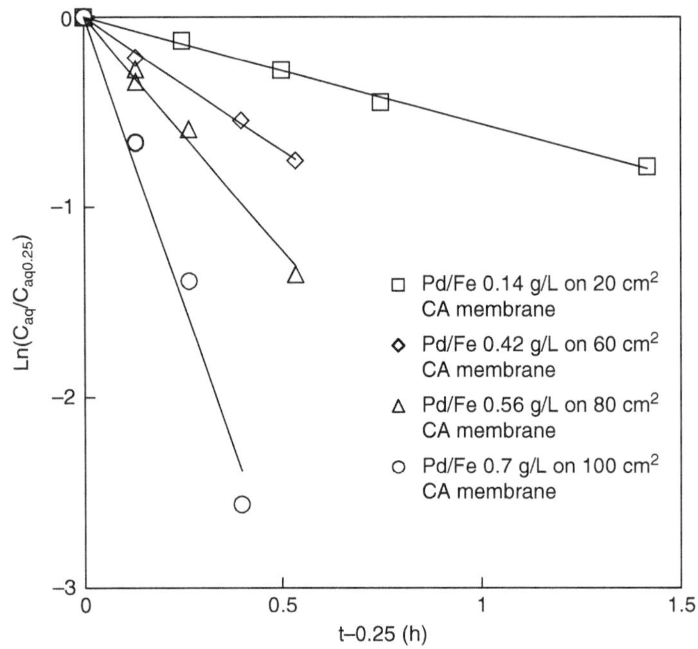

FIGURE 23.7

Representative results for dechlorination of trichloroethylene with membrane-supported palladium–iron nanoparticles.

[23], this is a strong endorsement for the use of palladium as a second metal to increase the activity of membrane-supported iron nanoparticles.

23.5.2 Absorption in support polymer

One significant advantage of supporting nanoparticles for water treatment is the potential for synergistic effects. For example, an organic support would have a high partition coefficient for the chlorinated organic. This is particularly true for elastomers. Data from the literature show that recycled tires can uptake nearly four times their mass in TCE [9]. Common membrane polymers, although not generally elastomers, are excellent absorbents for chlorinated solvents. This is because the solubility parameters for chlorinated solvents and these polymers are similar. Therefore, chlorinated organic contaminants will be absorbed at low concentration from water. Absorption data for TCE from water into PES is shown in Fig. 23.8. Notice that even at very low TCE concentration in the aqueous phase, absorption is measured in mg TCE/g polymer.

The data in Fig. 23.5 showed a rapid decrease in TCE solution concentration in the first 15 minutes of the batch experiments. This was observed in all cases,

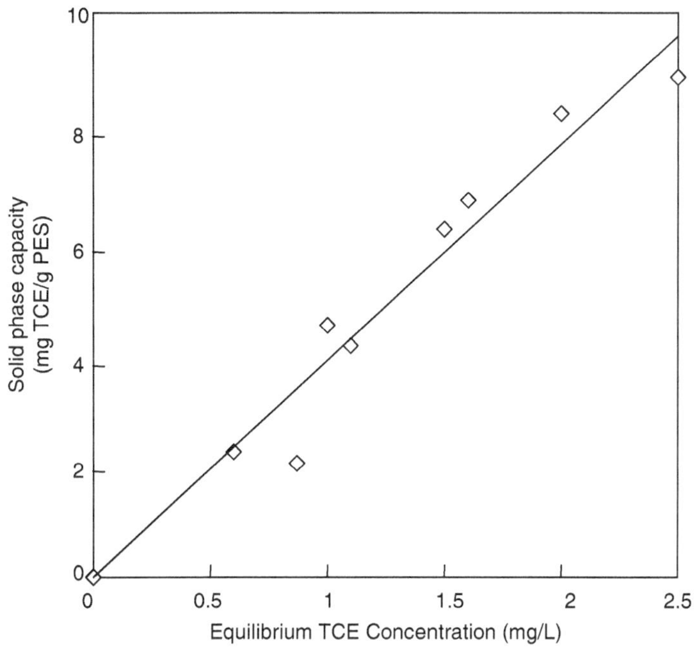

FIGURE 23.8

Trichloroethylene (TCE) absorption in polyethersulfone (PES) membrane.

including the case of a blank membrane. The blank did not contain any nanoparticles. However, its thickness and area was similar to the membrane-supported nanoparticles. It showed no decrease in TCE concentration after the initial period of absorption. The extent of absorption will depend on the amount of membrane and the chemistry of the polymer. The results in Fig. 23.5 were for CA. The more hydrophobic nature of PES would result in a higher amount of absorption. It should be noted that the solid phase concentration decreases as TCE is dechlorinated by the membrane-supported nanoparticles. The residual TCE concentration in solution and the total TCE in the system, however, will also decrease. Therefore, membrane-supported nanoparticles can rapidly absorb relatively intractable chlorinated species for immediate water treatment, and the supported nanoparticles can dechlorinate the organics over a longer timescale.

23.6 Conclusions

Membranes are an excellent support for nanoparticles when applied to water treatment. The membrane support is amenable to engineered water treatment processes, and does not seriously degrade dechlorination kinetics. The addition of a second metal, such as nickel, can improve reaction kinetics. However, we have observed that oxidized metals are lost to the environment. Palladium has a similar improvement at 10 percent of the loading required for nickel–iron nanoparticles. The use of membrane-supported nanoparticles allows for more controlled application of these nanoparticles to water treatment. The membrane has very high absorption of chlorinated organics such as TCE. Absorbed TCE is then dechlorinated by the nanoparticles over time.

References

[1] H.L Lien, W.X. Zhang, Nanoscale iron particles for complete reduction of chlorinated ethenes, Colloids Surf. A Physicochem. Eng. Asp. 191 (2001) 97–105.

[2] B. Schrick, J.L. Blough, A.D. Jones, T.E. Mallouk, Hydrodechlorination of trichloroethylene to hydrocarbons using bimetallic nickel–iron nanoparticles, Chem. Mater. 14 (2002) 5140–5147.

[3] S.M. Ponder, J.G. Darab, T.E. Mallouk, Remediation of Cr(VI) and Pb(II) aqueous solutions using supported, nanoscale zero-valent iron, Environ. Sci. Technol. 34 (2000) 2564–2569.

[4] N.E. Korte, J.L. Zutman, R.M. Schlosser, L. Liang, B. Gu, Q. Fernando, Field application of palladized iron for the dechlorination of trichloroethene, Waste Manag. 20 (2000) 687–694.

[5] L.J. Graham, G. Jovanovic, Dechlorination of p-chlorophenol on a Pd/Fe catalyst in a magnetically stabilized fluidized bed; implications for sludge and liquid remediation, Chem. Eng. Sci. 54 (1999) 3085–3093.

[6] C. Baker, S.I. Shah, S.K. Hasanain, Magnetic behavior of iron and iron-oxide nanoparticle/polymer composites, J. Magn. Magn. Mater. 280 (2004) 412–418.

[7] L. Wu, M. Shamsuzzoha, S.M.C. Ritchie, Preparation of cellulose acetate supported zero-valent iron nanoparticles for the dechlorination of trichloroethylene in water, J. Nanop. Res. 7 (2005) 469–476.

[8] L. Wu, S.M.C. Ritchie, Removal of trichloroethylene from water by cellulose acetate supported bimetallic Ni/Fe nanoparticles, Chemosphere 63 (2006) 285–292.

[9] P. Rangarajan, P. Sisk, D. Bhattacharyya, Novel applications of scrap tire for organic sorption/separations, Clean Prod. Process. 1 (1999) 199–209.

[10] C.B. Wang, W.X. Zhang, Synthesizing nanoscale iron particles for rapid and complete dechlorination of TCE and PCBs, Environ. Sci. Technol. 31 (1997) 2154–2156.

[11] A. Martino, M. Stoker, M. Hicks, C.H. Bartholomew, A.G. Sault, J.S. Kawola, The synthesis and characterization of iron colloid catalysts in inverse micelle solutions, Appl. Catal. A Gen. 161 (1997) 235–248.

[12] F. Li, C. Vipulanandan, K.K. Mohanty, Microemulsion and solution approaches to nanoparticle iron production for degradation of trichloroethylene, Colloids Surf. A Physicochem. Eng. Asp. 223 (2003) 103–112.

[13] J. Xu, A. Dozier, D. Bhattacharyya, Synthesis of nanoscale bimetallic particles in polyelectrolyte membrane matrix for reductive transformation of halogenated organic compounds, J. Nanop. Res. 7 (2005) 449–467.

[14] S.G. Dixit, A.R. Mahadeshwar, S.K. Haram, Some aspects of the role of surfactants in the formation of nanoparticles, Colloids Surf. A Physicochem. Eng. Asp. 133 (1998) 69–75.

[15] J. Xu, D. Bhattacharyya, Membrane-based bimetallic nanoparticles for environmental remediation: Synthesis and reactive properties, Environ. Prog. 24 (2005) 358–366.

[16] J. Xu, D. Bhattacharyya, Fe/Pd nanoparticle immobilization in microfiltration membrane pores: Synthesis, characterization, and applications in the dechlorination of polychlorinated biphenyls, Ind. Eng. Chem. Res. 46 (2007) 2348–2359.

[17] C.J. Lin, S.L. Lo, Y.H. Liou, Dechlorination of trichloroethylene in aqueous solution by noble metal-modified iron, J. Hazard. Mater. 116 (2004) 219–228.

[18] W.S.W. Ho, K.K. Sirkar (Eds.), Membrane Handbook, Chapman & Hall, New York, 1992.

[19] G.V. Lowry, K.M. Johnson, Congener-specific dechlorination of dissolved PCBs by microscale and nanoscale zerovalent iron in a water/methanol solution, Environ. Sci. Technol. 38 (2004) 5208–5216.

[20] W.X. Zhang, C.B. Wang, H.L. Lien, Treatment of chlorinated organic contaminants with nanoscale bimetallic particles, Catal. Today 40 (1998) 387–395.

[21] Y.H. Tee, E. Grulke, D. Bhattacharyya, Role of Ni/Fe nanoparticle composition on the degradation of trichloroethylene from water, Ind. Eng. Chem. Res. 44 (2005) 7062–7070.

[22] T.M. Vogel, C.S. Criddle, P.L. McCarty, Transformation of halogenated aliphatic compounds, Environ. Sci. Technol. 21 (1987) 722–736.

[23] X. Xu, H. Zhou, P. He, D. Wang, Catalytic dechlorination kinetics of p-dichlorobenzene over Pd/Fe catalysts, Chemosphere 58 (2005) 1135–1140.

[24] H.F. Mark, N.M. Bikales, C.G. Overberger, G. Menges (Eds.), Encyclopedia of Polymer Science and Engineering, second ed., Wiley, New York, 1985.

[25] J. Brandup, E.H. Immergut, E.A. Grulke (Eds.), Polymer Handbook, fourth ed., Wiley, New York, 1999.

[26] J.K. Gotpagar, E.A. Grulke, T. Tsang, D. Bhattacharyya, Reductive dechlorination of trichloroethylene using zero-valent iron, Environ. Prog. 16 (1997) 137–143.

CHAPTER

Synthesis of Nanostructured Bimetallic Particles in Polyligand-Functionalized Membranes for Remediation Applications

24

Jian Xu[1], Leonidas Bachas[2] and Dibakar Bhattacharyya[3]

[1]Coskata, Warrenville, IL, USA [2]University of Miami, Miami, FL, USA [3]Department of Chemical and Materials Engineering, University of Kentucky, Lexington, KY, USA

Street, Sustich, Duncan and Savage. Nanotechnology Applications for Clean Water, 2nd Edition.
 DOI: http://dx.doi.org/10.1016/B978-1-4557-3116-9.00024-X

24.1 Introduction

Nanosized metal particles have become an important class of materials due to their unique physical and chemical properties and high surface reactivity. Substantial studies of metal nanoparticles synthesis have been reported in the field of catalysis [1], optical [2], electronic [3], magnetic [4], biological devices [5,6], and pollution control. Recently, various studies have been reported on groundwater remediation for the degradation of toxic chlorinated organic compounds (COCs) with non-immobilized Fe0 based bimetallic nanoparticles (Fe/Ni, Fe/Pd) [7–10]. In this case, COCs are reduced to nontoxic hydrocarbons in the presence of the second catalytic metal (Pd or Ni) by substitution of chlorine with hydrogen. Compared to the single zero-valent Fe system such as iron filings, which has been used for decades for COCs degradation, this catalytic hydrodechlorination technology developed because of an enhanced reaction rate and elimination of toxic by-product formation due to the second, catalytic metal. The bimetallic nanoparticles used in most of the studies were synthesized in aqueous phase by reduction of metal cations with sodium borohydride [7–10]. The nanosized metals precipitated from solutions are extremely reactive due to the high surface energy, and they usually aggregate without the protection of their surface [11,12].

In order to avoid the agglomeration and aggregation, nanoparticles are usually stabilized by polymers or ligands in solution phase, or immobilized on solid supports [13–15]. Considerable attention has been given to the preparation of metal nanoparticles embedded in polymer films or membranes by a stepwise approach of ion-exchange/reduction [16–23]. In this method, ion-exchange ligands created in the thin films can bind metal cations from aqueous solution. Post reduction or precipitation produces nanoparticles from bound metal cation precursors. The advantage of this process is the creation of controllable nanostructure properties during nanoparticle synthesis by utilizing these ion-exchange ligands. The nanostructure properties include the particle size and distribution, particle concentration, and interparticle spacing. The amount of metal cations loaded is controlled by the amount of ligand sites and ion-exchange conditions such as the pH and competitive ions [17]. The distance between bound cations, which determines the final particle size, is also controlled by the space between ion-exchange ligands [24]. For example, Wang and coworkers reported that silver nanoparticles with various size and concentration have been produced in the polyelectrolyte multilayer film [24].

Various ligands and chelating groups can be used for the synthesis of metal nanoparticles by the ion-exchange/reduction method. Especially, polyelectrolytes [25] containing multifunctional chelating agents provide a great number of side functional ligands such as amines, carboxylic acids, amides, alcohols, aminoacids, pyridines, thioureas, iminos, and so on. These side functional groups can have strong interaction with metal ions to establish stable polymer-ion complexes. Attachment procedures of these chelating polymers and polypeptides on membrane internal pore surfaces have been extensively studied for metal ions sorption [26–30]. These functionalized membranes incorporating metal ions have the possibility to be the precursors for nanoparticle synthesis. Microfiltration (MF) and ultrafiltration (UF) membranes are no doubt the ideal base for functionalization due to the open structure and large pore size that are essential to attain high efficient utilization of available sites as well as the easy accessibility to the nanoparticles immobilized inside the membrane matrix.

Polyacrylic acid (PAA), a water soluble polyelectrolyte (polyligand), has been extensively used for metal capture and ion-exchange process because of the carboxylic acid group [25]. Generally, there are three different ways to functionalize MF/UF membrane with PAA: (i) dip or spin coating PAA on the membrane support; (ii) layer-by-layer assembly of polycations (polyallylamine hydrochloride (PAH), polyethyleneimine (PEI)) and PAA on membrane surface or pores; (iii) in situ polymerization of acrylic acid (PAA monomer) inside membrane pores. Literature results have also included PAA as the membrane selective layer to prepare pervaporation membrane (PV) [31–35] because of the highly preferential water permeation characteristics of PAA. PAA functionalization on MF/UF membrane (polyvinylidene fluoride (PVDF) polysulfone (PS), polyacrylonitrile (PAN), etc) is a well-known process for preparing PV, nanofiltration (NF) [36], and reverse osmosis (RO) membrane [37]. Incorporating the PAA coating layer with metal ions (Al^{3+}) as additives was investigated to increase the hydrophilicity of the layer and restrict the swelling of the membrane by forming cross-linking structure with metal ions [32,34,35]. However, few studies have been conducted for the immobilization of nanoparticles in microporous membranes modified with PAA. Microporous membranes functionalized with highly reactive metal nanoparticles are quite novel. The advantages of immobilization of metal nanoparticle in porous membrane in this work include prevention of nanoparticle agglomeration, control of particle, structure, and assembly, convective flow to eliminate the diffusion resistance, recapture of dissolved metal ions.

In our studies, we took a different approach based on the combination of ion exchange (with multiple COOH binding sites) and reduction to prepare metal nanoparticles in membranes. Polyvinylidene fluoride MF membrane provides an ideal support for functionalization due to the open structure and large porosity. In order to fully utilize the pore surface, PAA is synthesized inside PVDF support membrane by in situ polymerization of acrylic acid. Various metal ions can be introduced to the PAA domain as the nanoparticle precursor by the ion-exchange process. It is also possible to load two or more types of metal ions into the membrane for bimetallic or multimetallic nanoparticle synthesis. In addition to the new synthesis process, membrane and nanoparticles characterization is fully examined using various electron microscopy techniques. High resolution X-ray energy dispersive spectroscopy (EDS) mapping performed in scanning transmission electron microscopy (STEM) was employed to reveal the elemental distribution at nano scale. The reactivity of membrane-supported bimetallic nanoparticles was further investigated toward the reductive dechlorination of selected chlorinated organics (TCE and PCBs). The roles of the secondary metal, nanoparticle structure, composition, and reaction pathway were studied and correlated with the high resolution EDS mapping analysis.

24.2 Nanoparticle synthesis in functionalized membranes

24.2.1 Polyvinylidene flouride membrane functionalization with polyacrylic acid

The PVDF MF membranes functionalized with PAA were prepared by filling the membrane pores with the acrylic acid monomer solution, followed by literature reported (for polyethylene MF membrane) in situ free radical polymerization via thermal treatment [36]. The typical procedure is described in Fig. 24.1. The monomer solution was prepared by mixing the acrylic acid (30 wt%), benzoyl peroxide (0.5 wt%, initiator), 1,1,1-trimethylolpropane triacrylate (TMPTA) (1 wt%, cross-linking agent) in toluene. Benzoyl peroxide as the initiator was first dissolved

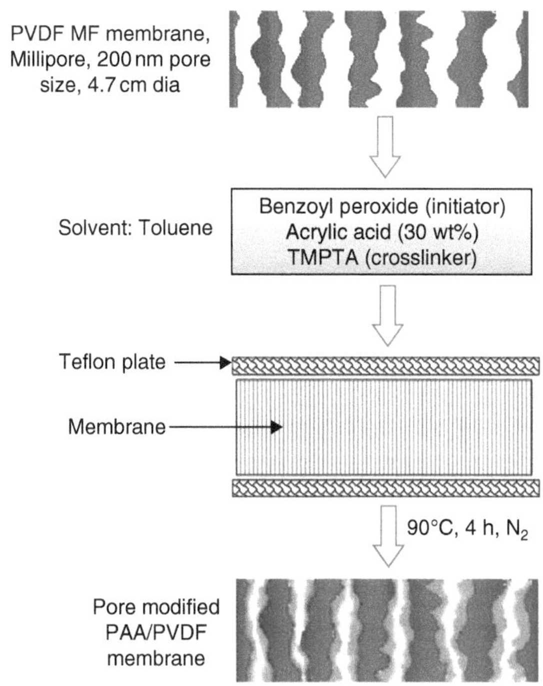

FIGURE 24.1

Schematic diagram of polyacrylic acid (PAA) functionalized polyvinylidene fluoride (PVDF) microfiltration (MF) membranes using in situ polymerization of acrylic acid via thermal treatment.

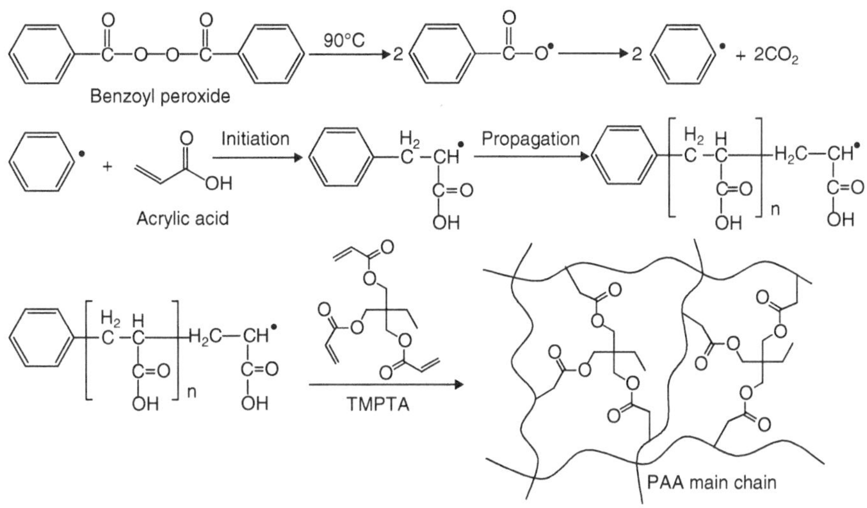

FIGURE 24.2

Thermal induced free radical polymerization reaction of acrylic acid using benzoyl peroxide as the initiator. Polyacrylic acid (PAA) is cross-linked during polymerization by TMPTA.

in toluene. The PVDF membranes were immersed into the monomer solution and quickly placed between two teflon plates that were subsequently clamped together. The membranes immobilized in the two teflon plates were then placed into the oven at 90 ± 2C with nitrogen purge. The TMPTA served as the cross-linking monomer due to the trifunctional double bonds (Fig. 24.2). After 4 hours, the membranes were released from teflon plates and washed in 200 mL ethanol to remove unreacted monomers. Finally, the PAA/PVDF membranes were rinsed with deionized water and kept in deionized water for nanoparticle synthesis.

24.2.2 Synthesis of fe-based bimetallic nanoparticles in polyacrylic acid layers

The flowchart for the membrane supported nanoparticle synthesis is described in Fig. 24.3. Briefly, the ferrous ions were first loaded into membranes by ion exchange, and Fe nanoparticles were formed followed by reduction of Fe^{2+} with $NaBH_4$. Prior to the ion exchange, the PAA/PVDF membrane was soaked in sodium hydroxide solution (0.1 M) overnight (12–14 hours) to convert PAA from hydrogen form (−COOH) to the sodium form (−COONa). After the excess sodium hydroxide was rinsed from the membrane with deionized water, the membrane was immersed in ferrous chloride solution at pH 4.8–5.0 with nitrogen purge for 12 hours. The membrane was then washed with deoxygenated deionized

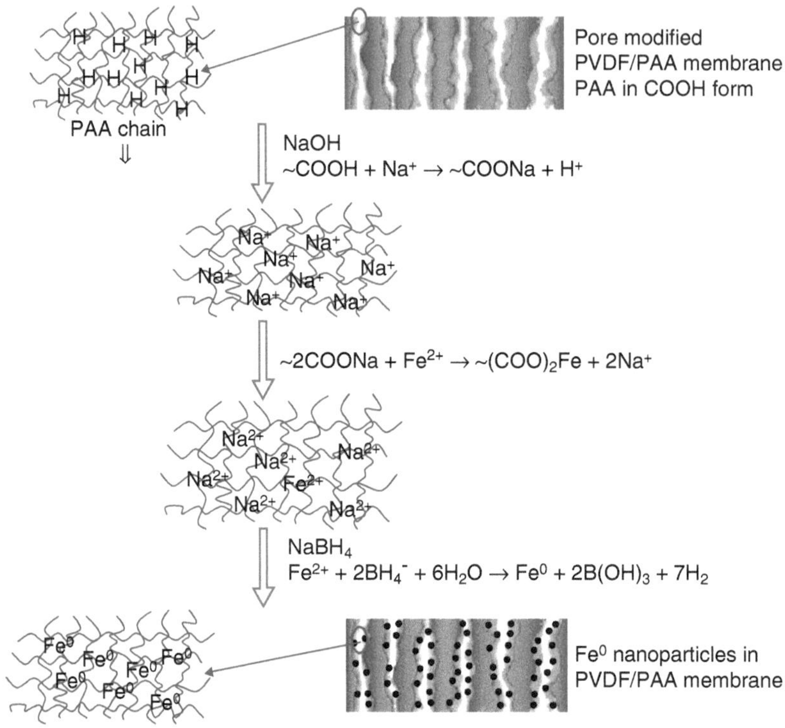

FIGURE 24.3

Schematic diagram of iron nanoparticles synthesized in polyacrylic acid (PAA)/ polyvinylidene fluoride (PVDF) membranes.

water. Subsequent immersion into sodium borohydride solution yielded Fe nanoparticles in the PAA/PVDF membrane. After rinsing with deoxygenated deionized water and ethanol sequentially, the membrane supported iron nanoparticles were soaked in 50 mL solution (90/10 vol.% ethanol/water) of K_2PdCl_4 (0.12 mM) for 30 minutes. This resulted in the deposition of Pd on the Fe surface through the redox reaction shown in Equation 24.1 [38]:

$$Pd^{2+} + Fe^0 \rightarrow Fe^{2+} + Pd^0 \tag{24.1}$$

Due to the high reactivity of Fe nanoparticles, 90 vol.% ethanol were used to minimize the Fe corrosion reaction and other side reactions. After rinsing with ethanol three times, the prepared Fe/Pd nanoparticles in PAA/PVDF membrane were stored in ethanol solution for further dechlorination study. Different Pd loading Fe/Pd nanoparticles in PAA/PVDF membrane can be prepared using various concentration of $K_{2P}dCl4$ in solution.

The core/shell Fe/Ni bimetallic nanoparticle were prepared by immersion of Fe^0 nanoparticle in PAA/PVDF membranes $NiSO_4 \cdot 6H_2O$ solution (0.07 wt %, pH = 4.8) for 10 minutes. Upon deprotonation of COONa in the PAA repeating unit, Ni^{2+} was ion-exchanged in the PAA domain. Subsequent immersion of the membrane into the $NaBH_4$ aqueous solutions (0.5 M) for 5 minutes results in nickel precipitation on the iron surface to form Fe/Ni nanoparticles. The alloy structure Fe/Ni nanoparticles were synthesized in PAA/PVDF membranes by adding $NiSO_4$ into $FeCl_2$ solution for ion-exchange process. Fe/Ni particles can be produced simultaneously from $NaBH_4$ (0.5 M) reduction with the membrane containing both iron and nickel ions.

24.3 Characterization of polyacrylic acid functionalized membranes

Figure 24.4 compares the scanning electron microscopy (SEM) surface images of the unmodified PVDF support membrane and the PAA functionalized PVDF membrane in the dry state. The PVDF MF support membrane (Fig. 24.4A) shows a highly porous microstructure. The pores are mostly circular in shape but highly nonuniform in size (0.2–2.0 μm). As expected, the modified membrane shows less porosity with small number and size of pores. This indicates that PAA has been filled into the pores to create smaller pores. The cross-section images of these membranes are shown in Fig. 24.5. By contrast, different regions of cross-sections clearly show the structure difference between PAA modified membrane and unmodified substrate membrane. As shown in Fig. 24.5C and D, the pores inside the PVDF substrate are filled with small grains, suggesting that PAA

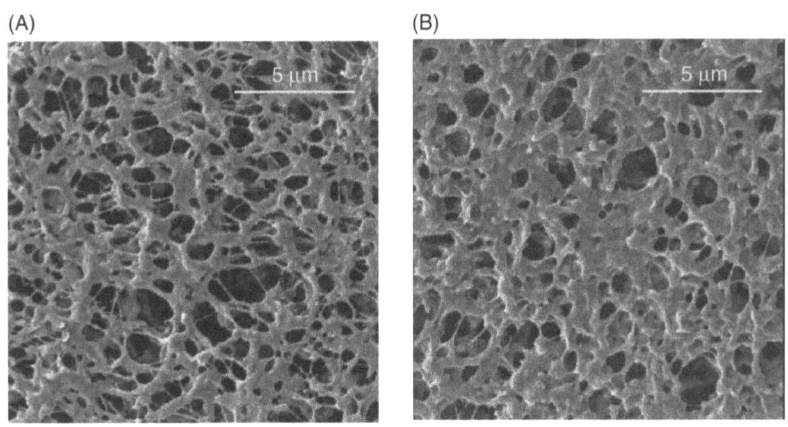

FIGURE 24.4

Scanning electron microscope images of membrane surface: (A) unmodified polyvinylidene fluoride (PVDF) support membrane; (B) pore-filled polyacrylic acid (PAA)/PVDF membrane.

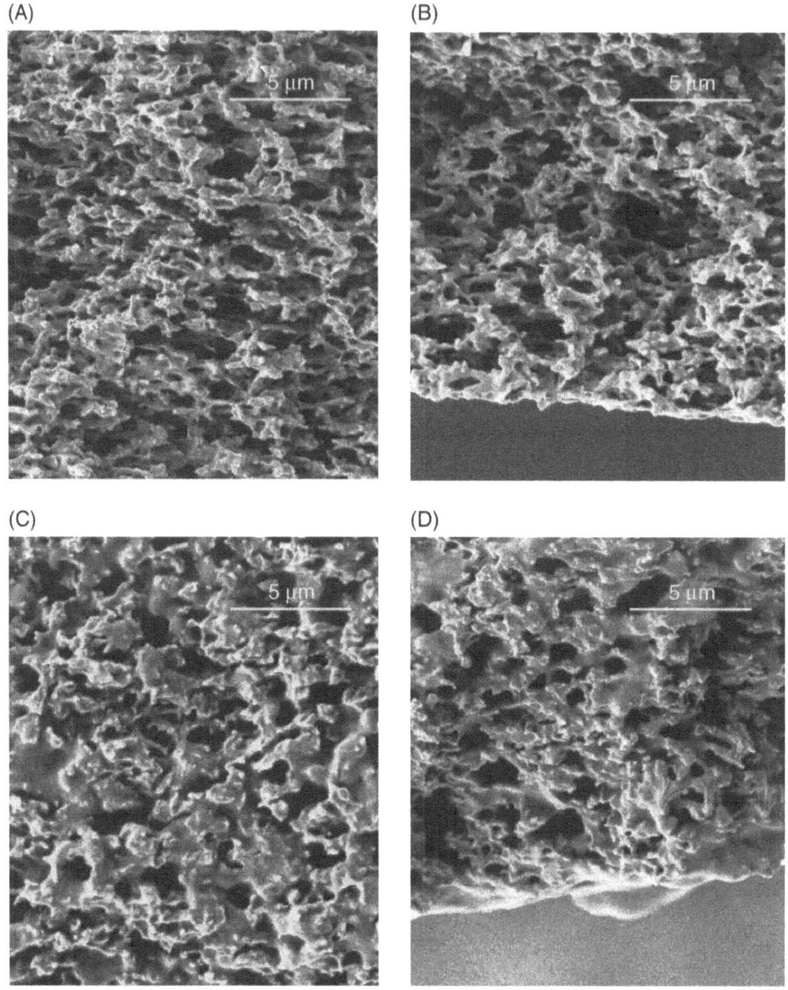

FIGURE 24.5

Scanning electron microscope images of membrane cross-sections. Unmodified polyvinylidene fluoride (PVDF) support membrane: (A) middle of the cross-section; (B) bottom of the cross-section. Pore-filled polyacrylic acid (PAA)/PVDF membrane: (C) middle of the cross-section; (D) bottom of the cross-section.

modification has taken place throughout the PVDF substrate. In order to further reveal the distribution of PAA inside membrane pores, EDS mapping analysis of oxygen atoms (from PAA) and fluorine atoms (from PVDF) were also performed using STEM. A region of the membrane cross-section image is shown in a STEM bright field image (Fig. 24.6A). The EDS mapping was performed at the selected area on the left STEM image (Fig. 24.6A) and reveals the position of elemental

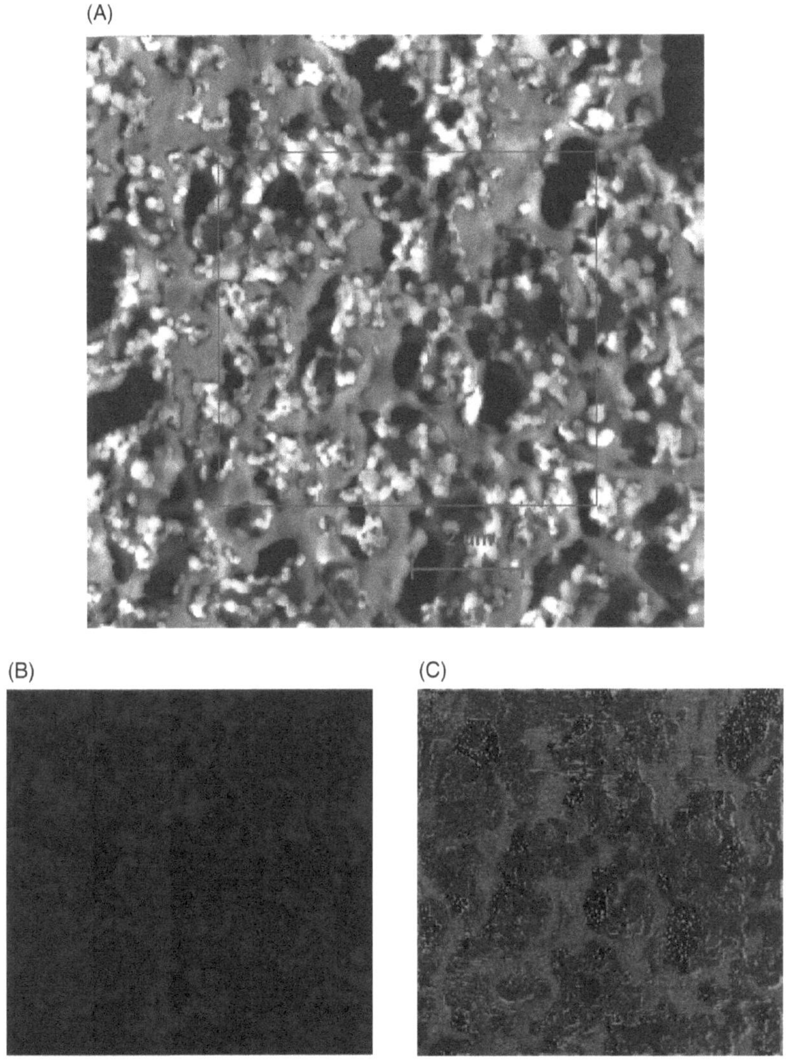

FIGURE 24.6

Scanning transmission electron microscopy (STEM)–energy dispersive spectroscopy (EDS) mapping of polyvinylidene fluoride (PVDF) membrane functionalized with polyacrylic acid (PAA). (A) STEM image of membrane cross-section; (B) F map from (A); and (C) O map from (A).

atoms of fluorine (F) and oxygen (O) in Fig. 24.6B and C, respectively. The map is generated by placing white dots on the image when an X-ray count of a particular element is received. As shown in the Fig. 24.6B and C, the dots for oxygen appear strongly in the map and oxygen atoms are mainly found in the regions where little fluorine atoms were identified. This indicates the presence of PAA inside the membrane because oxygen only comes from carboxylic acid group. And it also confirms the assumption that the small grains observed in the SEM cross-sections are PAA.

24.4 Characterization of nanoparticles in membranes

24.4.1 Chelation interaction between ferrous ions and polyacrylic acid

By using this PAA metal ions binding interaction (Fe^{2+} and COO^{-}) followed by reduction method, we created iron nanoparticles well dispersed inside membranes because the ferrous ions are bound and distributed along the PAA chains. The surrounding polymer chains also can prevent the ion migration and nanoparticle agglomeration, which plays a critical role in stabilizing nanoparticles and controlling the particle size [24]. Therefore, it is necessary to understand the binding interaction between ferrous ions and PAA. First, the mass balance of sodium and ferrous ions is calculated based on the ICP (inductively coupled plasma atomic emission spectroscopy) analysis of the $FeCl_2$ solution before and after loading of Fe^{2+} on the membrane. According to the ICP results, the atom ratio of sodium ions released from the PAA/PVDF membrane over the ferrous ions bound to the membrane is 1.9 ± 0.1. In theory, binding one ferrous ion from solution results in two sodium ions released from the membrane due to the charge balance. This indicates that the ferrous ions are well chelated with PAA (no physical adsorption). An elemental analysis was performed on the PAA/PVDF membrane loaded with $Fe2^{+}$, and the result is shown in Fig. 24.7. It has been found that the coordination number for PAA-divalent metal complex is 2 [39]. Therefore, one ferrous cation is satisfied with two carboxyl anions containing four oxygen atoms. As shown in Fig. 24.7, the EDS analysis gives the atom ratio of 3.5 (oxygen over iron). This value agrees well with the established PAA-metal binding stability constant.

Second, the bound ferrous ion proximity in the membrane plays an important role in controlling the size of nanoparticle [22,24]. It has been reported that the average particle size is larger when metal cation concentrations in membrane are higher [22,24]. This is due to the enhanced aggregation of Fe atoms because of the shorter distance between $Fe2^{+}$ at the higher loading density. Next, an EDS map was acquired in STEM to obtain a better understanding of the interaction between $Fe2^{+}$ and PAA. As shown in Fig. 24.8, all the iron atoms are associated with oxygen atoms and the Fe map matches perfectly with oxygen map. The black dots in the iron map are believed to be the gaps between chelated ferrous ions. It is important to point out that these images and maps were obtained under the completely dry state due to the requirement of sample preparation and TEM analysis. This can change the morphology of the membrane because PAA is an extremely swellable polyelectrolyte.

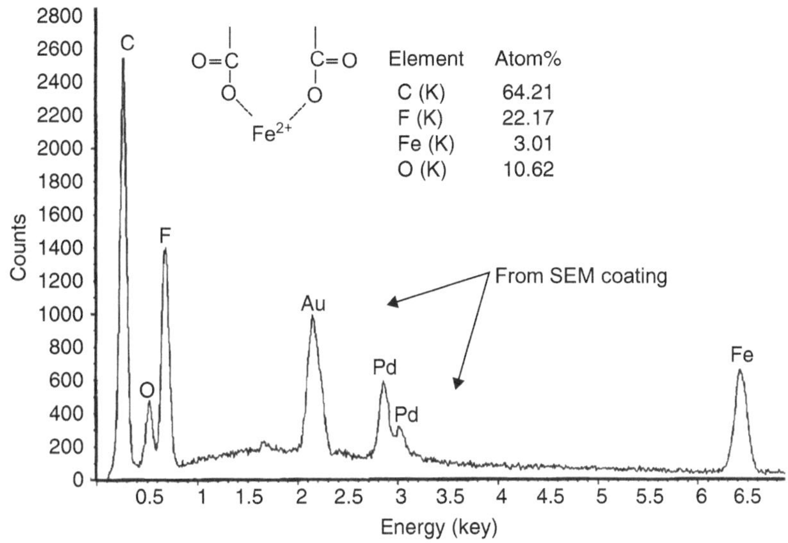

FIGURE 24.7

Scanning electron microscopy (SEM)—energy dispersive spectroscopy (EDS) spectra of polyacrylic acid (PAA) pore-filled polyvinylidene fluoride (PVDF) membranes loaded with ferrous ions.

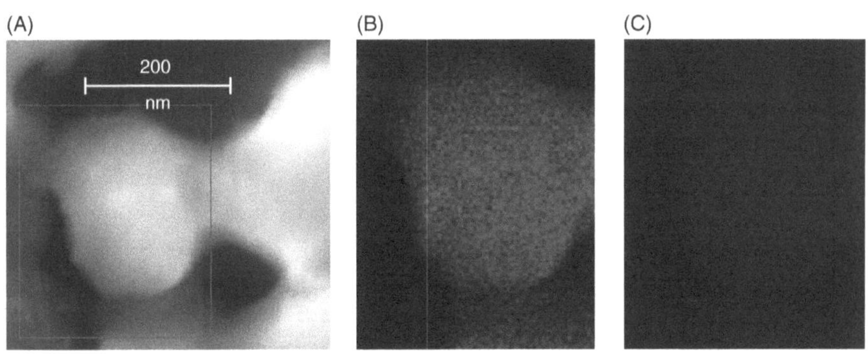

FIGURE 24.8

Scanning transmission electron microscopy (STEM)—energy dispersive spectroscopy (EDS) mapping of polyacrylic acid (PAA) chelation with $Fe2^{+}$. (A) STEM image of PAA domain chelated with $Fe2^{+}$; (B) Fe map from (A); and (C) O map from (A).

24.4.2 Fe/Pd nanoparticle characterization

Transmission electron microscope analysis at low magnification was performed to verify the nanoparticle formation and distribution inside PAA/PVDF membranes. As shown in Fig. 24.9, Fe/Pd nanoparticles in spherical shape were

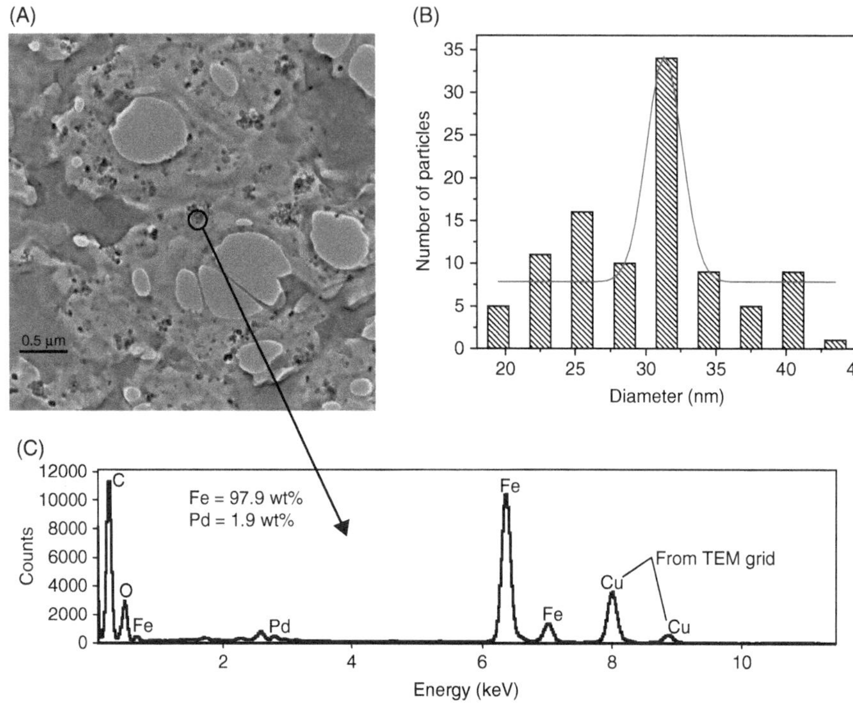

FIGURE 24.9

(A) Transmission electron microscope (TEM) image of polyacrylic acid (PAA)/polyvinylidene fluoride (PVDF) membrane cross-section containing Fe/Pd nanoparticles; (B) histogram from the left TEM image of 100 Fe/Pd nanoparticles; (C) energy dispersive spectroscopy (EDS) spectrum acquired from the nanoparticles in the TEM image.

homogeneously dispersed in the PAA phase over the membrane cross-section, whereas the regions containing no nanoparticles are believed to be the PVDF substrate phase. A statistical analysis of the image yielded an average particle size of 30 nm in diameter, with the size distribution standard deviation of 5.7 nm. Based on the mean diameter of 30 nm, the external surface area for nanoparticles was calculated to be approximately 25 m^2g^{-1}. The EDS analysis was also conducted during the TEM observation using a 2 nm electron beam spot to determine the elements present in the nanoparticles. The composition of nanoparticles identified in the TEM image was also quantified by EDS (Fig. 24.9C). The Pd content was found to be 1.9 wt%, which is consistent with the previous ICP analysis results. Boron as a light element at low content was not detected by EDS due to the low energy sensitivity.

Next, the nanostructures and element distribution of the Fe/Pd nanoparticles were observed by HRTEM and STEM-EDS mapping. Figure 24.10 shows a STEM bright field image and the elemental mapping images of the corresponding

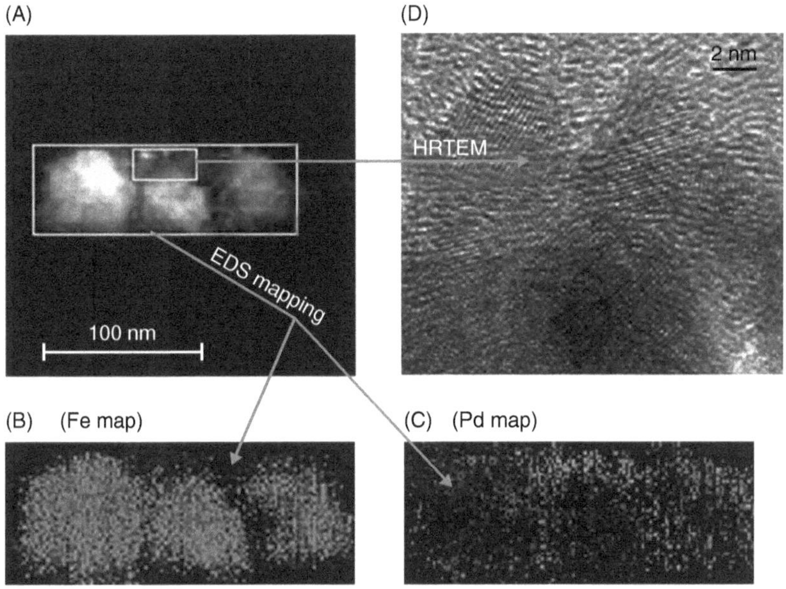

FIGURE 24.10

Characterization of Fe/Pd nanoparticles. (A) Scanning transmission electron microscopy image of Fe/Pd (Pd = 2.3 wt%) nanoparticles; (B) energy dispersive spectroscopy (EDS) mapping image of Fe; (C) EDS mapping image of Pd; (D) high-resolution transmission electron microscope image of Fe/Pd nanoparticles.

area for Fe and Pd. The probe size used was 1 nm in diameter. The mapping images clearly demonstrate a core/shell structure for the Fe/Pd nanoparticle with Fe in the core region and Pd in the shell region. This is as expected because Pd was post reduced by Fe^0 and deposited on the iron surface.

24.5 Reactivity of membrane-based nanoparticles

24.5.1 Catalytic hydrodechlorination of trichloroethylene

The reactive properties of the bimetallic Fe/Ni nanoparticles in PVDF membranes dip-coated with PAA were examined by reduction of TCE in water at room temperature. Trichloroethylene contamination of groundwater is widely reported in the literature [22]. Figure 24.11 shows the batch reaction (at pH 6) of TCE with Fe/Ni (Ni = 25 wt%) nanoparticles in PAA/PVDF membranes. The TCE transformation rate can be described by a simple pseudo-first-order model. Complete dechlorination of TCE was achieved within 2 hours. Ethane and Cl^- were formed as the only products in the headspace and aqueous phase, respectively. No chlorinated intermediates were detected and 92 percent carbon balance and 93 percent

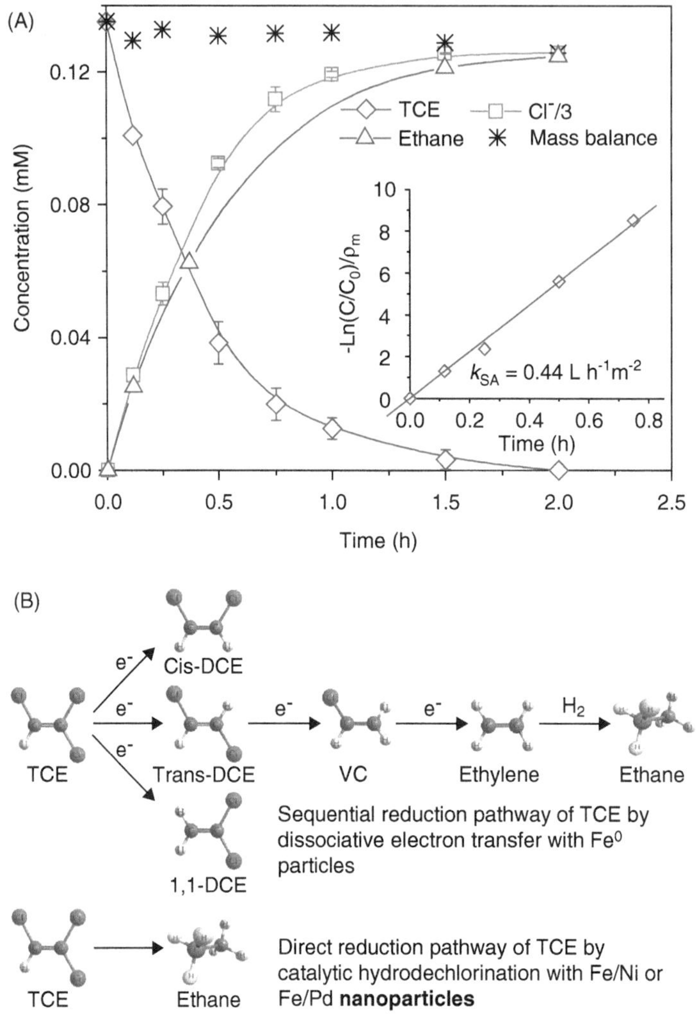

FIGURE 24.11

(A) Batch reaction of trichloroethylene (TCE) dechlorination and products formation (ethane and chloride) with Fe/Ni (Ni = 25 wt%, post coat Ni) nanoparticles in polyvinylidene fluoride (PVDF) membranes dip-coated with polyacrylic acid (PAA). $\rho_m = 0.2\ g\ L^{-1}$; (B): Schematic diagram for reductive dechlorination pathways of TCE with iron and Fe/Ni or Fe/Pd systems. From [22].

Cl balance were obtained in the whole system, indicating a direct reaction pathway from TCE to ethane for bimetallic Fe/Ni (Ni = 25 wt%) nanoparticles.

The complete conversion of TCE to ethane with bimetallic nanoparticles is totally different from the sequential reductive dechlorination with monometallic

Fe system. For the Fe system TCE is transformed to dichloroethylene (DCE) to vinyl chloride (VC) and finally to ethylene and ethane [40]. The presence of the secondary metal on nanosized Fe changes the reaction pathway dramatically [9,10]. In the monometallic Fe system, the dechlorination mechanism is preferably explained by dissociative electron transfer resulting in the formation of less chlorinated radicals as intermediates [41]. While in the bimetallic system, Fe is considered as the reductant for water to generate hydrogen and TCE is dechlorinated by catalytic hydrodechlorination [42] in the presence of Ni, resulting in the direct reduction to ethane. Another advantage of coating the secondary metal is to prevent the conversion of the Fe^0 to an oxide form (Fe_xO_y) that can deactivate the nanoparticle surface [41].

It has been reported in earlier studies that dechlorination reaction can be described by a pseudo-first-order kinetic model. The TCE degradation rate by Fe-based bimetallic nanoparticles is considered as first order in terms of both TCE concentration and the concentration of metals available in the solution. Therefore, the following Equation 24.2 has been used [43] to describe this pseudo-first-order reaction model.

$$\frac{dC}{dt} = -k_{SA} a_{SA} a_s p_m C \tag{24.2}$$

Regression of the kinetic data can be used to determine the surface area normalized reaction rate constant k_{SA}. Since k_{SA} is the characterized first-order reaction rate, it should be independent of variance of reaction conditions such as initial TCE concentration, metal mass and the volume of reaction system.

24.5.2 Effect of dopant material and nanoparticle structure

In order to further understand the catalytic dechlorination mechanism and bimetallic nanoparticle reactivity, TCE dechlorination experiments were conducted with bimetallic nanoparticles with different type of dopant metal and different structure. Figure 24.12 shows the normalized rate constant (k_{SA}) of TCE dechlorination using PAA/PVDF membrane based Fe/Cu nanoparticles, alloy Fe/Ni nanoparticles (simultaneous reduction of Fe and Ni), and core-shell Fe/Ni nanoparticles (post-coating Ni). The second metal composition was kept the same (25 wt%) for all the three reaction systems. As expected, the core-shell Fe/Ni nanoparticles exhibit higher TCE degradation rate than the alloy Fe/Ni nanoparticles. k_{SA} for core-shell nano Fe/Ni is about four times higher than that of alloy nano Fe/Ni. It has been demonstrated that most Ni atoms are located at the out side of iron surface for the core-shell structure. The alloy structure has a homogenous distribution of iron and nickel atoms inside particle, which results in a lesser amount of Ni atoms on the surface. Since the Ni is the active site for the catalytic hydrodechlorination reaction and the reaction only takes place at the particle surface, the lower k_{SA} for alloy Fe/Ni nanoparticles is due to the less active sites (Ni atoms) on the surface.

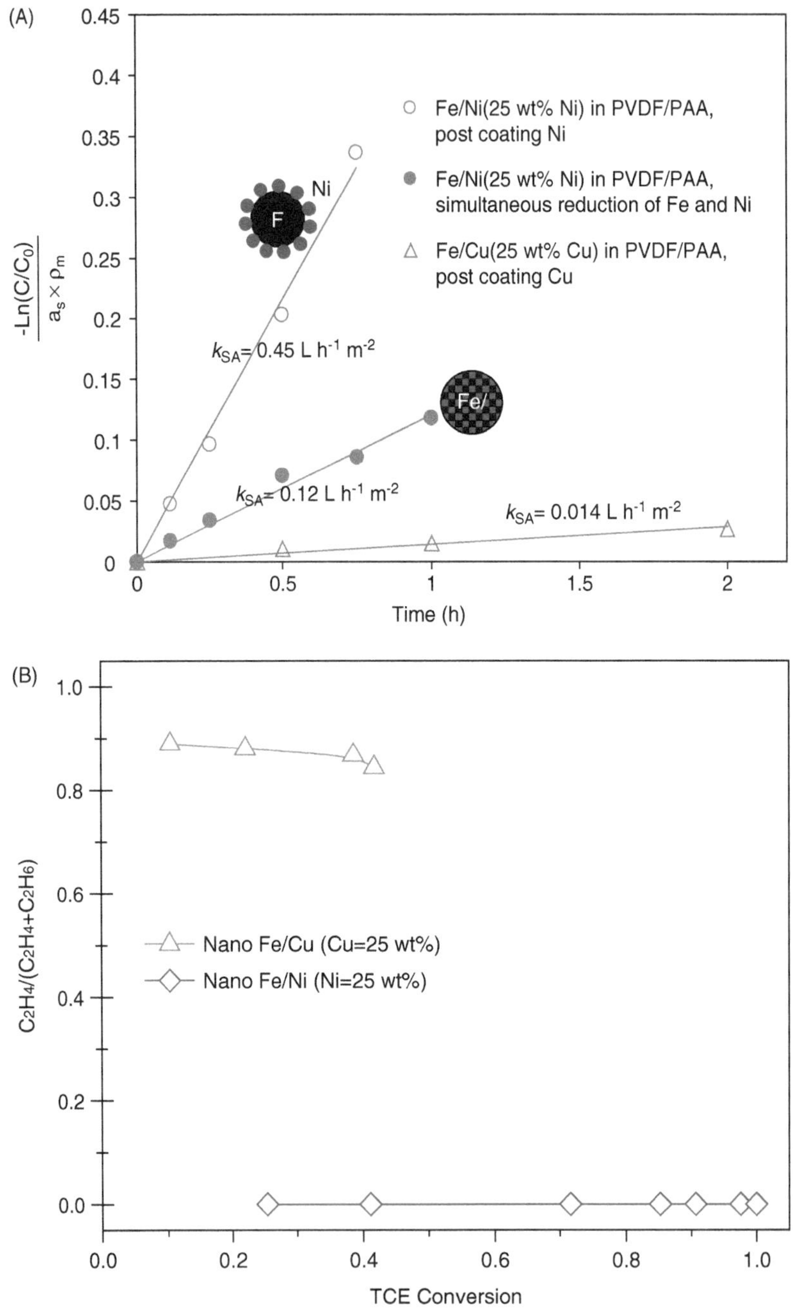

FIGURE 24.12

(*Continued*) caption for figure 24.12 is on page 385

Compared to the Fe/Ni nanoparticles, Fe/Cu nanoparticles show a much slower reaction rate. k_{SA} for core-shell Fe/Cu nanoparticles is about 30 times lower than core-shell Fe/Ni nanoparticles at the same dopant composition. In order to understand the reactivity difference between Cu and Ni, the product formation was measured for TCE degradation with Fe/Cu and Fe/Ni nanoparticles. As shown in Fig. 24.12, the dominating product for Fe/Cu system is ethylene, suggesting the main reduction pathway is by electron transfer. In contrast, ethane is the only product formed in the Fe/Ni system due to the catalytic hydrogenation mechanism. This observation is consistent with the result reported in the literature [44] for the *cis*-dichloroethylene (*cis*-DCE) dechlorination by Fe/Ni and Fe/Cu particles. Ni and Pd are proven the most active dopant for the reductive dechlorination reaction due to the high hydrogenation activity.

24.5.3 Catalytic hydrodechlorination of selected polychlorinated biphenyls

To investigate the catalytic properties of Fe/Pd nanoparticles synthesized in PAA/PVDF membranes, we studied the reductive hydrodechlorination of selected PCBs [23] using the bimetallic nanoparticles. PCBs are among the most important chlorinated aromatic compounds that cause a stringent environmental problem due to their hydrophobic nature and excellent chemical stability. The dechlorination mechanism and kinetic rates were investigated using membrane supported Fe/Pd nanoparticles. In order to understand and quantify the role of second dopant metal, we studied the dechlorination rates as a function of Pd content on Fe as well as the reaction temperature.

Figure 24.13 shows the dechlorination of 15.6 mg/L PCB77 in 65/35 vol.% ethanol/water solution with Fe/Pd (Pd: 2.3 wt%) immobilized inside PAA/PVDF membranes. High concentration of ethanol in the solution matrix was used because of the lower solubility of PCB77 in water. As shown in the figure, the membrane supported Fe/Pd nanoparticles exhibit extremely fast dechlorination rate. Complete degradation of PCB77 by Fe/Pd in PAA/PVDF membrane was achieved within 2 hours. Biphenyl was formed as the main dechlorination product. PCB77 was completely transformed to biphenyl after 2 hours. The degradation of PCB77 by Fe/Pd nanoparticles occurred in a sequential reduction pathway, which is indicated by the detected less chlorine intermediates. All the PCB intermediates were only identified in the low concentration level within 1 hour. It

◀ (A) Trichloroethylene (TCE) degradation with various bimetallic nanoparticles in polyacrylic acid (PAA)/polyvinylidene fluoride (PVDF) membranes: ((o) Fe/Ni (Ni = 25 wt%, post-coating Ni); ((•) Fe/Ni (Ni = 25 wt%, simultaneous reduction of Fe and Ni); (△) Fe/Cu (Cu = 25 wt%, post-coating Cu). (B) Ethylene and ethane formation from TCE dechlorination with Fe/Ni (Ni = 25 wt%, post-coating Ni) nanoparticles and Fe/Cu nanoparticles (Cu = 25 wt%, post-coating Cu).

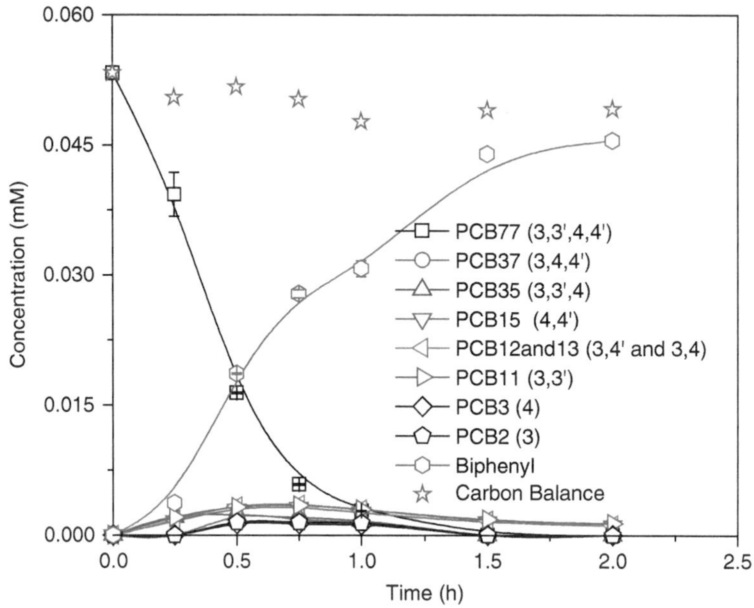

FIGURE 24.13

Batch reaction of PCB77 with Fe/Pd (Pd = 2.3 wt%) in pore-filled polyacrylic acid (PAA)/polyvinylidene fluoride (PVDF) membrane. Metal loading: 0.8 g L^{-1}. Initial PCB77 concentration: 15.6 mg L^{-1}.

has been proven in the literature [8] that non-ortho-chlorinated PCB congeners dechlorinate faster than the ortho-chlorinated isomers due to the effect of higher steric hindrance for ortho-position. The reactivity of the chlorine substitutents decreases in the order para $\approx$ meta > ortho [45]. Due to the increasing torsion angles with the increase of ortho substitution, non-ortho substituted congeners could adsorb in a closed planar position with nanoparticles, which is beneficial for the reductive dechlorination. The more positive reduction potentials measured in the literature [46] also support the increased reactivity of non-ortho substituted congeners.

24.5.4 Dechlorination efficiency of different polychlorinated biphenyls

In theory, there are 209 different PCB congeners. In order to understand and quantify different PCBs dechlorination reactions, degradation of PCB4, PCB44, PCB77 were conducted with same Fe/Pd nanoparticles immobilized in PAA/PVDF membrane. Figure 24.14 plots the yield and selectivity for biphenyl as a function of reaction time. It can be seen that biphenyl yield is about the same for all three PCBs, whereas the biphenyl selectivity varies significantly. Although the biphenyl

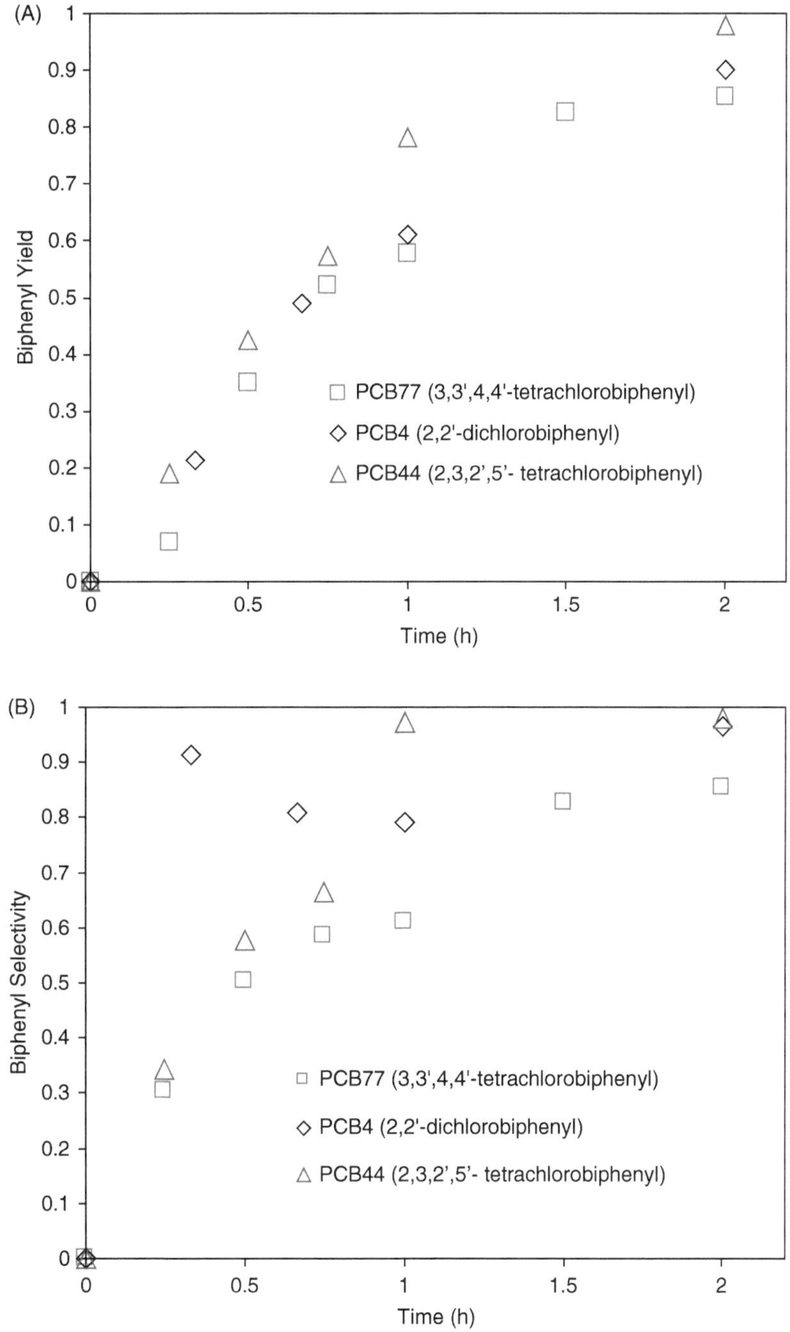

FIGURE 24.14

(*Continued*) caption for figure 24.14 is on page 388

selectivity for the three different PCBs reaches about the same value at the end of reaction, PCB4 shows much higher selectivity than the other two PCBs within 1 hour. The higher selectivity is due to the negligible formation of intermediate 2-chlorobiphenyl as the only intermediate is detected in trace level. The degradation of PCB77 and PCB44 has nine and five different intermediates, respectively. Each individual intermediate is detected in trace amount, but the total amount of intermediates are not negligible. This indicates that the degradation of PCB77 and PCB44 does not result in the correspondingly equal amount formation of biphenyl. Biphenyl is formed by sequential reduction pathway. The lower biphenyl selectivity is due to the complex reduction pathway for higher number chlorine substituted PCB congeners.

24.5.5 Catalytic activity as a function of palladium coating content

In order to understand the role of Pd as the second dopant, the batch dechlorination rates of DiCB (2,2'-dichlorobiphenyl) were measured as a function of Pd coating content (Fig. 24.15). The k_{SA} is 0.017, 0.068, and 0.166 L $h^{-1}m^{-2}$ for Fe/Pd nanoparticles with 0.6, 2.3, and 5.6 wt% Pd, respectively. In the bimetallic system, the role of Fe is to generate hydrogen by corrosion reaction, whereas Pd serves as the catalyst and the chlorine atom in DiCB is mainly replaced by hydrogen on the Pd surface [8]. Therefore the Pd atoms are considered as the surface reactive sites for the dechlorination of DiCB. The variation of the k_{SA} as a function of Pd content is due to the difference of reactive sites. By normalizing the k_{SA} in terms of Pd content (reactive sites), we found the same reaction rate of Fe/Pd nanoparticles. The modified reaction model shown in Equation 24.3, developed by Johnson, provided a better way to understand and quantify the effect of variation in reactivity of different metal system [23,43].

$$\frac{dC}{dt} = k_{SA}p_{m}a_{s}C = k_{2}\Gamma a_{s}p_{m}C \tag{24.3}$$

where k_2 is the second order rate constant at a particular type of site (L h^{-1} mol^{-1}) and Γ is the surface concentration of reactive sites (mol m^{-2}). In this model, k_{SA} is expressed as the product of k_2 and Γ, which is more reasonable

◀ Polychlorinated biphenyl 4 (PCB4), PCB44, and PCB77 dechlorination with Fe/Pd nanoparticles in polyacrylic acid (PAA)/polyvinylidene fluoride (PVDF) membranes. Metal loading is kept the same (0.8 g L^{-1}). (A) Biphenyl yield; (B) biphenyl selectivity. (Biphenyl yield = biphenyl formed/initial amount of parent PCB; biphenyl selectivity = biphenyl formed/parent PCB consumed.)

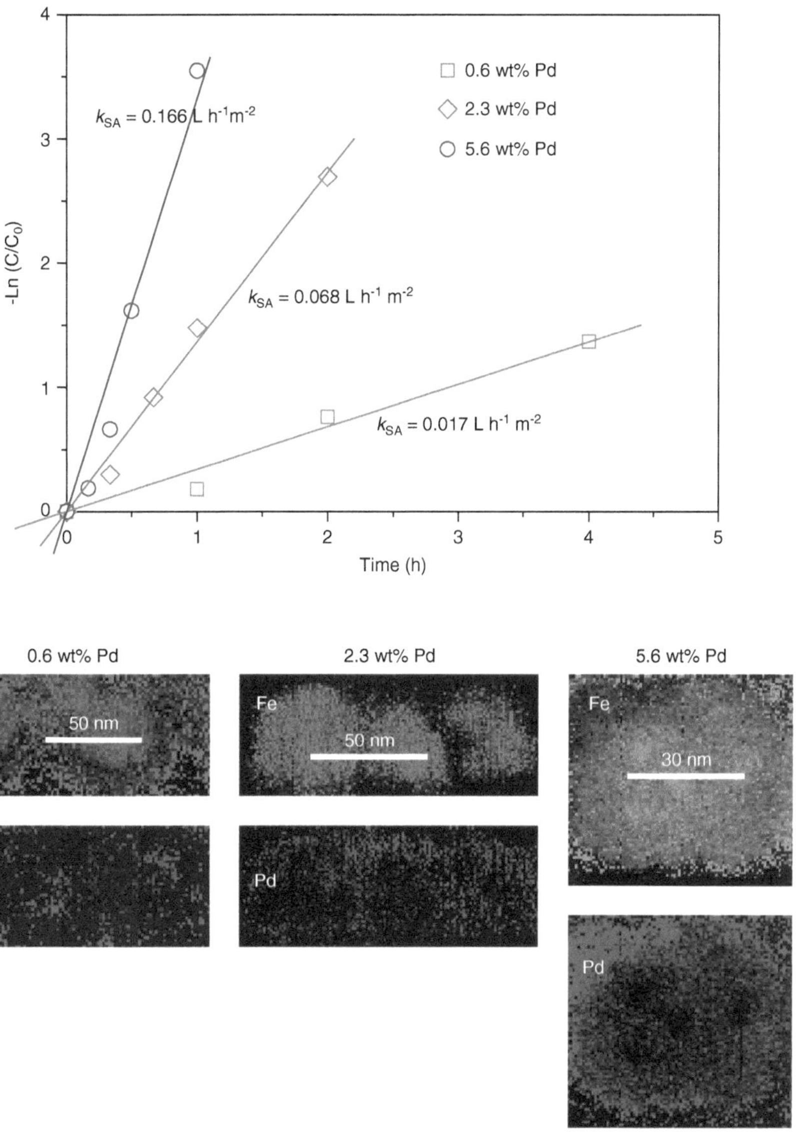

FIGURE 24.15

Best linear fit of k_{SA} for dechlorination of DiCB (2,2'- dichlorobiphenyl) with various Fe/Pd nanoparticles in polyacrylic acid (PAA)/polyvinylidene fluoride (PVDF) membranes. Metal loading: 0.8 g L^{-1}. Bottom images: high-resolution Scanning transmission electron microscopy (STEM)–energy dispersive spectroscopy (EDS) mapping of Fe and Pd in nanoparticles.

when dechlorination reaction preferentially occurs at the reactive catalytic surface sites (bimetallic system).

Based on the 30 nm average diameter of nanoparticles, we calculated the Pd coverage and surface Pd atoms for different Fe/Pd nanoparticles by using a Pd atom cross-sectional area of 0.0787 nm^2 [47]. Our calculations indicate that Fe/Pd nanoparticles with 0.6, 2.3, and 5.6 wt% Pd have 0.1, 0.4, and 0.97 layers of Pd atoms, respectively. Since maximum Pd coverage is less than one layer, all the Pd atoms are considered as surface reactive sites. Γ for the three different nanoparticles with 0.6, 2.3, and 5.6 wt% Pd is $2.20*10^{-6}$, $8.43*10^{-6}$, and $2.05*10^{-5}$ mol m^{-2}, respectively. It should be noted that total surface area was used in all k_{SA} calculations. By applying Γ in the Equation 24.3, k_2 was determined to be 7,727, 8,066, and 8,098 L h^{-1} mol^{-1}, respectively. The enhanced reaction rate (k_{SA}) is only due to the increase of surface Pd atoms.

High-resolution STEM-EDS mapping images were also acquired in Fig. 24.15 to compare the Pd atoms distribution for different Pd coating nanoparticles. The STEM-EDS mapping technique presents us a 2D image of 3D sample in transmission [48]. All the Fe/Pd nanoparticles show a core/shell structure with Fe rich in the core region and Pd rich in the shell region. More Pd atoms were deposited on the iron surface and the Pd shell layer became denser with the increased Pd content. In spite of the limited spatial resolution in the EDS mapping, the distribution of Pd atoms is still in qualitative agreement with the result based on the calculation. This result implies that a uniform Pd coating with controllable thickness can be obtained by post-reduction of Pd^{2+} with Fe nanoparticles immobilized in membrane phase.

24.6 Conclusions

This work has demonstrated successful in situ synthesis of highly reactive bimetallic nanoparticles with controllable size, distribution, and structure in the functionalized membrane matrix. The membrane immobilized nanostructured bimetallic materials (Fe/Ni and Fe/Pd) exhibited fast and complete reduction capability toward both chlorinated aliphatics and conjugated aromatics. The normalized reaction rate for TCE degradation by membrane supported Fe/Ni nanoparticles is about 5 times higher than the rate for Fe/Ni synthesized in solution, 200 times higher than the rate for nanoscale iron particles, and 400 times higher than the rate for bulk iron particles. With complete conversion to biphenyl, the normalized reaction rate for PCB dechlorination with membrane supported Fe/Pd nanoparticles in this study is 10,000-fold higher than that reported in the literature. Our work has led to significant improvement of current chlorinated organics degradation and advancement of highly reactive nanostructured materials development. High-resolution STEM-EDS mapping was used to quantify iron functionalization in the PAA domain, characterize bimetallic nanoparticle structure, and correlate it with reactivity.

Acknowledgments

This study was supported by the NIEHS-SBRP (P42ES007380) program, and by DOE-KRCEE (DE-FG05-03OR23032). We thank Dr. Alan Dozier for the assistance with TEM and STEM–EDS mapping analysis and John May, Tricia Coakley from UK Environmental Research and Training Laboratory (ERTL) for GC-MS and ICP-AES analytical support.

References

[1] R.W.J. Scott, O.M. Wilson, R.M. Crooks, Synthesis, characterization, and applications of dendrimer-encapsulated nanoparticles, J. Phys. Chem. B 109 (2005) 692–704.

[2] P.K. Jain, K.S. Lee, I.H. El-Sayed, M.A. El-Sayed, Calculated absorption and scattering properties of gold nanoparticles of different size, shape, and composition: applications in biological imaging and biomedicine, J. Phys. Chem. B 110 (2006) 7238–7248.

[3] S. Paul, C. Pearson, A. Molloy, M.A. Cousins, M. Green, S. Kolliopoulou, et al. Langmuir-blodgett film deposition of metallic nanoparticles and their application to electronic memory structures, Nano Lett. 3 (2003) 533–536.

[4] K. Yakushiji, F. Ernult, H. Imamura, K. Yamane, S. Mitani, K. Takanashi, et al. Enhanced spin accumulation and novel magnetotransport in nanoparticles, Nat. Mater. 4 (2005) 57–61.

[5] F. Patolsky, Y. Weizmann, I. Williner, Actin-based metallic nanowires as bio-nanotransporters, Nat. Mater. 3 (2004) 692–695.

[6] C. Jiang, S. Markutsya, Y. Pikus, V. Tsukruk, Freely suspended nanocomposite membranes as highly sensitive sensors, Nat. Mater. 3 (2004) 721–728.

[7] J. Feng, T.T. Lim, Pathways and kinetics of carbon tetrachloride and chloroform reductions by nano-scale Fe and Fe/Ni particles: comparison with commercial micro-scale Fe and Zn, Chemosphere 59 (2005) 1267–1277.

[8] G.V. Lowry, K.M. Johnson, Congener-specific dechlorination of dissolved PCBs by microscale and nanoscale zerovalent iron in a water/methanol solution, Environ. Sci. Technol. 38 (2004) 5208–5216.

[9] B. Schrick, J.L. Blough, A.D. Jones, T.E. Mallouk, Hydrodechlorination of trichloroethylene to hydrocarbons using bimetallic nickel-iron nanoparticles, Chem. Mater. 14 (2002) 5140–5147.

[10] W.X. Zhang, C.B. Wang, H.L. Lien, Treatment of chlorinated organic contaminants with nanoscale bimetallic particles, Catal. Today 40 (1998) 387–395.

[11] R.G. Freeman, K.C. Grabar, K.J. Allison, R.M. Bright, J.A. Davis, A.P. Guthrie, et al. Self-assembled metal colloid monolayers: an approach to SERS substrates, Science 267 (1995) 1629–1631.

[12] P. Raveendran, J. Fu, S.L. Wallen, Completely 'green' synthesis and stabilization of metal nanoparticles, J. Am. Chem. Soc. 125 (2003) 13940–13941.

[13] J.G. Worden, Q. Dai, A.W. Shaffer, Q. Huo, Monofunctional group-modified gold nanoparticles from solid phase synthesis approach: solid support and experimental condition effect, Chem. Mater. 16 (2004) 3746–3755.

[14] T. Fukasawa, T. Suetsuna, K. Harada, S. Suenaga, Catalytic performance of metal nanoparticles supported by ceramic composite produced by partial reduction of solid solution with dopant, J. Am. Ceram. Soc. 88 (2005) 2938–2941.
[15] Z. Liu, X.Y. Ling, X. Su, J.Y. Lee, Carbon-supported Pt and PtRu nanoparticles as catalysts for a direct methanol fuel cell, J. Phys. Chem. B. 108 (2004) 8234–8240.
[16] S. Ikeda, K. Akamatsu, H. Nawafune, T. Nishino, S. Deki, Formation and growth of copper nanoparticles from ion-doped precursor polyimide layers, J. Phys. Chem. B 108 (2004) 15599–15607.
[17] J. He, I. Ichinose, T. Kunitake, A. Nakao, In situ synthesis of noble metal nanoparticles in ultrathin TiO2-gel films by a combination of ion-exchange and reduction processes, Langmuir 18 (2002) 10005–10010.
[18] J. He, I. Ichinose, T. Kunitake, A. Nakao, Y. Shiraishi, N. Toshima, Facile fabrication of Ag–Pd bimetallic nanoparticles in ultrathin TiO2-Gel films: nanoparticle morphology and catalytic activity, J. Am. Chem. Soc. 125 (2003) 11034–11040.
[19] J.C. Pivin, M. Sendova-Vassileva, G. Lagarde, F. Singh, A. Podhorodecki, Optical activation of Eu3+ ions by Ag nanoparticles in ion exchanged silica-gel films, J. Phys. D: Appl. Phys. 39 (2006) 2955–2958.
[20] C. Damle, K. Biswas, M. Sastry, Synthesis of Au-core/Pt-shell nanoparticles within thermally evaporated fatty amine films and their low-temperature alloying, Langmuir 17 (2001) 7156–7159.
[21] J. Xu, A. Dozier, D. Bhattacharyya, Synthesis of nanoscale bimetallic particles in polyelectrolyte membrane matrix for reductive transformation of halogenated organic compounds, J. Nanoparticle Res. 7 (2005) 449–467.
[22] J. Xu, D. Bhattacharyya, Membrane-based bimetallic nanoparticles for environmental remediation: synthesis and reactive properties, Environ. Prog. Vol. 24 (2005) 358–366.
[23] J. Xu, D. Bhattacharyya, Fe/Pd nanoparticle immobilization in microfiltration membrane pores: synthesis, characterization, and application in the dechlorination of polychlorinated biphenyls, Ind. Eng. Chem. Res. 46 (2007) 2348–2359.
[24] T.C. Wang, M.F. Rubner, R.E. Cohen, Polyelectrolyte multilayer nanoreactors for preparing silver nanoparticle composites: controlling metal concentration and nanoparticle size, Langmuir 18 (2002) 3370–3375.
[25] B.L. Rivas, E.D. Pereira, I. Moreno-Villoslada, Water-soluble polymer-metal ion interactions, Prog. Polym. Sci. 28 (2003) 173–208.
[26] S. Konishi, K. Saito, S. Furusaki, T. Sugo, Binary metal-ion sorption during permeation through chelating porous membranes, J. Membr. Sci. 111 (1996) 1–6.
[27] D. Bhattacharyya, J.A. Hestekin, P. Brushaber, L. Cullen, L.G. Bachas, S.K. Sikdar, Novel poly-glutamic acid functionalized microfiltration membranes for sorption of heavy metals at high capacity, J. Membr. Sci. 141 (1998) 121–135.
[28] S.M.C. Ritchie, L.G. Bachas, T. Olin, S.K. Sikdar, D. Bhattacharyya, Surface modification of silica-and cellulose-based microfiltration membranes with functional polyamino acids for heavy metal sorption, Langmuir 15 (1999) 6346–6357.
[29] S.M.C. Ritchie, K.E. Kissick, L.G. Bachas, S.K. Sikdar, C. Parikh, D. Bhattacharyya, Polycysteine and other polyamino acid functionalized microfiltration membranes for heavy metal capture, Environ. Sci. Technol. 35 (2001) 3252–3258.
[30] J.A. Hestekin, L.G. Bachas, D. Bhattacharyya, Poly (amino acid)-functionalized cellulosic membranes: metal sorption mechanisms and results, Ind. Eng. Chem. Res. 40 (2001) 2668–2678.

[31] Y.F. Xu, R.Y.M. Huang, Pervaporation separation of ethanol-water mixtures using ionically crosslinked blended polyacrylic acid (PAA)-Nilon-6 membranes, J. Appl. Polym. Sci. 36 (1988) 1121.
[32] H.S. Choi, T. Hino, M. Shibata, Y. Negishi, H. Ohya, The characteristics of a PAA-PSF composite membrane for separation of water ethanol mixtures through pervaporation, J. Membr. Sci. 72 (1992) 259–266.
[33] J.W. Rhim, M.Y. Sohn, H.J. Joo, K.H. Lee, Pervaporation separation of binary organic-aqueous liquid mixtures using crosslinked PVA membrane. I. Characterization of the reaction between PVA and PAA, J. Appl. Polym. Sci. 50 (1993) 679–684.
[34] J.W. Rhim, H.K. Kim, K.H. Lee, Pervaporation separation of binary organic-aqueous liquid mixtures using crosslinked poly (vinyl alcohol) membranes. IV. Methanol-water mixtures, J. Appl. Polym. Sci. 61 (1996) 1767–1771.
[35] H. Ohya, M. Shibata, Y. Negishi, Q.H. Guo, H.S. Choi, The effect of molecular weight cut-off of PAN ultrafiltration support layer on separation of water-ethanol mixtures through pervaporation with PAA-PAN composite membrane, J. Membr. Sci. 90 (1994) 91–100.
[36] E.M. Gabriel, G.E. Gillberg, In situ modification of microporous membranes, J. Appl. Poly. Sci. 48 (1993) 2081–2090.
[37] J. Huang, Q. Guo, H. Ohya, J. Fang, The characteristics of crosslinked PAA composite membrane for separation of aqueous organic solutions by reverse osmosis, J. Membr. Sci. 144 (1998) 1–11.
[38] F. He, D. Zhao, Preparation and characterization of a new class of starch-stabilized bimetallic nanoparticles for degradation of chlorinated hydrocarbons in water, Environ. Sci. Technol. 39 (2005) 3314–3320.
[39] T. Tomida, K. Hamaguchi, S. Tunashima, M. Katoh, S. Masuda, Binding properties of a water-soluble chelating polymer with divalent metal ions measured by ultrafiltration. poly(acrylic acid), Ind. Eng. Chem. Res. 40 (2001) 3557–3562.
[40] W.S. Orth, R.W. Gillham, Dechlorination of trichloroethene in aqueous solution using Fe0, Environ. Sci. Technol. 30 (1996) 66–71.
[41] L.J. Matheson, P.G. Tratnyek, Reductive dehalogenation of chlorinated methanes by iron metal, Environ. Sci. Technol. Vol. 28 (1994) 2045–2053.
[42] Y. Liu, S.A. Majetich, R.D. Tilton, D.S. Sholl, G.V. Lowry, TCE dechlorination rates, pathways, and efficiency of nanoscale iron particles with different properties, Environ. Sci. Technol. 39 (2005) 1338–1345.
[43] T.L. Johnson, M.M. Scherer, P.G. Tratnyek, Kinetics of halogenated organic compound degradation by iron metal, Environ. Sci. Technol. 32 (1996) 2634–2640.
[44] D.M. Cwiertny, S.J. Bransfield, A.L. Roberts, Influence of the oxidizing species on the reactivity of iron-based bimetallic reductants, Environ. Sci. Technol. 41 (2007) 3734–3740.
[45] Y. Noma, M. Ohno, S. Sakai, Pathways for the degradation of PCBs by palladium-catalyzed dechlorination, Fresenius Environ. Bull. 12 (2003) 302–308.
[46] Q. Huang, J.F. Rusling, Formal reduction potentials and redox chemistry of polyhalogenated biphenyls in a bicontinuous microemulsion, Environ. Sci. Technol. 29 (1995) 98–103.
[47] M. Nutt, J.B. Hughes, M. Wong, Designing Pd-on-Au bimetallic nanoparticle catalysts for trichloroethene hydrodechlorination, Environ. Sci. Technol. 39 (2005) 1346–1353.
[48] D.B. Williams, C.B. Carter, Transmission Electron Microscopy, Kluwer Academic Pub., 1996.

CHAPTER

25 Magnesium-Based Corrosion Nano-Cells for Reductive Transformation of Contaminants

Shirish Agarwal[1], Souhail R. Al-Abed[2] and Dionysios D. Dionysiou[3]

[1]Department of Civil and Environmental Engineering, University of Cincinnati, Cincinnati, OH, USA [2]National Risk Management Research Laboratory, US Environmental Protection Agency, Cincinnati, OH, USA [3]Environmental Engineering and Science Program, University of Cincinnati, Cincinnati, OH, USA; Nireas–International Water Research Centre, University of Cyprus, Nicosia, Cyprus

25.1 Introduction

Enhanced corrosion-based palladium/magnesium (Pd/Mg) bimetallic systems were very effective in rapid and complete dechlorination of polychlorinated biphenyls (PCBs), a group of recalcitrant organic pollutants. Electron production for reduction of contaminants through Mg corrosion was accelerated by galvanic coupling of reactive Mg with inert Pd, also an excellent hydrogenation catalyst. Each Pd/Mg particle functioned as numerous nano-cells generating electrons that eventually led to reduction of the target compounds at the bimetallic interface. In

Street, Sustich, Duncan and Savage. Nanotechnology Applications for Clean Water, 2nd Edition.
 DOI: http://dx.doi.org/10.1016/B978-1-4557-3116-9.00025-1

general, bimetallic systems can remediate matrices contaminated with chlorinated organics (such as PCBs, trichloroethylene, DDT, chlorophenols) and inorganics (nitrates, metals). Pd/Fe is arguably the most widely studied bimetallic system. However, Fe corrodes spontaneously in air, necessitating surface pretreatment and synthesis and storage in anaerobic conditions. Candidature of Mg is based on its unique corrosion properties that afford synthesis and storage under ambient conditions and application-based advantages.

In Pd/Mg systems, doping Pd on a nanoscale is crucial as it provides enhanced catalytic activity owing to a higher fraction of available surface atoms and hence reduced requirements of expensive catalyst. Fine-tuning the synthesis procedure with respect to the Pd nanoparticle precursors, the synthesis medium, and the nature and amount of surfactants allows tailor-made bimetallic particles having small Pd islands enabling maximized reduction performance. Also, such nanosized islands are inherently more stable on the support compared to their micro-sized counterparts.

Thus, the unique corrosion properties of Mg that allow easy synthesis, storage, and application, when combined with precise procedures for optimized nanoscale doping, lead to bimetallic particles with maximized reduction capacities. Additionally, owing to the high oxidation potential of Mg (2.37V), a number of metals (Ni, Ag, Cu, etc.) can be coupled with it to provide an array of efficient corrosion nano-cells.

25.2 Magnesium-based bimetallic systems

In magnesium-based corrosion systems such as Pd/Mg, electron production for the reduction of contaminants through Mg corrosion is accelerated by galvanic coupling of reactive Mg with inert Pd, also an excellent hydrogenation catalyst, to drive the reduction of contaminants at the bimetallic interface [1]. A typical bimetallic corrosion cell has three constituents: (i) a sacrificial anode where the metal corrodes (Mg), releasing electrons in the process, (ii) an electrically conducting solution that forms the corrosion medium, and (iii) a cathode, which may also act as a hydrogenation catalyst (as in case of Pd). Similar bimetallic systems such as Pd/Fe and Ni/Zn have been used for diverse applications such as organic and inorganic syntheses, sample analyses, and environmental clean-up. In most studies with environmental application, palladium was chosen for its unique ability to intercalate and maintain a high surface concentration of hydrogen, a powerful reducing agent, making it superior for hydrogenation than Pt or Ni [2].

Pd/Fe systems have been known to completely dechlorinate trichloroethylenes [3] and PCBs [4]. However, Fe tends to corrode spontaneously requiring surface pretreatment and synthesis and storage under anaerobic conditions. Mg, on the other hand, is considered better-suited for such purposes, due to its unique corrosion properties (elaborated in the next section). In addition, it has an oxidation potential of 2.37V that is significantly higher than 0.44V of Fe [5], and thus a greater force to drive the reduction [6]. These advantages of Mg along with its natural abundance, low cost, and environment-friendly nature have led to growing interest in Pd/Mg-based systems [7,8]. In our studies, we found that Pd/Mg particles produce rapid and complete dechlorination (within 15 minutes) of PCBs, one of the most recalcitrant environmental contaminants in existence [9].

25.3 Unique corrosion properties of magnesium

In our first study, we traced the distinctive advantages Mg offers in terms of synthesis, storage, and application of bimetallic particles in environmental systems to its unique corrosion properties. The surface of elemental Mg in ambient conditions may show scattered corrosion but otherwise remains largely unaffected. Impure Mg forms a partially protective layer of magnesium hydroxide ($Mg(OH)_2$) by superficial corrosion. This corrosion film is weaker than that formed in aluminum but stronger than the flaky, non-adherent film in Fe [10]. On immersing in water, part of the oxide/hydroxide dissolves quickly saturating the water with sparingly soluble $Mg(OH)_2$ and raising the pH to 10.5 [10]. In Pd/Mg systems, the Pd islands act as cathodes facilitating accelerated localized corrosion, breaching the quasi-stable film further. Mg starts to corrode and generate OH^- in the process, favoring the formation of $Mg(OH)_2$, part of which goes into repairing the protective oxide layer [11]. This implies that corrosion prompts repair of the film making the system self-regulated. These unique corrosion properties of Mg have tremendous application-based implications. The largely pristine nature of exposed Mg particles allows doping without surface pretreatment. The metastable hydroxide film formed on thus doping Mg enables storage in ambient conditions without losing their dechlorination potential. Breaching of the partially protective film in water by corrosion at localized cathodes produces desired dechlorination at the bimetallic interface.

The well-established dechlorination mechanism in Pd/Fe particles [2] has been modified by including the self-limiting corrosion behavior of Mg to propose a mechanism for dechlorination in Pd/Mg systems based on Fig. 25.1 and Equations 25.1–25.6:

Corrosion of Mg, the anode, $E^o = -2.372$ V

$$Mg \rightarrow Mg^{2+}(aq) + 2e^- \tag{25.1}$$

Electrolysis of water, and intercalation of H_2 (g) in Pd lattice at cathode:

$$2H_2O + 2e^- \rightarrow H_2(g) + 2OH^- \tag{25.2}$$

$$H_2(g) \xrightarrow{pd} Pd \cdot H_2 \tag{25.3}$$

Reduction of organics at the bimetallic interface [12]:

$$Pd \cdot H_2 + Ar_2Cl \rightarrow Ar_2H + PdClH \tag{25.4}$$

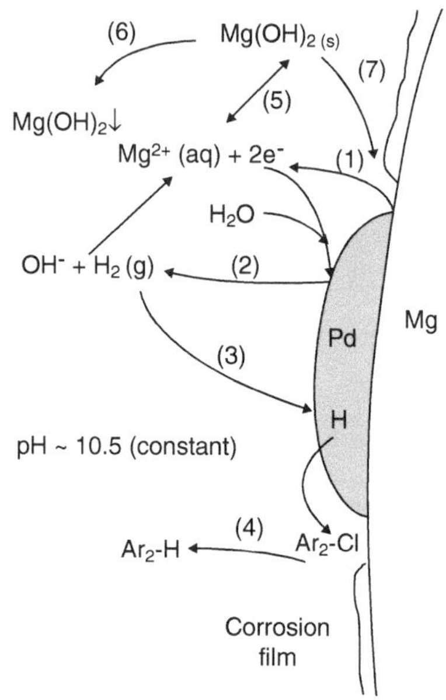

FIGURE 25.1

Proposed mechanism for reduction of chlorinated organics by Pd/Mg bimetallic systems [1].

Dynamic partitioning of $Mg(OH)_2$ in the bulk as soluble species, for anodic film repair and precipitate; K_{sp} of $Mg(OH)_2 = 10.92$ [13]:

$$Mg^{2+} + 2OH^- \leftrightarrow Mg(OH)_2(s) \tag{25.5}$$

$$Mg(OH)_2(s) \rightarrow Mg(OH)_2\downarrow + Mg(OH)_2(\text{Anodic repair}) \tag{25.6}$$

25.4 Doping nanoscale palladium onto magnesium—modified alcohol reduction route

Alcohol reduction of Pd salts is commonly used for synthesizing Pd nanoparticles [14] wherein the alcohol undergoes oxidization to aldehydes and other products. The Pd precursor, the organic medium of synthesis, and nanoparticle stabilizers are critical synthesis parameters that are commonly fine-tuned to customize nanoparticles. The effects of these parameters have been widely reported for synthesis of nanoparticles in a colloidal or suspended form. Introducing Mg during such synthesis for doping it with Pd dramatically alters the reaction environment as Mg is a reactive substrate and is known to reduce metal salts including that of Pd.

To obtain tailor-made bimetallic particles having small Pd islands and maximized reduction potential, the synthesis procedure must be fine-tuned with the presence of reactive Mg as an additional parameter influencing all others, namely, (i) the Pd nanoparticle precursors (different kinetics of reduction hence varying nanoparticle sizes), (ii) the synthesis medium (nanoparticle size control through their oxidation kinetics and diffusion coefficient), and (iii) nature, and (iv) amount of stabilizers (nanoparticles stabilization preventing aggregation).

For instance, the technique for Pd/Mg synthesis by reductive doping of elemental Pd onto Mg in ethanol from our previous study [1] was modified by using polyvinylpyrrolidone (PVP) to produce smaller Pd islands by limiting the agglomeration of nascent Pd particles [14]. In the modified procedure, a calculated amount of PVP (M.W. = 10,000) was added to 2L ethanol and stirred until dissolved. Palladium acetate was then added and stirred until dissolved. A ratio of 1:10 for palladium acetate and PVP was used. Mg particles (60 g,$<$40 μm) were then added to the flask and the resulting slurry stirred for 2 hours wherein elemental Pd was deposited onto Mg to form the bimetallic particles. The slurry was then vacuum filtered, the particles washed thrice with ethanol, and stored anaerobically.

This synthesis produced Pd/Mg particles active at a low Pd content of 0.02 percent. The reduction potential of these particles was evaluated by conducting dechlorination experiments with a 4 mg/L aqueous solution of 2-chlorobiphenyl (2-CB), a model PCB. It was found that particles synthesized with PVP with close to 30 times lower Pd content than those synthesized without PVP (0.02 percent vs. 0.58 percent Pd) exhibited faster dechlorination kinetics indicating drastically improved catalytic activity. This is because the site for these dechlorination reactions is the bimetallic interface [2], and a reduction in the size of the Pd islands through the protective agent during synthesis leads to an increased number of reactive islands and thereby improved dechlorination performance. Also, hydrogen production, which occurs as a side-reaction (Eqn 25.2), was seen at much lower levels in case of particles synthesized using PVP than in those without it. This indicates improved selectivity toward dechlorination in Pd/Mg synthesized through this route.

Unlike other studies where refluxing [15,16] or strong reducing agents such as hydrazine, hydroxylamine, sodium borohydride [17] have been employed to reduce Pd salts, this is a simple wet chemistry route for synthesis of highly reactive Pd/Mg particles with PVP as a protective agent and Mg in a dual role as an active substrate and a reducing agent for the Pd(II). Also, synthesizing Pd/Mg through this route comes with the advantages of the generic alcohol-reduction processes [18], namely, (i) simplicity and reproducibility of synthesis, (ii) small Pd islands with a narrow size distribution, (iii) tailoring Pd particles easily accomplished by altering choice of alcohol, reducing temperature, quantity and variety of stabilizing agent, nanoparticles precursor and concentrations, and (iv) Pd particles with high catalytic activity produced. In addition, such nanosized Pd islands are inherently more stable on Mg compared to their micro-sized counterparts.

25.5 Role of nanosynthesis in assuaging concerns from palladium usage

In the synthesis of Pd/Mg using nanotechniques, doping nanoscale Pd islands onto Mg substrates addresses two major concerns regarding use of Pd as a catalyst in these systems, namely, their possible toxicity and high cost. The nanosynthesis techniques produce highly reactive Pd islands even at very low Pd amounts (approximately 0.02 percent). In addition, such nanosized islands are inherently more stable on the support and selective compared to their micro-sized counterparts. This implies high catalytic activity with drastically reduced requirement of expensive catalyst and reduced bioavailability of Pd during the application of Pd/Mg systems. Other possible routes of addressing these issues include (i) use of alternate coupling metals such as iron, nickel, silver, and so on, made possible due to the high oxidation potential of Mg (2.37V) to provide an array of effective corrosion nano-cells. For instance Ni/Mg/Al [19], Ag/Mg [20], and Mg/Fe in our own studies have proven to be very effective in the reduction of a variety of contaminants including trichlorinated benzenes, chlorophenols, azo-dyes, and nitrates; ii) recycling Pd through one of the numerous processes for the recovery of supported Pd [21] after consumption of Mg; and (iii) using highly active Mg nanoparticles without any catalyst.

25.6 Challenges in nanoscaling magnesium

Magnesium, with its potential to reduce a variety of water contaminants, self-limiting corrosion behavior leading to long active life times, natural abundance, low cost and environmentally friendly nature promises to be a very effective technology, more so in the nanoscale. However, nanoparticles of Mg are difficult to produce by reduction of its salts in aqueous solutions owing to the low redox potential of Mg^{2+} [18]. Other procedures for synthesis of Mg nanoparticles have been reported, such as (i) heating magnesium oxide in an atmosphere of H_2 at 500°C and above [22]; (ii) reduction of Mg salts by alkali metal [23]; (iii) dehydrogenation of Mg hydride [24]; and iv) vaporization of Mg in the presence of tetrahydrofuran [25]. However, these techniques are very specialized and resource intensive. For environmental applications, the top−down synthesis pathway of breaking down micro-sized Mg to nanosized particles in a ball mill under an inert atmosphere may be more suitable. Reduction of MgO in an H_2 atmosphere at high temperatures may also be adopted for large-scale production.

Apart from synthesizing zero-valent Mg nanoparticles, the other challenge is their handling and application owing to the high reactivity of nano-Mg thus synthesized. A possible solution would be exposure of these particles to small amounts of O_2 to reduce their reactivity before bringing them out into the open air as is sometimes done for nanoscale iron [26]. The thin layer of magnesium oxide thus formed would slowly break down in contaminated waters producing the desired reduction of contaminants.

25.7 Other environmental applications

Magnesium-based corrosion systems are relatively new and not very widely explored. In reported studies so far, they have proven to be effective in the reductive transformation of a variety of organic contaminants in water such as PCBs [1,7], chlorophenols, DDT [6,27], azo dyes [28], as well as for inorganic pollutants such as nitrates and metals [29]. Pd/Mg systems are very robust and selectively dechlorinate organics even with interference from large quantities of ethanol and acetone. They also showed complete dechlorination of PCBs adsorbed on sediments [1,7].

Reductive bimetallic systems can be used in reactive barriers to treat highly chlorinated plumes in groundwater aquifers and submarine matrices where the redox environment favors reduction. Pd/Mg barriers in sediment beds can deplete the dissolved organics adjacent to the contaminated sediments causing their desorption into the aqueous phase and in so doing reduce their concentrations in the sediment. Pd may be contained in such barriers allowing it to be recycled while ensuring that it is not free to enter the natural waters. Reductive dechlorination by Pd/Mg can also be a pretreatment step before the application of oxidative technologies that may falter with highly chlorinated organics [30]. In natural waters, Pd/Mg particles will experience enhanced corrosion as chloride ions attack and break down the film locally [11]. Irrespective of corrosion conditions, the self-limiting nature of the process due to corrosion induced oxide layer repair implies that Pd/Mg particles will likely have a long active lifetime.

Upon corrosion during application to aqueous systems, Mg transforms to its oxides and hydroxides. Incidentally, MgO nanoparticles in slurries as well as dry forms have been reported to be anti-biofouling agents [31]. Also, nanocrystalline MgO is known to be an efficient adsorbent for acid gases, chlorocarbons, organophosphorus compounds, alcohols, and other organics due to its unique morphology and high surface areas [32,33]. MgO has been reported to destructively adsorb organics including acetone, ammonia, and benzaldehyde [34].

In broader environmental terms, Mg based composites are being explored for hydrogen storage toward sustainable solutions to the energy crisis. Mg, with its surface altered by small amounts of doped Pd, is a promising rechargeable hydrogen carrier [35,36]. Other composites such as Ni/Mg intermetallic compounds with and without Pd coating [37,38] have also shown immense promise in hydrogen storage and release.

Acknowledgments

This chapter has not been subjected to internal policy review of the US Environmental Protection Agency. Therefore, the research results presented herein do not, necessarily, reflect the views of the Agency or its policy. Mention of trade names or commercial products does not constitute endorsement or recommendation for use.

References

[1] S. Agarwal, S.R. Al-Abed, D.D. Dionysiou, Environ. Sci. Tech. 41 (2007) 3722–3727.

[2] I.F. Cheng, Q. Fernando, N. Korte, Environ. Sci. Tech. 31 (1997) 1074–1078.

[3] R. Muftikian, Q. Fernando, N. Korte, Water Res. 29 (1995) 2434–2439.

[4] C. Grittini, M. Malcomson, Q. Fernando, N. Korte, Environ. Sci. Tech. 29 (1995) 2898–2900.

[5] P. Vanysek, in: D.R. Lide (Ed.), CRC Handbook of Chemistry and Physics, 71st ed., Chemical Rubber Publishing Company, Boston, 1991, pp. 8-16–8-23.

[6] M.D. Engelmann, J.G. Doyle, I.F. Cheng, Chemosphere 43 (2001) 195–198.

[7] E. Hadnagy, L.M. Rauch, K.H. Gardner, J. Environ. Sci. Heal. A. 42 (2007) 685–695.

[8] B.R. Halle, K.M. Carvalho-Knighton, C.L. Geiger, C.A. Clausen, 229th Nat. Meeting Am. Chem. Soc. 45 (2005) 544–547.

[9] S. Agarwal, S.R. Al-Abed, D.D. Dionysiou, J. Environ. Eng. Sci. 133 (2007) 1075–1078.

[10] G.L. Song, A. Atrens, Adv. Eng. Mater. 1 (1999) 11–33.

[11] G. Song, A. Atrens, Adv. Eng. Mater. 5 (2003) 837–858.

[12] W. Wafo, S. Coen, A. Perichaud, M. Sergent, Analysis 24 (1996) 60–77.

[13] H.P.R. Frederikse, in: D.R. Lide (Ed.), CRC Handbook of Chemistry and Physics, 44th ed., Chemical Rubber Publishing Company, Boston, 1963.

[14] Y. Xiong, I. Washio, J. Chen, H. Cai, Z. Li, Y. Xia, Langmuir 22 (2006) 8563–8570.

[15] S. Ayyappan, R. Srinivasa Gopalan, G.N. Subbanna, C.N.R. Rao, J. Mater. Res. 12 (1997) 398–401.

[16] T. Teranishi, M. Miyake, Chem. Mater. 10 (1998) 594–600.

[17] H. Bönnemann, R.M. Richards, Eur. J. Inorg. Chem. (2001) 2455–2480.

[18] N. Toshima, T. Yonezawa, New J. Chem. 22 (1998) 1179–1201.

[19] Y. Cesteros, P. Salagre, F. Medina, J.E. Sueiras, D. Tichit, B. Coq, Appl. Catal. B-Environ. 32 (2001) 25–35.

[20] U. Patel, S. Suresh, J. Colloid Interf. Sci. 299 (2006) 249–259.

[21] R.V. Jasra, P.K. Ghosh, H.C. Bajaj, A.B. Boricha, Process for Recovery of Palladium from Spent Catalyst, US Patent Office, 20040241066, 2004.

[22] J. Cao, P. Clasen, W. Zhang, J. Mater. Res. 20 (2005) 3238–3243.

[23] R.D. Rieke, Acc. Chem. Res. 10 (1977) 301–306.

[24] E. Bartmann, B. Bogdanovic, N. Janke, S. Liao, K. Schlichte, B. Spliethoff, et al., Chem. Ber. 123 (1990) 1517–1528.

[25] H. Imamura, T. Nobunaga, M. Kawahigashi, S. Tsuchiya, Inorg. Chem. 23 (1984) 2509–2511.

[26] G.V. Lowry, K.M. Johnson, Environ. Sci. Tech. 38 (2004) 5208–5216.

[27] M.D. Engelmann, R. Hutcheson, K. Henschied, R. Neal, I.F. Cheng, Microchem. J. 74 (2003) 19–25.

[28] R. Patel, S. Suresh, J. Hazard. Mater. 137 (2006) 1729–1741.

[29] M. Kumar, S. Chakraborty, J. Hazard. Mater. 135 (2006) 112–121.

[30] Y. Liu, J. Schwartz, C.L. Cavallaro, Environ. Sci. Tech. 29 (1995) 836–840.

[31] P.K. Stoimenov, R.L. Klinger, G.L. Marchin, K.J. Klabunde, Langmuir 18 (2002) 6679–6686.
[32] R. Richards, W. Li, S. Decker, C. Davidson, O. Koper, V. Zaikovski, et al., J. Am. Chem. Soc. 122 (2000) 4921–4925.
[33] K.J. Klabunde, J. Stark, O. Koper, C. Mohs, D.G. Park, S. Decker, et al., J. Phys. Chem.-US 100 (1996) 12142–12153.
[34] A. Khaleel, P.N. Kapoor, K.J. Klabunde, Nanostruct. Mater. 11 (1999) 459–468.
[35] X. Xu, C. Song, Appl. Catal. A-Gen. 300 (2006) 130–138.
[36] X. Xu, J. Zheng, C. Song, Eng. Fuels 19 (2005) 2107–2109.
[37] R. Janot, A. Rougier, L. Aymard, C. Lenain, R. Herrera-Urbina, G.A. Nazri, et al., Alloy Compd. (2003) 356–357, 438–441.
[38] H. Shao, T. Liu, X. Li, L. Zhang, Scr. Mater. 49 (2003) 595–599.

PART

4

Stabilization Technologies

CHAPTER

26 Multifunctional Materials Containing Nanoscale Zerovalent Iron in Carbon Microspheres for the Environmentally Benign Remediation of Chlorinated Hydrocarbons

Jingjing Zhan[1], Bhanukiran Sunkara[1], Rubo Zheng[1], Pradeep Venkataraman[1], Sehinde Owoseni[1], Gary L. McPherson[1], Vijay T. John[1], Dick Brown[2] and David Culpepper[3]

[1]Department of Chemical and Biomolecular Engineering, Tulane University, New Orleans, LA, USA [2]ERM Corporation, Ewing, NJ, USA [3]NanoFex LLC., New Orleans, LA, USA

26.1 Introduction

In recent years, extensive efforts have been carried out to develop and synthesize nanomaterials with unique reactivity and functional characteristics for environmental applications [1–9]. For example, the use of nanoscale zerovalent iron (NZVI) particles represents a promising approach to the remediation of

 DOI: http://dx.doi.org/10.1016/B978-1-4557-3116-9.00026-3

chlorinated organic-contaminated groundwater such as trichloroethylene (TCE) [10–12], where the following redox reaction leads to conversion of TCE to innocuous gas phase products like ethane.

$$C_2HCl_3 + 4Fe^0 + 5H^+ \rightarrow C_2H_6 + 4Fe^{2+} + 3Cl^-$$

Compared to conventional micron-sized granular iron powders, the advantages of using NZVI particles include the potentially high reactivity as a consequence of high surface areas, and the fact that they can be colloidally stabilized, suspended as a slurry and injected into the subsurface [13–16]. However, the intrinsic ferromagnetism of NZVI particles leads to aggregation and there continues to be difficulties in developing efficient *in situ* technologies [17–19]. Several criteria need to be met in the design of effective systems for *in situ* degradation of chlorinated compounds. Such systems must be able to move through the subsurface with high mobility, show affinity toward hydrophobic TCE, and break down the contaminant efficiently. The mobility of colloids in the subsurface is determined by competitive mechanisms of Brownian motion, interception by soil and sediment grains, and sedimentation effects [20]. The Tufenkji–Elimelech model, which considers the effect of hydrodynamic forces and van der Waals interactions between colloidal particles and sediment grains, predicts that particles in the size ranging from 0.1 to 1.0 μm are likely to be the most mobile at typical groundwater flow conditions [21–23]. In addition, considering that hydrophobic organic contaminants are retained by soil grains via capillary forces and adsorption [22], it would be advantageous if these particles also reduce the concentration of dissolved TCE through a combination of sequestration by adsorption followed by degradation of TCE. Commonly, the mobility of NZVI particles can be enhanced by adsorption of hydrophilic or amphiphilic organic species, such as surfactants, vegetable oils, starch, or polyelectrolytes such as carboxymethyl cellulose (CMC) and poly(acrylic acid) (PAA), or triblock copolymers on the NZVI particle surface [18,22,24–28]. These adsorbed organics inhibit NZVI aggregation and enhance solution stability through steric hindrance and/or electrostatic repulsion [29,30]. Alternatively, NZVI immobilized onto support materials such as activated carbon granules (1–3 mm) is an effective way to inhibit aggregation of NZVI particles [31]. Composites with carbon introduce a strong adsorptive aspect into remediation technology as the carbon adsorbs chlorinated compounds, and these materials have been used in the development of adsorptive–reactive barriers [32,33]. In a pioneering paper on the use of carbon, Schrick et al. have shown that carbon black is a useful additive to zerovalent iron to prevent aggregation and facilitate reaction and transport [22]. More recently, Mackenzie et al. have incorporated iron onto activated carbon through traditional wetness impregnation procedures and have shown that the composite (termed Carbo-Iron), while somewhat limited in its reactivity, does adsorb and dechlorinate TCE [19].

In recent work from our laboratory, we have developed a multifunctional particulate system containing zerovalent iron that has the requisite characteristics of reaction, adsorption, and transport to effectively address the degradation of chlorinated compounds [34–36]. The particulate system is obtained from inexpensive precursors and through a semicontinuous method which allows large-scale synthesis necessary for eventual *in situ* application. The particulates contain NZVI supported on carbon microspheres and are synthesized through an aerosol route using inexpensive sugars as precursors. Prior studies have demonstrated that the aerosol-based technology is a simple approach to prepare particles in the submicron (typically 100–800 nm) size range and we have explored this concept in preparing NZVI supported on silica [23,37,38]. In our work, we have expanded the aerosol-based technology to produce carbon-based functional nanocomposites of zerovalent iron supported on carbon spheres. The postulates of the work are the following: (i) immobilization of NZVI onto carbon spheres may make the ZVI less prone to aggregation, while maintaining reactivity; (ii) carbon produced by an aerosol-based process serves as a strong adsorbent for TCE increasing local concentrations at the ZVI reaction sites, thereby enhancing the driving force of reaction; and (iii) the aerosol-based process is an efficient method to synthesize such multifunctional adsorptive-reactive materials in the optimal size range for transport through sediments. Finally, it is important to recognize that while the zerovalent iron particles are nanoscale, they are embedded in carbon microspheres that are larger than 100 nm size below which toxicity aspects of nanoscale materials may be of concern. As a result of such embedding, the composite particles might be intrinsically environmentally innocuous.

26.2 Materials synthesis

Figure 26.1A illustrates the aerosol process, with a system containing an atomizer (an inexpensive nebulizer used in inhalant therapies is all that is necessary) to generate the droplets, a furnace to convert the liquid precursors to the solid product, and a filter to collect the solids. Starting with a homogenous aqueous solution containing sucrose, iron chloride, sulfuric acid (optional at higher temperatures), the nebulizing unit atomizes the solution into droplets that undergo a heating and drying step, generating submicron particles that are collected on a filter. Figure 26.1B is a representation of the formation route of Fe/C composites. When the aerosol droplets pass through the heating zone, solvent evaporation and dehydration/carbonization of sucrose occur. The role of sulfuric acid (when used) is to accelerate the process of carbonization especially at lower furnace temperatures; at furnace temperatures over 450°C, this is completely optional. In addition, precipitation of solidified iron salt is concomitant with the dehydration of sucrose, generating a black powder of Fe salt/C composites. To obtain Fe/C composites, the collected powder is treated with sodium borohydride solution in excess to reduce ferric iron to zerovalent iron [10,38]. The final weight percentage of zerovalent iron in the Fe/C composites is typically between 20 and 30 wt%

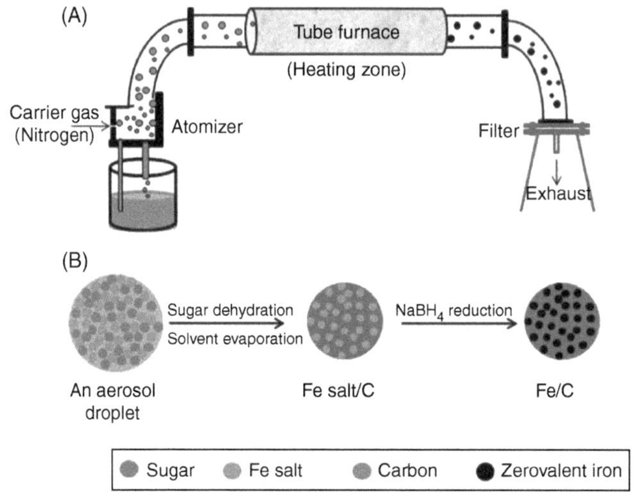

FIGURE 26.1

Schematic of (A) aerosol reactor for composite synthesis and (B) reaction in an aerosol droplet.

of the composite. It is noteworthy that operating conditions for synthesis of the aerosol-based Fe/C composites are adjustable. For instance, Fe/C composites can be obtained at temperatures as low as 350°C with dilute sulfuric acid added to the precursor solution to catalyze carbonization or at higher temperatures without any sulfuric acid. The reduced Fe/C composites are then tested with chlorinated hydrocarbons with TCE being the model reactant. For consistency in analysis, the TCE concentration in solution is kept at 20 ppm, and the reaction is monitored through headspace analysis using established procedures [39,40].

The microstructure and morphology of the multifunctional nanostructured particles were analyzed through transmission and scanning electron microscopy. Figure 26.2A shows the TEM image of aerosol-based bare carbon particles as the control. The particles are well-defined microspheres. For Fe/C composites as shown in Figure 26.2B, the presence of NZVI with higher electron contrast on carbon supports indicates distribution of nanoiron throughout the surface of carbon. The ZVI nanoparticles have a mean of 15.5 nm, and a standard deviation of 5 nm and the lack of aggregation to large clusters is noted in contrast to alternate methods of synthesizing NZVI for dechlorination [41]. The fact that NZVI particles are attached on the surface of carbon and are independent particles was further confirmed by the cut-section TEM image as shown in Figure 26.2C. Clearly, the cut-section TEM image shows a strong contrast between the dark edge and pale core, implying that zerovalent iron nanoparticles are attached to the surface rather than located in the interior. In agreement, from a typical SEM image (Figure 26.2D), it can be seen that all nanostructured Fe/C particles are spherical

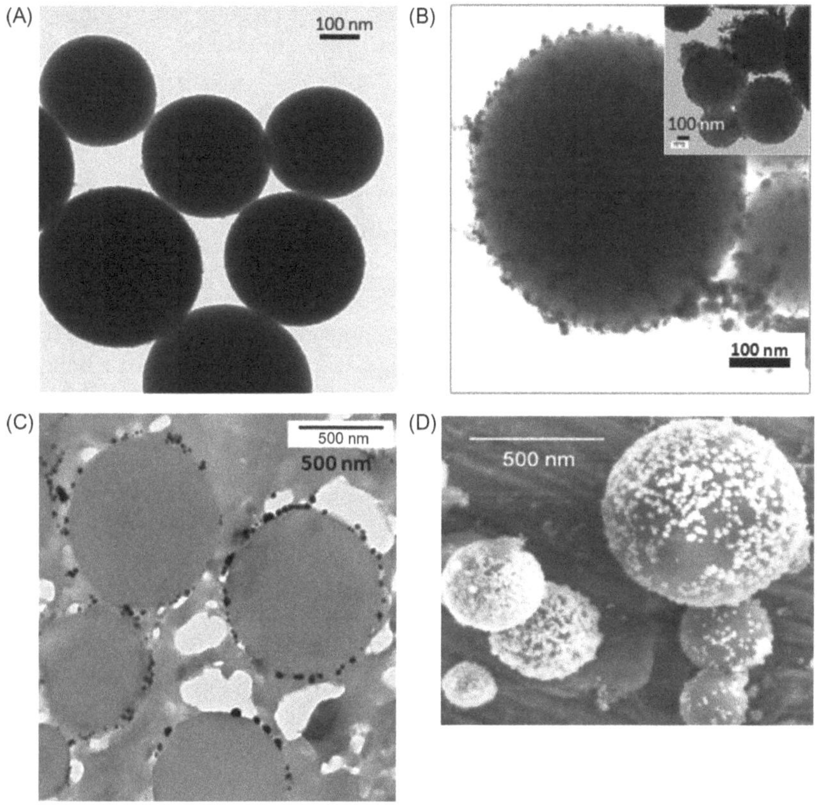

FIGURE 26.2

(A) TEM of carbon prepared by an aerosol-based process, (B) TEM, (C) cut-section TEM, and (D) SEM of Fe/C. The inset is the low magnification TEM of Fe/C.

with a size range of 100–800 nm and discrete zerovalent iron nanoparticles are decorated on the surface of carbon spheres. X-ray diffraction (XRD) and X-ray photoelectron spectroscopy (XPS) data further indicate the presence of zerovalent iron [36].

We have also realized a remarkable characteristic of the aerosol-based process, demonstrating that it is possible to vary the placement of iron nanoparticles on carbon microspheres either on the surface of the microsphere or in the interior [35]. Such a variation of metal placement can be accomplished simply through operation at various temperatures and are the result of modifying the relative rates of carbonization and metal precipitation in an aerosol droplet. At lower aerosolization temperatures, carbonization occurs first leading to iron salt precipitation on the external surface, while at higher temperatures interior placement occurs through concurrent iron salt precipitation and sucrose carbonization. Since both iron and carbon are widely used catalysts and catalyst supports, the simple process of modifying iron placement has significant potential applications in heterogeneous catalysis. Figure 26.3 illustrates the competition between sucrose carbonization and iron salt precipitation. Figure 26.4 illustrates single particle cut-section TEMs indicating the placement of NZVI on the surface at 300°C and 500°C transitioning to placement in the interior at 1000°C.

26.2.1 Adsorption and reactivity studies

TCE removal from solution and gas product evolution rates are shown in Figure 26.5, where the performance of Fe/C and bare carbon when contacted with dissolved TCE is compared. We note an immediate sharp decrease of the dissolved TCE concentration to 18% of its original value followed by a much slower decrease. The initial sharp decrease is due to TCE partitioning from solution to the carbon through strong adsorption. This is an important aspect to the design of these materials as the phenomenon leads to enhanced reactant concentrations in the vicinity of the reactive NZVI sites. The rate at which gas products evolve is indicative of the observed reaction kinetics of TCE. To prove the concept that the carbon is a strong adsorbent for TCE, we exposed bare aerosol carbon particles to TCE-containing solutions. As expected, there is an immediate and sharp reduction of solution concentration with an average adsorption of 0.66 mg of TCE/(gram of aerosol carbon) (or 82.5% of the total TCE) at the TCE concentration used in this study, but no further decrease in concentration due to reaction.

Since adsorption is rapid, reaction is the rate controlling step and it is possible to calculate a pseudo-first-order rate constant by following the evolution of the

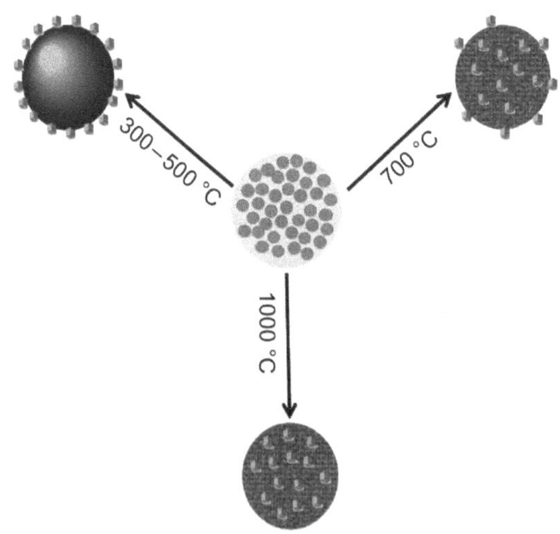

FIGURE 26.3

Schematic of the method to place NZVI either on the surface or in the interior of carbon microparticles based on the competition between carbonization and salt precipitation.

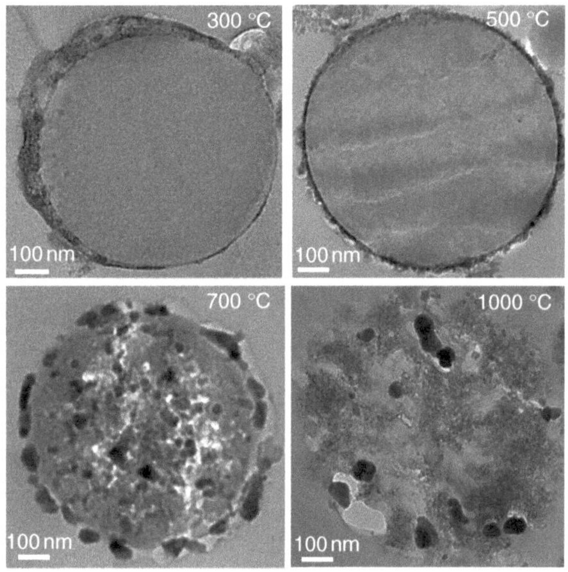

FIGURE 26.4

Cut-section TEM of composites at different heating zone temperatures. The evolution of iron nanoparticle placement from the external surface to the particle interior with an increase in the heating zone temperature from 300°C to 1000°C is noted.

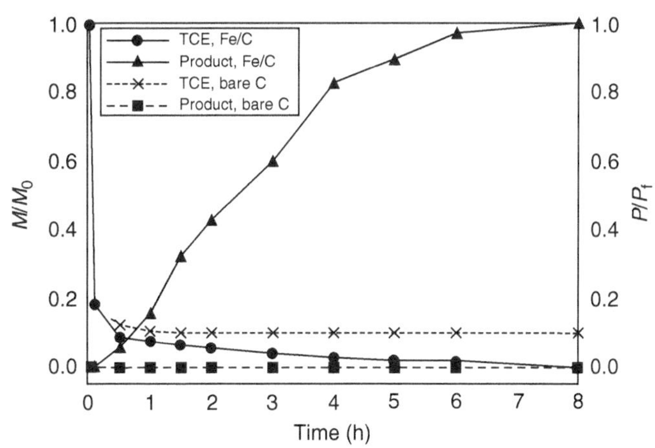

FIGURE 26.5

TCE removal from solution and gas product evolution rates for Fe/C composites. M/M_0 is the fraction of the original TCE remaining and P/P_f is the ratio of the gas product peak to the gas product peak at the end of 8 h.

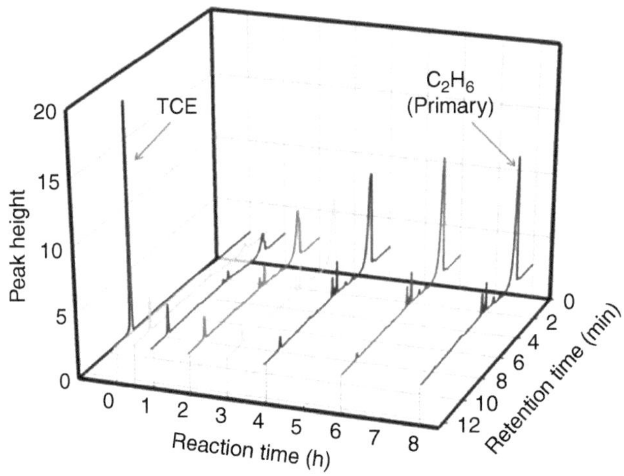

FIGURE 26.6

Representative GC trace of headspace analyses showing TCE degradation and reaction product evolution at various reaction times.

lumped gas phase products [39,40]. For Fe/C composites, the apparent reaction rate constant, k_{obs}, is approximately 0.47 h^{-1} with the mass-normalized reaction constant, k_m is 0.12 L/hg based on the mass of zerovalent iron. In contrast, commonly reported rate constants for NZVI for the remediation of TCE are 0.013 h^{-1} and 0.0026 L/hg, respectively [42]. The ~45-fold difference in k_m suggests that the application of Fe/C not only provides a strong sequestration mechanism but also greatly enhances the reactivity. The enhanced reactivity could be due to (a) the high surface area of nonaggregated NZVI and (b) the increased local concentrations of TCE due to adsorption. Figure 26.6 illustrates the time evolution of gas phase products. The chromatogram is illustrative in that it shows the sharp decrease of solution TCE level as soon as the reactive particles are added, a consequence of TCE adsorption. It is interesting to note that toxic intermediates, such as dichloroethane ($C_2H_2Cl_2$) and vinyl chloride (C_2H_3Cl), are not observed, and the entire product range is based on the light gases. We consider this is due to the strong adsorptive characteristics of the carbon to sequester intermediates till they are reacted away to the light gases, primarily ethane and ethylene, but including a small amount of butane and butene. Our continuing work with the individual intermediates has shown that this concept is indeed true and that all the intermediates are adsorbed to the carbon and subsequently reacted away.

Figure 26.7 compares adsorptive capacities of the aerosol-based Fe/C composites with humic acid (the major natural organic matter of soil) and commercially activated carbon. In all experiments, 20 mL of a 20 ppm TCE solution and 0.2 g of particles were used. Clearly, the adsorption of TCE on Fe/C (~85%) is higher than that of humic acid (~30%) and comparable to that on commercially available granular

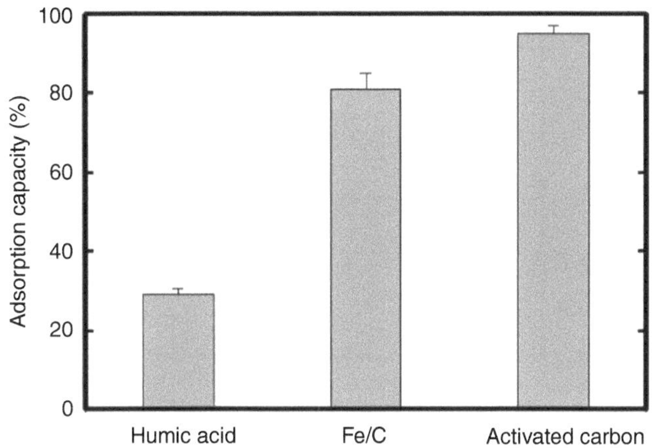

FIGURE 26.7

Comparison of adsorption capacity of humic acid, Fe/C from an aerosol-based process (1000°C), and commercial activated carbon. In all experiments, 20 mL of a 20 ppm TCE solution, 0.2 g of particles were used.

and irregularly defined activated carbons (~95%). The implication of the strong adsorption on the aerosol-based carbon is the ability to establish a driving force for chlorinated compounds to desorb from natural organic matter and partition to the carbon containing NZVI which leads to destruction of the TCE. This leads to highly effective remediation of contaminated sediments. We have calculated the partition coefficient for TCE adsorption on the aerosol-based Fe/C particles using the definition of Phenrat et al. [29] and fully discussed in the supporting information. The measured partition coefficient for TCE adsorption on humic acid is 85, while K_p for the adsorption of TCE on aerosol-based Fe/C composites is 1560, an 18.3-fold higher in adsorption capacity. The implication is that even in systems containing TCE adsorbed to natural organic materials, there will be a driving force to transfer TCE to the Fe/C composites where they will be reacted away.

26.3 Stability and transport characteristics

Figure 26.8 demonstrates that colloidal stability of Fe/C particles can be significantly enhanced by the addition of polyelectrolytes such as CMC, a well-studied additive for colloidal stabilization through both steric and electrostatic repulsion effects [43]. In the experiment, the initial concentration of Fe/C particles was maintained at 250 mg/L (0.01 g in 40 mL solution), and the content of CMC was 4% by weight. Sedimentation curves of suspensions were obtained by monitoring the turbidity of suspensions with a nephelometric turbidimeter (DRT100B, HF Scientific, Inc., Fort Myers, FL). The role of CMC in maintaining colloidal

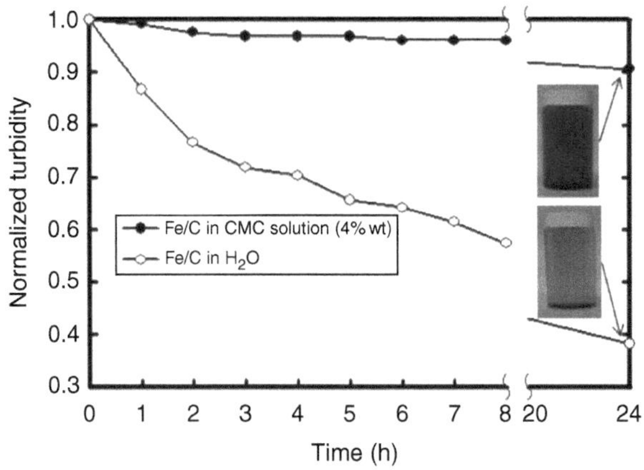

FIGURE 26.8

Sedimentation curves of Fe/C composites in 4% (w/w) CMC solution (solid circles) and water (open circles). The inset images are Fe/C composites in CMC solution and water after 24 h, respectively. The normalized turbidity is defined as the ratio of real-time turbidity to the initial turbidity of the colloidal suspension.

stability is clearly observed with over 90% of the particles remaining suspended after 24 h. Increased amounts of CMC enhance stability further (data not shown here). We contrast the findings with results in the literature that indicate that bare NZVI particles rapidly aggregate and precipitate from solution in less than an hour, indicating the necessity to functionalize the NZVI or add colloidal stabilizers [17,44,45]. In the technology described here, aggregation of NZVI is avoided by immobilization on carbon, and colloidal stability is brought about through the addition of an inexpensive polyelectrolyte.

Transport characteristics of these multifunctional materials are examined through capillary transport experiments. The capillary experiment is an effective and intuitive method to study particle transport through porous media and has been reported in our previous work [23]. As shown in Figure 26.9A, glass melting-point tubes with both ends open (1.5–1.8 mm i.d. × 100 mm length, Corning, NY) were used as capillaries. The capillary tubes were packed with wet Ottawa sand over a 3 cm length and were placed horizontally to simulate groundwater flow. A continuous water flow at a flow rate of 0.1 mL/min (Darcy velocity: 5 cm/min) was provided by a syringe pump. The exit point of the capillary was capped with a small glass wool plug. After 30 μL of CMC-stabilized Fe/C suspension was injected into the inlet of the capillary, water flushing was initiated and an inverted optical microscope was used to observe the pore-scale transport of the particles. Figure 26.9B illustrates photographs of the capillaries depicting the capillary containing Fe/C colloids before and after the water flush. The images indicate that carbon-supported NZVI particles readily transport through the packed capillaries and become captured in the glass wool. In contrast, bare NZVI particles

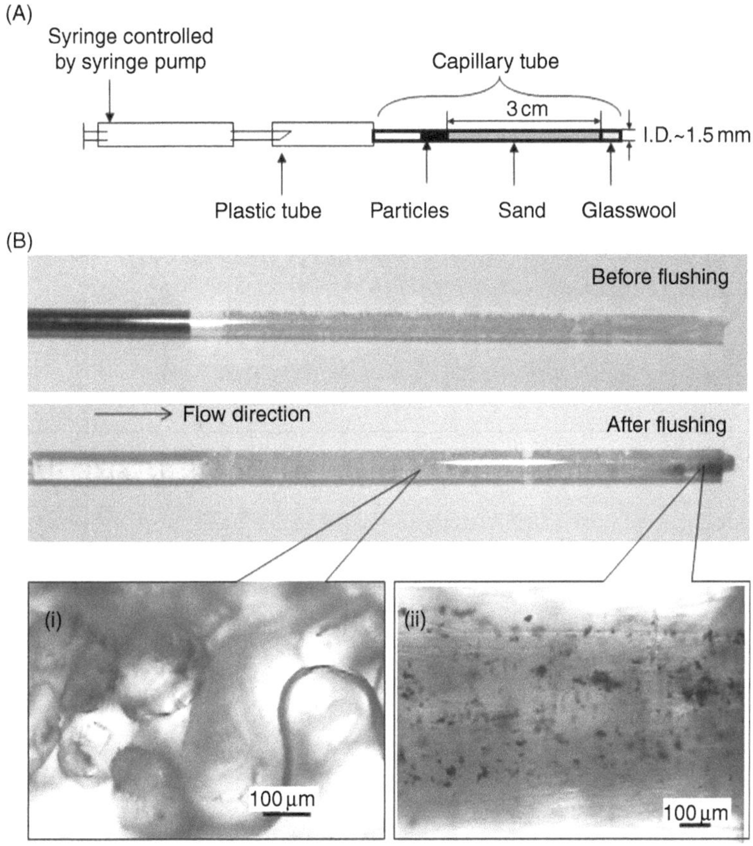

FIGURE 26.9

(A) Experimental setup to study transport in horizontal capillaries. (B) Photographs of before (top) and after (bottom) water flushing showing the characteristics of transport through packed capillaries. Panels showing optical micrographs of particles at various locations: (i) in the middle of the capillary and (ii) on glass wool at the end of the capillary after water flushing. Flow rate: 0.1 mL/min, sand length: 3 cm, and injected suspension volume: 30 μL.

agglomerate and do not transport through the capillary [23]. With over 97% of the particles being eluted through the capillary, the attachment coefficient denoting the attachment probability of particles to the sediment is calculated at 0.09.

26.4 Partitioning at TCE–water interfaces

We have also shown that these carbon microspheres can form Pickering emulsions of water-in-TCE that are stable over a period of 4 months. The carbon microspheres

FIGURE 26.10

Cryo-SEM of a water-in-TCE Pickering emulsion stabilized by carbon microspheres.

cause a reduction in the TCE–water interfacial tension and adsorb irreversibly to the TCE–water interface. Figure 26.10 is a cryo-SEM of such Pickering emulsions. The stability of the emulsion phase is attributed to the formation of a monolayer of carbon particles around the water droplets. The particle bilayer at the contact point of the water droplets offers a steric barrier to coalescence.

The results have broad implication to the remediation of chlorinated solvents. These solvents accumulate in pools in impervious bedrock fractures and slowly leach into groundwater. Treatment by carbon containing zerovalent iron nanoparticles is a potentially highly viable route as the carbons are strong adsorbents for TCE and the iron reduces the TCE to innocuous light hydrocarbons such as ethane. Our results indicate that carbon particles with zerovalent iron will accumulate at the water–TCE interface and gradually destroy the contaminant. The ability of the carbon particles to form water-in-TCE emulsions at the surface of a TCE pool is desirable from a remediation perspective since it allows access of water to zerovalent iron particles that are present on the carbon in the real application. The reverse case of TCE-in-water emulsions may be less desirable if it leads to mobilization of TCE droplets into groundwater flow. The results are also important for a variety of other technologies since they indicate the possibility of generating carbon materials with appropriate functionalization and treatment of the carbons [35]. Such materials have tremendous applications to energy technologies and their applications will be pursued in continuing research.

26.5 Summary

In conclusion, nanoscale zerovalent iron particles supported on carbon particles using an aerosol-based process followed by reduction have potential in the environmental remediation of chlorinated hydrocarbon. These composites are specifically designed for use in the *in situ* breakdown of chlorinated hydrocarbons such as TCE. The following are beneficial characteristics of these systems: (i) the presence of nanoscale zerovalent iron in the composites ensures efficient TCE remediation; (ii) the aerosol-based carbon strongly adsorbs TCE removing dissolved TCE rapidly and facilitates reaction by increasing TCE concentrations in the vicinity of the iron; (iii) the strongly adsorptive carbon prevents release of any toxic chlorinated intermediate; (iv) the particle size distribution is optimal for effective transport through soil; and (v) the composite particles are environmentally benign. Finally, the aerosol process is conducive to scale up as it can easily be designed for continuous operation.

There are additional generalizations that can be drawn. The novelty of the current work is the demonstration of the versatility of the aerosol-based process and the understanding of the competitive nature of carbonization of sucrose and iron salt precipitation during evaporation of an aerosol droplet. By a simple increase of heating zone temperatures, the relative increase in the iron precipitation rate to the carbonization rate results in iron placement in the interior of the carbon microsphere than on the external surface. Regardless of the environmental application described here, carbon is a ubiquitous catalyst support and iron is an equally ubiquitous catalytic material. Whether these concepts can be extrapolated to the preparation of other catalytic metals on carbon supports is the subject of further study. But one can imagine the possibilities of a facile method to control metal placement in catalytic supports. In highly diffusion limited reactions, external placement of the metal species may be preferred, whereas in cases where diffusion hinders transport of catalyst poisons or coke precursors, internal placement of the metal species would be preferred. The aerosol route to such placement modification is therefore of significant use to catalyst design.

Acknowledgments

Funding from the Environmental Protection Agency (EPA-GR832374), the National Science Foundation (Grant 0933734), and the Louisiana Board of Regents is gratefully acknowledged.

References

[1] G.A. Mansoori, T.R. Bastami, A. Ahmadpour, Z. Eshaghi, Environmental application of nanotechnology, in: G. Cao, C.J. Brinker (Eds.), Annual Review of Nanoresearch, vol. 2, World Scientific Publishing Company, 2008, pp. 439–493.

[2] P.G. Tratnyek, R.L. Johnson, Nanotechnologies for environmental cleanup, Nano Today 1 (2) (2006) 44–48.

[3] M.S. Mauter, M. Elimelech, Environmental applications of carbon-based nanomaterials, Environ. Sci. Technol. 42 (16) (2008) 5843–5859.

[4] K. Yoon, B.S. Hsiao, B. Chu, Functional nanofibers for environmental applications, J. Mater. Chem. 18 (44) (2008) 5326–5334.

[5] H. Zhang, G. Chen, D.W. Bahnemann, Photoelectrocatalytic materials for environmental applications, J. Mater. Chem. 19 (29) (2009) 5089–5121.

[6] W.A. Arnold, A.L. Roberts, Pathways of chlorinated ethylene and chlorinated acetylene reaction with Zn(0), Environ. Sci. Technol. 32 (19) (1998) 3017–3025.

[7] S.-S. Chen, H.-D. Hsu, C.-W. Li, A new method to produce nanoscale iron for nitrate removal, J. Nanopart Res. 6 (6) (2004) 639–647.

[8] X.G. Li, S. Takahashi, K. Watanabe, Y. Kikuchi, M. Koishi, Hybridization and characteristics of Fe and Fe–Co nanoparticles with polymer particles, Nano Lett. 1 (9) (2001) 475–480.

[9] L.B. Hoch, E.J. Mack, B.W. Hydutsky, J.M. Hershman, J.M. Skluzacek, T.E. Mallouk, Carbothermal synthesis of carbon-supported nanoscale zero-valent iron particles for the remediation of hexavalent chromium, Environ. Sci. Technol. 42 (7) (2008) 2600–2605.

[10] C.B. Wang, W.X. Zhang, Synthesizing nanoscale iron particles for rapid and complete dechlorination of TCE and PCBs, Environ. Sci. Technol. 31 (7) (1997) 2154–2156.

[11] E.K. Nyer, D.B. Vance, Nano-scale iron for dehalogenation, Ground Water Monit. Rem. 2 (2001) 41–46.

[12] W.X. Zhang, Nanoscale iron particles for environmental remediation: an overview, J. Nanopart Res. 5 (2003) 323.

[13] Y.P. Sun, X.Q. Li, J. Cao, W.X. Zhang, H.P. Wang, Characterization of zero-valent iron nanoparticles, Adv. Colloid Interface Sci. 120 (2006) 47.

[14] V. Sarathy, P.G. Tratnyek, J.T. Nurmi, D.R. Baer, J.E. Amonette, C.L. Chun, et al., Aging of iron nanoparticles in aqueous solution: effects on structure and reactivity, J. Phys. Chem. C 112 (7) (2008) 2286–2293.

[15] Y. Liu, H. Choi, D. Dionysiou, G.V. Lowry, Trichloroethene hydrodechlorination in water by highly disordered monometallic nanoiron, Chem. Mater. 17 (21) (2005) 5315–5322.

[16] B. Schrick, J.L. Blough, A.D. Jones, T.E. Mallouk, Hydrodechlorination of trichloroethylene to hydrocarbons using bimetallic nickel–iron nanoparticles, Chem. Mater. 14 (12) (2002) 5140–5147.

[17] T. Phenrat, N. Saleh, K. Sirk, R.D. Tilton, G.V. Lowry, Aggregation and sedimentation of aqueous nanoscale zerovalent iron dispersions, Environ. Sci. Technol. 41 (1) (2007) 284–290.

[18] F. He, D.Y. Zhao, Manipulating the size and dispersibility of zerovalent iron nanoparticles by use of carboxymethyl cellulose stabilizers, Environ. Sci. Technol. 41 (17) (2007) 6216–6221.

[19] K. Mackenzie, A. Schierz, A. Georgi, F.D. Kopinke, Colloidal activated carbon and carbo-iron novel materials for in-situ groundwater treatment, Glob. Nest J. 10 (1) (2008) 54–61.

[20] K. Yao, M. Habibian, C. O'Melia, Water and waste water filtration: concepts and applications, Environ. Sci. Technol. 5 (11) (1971) 1105–1112.

[21] N. Tufenkji, M. Elimelech, Correlation equation for predicting single-collector efficiency in physicochemical filtration in saturated porous media, Environ. Sci. Technol. 38 (2) (2004) 529.

[22] B. Schrick, B.W. Hydutsky, J.L. Blough, T.E. Mallouk, Delivery vehicles for zerovalent metal nanoparticles in soil and groundwater, Chem. Mater. 16 (11) (2004) 2187–2193.

[23] J. Zhan, T. Zheng, G. Piringer, C. Day, G.L. McPherson, Y. Lu, et al., Transport characteristics of nanoscale functional zerovalent iron/silica composites for in situ remediation of trichloroethylene, Environ. Sci. Technol. 42 (23) (2008) 8871–8876.

[24] N. Saleh, T. Phenrat, K. Sirk, B. Dufour, J. Ok, T. Sarbu, et al., Adsorbed triblock copolymers deliver reactive iron nanoparticles to the oil/water interface, Nano Lett. 5 (12) (2005) 2489–2494.

[25] D.S. Alessi, Z. Li, Synergistic effect of cationic surfactants on perchloroethylene degradation by zero-valent iron, Environ. Sci. Technol. 35 (18) (2001) 3713–3717.

[26] S.R. Kanel, H. Choi, Transport characteristics of polymer stabilized nano scale zero-valent iron in porous media, Water Sci. Technol. 55 (1–2) (2007) 157–162.

[27] J. Quinn, C. Geiger, C. Clausen, K. Brooks, C. Coon, S. O'Hara, et al., Field demonstration of DNAPL dehalogenation using emulsified zero-valent iron, Environ. Sci. Technol. 39 (5) (2005) 1309–1318.

[28] F. He, D. Zhao, Preparation and characterization of a new class of starch-stabilized bimetallic nanoparticles for degradation of chlorinated hydrocarbons in water, Environ. Sci. Technol. 39 (9) (2005) 3314–3320.

[29] T. Phenrat, Y. Liu, R.D. Tilton, G.V. Lowry, Adsorbed polyelectrolyte coatings decrease Fe(0) nanoparticle reactivity with TCE in water: conceptual model and mechanisms, Environ. Sci. Technol. 43 (5) (2009) 1507–1514.

[30] T. Phenrat, T.C. Long, G.V. Lowry, B. Veronesi, Partial oxidation ("aging") and surface modification decrease the toxicity of nanosized zerovalent iron, Environ. Sci. Technol. 43 (1) (2009) 195–200.

[31] H. Choi, S.R. Al-Abed, S. Agarwal, D.D. Dionysiou, Synthesis of reactive nano-Fe/Pd bimetallic system-impregnated activated carbon for the simultaneous adsorption and dechlorination of PCBs, Chem. Mater. 20 (11) (2008) 3649–3655.

[32] N. Lubick, Cap and degrade: a reactive nanomaterial barrier also serves as a cleanup tool, Environ. Sci. Technol. 43 (2) (2009) 235.

[33] H. Choi, S. Agarwal, S.R. Al-Abed, Adsorption and simultaneous dechlorination of PCBs on GAC/Fe/Pd: mechanistic aspects and reactive capping barrier concept, Environ. Sci. Technol. 43 (2) (2009) 488–493.

[34] J.J. Zhan, B. Sunkara, J.J. Tang, Y.Q. Wang, J.B. He, G.L. McPherson, et al., Carbothermal synthesis of aerosol-based adsorptive-reactive iron–carbon particles for the remediation of chlorinated hydrocarbons, Ind. Eng. Chem. Res. 50 (23) (2011) 13021–13029.

[35] B. Sunkara, J.J. Zhan, I. Kolesnichenko, Y.Q. Wang, J.B. He, J.E. Holland, et al., Modifying metal nanoparticle placement on carbon supports using an aerosol-based process, with application to the environmental remediation of chlorinated hydrocarbons, Langmuir 27 (12) (2011) 7854–7859.

[36] J.J. Zhan, I. Kolesnichenko, B. Sunkara, J.B. He, G.L. McPherson, G. Piringer, et al., Multifunctional iron–carbon nanocomposites through an aerosol-based process for

the in situ remediation of chlorinated hydrocarbons, Environ. Sci. Technol. 45 (5) (2011) 1949–1954.
[37] T. Zheng, J. Pang, G. Tan, J. He, G. McPherson, Y. Lu, et al., Surfactant templating effects on the encapsulation of iron oxide nanoparticles within silica microspheres, Langmuir 23 (9) (2007) 5143–5147.
[38] T. Zheng, J. Zhan, J. He, C. Day, Y. Lu, G.L. McPherson, et al., Reactivity characteristics of nanoscale zerovalent iron–silica composites for trichloroethylene remediation, Environ. Sci. Technol. 42 (12) (2008) 4494–4499.
[39] J. Zhan, B. Sunkara, L. Le, V.T. John, J. He, G.L. McPherson, et al., Multifunctional colloidal particles for in situ remediation of chlorinated hydrocarbons, Environ. Sci. Technol. 43 (22) (2009) 8616–8621.
[40] B. Sunkara, J. Zhan, J. He, G.L. McPherson, G. Piringer, V.T. John, Nanoscale zerovalent iron supported on uniform carbon microspheres for the in situ remediation of chlorinated hydrocarbons, ACS Appl. Mater. Interfaces 2 (10) (2010) 2854–2862.
[41] J.T. Nurmi, P.G. Tratnyek, V. Sarathy, D.R. Baer, J.E. Amonette, K. Pecher, et al., Characterization and properties of metallic iron nanoparticles: spectroscopy, electrochemistry, and kinetics, Environ. Sci. Technol. 39 (5) (2005) 1221–1230.
[42] H.L. Lien, W.X. Zhang, Nanoscale Pd/Fe bimetallic particles: catalytic effects of palladium on hydrodechlorination, Appl. Catal. B-Environ. 77 (1–2) (2007) 110–116.
[43] K.J. Cantrell, D.I. Kaplan, T.J. Gilmore, Injection of colloidal Fe^0 particles in sand with shear-thinning fluids, Anglais 123 (8) (1997) 786–791.
[44] F. He, D. Zhao, J. Liu, C.B. Roberts, Stabilization of Fe–Pd nanoparticles with sodium carboxymethyl cellulose for enhanced transport and dechlorination of trichloroethylene in soil and groundwater, Ind. Eng. Chem. Res. 46 (1) (2007) 29–34.
[45] F. He, D. Zhao, Hydrodechlorination of trichloroethene using stabilized Fe–Pd nanoparticles: reaction mechanism and effects of stabilizers, catalysts, and reaction conditions, Appl. Catal. B Environ. 84 (2008) 533–540.

CHAPTER

Water Decontamination Using Iron and Iron Oxide Nanoparticles

27

Kimberly M. Cross[1], Yunfeng Lu[1], Tonghua Zheng[2], Jingjing Zhan[2], Gary L. McPherson[2] and Vijay T. John[2]

[1]Department of Chemical & Biomolecular Engineering, University of California at Los Angeles, Los Angeles, CA, USA [2]Department of Chemical and Biomolecular Engineering, Tulane University, New Orleans, LA, USA

27.1 Introduction

Nanoscale materials are emerging as an important class of materials that are highly desired for a broad spectrum of applications due to their unique and superior properties. Among this material family, nanoparticles have great promise for use in many areas including catalyst, optical, biological, microelectronic, and environmental applications. This chapter focuses on the use of nanoparticles for water decontamination. Water pollutants, such as waterborne bacteria, toxic metals, and chlorinated organics, are introduced into the environment from both natural

Street, Sustich, Duncan and Savage. Nanotechnology Applications for Clean Water, 2nd Edition.
 DOI: http://dx.doi.org/10.1016/B978-1-4557-3116-9.00027-5

and anthropogenic sources [1]. Nanoparticles are an extremely versatile remediation tool, mainly due to their small size (1–100 nm) and enhanced reactivity. They can be easily incorporated within a slurry, injected under pressure and/or by gravity to the contaminated plume, or effectively transported within the contaminated sites by the flow of groundwater [2]. These unique capabilities provide numerous opportunities for both in situ and ex situ applications, such as the remediation of contaminated soils, sediments, and solid wastes. Alternatively, nanoparticles can be anchored onto solid supports (such as polymers, activated carbon, zeolites, and other porous supports), which help to avoid agglomeration and aggregation and enhance their remediation effects [2]. Compared with more conventional approaches, the use of nanoparticles enables a more rapid, cost-effective clean up process. Complete decontamination of water, however, is a highly sophisticated topic beyond the scope of this chapter. This chapter focuses on the remediation of contaminants in water, such as arsenic, chromium, phosphate, chloro-organics, and waterborne bacteria (e.g., *E. coli*), by adsorption or dechlorination methods using iron and iron oxide nanoparticles synthesized from various methods.

27.2 Synthesis and properties of iron and iron oxide nanoparticles

27.2.1 Iron nanoparticles

Iron is one of the most abundant elements on earth. Elemental iron has been used as an ideal candidate for remediation because it is inexpensive, abundant, easy to prepare and apply to a variety of systems, and devoid of any known toxicity induced by its usage. The concept of using metals, such as iron, as remediation agents is based on reduction–oxidation or "redox" reactions, in which a neutral electron donor (a metal) chemically reduces an electron acceptor (a contaminant). Nanoscale iron particles have surface areas significantly greater than larger-sized powders or granular iron, which leads to enhanced reactivity for the redox process. As a result, iron nanoparticles have been extensively investigated for the decomposition of halogenated hydrocarbons to benign hydrocarbons and the remediation of many other contaminants including anions and heavy metals [3].

The most commonly used method to synthesize nanoscale zero-valent iron nanoparticles is based on reduction of $FeCl_3$ using borohydride [4,5]. Transmission electron microscopy (TEM) micrographs have shown that iron nanoparticles synthesized using the previously mentioned method can range from 1 to 100 nm in size [4,5], specifically a study on perchlorate reduction reported iron particles with an average diameter of 57 ± 16 nm [6]. In most cases excessive borohydride is needed to accelerate the reaction and provide uniform growth of iron crystals [4,5]. Zero-valent iron nanoparticles are highly reactive and react rapidly with surrounding media in the subsurface [7]. A significant loss of reactivity can occur before the particles are able to reach the target contaminant.

In addition, zero-valent iron nanoparticles tend to flocculate when added to water, resulting in a reduction in effective surface area of the metal. Therefore, the effectiveness of a remediation depends on the accessibility of the contaminants to the nanoparticles; and the maximum efficiency of remediation will be achieved only if the metal nanoparticles can effectively migrate without oxidation to the contaminant or the water/contaminant interface. To overcome such difficulties, a commonly used strategy is to incorporate iron nanoparticles within support materials, such as polymers, porous carbon, and polyelectrolytes [8–10].

The presence of a secondary metal on iron nanoparticles leads to the formation of bimetallic nanoparticles with novel catalytic activity [11]. The deposition of the second metal can enhance the reactivity of iron nanoparticles by changing their surface electronic properties [12]. Such bimetallic nanoparticles are often prepared by coating iron particles with palladium or gold through a reduction and deposition process, experimentally completed by soaking freshly prepared iron particles with an ethanol solution containing 1 wt% of palladium acetate [4]. The synthesized nanoparticles are extremely reactive and usually aggregate if their surfaces are left unprotected [13,14]. Common approaches used for stabilization of bimetallic nanoparticles include protection by capping ligands, such as polymers or surfactants, and dispersion in solid supports, such as active carbon, metal oxides, zeolites, or polymer films [15]. However, immobilization of nanoparticles in the solid matrix may increase diffusion resistance [15]. To overcome this problem, it is best to synthesize and immobilize bimetallic nanoparticles in an open matrix. Micro-filtration membranes are of great interest for this purpose because of their open structure and large pore size (100–500 nm) [9]. Membranes functionalized with bimetallic nanoparticles offer great advantages for the catalytic reaction because diffusion limitation can be minimized under convective flow. For example, Fe/Pd bimetallic nanoparticles have been incorporated into a polyacrylic acid (PAA) and polyvinylidene fluoride (PVDF) membrane matrix by ion exchange and subsequent reduction [9]. The role of PAA as a chelating (ion exchange) polymer and their dechlorination behavior has also been investigated [9].

27.2.2 Iron oxide nanoparticles

Compared with zero-valent iron nanoparticles, iron oxide nanoparticles (Fe_3O_4) have super-paramagnetic properties and are capable of being separated from purified water by application of a magnetic field [16]. The representative synthesis method involves a one-pot synthesis from iron precursor, oleic acid, and 1-octadecene at high temperature (e.g., 320°C) [17,18]. The synthesized nanoparticles were then dispersed in water with the assistance of a surfactant, Brij 30, and sonication [17]. After successive washing with water, the Brij 30 was removed using ultra-high-speed centrifugation [17]. Using this method, monodispersed 12 nm iron oxide particles have been synthesized from iron (III) oxide monohydrate, FeO(OH). Similarly, surfactants and/or polymers have been used to stabilize the nanoparticles and prevent aggregation [19]. Enhanced stability can also be achieved by encapsulating iron oxide particles within porous silica support [19]. In this approach, additional functionality can be further provided by the silica surface, allowing for particles to be hydrophobic or hydrophilic, in order to stay sustained in an aqueous or organic phase [19]. Aerosol processing has been used to encapsulate Fe_3O_4 nanoparticles (20–40 nm) in silica microspheres [20]. The aerosol process initially atomizes a solution of iron and a silica precursor into droplets. The droplets are then dried and solidified forming nanoparticles that are collected into a filter [20].

Iron oxide has also been incorporated into membrane supports for various applications. Membrane technology is considered to be an effective alternative to conventional water treatment for the removal of particles, microorganisms, and organic matter [21]. Ceramic membranes combined with ozonation generate very high and stable permeate fluxes without causing membrane damage [22]. In addition they have been shown to achieve complete removal of fecal coliforms and *E. coli*, effectively disinfecting water [23]. Incorporating iron oxide with catalyzed ozonation and membrane filtration will improve inactivation and removal of bacteria from various water sources [17].

Besides the Fe_3O_4 nanoparticles, hydrated iron oxide nanoparticles are another important class of material candidates that are nontoxic, inexpensive, readily available, and chemically stable over a wide pH range. Compared to crystalline forms of iron oxyhydroxide (namely, goethite, feroxyhyte, and lepidocrocite), amorphous hydrated iron oxides have higher specific surface area. Since sorption sites reside primarily on the surface, amorphous hydrated iron oxides (referred to as HFO) offer the highest sorption capacity on a mass basis. Ligand sorption capacity can be greatly increased by dispersing hydrate iron oxide nanoparticles within polymeric anion exchangers. Ion exchange membranes, such as fibrous ion-exchanger (FIBAN) and hybrid anion exchanger (HAIX), have been used for this purpose. FIBAN is generally constructed from polypropylene (PP) industrial fibers radiochemically grafted with styrene-divinylbenzene copolymer [8]. Subsequent absorbance of Fe^{3+} and precipitation led to the supported HFO [8]. Similarly, commercially available strong-base anion exchange resin, IRA-900, was used to prepare HAIX, then Fe(III) ions from ferric chloride were impregnated onto HAIX [24]. HFO particles provide high sorption affinity toward dissolved contaminant species, whereas the fibrous polymer matrix provides durability, mechanical strength, and excellent hydraulic and kinetics characteristics of fixed beds [8]. Higher adsorption results have been shown on hydrated iron oxides in comparison to small nanoscaled dispersed iron oxide particles.

27.3 Removal of pollutants through sorption/dechlorination by iron/iron oxide nanoparticles

Zero-valent iron removes aqueous contaminants by reductive dechlorination, in the case of chlorinated solvents, or by reduction to an insoluble form, in the case of aqueous metal ions [2]. Increasing the surface area of zero-valent iron nanoparticles results in an increased rate of remediation. In general, chlorinated organics ($C_xH_yCl_z$) and iron in aqueous solutions can be expressed by Equation 27.1 [5]:

$$C_xH_yCl_z + zH^+ + zFe^0 \rightarrow C_xH_{y+z} + zFe^{2+} + zCl^- \quad (27.1)$$

Iron undergoes classical electrochemical/corrosion redox reactions in which iron is oxidized from exposure to oxygen and water (Equations 27.2–3) [2] :

$$2Fe^0_{(s)} + O_{2\ (g)} + 2H_2O \rightarrow 2Fe^{2+}{}_{(aq)} + 4OH^-{}_{(aq)} \quad (27.2)$$

$$Fe^0_{(s)} + 2H_2O_{(g)} \rightarrow Fe^{2+}{}_{(aq)} + H_{2(g)} + 2OH^-_{(aq)} \quad (27.3)$$

Fe(II) reacts to give magnetite (Fe_3O_4), ferrous oxide ($Fe(OH)_2$), and ferric hydroxide ($Fe(OH)_3$) depending on redox conditions and pH. For example chromium can be reduced by Fe(II) to the generic reaction scheme shown in Equation 27.4 [25]:

$$Cr^{+6} + 3Fe^{2+} \rightarrow Cr^{3+} + 3Fe^{3+} \quad (27.4)$$

27.3.1 Removal of arsenic in water

Due to its extreme toxicity, the practical and effective removal of arsenic from groundwater remains an important water treatment issue. The presence of arsenic species in drinking water, even in low concentrations, is a severe threat to human health. Effective January 23, 2006, the maximum permissible concentration (MPC) of arsenic in drinking water mediated by the World Health Organization and the US Environmental Protection Agency (US EPA) was decreased from 50 μg/L to 10 μg/L [26].

Arsenic occurs naturally in rocks, soils, water, plants, and animals. In nature, arsenic occurs in the form of arsenate, AsO_3 (As(III)), and arsenite, AsO_5 (As (V)). The distribution of As(III) to As(V) depends on pH and redox potential [27]. It has been reported that As(III) is four–ten times more soluble in water and ten times more toxic than As(V) [28]. Most sorption techniques available for arsenic removal have a greater efficiency for As(V) than for As(III) [17]. Arsenate is retained at low pH values, whereas increases in arsenite adsorption are observed with increasing pH, with a maximum adsorption at pH 9 [17].

The effect of particle size on the adsorption and desorption behaviors of arsenite and arsenate have been previously studied to determine the efficiency of arsenic removal from water. Iron oxide particles, Fe_3O_4, 12 nm in diameter, have been tested and compared with larger sized (20 nm and 300 nm) particles. The result showed that iron oxide particles are capable of achieving residual arsenic concentrations of less than 8 μm/L, removing over 98 percent of arsenite and arsenate in water samples [16]. Afterward, these nanoparticles were retrieved and reused by deactivating the magnetic field [16]. As seen in Fig. 27.1A, as the

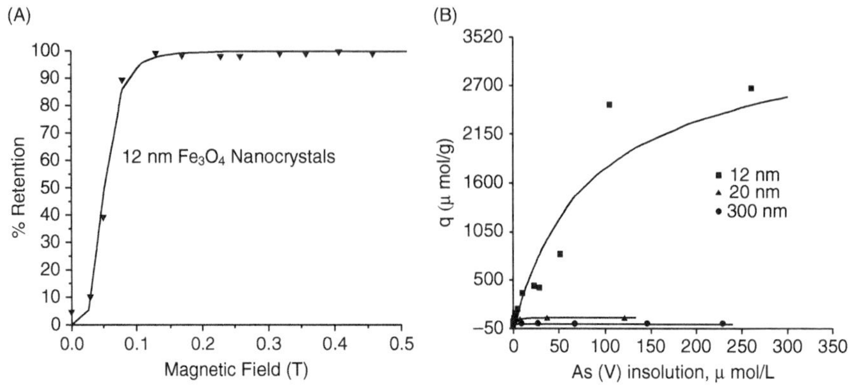

FIGURE 27.1

(A) The magnetic field dependence of 12 nm iron oxide on particle retention. (B) Arsenite adsorption on different iron oxide nanoparticles (12 nm, 20 nm, 300 nm). The solid lines are drawn using Langmuir isotherm equations.

The figures are adapted from [16].

Table 27.1 The Effect of Fe_3O_4 Size on Arsenic Removal Efficiency [16]

Particle Size (nm)	Arsenic Oxidation State	Initial As Concentration (μg/L)	Residual As Concentration (μg/L)	Removal (%)
12	As(III)	500	3.9	99.2
12	As(V)	500	7.8	98.4
20	As(III)	500	46.3	90.9
20	As(V)	500	17.2	96.5
300	As(III)	500	375	24.9
300	As(V)	500	356.4	24.2

magnetic field is increased, there is an increased quantity of nanocrystals retained [16]. From the data shown in Table 27.1, the removal strongly depends on the size of Fe_3O_4 particles [16]. Figure 27.1B shows an increase in the weight-based arsenite adsorption density with decreasing particle size, due to less aggregation and more sites being exposed to arsenic adsorption [16]. Thus, it has been shown that arsenic can be removed from groundwater through the optimization of particle size and the usage of magnetic separation systems at low field strength.

Adsorption of arsenic on hydrated iron oxide prepared via the electrochemical peroxide (ECP) process is extremely high, rapid, and independent of the initial pH of the contaminated water. The removal rate of arsenic has been reported to be as high as 99 percent [29]. Reports have been made that after only 3 minutes of generating an electric direct current in the contaminated water system,

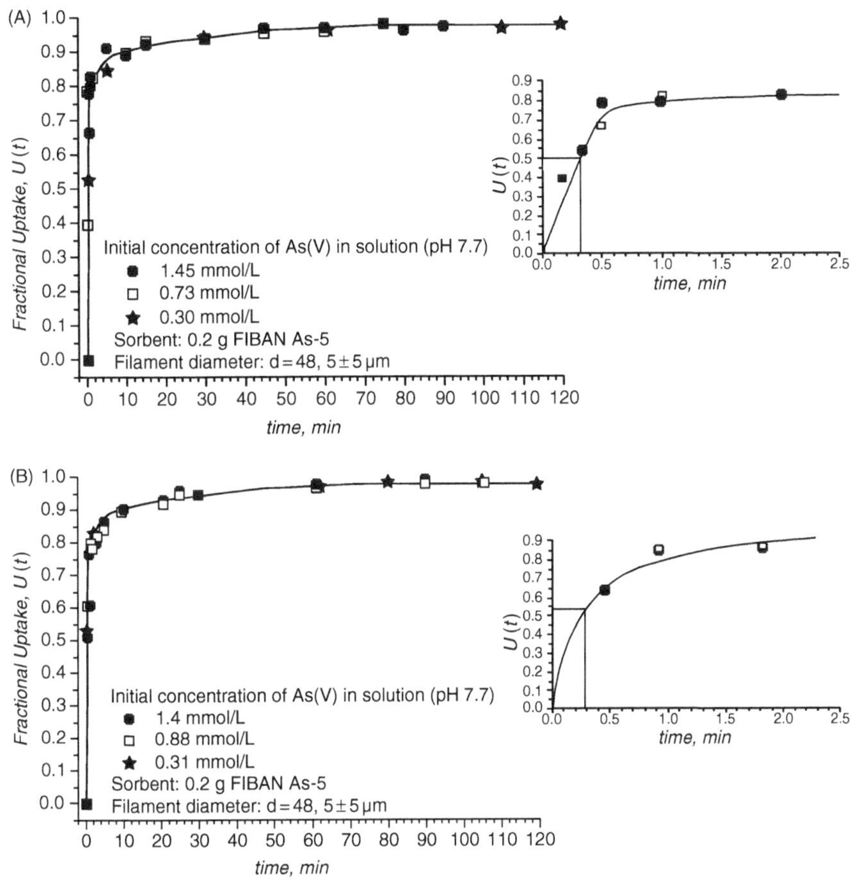

FIGURE 27.2

Kinetics curves of (A) arsenate and (B) arsenite sorption onto fibrous ion exchangers (FIBAN)—with hydrated iron oxide particles.

Adapted from [8].

approximately 4150 μg of arsenic out of an initial mean concentration of 4300 μg arsenate is retained by HFO particles [29]. After exposure to HFO, very small traces of the element, approximately 10.0 μl^{-1}, remain in solution [29]. The presence of hydrogen peroxide in the ECP process induces the adsorption of arsenite to lower pH values of 3.5–6.0. Thus, the ECP process provides promising results for the removal of arsenite at lower pH values [29].

A similar study with high selectivity toward arsenic was also completed using polymeric fibrous support containing hydrated iron oxides. Within a pH range of 6.5–8.5, arsenate and arsenite were both completely desorbed from contaminated water using crystalline and amorphous HFO nanoparticles adsorbed on FIBAN [30]. Fig. 27.2A and B shows that arsenate and arsenite sorption processes were completed within 1 hour, with more than 90 percent sorption achieved after 10 minutes [8]. Dixit and Hering found a maximum sorption density for As(III) on HFO of 0.31 mol As/mol Fe, which is 2.2 times the amount adsorbed per mole of iron on 12 nm particles [31]. In the pH range typical for drinking waters, FIBAN with HFO nanoparticles shows a high capacity for As(V) as well as As(III).

27.3.2 Removal of chromium in water

Contamination of both soil and groundwater by hexavalent chromium has been one of the most challenging environmental issues [32]. Processes with zero-valent iron have been effectively shown to reduce Cr(VI) to Cr(III), which is insoluble in water. Trivalent chromium is much easier to clean, since chromium is easier to filter in this state. Carboxymethyl cellulose (CMC) stabilized zero-valent nanoparticles are able to reduce and remove Cr(VI) in water. Figure 27.3A shows an approximate 53 percent reduction of Cr(VI) at equilibrium (36 hours after the reaction) with a 0.08 g/L Fe dose [25]. When the iron dose was increased from 0.04 g/L to 0.12 g/L the reduction of Cr(VI) in water increased from 24 to 90 percent, which is seen in Fig. 27.3B [25]. This study provided substantial evidence that stabilized zero-valent iron nanoparticles can lead to innovative remediation technology that is more cost effective and environment friendly.

27.3.3 Removal of phosphates in water

Phosphates are a growth-nutrient for microorganisms in water [33]. As a result of increased phosphorus concentration, an excessive growth of photosynthetic aquatic micro- and macro-organisms occurs and ultimately becomes a major cause of eutrophication, or extensive algae growth [33]. All parameters being equal, HAIX containing hydrated iron oxide nanoparticles was compared with a granular ferric hydroxide (GFH) without any ion exchange material. As seen in Fig. 27.4, HAIX provided significantly greater phosphate removal capacity [24]. Phosphate breakthrough with HAIX occurred after nearly four thousand bed volumes, while the commercially available GFH column from the US Filter Co. showed breakthrough after one thousand bed volumes [24]. HAIX is able to provide high phosphate selectivity in the presence of other common anions such as chlorides and sulfates, and is able to be used for multiple cycles of phosphate removal without the presence of physical deterioration. Phosphates can effectively be removed using HAIX, which provides a durable, cost-effective, and reusable remediation process.

In addition to using hydrate iron oxide as a source to remove phosphorous, laboratory studies have also been conducted using allophone nanoclay. Allophane nanoclay is a natural, inexpensive, and environment-friendly material, suitable for removing phosphorous from wastewater [34]. Clays have long been used as a flocculate and sorbent of suspended particles, disease-bearing organisms, and toxic compounds in water [35]. The unit particles of allophane consist of hallow

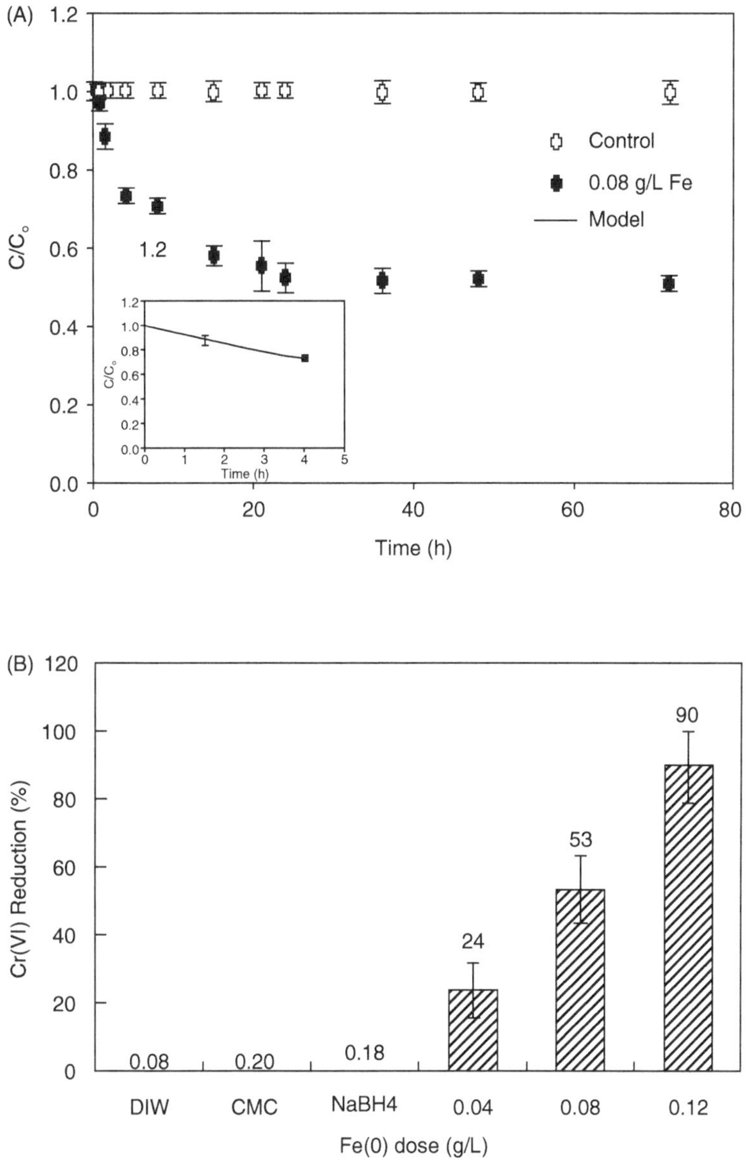

FIGURE 27.3

(A) Reduction of Cr(VI) in water by carboxymethyl cellulose (CMC) stabilized Fe nanoparticles, 0.8 g/L metal loading. (B) Equilibrium Cr(VI) reduction in the presence of CMC, borohydride, or various concentrations of CMC-stabilized Fe nanoparticles.

Adapted from [25].

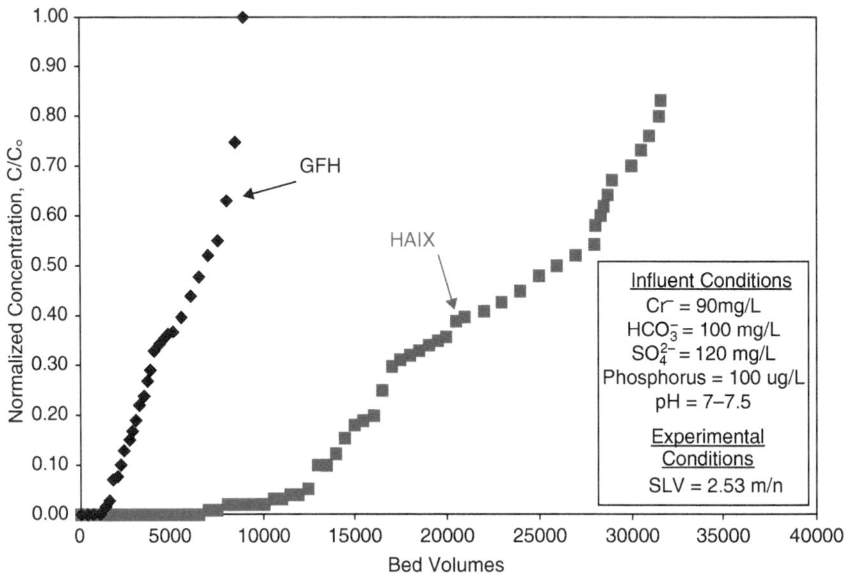

FIGURE 27.4

Comparison of effluent histories of two separate fixed-bed columns granular ferric hydroxide (GFH) obtained from US Filer Co. and the hybrid anion exchanger (HAIX) using an influent of identical composition.

Adapted from [24].

aluminosilicate nanoballs with an outer diameter of 3.5–5.0 nm, and hollow aluminosilicate tubules 2 nm in diameter [34]. Allophane is able to adsorb phosphate over a wide range of solution concentrations, and can be recovered from wastewater after use. Figure 27.5 shows the concentration of phosphorous in a flask after contacting the solution with allophane nanoclay for different periods of time. After 24 hours of contact, the phosphorous concentration decreased from 0.96 to 0.55 mg/L [34]. The phosphorous concentration continued to decrease as contact time increased [34]. Due to these features, the use of clay should be further investigated for water treatment purposes.

27.3.4 Removal of chloro-organics in water

Chloro-organics, including chlorinated methanes, ethanes, benzenes, and polychlorinated biphenyls, are a major class of contaminates and several nanomaterials have been used to aid in their remediation [36]. Trichloroethylene (TCE) and chloroform ($CHCl_3$) are toxic and carcinogenic containments found in water. A study was conducted in Research Triangle Park, NC, in which 1,600 gallons of 1.9 g/L iron nanoparticles were injected into a TCE plume (average concentration of about 14 mg/L) over a period of two days. The pollutants were

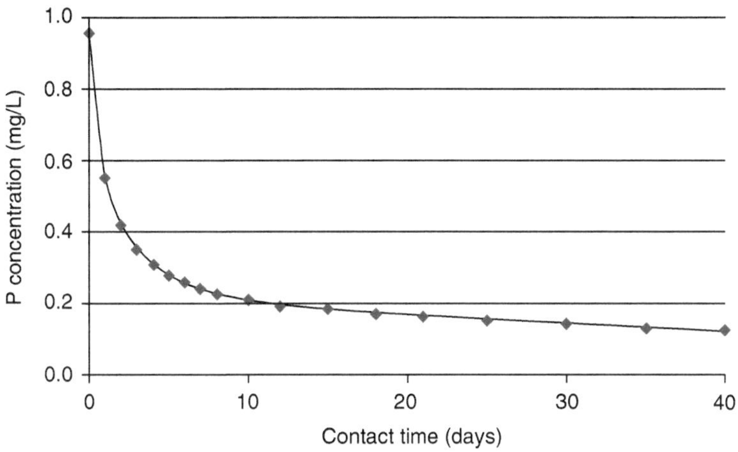

FIGURE 27.5

Concentration of phosphate (mg/L) in solution after adsorption by allophane nanoclay.

Adapted from [34].

completely removed within 20 days near the injection well and within 50 days 7.5 m downstream [37]. Similarly, a field demonstration was performed at a manufacturing site in Trenton, NJ, in which nanoscale palladium-coated iron particles were gravity-fed into groundwater contaminated by TCE and other chlorinated aliphatic hydrocarbons [38]. Approximately 1.7 kg of nanoparticles were introduced into the test area over a 2-day period, resulting in TCE reduction efficiencies up to 96 percent over a 4-week monitoring period [38].

Iron nanoparticles have been encapsulated with silica in order to increase stability and prevent aggregation. Nanocomposite particles synthesized from an aerosol process are shown in Fig. 27.6A, where the dark spots are zero-valent iron nanoparticles [20]. Examining the ratio of remaining TCE to original TCE content, Fig. 27.6B shows that the synthesized particles are effective in dehalogenation of TCE [20].

Currently the US EPA sets maximum contaminant levels (MCL) of 5 ppb for TCE and 70 ppb for aqueous chloroform ($CHCl_3$) [10]. Activated carbon fibers (ACFs) coated with 30–70 percent phenolic resin, which allows for a tailed porous structure, are superior in performance in comparison to granular activated carbon (GAC) [10]. Increased nanopore volumes (approximately 1.2 nm) and unsaturated surface chemistry interact favorably with chlorinated hydrocarbons [10]. Fig. 27.7A and B shows that tailored ACF filters resulted in undetectable (below 1 ppb) TCE and $CHCl_3$ concentrations [10]. In addition to using iron and iron oxide, chloro-organics have been effectively removed using carbon nanoporous fibers.

Polychlorinated biphenyls (PCBs), which were previously extensively used in coolants and lubricants in various electrical equipment, are a family of manmade

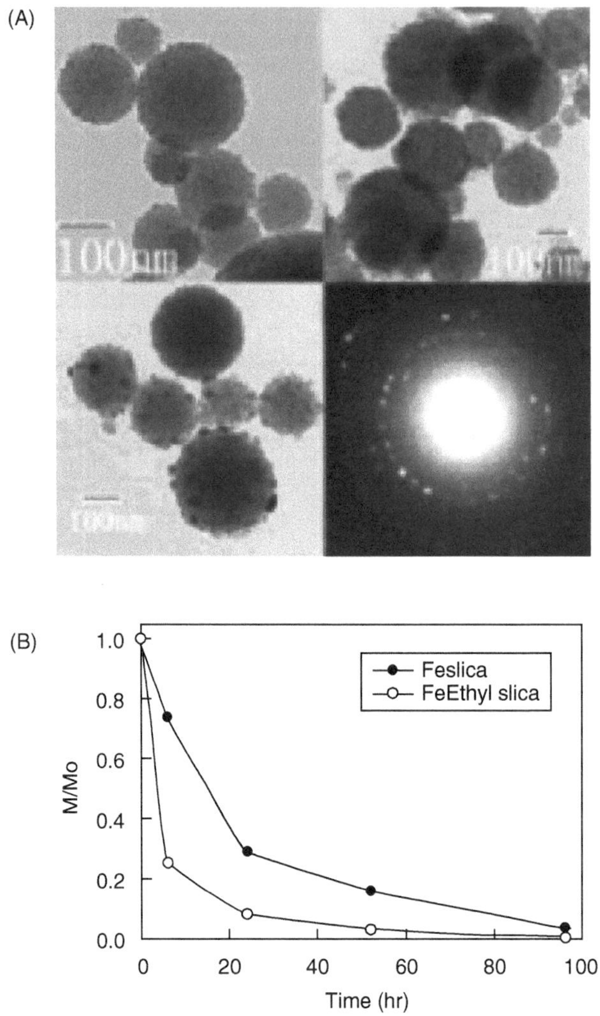

FIGURE 27.6

(A) Aerosolized silica particles containing iron nanoparticles [20]. (B) Characteristics of trichloroethylene (TCE) destruction using the silica–iron oxide composite nanoparticles.

Adapted from [20].

chemicals that contain 209 individual compounds with various toxicity levels [39,40]. The reductive hydro-dechlorination of 2,2′-dichlorbiphenyl (DiCB) has been completed with a high degradation rate using Pd/Fe supported by polymerized membranes, as shown in Fig. 27.8. More than 90 percent dechlorination of DiCB was achieved in 2 hours with a 0.8 g/L metal loading [9]. To examine the effect of particle size on reactivity and reaction pathway shift, dechlorination

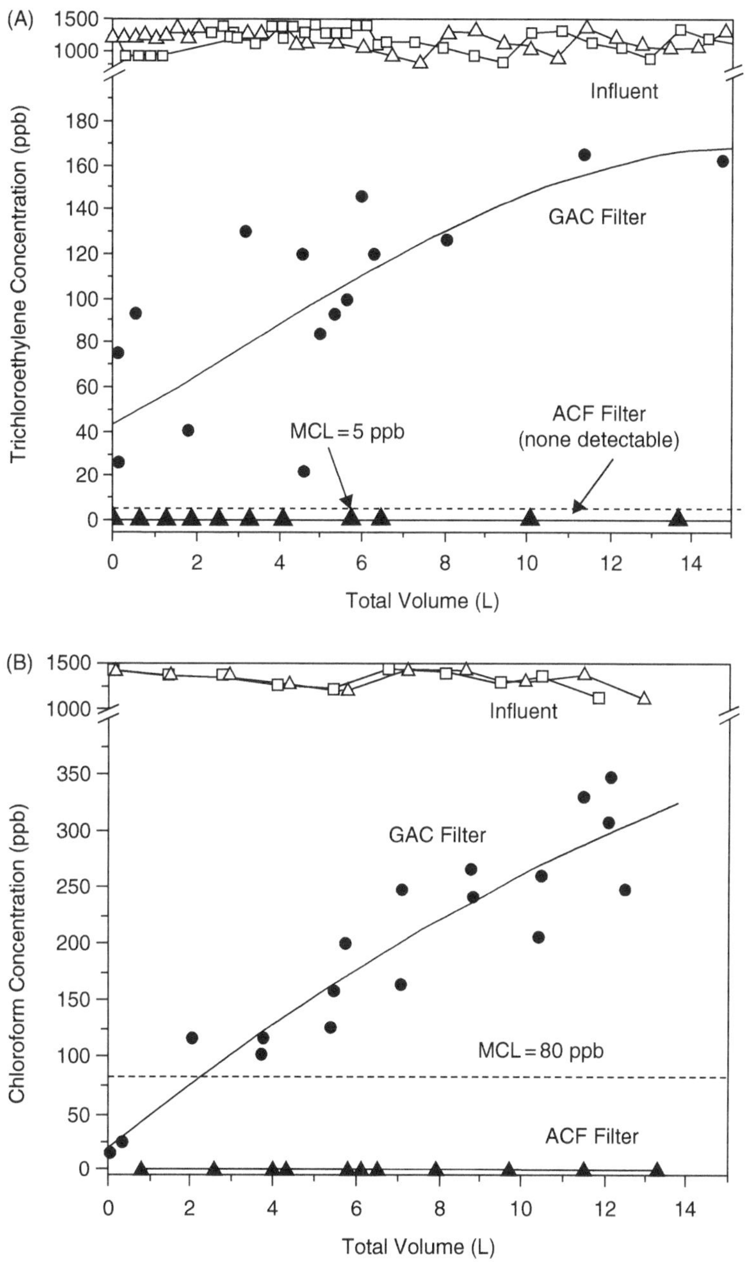

FIGURE 27.7

Breakthrough curves for (A) trichloroethylene (TCE) and (B) chloroform ($CHCl_3$) from both activated carbon filters (ACF) and granulated activated carbon (GAC) filters.

Adapted from [10].

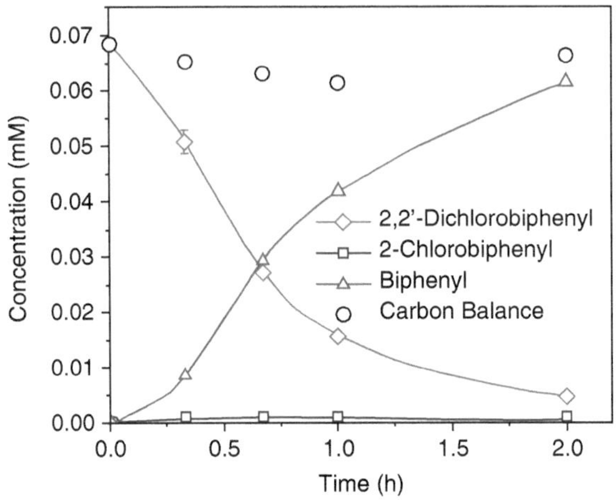

FIGURE 27.8

Batch reaction of 2,2'-dichlorobiphenyl with Fe/Pd in polyacrylic acid (PAA)/ polyvinylidene fluoride (PVDF) membrane at room temperature with 0.8 g/L metal loading.

Adapted from [9].

of DiCB with bulk Fe particles (approximately 120 μm) coated with Pd resulted in only a 10 percent dechlorination of DiCB, with a high metal loading of 87.5 g/L [9]. The surface-area-normalized rate constant k_{SA} calculated for the degradation of DiCB was only 0.00011 L/h m^2, which is over six hundred times lower than that obtained for the membrane supported Fe/Pd nanoparticles [9]. Enhanced reactivity is due to the increased number of catalytic reaction sites on the various facets, edges, corners and defects on the nanoparticles.

27.3.5 Removal of *E. coli* in Water

Iron oxide has been shown to retard the proliferation of bacteria. The incorporation of iron oxide catalyzed ozonation technology increases the retention of bacteria to the surface of membranes, resulting in improved remediation of water. Iron oxide catalyzed ozonation and membrane filtration will combine to improve inactivation and/or removal of bacteria [41]. The mortality of *E. coli* in the product water after treatment using the ozonation–membrane filtration process with iron coated nanoparticles 4–6 nm in diameter was 99 percent [42]. In Fig. 27.9A, a comparison of the mortality rate of *E. coli* is made between various membrane treatment processes [42]. The improved disinfection observed using the combined process is due to the catalytic decomposition of ozone at the iron oxide surface, which resulted in the formation of ^{-}OH or other radical species that inactivate the bacteria near the surface. In addition, assembled organic carbon (AOC) concentrations were reduced using the combined process (as shown in Fig. 27.9B), indicating that there is a reduced potential for regrowth in water treatment using the

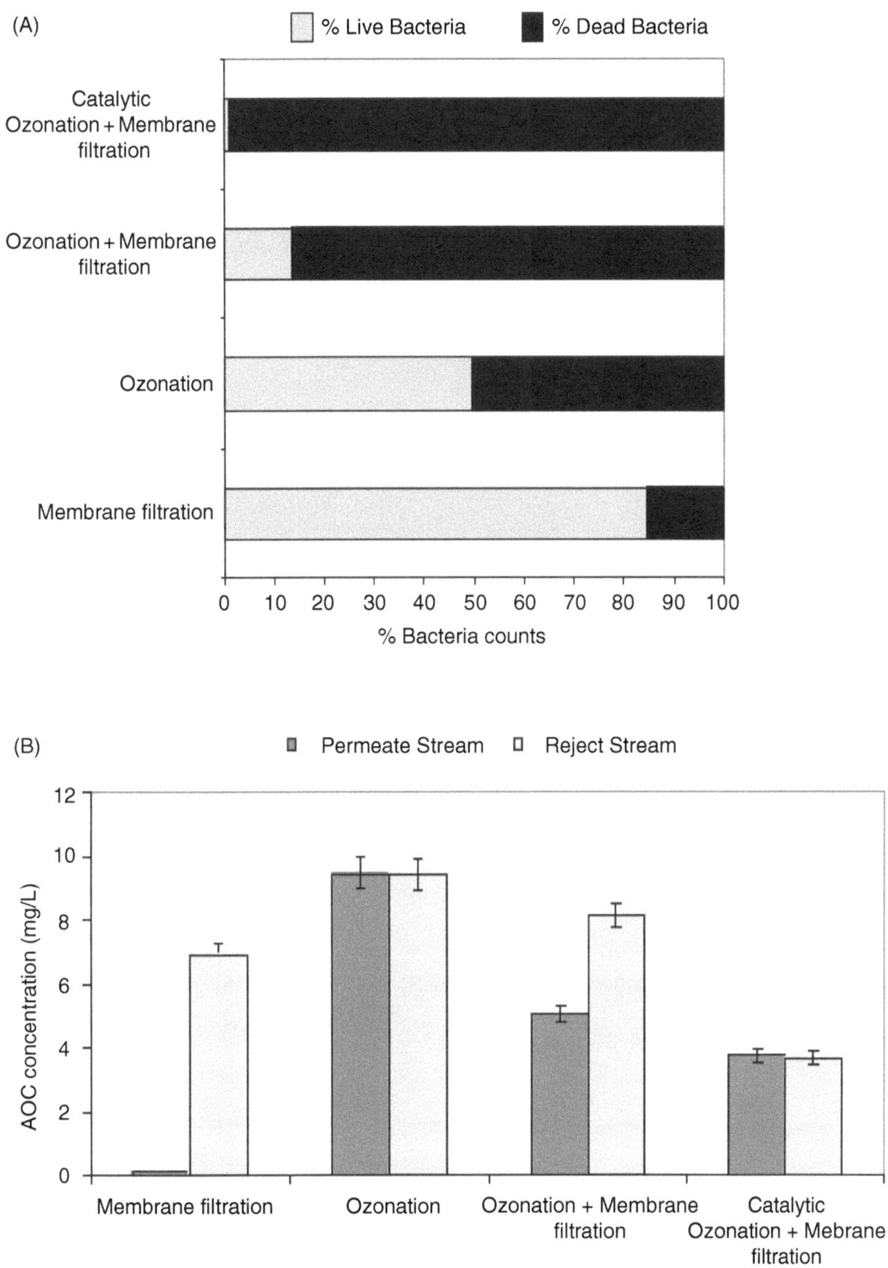

FIGURE 27.9

(A) Percent of live and dead bacteria in the permeate after different treatments.
(B) Assimilate organic carbon (AOC) concentration after different treatments for the permeate and reject streams.

Adapted from [42].

coated membranes [42]. Due to the inactivation of *E. coli* and the lowering of AOC concentrations, using iron oxide catalyzed ozonation technology is likely to be very effective to disinfect and control bacterial regrowth in water.

27.4 Conclusions

Nanotechnology has the potential to contribute to improved, more cost effective removal of drinking water contaminants, such as metals, toxic halogenated organic chemicals, suspended particulate matter, and pathogenic microorganisms. Current challenges include further research to determine methods that increase the stability of nanoparticles used in remediation, and the need to develop manufacturing techniques for mass production of these materials. Applications involving dispersive uses of nanomaterials in water have the potential for wide exposure to aquatic life and humans. More research is needed regarding the impact of nanoparticles and their fate once they have been transported into the environment.

References

[1] N. Savage, M.S. Diallo, J. Nanopart. Res. 7 (2005) 331–342.
[2] W. Zhang, J. Nanopart. Res. 5 (2003) 323–332.
[3] M. Cowell, T. Kibbey, J. Zimmerman, K. Hayes, Environ. Sci. Technol. 34 (8) (2000) 1583–1588.
[4] C. Wang, W. Zhang, Environ. Sci. Technol. 31 (7) (1997) 2154–2156.
[5] W. Zhang, C. Wang, H. Lien, Catal. Today 40 (4) (1998) 387–395.
[6] J. Cao, D.W. Elliot, W.-X. Zhang, J. Nanopart. Res. 7 (2005) 499–506.
[7] M.I. Pierce, C.B. Moore, Water Res. 15 (1982) 1247.
[8] O.M. Vatutsina, V.S. Soldatov, V.I. Sokolova, J. Johann, M. Bissen, A. Weissenbacher, React. Funct. Polym. 67 (2007) 184–201.
[9] J. Xu, D. Bhattacharyya, Ind. Eng. Chem. Res. 46 (2007) 2348–2359.
[10] Z. Yue, J. Economy, J. Nanopart. Res. 7 (2005) 477–487.
[11] S. Mandal, P.R. Selvakannan, R. Pasricha, M. Sastry, J. Am. Chem. Soc. 125 (2003) 8440–8441.
[12] B. Zhou, S. Hermans, G.A. Somorjai, Nanotechnology in Catalysis, Kluwer Academic/Plenum, New York, NY, 2003.
[13] R.G. Freeman, K.C. Grabar, K.J. Allison, R.M. Bright, J.A. Davis, A.P. Guthrie, et al., Science 267 (1995) 1629–1631.
[14] P. Raveendran, J. Fu, S.L. Wallen, J. Am. Chem. Soc. 125 (2003) 13940–13941.
[15] S. Kiadambi, M.L. Bruening, Chem. Mater. 17 (2005) 301–307.
[16] J.T. Mayo, C.T. Yavuz, S. Yean, L. Cong, H. Shipley, W.W. Yu, et al., 8, pp. 71–75, 2007.
[17] S. Yean, L. Cong, C.T. Yavuz, J.T. Mayo, W.W. Yu, A.T. Kan, et al., J. Mater. Res. 20 (12) (2005) 3255–3264.
[18] W.W. Yu, J. Falkner, B.S. Shih, V.L. Colvin, Chem. Mater. 16 (2004) 3318.

[19] T. Zheng, J. Pang, G. Tan, J. He, G.L. McPherson, Y. Lu, et al., Langmuir 23 (9) (2007) 5143–5147.

[20] Y. Lu, V.T. John, Progress Report: Decontamination of Chlorinated Compounds, EPA, National Center for Environmental Research, 2006.

[21] C.T. Cleveland, J. Am. Wat. Works Assoc. 91 (6) (1999) 10.

[22] B.S. Karnik, S.H. Davies, K.C. Chen, D.R. Jaglowski, M.J. Baumann, S.J. Masten, Water Res. 39 (4) (2005) 728–734.

[23] S. Benfer, B. Arki, G. Tomandl, Adv. Eng. Mater. 6 (7) (2004) 495–500.

[24] L.M. Blaney, S. Cinar, A.K. Sen Gupta, Water Res. 41 (2007) 1603–1613.

[25] Y. Xu, D. Zhao, Water Res. 41 (2007) 2101–2108.

[26] US EPA, Arsenic and Clarifications to Compliance and New Source Contaminants Monitoring Final Rule, Federal Register (66 FR 6976), January 22 2001.

[27] A. Vaseashta, M. Vaclavikova, S. Vaseashta, G. Gallios, P. Roy, O. Pummakarnchana., Sci. Technol. Adv. Mater. 8 (2007) 47–59.

[28] US EPA, EPA-542-S-02-002, Cincinnati, OH, Report No.: EPA Report-816-D-02-005, 2002.

[29] M. Arienzo, J. Chiarenzelli, R. Scrudato, J. Hazard. Mater. B87 (2001) 187–198.

[30] M.J. DeMarco, A.K. Sen Gupta, J.E. Greenleaf, Water Res. 37 (2003) 164.

[31] S. Dixit, J.G. Hering, Environ. Sci. Technol. 37 (2003) 4182.

[32] S. Guha, P. Bhargava, Water Environ. Res. 77 (2005) 411–416.

[33] R.H. Foy and P.J.A. Withers, The contribution of agricultural phosphorous to eutrophication. Proceedings of the Fertilizer Society No, 365, The Fertilizer Society, London, 1995. in Fertilizer Society Proceedings, 1995.

[34] G. Yuan, L. Wu, Sci. Technol. Adv. Mater. 8 (2007) 60–62.

[35] G.J. Churchman, W.P. Gates, B.K.G. Theng, G. Yuan, Clays and clay minerals for pollution control, in: F. Bergaya, B.K.G. Theng, G. Lagaly (Eds.), Handbook of Clay Science, Amsterdam, Elsvier Ltd., 2006, pp. 625–675.

[36] Y. Liu, H. Choi, D. Dionysiou, G.V. Lowry, Chem. Mater. 17 (2005) 5315–5322.

[37] R. Glazier, R. Venkatakrishnan, F. Gheorghiu, L. Wlata, R. Nash, W. Zhang, Civ. Eng. 73 (5) (2003) 64–69.

[38] D.W. Elliott, W.-X. Zhang, Environ. Sci. Technol. 35 (2001) 4922–4926.

[39] US EPA, National Air Toxic Information Report: Qualitative and quantitative carcinogenic risk assessment, in STAPPA/ALAPCO U-Ea, ed., pp. 11-1–11-6, 1987.

[40] M. Shikiya, W. Barcikowski, and M.I. Kahn, Analysis of ambient data from potential toxics ‘hotpoints’ in South Coast Air Basin, in District SCAQM, ed., South Coast Air Quality Management District, 1998.

[41] A. Bottino, C. Capannelli, A.D. Borghi, M. Colombino, O. Conio, Desalination 141 (2001) 75–79.

[42] B.S. Karnik, S.H. Davies, M.J. Baumann, S.J. Masten, Ozone: Sci. Eng. J. Int. Ozone Assoc. 29 (2007) 75–84.

CHAPTER

28 Nanotechnology for Contaminated Subsurface Remediation: Possibilities and Challenges

Denis M. O'Carroll

Department of Civil and Environmental Engineering, The University of Western Ontario, London, Ontario, Canada

28.1 Introduction

Groundwater is an important source of water throughout the world and as such, its protection is paramount. It is estimated that groundwater sources provide between 95 and 100 percent of potable water supplies in some developing countries (e.g., Oman and Cuba) whereas most regions of the world obtain 30–50 percent of their potable water supplies from groundwater sources [1,2]. Specifically, 70–90 percent of drinking water supplies come from groundwater sources in Central America and 70 percent from groundwater sources in China [2,3]. In arid regions such as Northern China where groundwater sources represents 87 percent of the total water consumed in 27 cities, this resource is vital [4]. Even in Canada, a country with abundant surface water resources, 30 percent of the population rely on groundwater for domestic use [5]. In the United States approximately one-quarter of consumed fresh water comes from groundwater sources [6]. Despite its importance throughout the developed and developing world,

Street, Sustich, Duncan and Savage. Nanotechnology Applications for Clean Water, 2nd Edition.
 DOI: http://dx.doi.org/10.1016/B978-1-4557-3116-9.00028-7

contamination of drinking water resources by natural and anthropogenic substances is common. For example it is estimated that 90 percent of groundwater in Chinese cities is contaminated to some extent and 61 percent of tested wells in Nicaragua were contaminated with pesticides above United States Environmental Protection Agency (US EPA) drinking water standards [2,3]. This contamination causes serious health and environmental problems. Furthermore, brownfield sites, abandoned properties with soil and groundwater contamination, severely limit redevelopment and reinvigoration of many urban cores. Contamination of water resources, therefore, has significant health, ecological, and economic societal impacts.

28.2 Sources of groundwater contamination and remediation costs

From the industrial revolution until the 1970s, vast amounts of hazardous waste and spent chemicals were disposed directly to the subsurface with the belief that the subsurface acted as a natural filter of disposed chemicals. In the United States it was not until an abnormally high number of birth defects and miscarriages were found in Love Canal, New York, that the public became aware of the impacts of improper contaminant disposal practices. Many of the chemicals historically disposed of to the subsurface are particularly persistent in the subsurface, frequently contaminating water sources for decades or centuries. As an example of the magnitude of the problem, Canada is home to an estimated thirty thousand brownfield sites, many of which exhibit contamination of soil and groundwater by hazardous industrial chemicals. The restoration of these brownfield sites has been identified as critical to the health and overall sustainability of Canada, by providing significant economic, environmental, and social benefits [7]. It is estimated that for every dollar spent on restoration and redevelopment, this investment yields a fourfold return in terms of economic benefit to the Canadian economy (direct and indirect), providing up to seven billion dollars annually in expected net national public benefit [8]. To put this benefit in context, it represents approximately 0.6 percent of Canada's gross domestic product [9]. A 1997 US National Research Council report suggests that there are between three hundred thousand and four hundred thousand contaminated sites in the United States requiring cleanup [10]. Furthermore it is estimated that there are over thirty thousand dense nonaqueous phase liquid (DNAPL) contaminated sites in the United States including twenty thousand sites at existing or former dry cleaner installations [11]. These DNAPL sites represent a subset of the larger number of brownfield sites in the United States and are often considered the most persistent and problematic sites. In 1996 approximately $9 billion were spent on environmental remediation [10]. To restore brownfield sites the US EPA Superfund Program in 2006 alone obligated $540 million for the construction and active remediation of contaminated sites where no other responsible parties could pay for remedial activities. Of this amount 45 percent went to only 14 sites, or an average of $17 million per year for each of these 14 sites. If remedial activities from the private sector were also included in this figure the total spent on remediation in the United States would increase considerably, as indicated by the 1996 figure.

Anthropogenic groundwater contaminants were often generated in various industrial processes prior to their disposal to the subsurface. For example chemical solvents were commonly used to clean cutting tools, entraining small heavy metal fragments from the cutting process and generating significant quantities of waste liquids. A survey of 91 US Department of Energy (DOE) wastes sites found that 20 percent of their sites were contaminated with this complex waste mixture [12]. Other industrial and commercial uses of common groundwater contaminants include the use of polychlorinated biphenyls (PCB) in electrical transformers, heavy metals as paint additives or in metal plating and smelting operations, pesticides for agriculture, and chlorinated solvents in dry cleaning installations [10]. All of these operations have caused subsurface contamination due to their improper disposal of hazardous chemicals. Common disposal practices included accidental releases (e.g., due to leaky underground storage tanks or compromised landfill liners) and intentional releases (e.g., underground storage systems designed to slowly leach liquids into the subsurface, subsurface injection wells, or land application of contaminants) [10,11,13]. These contaminants can cause serious health problems and even death. For example, PCBs, and the chlorinated solvent and DNAPL, trichloroethylene (TCE), are carcinogens [14,15]; pesticides have been found to cause birth defects [16], and lead can impair organ development in fetuses [17]. Unfortunately groundwater contamination of drinking water sources is widespread in the developing and developed world due to these poor disposal practices, causing significant health problems. As a result the development of innovative remediation technologies is an active area of research.

28.3 Remediation alternatives

Substantial advances in our understanding of the phenomena governing subsurface remediation have been made and a number of innovative remediation technologies have been developed, such as steam and brine flooding or density modified displacement [18–20]. A 2005 US National Research Council report grouped remediation technologies of DNAPL and chemical explosive contaminated sites into two classifications, extraction or transformation [11]. Many of these remediation alternatives could be used for a variety of other classes of subsurface contaminants. Extraction refers to the removal of the contaminant from the subsurface for subsequent aboveground treatment and disposal. Transformation refers to the conversion of hazardous subsurface contaminants to more benign daughter products or forms. Examples of extraction technologies include excavation and pump and treat, and examples of transformation technologies include chemical oxidation and enhanced bioremediation. This report concluded that, for the reviewed technologies, contaminant mass removal did not significantly reduce toxicity or mass flux from the contaminant source zone. Furthermore, the report found that no large-scale DNAPL contaminated site has been cleaned up such that the aquifer water meets drinking water standards. This report did not include the potential for nanotechnology-based remediation alternatives in its assessment. Based on this report and other similar studies, existing technologies for the remediation of more persistent contaminants (i.e., chlorinated solvents and heavy metals) are rarely able to clean up contaminated sites such that the water source meets drinking water objectives at the completion of remedial activities. The problem relates to the inability of existing remedial technologies to remove enough contaminant mass in the subsurface to significantly reduce dissolved aqueous phase concentrations. The decision to remediate contaminated sites is, therefore, still a source of considerable debate despite over two decades of active research and development [11,21,22]. The development of new and innovative remediation technologies is, therefore, crucial to achieve clean up goals at contaminated sites.

28.4 Contaminated site remediation via reactive nanomaterials

Nanoparticles are receiving widespread interest in a variety of fields due to their unique, beneficial chemical, physical, and mechanical properties. They have recently been proposed to address a variety of environmental problems including the treatment of surface water, groundwater, and industrial wastewaters containing a range of organic, inorganic, and microbial contaminants [23]. With regards to contaminated site remediation they have tremendous potential to remediate a wide variety of common subsurface contaminants. For example, carbon nanotubes have been found to have a heavy metal adsorption capacity five times greater than that of granular activated carbon [24]. Their ability to adsorb organic contaminants such as PAHs [25] and dichlorobenzene [26], both US EPA priority pollutants, has also been evaluated. Reactive nanometals, such as nanoscale zero valent iron (nZVI), are also the focus of significant research for the remediation of environmental contaminants (e.g., [27–35]). A number of studies have found that reaction rates are much faster for nanometer scale metals when compared to larger, micrometer or millimeter scale metals and that the nanometer scale metals can degrade a wider range of contaminants (e.g., [28,34,36]). Furthermore the incorporation of a second noble metal catalyst in the zero valent nanometal has been shown to significantly improve degradation rates beyond that of a single zero valent nanometal [36,37]. Contaminants that can be remediated using nanometals include PCBs, chlorinated ethenes (e.g., TCE), chlorinated ethanes [e.g., hexachloroethane (HCA), pentachloroethane (PCA)], and heavy metals (e.g., AS^{III}) [28,33,34,38,39]. However, there are still a number of common contaminants found at sites requiring remediation that, thus far, cannot be degraded using nanoparticles. For example, 1,1-dichloroethane and 1,2-dichloroethane, both EPA priority pollutants, to date have not been dechlorinated using reactive nanometals [39].

Much of the work investigating the applicability of nanoparticles for remediation has been completed in small bench scale experiments under ideal conditions (e.g., using DI water and dissolved phase single component contaminants) and low contaminant concentrations. At many sites, however, NAPLs were not disposed of as pure liquids, but in acidic or basic mixtures sometimes containing surface active compounds [12,13,40]. A review of Sloat [13] provides a good perspective of common disposal practices in the 1960s. Sloat [13] discusses, in great detail, the composition of waste liquids (e.g., fabrication oil comprised of 75 percent carbon tetrachloride and 25 percent lard oil) that were improperly disposed of to the subsurface at the Hanford Low Level Waste Management site. Another complicating factor that needs to be investigated regarding contaminated site remediation using nanoparticles is the presence of chemical heterogeneities in the subsurface (e.g., natural anions and cations dissolved in water as well as soil surface constituents). A limited amount of studies have been published related to the field application of reactive nanometals and their associated complexities for chlorinated hydrocarbon remediation (e.g., [30,41,42]). Each study reported significant chlorinated hydrocarbon reductions; however, significant questions remained at the completion of these field trials including the mobility of the nanoparticles in the field, the premature passivation of the reactive nanometal and the potential appearance of more toxic degradation products. The complexities associated with nanometal remediation under field conditions have also been the focus of controlled laboratory experiments [35,43]. For example Lien et al. [35] investigated the ability of nZVI to remediate a mixture of carbon tetrachloride and heavy metals. As discussed earlier, mixtures of chlorinated hydrocarbons and heavy metals are common at many DOE contaminated facilities [12]. Lien et al. [12] found that different heavy metals affected the dechlorination rate of carbon tetrachloride as well as the degradation products. In another study Liu et al. [43] found that aqueous phase anions commonly found in the field, such as nitrate, decrease TCE dechlorination rates due to surface passivation of nZVI. In their study Liu et al. [43] also investigated nZVI promoted dechlorination rates as TCE concentrations approached their solubility limit. TCE concentrations would be expected to approach the solubility limit near the contaminated source zone. They found that dechlorination rates were slightly reduced as TCE approached the solubility limit. Significant work is needed beyond these studies to fully understand remediation via reactive nanometals under the complex conditions found at the field scale to fully take advantage of the tremendous opportunities nanotechnology offers.

28.5 Example of contaminated site remediation via reactive nanometals

An understanding of the subsurface contaminant architecture is also important for the design of efficient remediation alternatives. One significant challenge faced by those developing site remediation plans is actually locating the contaminated source zone, if one exists. As an example, consider the migration of a chlorinated solvent, a DNAPL, from a leaking underground storage tank (Fig. 28.1). Chlorinated solvents are the most common groundwater contaminants at US hazardous waste sites [44]. The migration and ultimate entrapment of NAPLs are governed by gravitational, viscous, and capillary forces [45]. Chlorinated solvents are denser than water and thus migrate vertically downward through the unsaturated and saturated zones following their release to subsurface environments and only coming to rest once they have reached an impervious layer (e.g., clay or bedrock), as illustrated in Fig. 28.1A. However, a fraction of the disposed DNAPL may reside in high saturation pools on lenses of impervious soils above the impervious layer or may be retained in residual ganglia between the DNAPL release point and the impervious layer. Remediation strategies need to target both residual ganglia as well as pooled DNAPLs. The application of reactive nanoparticle remediation for contaminated DNAPL source zones such as this can be broken down into a series of steps: the transport of the nanoparticles, with the bulk aqueous phase (or some delivery fluid) (Fig. 28.1A), to the DNAPL contaminated zone, their partitioning to the DNAPL/aqueous phase interface (Fig. 28.1B) and their reaction with and degradation of the DNAPL while avoiding the generation of more toxic daughter products (Fig. 28.1C). It is clear that reactive nanoparticles have significant potential to complete step three of this process as the work completed to date investigating remediation using reactive nanometals has shown they can significantly reduce dissolved aqueous phase contaminant concentrations. It should be pointed out that they have, thus far, been unable to significantly reduce source zone free phase NAPL mass [46]. Significant reductions in source zone free phase NAPL mass are necessary to significantly reduce the risk to downstream groundwater receptors. Further work is necessary for the optimized delivery of nanoparticles to the source zone (step one) and their partition to the NAPL/water interface (step two).

Recent work has focused on the development of surface modified nanometals (e.g., [27,29,33,46–49]) for optimum transport of reactive nanometals to the contaminated source zone. Although reactive nanometals are quite reactive they are not stable in aqueous phase solutions without some modification (e.g., [27,46,49]). These studies suggest that it will be possible to develop surface modified reactive nanometals that can be transported, with the aqueous phase, to the contaminated source zone. Further work, however, is necessary as many of these studies were conducted in simplified systems and their transport through permeable media systems has not been adequately evaluated. For example controlled nanometal transport column experiments are typically on the decimeter scale

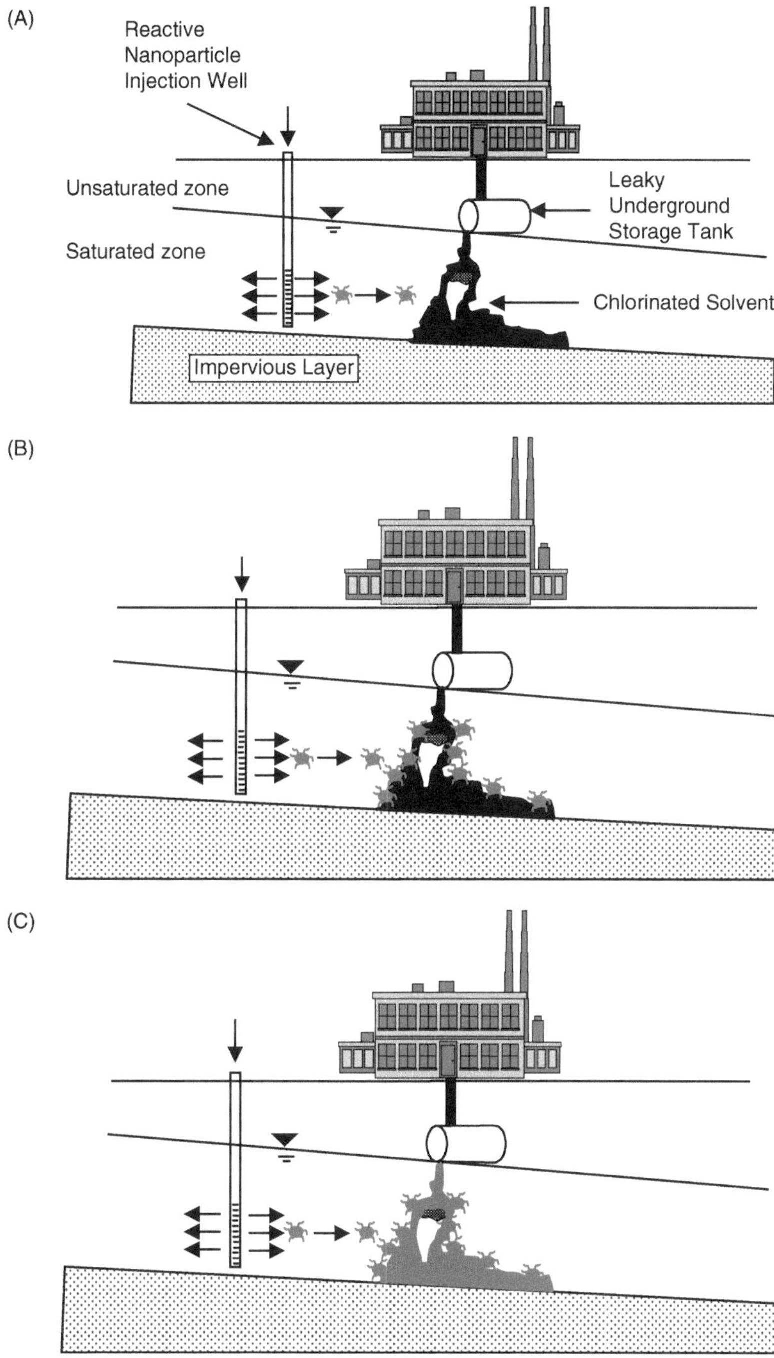

FIGURE 28.1

Schematic of dense nonaqueous phase liquid (DNAPL) remediation via reactive nanoparticles: (A) transport from the injection well to the DNAPL source zone; (B) partitioning to the water/NAPL interface; (C) reaction with the DNAPL.

(e.g., [27,48]) however nanometals will likely have to travel at least one or two orders of magnitude more at the field scale. Therefore even a small fraction of nanometal loss from the aqueous phase at the decimeter scale may become quite important at the field scale as it could translate into significant nanometal loss from the aqueous phase prior to reaching the contaminated source zone. In addition to experimental transport studies the ability to predict nanoparticle mobility is a prerequisite to the design of nanometal delivery systems for source zone remediation. The limited nanoparticle mobility studies published to date (e.g., [27,50]) have used predictive models originally developed for colloidal transport or filtration in porous media systems (e.g., [51,52]). Further study is necessary to determine if these models are appropriate for reactive nanometal transport at the field scale. Reactive nanometals may be removed from the aqueous phase due to deposition on the soil surface and straining (retention at grain/grain intersections). Traditional colloid filtration theory predicts colloid removal due to deposition on the solid surface but does not incorporate straining. A recent study by Saleh et al. [29] found that straining is an important nanometal removal mechanism and suggested that traditional colloid filtration theory may not be appropriate for the prediction of nanometal transport in the subsurface.

The second important step in the application of reactive nanoparticle remediation for contaminated DNAPL source zones, partitioning to the NAPL/water interface, has received much less attention than the other two steps. For reactive nanometals to realize their full DNAPL source zone remediation potential, they must either partition to the DNAPL/water interface or reside in the immediate vicinity. To address this issue Saleh et al. [46] synthesized nZVI on which surface active triblock copolymers were anchored. This structure facilitates nanoparticle transport to the NAPL/aqueous phase interface where the zero valent iron core can degrade the chlorinated solvent. In a subsequent study Saleh et al. [29] emplaced a NAPL in their column transport experiments and suggested that selection of a polymer with an appropriate hydrophobe/hydrophile ratio is vital for optimal partitioning. The encapsulation of nZVI in a nonionic surfactant sorbitan triolate emulsifier has also been used at a field trial to enhance partitioning to the NAPL/water interface [41]. In addition the emulsifier was designed to protect nZVI from reacting with groundwater constituents that would decrease its reducing capacity. To date our ability to precisely design and synthesize reactive nanometals such that they partition to the NAPL/water interface, while still being mobile and able to degrade the contaminants of choice, is lacking and further research is necessary.

Nanotechnology based remediation technologies offer tremendous potential for contaminated site cleanup. There are, however, challenges that need to be addressed prior to the widespread usage of nanotechnology based remediation technologies. As an example of the conditions that could be encountered at a DNAPL contaminated site the multiphase flow and contaminant transport numerical simulators M-VALOR [53,54] and MISER [55,56] were used to model a hypothetical DNAPL release, redistribution, and subsequent water flood. The water flood simulations give an estimate of the mean arrival time of an injected nanometal at the source zone. The reactive lifetime of nZVI is a concern as they can be rapidly corroded, thereby losing their reactivity (e.g., [37,57]). Furthermore nZVI reacts with natural groundwater constituents, decreasing its reducing capacity prior to reaching the source zone [41,43]. The selected DNAPL was tetrachloroethylene (PCE), a US EPA priority pollutant. PCE was released into a two-dimensional heterogeneous domain (7.925 m wide by 9.754 m deep) at equivalent rates of 0.08 L/d and 0.27 L/d (cases a and b, respectively) for 400 days followed by 300 days of redistribution. Soil properties for the simulations are based on a geostatistical representation of a surfactant enhanced aquifer remediation demonstration site in Oscoda, MI [58–62]. Soil permeabilities of the two-dimensional domain are presented in Fig. 28.2. As illustrated in this figure some soil layering exists at this site and soil characteristics at the site are only moderately heterogeneous. This site was subject to extensive site characterization with 14 vertical and angled cores. Grain size distributions of 167 subsamples, subdivided from the 14 core samples, were quantified and used to estimate soil sample permeability using the Carman–Kozeny equation [58,61,63]. Representative capillary pressure/saturation retention properties were estimated using the Haverkamp and Parlange method [64] and Brooks–Corey retention curve entry pressures [65] were estimated using Leverett scaling [66]. The numerical simulator M-VALOR [53,54] was used to model PCE infiltration and redistribution. In this example the PCE release and redistribution conditions were identical to those of Christ et al. [62] with the exception of the higher injection rate in case b.

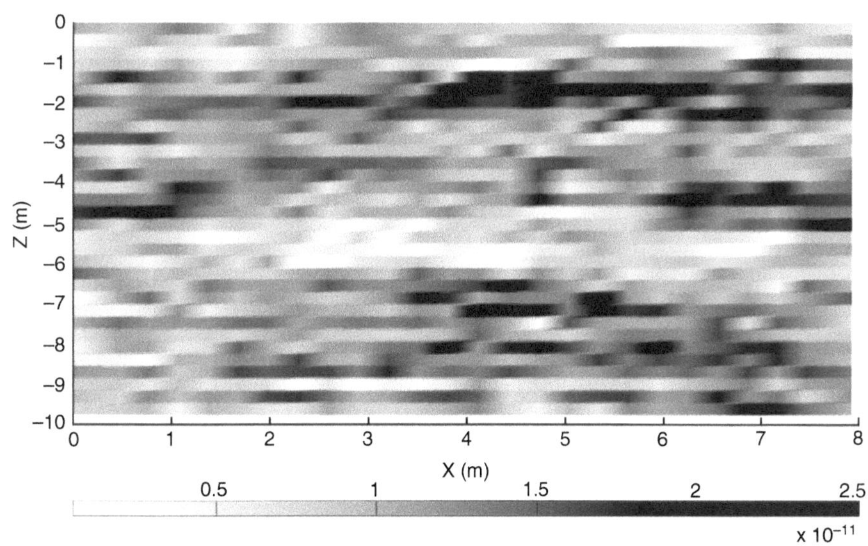

FIGURE 28.2

Vertical soil permeability (m^2) in the tetrachloroethylene (PCE) infiltration, redistribution, and water flood simulation.

Following 300 days of PCE redistribution, a water flood was initiated at an equivalent rate of 28.4 L/min (hydraulic gradient of 3.1 percent). These water flood conditions are similar to those during the surfactant flood and recovery activities at the field site [59]. The mean arrival times from the water flood simulations neglect nanometal removal from the aqueous phase due to deposition on the soil surface as well as removal due to straining. Furthermore, this analysis neglects any kind of dispersive flux. As such this simple analysis could be considered a best case scenario regarding nanometal arrival times. The numerical simulator MISER [55,56] was used to model the water flood.

Following 400 days of PCE infiltration and 300 days of redistribution, free phase PCE is present in pools, above low permeability lenses, and at residual ganglia saturations between these pools (Figs. 28.3 and 28.4). PCE contaminates the domain to a greater extent in the simulation with the higher PCE injection rate and PCE saturations are generally higher (i.e., NAPL saturations are generally higher in Fig. 28.4 when compared to Fig. 28.3). Maximum PCE saturations are 0.33 and 0.65 for cases a and b, respectively. Maximum PCE saturations would likely increase if the simulated site had a lower mean soil permeability and a larger variance in soil permeability [67]. Following DNAPL release and redistribution a waterflood was initiated, representing the injection of a nanometal slurry at the left side of the simulated domain. Mean arrival times for the reactive nanometals are shown in Figs. 28.5 and 28.6. Mean arrival times generally increase from the left of the model domain, where the nanometals are injected, to the right side of the model

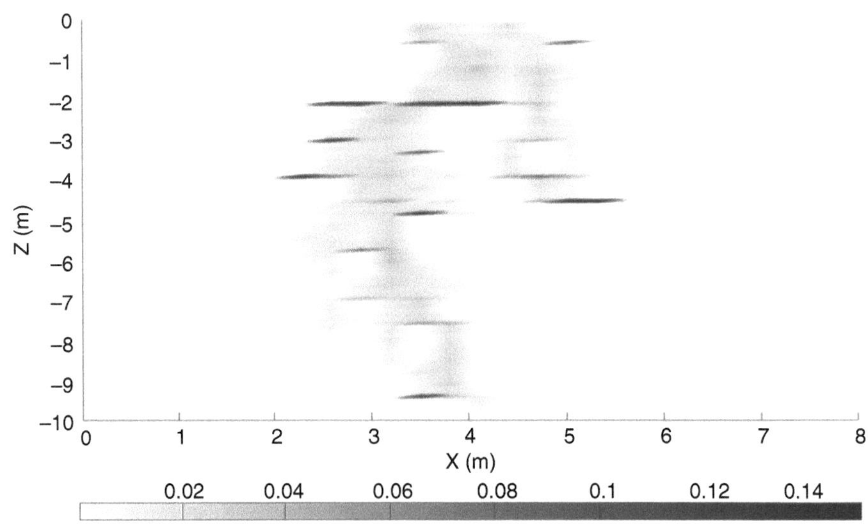

FIGURE 28.3

Case a—nonaqueous phase liquid (NAPL) saturations after 400 days of tetrachloroethylene (PCE) infiltration at an equivalent rate of 0.08 L/d followed by 300 days of PCE redistribution.

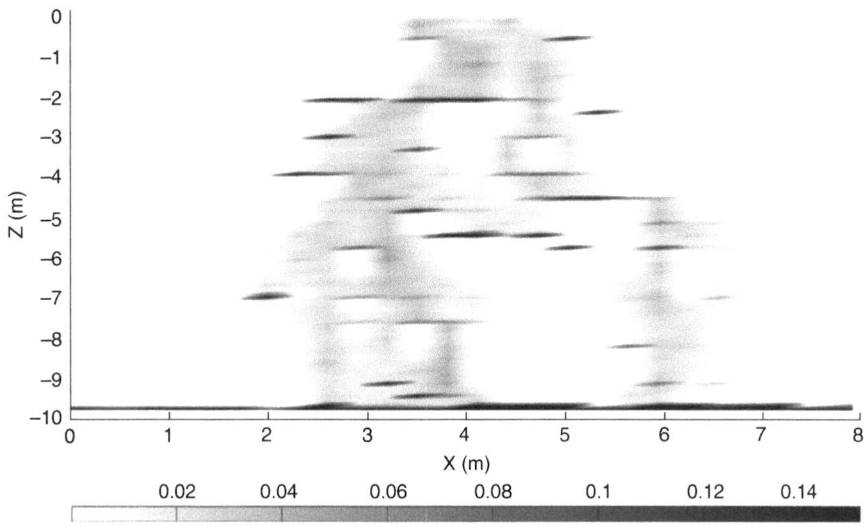

FIGURE 28.4

Case b—nonaqueous phase liquid (NAPL) saturations after 400 days of tetrachloroethylene (PCE) infiltration at an equivalent rate of 0.27 L/d followed by 300 days of PCE redistribution.

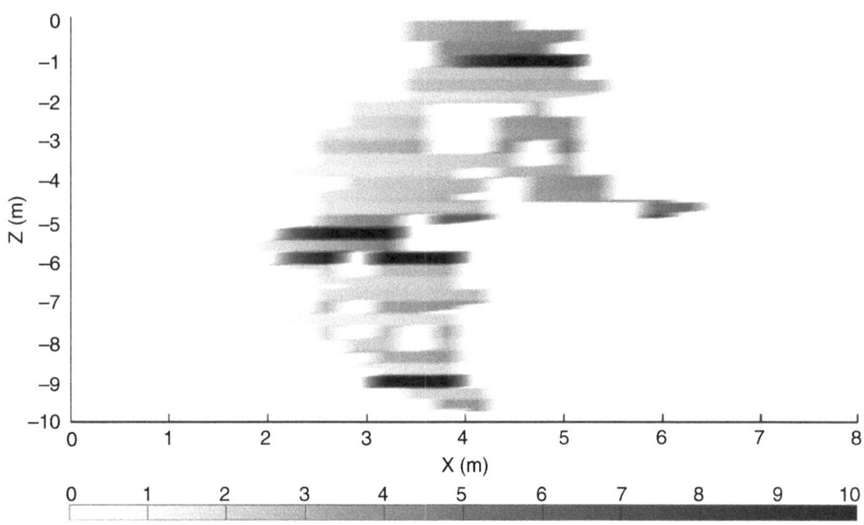

FIGURE 28.5

Case a—mean arrival time in days for reactive nanometals to reach tetrachloroethylene source zone.

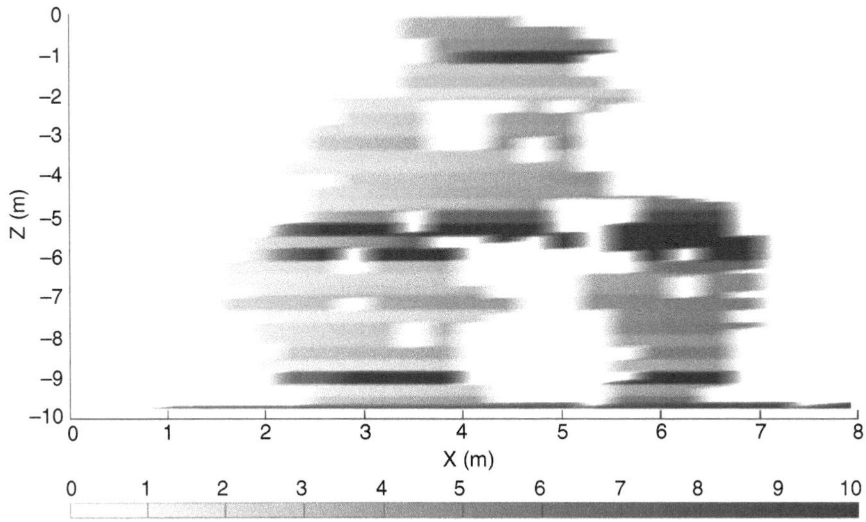

FIGURE 28.6

Case b—mean arrival time in days for reactive nanometals to reach the tetrachloroethylene source zone.

domain. The mean arrival times for the nanometals are generally less than 10 days in these simulations; maximum were 16 and 24 days for cases a and b, respectively. These arrival times are certainly within the expected reactive lifetime of nZVI under ideal conditions (e.g., [37,57]) however further work is necessary to determine reactive lifetimes under conditions found in the field. Due to the horizontal extent of the PCE contamination in case b, numerous pore volumes of nZVI slurries would have to be injected prior to reactive nZVI reaching the downstream PCE contamination, as the nZVI would first react with upstream PCE or natural groundwater constituents. Increasing PCE saturations would also increase the reactive nanometal arrival times due to decreased permeability to water in the PCE saturated zones. Nanometal arrival times may be much greater at field sites with lower permeability sands or soils. The sands at the site used in this example, with a mean permeability of 2.2×10^{-11} m^2, would certainly not be considered low permeability sands. For example, the well-known Borden fieldsite, with medium to fine grained sands, has a mean permeability of 7.6×10^{-12} m^2 [68]. Prior to the implementation of reactive nanometals for DNAPL remediation, an assessment of the reactive lifetime required to reach the source zone is, therefore, necessary on a site-specific basis.

28.6 Summary

A number of exciting, new, and innovative remediation technologies have been made possible through the advent of nanotechnology. These remediation alternatives hold significant promise but require further research and development. An additional hurdle that will require attention is the public and regulatory perception surrounding nanotechnology. For example, although carbon nanotubes have potential for the removal of metal contaminants their toxicity is unknown. Similarly, many of the noble metal catalysts used with nZVI are known contaminants, although the mass fraction of these catalysts is usually quite small. A risk assessment is needed to determine the benefit of using a known or potential contaminant to remediate another contaminant prior to the widespread implementation of nanotechnology-based remediation alternatives. Furthermore, the tremendous benefits of nanotechnology will need to be communicated to the public and regulators prior to its widespread use.

References

[1] Hydrogeologists Without Borders, cited January 4, 2008. Available from: <http://hydrogeologistswithoutborders.org/HWB percent20Global percent20Groundwater percent20 Situation.pdf>.

[2] D. Bethune, C. Ryan, M. Losilla, J. Krasny, Hydrogeology, in: J. Bundsheh (Ed.), Geology, Resource and Geo-hazards of Central America, Balkama, Netherlands, 2007.

[3] International Water and Sanitation Centre, cited January 3, 2008. Available from: <http://www.irc.nl/page/27848>.

[4] C. Mengxiong, Groundwater resources and development in China, Environ. Geol. 10 (3) (1987) 141–147.

[5] Environment Canada, cited January 3, 2008. Available from: <http://www.ec.gc.ca/Water/images/nature/grdwtr/a5f6e.htm>.

[6] USGS, cited January 3, 2008. Available from: <http://ga.water.usgs.gov/edu/wugw.html>.

[7] NRTEE, Cleaning up the Past, Building the Future, Govt. of Canada, 2003.

[8] Regional Analytics Inc., A Preliminary Investigation into the Economic Impact of Brownfield Redevelopment Activities in Canada, (2002).

[9] Statistics Canada, cited January 3, 2008. Available from: <http://www40.statcan.ca/l01/cst01/trade26.htm>.

[10] National Research Council, Innovations in Ground Water and Soil Cleanup: From Concept to Commercialization, National Academy Press, Washington, D.C., 1997.

[11] National Research Council, Contaminants in the Subsurface: Source Zone Assessment and Remediation, Committee On Source Removal of Contaminants in the Subsurface, National Academies Press, Washington, D.C., 2005.

[12] R.G. Riley, J.M. Zachara, Chemical contaminants on DOE lands and selection of contaminant mixtures for subsurface science research, Report No.: DOE-ER0547T. Battelle Pacific Northwest Labs, Richland, WA, 1992.

[13] R.J. Sloat, Hanford low level waste management reevaluation study. Atlantic Richfield Handford Company, Richland, WA, Report No.: ARH-231, 1967.

[14] National Research Council, Assessing the Human Health Risks of Trichloroethylene: Key Scientific Issues, National Academies Press, Washington, D.C., 2006.

[15] US EPA, cited January 4, 2008. Available from: <http://www.epa.gov/pcb/pubs/effects.htm>.

[16] B.B. Eskenazi, A.R. Castorina, Exposures of children to organophosphate pesticides and their potential adverse health effects, Environ. Health Perspect. 107 (3) (1999).

[17] National Research Council, Measuring Lead Exposure in Infants, Children, and Other Sensitive Populations, National Academies Press, Washington, D.C., 1993.

[18] C.A. Ramsburg, K.D. Pennell, Density-modified displacement for DNAPL source zone remediation: Density conversion and recovery in heterogeneous aquifer cells, Environ. Sci. Technol. 36 (14) (2002) 3176–3187.

[19] H.Y. She, B.E. Sleep, Removal of perchloroethylene from a layered soil system by steam flushing, Ground Water Monit. Remediat. 19 (2) (Spring 1999) 70–77.

[20] D.N. Johnson, J.A. Pedit, C.T. Miller, Efficient, near-complete removal of DNAPL from three-dimensional, heterogeneous porous media using a novel combination of treatment technologies, Environ. Sci. Technol. 38 (19) (2004) 5149–5156.

[21] T.C. Sale, D.B. McWhorter, Steady state mass transfer from single-component dense non-aqueous phase liquids in uniform flow fields, Water Resour. Res. 37 (2) (2001) 393–404.

[22] P.S.C. Rao, J.W. Jawitz, Comment on 'Steady state mass transfer from single-component dense nonaqueous phase liquids in uniform flow fields' by T.C. Sale and D.B. McWhorter, Water Resour. Res. 39 (3) (2003).

[23] N. Savage, M.S. Diallo, Nanomaterials and water purification: Opportunities and challenges, J. Nanopart. Res. 7 (4–5) (2005) 331–342.

[24] Y.H. Li, J. Ding, Z.K. Luan, Z.C. Di, Y.F. Zhu, C.L. Xu, et al., Competitive adsorption of Pb2 + , Cu2 + and Cd2 + ions from aqueous solutions by multiwalled carbon nanotubes, Carbon 41 (14) (2003) 2787–2792.

[25] K. Yang, L.Z. Zhu, B.S. Xing, Adsorption of polycyclic aromatic hydrocarbons by carbon nanomaterials, Environ. Sci. Technol. 40 (6) (2006) 1855–1861.

[26] X.J. Peng, Y.H. Li, Z.K. Luan, Z.C. Di, H.Y. Wang, B.H. Tian, et al., Adsorption of 1,2-dichlorobenzene from water to carbon nanotubes, Chem. Phys. Lett. 376 (1–2) (2003) 154–158.

[27] B. Schrick, B.W. Hydutsky, J.L. Blough, T.E. Mallouk, Delivery vehicles for zerovalent metal nanoparticles in soil and groundwater, Chem. Mater. 16 (11) (2004) 2187–2193.

[28] Y.Q. Liu, S.A. Majetich, R.D. Tilton, D.S. Sholl, G.V. Lowry, TCE dechlorination rates, pathways, and efficiency of nanoscale iron particles with different properties, Environ. Sci. Technol. 39 (5) (2005) 1338–1345.

[29] N. Saleh, K. Sirk, Y.Q. Liu, T. Phenrat, B. Dufour, K. Matyjaszewski, et al., Surface modifications enhance nanoiron transport and NAPL targeting in saturated porous media, Environ. Eng. Sci. 24 (1) (2007) 45–57.

[30] D.W. Elliott, W.X. Zhang, Field assessment of nanoscale bimetallic particles for groundwater treatment, Environ. Sci. Technol. 35 (24) (2001) 4922–4926.

[31] X.Q. Li, D.W. Elliott, W.X. Zhang, Zero-valent iron nanoparticles for abatement of environmental pollutants: Materials and engineering aspects, Crit. Rev. Solid State Mater. Sci. 31 (4) (2006) 111–122.

[32] W.X. Zhang, D.W. Elliott, Applications of iron nanoparticles for groundwater remediation, Remediation (Spring 2006) 7–21.

[33] F. He, D.Y. Zhao, J.C. Liu, C.B. Roberts, Stabilization of Fe-Pd nanoparticles with sodium carboxymethyl cellulose for enhanced transport and dechlorination of trichloroethylene in soil and groundwater, Ind. Eng. Chem. Res. 46 (1) (2007) 29–34.

[34] H. Song, E.R. Carraway, Reduction of chlorinated ethanes by nanosized zero-valent iron: Kinetics, pathways, and effects of reaction conditions, Environ. Sci. Technol. 39 (16) (2005) 6237–6245.

[35] H.L. Lien, Y.S. Jhuo, L.H. Chen, Effect of heavy metals on dechlorination of carbon tetrachloride by iron nanoparticles, Environ. Eng. Sci. 24 (1) (2007) 21–30.

[36] C.B. Wang, W.X. Zhang, Synthesizing nanoscale iron particles for rapid and complete dechlorination of TCE and PCBs, Environ. Sci. Technol. 31 (7) (1997) 2154–2156.

[37] B. Schrick, J.L. Blough, A.D. Jones, T.E. Mallouk, Hydrodechlorination of trichloroethylene to hydrocarbons using bimetallic nickel-iron nanoparticles, Chem. Mater. 14 (12) (2002) 5140–5147.

[38] S.R. Kanel, B. Manning, L. Charlet, H. Choi, Removal of arsenic(III) from groundwater by nanoscale zero-valent iron, Environ. Sci. Technol. 39 (5) (2005) 1291–1298.

[39] H.L. Lien, W.X. Zhang, Hydrodechlorination of chlorinated ethanes by nanoscale Pd/Fe bimetallic particles, J. Environ. Eng. 131 (1) (2005) 4–10.

[40] R.E. Jackson, V. Dwarakanath, Chlorinated degreasing solvents: Physical–chemical properties affecting aquifer contamination and remediation, Ground Water Monit. Remediat. 19 (1999) 102–110.

[41] J. Quinn, C. Geiger, C. Clausen, K. Brooks, C. Coon, S. O'Hara, et al., Field demonstration of DNAPL dehalogenation using emulsified zero-valent iron, Environ. Sci. Technol. 39 (5) (2005) 1309–1318.

[42] K.W. Henn, D.W. Waddill, Utilization of nanoscale zero-valent iron for source remediation—a case study, Remediation 6 (2) (2006).

[43] Y. Liu, T. Phenrat, G.V. Lowry, Effect of TCE concentration and dissolved groundwater solutes on NUI-promoted TCE dechlorination and H-2 evolution, Environ. Sci. Technol. 41 (22) (2007) 7881–7887.

[44] National Research Council, Alternatives for Ground Water Cleanup, National Academies Press, Washington, D.C., 1994.

[45] K.D. Pennell, G.A. Pope, L.M. Abriola, Influence of viscous and buoyancy forces on the mobilization of residual tetrachloroethylene during surfactant flushing, Environ. Sci. Technol. 30 (4) (1996) 1328–1335.

[46] N. Saleh, T. Phenrat, K. Sirk, B. Dufour, J. Ok, T. Sarbu, et al., Adsorbed triblock copolymers deliver reactive iron nanoparticles to the oil/water interface, Nano Lett. 5 (12) (2005) 2489–2494.

[47] S.R. Kanel, D. Nepal, B. Manning, H. Choi., Transport of surface-modified iron nanoparticle in porous media and application to arsenic(III) remediation, J. Nanopart. Res. 9 (5) (2007) 725–735.

[48] B.W. Hydutsky, E.J. Mack, B.B. Beckerman, J.M. Skluzacek, T.E. Mallouk, Optimization of nano- and microiron transport through sand columns using polyelectrolyte mixtures, Environ. Sci. Technol. 41 (18) (2007) 6418–6424.

[49] F. He, D.Y. Zhao, Preparation and characterization of a new class of starch-stabilized bimetallic nanoparticles for degradation of chlorinated hydrocarbons in water, Environ. Sci. Technol. 39 (9) (2005) 3314–3320.

[50] H.F. Lecoanet, M.R. Wiesner, Velocity effects on fullerene and oxide nanoparticle deposition in porous media, Environ. Sci. Technol. 38 (16) (2004) 4377–4382.

[51] N. Tufenkji, M. Elimelech, Correlation equation for predicting single-collector efficiency in physicochemical filtration in saturated porous media, Environ. Sci. Technol. 38 (2) (2004) 529–536.

[52] A. Amirtharajah, Some theoretical and conceptual views of filtration, J. Am. Water Works Assn. 80 (12) (1988) 36–46.

[53] L.M. Abriola, K.M. Rathfelder, M. Maiza, S. Yadav, VALOR code version 1.0: A PC code for simulating immiscible contaminant transport in subsurface systems: EPRI TR-101018. Report No.: EPRI TR-101018 (1992).

[54] D.M. O'Carroll, S.A. Bradford, L.M. Abriola, Infiltration of PCE in a system containing spatial wettability variations, J. Cont. Hydrol. 73 (2004) 39–63.

[55] T.J. Phelan, L.D. Lemke, S.A. Bradford, D.M. O'Carroll, L.M. Abriola, Influence of textural and wettability variations on predictions of DNAPL persistence and plume development in saturated porous media, Adv. Wat. Res. 27 (4) (2004) 411–427.

[56] K.M. Rathfelder, J.R. Lang, L.M. Abriola, A numerical model (MISER) for the simulation of coupled physical, chemical and biological processes in soil vapor extraction and bioventing systems, J. Contam. Hydrol. 43 (3–4) (2000) 239–270.

[57] Y.Q. Liu, G.V. Lowry, Effect of particle age (Fe-O content) and solution pH on NZVI reactivity: H-2 evolution and TCE dechlorination, Environ. Sci. Technol. 40 (19) (2006) 6085–6090.

[58] L.M. Abriola, C.D. Drummond, E.J. Hahn, K.F. Hayes, T.C.G. Kibbey, L.D. Lemke, et al., Pilot-scale demonstration of surfactant-enhanced PCE solubilization at the Bachman road site. 1. Site characterization and test design, Environ. Sci. Technol. 39 (6) (2005) 1778–1790.

[59] C.A. Ramsburg, K.D. Pennell, L.M. Abriola, G. Daniels, C.D. Drummond, M. Gamache, et al., Pilot-scale demonstration of surfactant-enhanced PCE solubilization at the Bachman road site. 2. System operation and evaluation, Environ. Sci. Technol. 39 (6) (2005) 1791–1801.

[60] L.D. Lemke, L.M. Abriola, J.R. Lang, Influence of hydraulic property correlation on predicted dense nonaqueous phase liquid source zone architecture, mass recovery and contaminant flux, Water Resour. Res. 40 (12) (2004).

[61] L.D. Lemke, L.M. Abriola, P. Goovaerts, Dense nonaqueous phase liquid (DNAPL) source zone characterization: Influence of hydraulic property correlation on predictions of DNAPL infiltration and entrapment, Water Resour. Res. 40 (1) (2004).

[62] J.A. Christ, L.D. Lemke, L.M. Abriola, Comparison of two-dimensional and three-dimensional simulations of dense nonaqueous phase liquids (DNAPLs): Migration and entrapment in a nonuniform permeability field, Water Resour. Res. 41 (1) (2005).

[63] J. Bear, Dynamics of Fluids in Porous Media, Elsevier Science, New York, 1972.

[64] R. Haverkamp, J.Y. Parlange, Predicting the water-retention curve from particle-size distribution:1. Sandy soils without organic-matter, Soil Sci. 142 (6) (1986) 325–339.

[65] R.H. Brooks and A.T. Corey, Hydraulic properties of porous media, Hydrology Papers No. 3, Colorado State University, Boulder, CO, (1964).

[66] M.C. Leverett, Capillary behavior in porous solids, Transactions of the American Institute of Mining and Metallurgical Engineers 142 (1941) 152–169.

[67] J.I. Gerhard, B.H. Kueper, Influence of constitutive model parameters on the predicted migration of DNAPL in heterogeneous porous media, Water Resour. Res. 39 (10) (2003).

[68] M.A. Turcke, B.H. Kueper, Geostatistical analysis of the Borden aquifer hydraulic conductivity field, J. Hydrol. 178 (1–4) (1996) 223–240.

CHAPTER

29 Green Remediation of Hexavalent Chromium Using Naturally Derived Flavonoids and Engineered Nanoparticles

Omowunmi A. Sadik, Veronica Okello, Francis Osonga, Marcells A. Omole, and Isaac K'Owino

Department of Chemistry, Center for Advanced Sensors and Environmental Systems, State University of New York at Binghamton, Binghamton, NY, USA

29.1 Introduction

Contaminated ground- and surface water supplies such as the municipalities, wells, lakes, and rivers exhibit various levels of contamination that can pose numerous health risks. The release of contaminants into the environment can occur during production, use, and disposal of chemicals, thereby leading to potential contamination of water supplies. Among the numerous sources of pollution is

Street, Sustich, Duncan and Savage. Nanotechnology Applications for Clean Water, 2nd Edition.
 DOI: http://dx.doi.org/10.1016/B978-1-4557-3116-9.00029-9

treated wastewater, which is reinjected into groundwater aquifers for indirect reuse. In order to reach drinking water quality standards, surface water typically requires both filtration and disinfection because of its exposure to the environment and the higher potential for contamination. Industry and municipalities use about 10% of the globally accessible runoff and generate a stream of wastewater that flows or seeps into rivers, lakes, groundwater, or the coastal seas [1]. These wastewaters contain numerous chemical compounds in varying concentrations. About 300 million tons of synthetic compounds used in industrial and consumer products partially find their way into natural waters. Additional pollution comes from diffuse sources such as agriculture, where 140 million tons of fertilizers and several million tons of pesticides are applied each year [2]. The input of 0.4 million tons of oil and gasoline components through accidental spills represents yet another important source of water pollution. Other notable sources of contamination include the intrusion of salty water into groundwater due to overexploitation of aquifers, the human-driven mobilization of naturally occurring geogenic toxic chemicals, including heavy metals and metalloids, and the biological production of toxins and malodorous compounds.

Groundwater is normally considered to be the purest source of water because it is naturally filtered as it passes through the layers of rock and sediments in an aquifer. However, the geology of the rocks in the aquifer may have a major impact on the quality of the groundwater and quite often, the technology required to remove these contaminants can be more complicated and expensive than surface water treatments. As a result of extensive regulations, guidelines, and water quality testing, drinking water supplies in the United States are among the cleanest and safest in the world, yet despite efforts to standardize the cleanliness and quality of drinking water nationwide, the quality of water served by individual systems varies over time due to changes in the water source from which it is drawn and the treatment it undergoes. To date, an effective and sustainable global strategy against this insidious and mostly unseen contamination of aquatic environments barely exists. Source controls and technical systems, such as wastewater treatment plants, only function as partial barriers, but major challenges remain. The source, behavior, and treatment of the relatively small number of pollutants [3], such as acids, salts, nutrients, and natural organic matter, occurring at μg/L to mg/L concentrations are relatively well understood. However, high nutrient loads can lead to increased primary production, oxygen depletion, and toxic algal blooms. In such cases, the challenges are to predict ecosystem responses, to optimize treatment technologies, and to develop integrated policies at the scale of river basins [4].

The effects of the thousands of synthetic and natural trace contaminants on the aquatic environment are difficult to assess at low to very low concentrations (pg/L to ng/L) [5]. These chemicals are ubiquitous in natural waters, not only in industrialized areas but in more remote environments as well. Some chemicals such as heavy metals are not degraded at all, while others such as persistent organic compounds (e.g., DDT, lindane, or polychlorinated biphenyls) are degraded only very slowly. These can therefore be transported via water or air to locations hundreds or even thousands of miles away from their source [6,7]. Some compounds that are less persistent and not prone to long-range transport may still be of concern if they are continuously emitted or form problematic (bio) transformation products [8]. Examples of this category include hormones and drugs, or persistent degradation products of surfactants such as nonylphenol. Therefore, assessing the impact of micropollutants in aquatic systems is a formidable task requiring improved analytical and modeling tools to probe the distribution, bioavailability, and biological effects of single compounds and/or chemical mixtures. Thus methods to classify existing and new chemicals on the basis of their potentials on human health and the environment must be continually refined. Moreover, mitigation technologies to reduce the impact of micropollutants as well as strategies to minimize their introduction into the environment require further development.

In the past 10 years, emerging technologies such as phytoremediation, bioremediation, and permeable reactive barriers have become popular new tools. These novel treatments have begun to compete with more established technologies such as solidification/stabilization, soil vapor extraction, and thermal desorption for soil, and pump and treat systems for groundwater remediation [9]. At the very forefront of these emerging technologies lies the development of nanotechnology. Currently a wide variety of potential remedial tools employing nanotechnology are being examined at the bench scale for use in wastewater and soil remediation. One emerging nanotechnology, nanosized zerovalent iron (ZVI) and its derivatives, has reached the commercial market for field-scale remediation and studies. One of the emerging compounds of concern is chromium. Chromium usually occurs in its compounds in the form of chromium (VI) or chromium (III), which are the most stable and common oxidation states of chromium. Compounds of chromium (VI) and chromium (III) have different solubilities and toxicities [10]. Chromium (III) is essential to animals and human beings. It is recommended that a daily uptake of 50–100 μg is helpful for human beings, without toxic effects observed even at a higher dosage [11]. In contrast, chromium (VI) has been proven to be highly toxic, and some chromates are considered as carcinogens [12]. Unfortunately, a large amount of chromium (VI)-containing wastes arise every day from tanning, electroplating, wood preservation, dying, and production of chromium chemicals. Therefore, many countries have adopted severe restrictions on disposals of those industrial wastes. Quite a few methods for the removal of chromium (VI) have been studied, such as biosorption, ion exchange, solvent extraction, nanofiltration, micelle-enhanced ultrafiltration, adsorption with inorganic sorbent materials, reduction, and precipitation [13].

We have developed new treatment approaches based on catalytic reduction of Cr(VI) to the benign form {(Cr(III)} using palladium nanoparticles (PdNPs). This chapter examines the most current information regarding metal contamination and the *in situ* remediation of chromium. This work focuses on the use of PdNPs for the catalytic conversion of Cr(VI) to Cr(III) using formic acid (FA) and sulfur.

Subsequent practical application in aqueous samples indicates a complete elimination of Cr(VI). Finally, this chapter also discusses the use of naturally occurring flavonoids as reducing agents for Cr(VI).

29.2 Nanotechnologies for site remediation and wastewater treatment

The development and implementation of water treatment technologies have been mostly driven by three primary factors: the discovery of new rare contaminants, the promulgation of new water quality standards, and cost [14]. For the first 75 years of this century, chemical clarification, granular media filtration, and chlorination were virtually the only treatment, processes used in municipal water treatment [15]. However, the past 20 years have seen a dramatic change in the water industry's approach to water treatment, in which water utilities have started to seriously consider alternative treatment technologies to the traditional filtration/chlorination treatment approach. Regardless of the water treatment technology employed to clean drinking water, most water systems add chlorine or other disinfectants to ensure water remains clean within the water distribution system. Disinfection of drinking water supplies was an important advance in public health and a major accomplishment of the twentieth century. The Safe Drinking Water Act (SDWA) directs the US EPA to establish national standards for contaminants in public drinking water supplies. Enforceable standards are to be set at concentrations at which no adverse health effects in humans are expected to occur and for which there are adequate margins of safety achievable with the use of the best technology available.

Nanotechnology, the science and art of manipulating matter at the atomic and molecular level, has the potential to substantially enhance environmental quality and sustainability through pollution prevention, treatment, and remediation. The nanotechnology industry is increasingly promoting nano as a "green" technology that will improve the environmental performance of existing industries, reduce consumption of resources and energy, and allow achievement of environmentally benign economic expansion. Cost-effective remediation techniques pose a major challenge for the EPA in the development of adequate hazard removal techniques that protect the public and safeguard the environment. The EPA supports research into new remediation approaches that are more productive and cost-effective in removing contamination than currently available techniques [16].

Metallic substances of significant concern in remediation of soils, sediment, and groundwater are arsenic, chromium, mercury, lead, and cadmium. Nanotechnology offers the possibility of more effective remediation due to the higher surface-to-volume ratios of nanomaterials, and the possibility of novel collection and separation protocols due to the unique physical properties of nanomaterials is feasible. Specific control and design of materials at the molecular level may impart increased affinity, capacity, and selectivity for pollutants, thereby reducing releases of such hazardous materials to the air and water, providing safe drinking water, and minimizing quantities and exposure to hazardous materials.

29.2.1 Bimetallic nanoparticles remediation approach

Bimetallic particles are those particles on which a thin layer of catalytic metal (e.g., Pd, Pt, which are not active in themselves) is doped onto the surface of the active (reducing) metal, Fe or Zn, as shown in Figure 29.1. Physically mixing the two metals does not increase the rate of reaction; the palladium must be doped onto the surface [9]. Doping palladium on the surface sets up a galvanic couple, which increases the rate of corrosion of the iron and hence increases the rate of oxidation and reduction. Palladium and nickel have also been found to significantly enhance the dechlorination of polychloroethylene (PCE) in a zerovalent silicon/water system [8]. Another advantage of these bimetallic particles is that they can add stability, as ZVI particles lose reactivity within a few days while Fe/Pd particles remain active for at least 2 weeks. This is regardless of the fact that the galvanic couple increases the iron corrosion rate during oxidation and reduction reactions. Doping the zinc surface with palladium can also prevent passivation occurring [9].

Bimetallic catalysts [17] are especially interesting for several reasons: combining two metals may provide control over the catalytic activity, selectivity, and

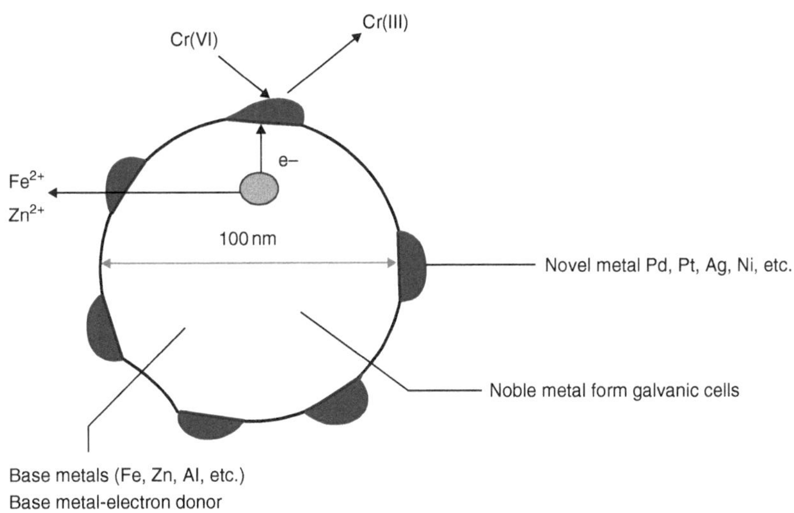

FIGURE 29.1

Schematic of bimetallic nanoparticle design.

stability, and some combinations may exhibit synergistic effects [18]. Moreover, by controlling the type of cluster synthesized, one can improve the "catalyst atom economy" [19]. The control of homogeneity, dispersion, and alloying extent has a profound influence on the surface properties which affect the catalytic activity and stability of the bimetallic NPs. The alloying extent in bimetallic NPs causes changes in atomic distribution in bimetallic NPs which has a strong influence on physicochemical properties of NPs [20]. Theory of nanoparticle catalysis, electrocatalysis, and modeling of these reactions involving simulations of the reaction kinetics on nm-supported catalyst particles based on electronic structure and chemisorption properties of supported metal clusters have been studied [19]. Particle size, support, and effects of electrochemical and chemical promotion on metal films and nanoparticles have been exploited for the design of novel nanostructured material based on transition metal compounds for electrocatalysis.

Palladium-mediated redox reactions are not new and pure Pd clusters have been shown to give lower catalytic activities (44%) compared to the alloy Ni–Pd (63%), showing Pd cluster to be less active than the Ni–Pd core–shell clusters. As all these catalysts contain the same amount of palladium, this indicates that the core–shell structure results in more Pd atoms on the surface. The total coordination number around Ni atoms in bimetallic clusters is usually higher than that around Pd [20], suggesting that the Pd atoms are located preferentially on the surface. The tendency of Pd to go to the surface may explain the difference in activity between the Pd clusters and the alloy Ni–Pd clusters noting that no reaction takes place when the Ni clusters or Ni(II) alone are used. Therefore, it is likely that only Pd is responsible for the catalysis in the case of the alloy and the core–shell clusters. The most important finding is that by combining palladium with another non-reactive metal (in this case Ni), we can increase the activity per Pd atom (segregated Pd clusters $<$ alloy Ni–Pd clusters $<$ core–shell Ni–Pd clusters).

Environmental applications of zerovalent metals (ZVMs) also overlap with the burgeoning field of nanotechnology. However, use of zerovalent single metals to reduce chlorinated organics has some drawbacks [21]. For example, even when nanoscaled ZVI particles are used, the metal mass normalized observed rate constant for dechlorination of TCE is still very low, in the order of 10^{-2} L/g h [21]. More important is that a hydroxide or oxide layer will form on the particle surface during the reaction or upon contact of the nanoparticles with air, significantly reducing their reactivity and decreasing the effective use of the metal particles. Efforts to improve the ZVM technique have led to the use of Ni/Fe and Pd/Fe particles to dechlorinate chlorinated organics [19]. Reports show that physical addition of Pd^0 or Ni^0 micron-sized powder could reactivate Fe^0 particles that have lost their surface activity [19]. It has been reported that the reduction of chlorinated organics by bimetallic particles happens via hydrodechlorination instead of electron transfer, in which Fe or Zn acts as the reducing agent, and Ni or Pd acts as a catalyst.

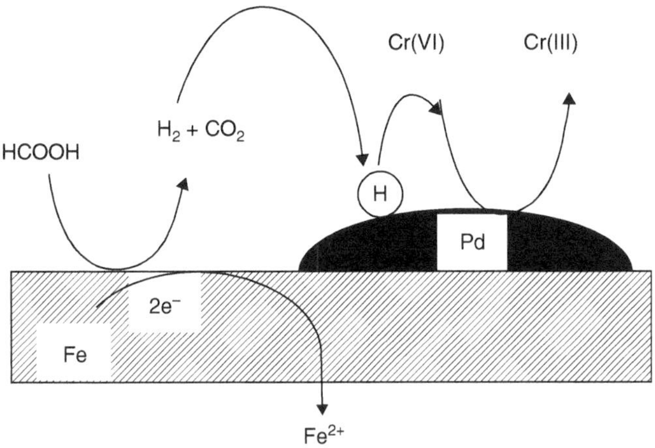

FIGURE 29.2

Depiction of ZVI-mediated degradation mechanism: the direct reduction model of Cr(VI) to Cr(III).

Figure 29.2 shows the applicability of this approach in FA reduction of hexavalent chromium. The latter are good hydrogenation catalysts and have a high ability to dissociate H_2 [21]. The introduction of a second metal not only increases the reactivity and reduces the accumulation of toxic by-products, but has been reported to make the particles more stable in air by inhibiting the oxidation in some cases [21].

29.2.2 Remediation of chromium using nanotechnology

Chromium (VI) is one of the major industrial contaminants in groundwater and a known human carcinogen. Hexavalent chromium (Cr(VI)) is believed to be the third most common pollutant at hazardous waste sites as well as the second most common inorganic contaminant after lead [4]. The chromate can readily cross cell membranes and be converted into reactive Cr(IV) and Cr(V) as well as stable Cr(III)–DNA adducts, causing mutation and cancer [5]. Thus, site remediation is often needed to reduce the risks it poses to humans and the ecosystem. Consequently, the reduction of Cr(VI) to Cr(III) in both aqueous and natural soil media is a prerequisite step in the eradication of environmental hazards associated with high levels of Cr(VI).

We report the feasibility of using Pd nanoparticles as innovative catalysts in the conversion of reducible contaminants from toxic to benign forms. Cr(VI) is a known carcinogen while the trivalent chromium salts are believed to be nontoxic. The ability of Pd nanoparticles to catalyze the rapid reduction of Cr(VI) to Cr(III) using reactive sulfur intermediates produced *in situ* has been demonstrated [22]. We used a microchamber set at 130°C as shown in Figure 29.3A and the

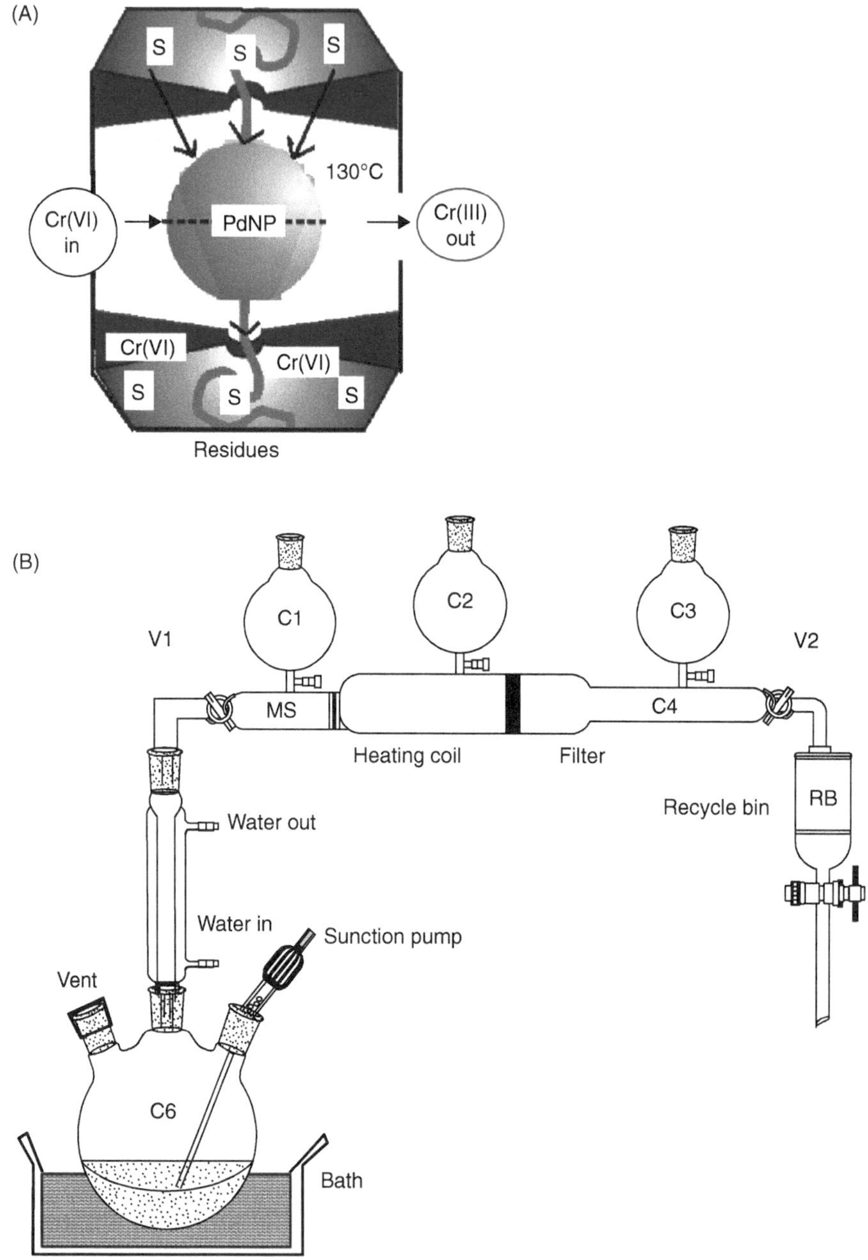

FIGURE 29.3

(A) Schematics of a nanoreactor designed for Cr(VI) remediation into Cr(III) using composite in the presence of Pd nanoparticles; (B) design and application of a nanoreactor designed for remediation of Cr(VI) into Cr(III) using FA in the presence of Pd nanoparticles.

reduction mixture consisted of PdNPs and sulfur (PdNPs/S), which generated highly reducing sulfur intermediates that effected the reduction of Cr(VI) to Cr (III). UV/visible spectroscopy and cyclic voltammetry were employed to monitor the reduction process. The results showed that 99.8% of 400 mM Cr(VI) was reduced to Cr(III) by PdNPs/S in 1 h compared to 2.1% by a control experiment consisting of sulfur only. The rate of Cr(VI) reduction was found to be dependent on temperature and pH and was greatly enhanced by the addition of PdNPs. Subsequent application of this approach in the reduction of Cr(VI) in soil and aqueous media was conducted. In contrast to the control experiments with and without PdNPs or sulfur, >92% conversion rate was obtained in the presence of PdNPs/S within 1 h. This represents over a 500-fold improvement in conversion rate compared to current microbial approaches. XPS analysis provided the confirmation regarding the oxidation states of Cr(VI).

29.2.3 Determination of Cr(VI) concentration

In order to determine the extent of the catalytic reduction of Cr(VI) due to PdNPs at a fixed temperature, three different experimental setups were used. These include (i) a buffered Cr(VI) solution involving neither the FA nor the PdNPs as controls; (ii) a buffered reaction mixture consisting of Cr(VI) and FA; and (iii) a buffered reaction mixture containing Cr(VI), FA, and PdNPs. For each experiment, the reaction mixture contained 5.0 mL of a 10.0 mM Cr(VI), 220 μL FA, 1.68 mL of 0.5 M buffer (acetate buffer pH 2.0), and 100 μL of colloidal PdNPs. The temperature of the 7.0 mL cocktail reaction mixture was raised to 45°C and the temperature maintained for a period of 5 min or more. Samples were periodically withdrawn from the reaction vessel and transferred immediately into an ice bath at 4°C to minimize any further reactions during the 3 min time lapse. Samples were then passed through a 0.2 μm membrane filter before carrying out spectrophotometric analysis on the filtrate. The residual Cr(VI) concentrations were determined after appropriate dilution of the resulting mixtures. The rate of change in Cr(VI) was determined as the difference in the initial and final Cr(VI) concentrations over time expressed as a percentage. To ascertain negligible Cr(VI) reduction at 4°C, tests were performed on possible Cr(VI) reduction at that same temperature for a period of 30 min.

Sadik and coworkers [23] further designed a nanoreactor shown in Figure 29.3B. The reactor employed colloidal Pd nanoparticles catalyst for the reduction of Cr(VI) to Cr(III) using FA as the reducing agent. The reaction kinetics of Cr(VI) reduction and the effects of other parameters such as the temperature, the FA concentration, the pH, Pd loading, and elemental hydrogen were studied. Based on the experimental results, this approach has shown that colloidal PdNPs enhanced the reduction rate Cr(VI) to Cr(III). The reduction of Cr(VI) to Cr(III) was dependent on temperature, pH, the amount of PdNPs, and FA concentration with optimum reduction at 45°C and over 99% reduction at low pH values.

29.2.4 Removal of Cr(VI) from complex aqueous media

We have tested the use of PdNPs/S mixture in the removal of Cr(VI) to Cr(III) in a complex aqueous media containing 400, 500, 800, and 1000 μM of Cr(VI). To achieve this, we mimicked the mineral composition of the natural environment by using m9k buffer. Figure 29.4 shows the results obtained. PdNPs/S reduced 400 μM of Cr(VI) at a rate of 99.8(± 0.2)% per hour compared with 2.1(± 0.9)% per hour and 65.22(± 3.3)% per hour by respective controls of sulfur only and PdNPs only. It is believed the drop in the residual amount of Cr(VI) in the presence of PdNPs only was consistent with the adsorption of Cr_2O_3 onto Pd surfaces and the probable retention of Cr(VI) by sample matrices during the filtration process [24–28]. In the process, drops in residual amounts of Cr(VI) that depended on the initial Cr(VI) concentration were obtained (Figure 29.4). This represents over 500-fold improvement in conversion rate compared to current microbial approaches [25–29]. Table 29.1 gives the comparison of the nanoremediation approaches for Cr using microbial, Bio-Pd, FeNPs, and PdNP methods. Using Pd/NPs, 99.8% removal of Cr(VI) in 336 μM concentration was achieved within 5 min as opposed to ~90% removal from 100 μM using Bio-Pd method [30].

29.3 Naturally occurring flavonoids as reducing agents for hexavalent chromium

Currently the demand for a safe and cleaner environment is the concern of scientists. Consequently the reduction of heavy metals from the ecosystem is one of the

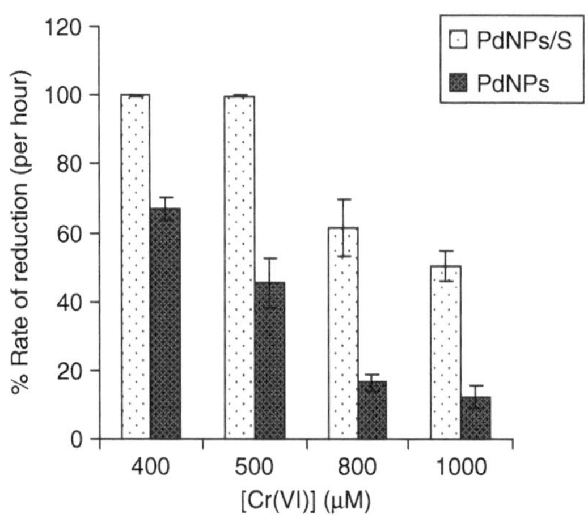

FIGURE 29.4

Variation of percentage rate of reduction of Cr(VI) by PdNPs/S with initial Cr(VI) concentration. Conditions: 1.6 mg/mL PdNPs; 10 mg/mL sulfur; 400, 500, 800, and 1000 μM Cr(VI). Controls consisted of all other reaction components except sulfur.

Table 29.1 Comparison of Nanoremediation of Chromium with Existing Approaches

Parameter	Microbial	Bio-Pd	Fe-NPs	TiO_2	PdNPs/FA	PdNPs/S
Time	6 days	~680 h	Over 2 months	2 h	5 min	1 h
Yield	~90% of 200 μM	90% of 100 μM	Not reported	93.7% of 336 μM	99.8% of 10 mM	99.83% of 400 μM
References	[31a]	[31b]	[31c]	[31d]	[19]	[20]

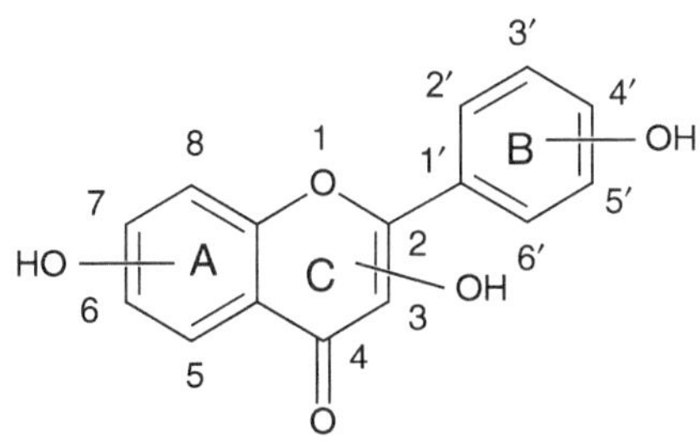

FIGURE 29.5

Generic structure of flavonoids.

most important research activities being pursued by most research groups, in order to find better ways to remove heavy metals from the environment. However, the use of green approach, particularly naturally occurring flavonoids, to reduce the levels of heavy metals is still rare. The overall goal of this section was to reduce toxic hexavalent chromium to its trivalent state using naturally occurring flavonoid, namely quercetin (QCR) and two of its derivatives (quercetin pentaphosphate (QPP) and quercetin sulfonic acid (QSA)). The specific objectives were to eliminate the use of organic solvents and to minimize the overall reaction time using PdNPs. These objectives were successfully achieved as highlighted below and a comparative study with previous studies from our laboratory was therefore carried out.

Flavonoids are a subclass of the polyphenols, which are characterized as containing two or more aromatic rings, each bearing at least one aromatic hydroxyl and connected with a carbon bridge [31,32]. For flavonoids, this bridge consists of three carbons that combine with an oxygen and two carbons of one of the aromatic rings (A ring) to form a third 6-member ring (C ring) (Figure 29.5). The flavonoids are further divided into subclasses based on the connection of the B ring to the C ring, as well as the oxidation state and functional groups of the C ring. Individual flavonoids are identified and characterized by hydroxylation and conjugation patterns of the B ring, as well as the conjugation patterns of hydroxyls on the A and C rings.

A number of these polyphenols have shown interesting biological activities [33] and lately, modification for applications in environmental remediation [34]. However, this class of compounds utility has been hampered by solubility difficulties.

Our group has been actively involved in the synthesis and characterization of flavonoid derivatives and reported the very first evidence of the removal of Cr (VI) from the environment using naturally derived flavonoids [35]. In this study, QCR and two successfully synthesized QCR derivatives named as QPP and QSA were used to reduce Cr(VI) from the environment. The derivatives were characterized using UV–vis spectroscopy, ^{1}H NMR, ^{13}C NMR, ^{31}P NMR, and LC–MS techniques. The solubility study conducted on QPP found its solubility to be 840 mg/mL. Furthermore the solutions of QPP and QSA were found to be stable for more than 1 year at 0–4°C and the lifetime could be extended by storing at −20°C. The application of QCR and its derivatives (QPP and QSA) to remove Cr(VI) was very successful, with the results showing that QCR had a higher reduction efficiency of 99.8% (30 min), closely followed by QPP/PdNPs mixture at 96.5% (60 min), and then finally QSA/PdNPs mixtures at 91.7%. This highest reduction of QCR could be due to its higher hydrophobicity compared to QPP and QSA hence depicts stronger affinity for Cr(III).

The study revealed that PdNPs catalyst increased the reduction efficiency of chromium (VI) by ~36.5% while a change in operating temperature from 25°C to 45°C improved the efficiency by ~46.8%.The reduction results were validated in environmental (Binghamton University (BU)) and standard reference material (buffalo river sediments (BRS)) soil samples. The results indicated that the analysis could be completed within 1 h. However, the efficiency was found to be higher in BU soil than in BRS soil by 16.1%. It was found that QPP recorded the highest % atom economy of 94.6% and hence indicated higher enhanced performance compared to bioremediation approach which requires several months to achieve ~90% reduction efficiency [35]. The study clearly demonstrated that Quercetin, the water-soluble QPP and QSA, provides a green approach for the reduction of chromium (VI) from the environment.

We have shown that the presence of PdNPs in the reaction media showed remarkable enhancement in the conversion of Cr(VI) to Cr(III) in all the cases studied. FA/PdNPs composite registered the lowest reaction time of 5 min with an overall efficiency of 99.8%. Derivatization of QCR tremendously improved its solubility. It was also observed that reduction of Cr(VI) was pH dependent with higher reduction efficiencies observed at $pH<3$. QCR registered the highest % efficiency in both soil and water samples (95.6% BU soil) followed by QPP (90.5%) and finally QSA (89.1%). The overall reduction efficiency of QPP and QSA in the presence of PdNPs at 45°C in water samples over a period of 1 h was found to be 96.5% and 91.2%, respectively. Over 98% efficiency was achieved for QCR water samples. All the reactions followed a first-order kinetics with respect to Cr(VI).

Table 29.2 Comparison of Reduction Efficiency of Cr(VI) Using PdNPs, Flavonoid Derivatives, and Other Reducing Agents

	[Cr(VI)] (mM)	Time (min)	Efficiency PdNPs (%)	Overall Efficiency (%)
FA/PdNPs	7.14	5, Temp 45°C	68.0	99.8
Sulfur	0.40	60, Temp 130°C	97.7	99.8
PAA/PdNPs	0.10	14, Temp 45°C	14	99.9
QPP/PdNPs	0.065	60, Temp 45°C	36.7	96.5
QSA/PdNPs	0.065	60, Temp 45°C	36.1	91.7
QCR	0.072	30, Temp 45°C	–	99.8
QSA	0.065	60, Temp 45°C	–	55.6
QPP	0.065	60, Temp 45°C	–	59.6

29.4 Conclusions

We have reviewed the use of nanoparticles for the remediation of environmental contaminants such as chlorinated organic solvents, trichloroethylene, heavy metals, and others. Nanomaterials offer the possibility of more effective remediation due to their higher surface-to-volume ratios and the possibility of novel collection and separation protocols. This chapter has focused on *in situ* remediation of inorganic contaminants, specifically chromium. The removal of 99.8% Cr(VI) from complex aqueous media using Pd/NPs was achieved within minutes as opposed to ~90% removal using Bio-Pd method. This chapter also discusses the use of naturally occurring flavonoids as reducing agents for Cr(VI). However, it should be noted that despite the short reaction time reported for PAA and FA reductants (Table 29.2), the use of flavonoids presents greener chemistry involving naturally occuring flavonoids, and further supports the importance of consuming flavonoids such as QCR and its derivatives to successfully reduce Cr(VI) to benign Cr(III) in acidic medium of human digestive system and the environment. The temperature at which the reduction was achieved also compares well with most of the other methods in Table 29.2, apart from sulfur which required very high temperatures of about 130°C.

Acknowledgments

The authors acknowledge the National Science Foundation CBET 1230189 for funding. N.D. also acknowledges the support of the IGERT fellowship of the National Science Foundation under Agreement # DGE-0654112, administered by the Nanobiotechnology Center at Cornell.

References

[1] United Nations Educational, Scientific, and Cultural Organization, World Water Assessment Program Water for People, Water for Life—the United Nations World Water Development Report. Berghahn Books, Barcelona, 2003.

[2] FAO, Statistical Database, Food and Agriculture Organization of the United Nations, Rome, 2006 <http://faostat.fao.org/>.

[3] M. Mengis, S. Bernasconi, R. Gächter, B. Wehrli, Nitrogen elimination in two deep eutrophic lakes, Limnol. Oceanogr. 42 (1997) 1530–1543.

[4] R.B. Jackson, S.R. Carpenter, C.N. Dahm, D.M. McKnight, R.J. Naiman, S.L. Postel, et al., Water in a changing world, Ecol. Appl. 11 (2001) 1027–1045.

[5] S. Jobling, M. Nolan, C.R. Tyler, G. Brighty, J.P. Sumpter, Widespread sexual disruption in wild fish, Environ. Sci. Technol. 32 (1998) 2498–2506.

[6] R.W. Macdonald, L.A. Barrie, T.F. Bidleman, M.L. Diamond, D.J. Gregor, R.G. Semkin, et al., Contaminants in the Canadian Arctic: 5 years of progress in understanding sources, occurrence and pathways, Sci. Total Environ. 254 (2000) 93–234.

[7] M.H.A. Kester, S. Bulduk, D. Tibboel, W. Meinl, H. Glatt, C.N. Falany, et al., Potent inhibition of estrogen sulfotransferase by hydroxylated PCB metabolites: a novel pathway explaining the estrogenic activity of PCBs, Endocrinology 141 (2000) 1897–1900.

[8] H. Iwata, S. Tanabe, N. Sakai, A. Nishimura, R. Tatsukawa, Geographical distribution of persistent organochlorines in air, water and sediments from Asia and Oceania, and their implications for global redistribution from lower latitudes, Environ. Pollut. 85 (1994) 15–33.

[9] R.W.J. Scott, H.C. Ye, R.R. Henriquez, R.M. Crooks, Synthesis, characterization, and stability of dendrimer-encapsulated palladium nanoparticles, Chem. Mater. 15 (2003) 3873.

[10] Y. Ding, Z. Ji, Production and Application of Chromium Compounds, Chemical Industry Press, Beijing, 2003, pp. 334

[11] K. Pohlantdt-Schwandt, Treatment of wood ash containing soluble chromate, Biomass Bioenergy 16 (1999) 447.

[12] G.S. Gupta, Y.C. Sharma, Environmental management of textile and metallic industrial effluents, J. Colloid Interface Sci. 168 (1994) 118.

[13] T. Wang, Z. Li, Removal of chromium from water and wastewater, Environ. Sci. Technol. 26 (2003) 85.

[14] W. Giger, Emerging Chemical Drinking Water Contaminants, in Identifying Future Drinking Water Contaminants, National Academy Press, Washington, DC, 1999, pp. 112–119

[15] W. Giger, et al., Setting Priorities for Drinking Water Contaminants, in Committee on Drinking Water Contaminants of the US Academy of Sciences (Ed.), National Academy Press, Washington, DC, 1999.

[16] US Environmental Protection Agency, Proceedings: EPA Nanotechnology and the Environmental: Applications and Implications STAR Progress Review Workshop, 2003. EPA Document Number: EPA/600/R-02/080.

[17] L. Guczi, Bimetallic nano-particles: featuring structure and reactivity, Catal. Today 101 (2005) 53.

[18] M. Takanori, T. Asakawa, Recently developed catalytic processes with bimetallic catalysts, Appl. Catal. A 280 (2005) 47.

[19] M.T. Reetz, M. Winter, R. Breinbauer, T. Thurn-Albrecht, W. Vogel, Size-selective electrochemical preparation of surfactant-stabilized Pd-, Ni- and Pt/Pd colloids, Chem. A Eur. J. 7 (2001) 1084.

[20] P. Lu, T. Teranishi, K. Asakura, M. Miyake, N. Toshima, Polymer-protected Ni/Pd bimetallic nano-clusters: preparation, characterization and catalysis for hydrogenation of nitrobenzene, J. Phys. Chem. B. 103 (1999) 9673.

[21] T. Miyake, T. Asakawa, Recently developed catalytic processes with bimetallic catalysts, Appl. Catal. A 280 (2005) 47.

[22] T. Teranishi, M. Miyake, Novel synthesis of monodispersed Pd/Ni nanoparticles, Chem. Mater. 11 (1999) 3414.

[23] I.O. K'Owino, M.A. Omole, O.A. Sadik, Tuning the surfaces of palladium nanoparticles for the catalytic conversion of Cr(VI) to Cr(III), J. Environ. Monit. 9 (2007) 657–665.

[24] M.A. Omole, I.O. K'Owino, O.A. Sadik, Palladium nanoparticles for catalytic reduction of Cr(VI) using formic acid, Appl. Catal. B: Environ. 76 (2007) 158–167.

[25] (a) Q. Zhang, H. Minami, S. Inoue, A. Ikuo, Preconcentration by co-precipitation of chromium in natural waters with Pd/8-quinolinol/tannic acid complex and its direct determination by solid-sampling atomic absorption spectrometry, Anal. Chim. Acta 402 (1999) 277–282; (b) H. Brim, S.C. McFarlan, J.K. Fredrickson, K.W. Minton, M. Zhai, L.P. Wackett, et al., Engineering Deinococcus radiodurans for metal remediation in radioactive mixed waste environments, Nat. Biotechnol. 18 (2000) 85–90; (c) B. Deng, T.A. Stone, Surface-catalyzed chromium (VI) reduction: reactivity comparisons of different organic reductants and different oxide surfaces, Environ. Sci. Technol. 30 (1996) 2484–2494.

[26] L. Hu, G. Xia, L. Qu, M. Li, C. Li, Q. Li, et al., The effect of chromium on sulfur resistance of Pd/HY–Al_2O_3 catalysts for aromatic hydrogenation, J. Catal. 202 (2001) 220–228.

[27] S. Khairrulin, B. Beguin, E. Garbowski, M. Primet, Catalytic properties of chromium–palladium loaded alumina in the combustion of methane in the presence of hydrogen sulfide, J. Chem. Soc., Faraday Trans. 93 (1997) 2217.

[28] J. Hu, I.M.C. Lo, G. Chen, Removal of Cr(VI) by magnetite, Water Sci. Technol. 50 (2004) 139–146.

[29] M.A. Schlautman, I. Han, Effects of pH and dissolved oxygen on the reduction of hexavalent chromium by dissolved ferrous iron in poorly buffered aqueous systems, Water. Res. 35 (2001) 1534–1546.

[30] M. Quilntana, G. Curutchet, E. Donati, Factors affecting chromium(VI) reduction by Thiobacillus ferrooxidans, Biochem. Eng. J. 9 (2001) 11–15.

[31] M. Clifford, A nomenclature for phenols with special reference to tea, Crit. Rev. Food. Sci. Nutr. 41 (Suppl.) (2001) 393–397.

[32] G.R. Beecher, Overview of dietary flavonoids: nomenclature, occurrence and intake, J. Nutr. 133 (10) (2003) 3248S–3254S.

[33] U. Nöthlings, S.P. Murphy, L.R. Wilkens, B.E. Henderson, L.N. Kolonel, Flavonols and pancreatic cancer risk: the multiethnic cohort study, Am. J. Epidemiol. 166 (8) (2007) 924.

[34] V.A. Okello, S. Mwilu, N. Noah, A. Zhou, J. Chong, T.M. Knipfing, et al., Reduction of hexavalent chromium using naturally-derived flavonoids, Environ. Sci. Technol. 46 (2012) 10743–10751.

CHAPTER

30

Physicochemistry of Polyelectrolyte Coatings that Increase Stability, Mobility, and Contaminant Specificity of Reactive Nanoparticles Used for Groundwater Remediation

Tanapon Phenrat[1] and Gregory V. Lowry[2]

[1]Department of Civil Engineering, Naresuan University, Phitsanulok, Thailand [2]Department of Civil and Environmental Engineering, Carnegie Mellon University, Pittsburgh, PA, USA

Street, Sustich, Duncan and Savage. Nanotechnology Applications for Clean Water, 2nd Edition.
 DOI: http://dx.doi.org/10.1016/B978-1-4557-3116-9.00030-5

30.1 Challenges of using reactive nanomaterials for in situ groundwater remediation

Nanocatalysts and redox active nanoparticles (10–500 nm) such as nanoscale zero-valent iron (NZVI), bimetallic NZVI, magnetite (Fe_3O_4), and titanium dioxide (TiO_2) have great potential for purification and remediation of contaminated water and groundwater [1–3]. The ability of NZVI and bimetallic NZVI particles to rapidly dechlorinate chlorinated organics [4–8] or to immobilize heavy metals [8–10] found in contaminated groundwater is well documented, and NZVI and bimetallic NZVI have already been applied at more than 20 sites for the in situ groundwater remediation in pilot- or full-scale operations [11]. Fe_3O_4 nanoparticles are demonstrated to be an effective adsorbent for arsenic removal [12]. The small size of nanomaterials results in an increasing fraction of atoms at the surface, excess surface energy, and high surface area. All these properties lead to higher contaminant degradation rates or increased adsorption capacity per mass of the remediation agents compared to bulk materials. The small size of nanomaterials also offers the potential for injection into the subsurface for in situ remediation. This feature is particularly attractive for treating deep contaminant source areas (e.g., dense nonaqueous phase liquid or DNAPL) which are costly to remediate, technically challenging, and difficult (if possible) to meet cleanup targets in a reasonable amount of time [13–16].

Significant research progress has been made toward synthesizing reactive nanomaterials [3,4,17,18] and various inexpensive reactive or highly adsorptive nanomaterials specifically engineered for environmental remediation and are commercially available. Nevertheless, the low mobility of bare nanomaterials in the subsurface due to particle aggregation and particle attachment to aquifer materials, and the inability of bare nanomaterials to target specific contaminants such as DNAPL limit the potential application of these materials. Enhancing colloidal stability and the mobility of nanomaterials in porous media together with maximizing their affinity to target contaminants of interest are needed to ensure the success of in situ groundwater remediation. Polymeric surface modification/functionalization is an attractive alternative to improve the performance of nanomaterials in these aspects. This chapter describes the state-of-the art in using polymeric surface modification with reactive nanoparticles for environmental remediation, and then explains the physicochemical reasons of how polymeric surface modification and functionalization can overcome the challenges—aggregation, deposition, and targeting. Finally, the remaining challenges and ongoing research and development opportunities are discussed.

30.2 Polymeric surface modification/functionalization

30.2.1 Definitions and materials

Polymeric surface modification has received a great deal of attention, especially in industrial applications, as a means of enhancing colloidal stability of particles. A polymer is a macromolecule consisting of a repetition of smaller chemical units, monomers [19]. The number of units in a given chain is called the degree of polymerization, D_p. A polymer of which some monomer units are ionized and charged is referred to as a polyelectrolyte. Polyelectrolytes often achieve a net negative charge from carboxylate, sulfate, or sulfonate groups ($-COO^-$, $-SO_4^-$, $-SO_3^-$, respectively) or a net positive charge from ammonium or protonated amines [19]. Homopolymers consist of the same monomer units in the macromolecule whereas block copolymers consist of one repeating monomer unit followed by one or more blocks of different repeating units. The chemical and structural complexity of different polymers influences their ability to stabilize nanoparticle dispersions, enhance transport, and afford contaminant targeting. Block copolymers have advantages over homopolymers in that different functionalities for different tasks can be built into different blocks [14,19,20]. Table 30.1

Table 30.1 Polymeric Modifiers Used to Modify Nanoparticles for Environmental Remediation and their Reported Ability to Enhance Colloidal Stability and Transport in Porous Media

Surface Modifier	Charge/ Stabilization Type	Performance	
		Stabilization	**Transport**
Polymers			
Polyethylene glycol (PEG)	Nonionic/steric	Good[a]	Poor[a,b]
Polyvinyl alcohol (PVA)	Nonionic/steric	Poor[a]	Suspected to be poor
Guargum	Nonionic/steric	Poor[a]	Suspected to be poor
Polyelectrolytes			
Triblock copolymers[c]			
$PMAA_{48}$–$PMMA_{17}$–PSS_{650} [14,15,21]	Anionic/electrosteric	Excellent	Excellent
$PMAA_{42}$–$PMMA_{26}$–PSS_{462} [14,15,21]	Anionic/electrosteric	Good	Excellent
Polystyrene sulfonate (PSS) [22]	Anionic/electrosteric	Excellent	Good
Polyaspartate (PAP) [22]	Anionic/electrosteric	Excellent	Good
Carboxymethyl cellulose (CMC) [22,23][d]	Anionic/electrosteric	Fair [22], Excellent [23]	Poor,[a] excellent (23)
Poly acrylic acid (PAA) [24,25]	Anionic/electrosteric	Excellent	Good
Mixture of PAA–PSS–bentonite [26]	Anionic/electrosteric	—	Excellent

[a]*Unpublished data from our laboratory.*
[b]*The poor transportability of PEG modified nanoparticles is attributed to the strong specific interaction between PEG and silica sand.*
[c]*Triblock copolymer-functionalized nanoscale zero-valent iron (NZVI) can form stable picking emulsions of dense nonaqueous phase liquid (DNAPL) (TCE) water.*
[d]*The difference in the performance of CMC between [22] and [23] is presumably due to different modification approaches.*

summarizes various polyelectrolytes used for surface modification/functionalization of nanoparticles for groundwater remediation. It should be noted that only negatively charged polyelectrolytes or uncharged polymers are used for this purpose because aquifer materials at neutral pH are normally negatively charged. The mobility of positively charged polyelectrolyte-modified nanoparticles would be limited due to the electrostatic attraction between nanoparticles and aquifer materials.

30.2.2 Nanoparticle surface modification approaches

Surface modification/functionalization of nanoparticles is typically achieved by: (i) grafting polyelectrolytes from the surface of pre-synthesized nanoparticles; (ii) physisorption of polyelectrolytes onto the surface of pre-synthesized nanoparticles; or (iii) incorporating polyelectrolytes into the nanoparticles during the particle synthesis. The first approach uses an advanced synthesis method such as atom transfer radical polymerization (ATRP) [20] to produce a dense polymer brush layer from initiators that are covalently bound to the nanoparticle surface. The dense polymer brush is theoretically predicted to enhance colloidal stability and transport of nanoparticles as will be further discussed. Adsorbing polymers onto the nanoparticles (the second approach) is a less time- and material-intensive procedure than growing polymers from the particles and is desirable for the application of a large quantity such as environmental remediation (if the resulting surface modified nanoparticles can provide sufficient adsorbed polymer layer thickness, density, and appropriate configuration to achieve the intended task). The last approach is a single-step synthesis procedure where a polymeric modifier interacts with the particle surface during nanoparticle synthesis. The physicochemical properties and structures of nanoparticles modified in this way are likely to be different from the reactive nanoparticles obtained by grafting or physisorption of polymeric modifiers on to the surface of pre-synthesized nanoparticles. For example, Fe–Pd nanoparticles synthesized in the presence of sodium carboxymethyl cellulose (CMC) are reported to be colloidally stable, have excellent transportability through a loamy-sand soil, and are more reactive than bare Fe–Pd nanoparticles [23]. In contrast, NZVI particles modified by physisorption of CMC become less reactive [27], have moderate colloidal stability [22], but poor transportability through porous media.

The mass and configuration of charged macromolecules adsorbed onto the particle surface is governed by the molecular weight, ionization, and charge density of the macromolecule, the charge density and polarity of the solid surface, the solvent quality, and ionic strength [19,28]. The mass adsorbed and the configuration of the adsorbed layer is dictated by a balance between electrostatic attraction to the surface and repulsions among neighboring ionized monomer units, a loss of chain entropy upon adsorption, and also nonspecific dipolar interactions among the macromolecule, the solvent, and the surface [19,28]. Homopolymers are normally sorbed onto the surface in the train–loop–tail

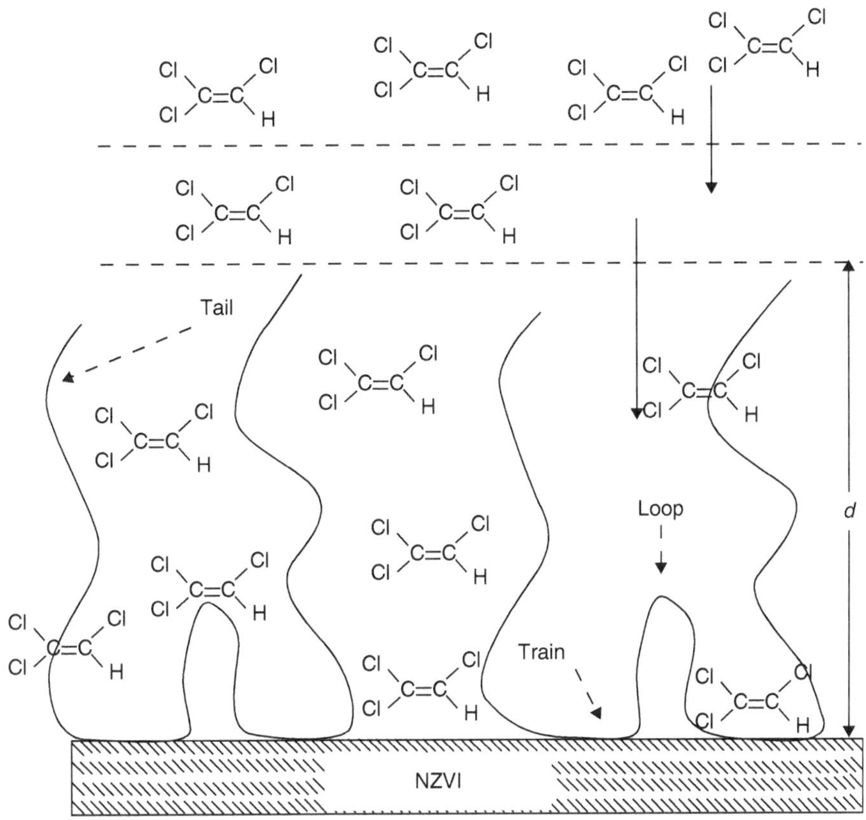

FIGURE 30.1

A schematic diagram illustrating train–loop–tail of orientation of the adsorbed homopolymers on the surface of nanoscale zero-valent iron (NZVI) and the site blocking effect and the mass transfer limitation on trichloroethylene (TCE) dechlorination due to trains, loops, and tails of adsorbed polyelectrolytes.

orientation (Fig. 30.1). Trains are sequences in contact with the sorbed surface. Loops and tails are attached but extend from the surface [19,29,30]. Block copolymers adsorb similarly, but they can be designed to anchor to the surface and theoretically can control the adsorbed layer configuration [19]. For example, poly (methacrylic acid) (PMAA) (Table 30.1), which has specific sorption affinity on the oxide surface, is used as an anchoring block of the trick block copolymer for NZVI surface modification [14]. The adsorbed layer properties, including adsorbed layer thickness (*d*) (Fig. 30.1) and adsorbed polymer mass per specific surface area of nanoparticles (surface excess, Γ), play an important role in stabilizing and enhancing the mobility of nanoparticles in the subsurface as will be discussed next.

30.3 Effect of surface modifiers on the mobility of nanomaterials in the subsurface

The sizes of pores in a groundwater aquifer are typically of the order of the sizes of aquifer materials themselves (tens to hundreds of micrometers), which are several orders of magnitude larger than the sizes of nanomaterials (10–500 nm) used for environmental remediation. Intuitively the transport of nanomaterials in the subsurface should be facile, however, recent studies have reported limited mobility of nanoparticles for environmental remediation in saturated porous media, that is, practical transport distances of only a few centimeters or less for bare nanoparticles [13,15,26]. In situ remediation with reactive nanomaterials typically involves injection of relatively high concentration particle dispersions being injected into the subsurface. Given this scenario, there are two physical phenomena limiting the transport of nanomaterials in water-saturated porous media. First, nanomaterials can be filtered from solution by deposition onto aquifer materials [31]. Second, aggregation or agglomeration can cause pore plugging that limits transport [15,32]. Both aggregation and deposition can be considered as two-step process, transport and attachment. Whether the particles will aggregate (or agglomerate, i.e., loose aggregation) or deposit onto a collector is controlled by attachment, which is governed by colloidal forces between two particles (aggregation) or particles and collectors (deposition) acting at a separation distance on the order of nanometers [31]. For bare nanoparticles, these colloidal forces are electrical double-layer repulsion, van der Waals attraction, magnetic attraction for magnetic nanoparticles (e.g., NZVI), hydration forces, and hydrophobic interactions [31,33]. For polyelectrolyte-coated nanomaterials (a common approach to inhibit aggregation) elastic and osmotic repulsive forces may also occur. Because of the importance of these forces on aggregation, deposition, and thus the mobility of nanomaterials in the subsurface environment, a brief overview of relevant physical chemistry of colloidal forces is provided here. More thorough reviews of this topic can be found elsewhere [31,33].

30.3.1 Colloidal forces and Derjaguin–Landau–Verwey–Overbeek theory

30.3.1.1 Bare nanoparticles

For bare nanoparticles, aggregation and deposition in aqueous environments are generally modeled using Derjaguin–Landau–Verwey–Overbeek (DLVO) theory [31,33]. According to classical DLVO theory, the van der Waals forces (V_{vdW}) are the primary attractive force, whereas repulsive forces are derived from the electrostatic double layer (V_{ES}) (Fig. 30.2A). The V_{vdW} attractive force between two spherical particles of radius R_1 and R_2 or a spherical particle of radius R and

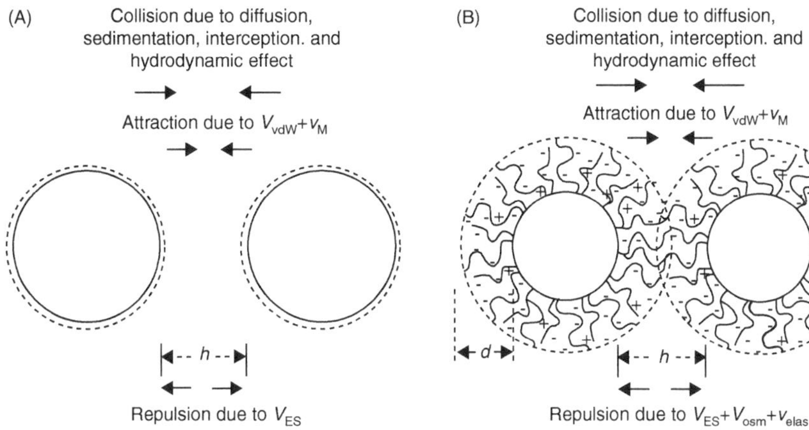

FIGURE 30.2

Schematic representation of particle–particle interaction forces acting on (A) bare charged nanoparticles and (B) electrosterically stabilized polyelectrolyte-modified nanoparticles including van der Waals attraction (V_{vdW}), electrostatic double layer repulsion (V_{ES}), magnetic attraction (V_M), osmotic repulsion (V_{osm}), and elastic-steric repulsion (V_{elas}).

a flat surface (a representative of a large collector grain relative to a nanoparticle) are given in Equations 30.1 and 30.2, respectively [34].

$$V_{\substack{\text{vdw}\\ \text{sphere}-\\ \text{sphere}}}(h) = -\frac{A}{6h}\frac{R_1R_2}{R_1+R_2} \tag{30.1}$$

$$V_{\substack{\text{vdw}\\ \text{sphere}-\\ \text{wall}}}(h) = -\frac{A\times R}{6h} \tag{30.2}$$

where A is the Hamaker constant, which is 1×10^{-19} and 3.7×10^{-20} N m for iron nanoparticles and titanium dioxide (anatase), respectively [35,36], and h(m) is separation distance between two interacting surfaces. The V_{vdW} decays with increasing h and increases in magnitude and extent as the particle size increases. This attractive energy promotes aggregation and deposition. Electrostatic repulsion between two identical particles of radius R and between a particle and a flat surface are given in Equations 30.3 and 30.4, respectively [37,38].

$$V_{\substack{\text{ES}\\ \text{sphere}-\\ \text{sphere}}} = 2\pi\varepsilon_r\varepsilon_o R\zeta_1^2 \text{In}[1+\text{e}^{-kh}] \tag{30.3}$$

$$V_{\substack{\text{ES}\\ \text{sphere}-\\ \text{wall}}} = \pi\varepsilon_r\varepsilon_o R\left[2\zeta_1\zeta_2\text{In}\left(\frac{1+e^{-kh}}{1-e^{-kh}}\right)+(\zeta_2^1+\zeta_2^2)\text{In}(1-e^{-2kh})\right] \tag{30.4}$$

where ε_r is the relative dielectric constant of the liquid, and ε_0 is the permittivity of the vacuum. ζ_1 and ζ_2 are the zeta potentials of a particle and a collector, respectively. κ is the inverse Debye length, which, for symmetrical ($z-z$) electrolytes, can be expressed as in Equation 30.5 [38]:

$$k = \sqrt{\frac{e^2 \sum n_i z_i^2}{\varepsilon_r \varepsilon_o k_B T}} \tag{30.5}$$

where e is the electron charge, n_i is the number concentration of ion i in the bulk solution, and z_i is the valence of ion i. The V_{ES} decays exponentially with separation distance (h) and decreases in magnitude as the Debye length ($1/\kappa$) decreases. In addition, V_{ES} decreases as particle size decreases. This repulsive energy inhibits aggregation and deposition. Because of their small size, the energy barrier for nanoparticles to resist aggregation and deposition may be less than that of micron-sized particles of the same surface charge. The distance from the particle surface over which V_{ES} acts to counter Van der Waal's attractive forces V_{vdW} is theoretically predicted to be similar to the Debye length ($1/\kappa$). As shown in Equation 30.5, the Debye length decreases with increasing ionic strength (n_i) and valency of ionic species present in the bulk solution. Therefore, the ionic strength and composition of groundwater will affect aggregation, deposition, and thus the mobility of nanoparticles [21].

Besides V_{vdW} and V_{ES}, other non-DLVO forces can affect aggregation and deposition in some circumstances. For example, NZVI particles that are single magnetic domain particle [13,32] have an intrinsic permanent magnetic dipole moment even in the absence of an applied magnetic field [35,39]. When particle dipoles are oriented in a head-to-tail configuration, the maximum magnetic attraction energy (V_M) is expressed as Equation 30.6 [32]:

$$V_M = \frac{-8\pi\mu_o M_s^2 R^3}{9\left(\frac{h}{R}+2\right)^3} \tag{30.6}$$

where μ_0 is the permeability of the vacuum, and M_s is the saturation magnetization of the particles. It should be noted that V_M is a longer-range attractive interaction compared to V_{vdW} and increases in magnitude with particle radius to the sixth power. Thus, the size of magnetic nanoparticles such as NZVI significantly affects aggregation and deposition [22].

Classical DLVO theory assumes that the total interaction energy between a pair of particles (aggregation) or particle–flat surface (deposition) is additive, that is, a combination of DLVO forces acting between two surfaces. Figures 30.2A and 30.3 show a schematic diagram of total energy of interaction between a bare nanoparticle and a flat surface. The sum of the interaction energy results in an attractive secondary minimum (Fig. 30.3) where particles can agglomerate or deposit [22,40]. Typically, the magnitude of this attractive secondary minimum well is relatively small. Therefore, this aggregation or deposition mode is reversible if Brownian

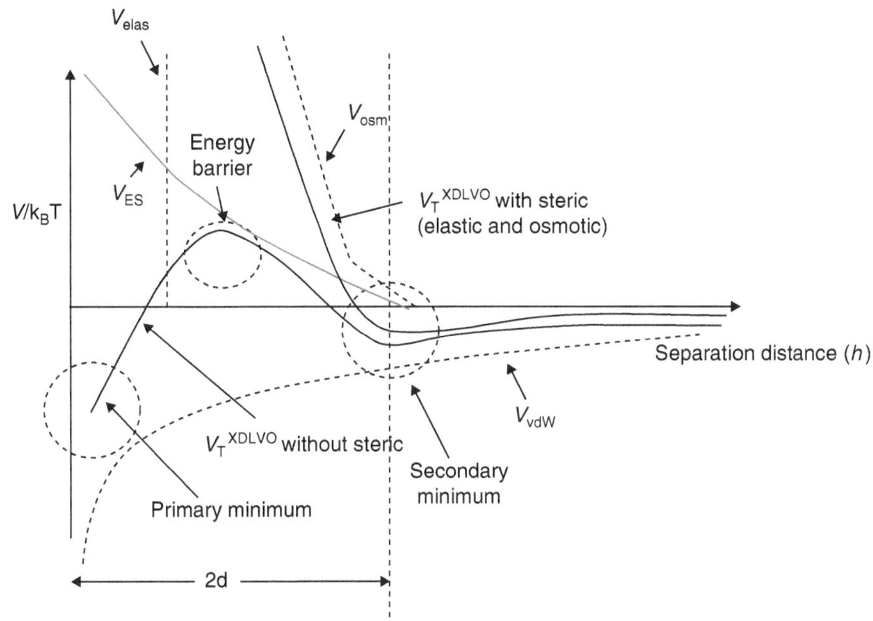

FIGURE 30.3

Hypothetical interaction energy profiles showing various components of the particle–collector/particle–particle interaction energy profile including total interaction energy with and without the electrosteric repulsive forces afforded by adsorbed or grafted polyelectrolytes (V_{osm} and V_{elas}).

energy ($1.5k_BT$) or external forces such as shear force due to fluid flow is large enough to get particles out of this attractive energy well [22,40,41]. Natural geochemical conditions that decrease the magnitude of the energy barrier, such as high ionic strengths and multivalent ions in groundwater, can promote aggregation and deposition under a primary minimum that would be much less reversible.

30.3.1.2 Polymer-modified nanoparticles

Polyelectrolyte surface modification can improve the mobility of nanomaterials in porous media by introducing an additional electrosteric repulsion to counter V_{vdW}, or magnetic attractive forces, that promote aggregation and deposition (Fig. 30.2B) [42–44]. Between two polymer-modified surfaces, the electrosteric repulsion consists of: osmotic repulsion (V_{osm}) and elastic–steric repulsion (V_{elas}). Overlap of the polyelectrolyte layers on two approaching surfaces increases the local polymer segment concentration and thus increases the local osmotic pressure in the overlap region (V_{osm}). Any compression of the adsorbed polyelectrolyte layers below the thickness of the unperturbed layer (*d*) leads to a loss of entropy and gives rise to the elastic repulsion (V_{elas}) [42,44]. The range and magnitude of the electrosteric repulsion between a pair of particles or a particle and a collector depends on the surface concentration of adsorbed polyelectrolyte, and extension and charge density of the adsorbed polyelectrolyte layer. The conformation of the adsorbed polymer can also contribute to the magnitude of the electrosteric repulsion [42]. However, the range of electrosteric repulsion is solely controlled by the adsorbed layer thickness. As shown in Fig. 30.3 (V_T^{XDLVO} with steric), for separation distances beyond $2d$, V_{vdW} attractive forces are dominant and result in the net attractive potential. However, when the polyelectrolyte-modified particles approach one another to a distance less than $2d$, there is a large energy barrier due to the adsorbed layer. In fact, the polymer coatings make it very difficult for these particles to aggregate in a primary minimum, so all agglomeration or deposition must occur under a shallower secondary minimum. This has very important consequences regarding the dispersion stability and mobility of nanoparticles in the subsurface because agglomeration and attachment under a secondary minimum are subjected to disaggregation and detachment due to shear force and Brownian energy. For this reason, aggregation and mobility of surface modified nanoparticles should be largely controlled by the adsorbed polyelectrolyte layer properties. A recent study [22] correlated the adsorbed polyelectrolyte layer properties (d estimated using Ohshima's soft particle analysis, to be discussed next) and the adsorbed polyelectrolyte mass (Γ) on NZVI to their dispersion stabilities (Fig. 30.4A). The higher the adsorbed polyelectrolyte mass and the more extended the layer thickness, the larger the magnitude and extent of electrosteric repulsion and the smaller the secondary minimum well for agglomeration. For similar reason,

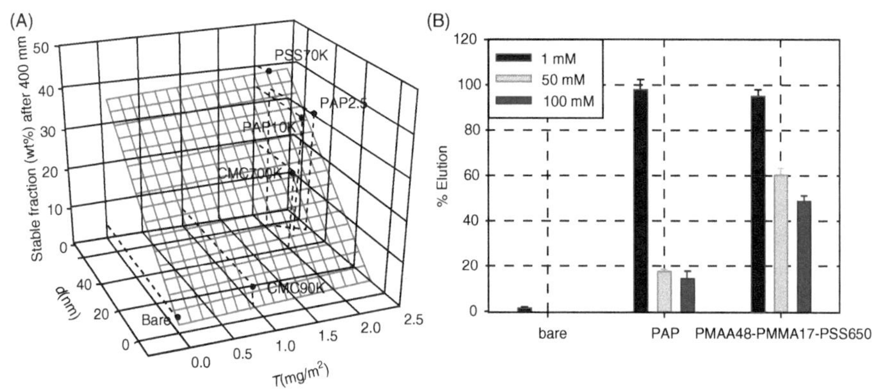

FIGURE 30.4

(A) Correlation between the colloidally stable fraction (wt %) of nanoparticles and the measured surface excess (Γ, mg/m^2) and layer thickness (d, nm) of each adsorbed polyelectrolyte [22]. (B) Effect of ionic strength on the elution of bare and modified reactive nanoscale iron particle (RNIP) through a 12.5-cm column with water-saturated silica sand having a porosity of 0.33. Particle mass concentration is 3 g/L. All samples also contained 1 mM $NaHCO_3$ to control pH at 7.4. The approach velocity was 9.5 m/d [15].

considering the particle–collector interaction under the influence of adsorbed polymer layers, collector ripening, and pore plugging are not expected for the transport of colloidally stable polyelectrolyte modified nanoparticles in saturated porous media. Figure 30.4B illustrates the enhanced mobility of NZVI particles modified by various surface modifiers compared to bare NZVI particles. The influence of electrosteric stabilization on the enhancement of dispersion stability and nanoparticle transport is also generally observed as summarized in Table 30.1.

Another unique feature of electrosteric stabilization with implications on the subsurface transport of nanoparticles is the ability to maintain the strong repulsive forces at high ionic strength and in the presence of multivalent ionic species. Aggregation and deposition of electrostatically stabilized nanoparticles are sensitive to ionic strength and divalent cations due to charge neutralization and shrinking of the Debye length. Electrosterically stabilized nanoparticles are much less sensitive to ionic strengths and types of ionic species [21,42] (Fig. 30.4B).

30.3.2 Adsorbed layer characterization

The adsorbed layer properties including Γ, d, and charge density of the adsorbed layers correlate with dispersion stability and mobility of nanoparticles in the subsurface. Thus, characterization of the adsorbed layer properties is important for modeling and design purposes. While Γ can be measured using well-established solution depletion methods [22], the measurement of d and charge densities are more complicated. Ellipsometry and atomic force microscopy (AFM) can be used to measure the polymer layer thickness adsorbed on a flat surface; however, they cannot be used to measure the layer thickness on the surface of nanoparticles, which are not atomically flat. Dynamic light scattering (DLS) can be used to measure the hydrodynamic layer thickness of surface modified nanoparticles. However, the hydrodynamic layer thickness measured by this technique is very sensitive to particle polydispersity making it unattractive for most nanoparticles that are typically polydisperse [22]. A recent study [22] applied Ohshima's soft particle theory [45], which is much less sensitive to particle polydispersity, to estimate adsorbed polyelectrolyte layer properties by interpreting electrophoretic mobility (EM) measurements as a function of ionic strength.

Ohshima's approach assumes that the electrical potential around a polyelectrolyte adsorbed onto a solid surface consists of two separate electrical potentials in two different regions: the Donnan potential (ψ_{DON}) inside the adsorbed layer of thickness d, and the surface potential (ψ_0) at the boundary between the adsorbed layer and the surrounding solution [45]. This model assumes that the polyelectrolyte segments act as resistance centers that exert frictional forces on the liquid given by $-\Gamma u$, where Γ is the friction coefficient and u is the local liquid velocity inside the polyelectrolyte layer. It is assumed that the ionized groups each contribute a charge Ze (Z, valence of the ionized groups on the polyelectrolyte; e, the electron charge), that they are uniformly distributed within the adsorbed layer at a number density N (m^{-3}), and that the aqueous phase contains a symmetrical electrolyte of valence z and bulk concentration n (m^{-3}). Considering the frictional force of the adsorbed layer in the Navier–Stokes equation, a modified expression for polyelectrolyte-coated nanoparticles that relates their electrophoretic mobility u_e to the characteristics of a charged adsorbed layer on the charged surface of a particle of radius a is given by Equation 30.7 [45]:

$$u_e = \frac{\varepsilon}{\eta}\frac{\psi_o/k_m + \psi_{DON}/\lambda}{1/k_m + 1/\lambda} f\left(\frac{d}{a}\right) + \frac{ZeN}{\eta\lambda^2} + \frac{8\varepsilon k_B T}{\eta\lambda ze} \cdot \tanh\frac{ze\zeta}{4k_B T} \cdot \frac{e^{-\lambda d}/\lambda - e^{-k_m d}/k_n}{1/\lambda^2 - 1/k_m^2} \tag{30.7}$$

where ε is the electric permittivity of the liquid medium, η is its viscosity, λ is a frictional parameter given by $(\gamma/\eta)^{1/2}$, and κ_m is the effective Debye–Hückel parameter of the surface hydrogel layer, which includes the contribution of the fixed charge ZeN [45]. ζ is the apparent zeta potential of the bare particles calculated from EM measurements using Smoluchowski's formula. The function $f(d/a)$ varies between one for a thin adsorbed layer relative to radius of the core particle (a), to two-thirds for a thick layer. Equation 30.7 is valid when λd and $\kappa d > 1$ [45]. The corresponding expressions for ψ_{DON}, ψ_0, $f(d/a)$, and κ_m are given in Equations 30.8–11 [45]:

$$\psi_{DON} = \frac{k_B T}{ze}\sinh^{-1}\left(\frac{ZN}{2zn}\right) \tag{30.8}$$

$$\psi_o = \psi_{DON} - \frac{k_B T}{ze}\tanh\left(\frac{ze\psi_{DON}}{2k_B T}\right) + \frac{4k_B T}{ze} \cdot e^{-k_m d}\tanh\frac{ze\zeta}{4k_B T} \tag{30.9}$$

$$f\left(\frac{d}{a}\right) = \frac{2}{3}\left[1 + \frac{1}{2(1+d/a)^3}\right] \tag{30.10}$$

$$k_m = k\left[\cosh\left(\frac{ze\psi_{DON}}{k_B T}\right)\right]^{1/2} \tag{30.11}$$

where κ_B is Boltzmann's constant, T is absolute temperature, and κ is the Debye–Hückel parameter of the solution. Use of the Ohshima method requires data for the electrophoretic mobility for both the bare particles and for the polyelectrolyte-coated particles as a function of the bulk solution ionic strength.

The procedure for extracting the characteristics of the polyelectrolyte layer from EM data involves a nonlinear fit of Equation 30.7, with terms defined as in Equations 30.8–30.11, to the experimental electrophoretic mobility versus ionic strength of a symmetrical electrolyte to obtain the best fit N, λ, and d [45].

30.4 Contaminant targeting of polymeric functionalized nanoparticles

Polymeric surface modification/functionalization offers a wealth opportunity to specifically target contaminants in situ. For example, novel polymeric surface modified nanoparticles offer the potential for in situ targeting and remediation of DNAPL source zone. A recent study [14] modified NZVI particles with triblock copolymers of three different blocks designed to provide three different functionalities for three different tasks. The PMAA block serves to anchor the triblock copolymers to the NZVI particles. Hydrophobic attractions arising from the PMMA block provide the strong affinity to the NAPL and create a low polarity region that hinders water access to the NZVI particles with the goal of minimizing ZVI oxidation during transport in the soil before it reaches the NAPL. The strongly charged PSS block serves to provide strong electrosteric interparticle repulsions that promote colloidal stability and electrosteric repulsion from the negatively charged surfaces that are predominant in the subsurface at near neutral pH. In water, the PMMA block is collapsed, whereas the extended PSS block provides electrosteric protection. At the DNAPL/water interface, the PMMA block swells in the organic solvent and anchors the particle at the interface (Fig. 30.5A). PSS solubility in the organic phase is too weak to allow full passage of the nanoparticle through the interface into the bulk NAPL phase. NZVI nanoparticles modified by triblock copolymers with this architecture and chemical composition successfully anchored to DNAPL–water interfaces in the laboratory (Fig. 30.5B) [14]. This process required significant input of energy (ultrasonication) to anchor the polyelectrolyte particles to the interface, and the ability for such particles to attach to entrapped DNAPL without this input of energy is unlikely. The use of DNAPL-soluble microemulsions containing reactive nanoparticles is another potential approach to deliver reactive NZVI to

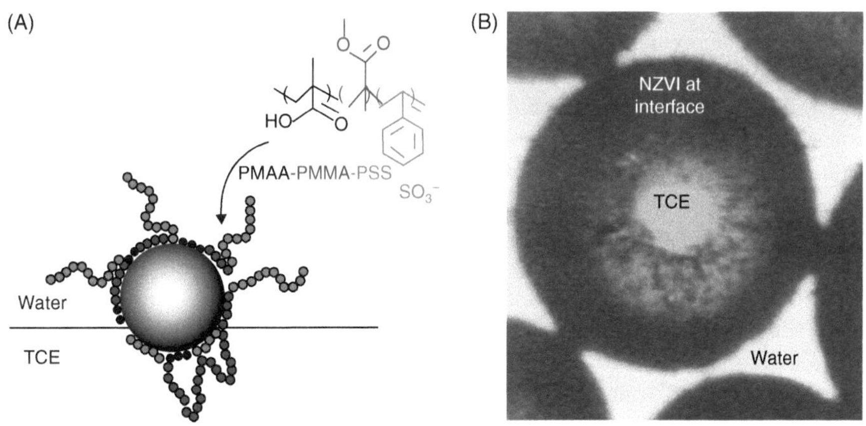

FIGURE 30.5

(A) A schematic diagram illustrating the proposed polyelectrolyte response used to anchor particles at the dense nonaqueous phase liquid (DNAPL)/water interface [15] and (B) micrographs of emulsified TCE droplets in water stabilized by $PMAA_{42}$–$PMMA_{26}$–PSS_{462} triblock copolymer-modified nanoscale zero-valent iron (NZVI) nanoparticles at the C2HCl3 (TCE)–water interface [14].

DNAPL. The iron nanoparticles and water contained in the emulsions coalesce with the DNAPL, bringing reactive NZVI and water (needed for the dechlorination reaction) into the DNAPL.

Beside polymeric surface functionalization, other approaches can be used to functionalize nanoparticles to provide contaminant specificity. A chelating agent such as an ethylenediamine (EDA) was used to functionalize TiO_2 nanoparticles. EDA forms coordination compounds with divalent metal ions such as Cu^{2+} and Pb^{2+}. These EDA-functionalized TiO_2 nanoparticles are designed for the injection into the subsurface to form semipermeable barriers that can immobilize metal ions in situ [46]. Similarly, hydrophobic contaminants such as polychlorinated biphenyls (PCBs) and polyaromatic hydrocarbons (PAHs), which strongly adsorb to soil and sediment, can also be remediated using functionalized hydrophobic nanoparticles that concentrate such contaminants onto them; then, the nanoparticles together with the contaminants can be removed from the environment [47].

30.5 Effect of surface modification/functionalization on contaminant degradation

Polymeric surface modification is needed to increase nanoparticle mobility and may offer some affinity for specific subsurface contaminants. However, because the reactions are heterogeneous (the contaminant must contact the particle surface to be adsorbed or degraded) surface modification by physisorption of polymers can decrease the reactivity of nanoparticles. For example, the polymeric modified NZVI is covered with a layer of sorbed macromolecules (Fig. 30.1). According to the well-defined Scheutjens and Fleer conceptual model for homopolymer sorption [19], charged homopolymers are normally adsorbed onto the surface in the train–loop–tail orientation. Trains are segments of polymer directly attached to the particle surface, whereas loops and tails form an extended polyelectrolyte brush away from the surface. For electron transfer to occur for TCE dechlorination, TCE must be on the NZVI surface. Therefore, TCE in the bulk aqueous phase must diffuse through the adsorbed polyelectrolyte layer to reach the surface. The hydrodynamic resistance and partitioning of TCE into the hydrophobic portion (carbon backbone) of this layer might inhibit diffusion of TCE to the particle surface. In addition, trains of these macromolecules cover the NZVI surface, stopping or inhibiting electron transfer from NZVI to TCE, which can decrease reactivity. A reported study on TCE degradation by triblock copolymer-modified NZVI supports this hypothesis in that the TCE dechlorination rates decrease two–nine times compared to bare NZVI depending on the type and molecular weight of polyelectrolyte modifiers used. Ultimately, there will be a trade-off between mobility and reactivity of nanoparticles used for environmental remediation. There is significant opportunity to develop novel surface coatings that enhance mobility and minimize the effect on reactivity. It may also be possible to use surface modification to control the reaction rate and hence the reactive lifetime of reactive nanomaterials.

30.6 Remaining challenges and ongoing research and development opportunities

Polymeric surface modification/functionalization is a promising approach to increase the mobility of reactive nanoparticles in the subsurface and may allow for efficient emplacement of particles for in situ remediation. Significant progress has been made in this area. However, there are many remaining challenges and ongoing research and development opportunities to improve the effectiveness of this technology. First, effective subsurface remediation requires the ability to emplace reactive nanomaterials in the contaminated zone. Although several field demonstrations and applications of reactive nanoparticles for in situ remediation have been conducted, the controlled or designed transport and emplacement of nanoparticles in the subsurface was not achieved because of the absence of fundamental, quantitative understanding of the transport of concentrated polyelectrolyte modified nanoparticles in the subsurface. Transport and deposition of nanoparticles in saturated porous media has typically been quantitatively modeled using a conventional clean bed filtration model based on the advection–dispersion equation for colloid transport [31]. However, the constraints on this model (e.g., the assumption of irreversible deposition, no aggregation, and perfect sink collectors) make it inappropriate for modeling the injection of high particle concentration dispersions used in remediation. An alternative approach to predicting the transport of polymer modified nanoparticles at high concentration is needed. This approach must take into account the effect of the adsorbed layer properties, geochemical–physical conditions, hydrodynamics, and particle concentration in order to predict the transport and emplacement of polymeric modified nanoparticles using various surface modifiers in different geochemical–physical and hydrodynamic conditions.

Second, the nanomaterials used for remediation are often more expensive than their bulk counterparts; hence, methods to improve the selectivity and efficiency of injected materials are still needed. This can be achieved through synthesis of reactive materials that specifically adsorb heavy metals [46], or have reactive sites specific for the target contaminants. Surface modifiers that can serve a dual role, that is, improve transport and improve selectivity, are needed. For example, surface modifiers that inhibit reactivity until the nanoparticles are emplaced in the contaminant source zone could also increase the effectiveness of nanoparticle remediation strategies by maintaining high reactivity until they are in a location where the contaminant concentrations are the highest. Since the mobility of reactive nanomaterials in the subsurface depends on the type of surface modification used, appropriate surface modification can allow one to select the transport distance, minimize the potential for "run away" particles, and decrease the potential for negative effects of nanoparticles on the environment and human health.

References

[1] N. Savage, M.S. Diallo, Nanomaterials and water purification: opportunities and challenges, J. Nanopart. Res. 7 (4–5) (2005) 331–342.

[2] M.R. Wiesner, J.-Y. Bottero (Eds.), Environmental Nanotechnology: Applications and Impacts of Nanomaterials, first ed., The McGraw-Hill, New York, 2007.

[3] G.E. Fryxell, G. Cao (Eds.), Environmental Applications of Nanomaterials: Synthesis, Sorbents and Sensors, Imperial College Press, 2007.

[4] Y. Liu, S.A. Majetich, R.D. Tilton, D.S. Sholl, G.V. Lowry, TCE dechlorination rates, pathways, and efficiency of nanoscale iron particles with different properties, Environ. Sci. Technol. 39 (5) (2005) 1338–1345.

[5] Y. Liu, G.V. Lowry, Effect of particle age (Fe0 content) and solution pH On NZVI reactivity: H2 evolution and TCE dechlorination, Environ. Sci. Technol. 40 (19) (2006) 6085–6090.

[6] Y. Liu, H. Choi, D. Dionysiou, G.V. Lowry, Trichloroethene hydrodechlorination in water by highly disordered monometallic nanoiron, Chem. Mater. 17 (21) (2005) 5315–5322.

[7] G.V. Lowry, K.M. Johnson, Congener-specific dechlorination of dissolved PCBs by microscale and nanoscale zerovalent iron in a water/methanol solution, Environ. Sci. Technol. 38 (2004) 5208–5216.

[8] W. Zhang, Nanoscale iron particles for environmental remediation: an overview, J. Nanopart. Res. 5 (2003) 323–332.

[9] S.M. Ponder, J.G. Darab, T.E. Mallouk, Remediation of Cr(VI) and Pb(II) aqueous solutions using supported, nanoscale zero-valent iron, Environ. Sci. Technol. 34 (2000) 2564–2569.

[10] S.R. Kanel, B. Manning, L. Charlet, H. Choi, Removal of arsenic(III) from groundwater by nanoscale zero-valent iron, Environ. Sci. Technol. 39 (5) (2005) 1291–1298.

[11] M.R. Wiesner, G.V. Lowry, P. Alvarez, D. Dionysiou, P. Biswas, Assessing the risks of manufactured nanomaterials, Environ. Sci. Technol. A. 40 (14) (2006) 4336–4337.

[12] C.T. Yavuz, J.T. Mayo, W.W. Yu, A. Prakash, J.C. Falkner, S. Yean, et al., Low-field magnetic separation of monodisperse Fe_3O_4 nanocrystals, Science 314 (2006) 964–967.

[13] G.V. Lowry, Nanomaterials for groundwater remediation, in: M.R. Wiesner, J.-Y. Bottero (Eds.), Environmental Nanotechnology: Applications and Impacts of Nanomaterials, McGraw-Hill, New York, 2007.

[14] N. Saleh, T. Phenrat, K. Sirk, B. Dufour, J. Ok, T. Sarbu, et al., Adsorbed triblock copolymers deliver reactive iron nanoparticles to the oil/water interface, Nano. Lett. 5 (12) (2005) 2489–2494.

[15] N. Saleh, K. Sirk, Y. Liu, T. Phenrat, B. Dufour, K. Matyjaszewski, et al., Surface modifications enhance nanoiron transport and NAPL targeting in saturated porous media, Environ. Eng. Sci. 24 (1) (2007) 45–57.

[16] A. Leeson, M.C. Kavanaugh, J.A. Marqusee, B. Smith, H. Stroo, M. Unger, et al., Remediating chlorinated source zones, Environ. Sci. Technol. 37 (11) (2003) 224A–230A.

[17] L. Wu, M. Shamsuzzoha, S.M.C. Ritchie, Preparation of cellulose acetate supported zero-valent iron nanoparticles for the dechlorination of trichloroethylene in water, J. Nanopart. Res. 7 (2005) 469–476.

[18] C.B. Wang, W.X. Zhang, Synthesizing nanoscale iron particles for rapid and complete dechlorination of TCE and PCBs, Environ. Sci. Technol. 31 (7) (1997) 2154–2156.

[19] G.J. Fleer, M.A. Cohen Stuart, J.M.H.M. Scheutjens, T. Cosgrove, B. Vincent, Polymers at Interfaces, Chapman & Hall, London, 1993.

[20] K. Matyjaszewski, J. Xia, Atom transfer radical polymerization, Chem. Rev. 101 (9) (2001) 2921–2990.

[21] N. Saleh, H.-J. Kim, K. Matyjaszewski, R.D. Tilton, G.V. Lowry, Ionic strength and composition affect the mobility of surface-modified NZVI in water-saturated sand columns, Environ. Sci. Technol. Change in press to 42 (9) (2008) 3349–3355.

[22] T. Phenrat, N. Saleh, K. Sirk, H.-J. Kim, R.D. Tilton, G.V. Lowry, Stabilization of aqueous nanoscale zerovalent iron dispersions by anionic polyelectrolytes: adsorbed anionic polyelectrolyte layer properties and their effect on aggregation and sedimentation, J. Nanopart. Res. 10 (2008) 795–814.

[23] F. He, D. Zhao, J. Liu, C.B. Roberts, Stabilization of Fe–Pd nanoparticles with sodium carboxymethyl cellulose for enhanced transport and dechlorination of trichloroethylene in soil and groundwater, Ind. Eng. Chem. Res. 46 (1) (2007) 29–34.

[24] B. Schrick, B.W. Hydutsky, J.L. Blough, T.E. Mallouk, Delivery vehicles for zerovalent metal nanoparticles in soil and groundwater, Chem. Mater. 16 (11) (2004) 2187–2193.

[25] G.C.C. Yang, H.-C. Tu, C.-H. Hung, Stability of nanoiron slurries and their transport in the subsurface environment, Sep. Purif. Technol. 58 (2007) 166–172.

[26] B.W. Hydutsky, E.J. Mack, B.B. Beckerman, J.M. Skluzacek, T.E. Mallouk, Optimization of nano- and microiron transport through sand columns using polyelectrolyte mixtures, Environ. Sci. Technol. 41 (18) (2007) 6418–6424.

[27] T. Phenrat, Y. Liu, R.D. Tilton, and G.V. Lowry. Effect of adsorbed polyelectrolytes on TCE dechlorination, product distribution, and h2 evolution by nanoscale zerovalent iron particles, Environ. Sci. Technol., under preparation.

[28] K. Holmberg, B. Jonsson, B. Kronberg, B. Lindman, Surfactants and Polymers in Aqueous Solution, second ed., John Wiley & Sons, Ltd., West Sussex, 2003.

[29] J.M.H.M. Scheutjens, G.J. Fleer, Statistical theory of the adsorption of interacting chain molecules. 1. Partition function, segment density distribution, and adsorption isotherms, J. Phys. Chem. 83 (1979) 1619–1635.

[30] J.M.H.M. Scheutjens, G.J. Fleer, Statistical theory of the adsorption of interacting chain molecules. 2. Train, loop, and tail size distribution, J. Phys. Chem. 84 (1980) 178–190.

[31] M. Elimelech, J. Gregory, X. Jia, R. Williams, Particle deposition and aggregation: Measurement, Modeling, and Simulation, Butterworth-Heinemann, Boston, 1995.

[32] T. Phenrat, N. Saleh, K. Sirk, R.D. Tilton, G.V. Lowry, Aggregation and sedimentation of aqueous nanoscale zerovalent iron dispersions, Environ. Sci. Technol. 41 (1) (2007) 284–290.
[33] D.F. Evans, H. Wennerstrom, The Colloidal Domain; Where Physics, Chemistry, Biology, and Technology Meet, Wiley-VCH, New York, 1999.
[34] J.N. Israelachvili, Intermolecular and Surface Forces, Second Edition: With Applications to Colloidal and Biological Systems, second ed, Academic Press, 1992.
[35] R.E. Rosensweig, Ferrohydrodynamics, Cambridge University Press, New York, 1985.
[36] A.I. Gómez-Merino, F.J. Rubio-Hernández, J.F. Velázquez-Navarro, F.J. Galindo-Rosales, P. Fortes-Quesada, The Hamaker constant of anatase aqueous suspensions, J. Colloid Interface Sci. 316 (2) (2007) 451–456.
[37] J. de Vicente, A.V. Delgado, R.C. Plaza, J.D.G. Durán, F. González-Caballero, Stability of cobalt ferrite colloidal particles: Effect of pH and applied magnetic fields, Langmuir 16 (2000) 7954–7961.
[38] J. Brant, J. Labille, J.-Y. Bottero, M.R. Wiesner, Nanoparticle transport, aggregation, and deposition, in: M. Wiesner, J.-Y. Bottero (Eds.), Environmental Nanotechnology: Applications and Impacts of Nanomaterials, 1st ed., McGraw-Hill, New York, 2007.
[39] R.A. McCurrie, Ferromagnetic Materials: Structure and Properties, Academic Press, London, 1994.
[40] A. Franchi, C.R. O'Melia, Effects of natural organic matter and solution chemistry on the deposition and reentrainment of colloids in porous media, Environ. Sci. Technol. 37 (6) (2003) 1122–1129.
[41] S. Torkzaban, S.A. Bradford, S.L. Walker, Resolving the coupled effects of hydrodynamics and DLVO forces on colloid attachment in porous media, Langmuir 23 (2007) 9652–9660.
[42] D.H. Napper, Polymeric Stabilization of Colloidal Dispersions, Academic Press, New York, 1983.
[43] M.S. Romero-Cano, A. Martín-Rodríguez, F.J. de las Nieves, Electrosteric stabilization of polymer colloids with different functionality, Langmuir 17 (2001) 3505–3511.
[44] G. Fritz, V. Schadler, N. Willenbacher, N.J. Wagner, Electrosteric stabilization of colloidal dispersions, Langmuir 18 (2002) 6381–6390.
[45] H. Ohshima, Electrophoresis of soft particles, Adv. Colloid Interface Sci. 62 (1995) 189–235.
[46] S.V. Mattigod, G.E. Fryxell, K. Alford, T. Gilmore, K. Parker, J. Serne, et al., Functionalized TiO_2 nanoparticles for use for in situ anion immobilization, Environ. Sci. Technol. 39 (2005) 7306–7310.
[47] W. Tungittiplakorn, L. Lion, C. Cohen, J. Kim, Engineered polymeric nanoparticles for soil remediation, Environ. Sci. Technol. 38 (5) (2004) 1605–1610.

CHAPTER

31 Stabilization of Zero-Valent Iron Nanoparticles for Enhanced In Situ Destruction of Chlorinated Solvents in Soils and Groundwater

Feng He[1], Dongye Zhao[2], and Chris Roberts[3]

[1]*Oak Ridge National Laboratory, Oak Ridge, TN, USA*

[2]*Environmental Engineering Program, Department of Civil Engineering, Auburn University, Auburn, AL, USA* [3]*Department of Chemical Engineering, Auburn University, Auburn, AL, USA*

31.1 Introduction

Chlorinated hydrocarbons such as trichloroethylene (TCE) and tetrachloroethylene (PCE) are among the most widely detected contaminants in soils at thousands of priority sites in the United States. The economic impact of chlorinated hydrocarbons in soils and groundwater is enormous. The US government estimated that it will cost $750 billion (or approximately $8,000 per American family) over the next 30 years to clean up the nation's contaminated groundwater [1].

A variety of technologies have been explored to remediate soils contaminated with chlorinated hydrocarbons, including bioremediation, pump-and-treat, thermal treatment, and permeable reactive barriers (PRB) [2]. Yet, it remains highly

Street, Sustich, Duncan and Savage. Nanotechnology Applications for Clean Water, 2nd Edition.
 DOI: http://dx.doi.org/10.1016/B978-1-4557-3116-9.00031-7

challenging to remove or destroy these dense nonaqueous phase liquids (DNAPLs) in a timely and cost-effective manner due to the magnitude and complexity of the problem [3]. In fact, current remediation practices remain, to a great extent, reliant upon excavation and subsequent landfill, which is costly and environmentally disruptive [4].

In the past decade or so, abiotic dechlorination using granular zero-valent iron (ZVI) particles has elicited great interest [5–7]. By 2003, approximately 70 PRBs employed commercial granular iron to degrade chlorinated hydrocarbons [7]. However, the dechlorination rate using granular iron particles is very slow, with a half-life for TCE reduction being in the order of days or longer [8,9]. As a result, toxic intermediate by-products such as vinyl chloride (VC) are often produced [10,11]. When applied to more persistent chemicals such as polychlorinated biphenyls (PCBs), these ZVI particles are not effective unless under rather extreme conditions (e.g., >300°C) [12,13]. Physically, granular iron particles are too big to be delivered into soils and are not transportable in subsurface. As a result, they can only be used in a passive process such as in the "funnel and gate" or PRB.

Various modifications have been explored to improve the performance of iron-based particles. One strategy is to reduce the particle size to the nanoscale, thereby increasing the surface area and enhancing the degradation rate [6]. Coating iron particles with a second catalytic metal such as Pd or Ni can also accelerate the dechlorination process [6,14,15]. It was reported [6] that reducing Pd-coated iron particle size from mm to nm (10–100 nm) increased TCE degradation rate by 10–100 times. In addition, truly nanoscale iron particles offer an unrivaled advantage in that they can be highly dispersible, and thus can be applied in situ to proactively attack the contaminant sources.

However, this promising technology was confined with a key technical barrier in that ZVI nanoparticles tend to rapidly agglomerate in water and grow to micron-scale or larger in a few minutes, thereby losing their soil mobility and chemical reactivity rapidly [16–18]. Conventionally, ZVI powders were prepared by reduction of Fe^{2+}/Fe^{3+} salts using borohydride (BH_4^-) in the aqueous phase. The earliest studies on this preparation method can be dated back to 1953 [19] and 1962 [20]. However, detailed studies on the synthetic chemistry were not reported until 1990 [21] and then 1995 [22]. Although other solvent-based methods such as micro-emulsion-based methods, sonication-assisted methods, and sol–gel methods have been reported, the water-based approach appears most suitable for environmental applications for its minimal use of organic solvents or chemicals. However, due to their extremely large area-to-volume ratio and high surface reactivity, ZVI nanoparticles tend to react rapidly with the surrounding media (dissolved oxygen (DO) and water) and interact with other particles, resulting in the formation of micron- to mm-scale aggregates.

31.2 Stabilization of zero-valent iron nanoparticles using polysaccharides

To prevent nanoparticle agglomeration, various strategies have been explored to stabilize various metal nanoparticles. To stabilize nanoscale iron oxides, various stabilizers have been found to be effective, including thiols [23], carboxylic acids [24], silica [25], surfactants [26], polymers, including some water soluble polysaccharides [27–29], copolymers of acrylic acids [30,31], styrenesulfonic acids [30], vinylsulfonic acid [30], and long-chain alcohols [32]. However, for environmental uses, not all these stabilizers are applicable to stabilizing ZVI nanoparticles. For example, thiols and carboxylic acids may be reduced by ZVI, some polymers may not function properly in water [32], and some stabilizers themselves are either environmentally harmful or cost-prohibitive.

Stabilization of ZVI nanoparticles was also explored recently. Mallouk and coworkers employed carbon nanoparticles and poly(acrylic acid) (PAA) for stabilizing and/or delivering iron-based nanoparticles [33,34]. Lowry et al. [35] prepared sorptive nanoparticles using so-called block copolymer shells consisting of a hydrophobic inner shell surrounded by a hydrophilic outer shell for dechlorination of DNAPLs. Bhattacharyya et al. [36] synthesized iron nanoparticles with the aid of cellulose acetate membrane supports and observed improved reaction rates for TCE degradation. Using a micro-emulsion method, Li et al. [37] synthesized iron nanoparticles that degraded TCE 2.6 times faster than non-stabilized particles. Sun and Zhang [38] reported that polyvinyl alcohols can reduce the size of ZVI nanoparticles from 60 to 7.9 nm.

Zhao and coworkers developed a particle stabilization strategy that employs low-cost, food-grade polysaccharides (starch or cellulose) as stabilizers [16,17, 39,40]. These polyhydroxylated and/or polycarboxylated macromolecules possess some novel features, which may prove highly useful for stabilizing ZVI nanoparticles. First, they can serve as molecular level capsules to shield formed nanoparticles from agglomeration. Second, they are much cheaper than virtually all other stabilizers tested so far. Third, they are environmentally benign and biodegradable. And fourth, they are mobile in soils, and hence, suitable for injection uses.

Polysaccharides are the most abundant natural biopolymers, and are composed of interconnected glucose and xylose subunits. Starch and cellulose are the most abundant polysaccharide members. Starch was recently used in surface modification of superparamagnetic nanoparticles [41,42]. Under the protection of starch, Ag nanoparticles remained suspended in water for months [42]. Recently, water-soluble starches were also used to disperse Au nanoparticles in water [43], and used as a morphology-directing agent for tellurium nanowires [44].

Similar to starch, cellulose also consists of anhydroglucose subunits, which however, are joined by so-called beta linkages to form a linear chain structure. Although it is linear and nominally thermoplastic, native cellulose is not water-soluble. However, cellulose (and starch) can be easily converted to water-soluble derivatives with desired features. Numerous modified celluloses are commercially available, many of which may serve as novel stabilizers for ZVI nanoparticles. Carboxymethyl cellulose (CMC) was used as a stabilizer in the preparation of superparamagnetic iron [45] and Ag nanoparticles [46]. The NaCMC-stabilized Ag(0) nanoparticles remained highly stable in water after 2 months. Zhang et al. stabilized Se nanoparticles with modified polysaccharides, including chitosan, konjac glucomannan, acacia gum, and CMC [47]. Chitosan was also reported to be an efficient stabilizer for Ag, Au, Pd, and Pt nanoparticles [48,49]. Moreover, heparin, dextran, and dextrin were used as stabilizing and reducing agents for Au and Ag nanoparticles [50,51]. More recently, sodium alginate was used for preparing stabilized Au nanoparticles [52].

For environmental remediation uses, a stabilizer should possess some critical attributes: (i) it can effectively facilitate dispersion, that is, prevent agglomeration, of nanoparticles; (ii) it must not cause any harmful environmental effect; (iii) it will not significantly alter the conductivity of soils; and (iv) it must be cost-effective.

He et al. [16,17,39,40] demonstrated that starch, cellulose, and their derivatives can serve as effective stabilizers to yield ZVI nanoparticles suitable for in situ destruction of chlorinated hydrocarbons. Figure 31.1 depicts the schematic of this modified water-based procedure for synthesizing stabilized nanoparticles. The preparation is conducted in flasks attached to a vacuum line. In Step 1, a stabilizer solution is prepared at pH 6.7–6.9. In Step 2, an Fe^{3+}/Fe^{2+} stock solution is added to the stabilizer solution to yield a solution with desired Fe and stabilizer concentrations. In this step, the molar ratio of stabilizer to Fe and type of a stabilizer can be varied to yield nanoparticles of various particle sizes. In Step 3, Fe^{3+}/Fe^{2+} is reduced to Fe(0) using approximately 1.5 times stoichiometric amounts of borohydride ($NaBH_4$). To ensure efficient use of the reducing agent and to preserve the reactivity of the resultant Fe(0) nanoparticles, the reactor

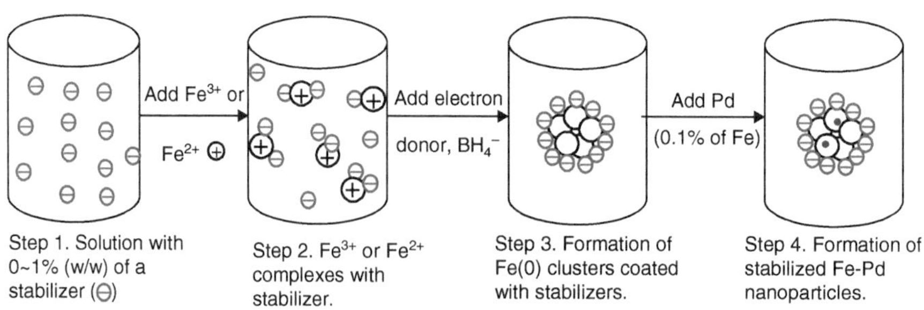

FIGURE 31.1

An innovative procedure for synthesizing stabilized zero-valent iron (ZVI) nanoparticles.

system should be operated under anaerobic conditions. Ferric iron is reduced by borohydride through the reaction shown in Equation 31.1:

$$Fe(H_2O)_6^{3+} + 3BH_4^- + 3H_2O \rightarrow Fe^0 + 10\tfrac{1}{2}H_2 + 3B(OH)_3 \qquad (31.1)$$

The resultant ZVI nanoparticles can be either used as a mono-metallic agent, or in Step 4, loaded with trace amounts (approximately 0.1 percent of Fe, w/w) of a second metal (e.g. Pd) as a catalyst to yield Fe–Pd bimetallic nanoparticles. It was shown that the addition of trace amounts of Pd substantially enhanced the dechlorination reactivity.

Figure 31.2 shows pictures of various iron particles. Whereas non-stabilized iron particles form agglomerates and precipitate in a few minutes, the starch- or cellulose-stabilized nanoparticles remain fully dispersible in water after a month. This observation suggests that the presence of a stabilizer can prevent ZVI nanoparticles from agglomeration. Fourier transform infrared spectroscopy (FTIR) study [17] revealed that the stabilization was facilitated by the sorption CMC molecules onto the surface of the ZVI nanoparticles.

Figure 31.3 compares transmission electron microscope (TEM) images of Fe–Pd nanoparticles prepared without a stabilizer (A) and with a soluble starch (B) or NaCMC (C) [16,17]. Figure 31.3A shows that the non-stabilized Fe–Pd particles appeared as much bulkier dendritic aggregates. In contrast, the starch or NaCMC-stabilized nanoparticles appeared as discrete and much finer nanoparticles even after 1 day of aging. The mean particle size was 14.1 nm for starched

FIGURE 31.2

(A) Commercial (Fisher Scientific) iron powder (labeled as 100 nm); (B) Fisher "nanoparticles" precipitate immediately in water; (C) 0.1 g/L iron nanoparticles stabilized with 0.2 percent (w/w) of a starch prepared in our lab—nanoparticles remain fully dispersed in water after a month; and (D) 0.1 g/L iron particles without a stabilizer prepared in our lab—non-stabilized iron particles agglomerate in minutes and precipitate in water.

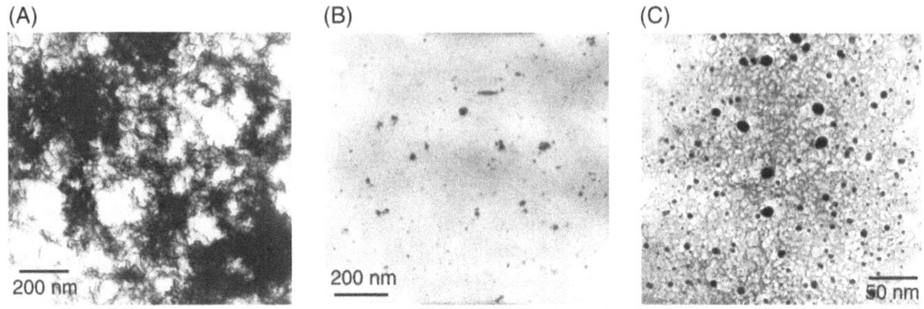

FIGURE 31.3

Freshly prepared zero-valent iron (ZVI) particles (A) without a stabilizer, (B) with 0.2 percent (w/w) of a starch, or (C) stabilized with 0.2 percent (w/w) CMC90K and after 1 day of aging (i.e., storage in a sealed vial at 4°C).

particles and 8.1 nm for NaCMC-stabilized particles. Evidently, the presence of the stabilizers prevented the agglomeration of the resultant iron particles.

Figure 31.4A shows that as the stabilizer-to-Fe molar ratio increases from 0.0025 to 0.003, the size distribution of the ZVI nanoparticles shifted substantially from primarily 37.2 and >200 nm particles to primarily 22.8 nm particles. Further increase in CMC/Fe ratio resulted in further decrease in particle size and narrower size distribution although not as significant. This observation indicates that ZVI nanoparticles of different size and size distribution can be prepared by varying the stabilizer-to-Fe molar ratio.

Figure 31.4B shows that based on an equal CMC concentration of 0.2 percent (w/w), a CMC with greater M.W. results in smaller nanoparticles. To test the effect of DS (degree of substitution), three types of CMC with the same M.W. of 90 k but with different DS (CMC90K with a DS of 0.7, EP-ML with a DS of 1.2, and 2-hydroxyethyl cellulose with no COO^- groups) were compared. (The higher DS indicates more –OH groups in CMC are substituted by carboxymethyl groups). Figure 31.4B shows that EP-ML resulted in smaller nanoparticles (16.0 vs. 18.6 nm) than CMC90K.

31.3 Reactivity of starch- or carboxymethyl-cellulose-stabilized zero-valent iron nanoparticles

In addition to improved and tunable soil transport, the stabilized ZVI nanoparticles displayed remarkably greater reactivity than non-stabilized particles when used for degradation of TCE and PCBs [16,17]. Zhao and coworkers [16,17] showed that at a modest Fe dose of 0.1 g/L and with a trace amount (0.1 percent of Fe) of Pd as a catalyst, the starch- and cellulose-stabilized nanoparticles destroyed, respectively, approximately 73 percent and 99 percent of TCE within

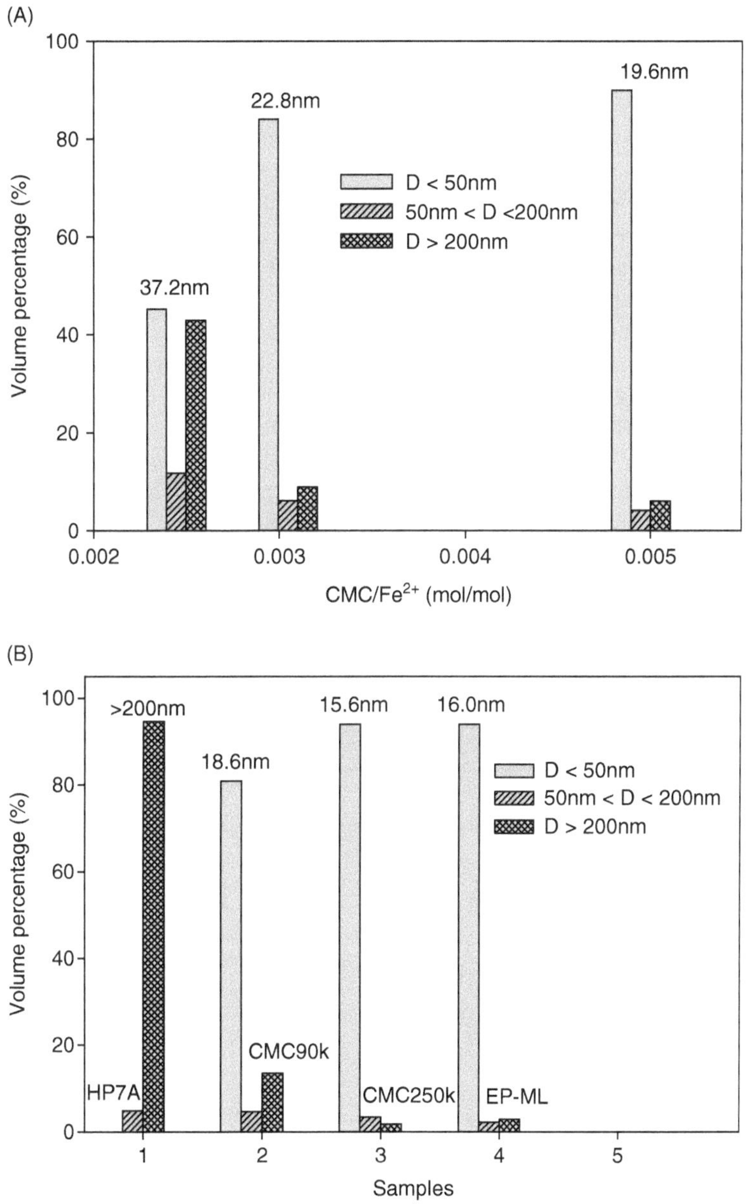

FIGURE 31.4

(A) Size distribution of zero-valent iron (ZVI) nanoparticles synthesized at different carboxymethyl cellulose (CMC)/Fe^{2+} ratios and at an initial Fe^{2+} concentration of 1 g/L (stabilizer: CMC90k, temp: 22°C). (B) Size distribution of Fe nanoparticles stabilized with CMC of various molecular weight and degree of substitution (Fe: 0.1 g/L, CMC90K: 0.2 percent (w/w), temp: 22°C).

40 minutes, compared to only approximately 25 percent for non-stabilized particles. In addition, He and Zhao [16] observed that chloride production is nearly stoichiometrically coupled with TCE degradation during the degradation of TCE with the stabilized ZVI nanoparticles, indicating that no intermediate products (e.g., dichloroethylene (DCE) or vinyl chloride (VC)) were detected.

The application of different stabilizers can also result in nanoparticles of different reactivity. At a fixed CMC-to-Fe^{2+} molar ratio of 0.0124, the use of three CMC stabilizers HP5A (M.W. = 13 k), CMC90k (M.W. = 90 k), and CMC250k (M.W. = 250 k) resulted in ZVI nanoparticles (0.1 g L^{-1}) of >200 nm, 18.6 nm, and 15.6 nm, respectively. The three stabilizers bear homologous chain-like molecular structures but with markedly different molecular size (length). Figure 31.5 shows that the smaller nanoparticles (stabilized with CMC of greater M.W.) exhibited greater reactivity for TCE degradation. FT-IR (17) studies indicated that the CMC molecules are adsorbed on the ZVI nanoparticle surface via the $-COO^-$ and $-OH$ groups. The larger the CMC molecules, the fewer molecules are adsorbed to the nanoparticles (i.e., larger CMC molecules are more effective stabilizers). Therefore, larger CMC molecules occupy fewer reactive sites on Fe surface, leaving more sites available for reactions. From a mass transfer viewpoint, the sorption of larger CMC molecules results in a bulkier and loosely assembled layer of CMC molecules on the particles (as opposed to a more compact layer for smaller CMC molecules), resulting in less mass transfer resistance of TCE to the Fe–Pd surface.

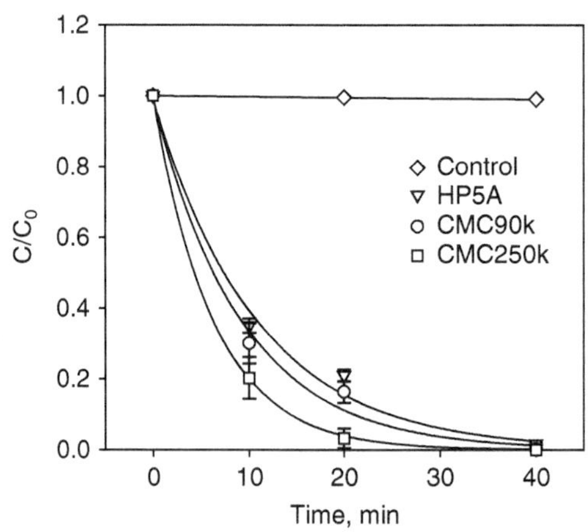

FIGURE 31.5

Hydrodechlorination of trichloroethylene (TCE) using Fe–Pd nanoparticles stabilized with carboxymethyl celluloses (CMCs) of various M.W. Iron dose = 0.1 g L^{-1} Fe, C_0 = 20 mg L^{-1} TCE, CMC = 0.2 wt. percent for all cases, Pd:Fe molar ratio = 1.0 mg Pd/g Fe, reaction pH = 8.3 ± 0.2. Symbols: mean of experimental duplicates with standard deviation; lines: the first-order reaction rate model fittings.

References

[1] T. Moran, New technology revolutionizing groundwater clean-up, E-Wire Press Release, 2004, <http://www.ewire.com>.

[2] J.H.C. Wong, C.H. Lim, G.L. Nolen, Design of Remediation Systems, CRC Press/Lewis Publishers, Boca Raton, FL, 1997.

[3] NASA, Groundwater remediation technologies: emulsified zero-valent iron, Kennedy Space Center NASA Kennedy Space Center, 2004, <http://pubdevelopment.rti.org/nasa/ksc/Remediation/ezvi.cfm>.

[4] W.R. Berti, S.D. Cunningham, In-place inactivation of Pb in Pb-contaminated soils, Environ. Sci. Technol. 31 (1997) 1359–1364.

[5] L.J. Matheson, P.G. Tratnyek, Reductive dehalogenation of chlorinated methanes by iron metal, Environ. Sci. Technol. 28 (1994) 2045–2053.

[6] C.B. Wang, W.X. Zhang, Synthesizing nanoscale iron particles for rapid and complete dechlorination of TCE and PCBs, Environ. Sci. Technol. 31 (1997) 2154–2156.

[7] R.W. Gillham, Discussion of nano-scale iron for dehalogenation, Ground Water Monit. Remed. 23 (1) (2003) 6–8.

[8] G.D. Sayles, G. You, M. Wang, M.J. Kupferle, DDT, DDD, and DDE dechlorination by zero-valent iron, Environ. Sci. Technol. 31 (1997) 3448–3454.

[9] T.L. Johnson, M.M. Scherer, P.G. Tratnyek, Kinetics of halogenated organic compound degradation by iron metal, Environ. Sci. Technol. 30 (1996) 2634–2640.

[10] S.W. Orth, R.W. Gillham, Dechlorination of trichloroethene in aqueous solution using Fe^0, Environ. Sci. Technol. 30 (1996) 66–71.

[11] W.A. Arnold, A.L. Roberts, Pathways and kinetics of chlorinated ethylene and chlorinated acetylene reaction with Fe(0) particles, Environ. Sci. Technol. 34 (2000) 1794–1805.

[12] F. Chuang, R.A. Larson, M.S. Wessman, Zero-valent iron-promoted dechlorination of polychlorinated biphenyls, Environ. Sci. Technol. 29 (1995) 2460–2463.

[13] V.S. Magar, PCB treatment alternatives and research directions, J. Environ. Eng. 129 (2003) 961–965.

[14] W.X. Zhang, C.B. Wang, H.L. Lien, Treatment of chlorinated organic contaminants with nanoscale bimetallic particles, Catal. Today 40 (1998) 387–395.

[15] B. Schrick, J.L. Blough, A.D. Jones, T.E. Mallouk, Hydrodechlorination of trichloroethylene to hydrocarbons using bimetallic nickel-iron nanoparticles, Chem. Mater. 14 (2002) 5140–5147.

[16] F. He, D. Zhao, Preparation and characterization of a new class of starch-stabilized bimetallic nanoparticles for degradation of chlorinated hydrocarbons in water, Environ. Sci. Technol. 39 (2005) 3314–3320.

[17] F. He, D. Zhao, J. Liu, C. Roberts, Stabilization of Fe-Pd bimetallic nanoparticles with sodium carboxymethyl cellulose for enhanced degradation of TCE in water, Ind. Eng. Chem. Res. 46 (2007) 29–34.

[18] B. Schrick, B.W. Hydutsky, J.L. Blough, T.E. Mallouk, Delivery vehicles for zerovalent metal nanoparticles in soil and groundwater, Chem. Mater. 16 (2004) 2187–2193.

[19] H.I. Schlesinger, H.C. Brown, A.E. Finholt, J.R. Gilbreath, H.R. Hoekstra, E.K. Hyde, Sodium borohydride, its hydrolysis and its use as a reducing agent and in the generation of hydrogen, J. Am. Chem. Soc. 75 (1) (1953) 215–219.

[20] H.C. Brown, C.A. Brown, Simple preparation of highly active platinum metal catalysts for catalytic hydrogenation, J. Am. Chem. Soc. 84 (8) (1962) 1493–1494.

[21] A. Corrias, G. Ennas, G. Licheri, G. Marongiu, G. Paschina, Amorphous metallic powders prepared by chemical reduction of metal ions with potassium borohydride in aqueous solution, Chem. Mater. 2 (1990) 363–366.

[22] G.N. Glavee, K.J. Klabunde, C.M. Sorensen, G.C. Hadjipanayis, Chemistry of borohydride reduction of iron(II) and iron(III) ions in aqueous and nonaqueous media. Formation of nanoscale Fe, FeB, and Fe_2B powders, Inorg. Chem. 34 (1995) 28–35.

[23] G. Kataby, T. Prozorov, Y. Koltypin, H. Cohen, C.N. Sukenik, A. Ulman, et al., Self-assembled monolayer coatings on amorphous iron and iron oxide nanoparticles: thermal stability and chemical reactivity studies, Langmuir 13 (1997) 6151–6158.

[24] G. Kataby, M. Cojocaru, R. Prozorov, A. Gedanken, Coating carboxylic acids on amorphous iron nanoparticles, Langmuir 15 (1999) 1703–1708.

[25] B.J. De Gans, C. Blom, J. Mellema, A.P. Philipse, Preparation and magnetisation of a silica-magnetite inverse ferrofluid, J. Magn. Magn. Mater. 201 (1999) 11–13.

[26] S. Sun, H. Zeng, Size-controlled synthesis of magnetite nanoparticles, J. Am. Chem. Soc. 124 (2002) 8204–8205.

[27] D.K. Kim, M. Mikhaylova, Y. Zhang, M. Muhammed, Protective coating of superparamagnetic iron oxide nanoparticles, Chem. Mater. 15 (2003) 1617–1627.

[28] J. Chatterjee, Y. Haik, C.J. Chen, Polyethylene magnetic nanoparticle: a new magnetic material for biomedical applications, J. Magn. Magn. Mater. 246 (2002) 382–391.

[29] H. Pardoe, W. Chua-anusorn, T.G St. Pierre, J. Dobson, Structural and magnetic properties of nanoscale iron oxide particles synthesized in the presence of dextran or polyvinyl alcohol, J. Magn. Magn. Mater. 225 (2001) 41–46.

[30] A. Ditsch, P.E. Laibinis, D.I.C. Wang, T.A. Hatton, Controlled clustering and enhanced stability of polymer-coated magnetic nanoparticles, Langmuir 21 (2005) 6006–6018.

[31] D. Rabelo, E.C.D. Lima, A.C. Reis, W.C. Nunes, M.A. Novak, V.K. Garg, et al., Preparation of magnetite nanoparticles in mesoporous copolymer template, Nano. Lett. 1 (2001) 105–108.

[32] G. Kataby, A. Ulman, R. Prozorov, A. Gedanken, Coating of amorphous iron nanoparticles by long-chain alcohols, Langmuir 14 (1998) 1512–1515.

[33] S.M. Ponder, J.G. Darab, T.E. Mallouk, Remediation of Cr(VI) and Pb(II) aqueous solutions using supported, nanoscale zero-valent iron, Environ. Sci. Technol. 34 (2000) 2564–2569.

[34] S.M. Ponder, J.G. Darab, J. Bucher, D. Caulder, I. Craig, L. Davis, et al., Surface chemistry and electrochemistry of supported zero-valent iron nanoparticles in the remediation of aqueous metal contaminants, Chem. Mater. 13 (2001) 479–486.

[35] G.V. Lowry, S. Majetich, D. Sholl, R.D. Tilton, K. Matyjaszewski, Developing functional Fe0-based nanoparticles for in situ degradation of DNAPL chlorinated organic solvents, Abstracts of Papers, 227th ACS National Meeting, Anaheim, CA, United States, March 28–April 1, 2004.

[36] D. Bhattacharyya, L.G. Bachas, J. Xu, D. Meyer, Y. Tee, Membrane-based synthesis of metal nanoparticles and applications to organic dechlorination, Abstracts of Papers, 227th ACS National Meeting, Anaheim, CA, United States, March 28–April 1, 2004.

[37] F. Li, C. Vipulanandan, K.K. Mohanty, Microemulsion and solution approaches to nanoparticle iron production for degradation of trichloroethylene, Colloid Surf. A 223 (1–3) (2003) 103–112.
[38] Y.P. Sun, W.X. Zhang, Dispersion of zero-valent iron nanoparticles, Abstracts of Papers, 229th ACS National Meeting, San Diego, CA, United States, March 13–17, 2005.
[39] J. Liu, F. He, D. Zhao, C. Roberts, Sugar-stabilized Pd nanoparticles exhibiting high catalytic activities for hydrodechlorination of environmentally deleterious trichloroethylene, Langmuir 24 (1) (2007) 328–336.
[40] F. He, D. Zhao, Manipulating the size and dispersibility of zero-valent iron nanoparticles by use of carboxymethyl cellulose stabilizers, Environ. Sci. Technol. 41 (17) (2007) 6216–6221.
[41] M. Mikhaylova, D.K. Kim, N. Bobrysheva, M. Osmolowsky, V. Semenov, T. Tsakalakos, et al., Superparamagnetism of magnetite nanoparticles: dependence on surface modification, Langmuir 20 (2004) 2472–2477.
[42] P. Raveendran, J. Fu, S.L. Wallen, Complete "green" synthesis and stabilization of metal nanoparticles, J. Am. Chem. Soc. 125 (2003) 13940–13941.
[43] T.K. Sarma, A. Chattopadhyay, Reversible encapsulation of nanometer-size polyaniline and polyaniline-au-nanoparticle composite in starch, Langmuir 20 (2004) 4733–4737.
[44] Q. Lu, F. Gao, S. Komarneni, A green chemical approach to the synthesis of tellurium nanowires, Langmuir 21 (2005) 6002–6005.
[45] S. Si, A. Kotal, T. Mandal, S. Giri, H. Nakamura, T. Kohara, Size-controlled synthesis of magnetite nanoparticles in the presence of polyelectrolytes, Chem. Mater. 16 (2004) 3489–3496.
[46] S. Magdassi, A. Bassa, Y. Vinetsky, A. Kamyshny, Silver nanoparticles as pigments for water-based ink-jet inks, Chem. Mater. 15 (2003) 2208–2217.
[47] S.Y. Zhang, J. Zhang, H.Y. Wang, H.Y. Chen, Synthesis of selenium nanoparticles in the presence of polysaccharides, Mater. Lett. 58 (2004) 2590–2594.
[48] M. Adlim, M.A. Bakar, K.Y. Liew, J. Ismail, Synthesis of chitosan-stabilized platinum and palladium nanoparticles and their hydrogenation activity, J. Mol. Catal. A-Chem. 212 (2004) 141–149.
[49] H. Huang, Q. Yuan, X. Yang, Preparation and characterization of metal-chitosan nanocomposites, Colloid Surf. B 39 (2004) 31–37.
[50] H. Huang, X. Yang, Synthesis of polysaccharide-stabilized gold and silver nanoparticles: a green method, Carbohyd. Res. 339 (2004) 2627–2631.
[51] J. Cao, M. Zheng, S. Deng, X. Ma, B. Fang, J. Tao, Synthesis of silver nanoparticles by polysaccharides, Abstracts of Papers, 229th ACS National Meeting, San Diego, CA, United States, March 13–17, 2005.
[52] A. Pal, K. Esumi, T. Pal, Preparation of nanosized gold particles in a biopolymer using UV photoactivation, J. Colloid Interf. Sci. 288 (2005) 396–401.

CHAPTER

Reducing Leachability and Bioaccessibility of Toxic Metals in Soils, Sediments, and Solid/Hazardous Wastes Using Stabilized Nanoparticles

32

Yinhui Xu, Ruiqiang Liu and Dongye Zhao

Environmental Engineering Program, Department of Civil Engineering, Auburn University, Auburn, AL, USA

32.1 Reductive immobilization of chromate in soil and water using stabilized zero-valent iron nanoparticles

32.1.1 Introduction

Chromium has been widely detected in groundwater and soils. To reduce human exposure, the US Environmental Protection Agency (EPA) set a maximum contaminant level (MCL) of 0.1 mg/L for total chromium in drinking water.

Street, Sustich, Duncan and Savage. Nanotechnology Applications for Clean Water, 2nd Edition.
 DOI: http://dx.doi.org/10.1016/B978-1-4557-3116-9.00032-9

Conventionally, Cr(VI) is removed from water through reduction of Cr(VI) to its less toxic form, Cr(III), followed by precipitation [1]. Researchers have demonstrated that Cr(VI) can be effectively reduced by Fe(II) according to the following generic reaction scheme, Equation 32.1 [2,3]:

$$\mathrm{Cr(VI)} + 3\mathrm{Fe(II)} \rightarrow \mathrm{Cr(III)} + 3\mathrm{Fe(III)} \tag{32.1}$$

Reduction of Cr(VI) to Cr(III) by powder or granular zero-valent iron (ZVI) particles and non-stabilized or aggregated ZVI nanoparticles has been investigated in laboratory and field studies [4–7]. The reduction was reported to follow the general pseudo-first order rate law shown in Equation 32.2 [7],

$$v = kA_s[Me] \tag{32.2}$$

where v is the reaction rate, k is the rate constant ($M^{-1}\ m^2\ s^{-1}$), $[Me]$ is the metal ion concentration (M), and A_s is the specific surface area of the iron particles (m^2/g). Equation 32.2 indicates that the reaction rate is directly proportional to the specific surface area of the ZVI particles. Consequently, reducing particle size is expected to greatly enhance the reaction rate exponentially. Cao and Zhang [8] reported that the surface-area-normalized reaction rate constant of Cr(VI) reduction by non-stabilized nanoparticles was about 25 times greater than that by iron powders (100 mesh). However, agglomerated ZVI particles are often in the range of micron scale. As a result, they are essentially not deliverable and cannot be used for in situ applications in soils.

To control nanoparticle agglomeration, He et al. [9,10] developed a technique for preparing stabilized ZVI nanoparticles by applying low concentrations of a starch or CMC as a stabilizer. The stabilized nanoparticles exhibited markedly improved soil mobility and greater reactivity when used for dechlorination of TCE.

32.1.2 Reduction and removal of Cr(VI) in water

Chromate is highly water soluble and mobile. In the subsurface environment, Cr (VI) will distribute between water and soil. Therefore, immobilization of Cr(VI) requires removal of chromate in the aqueous phase. Xu and Zhao [11] showed that at an Fe dose of 0.08 g/L (approximately 2.3 times the stoichiometric amount), about 53 percent of 34 mg/L of Cr(VI) was reduced at equilibrium, which was reached after roughly 36 hours of reaction.

It has been proposed that elemental Fe reduces Cr(VI) to Cr(III) following the stoichiometry in Equation 32.3 [12],

$$\mathrm{Fe^0} + \mathrm{CrO_4^{2-}} + 4\mathrm{H_2O} = \mathrm{Cr(OH)_3(s)} + \mathrm{Fe(OH)_3(s)} + 2\mathrm{OH^-} \tag{32.3}$$

In the absence of a stabilizer, the resultant $Cr(OH)_3$ is a sparingly soluble precipitate ($K_{sp} = 6.3 \times 10^{-31}$). Cr(III) can also be precipitated via the formation of Fe(III)–Cr(III) hydroxide according to Equation 32.4 [12],

$$x\mathrm{Cr^{3+}} + (1-x)\mathrm{Fe^{3+}} + 3\mathrm{H_2O} \Leftrightarrow (\mathrm{Cr}_x\mathrm{Fe}_{1-x})(\mathrm{OH})_3(\mathrm{s}) + 3\mathrm{H^+} \tag{32.4}$$

where x is equal to 0.75. The solubility of $Cr_xFe_{1-x}(OH)_3$ is lower than that of Cr $(OH)_3$. Alternatively, Cr(III) may also precipitate in the form of $Cr_xFe_{1-x}OOH$ [8]. In the presence of CMC, the particle agglomeration and precipitation may be somewhat inhibited. However, CMC is vulnerable to biodegradation and/or hydrolysis. Once decomposed, its particle stabilizing ability is ceased. Consequently, any residual fine precipitates will end up in the soil matrix through sorption and/or filtration effect.

Assuming complete mixing, the initial (<4 hours) reduction rate of Cr(VI) can be described by a pseudo-first-order kinetic model in Equation 32.5 [4,7]:

$$\frac{d[C]}{dt} = -k_{obs}[C] \tag{32.5}$$

where C is the concentration of Cr(VI) in water (mg/L), t the time (h), and k_{obs} the observed first-order rate constant (h^{-1}). The value of k_{obs} was determined to be 0.08 h^{-1} for CMC-stabilized ZVI nanoparticles [11]. Earlier, Ponder et al. [7] reported a k_{obs} value of 1.18 h^{-1} for resin-supported ZVI at an Fe-to-Cr molar ratio of 8:1, and they reported that the rate constant increases linearly with increasing Fe-to-Cr molar ratio [4,7]. In the subsurface environment, mixing is rather limited; thus, the overall reduction rate is likely controlled by mass transfer and the actual reaction rate can be much slower.

The reducing power of the ZVI nanoparticles can also be consumed by side reactions. For example, ZVI nanoparticles can also react with water via Equation 32.6 [13]:

$$Fe^0_{(s)} + 2H_2O \rightarrow Fe^{2+}_{aq} + H_{2(g)} + 2OH^-_{(aq)} \tag{32.6}$$

Although the resultant hydrogen and Fe^{2+} are fairly strong reducing agents, they are not strong enough to reduce Cr(VI) under the experimental conditions [11]. Xu and Zhao (11) observed that as the Fe dosage was increased from 0.04 g/L to 0.12 g/L, the equilibrium percentage removal of Cr(VI) increased from 24 to 90 percent. Evidently, at an Fe dosage of approximately 3.4 times the stoichiometric amount, the stabilized nanoparticles can reduce about 90 percent Cr(VI) under ambient conditions, which is 20 and 3 times, respectively, more effective than the resin-supported ZVI nanoparticles [7] and the non-stabilized ZVI nanoparticles [8].

32.1.3 Reduction and immobilization of Cr(VI) sorbed in soil

It was demonstrated that amending Cr(VI)-laden soil with CMC-stabilized ZVI nanoparticles can substantially reduce the chromate leachability [11]. Figure 32.1 shows the transient release of total Cr or Cr(VI) when 1.5 g of a Cr(VI)-laden soil sample was mixed with 15 mL of the nanoparticle suspension containing 0.08 g/L Fe and at an initial pH of 9.0. For comparison, Cr(VI) desorption kinetic data in DI water at pH 9.0 are also superimposed. At equilibrium, roughly 36 percent of preloaded Cr(VI) was desorbed from the soil when the nanoparticles were absent.

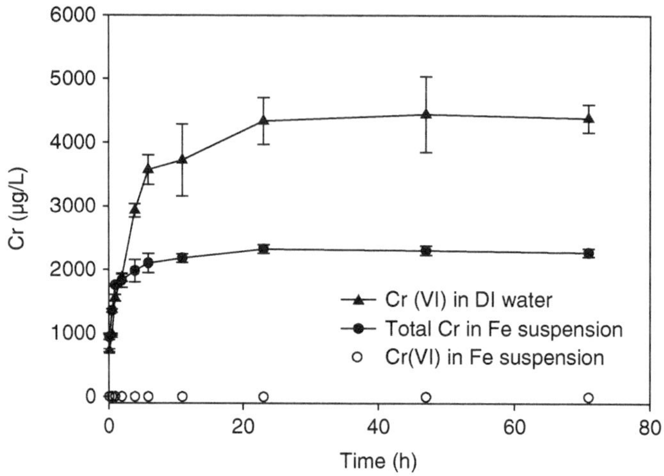

FIGURE 32.1

Leaching of Cr or Cr(VI) from a contaminated sandy loam soil with nanoparticle suspension (Fe = 0.08 g/L) or DI water. (Solution volume = 15 ml; Soil = 1.5 g; Initial Cr in soil = 83 mg/kg).

In contrast, when 0.08 g/L Fe nanoparticles were present, approximately 18 percent of the preloaded Cr(VI) was released, with no Cr(VI) detected in the aqueous phase. This observation indicates that a small dose of the stabilized nanoparticles was able to not only reduce the Cr(VI) leachability, but also completely transform all leached Cr(VI) to Cr(III).

Solution pH can affect both chromate speciation and the surface electric potential of soil, and thus, Cr(VI) reduction and immobilization. Xu and Zhao [11] tested Cr leachability from a sandy soil at the initial pH 9.0, 7.0, and 5.0 and in the presence or absence the Fe nanoparticles (soil = 1.5 g; solution = 15 ml). As the solution pH was decreased from 9.0 to 5.0, the DI-water desorbed Cr(VI) was reduced from 30 to 20 percent because at higher pH soil sorption sites become more negative and OH^- ions compete more fiercely with CrO_4^{2-} for the binding sites. However, when ZVI nanoparticles (0.08 g/L) were present, the total leachable Cr was reduced to <12 percent over the pH range of 5.0–9.0, and all desorbed Cr was detected as Cr(III). In addition, the total Cr release was much less pH dependent due to the added sorption capacity from Fe addition.

Figure 32.2 shows the chromium elution histories during two separate column runs when 0.06 g/L ZVI nanoparticle suspension at pH 5.60 or DI-water was pumped through a Cr-loaded sandy soil bed under otherwise identical conditions. As shown in Fig. 32.2A, the elution of total Cr with DI water displayed a much higher and broader peak and a longer tailing than with the ZVI nanoparticle suspension. Mass balance calculation revealed that DI water eluted a total of approximately 12 percent of the pre-sorbed Cr(VI), whereas the Fe suspension leached only about 4.9 percent,

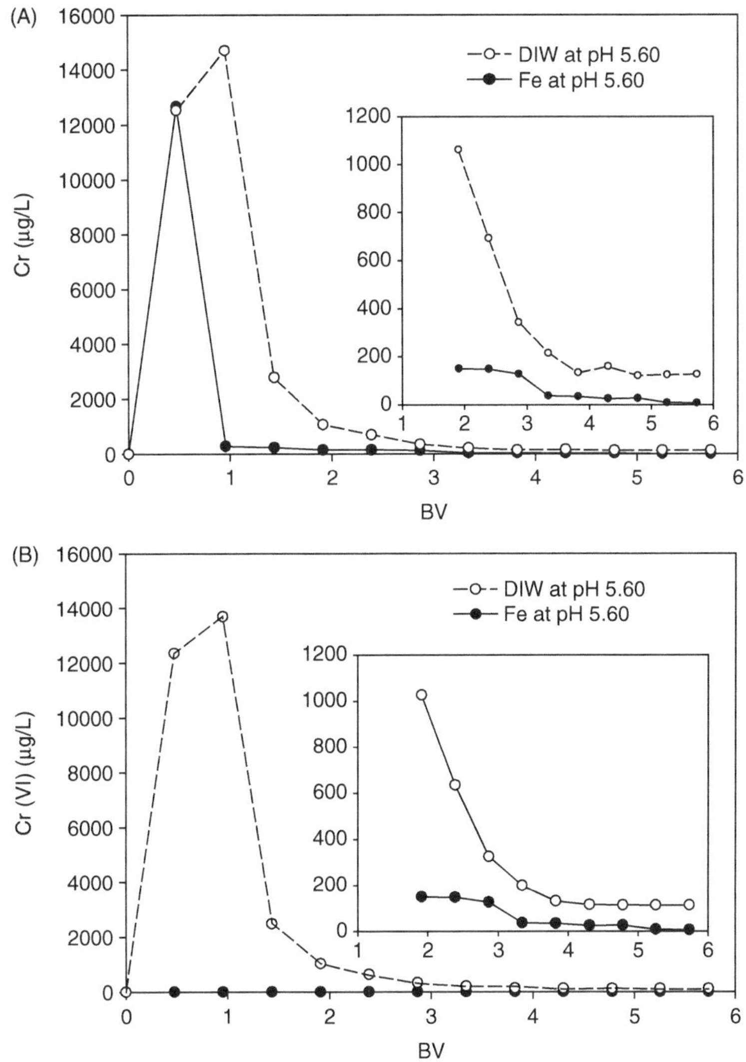

FIGURE 32.2

Cr elution histories during two separate column runs using nanoparticle suspension (Fe = 0.06 g/L) or DIW at an influent pH 5.6: (A) Total Cr; (B) Cr(VI); Insets: close-up of Cr elution histories after 1.9 BV.

a 59 percent reduction. When plotted as Cr(VI) (Fig. 32.2B), all of the leached Cr (VI) was converted to Cr(III) during the treatment. In both cases, the effluent pH was in the range of 5.2 to 5.7. In the presence of the ZVI Fe nanoparticles, the peak concentration dropped abruptly to <0.28 mg/L at 1 BV and to <0.007 mg/L after 5 BV of the nanoparticle suspension was passed. This observation indicates that the transformed Cr(III) is not only much less toxic but also much less mobile than Cr(VI). The transformed Cr(III), either in the form of fine precipitates or associated with the oxidized Fe nanoparticles, was subject to natural filtration effect as it travels through the soil [11].

The nanoparticle treatment also reduces the TCLP and the California Waste Extraction Test (WET) extractability of Cr(VI). Xu and Zhao [11] observed that the equilibrium Cr concentration in the TCLP extractant was 0.4 mg/L for the untreated soil, compared to only 0.04 mg/L when the same soil was treated with approximately 5.7 BV of the ZVI nanoparticle suspension at pH 5.60, and all TCLP-leached Cr for the treated soil was present in the form of Cr(III). Upon the same brief nanoparticle treatment, the WET-leached Cr concentration was reduced from 1.2 mg/L to 0.28 mg/L.

For in situ remediation uses, the nanoparticles must be highly mobile in soil to ensure delivery of the nanoparticles to the targeted locations. The soil transportability of the CMC-stabilized nanoparticles was tested by measuring the breakthrough curve of the Fe nanoparticles when a ZVI suspension (Fe = 0.06 g/L) was passed through a packed sandy soil bed [11]. More than 81 percent of ZVI introduced broke through rapidly in less than 1 BV, indicating that the nanoparticles were highly mobile through the soil bed. Approximately 19 percent of Fe introduced was retained in the soil bed after 3 BV of the suspension was passed and more Fe was retained in the soil thereafter. Under the subsurface conditions, the stabilized ZVI nanoparticles will be converted to iron minerals in weeks, and the minerals will eventually be incorporated in the ambient geo-media.

32.2 In situ immobilization of lead in soils using stabilized vivianite nanoparticles

Lead is a widespread contaminant and has been ranked the second most hazardous substance in the United States. Current remediation technologies rely largely on excavation and landfills, which are rather costly and often environmentally disruptive.

In recent years, in situ immobilization of Pb^{2+} in contaminated soils with phosphate-based amendments has elicited a great deal of attention [14–16]. Generally, this approach reduces the Pb^{2+} mobility, and thus toxicity, by transforming the labile form of Pb^{2+} in soils to the geochemically more stable pyromorphites ($Pb_5(PO_4)_3X$, where X = F, Cl, Br, OH). Pyromorphites are considered the most stable forms of Pb^{2+} under a wide range of environmental conditions, and are over 44 orders of magnitude less soluble than other common Pb^{2+} minerals in contaminated soils [17]. Phosphate has been applied to soils in its soluble forms [18,19], or in solid forms such as synthetic apatite ($Ca_5(PO_4)_3$(OH, F, Cl)) [14], natural phosphate rocks [20]. Among those additives, phosphoric acid was regarded as the most effective amendment for its easy delivery and superior ability to dissolve Pb^{2+} from minerals and transform it to pyromorphites [21]. Amendment dosage of 3 percent PO_4^{3-} by weight for soils has been proposed and applied by the EPA [21].

However, adding large amounts (e.g., the 3 percent PO_4^{3-} dosage) of phosphoric acid or phosphate salts into the subsurface is limited by not only the material cost but the secondary contamination problems associated with the excessive nutrient input. To avoid phosphate leaching, solid phosphate (e.g., rock phosphate) was also studied [14,20]. However, effectiveness of solid phosphate is hindered by the large size of the particles. In fact, even fine-ground solid phosphate particles are not mobile in soils, preventing solid phosphate from being delivered to the lead-affected zone and from reaching and reacting with Pb^{2+} sorbed in soils.

In a recent study [22], Liu and Zhao synthesized and tested a new class of nanoscale iron phosphate (vivianite) particles with CMC as a stabilizer. Batch test results showed that the CMC-stabilized nanoparticles can effectively reduce the TCLP leachability and PBET bioaccessibility of Pb^{2+} in three representative soils (calcareous, neutral, and acidic). When the soils were treated for 56 days at a dosage ranging from 0.61 to 3.0 mg/g-soil as PO_4^{3-}, the TCLP leachability of Pb^{2+} was reduced by 85–95 percent, and the bioaccessibility by 31–47 percent. Results from a sequential extraction procedure showed a 33–93 percent decrease in exchangeable Pb^{2+} and carbonate-bound fractions, and an increase in residual-Pb^{2+} fraction when Pb^{2+}-spiked soils were amended with the nanoparticles. The addition of chloride in the treatment further decreased the TCLP leachable Pb^{2+} in soils, suggesting formation of chloro-pyromorphite minerals. Compared to soluble phosphate used for in situ metal immobilization, application of the nanoparticles results in approximately 50 percent reduction in phosphate leaching into the environment.

32.3 Mechanisms of nanoparticle stabilization by carboxymethyl cellulose

As discussed earlier, the use of CMC as a stabilizer was able to prevent agglomeration of the nanoparticles and facilitate effective delivery of the nanoparticles into contaminated subsurface zones. He et al. [10] studied the CMC–nanoparticle interactions via Fourier Transform Infrared Spectroscopy (FTIR), and claimed that the stabilization of the ZVI nanoparticles is attributed to the adsorption of CMC molecules to the surface of the nanoparticles. FTIR results suggested that CMC molecules were adsorbed to iron nanoparticles primarily through the carboxylate groups through monodentate complexation. In addition, –OH groups in CMC were also involved in interacting with iron particles. The adsorption process results in the encapsulation of the nanoparticles with a thin layer of negatively charged CMC, which suppresses the growth of the iron nanoparticles and prevents the particles from agglomeration through the electrostatic repulsion and/or steric hindrance between the CMC-coated nanoparticles. In a recent study on the aggregation of agitate-coated nanoparticles [23], researchers observed that electrostatic or Derjaguin–Landau–Verwey–Overbeek (DLVO) interactions were the primary mechanism for particle stabilization/aggregation.

In addition, a recent study [24] reveals that the size and mobility of CMC-stabilized nanoparticles may be manipulated by varying the type and concentration of CMC and/or other synthesizing conditions such as temperature, pH, and cations.

32.4 Conclusions

Contamination of soil and groundwater by toxic metals has been one of the most challenging environmental issues. Because of the extent and scope of the contamination legacy and because of the high contaminant mobility, it is virtually impossible to cost-effectively remove the metals from many contaminated sites. The stabilized ZVI nanoparticles appear promising for developing more cost-effective in situ remediation technologies for these sites. Because of the small particle size, large surface area, easy soil deliverability, and high reactivity, these nanoscale materials can offer some unique advantages over conventional techniques for remediation of contaminated soil and groundwater. The nanoparticles can be applied in situ and may turn out particularly advantageous for contaminant zones that are deep or unreachable with normal technologies. The stabilized nanoparticles can effectively reduce leachability and bioaccessibility of chromate, lead, and arsenate in water, soil, and other porous media. However, detailed information on the transport and environmental fate of the stabilized nanoparticles is needed for field implementation of the technology. The particle stabilization strategy using polysaccharides offers a simple and effective means for controlling the size and transport behavior of the nanoparticles.

References

[1] S. Guha, P. Bhargava, Removal of chromium from synthetic plating waste by zero-valent iron and sulfate-reducing bacteria, Water Environ. Res. 77 (4) (2005) 411–416.

[2] L. Legrand, A. El Figuigui, F. Mercier, A. Chausse, Reduction of aqueous chromate by Fe(II)/Fe(III) carbonate green rust: kinetic and mechanistic studies, Environ. Sci. Technol. 38 (17) (2004) 4587–4595.

[3] M. Pettine, L. D'ottone, L. Campanella, F.J. Millero, R. Passino, The reduction of chromium (VI) by iron (II) in aqueous solutions, Geochim. Cosmochim. Acta 62 (9) (1998) 1509–1519.

[4] M.J. Alowitz, M.M. Scherer, Kinetics of nitrate, nitrite, and Cr(VI) reduction by iron metal, Environ. Sci. Technol. 36 (3) (2002) 299–306.
[5] D.W. Blowes, C.J. Ptacek, J.L. Jambor, In-situ remediation of Cr(VI)-contaminated groundwater using permeable reactive walls: laboratory studies, Environ. Sci. Technol. 31 (12) (1997) 3348–3357.
[6] N. Melitas, O. Chuffe-Moscoso, J. Farrell, Kinetics of soluble chromium removal from contaminated water by zerovalent iron media: corrosion inhibition and passive oxide effects, Environ. Sci. Technol. 35 (19) (2001) 3948–3953.
[7] S.M. Ponder, J.G. Darab, T.E. Mallouk, Remediation of Cr(VI) and Pb(II) aqueous solutions using supported, nanoscale zero-valent iron, Environ. Sci. Technol. 34 (12) (2000) 2564–2569.
[8] J. Cao, W.X. Zhang, Stabilization of chromium ore processing residue (COPR) with nanoscale iron particles, J. Hazard. Mater. 132 (2–3) (2006) 213–219.
[9] F. He, D. Zhao, Preparation and characterization of a new class of starch-stabilized bimetallic nanoparticles for degradation of chlorinated hydrocarbons in water, Environ. Sci. Technol. 39 (9) (2005) 3314–3320.
[10] F. He, D. Zhao, J. Liu, C.B. Roberts, Stabilization of Fe/pd bimetallic nanoparticles with sodium carboxymethyl cellulose to facilitate dechlorination of trichloroethene and soil transportability, Ind. Eng. Chem. Res. 46 (2006) 29–34.
[11] Y. Xu, D. Zhao, Reductive immobilization of chromate in soils and groundwater by stabilized zero-valent iron nanoparticles, Water Res. 41 (2007) 2101–2108.
[12] T. Astrup, S.L.S. Stipp, T.H. Christensen, Immobilization of chromate from coal fly ash leachate using an attenuating barrier containing zero-valent iron, Environ. Sci. Technol. 34 (19) (2000) 4163–4168.
[13] S.M. Ponder, J.G. Darab, J. Bucher, D. Gaulder, Surface chemistry and electrochemistry of supported zerovalent iron nanoparticles in the remediation of aqueous metal contaminants, Chem. Mater. 13 (2001) 479–486.
[14] Q.Y. Ma, T.J. Logan, S.J. Traina, Lead immobilization from aqueous solutions and contaminated soils using phosphate rocks, Environ. Sci. Technol. 29 (1995) 1118–1126.
[15] P. Zhang, J.A. Ryan, Formation of pyromorphite in anglesite-hydroxyapatite suspensions under varying pH conditions, Environ. Sci. Technol. 32 (1998) 3318–3324.
[16] X. Cao, L.Q. Ma, M. Chen, S.P. Singh, W.G. Harris, Impacts of phosphate amendments on lead biogeochemistry at a contaminated site, Environ. Sci. Technol. 36 (2002) 5296–5304.
[17] M.V. Ruby, A. Davis, A. Nicholson, In situ formation of lead phosphates in soils as a method to immobilize lead, Environ. Sci. Technol. 28 (1994) 646–654.
[18] T.T. Eighmy, B.S. Crannell, L.G. Butler, F.K. Cartledge, E.F. Emery, D. Oblas, et al., Heavy metal stabilization in municipal solid waste combustion dry scrubber residue using phosphate, Environ. Sci. Technol. 31 (1997) 3330–3338.
[19] R. Stanforth, J. Qiu, Effect of phosphate treatment on the solubility of lead in contaminated soil, Environ. Geol. 41 (2001) 1–10.
[20] L.Q. Ma, G.N. Rao, Aqueous Pb reduction in Pb-contaminated soils by phosphate rocks, Water Air Soil Poll. 110 (1999) 1–16.
[21] US Environmental Protection Agency Region 10, Consensus plan for soil and sediment Studies: Coeur d'Alene River soils and sediments bioavailability studies, URS DCN: 4162500.06161.05.a. EPA:16.2, 2001.

[22] R. Liu, D. Zhao, Reducing leachability and bioaccessibility of lead from soils using a new class of stabilized iron phosphate nanoparticles, Water Res. 41 (2007) 2491–2502.
[23] K.L. Chen, S.E. Mylon, M. Elimelech, Aggregation kinetics of alginate-coated hematite nanoparticles in monovalent and divalent electrolytes, Environ. Sci. Technol. 40 (2006) 1516.
[24] F. He, D. Zhao, Manipulating the size and dispersibility of zerovalent iron nanoparticles by use of carboxymethyl cellulose stabilizers, Environ. Sci. Technol. 41 (2007) 6216–6221.

PART

5

Societal Issues

CHAPTER

Introduction to Societal Issues: The Responsible Development of Nanotechnology for Water

33

Jeremiah S. Duncan[1], Nora Savage[2] and Anita Street[3]

[1]*Department of Atmospheric Science and Chemistry, Plymouth State University, Plymouth, NH*

[2]*Office of Research and Development, US Environmental Protection Agency, Washington, DC*

[3]*US Department of Energy, Washington, DC*

Globally, the issues with water quality and quantity are approaching a critical stage. Sections 1–4 of this book present a variety of promising research areas where advances in nanotechnology may provide some much needed technical solutions to a host of water-related problems—drinking water, treatment and reuse, remediation, and detection of pollutants (sensors). As a complement to these sections, Section 5 considers the nontechnical issues related to the use and acceptance of nanotechnology to improve water quality.

In an insightful essay on the relationships between society and the development of nanotechnology [1], Keller has considered the global successes and failures of nuclear energy and genetically modified foods, arguing that the success of a technology does not depend entirely on its scientific merits. To the contrary, the failure of various technologies may be attributed largely to a host of more complicated and difficult-to-predict societal pressures, including public opinion and acceptance, appropriate oversight and governance, demand (or lack thereof) for a solution to a given problem, or failure to study and manage risk at an early enough stage. It is imperative, therefore to properly consider these nontechnical (i.e., "societal") issues, lest the potential of the technical solutions be wasted due to lack of foresight and/or understanding.

In other words, there are several questions that ought to be considered with respect to public acceptance of nanotechnology: (1) what is the public's

Street, Sustich, Duncan and Savage. Nanotechnology Applications for Clean Water, 2nd Edition.
 DOI: http://dx.doi.org/10.1016/B978-1-4557-3116-9.00033-0

perception (bearing in mind this often has little to do with personal knowledge or level of education)? (2) What is the level of scientific literacy (separating science fact from science fiction)? (3) What is and should be the level of public participation in science decision making? (4) Who owns the technology, what will be developed, for what purpose and for whose benefit? (5) What is the scientific community's obligation (ethically or morally) to actively engage the public in the pursuit of science when the research is funded with public monies? (6) How should science balance market-driven research agendas with societal needs? and (7) How and when can the public contribute most meaningfully to encourage responsible development of nanotechnology? In this section, these issues and questions are examined to provide some food for thought as the development of nanotechnology—especially in the area of water quality—is developed.

Historically, technologies have been allowed to develop and mature largely without societal influence, at least in the early stages of research. In the words of Vannevar Bush, "Scientific progress on a broad front results from the free play of free intellects, working on subjects of their own choice, in the manner dictated by their curiosity for exploration of the unknown" [2]. Although largely successful, as evidenced simply by the historical advances in technology, particularly in the last century, this "technology push" model has meant that ethical, societal (or "social," depending upon your background), and even environmental considerations have often been overlooked or minimized during the introduction of novel technologies. On the other hand, the "societal pull" model has been used infrequently but has also been very successful (consider for example the space program's race to the moon in the 1960s, or to a lesser extent AIDS and cancer research). However, as Keller points out, these "push" and "pull" interactions are, in fact, bidirectional: on the one hand, new technologies influence a society's economic and political structures and often raise issues related to the society's values and culture. On the other hand, the way society structures its policies and institutions for supporting, regulating, and judging the safety of technologies has a strong influence on the pace and direction of their development [1].

Certainly, the early development of nanotechnology was no exception to the standard model, being more influenced by the "technology push," though concerns were raised early, e.g., by Drexler [3], about the long-term potential for negative consequences. However, a "societal pull" was begun with the initiation of nanotechnology as a national initiative in the United States, by the passing of the 21st Century Nanotechnology Development Act (Public law 108-153) in 2003, which specifically included "ethical, legal, environmental, and other appropriate societal concerns." Nanotechnology is arguably the first major technological advancement to begin addressing these issues on a large scale at such an early stage in its development. Continuing on this path will be critical as nanotechnology enters, in the words of Mihail Roco, its "second generation" [4]. In this stage, nanotechnology will go beyond simple material modification to the development of "smart" or active materials, introducing a host of new societal issues. Gorman et al. (Chapter 36) describe a social framework in which these questions may be addressed, and the seemingly conflicting pressures of technological push and societal pull can be resolved by bringing together historically separated groups of people (e.g., technical and social scientists), while Rejeski and Michelson (Chapter 37) look at more formal governance structures and policies that may already be acting as obstacles to innovation (i.e., technical push) and adoption of beneficial technologies (i.e., societal pull).

In Chapter 34, Street et al. further expand on some of the larger societal issues with respect to nanotechnology and water. The need to consider the wisdom of introducing and using a technology because we have the ability to do so is at the heart of the matter. Even in circumstances where there is a critical need—such as in the provision of clean water—and only benefits are foreseen, there is an obligation to ask and at least attempt to answer questions about potential risks to society and the environment. The understanding and balancing of these risks and benefits must be grounded in the best science available—a task, in relation to nanomaterials, that has only begun in earnest in the last few years. Street et al., therefore, also raise issues related to the environmental, health, and safety aspects of nanotechnology. But simply understanding the risks is not enough—they also need to be balanced against the benefits. As an illustration of the need to, as well as the difficulty of, understanding and balancing risk and benefit, Sengul and Theis (Chapter 41) present the results of a life cycle inventory of cadmium–selenide quantum dots—one of the few studies of its kind.

While perhaps at a more critical point than ever before, the need for clean water is not new. There have historically been a host of things competing for or preventing the supply of clean water, including technical challenges, pollution, political or cultural conflicts, and corruption and mismanagement. Chapter 35 by Duncan et al. considers some of these competitors for clean water, as well as several that may just be emerging.

Finally, as part of balancing benefits and risks, and ensuring the success of beneficial technologies, the public must be engaged. The need for and potential ways of achieving this public engagement is discussed by Berube (Chapter 38) and Street et al. (Chapter 34), as well as more specifically in the context of the developing world by Grimshaw et al. (Chapter 39), and Hlophe and Hillie (Chapter 40).

References

[1] K.H. Keller, Nanotechnology and Society, J. Nanopart Res. 9 (1) (2007) 5.

[2] V. Bush, Science: the endless frontier: Office of Scientific Research and Development, United States Government Printing Office, Washington DC, USA, 1945.

[3] E. Drexler, Engines of Creation: The Coming Era of Nanotechnology, Anchor books, New York, USA, 1986.

[4] M.C. Roco, W.A. Goddard III, D.W. Brenner, S.E. Lyshevski, G.J. Iafrate, National nanotechnology initiative – past, present, future, Handbook of Nanoscience, Engineering, and Technology, second ed. CRC Press, Taylor and Francis Group, LLC, Boca Raton, FL, USA.

CHAPTER

Nanotechnology in Water: Societal, Ethical, and Environmental Considerations

34

Anita Street[1], Jeremiah S. Duncan[2] and Nora Savage[3]

[1]US Department of Energy, Washington, DC, USA [2]Department of Atmospheric Science and Chemistry, Plymouth State University, Plymouth, NH, USA [3]Office of Research and Development, US Environmental Protection Agency, Washington, DC, USA

34.1 Introduction

Clean water is a necessity of life, not just for humans, but for every ecosystem on earth. Unfortunately, it is not a readily accessible resource everywhere. For this reason, any research done with the goal of purifying water may have dramatic social, ethical, and environmental consequences.

Worldwide, over one billion people lack reliable access to clean water, and 2.3 billion people or 41 percent of the global population live in water-stressed areas, a number that will increase 52 percent by 2025 [1,2]. Furthermore, water contamination causes over 2 million deaths annually [2,3]. This is arguably the most preventable of the major causes of death in the world—that is, the supply of clean water and sanitation taken for granted by most in the developed world. This gives rise to ethical considerations that will be discussed later in the chapter. In many cases, technological barriers to clean water have already been overcome,

Street, Sustich, Duncan and Savage. Nanotechnology Applications for Clean Water, 2nd Edition.
 DOI: http://dx.doi.org/10.1016/B978-1-4557-3116-9.00034-2

but political, social, financial, or educational barriers remain. In these cases, an argument can be made for furthering social sciences research to create solutions to nontechnical barriers. However, it is possible that basic research may result in solutions to these barriers as well by providing, for example, cheaper, simpler, more locally adaptable technologies. It is important, therefore, for researchers and policymakers to work together and carefully consider strategies for addressing problems and allocating the necessary resources to bring these technologies to the fore.

The social and ethical considerations do not apply only to providing clean water in developing countries. Developed nations with well-established infrastructures are also likely to be affected by breakthroughs in nanotechnology that could bring about revolutionary changes in the ways that water is treated, used, and provided. Water is of unparalleled importance and value to all life forms and, therefore, it is vital that new developments that affect quality and quantity be given proper consideration from a social and ethical standpoint.

The fact that environmental and health impacts are implied or inherent aspects of these broader societal and ethical concerns should not be taken for granted. Although the technologies previously discussed hold great promise for improving the availability of fresh water for people everywhere, there are uncertainties with regard to potential risks that need to be explored in greater depth. These potential risks should be communicated in ways that result in a constructive, respectful dialogue between scientists and affected communities.

34.2 Responsible development: ethical, social, and environmental concerns

The consideration of ethical, social, and environmental implications as part of a nanotechnology research and development agenda is termed collectively "responsible development." The meaning may vary slightly depending upon the source, but generally it denotes benign development, equitable deployment, and harmless end-of-life disposal of materials or products. Careful consideration of the complete life cycle of materials and products is needed to help eliminate or minimize deleterious human health and environmental consequences. Responsible development is not limited to risk considerations, but goes beyond the technical arena to include ethical concerns, equity, and other social and cultural issues [4].

The vision of the US National Nanotechnology Initiative (NNI) is one that "leads to a revolution in technology and industry that benefits society" [5]. This raises the question of whether nanotechnology as a whole or in part is benefiting society or will do so in the foreseeable future. Just because a scientific or technical breakthrough is achieved, an overall benefit to society from its development is not a foregone conclusion. Indeed, it is very likely that there may be risks, as history has shown with the development of other potentially disruptive and transformative technologies. Nor is it necessarily true that a nanotechnology with proven benefits is the best solution to pursue from a larger societal perspective—different technological options may be less costly or more socially acceptable, for example. The goal, then, of ensuring that the revolution of nanotechnology will truly become a public good can only be met by considering a number of issues beyond scientific and technical feasibility.

Nanotechnology is in a unique place on the development continuum. Although it is still in the early stages, broader societal issues are already being considered in conjunction with the research and development agenda. The 21st Century Nanotechnology Research and Development Act (Public law 108-153), passed in the United States in 2003, legislatively established the NNI. The act specifically mandates "that ethical, legal, environmental, and other appropriate societal concerns ... are considered during the development of nanotechnology." One of the first workshops organized soon after passage of this bill was devoted to the broader societal concerns related to the development and commercialization of nanotechnology [6]. Partly taking its cue from bioethics as well as the Ethical, Legal, and Societal Implications (ELSI) approach used by the Human Genome Project, the NNI is breaking new ground by emphasizing these issues at the early stages of development. As a result, specialized fields in ethics (nanoethics) and law have already emerged, as evidenced by the establishment of journals such as Nanoethics and Nanotechnology Law & Business.

In the premier issue of the journal Nanoethics, Deborah Johnson makes a convincing case that defines the role ethics must play in the development of nanotechnology [7], while pointing out that, at this point, the rather nascent field of nanoethics is heavily challenged "to examine the ethical implications of something that is not yet 'a something.'" In order for ethics to genuinely influence the development of nanotechnology, Johnson calls for a reconceptualization of technology as a "sociotechnical system" defined as a complex but inextricable linkage of artifacts (instruments, tools, materials, institutions), social practices, social arrangements, and relationships and systems of knowledge—people and things. Johnson sees this concept as a way to effectively bridge the gap between the philosophical and the material. With the multiplicity of actors involved in this evolving sociotechnical system, the challenge and opportunity for nanoethics is, then, to identify the values all parties bring to the system and forge a path forward that embraces these values and leads to positive outcomes, while laying the foundation for the responsible development of nanotechnology.

34.2.1 Access, parity, and effects of technology deployment

The responsible development of nanotechnology must involve consideration and understanding of the philosophical and social aspects of this emerging technology. Such issues include considering the questions of whether or not a technology should be deployed simply because it can be deployed. There may be instances where the general public does not believe a certain technology should be used or commercialized. It is all too tempting for scientists and engineers to believe that a developed technology will be beneficial to society and to dismiss as ignorant or uninformed any alternative views. Society as a whole should debate and determine what is best for it. The fact that select members of the society have technical expertise to develop a technology does not imply that they are the sole holders of the knowledge concerning questions regarding its usage in society. For example, a legal education is neither a requirement to vote on initiatives nor to be an elected official responsible for governing cities, states, or countries. Certain aspects of the law may require explanation but, given the opportunity, one can make a statement that reflects a personal value, regardless of legal knowledge.

Responsible development of nanotechnology should also involve taking a closer look at distributive equity of emerging technologies—the tendency for the widening rather than narrowing of the cultural and economic divide between the wealthy and the poor. Will nanotechnology-based water tools become luxuries that only wealthy nations and/or individuals can obtain? Will availability of these technologies invoke a social dynamic of environmental elitism, excluding those without the economic means to afford them? Further, for developing countries, the availability of such technologies could result in improved quality of life and new opportunities for economic growth. With increased access to more and better quality water, focus can be placed on other pressing societal needs, thereby conferring additional advantages. However, it is debatable whether solving problems (such as water quality/quantity) through nanotechnology will necessarily lead to economic growth. Ultimately, the heart of responsible development is ensuring that those with limited resources can also reap the benefits of nanotechnology.

Prevailing societal norms should also be considered, especially when technology developed in one country may be introduced to another. Determining whether or not a technology will clash with cultural norms or societal values is a reasonable, if not necessary action. These considerations are also useful for commercialization plans. Understanding the cultural and societal values within the intended consumer base will result in more effective deployment (see Grimshaw, Chapter 39 and Gorman, Chapter 36).

The deployment of a nanotechnology has environmental aspects as well. Potential environmental impacts of emerging technologies have historically been examined at the manufacturing or use stage (see the introduction to Part 5, Chapter 33), but consensus in academia and industry is growing that an assessment of the complete life cycle must be performed. For example, the assumption that engineered titanium dioxide used in sunscreens does not penetrate skin and will therefore have minimal negative impact fails to consider the washing off in water (both during personal cleaning and during swimming) or the disposal of unused portions. The entry into the environment and resulting unanticipated injury due to pharmaceutical products is a recent example of such incomplete assessments [8].

This is especially critical for the nations, states, and territories facing reduced water resources and increasing demands. Water and energy may likely prove to be two resources at the heart of major conflicts in the future. We should begin now to develop and deploy nano-enabled technologies accessible to all. Accomplishing this goal will necessitate international dialogue, including not just physical scientists and engineers but social scientists, policymakers, mathematicians, and legal scholars. Such an open and inclusive dialogue will result in innovative ways to reward the developer while providing equal access to all (see Gorman, Chapter 36).

34.2.2 Human health and environmental effects

The term "environmental health and safety" (EHS) is often used to frame the issues concerning the implications of nanotechnology. This term is misleading in that it is most commonly associated with occupational health—assessing the environmental conditions workers are exposed to and ensuring their health and safety. "Human health and environmental effects" may be a more appropriate term to indicate that exposure to nanomaterials can occur well beyond the occupational setting.

In the wake of what is frequently touted as the Third Industrial Revolution, nanotechnology continues to evolve and develop at breakneck speed. Many nano-based applications are already available commercially but little is known about the human health and environmental effects of exposure. This rapid surge in worldwide research and development and the potential large-scale use of nanomaterials in consumer products point toward the need for more definitive information regarding environmental health and safety impacts from the manufacturing process, disposal (industrial or personal), remediation, and treatment applications. As indicated in the previous four sections, the applications described show enormous promise for improving public health and the environment—specifically water quality—but cautious optimism may be warranted given the limited environmental and health risk information currently available [4]. However, there is information to be gleaned from the medical field, through research on medical applications, about effects of these materials and particles on the human body.

The very characteristics that make nanomaterials so unique and special—size and reactivity, for example—are the same characteristics giving rise to concerns about exposure to these materials, such as those used for remediation or water treatment. These concerns are often related to fate and transport issues: because of their exceedingly small size, the assumption is that they will be extremely mobile in porous media, thereby increasing the likelihood of human exposure due to dispersion and potential persistence in the environment [9]. Yet, how reactive such materials could be in the environment, what residual compounds are formed during degradation, and where and how these materials partition to various environmental and biological media are crucial but largely unknown. How these materials move from one medium to another, from one organism or ecosystem to another, and from organisms to the environment and vice versa will be critical for understanding and implementing proper manufacturing, usage, and recycling/disposal options that are most protective of human health and the environment [4]. In order to effectively assess these impacts, a full life-cycle perspective (impact of a product from the accumulation of starting materials to the development, manufacture, use, and eventual disposal or reuse of the item or portions thereof) of the various constituents and end products is an important component of a research framework (see Sengul and Theis, Chapter 41). Although complete, robust life-cycle analyses are difficult, and potentially impossible, when data is limited, and this should not be an excuse for ignoring the full life cycle in any risk analysis.

Considering all these unanswered questions, a relatively small amount of funding, compared to overall nanotechnology research and development spending, is dedicated to understanding environmental health and safety issues. Under the auspices of the NNI, the US government is assuming a leadership role in setting the directions for research for both the environmental applications and implications of nanotechnology [10]. Since 2001, approximately $12.2 million of federal grants have been awarded for the study of applications of nanotechnology to solve environmental problems, and $17.4 million has been awarded for the study of ecological and human health implications. Results of the research funded under these programs have helped to clarify important questions, principally in the use of nanomaterials for environmental cleanup/remediation and in better understanding the reactive nature of nanomaterials once they are introduced into biological systems. These questions relate to the environmental and health effects from exposure to materials that may be persistent and highly reactive; how those materials move through and interact in the various media—water, soil, and air—and how to utilize nanotechnology to better measure pollutants in the environment (see, e.g., grants awarded by the US Environmental Protection Agency, http://es.epa.gov/ncer/nano/index.html). There is, however, a growing recognition on the part of lawmakers that these important human health and environmental issues need to be addressed if nanotechnology is to continue to flourish.

In addition to the government, academics and industry have promoted the need for EHS research. For example, Maynard et al. proposed a framework comprised of five "Grand Challenges" for evaluating the human health and environmental health risks associated with nanotechnology. The hope is for the proffered research strategy to be adopted by the global science community and for fostering the responsible development of nanotechnology [11]

The international community is equally active in the quest to better understand the EHS implications of nanomaterials. For example, the Organisation for Economic Co-operation and Development (OECD, see http://www.oecd.org/sti/nano) established the Working Party on Manufactured Nanomaterials (WPMN) under the Joint Chemicals Committee to promote international cooperation and assist in the development of rigorous safety evaluation of nanomaterials. The WPMN is developing a research strategy based on the knowledge that large sums of money are being devoted to R&D for future applications of nanotechnology whereas, by comparison, relatively small sums are devoted to human health and environmental safety research. The objectives are to strengthen the international cooperation on safety research related to manufactured nanomaterial through: (i) identifying priority research areas; (ii) considering mechanisms for cooperative international research; and (iii) drawing up recommendations on research priorities for the short, medium, and longer term. The WPMN has also developed a comprehensive list of research themes on environment and human health safety.

The Working Party on Nanotechnology (WPN), somewhat distinct from the WPMN, was also established by OECD's Committee for Science and Technology Policy to advise the OECD on emerging policy-relevant issues in science, technology, and innovation related to the responsible development of nanotechnology. The WPN also promotes international cooperation that facilitates research, development, and the responsible commercialization and utilization of nanotechnology. It has identified the following six program areas as priorities for research:

- statistics and measurement;
- impacts on companies and the business environment;
- international R&D collaboration;
- communication and public engagement;
- policy dialogue; and
- global challenges: nanotechnology and water.

Thus, complementary efforts exist in the United States, Europe, and Asia. Many countries are beginning to consider the economic development opportunities associated with nanotechnology and are consulting more mature programs (e.g., NNI and OECD) to establish responsible nanotechnology research agendas.

34.3 Public engagement: what role should the public have?

There is an undeniable growing interest in providing the public with opportunities to voice their opinions about emerging technologies through the earlier engagement of the public in science and technology decision-making—largely in the wake of such controversies as genetically modified foods (GM) and stem cell research. Whereas some observers of nanotechnologies are calling for a more meaningful dialogue with the public, questions of how (and when) to best solicit and incorporate public values in the innovation process without stifling the freedom of the scientific enterprise still need to be addressed (see Berube, Chapter 38). This calls into question the existing paradigm of the conventional "black-boxed approach to popular science" that is normally practiced [12].

Macnaughten et al. and Demos, a U.K. think tank, have written convincingly on the subject of "upstream engagement" as it pertains to nanotechnology [13]. The concept of upstream engagement is rooted in the notion of engaging the public before technological innovations become fixed without benefit of social context. This is achieved through surveys, "nanodialogues," and focus groups to gauge public reaction to a technology. This is not unlike what happens in the marketing world when a manufacturer wants to develop a marketing strategy.

Moreover, gaining and maintaining public trust and support will be critical if the societal benefits of nanotechnology are to be fully realized. Studies in the United States and other parts of the developed world indicate that the general public perceives science as a public good. And, paradoxically, the same surveys show science literacy among laypersons is extremely low and highly contextual [14]. In other words, public perception of science can be heavily influenced by folk mythology and conspiracy theories. Thus, there is a complex relationship between knowledge and support. The public's pragmatic visions of the societal benefits of a technology may not be in sync with the loftier pursuits of research and scientific inquiry [14]. Improving how science is communicated can go a long way to engendering trust and paving the road to public acceptance. This includes using such techniques as "framing" or tailoring messages to address different publics.

Given what is at stake, if nanotechnology is to succeed in general, scientists must learn to communicate their research in a relatable and accessible way. Framing and upstream engagement approaches are just one such tool, but there is room for more interactive modes of communication and education to provide experiential learning opportunities to go beyond public meetings or lectures.

Enter the World Wide Web. The internet has virtually transformed how we work, how we shop, how we play, and how we communicate. There is virtually no aspect of our lives that has not changed due this technology. The next generation Web, or Web 2.0, has introduced collaborative technologies that could very well transform the scientific endeavor, scientist-to-scientist interactions, and interactions with the lay community. In a recent Scientific American article, Mitch Waldrop describes his concept of "Science 2.0" that features a number of scientists using Web 2.0 and other collaborative tools such as blogs and wikis to share their research and collaborate. Another such collaborative approach can be found through virtual social networking sites such as Second Life—a virtual networking site hosting more than 10 million registered residents worldwide. Although some scientists may be skeptical of these open sources, a small but growing number of scientists are beginning to utilize these tools and are exploiting them to their best advantage [15]. Web 2.0 technologies offer a conjunctive approach to education and public engagement as innovative as the nanotechnology itself.

In Chapter 38, Berube discusses in some detail several domestic water treatment scenarios that involve public engagement. In Chapter 39, Grimshaw describes an effort to engage citizens in a series of "upstream nanodialogues." This chapter discusses those engagements held in Zimbabwe involving scientists and representatives of two communities that experience problems with the supply of clean drinking water. The issue of public engagement is fraught with complications but its power is neither insignificant nor should it be underestimated.

34.4 Conclusions

So what does all this portend for the future of nanotechnology, and more specifically the future of water quality? For the poorest of the poor, advances in nanotechnology could potentially alleviate water access issues. If nothing else, the current state of the nanoscience argues for a contemporaneous social science research strategy to gain better understanding of early adoption, motivations for acceptance, or rejection of technology and the public's perception of science [7]. Nanotechnology will likely play a role in four key water industry areas: monitoring, desalinization, purification, and wastewater treatment [16] Additional funding is needed to better understand the risks associated with manufactured nanomaterials and their applications. It is clear that nanotechnology shows incredible promise to serve the growing demands for potable water and the quality of life for millions of people in the most underserved areas of the world. There is also a clear opportunity to reinvent the relationship between the science community and society at large—a new and improved social compact based on mutual trust and respect.

References

[1] UNESCO, Water: A Shared Responsibility, United Nations Educational, Scientific and Cultural Organization, 2006.
[2] P.H. Gleick, H. Cooley, The World's Water, 2006–2007: The Biennial Report on Freshwater Resources, Island Press, Washington, D.C., 2006.
[3] WHO, Water, sanitation and hygiene links to health, World Health Organization, 2004.
[4] N. Savage, A. Street, Commentary: Chapter 8, Sustainability and the environment, in: D. Bennett-Woods (Ed.), Nanotechnology: Ethics and Society, Taylor and Francis, London, 2008.
[5] The Subcommittee on Nanoscale Science, Engineering, and Technology, Committee on Technology, and National Science and Technology Council, The National Nanotechnology Initiative: Strategic Plan 2007, Washington, D.C., 2007.
[6] M.C. Roco, W.S. Bainbridge, Nanotechnology: Societal Implications, Springer, 2006.
[7] D. Johnson, Ethics and technology 'in the making': an essay on the challenge of nanoethics, NanoEthics 1 (1) (2007) 21.
[8] B. Halling-Sørensen, S.N. Nielsen, P.F. Lanzky, F. Ingerslev, H.C.H. Lützhøft, S.E. Jørgensen, Occurrence, fate and effects of pharmaceutical substances in the environment. A review, Chemosphere 36 (2) (1998) 357.
[9] J. Theron, J.A. Walker, T.E. Cloete, Nanotechnology and water treatment: applications and emerging opportunities, Crit. Rev. Microbiol. 34 (2008) 43–69.
[10] NNI, National Nanotechnology Initiative: Strategy for Nanotechnology-Related Environmental, Health, and Safety Research. Subcommittee on Nanoscale Science, Engineering, and Technology, Washington, D.C., 2008.

[11] A.D. Maynard, R.J. Aitken, T. Butz, V. Colvin, K. Donaldson, G. Oberdöster, et al., Safe handling of nanotechnology, Nature 444 (2006) 267.

[12] D. Dickson, The need to increase public engagement in science, (2004) [cited May 1, 2008]; Available from: <http://www.scidev.net/Editorials/index.cfm?fuseaction=readeditorials&itemid=138&language=1>.

[13] P. Macnaughten, M.B. Kearnes, B. Wynne, Nanotechnology, governance, and public deliberation: What role for the social sciences? Sci. Commun. 27 (2) (2005) 268–291.

[14] B. Lewenstein, Public perceptions of the connection between scientific research and social progress, in: A.M. Cetto (Ed.), World Conference on Science, United Nations Educational, Scientific, and Cultural Organization, Budapest, Hungary, 1999, p. 547.

[15] M.M. Waldrop, Science 2.0: is open access science the future? Sci. Am. (2008).

[16] J. Loncto, W. Marlan, F. Lynn, Nanotechnology in the water industry, Nanotech. Law Bus. 4 (2) (2007) 157–159.

CHAPTER

Competition for Water

35

Jeremiah S. Duncan[1], Nora Savage[2] and Anita Street[3]

[1]Department of Atmospheric Science and Chemistry, Plymouth State University, Plymouth, NH, USA [2]Office of Research and Development, US Environmental Protection Agency, Washington, DC, USA [3]US Department of Energy, Washington, DC, USA

Water, not unlike religion and ideology, has the power to move millions of people. Since the very birth of human civilization, people have moved to settle close to water. People move when there is too little of it; people move when there is too much of it. People move on it. People write and sing and dance and dream about it. People fight over it. And everybody, everywhere and every day, needs it. We need water for drinking, for cooking, for washing, for food, for industry, for energy, for transport, for rituals, for fun, for life. And it is not only we humans who need it; all life is dependent upon water for its very survival.

Mikhail Gorbachev

Street, Sustich, Duncan and Savage. Nanotechnology Applications for Clean Water, 2nd Edition.
 DOI: http://dx.doi.org/10.1016/B978-1-4557-3116-9.00035-4

35.1 Introduction

Water is essential to all life. It is used in virtually every aspect of human activity, from food production and personal hygiene to economic development and, as such, has deep-seated connections to social, cultural, and religious customs.

It may seem there is an abundance of water, especially to anybody living in a developed nation where water flows freely and cleanly from the kitchen tap, but scarcity is already a challenge in many areas of the world [1]. Further, it is predicted to become increasingly worse as the world population grows and climate change affects natural precipitation and evaporation patterns [1–4]. Indeed, over the next 20 years, the available freshwater resources are likely to dwindle by 30 percent [5].

Though water covers some 70 percent of the earth's surface, totaling roughly 1,400 million km^3, it is estimated that only 2.5 percent (35.2 million km^3) is fresh water, and much of this is inaccessible in the form of polar ice, and so on [1,6]. Indeed, competition for water resources is already occurring, with about 40 percent of the world's people living in regions that directly compete for shared water [7]. The average human requires an absolute minimum of 3–5 liters of clean water per day for drinking, and it has been suggested that we have a right to a minimum of 50 liters for drinking, cooking, bathing, and sanitation [8]. Yet over 2.6 billion people—a little over 40 percent of the world's population—lack basic sanitation, and more than 1 billion people still drink unsafe water (Fig. 35.1) [1,9]. Approximately 1.8 million people die each year from diarrheal diseases and millions more suffer from water-related illnesses due to lack of access to clean water and proper sanitation [8,10].

Understandably, access to clean water is an issue fraught with complications that may evoke tensions pushing grieving parties into armed conflict (see, e.g., Gleick's ongoing record of water-related conflicts [6]). Increasingly, we are looking to advances in science, such as nanotechnology, to provide water to people and areas in need. Many of the well-known causes of clean water scarcity—industrial pollution, biological and heavy metal contamination, increasing demand from expanding industry—are technical in nature, and Parts 1–4 of this book presented a number of advances in nanotechnology being brought to bear on these issues. However, there are clearly a number of nontechnical issues, such as population growth, climate change, misuse, corruption, mismanagement, and overconsumption, that result in scarcity by directly or indirectly competing for water. Additionally, advances in technology itself may soon contribute in ways

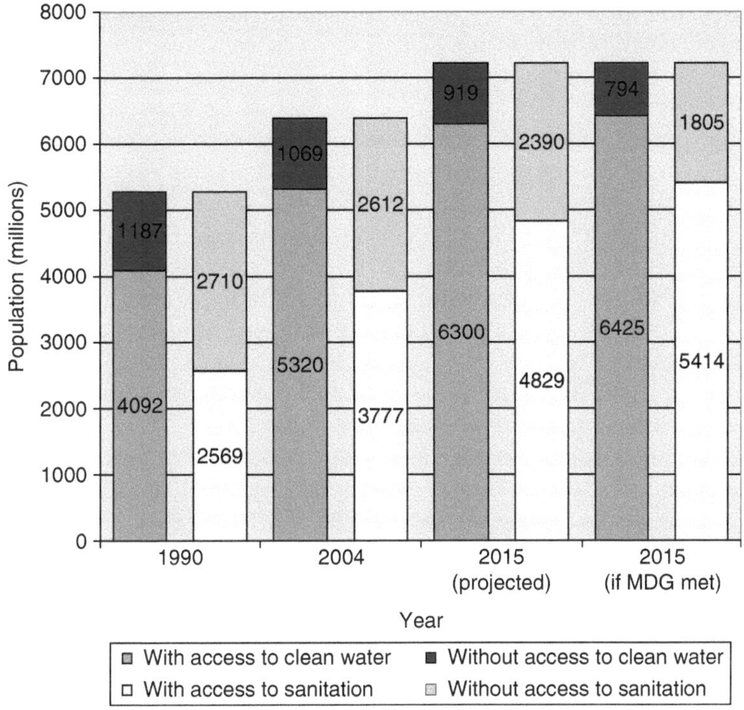

FIGURE 35.1

Global access to clean water and sanitation.

previously unknown, such as from increased agricultural production for biofuels. All of these are equally important to address if we are to solve this ever-increasing crisis in the supply of fresh water throughout the world. This chapter will explore these issues and consider ways that advances in nanotechnology might help.

35.2 Population and technological impacts on water

The field of industrial ecology has presented what has been termed "the master equation" for linking environmental impacts with population (Equation 35.1) [11]:

$$\text{Environmental impact} = \text{Population} \times \text{GDP}/\text{Population} \times \text{Impact}/\text{GDP} \quad (35.1)$$

This is an interesting framework through which to consider specifically demands for and impacts on water. The United Nations has estimated the world population will grow to over 9 billion by 2050, at an initial rate of roughly 1.1 percent, or about 72 million per year [12]. In other words, for the foreseeable future, the first term of the equation is expected to increase, and, to a first approximation, a larger population has an increased demand for water, if only for the 50 liters each person demands for basic living, without even considering the second and third terms of the equation.

Although the greatest increases in population will be in urban areas, rural areas suffer the most with the least amount of coverage. For instance, coverage (for sanitation) in rural Africa, Asia, Latin America, and the Caribbean is less than one-half that of their respective urban areas. In those three regions, about 2 billion dwellers in rural areas lack access to sufficient sanitation, and about 1 billion lack access to a reliable water supply. Indeed, water scarcity will likely be a motivator for rapid migration from rural areas, putting additional pressure on resources and infrastructure in swelling urban areas.

Likewise, the second term, GDP/population (a measure of personal wealth and to some extent, standard of living) is also increasing as (i) the natural human tendency is to strive for a higher quality of life, and (ii) large, developing nations, particularly in Asia, are showing potential for attaining Western-style standards of living.

As for the third term, there is a strong tendency for poorer countries to use less water, whereas more developed countries use more. Table 35.1 lists the GDP per capita and the average water withdrawals for a number of countries, including the 25 poorest (as defined by GDP/capita). This tendency can be seen in these data, although the correlation is imperfect (e.g., Luxembourg, with the highest GDP/capita, withdraws water at a rate less then one-tenth that of the United States). Nonetheless, Hoekstra and Chapagain, in studying the water footprint of various nations using the concept of "virtual water" used to make a product, have shown that more developed countries frequently have a higher water footprint [13]. Anecdotally, it may also be considered that more developed nations tend to be more industrialized, which has historically resulted in greater impacts on water.

With population and GDP/population trending upward, at least in the foreseeable future, only the third term, impact/GDP, is left to manipulate if we are to reduce demands and impacts on water. This, then, is where conservation and technological advances that reduce water usage and contamination can play a significant role.

35.3 Water access

Table 35.1 also lists the percentage of the populations with access to clean water, as reported by the UNICEF/WHO Joint Monitoring Program [16]. It is not surprising that all of the 25 poorest countries have limited access to clean water.

Table 35.1 GDP per Capita, Average Water Withdrawals, and Population with Access to Clean Water for Select Countries

	GDP per Capita ($US) [14]	Water Withdrawals [m^3/(capita-yr)] [6]	Population with Access to Clean Water (%) [15]
25 Lowest by GDP			
Afghanistan	1,000	779	39
Djibouti	1,000	25	73
Eritrea	1,000	68	60
Guinea	1,000	161	50
Tokelau	1,000	—	88
Rwanda	1,000	17	74
Madagascar	1,000	804	46
Kiribati	1,000	—	65
Mozambique	900	32	43
Togo	900	28	52
Burundi	800	38	79
Gambia, The	800	20	82
Sierra Leone	800	69	57
Malawi	800	78	73
Central African Republic	700	7	75
Ethiopia	700	72	22
Niger	700	156	46
Solomon Islands	600	20	70
Comoros	600	13	86
Somalia	600	400	29
Guinea-Bissau	600	113	59
Liberia	500	34	61
Zimbabwe	500	324	81
Congo, Democratic Republic	300	6	46
Others			
United States of America	46,000	1,600	100
Luxembourg (highest GDP)	80,800	121	100
Turkmenistan (largest withdrawal)	9,200	5,104	72

Indeed, people in impoverished places are the most vulnerable to water issues, particularly those who depend directly on water and other natural resources for their subsistence. Viewing these countries along with those deemed landlocked (Table 35.2), as ranked by the Central Intelligence Agency (CIA), it is notable that Afghanistan, Burundi, Malawi, and Zimbabwe (in bold font) appear in both tables. These countries are especially plagued with a critical need for clean, potable water and would most benefit from advances in nanotechnology for improving water quality. However, they are also the most likely not to be among those producing or developing the technology. It is morally imperative (see Chapter 34) that the specific problems of these countries be addressed, and that they be given open access to developing technologies.

35.4 Corruption, mismanagement, and overconsumption

According to the United Nations "there is enough water for everyone. The problem we face today is largely one of governance: equitably sharing this water while ensuring the sustainability of natural ecosystems" [1]. In many parts of the developing world, about 40 percent of the water is unaccounted for because of leaks and illegal connections [1]. Governance issues and the lack of a basic legal infrastructure to resolve international water disputes also present serious challenges to addressing management issues. Moreover, water is intrinsically undervalued as a resource and commodity in the marketplace, often leading to unsustainable practices and misuse. The illusion of abundance coupled with an inadequate pricing structure has made possible the overexploitation of freshwater

Table 35.2 Land-Locked Countries [14]

Afghanistan	Hungary	Paraguay
Andorra	Kazakhstan	Rwanda
Armenia	Kosovo	San Marino
Austria	Kyrgyzstan	Slovakia
Azerbaijan	Laos	Swaziland
Belarus	Lesotho	Switzerland
Bhutan	Liechtenstein	Tajikistan
Bolivia	Luxembourg	Turenistan
Botswana	Macedonia	Uganda
Burkina Faso	**Malawi**	Uzbekistan
Burundi	Mali	Vatican City
Central African Republic	Moldova	West Bank
Chad	Mongolia	Zambia
Czech Republic	Nepal	**Zimbabwe**
Ethiopia	Niger	

resources. Arguably then, one of the largest obstacles to the provision of clean water throughout the world is nontechnical, raising the intriguing question of whether advances in nanotechnology are able to address it. One possibility is for networked sensors to allow for better monitoring of leaks and overconsumption. Another is the possibility to enable point-of-use or other small water purification systems that would allow small groups of people to not have to depend on large, mismanaged distribution systems. Further, community- or individual-based water systems may help to illuminate the real cost of water for those individuals, subsequently discouraging waste and overconsumption.

35.5 Climate change and global warming

It is increasingly becoming apparent that climate change, which impacts environmental quality and quantity of natural resources, is a potential source of conflict and threat to national and, in turn, global security. As climate change threatens food supplies and economic development, even local and regional problems can have serious global implications.

According to Stephen Schneider, editor of the journal Climatic Change and a lead author for the authoritative Intergovernmental Panel on Climate Change (IPCC), "global warming will intensify drought. And it will intensify floods." This means areas such as southern Europe, the Mideast, North Africa, South Australia, Patagonia, and the US Southwest will dry out. Such severe conditions can compromise supplies of fresh water that are available for agriculture, and industrial and public use.

35.6 Patents: parity and access issues

One of the parity issues arising from the development and deployment of nanotechnology as a tool to increase both the quantity and quality of water resources involves power struggles, both political and economical. The employment of patents for protection of technological advances may provide a more level playing field for access to the technology. However, access to patents developed in the United States and Europe for third world countries is often fraught with difficulties and high costs. In addition, countries with limited financial resources often find obtaining access to newly developed technology problematic.

There are also countries where foreign patent rights are sometimes viewed without strict adherence to protection of those rights. In these situations, patents can be violated and the technology used without either notifying the patent holder or paying for access. The backward reconfiguration of the technology by close examination of the device or design is another way in which novel technologies are adopted. Such technologies can be inferior to the original design, and may even prove faulty.

However, with or without patent rights, companies that are economically strong can edge out competing companies by virtue of both the level of resources available for advancing research and the testing and deploying of new technologies. Such major corporations are also able to price new technologies low to attract more buyers, thus effectively destroying competing new technologies by smaller firms unable to reduce the price. By the same token, wealthy countries with huge financial and personnel resources to bear upon the development of novel technologies can gain the upper hand competitively. These countries are able to provide economic incentives for novel ideas, able to enhance the commercialization, and able to subsidize for a time such technologies. Such advantages can retard and even stifle competing companies and countries with fewer resources. The power such advantages confer upon nations and companies can result in the failure of a better technology to be developed and deployed. It is tempting to attempt to control the market through such means, however, and few in the position of being presented with the opportunity can resist.

The crucial need is to minimize the wielding of this power, especially when it begins to destroy healthy competition or when it results in unfair trade practices. This can be achieved through increased collaborations and consortia between and among industries and countries. Within the field of nanotechnology, there are a variety of such efforts, both national and international in scope, including efforts within the Organisation for Economic Co-operation and Development (OECD) and those within the National Nanotechnology Initiative.

35.7 Political demands

Political power can also result in the imposition of the needs of a more powerful nation subverting those of smaller ones. In addition, the political realities of nations that are land-locked or otherwise water poor may mean that research and deployment options for such nations are severely limited. Alternatively, it can also mean hostile relations between such neighboring nations faced with heavy water needs. Hostility can quickly deteriorate into conflicts that further strain the resources of these nation-states. One possible solution is to work together as partners more and to work as adversaries less (see Chapter 36). This would require sufficient water resources or the equal access to novel water technologies that would enable everyone to partake in the benefits without necessitating that some be left without.

35.8 Conflict

With population and development comes competition for available water. Water scarcity is an issue that affects diverse populations in all parts of the world, whether in the Southeastern United States or Sub-Saharan Africa. As a result scientists predict higher levels of conflict in already tense regions such as the Middle East.

Water has an obvious and tangible geopolitical aspect to it. The Middle East is a prime example, where ongoing tensions may be complicated by disputes over limited water resources. Water has become an issue of major political concern in the region in recent years to the extent that various peace agreements that have been proposed or signed have all included water. There is a worry on the parts of the global collective of intelligence communities that water—or the lack thereof—may spark regional flashpoints for international conflict. But whether that conflict will escalate to armed violence is the question. There are those observers who argue that the prospect of actual violence is unlikely whereas there is probably an equal number who assert that the probability of conflict is likely and growing.

In addition to the Middle East, Africa provokes the greatest concern about water shortage: by 2025, 40 countries are expected to experience water stress or scarcity [17]. Water scarcity is a function of supply and demand. Demand is increasing at an alarming rate in some regions, through population growth and increasing per capita use. In many water-scarce countries, such as the Darfur region of Sudan, Jordan, and Israel, there is no obvious and inexpensive way to increase water supply, which serves to fuel the potential for conflict. Recently in Darfur, a fossilized lake was discovered but to extract the water is costly and may not be practical in such an environmentally hostile region.

35.9 Biofuels

35.9.1 Biofuels introduction

Concerns about greenhouse gases, a desire to move away from foreign dependence on oil, and rising prices of oil are increasing attention on biofuels [18] However, an increased demand for biofuels is already putting pressure on stressed water sources—an example that a solution to one problem can potentially exasperate another one.

Simply, a biofuel is any fuel—solid, liquid, or gas—derived from a renewable, biological source, such as plants or organic waste matter. Organic material is converted to biofuels through biological and/or thermochemical processes [19–24]. Enzymes and/or microbes are used in biological processes, such as fermentation of sugars and starches to make bio-ethanol or digestion of waste material to make biogas (the equivalent of natural gas). Esterification of vegetable oils with methanol is a common chemical process used to produce bio-diesel, whereas gasification and the use of catalysts can provide a variety of biofuels in a manner similar to the Fischer–Tropsch process for converting coal into liquid fuel. Research is rapidly advancing on a variety of biofuel production processes, promising to make them more efficient and able to generate a wider variety of fuels (e.g., [23,25–31]).

Because petroleum supplies 40 percent of the US and 36 percent of the world's energy needs, and nearly two-thirds of it goes to transportation [32,33], current interest is largely on the production of liquid fuels for transportation. The two most commonly produced liquid biofuels are bio-ethanol and bio-diesel [18,21,34,35] and there are a variety of sources for each. Bio-ethanol typically comes from food crops high in sugar (e.g., sugarcane, sugar beet) or starch (e.g., corn, wheat, potatoes, cassava). Bio-diesel requires food crops high in oil (e.g., soybeans, oil seed rape, palm, sunflower) and can also be made from used cooking oil or animal fat [23,36]. Another potential source of biofuels is cellulose, an abundant material found in the cell walls of plants, bacteria, and algae. Being a polysaccharide, it can be broken down into simple sugars—in fact many bacteria and fungi use it as a food source by doing just that—that are suitable for conversion to bio-ethanol. Cellulose, therefore, represents the largest potential feedstock for biofuel production, and the ability to use it as such would overcome a number of hurdles to a viable biofuel industry [34,37]. Currently, however, there are technical and financial challenges to converting cellulose to bio-ethanol on an industrial scale [38], and both thermochemical and biological methods are being heavily pursued [37,39,40]. It is widely believed that the required technology will be available in the coming decade.

Bio-ethanol is the most widely produced biofuel. In 2006, roughly 38.2 billion liters[1] (GL) of bio-ethanol were produced worldwide [41][2] compared to 6.1 GL of biodiesel [42] (see Table 35.3). In 2006, the largest producers of bio-ethanol in the world were Brazil (16.7 GL) [41] and the United States (18.5 GL) [43], whereas the European Union produced the most bio-diesel (5.6 GL), mostly in Germany [44]. By comparison, in the same year, worldwide production of petroleum was 4857 GL [32], of which the United States consumed 812 GL for transportation [33].

35.9.2 Worldwide biofuels policy

Increased demand for biofuels has resulted largely from a variety of national policies (Table 35.4). For example, the US Energy Policy Act of 2005 calls for 7.5 billion gallons (28 GL) per year of biofuels to be used in fuel transportation by 2012 [48], and President Bush indicated a desire to increase that number to 35 billion gallons by 2017 in his 2007 State of the Union Address [49]. In Europe, the Biofuels Directive 2003/30/EC set standards of 2 and 5.75 percent of transportation fuels to be renewable by the end of 2005 and 2010, respectively [50], and there is interest in setting a target of 10 percent by 2020 [51]. A long-standing national policy in Brazil has promoted bio-ethanol such that, in recent years, it has comprised 35–40 percent of the automotive fuel market [45,52]. In 2005, China passed a Renewable Energy Law, and has begun to use biofuels as its demand for petroleum continues to increase [53,54].

Table 35.3 Production of Bio-Ethanol and Bio-Diesel (Millions of Liters)

		2002	2003	2004	2005	2006	2007
Bio-ethanol	Australia†				75	605‡	
	Brazil†	12,638	14,798	15,397	15,800	17,860	20,450‡
	China		25	380	1,166	1,648	1,838
	E.U.					1,708	2,928
	India†	180	90	20	100	250	550‡
	US	8,101	10,615	12,887	14,780	18,512	24,372‡
	World	26,091	30,022	31,644	34,508	38,285	
Bio-diesel	Australia†				105	524‡	
	Brazil†				0.7	155	730
	E.U.				3,361	6,016	9,453
	US	57	76	95	284	946	1,230‡
	World			2,100	3,900		

Note: Production of biofuels by a selection of the top producing countries. All data except for United States and World are from Global Agriculture Information Network reports [45] by the Foreign Agriculture Service, US Department of Agriculture. These are not considered official data, and are provided only for the purpose of comparison and indication of trends. Although better data may exist for some countries and some years, these sources were chosen to provide the best consistency in reporting. US bio-ethanol data are from the Energy Information Administration, US Department of Energy [33,43]; bio-diesel data are from the National Biodiesel Board [46]. Worldwide bio-ethanol estimates from F.O. Licht [41]; bio-diesel data are from Renewable Energy Policy Network for the 21st Century [47]. China and India have not produced significant quantities of bio-diesel.

†Dates are Marketing Years (e.g., "2005" = 2005/2006) for Australia (unspecified), Brazil (May–April), and India (October–September). Bio-diesel data for United States are reported sales and as such may be based on fiscal year.

‡Forecast.

Table 35.4 Overview of Biofuels Policies in Selected Countries

Australia	350 million liters target for bio-fuels production by 2010. Biofuels have no excise tax until 2011.
Brazil	Mandated 20–25 percent ethanol fuel content; 2 percent bio-diesel rising to 5 percent in 2013. Tax incentives for biofuel use.
China	Considering targets.
European Union	Mandated 5.75 percent biofuels in transport fuel by 2010. Targeting 10 percent by 2020.
India	Mandated 5 percent ethanol blending, increasing to 10 percent in the future. Developing program for bio-diesel from Jatropha curcas.
United States	7.5 billion gallons biofuels production by 2012. Considering increase to 35 billion gallons by 2017. Tax incentives.

Sources: Global Agriculture Information Network reports [45] by the Foreign Agriculture Service, US Department of Agriculture; Renewable Fuels Association [55]; and Renewable Energy Policy Network for the 21st Century [47].

35.9.3 Biofuels: solution to or creation of a problem?

Figure 35.2 shows historical and projected bio-ethanol production in the United States [33,43]. The data for 2007 and beyond are from a US Department of Energy (DOE) analysis that projects by considering only current policy and production trends; because predictions are not made about future policy changes, advances in technology, or societal/market forces, these may be considered very conservative. Also indicated are the congressionally mandated levels of all biofuels from the Energy Policy Act. It should be noted that even the conservative DOE projections indicate the 2012 goal from the Act will be exceeded by the end of 2008.

As noted earlier, there are a variety of motivating factors contributing to the increased production of biofuels. One of these is the potential to reduce greenhouse gases owing to the fact that biofuels, in theory, could be carbon-neutral sources of energy: CO_2 is captured during the growing phase of the plants and released in equal amounts when the biofuel is burned. A second is the ability to reduce dependence on foreign petroleum, since anywhere that a feedstock can be grown, biofuels can be produced. In addition, such a potentially self-sustainable energy supply would be beneficial to developing countries and remote populations. Finally, by generating a high demand for agricultural products, it is conceivable

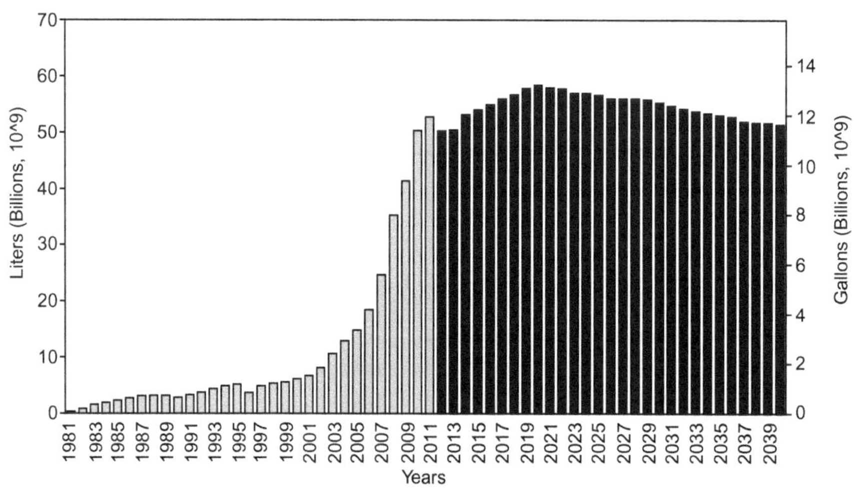

FIGURE 35.2

Bio-ethanol in the United States. Data for 1981–2005 from the US Dept. of Energy "Annual Energy Review 2006" [33]. Data for the 2006–2020 bar graph from US Dept. of Energy "Annual Energy Outlook 2008 with Projections to 2030" [43]. Data for the 2006–2012 line graph from mandated values in US Energy Policy Act, 2005, and includes biodiesel.

that a biofuels economy would alleviate poverty for many underdeveloped, agrarian-based societies throughout the world.

However, there are equally a number of potential negative implications of the blossoming biofuels industry, and before benefits can be touted unequivocally, the full life cycle of the fuel production—so-called "Well-to-Wheels" analyses—must be considered [56]. There are four primary areas: (i) Agricultural production of biomass, (ii) industrial conversion of biomass to biofuel, (iii) transportation of raw materials and final product, and (iv) use (generally combustion).

A number of analyses and several reviews looking at the net benefits of biofuels have been published [34,36,56–60]. For example, a 2007 review by von Blottnitz focused on bio-ethanol for transportation and found at least 47 related reports [36]. As the large number of reports might suggest, there is a considerable amount of debate on the benefits of biofuels. In particular, the focus has been on energy and carbon balances [23,59–71] because they relate to the ability to reduce greenhouse gases and dependence on petroleum. Specifically, on the issue of energy of corn-derived bio-ethanol, even very recent studies are in disagreement. For example, Pimental et al. have estimated that a liter of corn-ethanol requires 43% more energy to produce than it contains [66] whereas Shapouri et al. reported a positive energy return of 67 percent [72]. The discrepancies are largely due to disagreements over system boundaries and credits for co-products, and a recent meta-analysis indicates that corn-ethanol has a modest positive energy gain of 10–25 percent [59,60]. Interestingly, corn is among the least efficient crops currently being used to produce biofuels [36,61,62]. For example, sugarcane-derived bio-ethanol in Brazil has been estimated to return 214–287 percent more energy than it takes to produce [61], and soybean-bio-diesel was calculated to yield a 93 percent energy gain [62].

In addition to the energy balance issue, another concern is the amount of resources required to produce bio-energy crops. Table 35.5 lists approximate production capacities per hectare of biofuels from several crops. Consider the example of corn grown in the United States, where it is by far the largest source of biofuels. Figure 35.3 shows the US historical and projected utilization of corn for all domestic purposes and for bio-ethanol production. Similar to the DOE projections used in Fig. 35.2, these are from a US Department of Agriculture (USDA) analysis based on current policy and production trends, and may therefore be considered conservative. Shown with this data in bar graph form is the calculation of the percentage of corn used for bio-ethanol. It is notable that roughly in 2008, when the 2012 goal of 28 GL of biofuels is met, fully one-third of the US domestic use of corn will be going to this effort. Using the average of the most common estimates of ethanol production, 2800 L/ha (Table 35.5), this represents 10 Mha of corn harvested solely for bio-ethanol use but only 3.3 percent of the 2008 projected US gasoline consumption.[3] Figure 35.4 shows the estimated amount of land needed for additional corn production to meet various goals of gasoline replacement in 2008, indicating that every 10 percent substitution of bio-ethanol for gasoline will require roughly 30 Mha of corn to be harvested.

Table 35.5 Annual Agricultural Chemical Application and Bio-fuel Production for Various Crops[ξ]

Crop	Fertilizer (kg/ha)		Insecticide (g/ha)	Herbicide (kg/ha)	Biofuel Productivity[‡] (L/ha)	Net Energy Balance
	N	P				
Corn†	120–185	45–65	70–180	1–2	2800–3632 [62,64]	1.1–1.25 [59]
Soybean†	4–7	15–20	10–50	2–3	544 [62]	1.93 [62]
Sugarcane (US)	—	—	—	—	1760–2630 [83]	—
Sugarcane (Brazil) [61]	65	52	500	3	5220–7740 [41,61]	3.14–3.87 [61]
Switchgrass‡‡, [84]	50–200	0	0	0	3300–6600* [84]	4.43–6.40 [85,86]
Diverse grassland‡‡, [87]	0	12	75	0	1530**	5.44

Note: Conversion factors: 1 lb = 2.2046 kg, 1 gal = 3.785 L, 1 ha = 2.471 acres. Numbers have been averaged and rounded and are given only for comparison.
†Rates of chemical application for corn and soybeans from National Agricultural Statistics Service reports from 2000–2007 [88] calculated by total chemical applied divided by total acres harvested.
‡ Bio-diesel for soybean, bio-ethanol for all others.
†† The lower value of 2800 L/ha is an average of a number of similar estimates [64] and is probably more realistic.
‡‡Switchgrass and other perennials comprising diverse grassland are not well-established commodity crop; values will likely vary greatly in the future. A review [84] concluded that an average of 50 kg/ha of nitrogen fertilizer would be optimal for producing about 15 Mg/ha of switchgrass, though reviewed reports frequently used 100–200 kg/ha and had yields of 10–20 Mg/ha. Fertile diverse grasslands yields estimated at 6 Mg/ha [87].
**Based on estimated 0.330 L/kg conversion of dry switchgrass to ethanol [69].*
***Based on estimated 0.255 L/kg conversion of dry biomass to ethanol [68,87], though this seems conservative.*

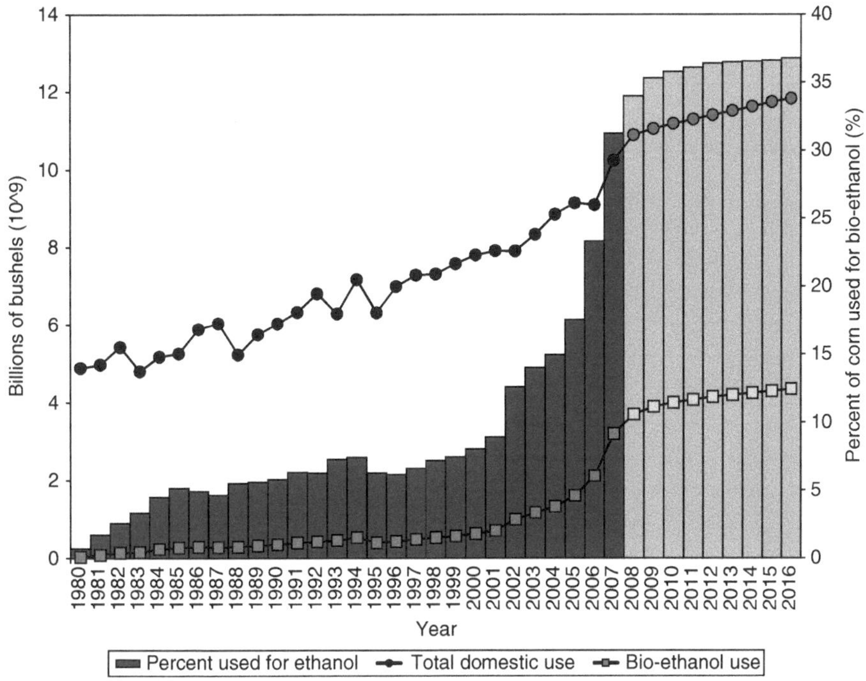

FIGURE 35.3

US corn used for bio-ethanol production compared to total supply. Data for 1980–2007 from http://www.ers.usda.gov/Data/Feedgrains/FeedGrainsQueriable.aspx using the following parameters: Data type = Supply and use; Attribute = Utilization, domestic total/ Utilization in alcohol for fuel; Commodity = corn; Frequency = Market Year; Location = United States of America; Market Years = 1980–2007. Data for 2008–2016 from USDA baselines projections [75].

In comparison, the total amount of agricultural land in the United States is 175.7 Mha [73], and the 5-year average of domestically used corn-acreage, less bio-ethanol production, was 19.5 Mha (see Fig. 35.4). To further illustrate, using 2005 numbers, Hill et al. estimated the conversion of the entire US corn and soybean crops to bio-ethanol and bio-diesel would account for only 12 percent and 6 percent of current petroleum products, respectively [62], and Anthrop calculated that replacing 15 percent of US gasoline demand with corn-derived bio-ethanol would consume 13.6 billion bushels of corn, or 130 percent of the US corn crop [65]. Similarly, Johnston and Holloway estimated the maximum bio-diesel production potential worldwide at 51 GL annually, if all commercially exported lipid feedstocks (i.e., not consumed internally) from existing agricultural lands were used [74].

In the context of these scales of agricultural and industrial production, it is clear that any negative impacts of biofuel production could be potentially huge.

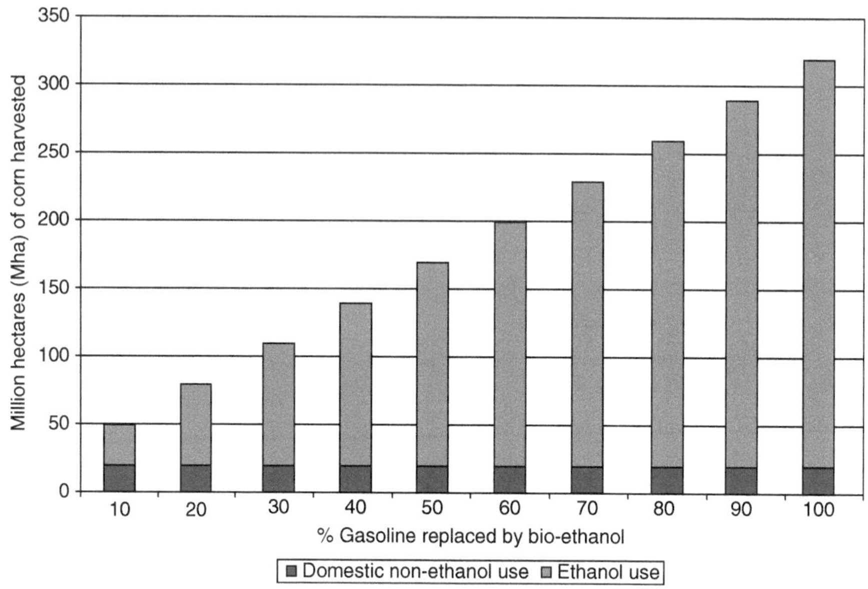

FIGURE 35.4

Farm land required to substitute gasoline with corn-based bio-ethanol. Calculated with data from http://www.ers.usda.gov/Data/Feedgrains/FeedGrainsQueriable.aspx using the following parameters: Data type = Supply and use; Attribute = Production/Area harvested for grain/Utilization in alcohol for fuel; Commodity = corn; Frequency = Market Year; Location = United States of America; Market Years = 2003–2007. Five-year average of acres producing domesitically used corn, not including use for bio-ethanol, was calculated as 19.5 Mha. On an energy basis, 839 GL/yr of bio-ethanol would be needed to completely replace the predicted US gasoline consumption in 2008 [48]. The lower value of 2800 L/ha was used (Table 35.5).

Unfortunately, in contrast to the debate about energy balances and greenhouse gas emissions, broader environmental implications have been far less examined [36,58,66,76–79]. Nonetheless, studies considering water in their analysis have been in general agreement: increased production of biofuels with current technology will likely have significant negative impacts on water supply and quality [57,58,65,79–81]. A study by Berndes calculated that biofuel production in the United States would more than double the per-capita water withdrawal by 2075 and would result in several countries, including China and India, having serious water shortages [81].

Growing crops requires water, with significant amounts coming from irrigation. Additionally, the industrial processing of biomass to biofuel is locally water-intensive, requiring large amounts in the distillation process. Pimentel et al. calculated that a liter of bio-ethanol produced from corn requires a total of 1,700 liters of water [66]. Though most of this is rainwater in the agricultural phase, it should be noted that an average of 13.9 percent of land used to grow corn over the last 10 years in the United States was irrigated [82], and it is reasonable to assume that increased crop production will eventually be done on marginal lands requiring more irrigation and agricultural chemical application. In contrast, a relatively small amount of the water, about 4 liters, is used in the industrial/refining phase [57], but the requirements can be locally substantial. For example, a National Academies report suggested that a bio-refinery producing "100 million gallons of ethanol per year would use the equivalent of the water supply for a town of about 5,000" [57]. The industrial processing of soybeans into bio-diesel requires about 1 liter of water per liter of fuel, while cellulose-derived bio-ethanol is projected to require 2–6 liters of water [57].

In addition to having a high demand on water, the agricultural production of bio-energy crops may require substantial amounts of fertilizers, herbicides, and pesticides, resulting in polluted water downstream. Table 35.5 lists the typical amounts of fertilizer, insecticide, and herbicide used to grow three common crops used in the production of bio-ethanol and bio-diesel.

35.9.4 Possible ways forward for biofuels

It seems clear that demand for petroleum-based energy cannot be met using current methods of producing biofuels from food sources such as corn and soybeans. As stated by Hill, "making biofuels that will be both environmentally superior to fossil fuels and displace significant quantities of fossil fuel use will require exploration of plant resources other than those that have been domesticated and bred primarily for their food, feed, or forage value" [34]. Technological breakthroughs, enabling the use of cellulosic materials as inputs for biofuels, will certainly go a long way [34,37,38]. Not only will this allow the extraction of energy from nonedible parts of the plants currently considered "waste," but it will also allow the use of wholly nonfood feedstocks as diverse as algae [89,90], trees, and naturally occurring perennial grasses. Because species such as switchgrass (Panicum virgatum) and miscanthus (Miscanthus sinensis) have evolved to be particularly well suited for their native climates and soils, they may require less fertilizer and be more resistant to pests, disease, and drought compared to food crops [84]. In fact, the US Department of Energy has emphasized switchgrass in its biofuels research program [85]. In addition to monocultures, highly diverse grasslands are also being considered [87], and appear to extend these benefits. For example, application of nitrogen fertilizer can be reduced, depending instead on naturally occurring legumes to fix nitrogen. Herbicides are also unnecessary, since the "weeds" are welcome additions. Perennials do not require re-seeding each year, naturally prevent erosion, can revive marginal lands, and overall have very low energy input requirements, especially compared to crops such as corn and soybeans [37,84,85,87,91]. Though current yields of bio-ethanol produced per hectare are not substantially better than corn (Table 35.5), the more favorable energy balance, lower use of agricultural chemicals, and erosion prevention potential make cellulosic ethanol from perennial grasses an attractive option, particularly from a water use and quality standpoint.

Meeting biofuel requirements with cellulosic bio-energy crops would still require large amounts of land, though it may be considerably more sustainable than using food-crops. A report from the Oak Ridge National Laboratory estimated that the United States could sustainably produce 1.3 billion metric tons of dry biomass per year using its agricultural and forest resources, while still meeting its food, feed, and export demands [92]. At a conservative conversion rate of 0.255 L/kg of dry biomass [68,87], this would be equivalent to 218 GL of gasoline.[2]

In the future, biofuel production should become more efficient, as the increased attention also spurs research in areas throughout the life cycle. In the agricultural phase, improved farming techniques and advances in biotechnology will likely improve crop and biofuel production. For example, plants will be bred or genetically modified to be better biofuel sources by increasing photosynthetic efficiency, CO_2 capture, nutrient use, and overall yield [23,57,93,94]. Major hurdles will likely be overcome when technology enables the use of cellulose as a feedstock. Industrial processes could be more efficient, for example, through better reuse of wastes, including heat, water, and unprocessed materials. Even efficiency in the use stage could be improved by designing vehicles more suited to biofuels, including using fuel cells instead of internal combustion engines. Interestingly, nanotechnology could play a role in many of these areas.

In the agricultural phase, nanotechnology could help boost efficiency and reduce impacts to water use and quality in several potential ways. For example, nano-enabled sensors providing real-time data on soil moisture and nutrient content would allow more efficient application of water and fertilizers. Additionally, nanomaterials could be used to direct pesticides and fertilizers to their targets and only be released when needed, decreasing the overall use of these chemicals.

Industrially, there may be a number of ways for nanotechnology to improve biofuel production. Particularly in the thermochemical production of biofuels, catalysts are being pursued for more efficient processes, as well as a wider variety of products. For example, though not specifically stated as being nanosized, two recently developed catalysts are indicative of the potential in this area: a Pt/Re-C catalyst used in the synthesis of fuel alkanes from bio-derived glycerol [25] and a CuRu/C catalyst used to produce 2,5-dimethylfurn (DMF)—a potentially superior biofuel as it has a 40 percent higher energy density than ethanol—from biomass [95]. Additionally, a Rh-based nano-catalyst loaded inside carbon nanotubes was found to improve ethanol production from syngas (CO and H2) over an order of magnitude from catalysts outside the tubes [96]. On another front, the distillation of ethanol stands out as an area requiring relatively large amounts of energy and water that could use the type of revolutionary breakthrough nanotechnology has provided in many other areas.

An indirect way for nanotechnology to mitigate potential impacts of biofuels is by enabling other renewable energy sources and reducing the need for biofuels.

Patzek has pointed out that plants only convert about 1 percent of the energy they receive from the sun into biomass [58]. Considering a 120-day growing season in temperate climates, this translates to a theoretical maximum of about 0.3 percent of the solar energy irradiating a given hectare in a year ultimately being converted to biofuel-energy. By contrast, commercially available solar cells have efficiencies of 5–17 percent [97] and can collect sunlight all year long. Nanotechnology-enabled plastic solar cells have lagged behind traditional silicon-based ones in terms of efficiency, but they offer considerable cost- benefits, for example, by using cheaper materials such as organics and polymer [98], and recently, efficiencies up to 6 percent have been reported [99].

The future path for biofuels—which will be most important based on efficiency of conversion processes, availability of feedstock, demand for particular products, advances in technology, and sustainability concerns—is difficult to predict. What is certain, however, is that global demand for energy continues to increase while petroleum reserves are being depleted. Biofuels are being sought as a solution, and the subsequently expanding activity in the agricultural and industrial sector will in turn impact water resources. Nanotechnology may have a role to play in enabling sustainable biofuels.

35.10 Bottled water

Over 40 billion gallons of bottled water are consumed each year worldwide [100], and in recent years this number has been increasing by nearly 10 percent per year as its popularity skyrockets in countries such as the United States, China, and India [100,101]. This popularity is despite the fact that bottled water is often more than a thousand times the cost of municipal water [101]. The willingness to pay so much more is often attributed to perceptions of improved taste, higher purity, convenience, or social trends. Contrast this to a recent survey of residents in Manaus, Brazil, on their willingness to pay to have fresh water piped directly into their house [102]. Manaus is an impoverished city of 1.5 million on the banks of the Amazon River. Although it is near one of the world's largest freshwater supplies, the municipal water system is adequate for less than 10 percent of the population. Despite having such an inadequate water system, and the fact that fully one-fifth of those surveyed had suffered from a waterborne illness in the last year, the survey found that citizens of Manaus were willing to pay an average of only US$6.12 per month to improve their water system. In comparison, they were already paying 50 percent more than this per month for their electricity [102].

Clearly, the perception of the value of water varies widely from person to person, throughout the world, and may in fact have little reflection on its actual value. For bottled water, this arguably skewed valuation of water is driving an enormous commercial market in developed countries. On the other hand, in many underdeveloped regions of the world, bottled water may be the best or only source of clean water. In either case, there are concerns with widespread distribution and dependence on bottled water.

First, bottled water must necessarily be bottled in a container at a central location. In underdeveloped regions, this centralized processing and use of sealed containers can help to ensure the sanitation of the water. However, in these regions and in developed areas, it also means the potentially unnecessary use of nonrenewable resources in the form of plastic bottles and petroleum use in transporting the product to its point-of-sale. Many of these bottles are not recycled—a problem that is exasperated in developing areas without facilities for recycling or often even for proper trash disposal. These inputs for bottled water are substantial. It has been estimated that every liter of bottled water consumed requires an additional two liters to produce, and that in the United States alone in 2006, 17 million barrels (2.7×109 liters) of petroleum went into the production of plastic for bottled water[4] [104].

A second area of concern is the safety of bottled water, and a number of studies have been published on the concentrations of nutrients, pollutants, and bacteria in commercially available bottled water throughout the world [105,106,107,108]. Although bottled water may be the best source of clean drinking water in many developing areas of the world, regulations regarding its safety are limited not only in these areas, but also in developed nations [101]. For example, regulations typically require regular testing of municipal water, but little if any such testing is required for bottled water. Unsurprisingly, bottled water is often found to contain unhealthy levels of bacteria [105,106,108], and a number of recalls have occurred [101]. Furthermore, bottled water is often lacking in essential nutrients, such as calcium, magnesium, potassium, and fluoride [105,106], which could be a health hazard if it is the sole source of water. Finally, one study found that bacterial growth in opened bottles can be substantial after 8 hours at room temperature, whereas tap water under the same conditions had minimal growth [107].

Despite considerations of cost, unnecessary use of resources, and potential health issues, bottled water use is growing rapidly throughout the world. In the developed world, it may be preferred for its taste and convenience, whereas in the developing world, it may be the only source of clean water. Advances in nanotechnology discussed throughout this book may provide viable cost and resource-saving alternatives in both cases by, for example, enabling point source options that not only sanitize drinking water, but also remove foul tastes and odors and are as portable as your typical bottle of water.

35.11 Future trends

Given the importance of access to clean water and basic sanitation for mitigating or eradicating poverty and improving people's livelihoods, The United Nations has declared 2005–2015 the "Water for Life Decade" and set a Millennium

Development Goal (MDG) of reducing by half the percentage of the world's population without sustainable access to clean drinking water and basic sanitation [9]. Although great strides are being made to provide clean water and sanitation, the goal is unlikely to be met on the current trajectory, which is likely to lead to more strife and conflict in the most sensitive regions.

By the year 2025, an estimated 2.7 billion people worldwide will be living in areas experiencing severe water scarcity. Although population is not the single driver affecting supplies, it affects the overall demand by all sectors competing for finite resources—potentially leading to regional conflicts in more dry arid areas (e.g., the Middle East, sub-Saharan Africa, parts of India and China). Africa, Asia, and the Middle East will account for 87 percent of the projected population growth over the next 30 years. These regions, collectively, are expected to grow by 2.3 billion people making them particularly vulnerable to water stress and food shortages. In addition the effects of climate change are likely to further exacerbate problems, whether it be too much water in some places or severe drought conditions in others.

Experts predict that increases in water demand will occur in developing countries where population growth and industrial expansion will be greatest. Some speculate that the wars of this century will not be fought over just politics and religion but over water rights, likening the current situation to the oil crisis of the 1970s.

35.12 Conclusions

Many of the issues discussed here—corruption, mismanagement, conflict, pollution, and so on—are not new to the water arena, but nanotechnology offers the promise of inexpensive, easily deployed, widely available novel technologies to increase both the quality and quantity of potable water, thereby eliminating many of the sources of hostility between nations and companies competing for a limited resource.

Examples of such technologies are discussed within this book and include nanomembranes, nanocatalysis, and nanoparticles. Membranes may be composed of carbon nanotubes or other engineered nanomaterials that provide improved techniques for water purification. Nanofiltration techniques can offer enhanced purification and removing even trace concentrations of pollutants is possible. These techniques may be utilized to enable the cost-effective desalination of water. Nanocatalysts can significantly reduce energy production, and usage. Engineered nanoparticles can be used for the removal of contaminants from aqueous streams, for the treatment of subsurface contaminated ground water sites, for the disinfection of water without the production of disinfection by-products, and for the addition of nutrients and other desired compounds into potable water. Overall, the potential benefits of engineered nanoparticles versus conventional techniques include reduced costs, energy, and material usage, in addition to revolutionary solutions to particularly challenging issues.

However, technology is not the only solution to the freshwater problem. For instance, water withdrawn from the Colorado River in several states for irrigation and other purposes drains the river nearly dry as it winds its way to the Sea of Cortes, Mexico. No technology exists that can double the flow of the Colorado River. Similarly, shrinking aquifer levels cannot be replaced by human technology. Obviously there is a critical need for extensive water conservation and proper water management. But where technology can help, nanotechnology is well positioned to provide solutions to many of the difficult water-related problems, thereby improving the quality of life for potentially millions, if not billions, of people worldwide.

Notes

1. For ease of comparison, numbers have been converted to liters per year unless reported otherwise. Slight errors may occur due to rounding. Conversion units: 1 year = 365 days, 1 barrel = 158.987 liters, 1 gallon = 3.7854 liters, density of ethanol = 0.789 g/mL, density of bio-diesel = 0.88 g/mL.
2. Numerous studies cite F.O. Licht [41] as the most authoritative source for bioethanol production statistics, though reported numbers are considerably higher than those here. A closer analysis of F.O. Licht's data suggests it is total ethanol production, including industrial and beverage use. F.O. Licht reports that for 2006, 76.9 percent of the total ethanol produced was fuel ethanol, so the data represented here have been adjusted accordingly.
3. On an energy basis. Projected gasoline consumption in 2008 is 552 GL/yr [43], and the lower heating values for ethanol and gasoline are 21.1 and 32.1 MJ/L.
4. These estimates are from The Pacific Institute's website, and although they seem reasonable they are not published, peer reviewed data. To our knowledge, life-cycle analyses of bottled water are limited [103] though as this discussion implies, there is a pressing need for detailed studies.

References

[1] UNESCO, Water: A Shared Responsibility, United Nations Educational, Scientific and Cultural Organization, 2006.
[2] D. Pimentel, H. Xuewen, C. Ana, P. Marcia, Impact of population growth on food supplies and environment, Popul. Environ. 19 (1) (1997) 9.
[3] D. Struck, Warming will exacerbate global water conflicts, Washington Post, Sect. A8, 2007.

[4] M. Kiparsky, P.H. Gleick, Climate change and California water resources, The World's Water, 2004–2005, The Biennial Report on Freshwater Resources, Island Press, Washington, D.C., 2004, pp. 157–188.
[5] P. Aldhous, The world's forgotten crisis, Nature 422 (6929) (2003) 251.
[6] P.H. Gleick, H. Cooley, The World's Water, 2006–2007, The Biennial Report on Freshwater Resources, Island Press, Washington, D.C., 2006. p. 368.
[7] A. de Sherbinin, Water and population dynamics: local approaches to a global challenge, Water and Population Dynamics: Case Studies and Policy Implications, American Association for the Advancement of Science, 1996.
[8] P.H. Gleick, The human right to water, Water Pol. 1 (5) (1998) 487.
[9] JMP, Meeting the MDG Drinking-Water and Sanitation Target: The Urban and Rural Challenge of the Decade, WHO/UNICEF Joint Monitoring Programme for Water Supply and Sanitation, 2006.
[10] WHO, Water, Sanitation and Hygiene Links to Health, World Health Organization, 2004.
[11] B.R. Allenby, Industrial Ecology: Policy Framework and Implementation, Prentice Hall, 1998.
[12] M.J. Alcalá, A Passage to Hope: Women and International Migration, United Nations Population Fund, 2006.
[13] Chapagain Hoekstra, Water footprints of nations: Water use by people as a function of their consumption pattern, Water Resour. Manag. 21 (1) (2007) 35.
[14] CIA, The 2008 World Factbook, Central Intelligence Agency, 2008.
[15] JMP, Joint monitoring programme for water supply and sanitation: Water data, Data Query [cited April 15, 2008]. Available at ⊲http://www.wssinfo.org/en/watquery.html⊳.
[16] C.N. Hamelinck, A.P.C. Faaij, Outlook for advanced biofuels, Energ. Policy 34 (17) (2006) 3268.
[17] S. Lonergan, Water and war, Our Planet 15 (4) (2005) 27.
[18] C. Somerville, Biofuels, Curr. Biol. 17 (4) (2007) R115–R119.
[19] G.W. Huber, S. Iborra, A. Corma, Synthesis of transportation fuels from biomass: chemistry, catalysts, and engineering, Chem. Rev. 106 (9) (2006) 4044.
[20] A.V. Bridgwater, Renewable fuels and chemicals by thermal processing of biomass, Chem. Eng. J. 91 (2–3) (2003) 87.
[21] M. Balat, Global bio-fuel processing and production trends, Energ. Explor. Exploit. 25 (3) (2007) 195.
[22] L.C. Meher, D. Vidya Sagar, S.N. Naik, Technical aspects of biodiesel production by transesterification—a review, Renew. Sust. Energ. Rev. 10 (3) (2006) 248.
[23] F. Ma, M.A. Hanna, Biodiesel production: a review, Bioresour. Technol. 70 (1) (1999) 1.
[24] Y. Lin, S. Tanaka, Ethanol fermentation from biomass resources: current state and prospects, Appl. Microbiol. Biot. 69 (6) (2006) 627.
[25] D.A. Simonetti, J. Rass-Hansen, E.L. Kunkes, R.R. Soares, J.A. Dumesic, Coupling of glycerol processing with Fischer-Tropsch synthesis for production of liquid fuels, Green Chem. 9 (10) (2007) 1073.
[26] R.W.R. Zwart, H. Boerrigter, High efficiency co-production of synthetic natural gas (SNG) and Fischer–Tropsch (FT) transportation fuels from biomass, Energ. Fuel 19 (2) (2005) 591.

[27] L. Bournay, D. Casanave, B. Delfort, G. Hillion, J.A. Chodorge, New heterogeneous process for biodiesel production: a way to improve the quality and the value of the crude glycerin produced by biodiesel plants, Catal. Today 106 (1–4) (2005) 190.
[28] S.V. Ranganathan, S.L. Narasimhan, K. Muthukumar, An overview of enzymatic production of biodiesel, Bioresour. Technol. (2007).
[29] G. Madras, C. Kolluru, R. Kumar, Synthesis of biodiesel in supercritical fluids, Fuel 83 (14–15) (2004) 2029.
[30] H. Xu, X. Miao, Q. Wu, High quality biodiesel production from a microalga chlorella protothecoides by heterotrophic growth in fermenters, J. Biotechnol. 126 (4) (2006) 499.
[31] J.M. Marchetti, V.U. Miguel, A.F. Errazu, Possible methods for biodiesel production, Renew. Sust. Energ. Rev. 11 (6) (2007) 1300.
[32] BP, BP Statistical Review of World Energy 2007, British Petroleum, 2007.
[33] EIA, Annual Energy Review 2006, Energy Information Administration, US Department of Energy, 2007.
[34] J. Hill, Environmental costs and benefits of transportation biofuel production from food- and lignocellulose-based energy crops, a review, Agron. Sustain. Dev. (2007) 1.
[35] A. Demirbas, Importance of biodiesel as transportation fuel, Energ. Policy 35 (9) (2007) 4661.
[36] H. Von Blottnitz, M.A. Curran, A review of assessments conducted on bio-ethanol as a transportation fuel from a net energy, greenhouse gas, and environmental life cycle perspective, J. Clean. Prod. 15 (7) (2007) 607.
[37] C. Schubert, Can biofuels finally take center stage? Nat. Biotech. 24 (7) (2006) 777.
[38] C.E. Wyman, What is (and is not) vital to advancing cellulosic ethanol, Trends Biotechnol. 25 (4) (2007) 153.
[39] M.J. Taherdazeh, K. Karimi, Acid-based hydrolysis processes for ethanol from lignocellulosic materials: a review, Bioresources 2 (3) (2007) 472.
[40] M.J. Taherdazeh, K. Karimi, Enzyme-based hydrolysis processes for ethanol from lignocellulosic materials: a review, Bioresources 2 (4) (2007) 707.
[41] F.O. Licht, World Ethanol Markets: The Outlook to 2015, FO Licht, 2006.
[42] Analysis FOLC, World Biodiesel Markets: The Outlook to 2010, F.O. Licht, 2007.
[43] EIA, Annual Energy Outlook 2008 with Projections to 2030 (Early Release), Energy Information Administration, US Department of Energy, 2007.
[44] EBB, Statistics: The EU Biodiesel Industry [cited January 7, 2008]. Available at <http://www.ebb-eu.org/stats.php>.
[45] FAS, Global Agriculture Information Network Reports, USDA Foreign Agricultural Service, (2007). Report No.: AS6043, BR6008, BR7011, BR7012, CH7039, E47051, IN6047, IN7047.
[46] NBB, Biodiesel sales [cited December 15, 2007]. Available at <http://www.biodiesel.org/pdf_files/fuelfactsheets/Biodiesel_Sales_Graph.pdf>.
[47] REN21, Renewables Global Status Report, 2006 Update: Renewable Energy Policy Network for the 21st Century, 2006.
[48] US Energy Policy Act of 2005, p. 551, 2005.
[49] G. Bush, State of Union Address, [cited December 17, 2007]. Available at <http://www.whitehouse.gov/news/releases/2007/01/20070123-2.html>, 2007.

[50] Directive 2003/30/EC of the European Parliament and of the Council of 8 May 2003 on the promotion of the use of biofuels or other renewable fuels for transport, p. 5, 2003.
[51] EC, Renewable Energy Road Map. Renewable Energies in the 21st Century: Building a More Sustainable Future, Commission of the European Communities, 2006.
[52] L.L. Nass, P.A.A. Pereira, D. Ellis, Biofuels in Brazil: an overview, Crop. Sci. 47 (6) (2007) 2228.
[53] B. Yang, C.E. Wyman, Advancing cellulosic ethanol technology in China, Prog. Chem. 19 (7–8) (2007) 1072.
[54] S. Fan, B. Freedman, J. Gao, Potential environmental benefits from increased use of bioenergy in China, Environ. Manage. 40 (3) (2007) 504.
[55] RFA, Ethanol Industry Outlook 2007: Building New Horizons, Renewable Fuels Association, 2007.
[56] N. Brinkman, M. Wang, T. Weber, T. Darlington, Well-to-Wheels Analysis of Advanced Fuel/Vehicle Systems—A North American Study of Energy Use, Greenhouse Gas Emissions, and Criteria Pollutant Emissions, Argonne National Lab, 2005.
[57] Committee on Water Implications of Biofuels Production in the United States NRC, Water Implications of Biofuels Production in the United States, National Research Council of the National Academies, Washington, D.C., 2007.
[58] T.W. Patzek, Thermodynamics of the corn-ethanol biofuel cycle, Crit. Rev. Plant Sci. 23 (6) (2004) 519–567.
[59] A.E. Farrell, R.J. Plevin, B.T. Turner, A.D. Jones, M. O'Hare, D.M. Kammen, Ethanol can contribute to energy and environmental goals, Science 311 (5760) (2006) 506.
[60] A.E. Farrell, R.J. Plevin, B.T. Turner, A.D. Jones, M. O'Hare, D.M. Kammen, Energy returns on ethanol production—response, Science 312 (5781) (2006) 1747.
[61] M.E.D. De Oliveira, B.E. Vaughan, E.J. Rykiel Jr, Ethanol as fuel: energy, carbon dioxide balances, and ecological footprint, Bioscience 55 (7) (2005) 593.
[62] J. Hill, E. Nelson, D. Tilman, S. Polasky, D. Tiffany, Environmental, economic, and energetic costs and benefits of biodiesel and ethanol biofuels, Proc. Natl. Acad. Sci. USA 103 (30) (2006) 11206.
[63] D. Pimentel, Ethanol fuels: energy balance, economics, and environmental impacts are negative, Nat. Resour. Res. 12 (2) (2003) 127.
[64] H. Shapouri, A.D. James, W. Michael, The Energy Balance of Corn Ethanol: An Update, US Department of Agriculture, 2002.
[65] D.F. Anthrop, Analysis highlights limits on energy promise of biofuels, Oil Gas J. 105 (5) (2007) 25.
[66] D. Pimentel, T. Patzek, G. Cecil (Eds.), Ethanol Production: Energy, Economic, and Environmental Losses, 2007.
[67] J. Sheehan, V. Camobreco, J. Duffield, M. Graboski, H. Shapouri, Overview of Biodiesel and Petroleum Diesel Life Cycles, US Department of Energy, US Department of Agriculture, 1998.
[68] J. Sheehan, A. Aden, K. Paustian, K. Killian, J. Brenner, M. Walsh, et al., Energy and environmental aspects of using corn stover for fuel ethanol, J. Ind. Ecol. 7 (3) (2003) 117.

[69] S. Spatari, Y. Zhang, H.L. MacLean, Life cycle assessment of switchgrass- and corn stover-derived ethanol-fueled automobiles, Environ. Sci. Technol. 39 (24) (2005) 9750.
[70] P.R. Adler, S.J. Del Grosso, W.J. Parton, Life-cycle assessment of net greenhouse-gas flux for bioenergy cropping systems, Ecol. Appl. 17 (3) (2007) 675–691.
[71] T. Searchinger, R. Heimlich, R.A. Houghton, F. Dong, A. Elobeid, J. Fabiosa, et al., Use of US croplands for biofuels increases greenhouse gases through emissions from land-use change, Science 319 (5867) (2008) 1238.
[72] H. Shapouri, D. James, M. Andrew, The 2001 net energy balance of corn ethanol, in Corn Utilization and Technology Conference, Indianapolis, IN, p. 6, 2004.
[73] NASS, 2002 Census of Agriculture, National Agricultural Statistics Service, US Department of Agriculture, 2004.
[74] M. Johnston, T. Holloway, A global comparison of national biodiesel production potentials, Environ. Sci. Technol. 41 (23) (2007) 7967.
[75] Interagency Agricultural Projections Committee, USDA Agricultural Baseline Projections to 2016, US Department of Agriculture, 2007.
[76] L.P. Koh, Potential habitat and biodiversity losses from intensified biodiesel feedstock production, Conserv. Biol. 21 (5) (2007) 1373.
[77] S. Kim, B.E. Dale, Life cycle assessment of various cropping systems utilized for producing biofuels: bioethanol and biodiesel, Biomass. Bioenerg. 29 (6) (2005) 426.
[78] U.N. Energy, Sustainable Bioenergy: A Framework for Decision Makers, United Nations, 2007.
[79] J.F. Kreider, P.S. Curtiss, Comprehensive evaluation of impacts from potential, future automotive fuel replacements, Proceedings of Energy Sustainability 2007, ASME, Long Beach, CA, 2007.
[80] T. Simpson, P. Jim, M. Beth, S. Matt, K. Ron, V. Jake, Biofuels and Water Quality: Meeting the Challenge & Protecting the Environment, The Mid-Atlantic Regional Water Program, USDA-CSREES, 2007.
[81] G. Berndes, Bioenergy and water—the implications of large-scale bioenergy production for water use and supply, Global Environ. Change 12 (4) (2002) 253.
[82] ERS, Agricultural Resource Management Survey: Farm Business and Household Survey Data, Economic Research Service, US Department of Agriculture.
[83] H. Shapouri, M. Salassi, The Economic Feasibility of Ethanol Production from Sugar in the United States, US Department of Agriculture, 2006.
[84] D.J. Parrish, J.H. Fike, The biology and agronomy of switchgrass for biofuels, Crit. Rev. Plant Sci. 24 (5/6) (2005) 423.
[85] S.B. McLaughlin, M.E. Walsh, Evaluating environmental consequences of producing herbaceous crops for bioenergy, Biomass Bioenerg. 14 (4) (1998) 317.
[86] M.R. Schmer, K.P. Vogel, R.B. Mitchell, R.K. Perrin, Net energy of cellulosic ethanol from switchgrass, Proc. Natl. Acad. Sci. USA 105 (2) (2008) 464.
[87] D. Tilman, J. Hill, C. Lehman, Carbon-negative biofuels from low-input high-diversity grassland biomass, Science 314 (5805) (2006) 1598.
[88] NASS, Agricultural Chemical Usage, Field Crops Summary (2001, 2002, 2004, 2005, 2006), National Agricultural Statistics Service, US Department of Agriculture, 2005.
[89] J. Cunningham, Biofuel joins the jet set, Prof. Eng. 20 (10) (2007) 32.

[90] A.H. Scragg, A.M. Illman, A. Carden, S.W. Shales, Growth of microalgae with increased calorific values in a tubular bioreactor, Biomass Bioenerg. 23 (1) (2002) 67.

[91] D. Tilman, P.B. Reich, J.M.H. Knops, Biodiversity and ecosystem stability in a decade-long grassland experiment, Nature 441 (7093) (2006) 629.

[92] R.D. Perlack, Biomass as Feedstock for a Bioenergy and Bioproducts Industry: The Technical Feasability of a Billion-Ton Annual Supply, US Dept. Energy Technical Report, 2005, p. 78.

[93] A.J. Ragauskas, C.K. Williams, B.H. Davison, G. Britovsek, J. Cairney, C.A. Eckert, et al., The path forward for biofuels and biomaterials, Science 311 (5760) (2006) 484.

[94] M.B. Sticklen, Feedstock crop genetic engineering for alcohol fuels, Crop Sci. 47 (6) (2007) 2238.

[95] Y. Roman-Leshkov, C.J. Barrett, Z.Y. Liu, J.A. Dumesic, Production of dimethylfuran for liquid fuels from biomass-derived carbohydrates, Nature 447 (7147) (2007) 982.

[96] X. Pan, Z. Fan, W. Chen, Y. Ding, H. Luo, X. Bao, Enhanced ethanol production inside carbon-nanotube reactors containing catalytic particles, Nat. Mater. 6 (7) (2007) 507.

[97] C. Honsberg, FAQ List for Solar Cells, [cited March 1, 2008]. Available at <http://www.solar.udel.edu/faq.html>.

[98] S.E. Gledhill, B. Scott, B.A. Gregg, Organic and nano-structured composite photovoltaics: an overview, J. Mater. Res. 20 (12) (2005) 3167.

[99] K. Kim, J. Liu, M.A.G. Namboothiry, D.L. Carroll, Roles of donor and acceptor nanodomains in 6% efficient thermally annealed polymer photovoltaics, Appl. Phys. Lett. 90 (16) (2007) 163511.

[100] International Bottled Water Association, Beverage Marketing's 2006 Market Report Findings, 2006.

[101] P.H. Gleick, The Myth and Reality of Bottled Water. The World's Water, 2004–2005, The Biennial Report on Freshwater Resources, Island Press, Washington, D.C., 2004, pp. 17–43.

[102] J.F. Casey, J.R. Kahn, A. Rivas, Willingness to pay for improved water service in Manaus, Amazonas, Brazil, Ecol. Econ. 58 (2) (2006) 365.

[103] K. Homaki, P.H. Nielsen, A. Sathasivan, E.L.J. Bohez, Life cycle assessment and environmental improvement of residential and drinking water supply systems in Hanoi, Vietnam, Int. J. Sust. Dev. World 10 (1) (2003) 27.

[104] The Pacific Institute [cited February 10, 2008]. Available at <http://www.pacinst.org/topics/water_and_sustainability/bottled_water/bottled_water_and_energy.html>.

[105] R.K. Mahajan, T.P.S. Walia, B.S. Lark, Analysis of physical and chemical parameters of bottled drinking water, Int. J. Environ. Heal. R. 16 (2) (2006) 89.

[106] J.A. Lalumandier, L.W. Ayers, Fluoride and bacterial content of bottled water vs tap water, Arch. Fam. Med. 9 (3) (2000) 246.

[107] S.D. Raj, Bottled water: how safe is it? Water Environ. Res. 77 (7) (2005) 3013–3018.

[108] L. Fewtrell, D. Kay, M. Wyer, A. Godfree, G. O'Neill, Microbiological quality of bottled water, Water Sci. Technol. 35 (11–12) (1997) 47–53.

CHAPTER

A Framework for Using Nanotechnology to Improve Water Quality 36

Michael E. Gorman[1], Ahson Wardak[2], Emma Fauss[3] and Nathan Swami[3]

[1]Department of Science, Technology and Society, University of Virginia, Charlottesville, VA, USA
[2]Department of Systems and Information Engineering, University of Virginia, Charlottesville, VA, USA [3]Department of Electrical and Computer Engineering, University of Virginia, Charlottesville, VA, USA

36.1 Introduction

If nanotechnology is to represent societal as well as technical progress, it will have to contribute to the solution of global problems like water quality. This chapter describes a framework for guiding nanotechnology away from risks and toward benefits, and applies it to the problem of water quality. The framework includes five major components; the first three are necessary conditions for accomplishing the last two.

Street, Sustich, Duncan and Savage. Nanotechnology Applications for Clean Water, 2nd Edition.
 DOI: http://dx.doi.org/10.1016/B978-1-4557-3116-9.00036-6

- Superordinate goals: Ensuring an adequate global supply of water is a superordinate goal because it affects the survival of the human race and therefore should trump the differences that separate us. From our viewpoint, nanotechnology is a set of solutions and capabilities in search of problems. Water quality could serve as a superordinate goal, linking nanotechnology researchers from different institutions and disciplines with those who have a compelling need.
- Trading zones: This linkage will require setting up a series of trading zones involving multiple stakeholders for exchanging ideas, resources, and solutions across different communities and interests. In order to trade, participants will have to develop one or more of the following:
  - A reduced common language, or creole.
  - A boundary object or technology like a database that participants in the zone share.
  - Interactional expertise on the part of one or more participants who will serve a role similar to trade agents, facilitating exchanges of ideas, and resources.
- Moral imagination: Differences in values can prevent adoption of a superordinate goal. Moral imagination is the equivalent of interactional expertise concerning values. In order to care about someone else's water supply, you have to see their need as if it were your own and also understand enough about their culture to coevolve appropriate solutions.

Developing the three capabilities above will permit:

- Adaptive management: Adaptive management is both difficult and essential in tightly coupled human–technological–natural systems. Participants in trading zones around nanotechnology and water quality will have to be in constant dialogue with each other and with the system they are managing, adapting strategies to new data.
- Anticipatory governance: Adaptive management can be done by multiple stakeholders forming trading zones that cut across traditional governance structures, but global solutions will require the involvement of policy institutions. Regulatory systems need to be more anticipatory and adaptive, steering a course between the precautionary principle and reactive risk management. Expert elicitation is one method for implementing adaptive management and anticipatory governance. Another set of methods or tools are provided by life cycle analyses that incorporate not only material flows but also social and behavioral interactions.
- Life cycle assessment: Another set of methods or tools are provided by life cycle analyses that incorporate not only material flows but also social and behavioral interactions.

In the following sections, we will describe each of these components and combine them into a framework, showing how they could be applied to nanotechnology and water.

36.2 Superordinate goals

The social psychologist Muzafer Sherif grew up in Turkey during the decline of the Ottoman Empire and saw firsthand how members of groups would exhibit compassion toward their own and hostility toward members of other groups. He decided to devote a career to understanding this phenomenon [1]. He and his wife Carolyn created a summer camp for boys in Oklahoma in which each boy was randomly assigned to one of two groups, and competition between the groups was encouraged [2]. The Sherifs were surprised at how quickly intergroup hostility developed.

They found the best way to bring the groups together was to introduce what the Sherifs called a superordinate goal. He faked a problem with the camp's water supply, one that the two groups had to work together to solve—they had to hike out to the water tank, inspect it, and finally discover a clogged pipe, which they had to fix. The water challenge broke the ice and began to get the groups working together. Another challenge involving starting the truck to get lunches used food as a superordinate goal, and here the groups worked together to drag the truck until it started.

The clogged pipe is an overly simplified analogy to the problem of water, worldwide—one that requires a lot more than unclogging. The World Health Organization (WHO)'s 2000 Water and Sanitation Report remarks that there are 2.1 billion people without access to improved water supply and 2.4 billion without access to improved sanitation service worldwide [3]. Though overwhelmingly problems of lower income countries, the lack of access to improved water supply and sanitation services affects poor communities in both lower and higher income countries.

The lack of access to these basic municipal sanitation services has impacts on human health, the environment, and the economies of the affected regions. Health effects include infectious diarrhea, trachoma, and hookworm [4], and have dramatic consequences on the population of low-income countries. Each year, more than 3 million people die from diarrheal disease, and some 1.5 million from malaria [3]. These consequences among others are principally due to poor personal and domestic hygiene, unsafe drinking water, and poor water and sanitation management. The United Nations Children's Fund (UNICEF) estimates that annual investments of approximately $25 billion are needed for municipal sanitations systems worldwide. However, the health costs associated with waterborne diseases and the increased costs for water treatment due to contaminated groundwater greatly exceed this amount [5].

Water should be a global superordinate goal, and indeed it is highlighted as one of three subgoals in the UN's Millennium Development Goal 7, Ensure Environmental Sustainability: "Reduce by half the proportion of people without sustainable access to safe drinking water."[1]

[1]Find more information about the UN's Millennium Development Goals at <http://www.un.org/millenniumgoals/>

Sherif's campers were from homogeneous backgrounds and were randomly separated into their groups. They therefore did not mimic the long-term divisions between religious, ethnic, and income groups that make intergroup hostility different to overcome. Access to water is a chronic problem for millions of human beings, but not for all, and those who have abundant supplies do not see access to water as a superordinate goal. Even if one does care enough to work on someone else's water problem, the appropriate technology needs to be adopted. One way to get different expertise communities to work together with stakeholders is to form trading zones.

36.3 Trading zones

The historian and philosopher of science Thomas Kuhn proposed that science evolves by discarding old paradigms in favor of new ones [6]. A paradigm is both a way of thinking and a way of doing; all the practitioners in a domain know which problems are worth solving and what methods to use in solving them. Much of this knowing is tacit. It is "obvious" to those within the paradigm, but opaque to those outside. Kuhn referred to this disjoint between those who hold different paradigms as the problem of incommensurability. During a paradigm shift, those who still operated in the old paradigm could not understand key aspects of the new one. For example, Einstein's special relativity created a new paradigm in which the distinctions between space and time and mass and energy disappeared, which made it hard for those in the old paradigm to understand Einstein's universe—indeed, it took about 5 years for signs of acceptance to emerge in the relevant physics community [7].

Galison noted that physicists and engineers working together on the development of radar during World War II came from apparently incommensurable paradigms and yet were able to work together because they formed trading zones in order to exchange ideas and resources. The key to the success of such a zone, in Galison's view, is gradual evolution of a common language, from a shared jargon to a pidgin to a creole. The creole can eventually grow into a full-fledged language, leading to a new specialist community. No doubt new fields will emerge out of nanotechnology trading zones, especially where nanotechnology intersects with biotechnology, information technology, and cognitive science [8,9].

In the case of nanotechnology and water, trading zones are crucial. Consider water quality in the developing world. In order for nanotechnologies to be a solution, scientists and engineers will have to coordinate with seekers from a variety of local situations around the world [10]. Seekers are those looking for a solution, who are primed to work with an outsider or anyone who comes in with a good idea, and is willing to work with them. But the seeker and the nanoscientist will come from different cultures—hence, the need for a trading zone, in which exchanges of knowledge and resources become more fluent over time.

Development of a creole is one way of facilitating exchanges across a trading zone. There are two other ways.

36.3.1 Interactional expertise

Trading zones are often facilitated by agents [11]. The best agent would not only know enough of the languages of both cultures to act as an interpreter, she or he would understand enough of their world views or paradigms to encourage them to trade.

But how does one achieve sufficient understanding of the different cultures in a trading zone to act as an effective agent? Collins and Evans [12] have solved this problem by identifying and studying a new type of expertise they refer to as interactional. Interactional expertise is the ability to adopt the language and concepts of an expertise community sufficiently to pass as a member, without being able to conduct the research. Collins' discovery of this kind of category emerges from his own experience as a sociologist of science, where he had to acquire sufficient expertise to study scientific expertise communities by interacting with the members [13,14].

One example of a trading zone involving water that includes both a creole and interactional expertise comes from water management in Arizona. To make a decision about dams, social scientists were recruited to develop a set of metrics that would allow Bureau of Reclamation engineers, Yavapai Indians, and other stakeholders to communicate about the value, use, and need given to water resources. We could say that the social scientists were engaged in two simultaneous projects: trying to become interactional experts in the worlds of the various negotiating parties and trying to build an interlanguage—in this case a common language of measurement—that both parties would share. The social scientists failed to build the interlanguage; both the Bureau engineers and the Yavapai felt that the metrics developed by the social scientists failed to capture important parts of their separate beliefs and values. The interactional expertise part of the enterprise worked, however. A compromise was reached that avoided the building of a new dam, preserved Yavapai land, and still permitted water management. Cultures that remained incommensurable traded their way to a solution with help from social scientists who acquired interactional expertise [15,16].

Similarly, interactional expertise will be essential to the development of trading zones linking nanotechnology to water needs around the globe. Interactional expertise is not limited to social scientists—any expert who enters into the language and customs of another domain can be an interactional expert.

36.3.2 Boundary object

Participants in a trading zone can also coordinate activities around a boundary object or system whose meaning is only partly shared by those involved in exchanges. "Boundary objects are objects that are both plastic enough to adapt to local needs and constraints of the several parties employing them, yet robust enough to maintain a common identity across sites" [17, p. 393]. "Sites" refers to the different communities involved in exchanges. Just as coordination of activities across trading zones can be facilitated by development of common languages and/or by individuals with shared expertise, so can coordination also be facilitated by technologies like software, maps, design specifications, and decision aids.

Consider an example. The capacity factors approach developed by Garrick Louis at the University of Virginia is used to assess community resource levels and technology management levels for the selection and implementation of a solution for water supply and sanitation systems in developing communities. The capacity factors are categorized under eight general headings: institutional, human resources, technical, economic/financial, environmental/natural resources, energy, sociocultural, and service [18]. The community is first assessed for its ability to sustain and implement a technology across all factors and is then given a composite score (technology management level). Then, each alternative technology is assessed for its needs from the community across all factors and given a composite score (community resource level). As long as the community resource level is equal to or lower than the technology management level, then the technology is viable.

The systems engineer's representation of community capacity may be very different from that of the locals. The engineer's representation is quantitative: typically, the options are scored on a one to five scale and put in a matrix for further comparison. The local view of capacities will likely be more tacit and holistic, reflecting a different paradigm or way of life. If capacity factors are treated as a boundary object, then the participants can gradually evolve a better representation for working together. If capacity factors are treated as a rational, quantitative solution into which the local problem must be fit, then there is no real trading zone and the end result will be another failure to truly address the concerns of local seekers.

This approach could be applied to nanotechnology and the water. An engineer or scientist with a promising nanotechnology could work with a community to assess whether any potential nanotechnology solution was a good fit with the community's resources and need. The capacity factors would constitute a boundary system because these would need constant adjustment and re-prioritization based on input from both the technology experts and the local stakeholders, who know their situations. The capacity factors approach is not an algorithm that can be applied *a priori*—instead, it is the basis for organizing a dialogue, and in that way, serves a function similar to a shared language. An interactional expert would be helpful at facilitating the dialogue; she or he could, for example, communicate the capacity factors, their rationale, and their scoring to the local community—and understand the community's responses well enough to translate them into the capacity matrix, modifying the matrix itself if necessary. Not all problems and situations can be nicely characterized by a matrix, which is why capacity factors must constitute a boundary object that serves the heuristic function of creating a space for dialogue over what a community really needs. The end result may be that the current state of nanotechnology can provide no solutions for a particular situation, but the scientists and engineers will leave with a clearer understanding of the local problem, which will help the evolution of future nanotechnologies.

A database can also serve as a boundary object, linking participants by providing information they can use as a kind of creole [19]. For example, Fauss has created a University of Virginia database for those interested in where and how silver nanoparticles are being used in products. This database is based primarily on Internet searches for information about products which incorporate silver nanotechnology; it provides an overview of how silver nanotechnology is being used in the marketplace today and what types of exposures are most likely.[2] Databases like this provide a common format for storing information and a public repository, making it easier for participants in nanotechnology trading zones around water quality to share and update information. Again, the database sits on the boundary between stakeholders, serving a function similar to a creole as it evolves based on multiple inputs.

Nanotechnology solutions for water quality will have to involve trading zones, facilitated by a combination of shared language, interactional expertise, and/or a boundary object.

36.4 Moral imagination

Exchanges across trading zones on vital issues like water can be limited by incommensurable values that keep participants from seeing that they have the same overall goal and that they may not survive if they do not work together. Suppose each group thinks its main goal is to dominate or even eliminate the other? A common language is not enough to overcome values differences, nor is the presence of a boundary object.

A kind of interactional expertise at the level of values may be helpful, however. This capability is known as moral imagination, which includes "the awareness of various dimensions of a particular context as well as the operative framework and narratives. Moral imagination entails the ability to understand that context or set of activities from a number of different perspectives, the actualizing of new possibilities that are not context-dependent, and the instigation of the process of evaluating those possibilities from a moral point of view" [20, p. 5].

The central tenet of moral imagination is that we learn practical ethics from stories [21], which become mental models for virtuous behavior [22]. "Mental models are deeply ingrained assumptions, generalizations, or even pictures or

[2]The database will be available at the Project on Emerging Nanotechnologies website, www.nanotechproject.org.

images that influence how we understand the world and how we take action. Very often, we are not consciously aware of our mental models or the effects they have on our behavior" [23]. The term "mental model" reflects the fact that human beings have mental representations of their experiences that they can "run" in their "mind's eyes" to make observations and decisions [24].

Those who hold incommensurable ideologies see their values as truths and other values as heresy. The first step in moral imagination is recognizing that these apparent truths are perspectives.

The second step is the interactional expertise move: trying to understand the perspective of the other culture or community or discipline (these terms are not mutually exclusive) involved in a potential exchange. Understanding another group's mental model is not the same as agreeing with it. Instead, it allows the interactional expert to start a deep dialogue that might lead all participants to adopt an improved mental model. "Each person's view is a unique perspective on a larger reality. If I can "look out" through your view and you through mine, we will each see something we might not have seen alone" [23, p. 248].

Once participants in a potential trading zone see that water quality is a problem of sufficient urgency so that old habits and values will have to be adjusted, then the third step is to creatively envision new alternatives for reaching the superordinate goal. These alternatives will have to be evaluated, which brings us to the governance and management parts of our framework.

36.5 Adaptive management

Mike Roco, one of the architects of the National Nanotechnology Initiative, calls for the application of adaptive management to nanotechnology [25]. Adaptive management is a technique first developed to restore salmon runs in the Pacific northwest, where multiple stakeholders and different value systems needed to cooperate in order to preserve a resource used and valued by all [26]. In adaptive management, "policies become hypotheses, and management actions become the experiments to test those hypotheses" [27, p. 447]. Falsifying a policy hypothesis is not a failure if it is transformed into an opportunity to learn.

So any application of nanotechnology to improving water quality should be regarded as an experiment, involving the collection of data on the intervention, including effects on culture and relationships as well as on water quality. Ideally, a nanotechnology intervention should be compared with alternatives. Note that the experiment will not be of the traditional laboratory sort, with perfect controls and procedures; instead, it will be a field experiment, in which one looks at how the nanotechnology is implemented by people and convenes the stakeholders periodically to discuss. Adaptive management requires a trading zone, where multiple parties evaluate the ongoing results, iterate, and improve—and link to other management experiments elsewhere. One requirement is that the process be documented for comparison with other attempts.

Management experiments have to be reversible, following one of the core principles of Earth Systems Engineering Management (ESEM) [28]. If an experiment leads to a semipermanent system change, then instead of using the result to design a new experiment, one is forced to develop a new management strategy for the changed system. Therefore, adaptive management requires that one be able to return the system to the state before the experiment. In practice, this may be quite difficult, especially in complex systems, where effects are nonlinear.

Two strategies for dealing with complexity are the use of multiple research methods [29] and the use of a variety of models that have the potential to signal systems change [30]. Choosing a combination of methods and models requires a trading zone, because more than one discipline has to be involved. The models and methods themselves will have to be adapted over time.

In order to succeed at the systems level, adaptive management will require moral imagination, because the environmental experiments are also social ones, involving multiple stakeholders with different values. What looks like a positive impact to one stakeholder may violate another's fundamental preconceptions.

36.6 Anticipatory governance

Trading zones emerge either outside of formal governmental and institutional structures or as a way of working around them—of cutting across the layers of bureaucracy, of "flying under the radar." Bureaucratic rules and institutions can emerge out of a trading zone, in order to enforce what participants agreed to—e.g., regulations in the United States depend on democratic institutions and involve consultation with multiple stakeholders. But regulations, once in place, are hard to change, and are also based on risk assessment, which requires that the costs and benefits of all possible scenarios be quantified: risk has to be proven rigorously. At the earliest stages of technological development, the risks and benefits are not known or quantifiable, and the law of unexpected consequences rules.

An alternative more favored in Europe than the United States is the precautionary principle, which comes in many forms [31], but the essential message is: "if you don't know, you don't go." Whereas risk methodology requires that harm be proved, the precautionary principle requires that safety be proved—impossible in general, and certainly impossible for emerging technologies. A premature moratorium not only blocks potential risks, it also blocks potential improvements in water quality.

When considering the impact of nanotechnology on water quality, neither risk analysis nor the precautionary principle is adequate. One way of going forward is anticipatory governance: "between adapting to a common revolution and halting development exists an array of government options...Anticipatory governance seeks to lay the intellectual foundation for (any of) these approaches early enough to be effective" [32, p. 992]. These options will include trading zones that link government institutions with NGOs, scientists and seekers who understand water quality issues in a local context. It is important to emphasize that "these network structures do not replace the accountability of existing hierarchical bureaucracies but operate within and complement them" [27, p. 450].

One of the best ways of avoiding unacceptable risks is to steer toward beneficial, "win–win" outcomes—which will have risks associated with them, as all human activity does, but where the corresponding benefits are identified and tracked from the beginning, and where multiple stakeholders have input into the management response. Consider the emergence of green nanotechnology [33], which was promoted by the EPA in its recent "Pollution Prevention through Nanotechnology" conference (<http://www.epa.gov/oppt/nano/nano-confinfo.htm>). Here is an example of how government could move toward anticipatory governance, if strategies developed at the conference encourage the development of trading zones capable of implementing them.

36.6.1 Expert elicitation as a method for facilitating anticipatory governance

In risk assessment situations where information is lacking, expert elicitation is often used to fill the gaps [34,35]. This method involves asking experts to estimate the potential risks of a nanomaterial, in the light of exposure scenarios. The experts may say there is too little research to be certain, in which case they have identified a research gap. Experts may also disagree on potential hazards, especially if experts come from more than one specialization.

We are conducting an expert elicitation on the potential impact of silver nanoparticles, beginning with how they are used in products now. The nine experts came from a diversity of backgrounds, including regulation, toxicology, industry, bioengineering, and polymer chemistry. We supplied these experts with information on the products from a database created by Fauss while working at the Project on Emerging Nanotechnologies (<http://www.nanotechproject.org/>). Eighty-eight percent of the 240 products currently listed in the UVA Silver Nanotechnology Database make some sort of antibacterial claim.

We asked the experts to consider both hazards of the material involved, based on its properties, and also exposure scenarios. Each interview created a platform for the next. Through this process, the interviewer (Fauss) became an interactional expert. The increased surface area of nano as opposed to bulk silver means the former will release more silver ions than the latter, which will increase toxic effects—good if the goal is to kill harmful bacteria in water, bad if one wishes to preserve the aquatic food chain. Experts did not agree on the probable aquatic effect, in part because of a difference of opinion over whether the silver nanoparticles would quickly become entrapped in the sediment or whether they could become lodged in the respiratory system of aquatic species, releasing silver ions into the gills. The experts noticed that the cause of the toxic effects of silver nanoparticles is not determined.

Expert elicitation at the earliest stages of nanotechnology research can complement efforts to address and close regulatory gaps regarding nanotechnology [36]. Consider the EPA's proposed regulation of any products containing silver nanoparticles that claim to kill germs [37]. The regulation is worded so that if a producer does not claim any antimicrobial benefits, the silver nano will escape regulation—because the EPA can only regulate the pesticide function of silver nanotechnology. The problem with the nanotechnology frontier is that products cut across existing agency boundaries, suggesting the need for new regulatory structures, ones that can adapt to potential risks. Expert elicitation is one method for anticipating regulatory gaps.

Ideally, the end result of a set of interviews would be information that could be used in a discussion, in which all the experts were connected to talk about the results from the interviews and decide on future research and governance priorities. This kind of a gathering could lead to formation of a trading zone, especially if other stakeholders were gradually added.

The same methodology could be used to steer nanotechnology research in directions most likely to benefit societies worldwide. Early state research at Rice University is one example, where iron nanoparticles and an electromagnet are used to filter arsenic out of running water [38]. The thoughtful application of this technology in countries, like Bangladesh, India, and China, could have positive impacts, but societal dimensions of such applications should be considered at the earliest stages to make certain a nanotechnology solution will work in the cultural context. The group of experts would have to include not only scientists and engineers familiar with the technology but also those who knew a variety of cultural contexts into which the technology could be introduced. Louis' capacity factors might serve as a boundary object for such an exchange.

Here the central ethical issue moves from avoiding harm to doing good. But of course that "doing good" can be misguided imperialism in the developing world—"the white man's burden" [10]. Consider the Rice solution. A capacity factor approach should be applied to determine when it is more effective to test for arsenic and dig wells in safer areas (<http://www.earth.columbia.edu/about/aboutcase01.html>) than to mitigate existing sites with nanotechnology.

36.6.2 Potters for peace

An example of a promising approach is the potential use of nano silver in ceramic filters. Potters for Peace (<http://www.pottersforpeace.org/>) is a US-based NGO that partnered with grassroots seekers in Nicaragua to develop, manufacture, and distribute filters that look like ceramic pots and sell for between $5 and $15 each. A sociologist/potter named Ron Rivera provided interactional expertise that helped facilitate the development of manufacturing in Managua.

The filters are typically made out of materials like clay and water mixed with combustible organic materials: sawdust, flour, or rice husks. The sawdust, flour, or rice husks combust when the materials are fired in a kiln, making the resulting ceramic material porous and therefore permeable to water. Typically, silver is added after the filter is fired. This silver contains zero-valent bulk particles and nanoparticles that lodge in the pores of the filter and presumably act as a disinfectant to deactivate pathogenic microorganisms.

Oyandel–Craver and Smith conducted the first study evaluating the efficacy of the filters and the added silver. They found that even filters without the silver treatment removed at least 98% of the *Escherichia coli* (*E. coli*), and the addition of silver removed almost 100% [39]. This example shows how science can add value to a trading zone composed of indigenous seekers and NGO experts. Future work needs to focus on estimating the proportion of this silver that is nano, in a typical preparation, and whether using nano silver has important advantages over an uncontrolled mixture of bulk and nano.

This kind of technology ranks well on Louis' capacity factors because it can be made from local materials, excepting only the silver, which can be purchased in bulk by an entrepreneur and used as a basis for a business. Where would such a local entrepreneur obtain a more precisely controlled mix of bulk and nano silver? Would it add sufficient value, when the current technology is close to 100% effective at eliminating *E. coli*? Life cycle scenarios that incorporate social and behavioral factors are a useful tool for anticipating different development use and disposal trajectories these filters might take, especially considering potential changes in the socio-technical system [40].

Nanotechnology research should be targeted at making sustainable improvements in mechanisms for ensuring water quality in the developing world and sharing the results widely. One potential barrier to such sharing is intellectual property rights. Certainly, innovators should be able to make a profit on nanotechnology solutions to water quality, but those with the greatest need are often least able to afford new technologies, as in the case of life-enhancing pharmaceuticals [41]. Governance mechanisms need to be put into place that will facilitate sharing solutions and getting them to those who need them the most.

36.7 Conclusions

A major goal of ESEM is to increase system resilience. Regarding water quality, resilience implies having a supply that is robust with respect to local and system disruptions due to droughts, natural disasters, and human impacts. It also implies having robust mechanisms for ensuring that every human being on the planet has access to potable water.

The systems-level question for nanotechnology is how it can help achieve this sort of resilience. A first step is to establish a trading zone that brings together a small group of nano scientists, engineers, policy makers, and seekers from around the world. These seekers understand the problem and also the local context in which it has to be solved, and they are eager to work on solutions. The scientists, engineers, and policy makers could think about how to direct future nanotechnology research in ways that would help these and other seekers.

The trading zone could begin with a workshop, preferably held in a location that illustrated a pressing water quality problem. The goal of the workshop would be to develop collaborations among the participants; the end result would be specific projects. All participants would commit to documenting their processes, the successes, as well as failures. This kind of documentation can turn reversible mistakes into beneficial learning experiences; it is important to know what technologies and processes fail in specific situations, not just which ones succeed. "A collective memory of experiences with resource and ecosystem management provides context for social responses and helps the socio-ecological system prepare for change. If experience embedded in institutions and organizations provides a context for the modification of management policy and rules, people can act adaptively in the face of surprise" [27, p. 453].

Therefore, attempts to apply nanotechnology to improve water quality need to be documented and shared, worldwide. Some of the technologies may be proprietary, especially in the developed world, but at least research on application should be conducted rigorously and shared. For example, the database of silver nanotechnology products should eventually include information from independent laboratories on whether and in what proportion the products actually contain silver nano, and on the impact of the products on a standard preparation of microbes and aquatic organisms. This kind of research could also help us gain a deeper understanding of how silver nanoparticles can be turned into nanotechnologies, where their antimicrobial function can be controlled precisely.

The same kind of sharing will be required for developing world solutions like the ceramic filters. In what locations are which kinds of filters most useful—and least? The filters are a kind of end-of-pipe solution: the water is contaminated, but the user can eliminate the contamination just before drinking. Will nanotechnology provide promising applications for larger-scale water treatment? Suites of nanotechnologies will eventually be linked into nano systems that should address water quality at a global level, with subsystems providing solutions adapted to local conditions. To see whether and how nanotechnology can be applied to water quality, we will have to encourage trading zones that bring participants from multiple disciplines and stakeholder perspectives together. All parties in the zone will have to see that water quality is a superordinate goal requiring their cooperation. These trading zones will produce new research and application collaborations that will have to be continually assessed to see whether they are working, and how lessons can be generalized.

For these collaborations to have an impact on the system, they will need support from governance structures. In particular, governance can reduce the need for constant meetings among stakeholders by imposing a structure and rules for mediating disagreements. Those focused on governance should take a system's perspective, keeping alignment across trading zones that are dealing only with parts of the overall problem. Gaps in our current governance structures can be anticipated now, and steps taken to make them more adaptive.

This whole process is iterative: unexpected results from a management experiment may force participants to reconvene in a trading zone, refreshing their alignment around the superordinate goal, and developing new collaborations. The process of trading zone formation should itself be subject to adaptive management; it should be studied as it is implemented and improved by lessons learned.

The natural and technological barriers to improved water quality are lower than the interpersonal ones, though all are connected. As Walt Kelly[3] famously said, "We have met the enemy and he is us." Kelly also said, "We are surrounded by insurmountable opportunities." Applying nanotechnology to water quality represents such an opportunity; the framework outlined in this chapter shows us how to take advantage of it.

Acknowledgments

The authors would like to thank Garrick Louis and James Smith for their comments and assistance.

References

[1] M.E. Gorman, Pre-war conformity research in social psychology: the approaches of Floyd H. Allport and Muzafer Sherif, J. Hist. Behav. Sci. 17 (1981) 3–14.

[2] C.W. Sherif, Orientation in Social Psychology, Harper & Row, New York, 1976.

[3] World Health Organization, UNICEF. Global Water Supply and Sanitation Assessment 2000 Report. Geneva, Switzerland. New York; 2000 Contract No.: Document Number.

[4] A. Pruss, D. Kay, L. Fewtrell, J. Bartram, Estimating the burden of disease from water, sanitation, and hygiene at a global level, Environ. Health Perspect. 110 (5) (2002) 537–542.

[5] S. Annamraju, B. Calaguas, E. Gutierrez, Financing Water and Sanitation Policy Briefing Paper, WaterAid, London, UK, 2001.

[6] T.S. Kuhn, The Structure of Scientific Revolutions, University of Chicago Press, Chicago, 1962.

[7] G. Holton, Thematic Origins of Scientific Thought, Harvard University Press, Cambridge, 1973.

[8] M.E. Gorman, Combining the social and the nano: a model for converging technologies, in: M.C. Roco, W.S. Bainbridge (Eds.), Converging Technologies for Improving Human Performance: Nanotechnology, Biotechnology, Information Technology and Cognitive Science, Kluwer, Dordrecht, Netherlands, 2003, pp. 367–373.

[3] Author of the Pogo comic strip (<http://www.pogopossum.com/>).

[9] M.E. Gorman, Expanding the trading zones for convergent technologies, in: M.C. Roco, W.S. Bainbridge (Eds.), Converging Technologies for Improving Human Performance: Nanotechnology, Biotechnology, Information Technology and Cognitive Science, Kluwer, Dordrecht, Netherlands, 2003, pp. 424–428.

[10] W.R. Easterly, The white man's burden: why the West's efforts to aid the rest have done so much ill and so little good. New York, Penguin Press, 2006.

[11] M. O'Leary, W. Orikowski, J. Yates, Distributed work over the centuries: trust and control in the Hudson's Bay Company, 1670–1826, in: P. Hinds, S. Kiesler (Eds.), Distributed Work, MIT Press, Cambridge, MA, 2002.

[12] H.M. Collins, R. Evans, The third wave of science studies, Soc. Stud. Sci. 32 (2) (2002) 235–296.

[13] H. Collins, Gravity's Shadow : The Search for Gravitational Waves, University of Chicago Press, Chicago, 2004.

[14] H.M. Collins, R. Evans, R. Rieiro, M. Hall, Experiments with interactional expertise, Stud. Hist. Philos. Sci. 37 (A4) (2006) 656–674.

[15] W.N. Espeland, The Struggle for Water: Politics, Rationality, Identity and the American Southwest, University of Chicago Press, Chicago, 1998.

[16] H. Collins, R. Evans, M. Gorman, Trading zones and interactional expertise, Stud. Hist. Philos. Sci. 39 (1) (2007) 657–666.

[17] S.L. Star, J.R. Griesemer, Institutional ecology, 'translations' and boundary objects: amateurs and professionals in Berkeley's museum of vertebrate zoology, 1907–39, Soc. Stud. Sci. 19 (3) (1989) 387–420.

[18] G.E. Louis, L. Magpili, A life-cycle capacity-based approach to allocating investments in municipal sanitation infrastructure, Struct. Infrastructure Eng. 3 (2) (2007) 121–131.

[19] T. Rached, Values and data in a boundary object: the treatment of risk assessments and chemical hazard data in the CleanGredients database [Masters]. Charlottesville, University of Virginia, 2006.

[20] P.H. Werhane, in: R.E. Freeman (Ed.), Moral Imagination and Management Decision Making, Oxford University Press, Oxford, 1999.

[21] M. Johnson, Moral Imagination, University of Chicago Press, Chicago, 1993.

[22] M.E. Gorman, M.M. Mehalik, P.H. Werhane, Ethical and Environmental Challenges to Engineering, Prentice Hall, Englewood Cliffs, N.J., 2000.

[23] P. Senge, The Fifth Discipline: The Art & Practice of Learning Organizations, Currency Doubleday, New York, 1990.

[24] M.E. Gorman, in: T. Gieryn (Ed.), Simulating Science: Heuristics, Mental Models and Technoscientific Thinking, Indiana University Press, Bloomington, 1992.

[25] O. Renn, M.C. Roco, Nanotechnology and the need for risk governance, J. Nanopart. Res. 8 (2006) 153–191. Available from: http://dx.doi.org/10.1007/s11051-006-9092-7.

[26] B.P. Hooper, C. Lant, Integrated, adaptive watershed management, in: D.S. Slocombe, K.S. Hanna (Eds.), Integrated Resource and Environmental Management: Concepts and Practice, Oxford University Press, Don Mills, Ont., 2007, pp. 97–118.

[27] C. Folke, T. Hahn, P. Olsson, J. Norberg, Adaptive governance of social-ecological systems, Annu. Rev. Environ. Resour. 30 (2005) 441–473.

[28] B.R. Allenby, Earth systems engineering and management, Technol. Soc. 19 (4) (2000/2001) 10–24.

[29] R.K. Plowright, S.H. Sokolow, M.E. Gorman, P. Daszak, J.E. Foley, Causal inference in disease ecology: investigating ecological drivers of disease emergence, Front. Ecol. Environ. 6 (2008) 420–429.

[30] H. Cabezas, C.W. Pawlowski, A.L. Mayer, N.T. Hoagland, Sustainability: ecological, social, economic and systems perspectives, Clean Technol. Environ. Policy (2003) 167–180.

[31] G. Marchant, From general policy to legal rule: aspirations and limitations of the precautionary principle, Environ. Health Perspect. 111 (14) (2003) 1799–1802.

[32] D. Barben, E. Fisher, C. Selin, D.H. Guston, Anticipatory governance of nanotechnology: foresight, engagement, and integration, in: J. Edward, O.A. Hackett, Michael Lynch, Judy Wajcman (Eds.), The Handbook of Science and Technology Studies, third ed., MIT Press, Cambridge, MA, 2007, pp. 979–1000.

[33] K.F. Schmidt, Green Nanotechnology: It's Easier than you Think, Project on Emerging Nanotechnologies, Washington, DC, 2007. <http://eprints.internano.org/68/1/GreenNano_PEN8.pdf>.

[34] M. Kandlikar, G. Ramachandran, A. Maynard, B. Murdock, W.A. Toscano, Health risk assessment for nanoparticles: a case for using expert judgment, J. Nanopart. Res. 9 (2006) 137–156.

[35] K. Morgan, Development of a preliminary framework for informing the risk analysis and risk management of nanoparticles, Risk Anal. 23 (6) (2005) 1621–1635.

[36] A. Wardak, M.E. Gorman, N. Swami, D. Rejeski, Environmental Regulatory Implications for Nanomaterials under the Toxic Substances Control Act (TSCA), IEEE Technol. Soc. (2007).

[37] R. Weiss, EPA to regulate nanoproducts sold as germ-killing. The Washington Post. 2006 November 23, Sect. A01, A7.

[38] C.T. Yavuz, J.T. Mayo, W.W. Yu, A. Prakash, J.C. Falkner, S. Yean, et al., Low-field magnetic separation of monodisperse Fe3O4 nanocrystals, Science 314 (5801) (2006) 964–967.

[39] V.A. Oyandel-Craver, J.A. Smith, Sustainable colloidal-silver-impregnated ceramic filter for point-of-use water treatment, Environ. Sci. Technol. (2008).

[40] N. Savage, M.E. Gorman, A. Street (Eds.), Emerging Technologies: Socio-Behavioral Life-Cycle Approaches, Pan Stanford Publishing, Singapore, 2013.

[41] P.H. Werhane, M. Gorman, Intellectual property rights, access to life-enhancing drugs, and corporate moral responsibilities, in: M.A. Santoro, T.M. Gorrie (Eds.), Ethics and the Pharmaceutical Industry, Cambridge University Press, Cambridge, 2005, pp. 260–281.

CHAPTER

International Governance Perspectives on Nanotechnology Water Innovation

37

David Rejeski[1] and Evan S. Michelson[2]

[1]*Science and Technology Innovation Program, Woodrow Wilson International Center for Scholars, Washington, DC, USA* [2]*Alfred P. Sloan Foundation, New York, NY, USA*

37.1 Introduction

As nanotechnology applications are being developed to provide novel solutions to many of the world's water problems—from developing improved desalination methods to cleaning up emerging pollutants—the international community has a unique opportunity to develop and implement new kinds of governance systems that will ensure that these applications can reach the market quickly, efficiently, and successfully. National and international regulatory bodies, from the Environmental Protection Agency (EPA) in the United States to the United Nations Water Program (UN-Water), can address this challenge of applying nanotechnology to improving water quality by adopting methodologies that spur innovation for development early in the research process, focusing on new ways of disseminating information about nanotechnology water applications, and considering the full life cycle of nanotechnology water applications.

This chapter will investigate some of the ways that barriers to collaboration around nanotechnology water applications in the international arena can be

Street, Sustich, Duncan and Savage. Nanotechnology Applications for Clean Water, 2nd Edition.
 DOI: http://dx.doi.org/10.1016/B978-1-4557-3116-9.00037-8

overcome. It begins by diagnosing the need for policy interventions with respect to nanotechnology and water and then offers a series of recommendations for approaches that may successfully address these challenges. The main purpose is to present an emerging set of policy options that can accelerate the application of nanotechnology toward improving water quality. However, without appropriate supportive policy options that can advance innovation responsibly, there is a potential that these benefits could be lost due to delay, lack of commercialization opportunity, and poor public outreach and communication.

37.2 Diagnosing the need

With a number of potentially groundbreaking nanotechnology water purification products entering the market—such as LifeStraw, a personal, portable water filtration product aimed at improving access to clean water in the developing world—the application of nanotechnology to water has the potential to become a major industry over the next 10–15 years [1]. In their comprehensive report "Nanotechnology, Water and Development," Hillie, Munasinghe, Hlope, and Deraniyagala [2] conclude that "nanotechnology applications for water treatment are not years away; they are already available and many more are likely to come on the market in the coming years." The promise of such nanotechnologies is also well documented, ranking high on a list of potential applications for the developing world [3]. However, it is anticipated that the realization of such benefits from nanotechnology could be hindered by challenges facing other, more conventional technologies that have attempted to solve such development-related problems. For example, a study comparing the use of conventional and nanotechnology water treatment and filtration technologies notes that each faces a range of access, ownership, social, economic, and environmental barriers to success [4–6]. However, in addition to these broader issues involving the application of new technologies to international development, there is an additional set of scientific and policy barriers more specifically related to the field of nanotechnology that could interfere with the long-term success of these water applications. Without addressing such obstacles, it remains an open question whether nanotechnology water applications will be able to surpass their conventional counterparts in terms of effectiveness, reliability, and ease of diffusion.

One of the first challenges facing the application of nanotechnology to water is addressing a number of critical, yet underlying, research areas as the field advances. In NanoFrontiers: Visions for the Future of Nanotechnology, author Karen Schmidt [7] reports that a discussion among leading scientists, engineers, and policy analysts about applying nanotechnology to long-term, global problems led to the articulation of a need for a set of information management, measurement, and communication tools that will allows researchers to share vast kinds of information quickly and efficiently. With multiple kinds of nanotechnology water treatment options in the pipeline, from carbon nanotube membranes to nanoporous ceramics to nanoscale zero-valent iron [4], such broad research tools must be developed to help organize and distribute a wealth of information that will emerge from laboratories and companies over the ensuing decades. Without such close-knit sharing of results, testing procedures, and standard material by way of databases, interdisciplinary collaborations, and other methods, there is a real risk that such efforts will fall prey to the drawbacks that other development-related technologies have encountered.

Second, a clear and transparent oversight system for nanotechnology is needed, one that demonstrates a vision, risk management principles, and a commitment to investigating and anticipating risks early in the research process. Without such a consistent regulatory approach from governments, both in the United States and around the world, even the development of promising nanotechnology water applications runs the risk of being hampered by distrust from the public at large. This may be one lesson that the introduction of nanotechnology can learn from the uneven and often resisted introduction of genetically modified foods: that initial lack of trust due to perceived secrecy, or concerns that potentially negative health and environmental impacts are not being addressed, can be difficult to overcome, if not insurmountable, in some cases.

As public perception research from the Project on Emerging Nanotechnologies and others indicates [8–11], the good news is that surveys and focus groups have shown a high degree of consistency in terms of what it takes to increase public confidence in government and industry involved with nanotechnologies: first, disclosure and transparency concerning the risk and benefits of nanotechnologies; second, more pre-market testing of products; and, third, testing done by trusted, third-party entities. This may be particularly important in the realm of nanotechnology water applications, where a dynamic could emerge that has companies from the developed world creating and marketing products for the developing world. Over the next few years, the social contract between government, industry, and the public around nanotechnology water applications will be defined, and creating trust will become a critical and essential factor in creating value and commercialization opportunities.

Closely tied to the issues of trust and transparency is the need for extensive risk research to determine how nanotechnology water applications might negatively impact human health and environmental well-being. To date, such risk research is generally scarce and offers little indication about how nanomaterials such as carbon nanotubes or nano-engineered silver, that may be used for environmental remediation or water filtration, could cause ecotoxicity, dispersion through the aquatic system, or contamination of the food chain. For example, an inventory of ongoing nanotechnology risk research projects maintained by the Project on Emerging Nanotechnologies [12] reveals that a disproportionately low amount of funding is being directed into looking into these environmental impact questions. Affected populations will begin to ask if such materials lead to uptake in the drinking water or persist in soil for extended periods of time.

For some nanomaterials, such as silver, there still remains uncertainty as to how government regulators will respond to its use in a variety of applications [13], and there are also concerns from environmental groups that the same properties that make nanoscale silver beneficial in improving water quality—because its enhanced properties are more effective at killing off bacteria and microbes—will be the same properties that reduce the effectiveness of municipal water treatment facilities that rely on the action of bacteria and microbes to purify sewage and wastewater [14–15]. As Breggin and Pendergrass [16] note, there is even growing concern that certain classes of nanomaterials may, in the future, be considered hazardous waste due to their as-of-yet unknown toxicological properties, creating potential legal liabilities for manufacturers, investors, and insurers. Addressing such uncertainties early in the development process would be beneficial and could help avoid health, environmental, and legal problems in the future.

Finally, nanotechnology faces the problem of waiting for a "killer application"—an indispensable, high-profile application that transforms the industry from its nascent stages of research to a more mature stage of commercialization—that has yet to arrive. Certainly, a nanotechnology water application, whether it is used for desalinization, purification, or recycling, could serve as such a visible use for the technology. However, without such a "got to have it" product or set of products, the presumed nanotechnology revolution risks becoming out of date in today's rapidly advancing technological landscape. There are already signs of "nano fatigue," with scientists, policymakers, and media outlets beginning to focus on emerging fields of synthetic biology, advanced climate change, and next-generation robotics. Whereas transformative breakthroughs take time, advancing nanotechnology water applications will require that a range of stakeholders, including government, identify long-term goals and develop a well-articulated strategy for reaching them. Although such roadmaps are available from a variety of organizations, such as the Foresight Nanotech Institute [17], governments needs an improved process of searching for new, "game changing" ideas for improving water quality and helping to transform them into revolutionary products and services that would benefit people in the developed and developing world.

37.3 The role for policy

Given these challenges, there is clearly a role for policymakers to play in advancing the application of nanotechnology toward improving water quality. It is anticipated that a suite of policy actions is needed that are both coordinated and integrated across a range of disciplinary boundaries and local, state, national, and international actors. One theoretical approach useful in conceptualizing such governance options is the "Frame One" and "Frame Two" context developed in a White Paper from the International Risk Governance Council [18] and a conference report from the Swiss Re Centre for Global Dialogue. The notion is that addressing nanotechnology's oversight and public risk perception may shift from a situation where "existing risk management approaches are directly applicable" (Frame One) to a situation where "a set of new risks could emerge through the profound shift in technical capabilities that nanotechnology offers" (Frame Two) [19].

Such changes in governance strategies are critical to adequately respond to the complex and interrelated impacts of these nanotechnology applications. As Olson and Rejeski [20] note, "traditional policy approaches based on hierarchical systems of command and control and market interventions will need to be complemented by the use of networks to steer change." In practice, this change of mindset has started to occur through the proliferation of voluntary codes of conduct—for example, the Responsible Nano Code in the United Kingdom [21]—and industry and nongovernmental risk management frameworks—for example, the Nano Risk Framework created by Environmental Defense and DuPont Corporation in the United States [22]. Such oversight experimentation will work to ensure that there is a commitment toward pursuing nanotechnology water applications in a coherent and sustainable manner. Although the options presented here are not the only ones available that could jump-start innovation aimed at addressing such long-term problems, they do center on actions that could be undertaken in the short term and that could foster cooperation among a range of interested stakeholders.

One way to address the challenges outlined earlier is to offer innovation inducement awards and prizes—a topic that has gained considerable attention from policy analysts over the past few years in reports from Kalil [23] and The National Academies [24]—in the area of nanotechnology applications for improving water quality. Such a prize could establish key scientific and technical benchmarks that would need to be achieved in order to receive the monetary amount of the prize or the award. As Schmidt states in Green Nanotechnology: It's Easier Than You Think [25], such a prize would fit well under the concept of green nanotechnology, an approach to risk mitigation that encompasses three complementary goals of advancing the development of clean technologies that use nanotechnology, minimizing potential environmental and human health risks associated with the manufacture and use of nanotechnology products, and encouraging the replacement of existing products with new nanotechnology products that are more environmentally friendly throughout their life cycles.

Such a "GreenNano Water Award" could help elevate green nanotechnology's visibility in a number of ways and, in turn, stimulate further innovation. For instance, recognizing innovative approaches to improving water quality based on nanotechnology would reward scientists and engineers working in this emerging area, may attract more scientists to the field, and help retain them over the course of their careers. Offering a financial award could help researchers and developers commercialize their green nanotechnology innovations and make green nanotechnology a visible national and international priority. An award program could also increase knowledge on efforts in green nanotechnology by consolidating and, in a sense, creating an inventory of ongoing activities. Anastas and Zimmerman [26] note in their report "Green Nanotechnology: Why We Need a Green Nano Award & How to Make It Happen" that such a prize would take advantage of the unprecedented opportunity to "green" the wider emerging nanotechnology production infrastructure and, in terms of applications that improve water quality, would have the opportunity to positively shape investments in environment-friendly facilities, foster open intellectual property arrangements, and create mutual responsibility across supply-chain relationships. In short, an award that recognizes green nanotechnology water applications would significantly influence key production choices that will become "locked-in" over the next 5–10 years.

Second, funding nanotechnology water research and applications could become a strategic investment goal of government agencies in the United States, including the EPA, the National Science Foundation (NSF), the Small Business Innovation Research (SBIR) Program, and the Small Business Technology Transfer (STTR) Program. Ideally, such funding efforts could be conducted in collaboration with international partners, in the European Union (EU) and East Asia, and through organizations such as UN-Water and the Organisation for Economic Co-operation and Development (OECD). Such joint funding projects on a particular topic, such as nanotechnology and water, would be a novel way to stimulate the creation of international research networks and share technical, logistical, and commercialization expertise across leading developing countries, such as India and China [27]. In his report EPA and Nanotechnology: Oversight for the 21st Century [28], J. Clarence Davies noted that a renewed emphasis on international efforts, such as the ones described earlier, would be necessary to help the United States maintain and continue its leadership in nanotechnology over the next 2–5 years. Such collaborations are rapidly occurring between developing countries [29] and between countries in the developed and developing world [30–31] and could easily be focused around an organizing topic such as nanotechnology and water. However, without policy changes and a renewed emphasis on technical assistance programs, diffusion of such cutting-edge technologies to the developing world will fall short and the problem of poor water quality will persist. Therefore, formal collaborations are needed at the micro-level (between individual researchers), meso-level (between individual universities or companies), and macro-level (between nations or groups of nations), and they could culminate in an International Year of Water Nanotechnology that brings increased attention to nanotechnology's environmental applications.

This concept of an "International Year" of nanotechnology leads to a third policy action, which centers on developing an advanced outreach and communication strategy for nanotechnology water applications. Although viewing communications as a policy tool may not appear, at first glance, to be a worthwhile endeavor, it is evident that the lack of public awareness about nanotechnology's potential applications can hinder its growth and potentially lead to backlash or rejection of the technology [9–10]. Disseminating information about such research and potential applications can be a powerful advancement tool, particularly when using interactive, new media outlets, such as podcasts, video links, blogs, and video games. For example, a podcast on nanotechnology and clean water applications—"Plenty of Clean Water at the Nanofrontier" [32], featuring researcher Eric Hoek (see also Hoek and Ghosh, Chapter 9)—is available online at the Project on Emerging Nanotechnologies website. There may also be other strategies—such as a word-of-mouth information campaign focusing on nanotechnology and water—that use the power of personal conversation to spread information about this emerging area of research. Such a word-of-mouth campaign could help diffuse knowledge of new ideas by targeting individual trendsetters that can inform larger groups of people through their own influential networks. The advantage of launching an informational nanotech campaign on the topic of water applications using word-of-mouth is that the communication infrastructure, know-how, and evaluation systems are already in place, and are rapidly improving. An innovative word-of-mouth campaign could place nanotechnology water applications into the world of everyday conversation, where messages are built on trust and understanding rather than on hype and jargon.

Finally, concerns about potential health and environmental impacts of nanotechnology water applications can be addressed by encouraging companies to undertake robust life cycle assessments (LCAs) of their products before they enter the market. A cradle-to-grave look at the health and environmental impact of a material, chemical, or product, LCAs can be essential tools for ensuring the safe, responsible, and sustainable commercialization of nanotechnology, provide the advantage of making potential problems known early in the innovation process, and encourage confidence in the consumer that companies have practiced due diligence and foresight. In particular, LCAs conducted in partnership between government and industry—or by independent, third parties—have the power of presenting a degree of objectivity about the scientific and technical findings. Such public–private partnerships also encourage the sharing of information among participants, with government gaining early information about new kind of products and with industry gaining experience in responding to and addressing critical questions about environmental safety and health.

Nanotechnology products designed to improve water quality are natural candidates for LCA analysis because they could potentially have long-term effects across multiple stages of use, from generation to consumption to disposal. A report from a workshop on this topic, "Nanotechnology and Life Cycle Assessment: A Systems Approach to Nanotechnology and the Environment" [33], points out that wisely implemented assessment tools, such as LCA, can help enable governments, industry, and consumers to compare the environmental performance of a novel nanotech product with that of conventional products already on the market. However, the report also points out that major future efforts related to data gathering, protocol implementation, and practical measurement methodologies are needed if potential risks are to be fully addressed by LCAs. Options are available to fill in these gaps, through the undertaking of LCA case studies of representative materials and the adoption of standardized LCA reporting mechanisms and terminologies, but action is needed soon if such information is going to significantly impact early stage innovation.

37.4 Conclusions

In the end, only a concerted effort to think ahead about nanotechnology water applications on a global level will ensure that their full potential will be realized. Clearly, challenges remain, from ensuring that appropriate research tools are widely available to addressing concerns about environmental health and safety risks to implementing a clear and transparent oversight system. Responding to these problems will require an integrated set of forward-looking policy solutions that combine high-profile incentive awards for innovation; targeted, coordinated, and strategically planned investment at the international level; a renewed focus on public outreach and communication; and reliance on life cycle assessments to identify long-term risks. Undertaking these actions will require intellectual, financial, human resource, and time commitments from a range of stakeholders. This process must begin to move forward at a rapid pace to match the speed of nanotechnology innovation.

References

[1] M. LaMonica, For disruptive technologies, look to material science, CnetNews.com [Internet]; October 15, 2007 [cited March 14, 2008]. Available at <http://www.news.com/8301-10784.3-9797268-7.html>.

[2] T. Hillie, M. Munasinghe, M. Thembela, Hlope, Y. Deraniyagala, Nanotechnology, Water & Development, Meridian Institute, Washington, 2007 [cited March 14, 2008]. Available at <http://www.merid.org/nano/waterpaper/NanoWaterPaperFinal.pdf>.

[3] F. Salamanca-Buentello, D.L. Persad, E.B. Court, D.K. Martin, A.S. Daar, P.A. Singer, Nanotechnology and the developing world, PLoS Med. 2 (2005) 383–386 [cited March 14, 2008]; available at <http://medicine.plosjournals.org/perlserv/?request=get-document&doi=10.1371/journal.pmed.0020097>.

[4] Meridian Institute, Overview and comparison of conventional and nano-based water treatment technologies, Washington, Meridian Institute, 2006 [cited March 14, 2008]. Available at <http://www.merid.org/nano/watertechpaper/watertechpaper.pdf>.

[5] J. Loncto, M. Walker, L. Foster, Nanotechnology in the water industry, Nano Law and Bus. 4 (2007) 157–159.

[6] M. Berger, Water, nanotechnology's premise, and economic reality, Nanowerk [Internet], August 15, 2007 [cited March 14, 2008]. Available at <http://www.nanowerk.com/spotlight/spotid=2372.php>.

[7] K.F. Schmidt, Nanofrontiers: visions for the future of nanotechnology, Washington, Project on Emerging Nanotechnologies, (2007), PEN 6 [cited March 14, 2008]. Available at <http://www.nanotechproject.org/file.download/files/PEN6.NanoFrontiers.pdf>.

[8] J. Macoubrie, Informed public perceptions of nanotechnology and trust in government, Washington, Project on Emerging Nanotechnologies, (2005), PEN 1 [cited March 14, 2008]. Available at <http://www.nanotechproject.org/process/files/2662/informed.public.perceptions.of.nanotechnology.and.trust.in.government.pdf>.

[9] Hart Research Associates, Awareness and attitudes toward nanotechnology: report findings, Washington, Project on Emerging Nanotechnologies, (2006) [cited March 14, 2008]. Available at <http://www.nanotechproject.org/file.download/files/HartReport.pdf>.

[10] Hart Research Associates, Awareness and attitudes toward nanotechnology and federal regulatory agencies, Washington, Project on Emerging Nanotechnologies, (2007) [cited March 14, 2008]. Available at <http://www.nanotechproject.org/process/files/5888/hart.nanopoll.2007.pdf>.

[11] D.M. Kahan, P. Slovic, D. Braman, J. Gastil, G.L. Cohen, and D. Kysar, Biased assimilation, polarization, and cultural credibility: an experimental study of nanotechnology risk perceptions, Washington, Project on Emerging Nanotechnologies, (2008) [cited March 14, 2008]. Available at <http://www.nanotechproject.org/process/files/5960/brief2kahan.final.pdf>.

[12] Nanotechnology Consumer Product Inventory [Internet], [cited March 14, 2008]. Available at <http://www.nanotechproject.org/inventories/consumer>.

[13] L.L. Bergeson, EPA clarifies position on ion-generating equipment, Chem. Process. [Internet], (2007) [cited March 14, 2008]. Available at <http://www.chemicalprocessing.com/articles/2007/205.html>.

[14] J. Sass and M.C. Wu. Registration of nanosilver as a pesticide under FIFRA [Internet], November 2006 [cited March 14, 2008]. Available at <http://www.nrdc.org/media/docs/061127.pdf>.

[15] R.L. Rundle, This war against germs has a silver lining, The Wall Street Journal [Internet], June 6, (2006) [cited March 14, 2008]. Available at <http://online.wsj.com/article/SB114955908525572199.html>.

[16] L.K. Breggin, J. Pendergrass, Where does the nano go? End-of-life regulation of nanotechnology, Washington, Project on Emerging Nanotechnologies, (2007), PEN 10 [cited March 14, 2008]. Available at <http://www.nanotechproject.org/process/files/2699/208.nanoend.of.life.pen10.pdf>.

[17] Foresight Nanotech Institute, Productive nanosystems: a technology roadmap, Menlo Park, CA, Foresight Nanotech Institute, (2007) [cited March 14, 2008]. Available at <http://www.foresight.org/roadmaps/Nanotech.Roadmap.2007.main.pdf>.

[18] International Risk Governance Council, White paper on nanotechnology risk governance, (2006) [cited March 14, 2008]. Available at <http://www.irgc.org/IMG/pdf/IRGC.white.paper.2.PDF.final.version-2.pdf>.

[19] Swiss Re Centre for Global Dialogue, The risk governance of nanotechnology: recommendations for managing a global issue, (2006), 7 [cited March 14, 2008]. Available at <http://www.ruschlikon.net/INTERNET/rschwebp.nsf/vwPagesIDKeyWebLu/GLBH-743CKG/$FILE/Nanotech.Report.2006.pdf>.

[20] R. Olson, D. Rejeski, Introduction: another chance, in: R. Olson, D. Rejeski (Eds.), Environmentalism and the Technologies of Tomorrow: Shaping the Next Industrial Revolution, Island Press, Washington, 2005, pp. 1–74.

[21] Responsible NanoCode [Internet], [cited March 14, 2008]. Available at <http://www.responsiblenanocode.org/index.html>.

[22] Environmental Defense and Dupont Corporation, Nano risk framework, (2007) [cited March 14, 2008]. Available at <http://www.edf.org/documents/6496.Nano%20Risk%20Framework.pdf>.

[23] Kalil T, Prizes for technological innovation, Washington, The Brookings Institution, (2006) [cited March 14, 2008]. Available at <http://www.brookings.edu/~/media/Files/rc/papers/2006/12healthcare.kalil/200612kalil.pdf.

[24] The National Academies, Innovation inducement prizes at the National Science Foundation, The National Academies, Washington, 2007.

[25] K.F. Schmidt, Green nanotechnology: it's easier than you think, Washington, Project on Emerging Nanotechnologies, (2007), PEN 8 [cited March 14, 2008]. Available at <http://www.nanotechproject.org/file.download/files/GreenNano.PEN8.pdf>.

[26] P. Anastas, J. Zimmerman, Green nanotechnology: why we need a green nano award & how to make it happen, Washington, Project on Emerging Nanotechnologies, (2007) [cited March 14, 2008]. Available at <http://www.nanotechproject.org/file.download/206>.

[27] L. Yeung, E.S. Michelson, China, nanotechnology, and the environment, in: J.L. Turner (Ed.), China Environment Series, Woodrow Wilson International Center for Scholars, Washington, 2006, pp. 82–84 [cited March 14, 2008]; available at <http://www.wilsoncenter.org/topics/pubs/CEF.Feature.4.pdf>

[28] J.C. Davies, EPA and nanotechnology: oversight for the 21st century, Washington, Project on Emerging Nanotechnologies, (2007), PEN 9 [cited March 14, 2008]. Available at <www.nanotechproject.org/file.download/197>.

[29] T.V. Padma India, Brazil, and South Africa discuss joint research, SciDev.com [Internet]; October 24, 2004 [cited March 14, 2008]. Available at <http://www.scidev.net/News/index.cfm?fuseaction=readNews&itemid=1693&language=1>.

[30] Finland and China team up on nanotechnology research, Azonano [Internet], December 7, 2007 [cited March 14, 2008]. Available at <http://www.azonano.com/news.asp?newsID=5489>.

[31] Nanotechnology collaboration between Europe and India, Nanotechnology Now [Internet], August 18, 2006 [cited March 14, 2008]. Available at <http://www.nanotech-now.com/news.cgi?story.id=16916>.

[32] Plenty of clean water at the nanofrontier [podcast], Washington, Project on Emerging Nanotechnologies, 2007, 3 [cited March 14, 2008]. Available at <http://www.penmedia.org/podcast/nano/Podcast/Entries/2007/8/7.Episode.3.-.Plenty.of.Clean.Water.on.the.NanoFrontier.html>.

[33] Nanotechnology and life cycle assessment: a systems approach to nanotechnology and the environment, Washington, Project on Emerging Nanotechnologies, (2007) [cited March 14, 2008]. Available at <http://www.nanotechproject.org/file.download/168>.

CHAPTER

Nanoscience and Water: Public Engagement at and Below the Surface 38

David M. Berube
North Carolina State University, Raleigh, NC, USA

38.1 Introduction

The subject of nanotechnology and water offers an opportunity to discuss some of the major societal issues associated with emerging technologies generally and nanotechnology specifically. Water is a staple and the subject of potable water is an emotional one. The public's reaction to applications of nanoscience for producing drinkable water either in the developed or in the developing world should be very positive. The alternative is disease and sometimes death. Moreover, fear of death tends to skew opinion formation positively.

At the same time, however, most of the public generally feel decisions about drinking water are out of their control and any exposure to nanoparticles in

Street, Sustich, Duncan and Savage. Nanotechnology Applications for Clean Water, 2nd Edition.
 DOI: http://dx.doi.org/10.1016/B978-1-4557-3116-9.00038-X

drinking water will be considered an involuntary risk, a variable with a negative valence. For example, there have been few efforts to include the public in decisions about water treatment and purification. For most consumers, water is provided by large public and impersonal utilities. They hear from them monthly when they receive their bills, bills that are often bundled with trash collection and other community services.

In addition, it is likely that Western public reactions to nanotechnology producing clean water for non-Westerners will be gauged differently than for themselves. It is likely these reactions would differ if the treatment strategy is for a different racial or socioeconomic group within the West. This is due to the unfortunate tendency to discount morbidity and mortality values when the groups involved are different from us. Self-interests counterbalanced against altruism have always played powerful roles in deciding how public resources are expended and the values associated with public welfare projects.

Although there may not be any moral or ethical duty, requirement, or obligation for government and industry to engage the public before adopting and marketing a technology, there are many pragmatic reasons to do so. Research and development is expensive and time-consuming with multiple opportunity trade-off costs. Unless public resources are shifted to purchase, install, and maintain new public technologies such as water treatment, they can be cost prohibitive for many markets. In addition, public monies often track public sentiment. Finally, the public functions as consumers *per se*, shareholders of industries in the business of providing potable water. Public support for high-cost facilities involves public contracts and bonds. Neutral attitudes, if not palpable opposition, should be serious concerns to public service providers.

The following draws heavily from my graduate work studying the miscommunication between Amerindian Native American gens or tribes and the Departments of Interior and War during the late nineteenth century and an unpublished manuscript in preparation on events leading to the assassinations of Sitting Bull and Big Foot and the Wounded Knee Creek massacre. Cultural anthropologists have studied the interaction between deployments of Western technologies and developing cultures for many years. Simply foisting a Western technology on a non-Western culture regardless of its utility can be counterproductive if not a recipe for disaster [1].

38.2 Water and the public

WHO estimates that approximately 1 billion people do not have access to a reliable water supply and 3.4 million die annually from water-related diseases. This means 42,000 people die from diseases related to low-quality drinking water each week. Over 90% of diarrheal diseases in the developing world today occur in children under 5 years of age. In 2002, 230 million school-age children were without a reliable water supply [2]. There are 525 million small farms in the world with over 2.5 billion people living off the land. Whereas water availability increases by about 0.5% annually, the demand is increasing by 10.5% [3].

Extrapolating from population data, the demand for drinking water and water for agriculture and industrial uses is expected to increase by as much as 70% over the next 25 years in the United States alone. Richard Sustich from the Center of Advanced Materials for Purification of Water with Systems at the University of Illinois at Urbana–Champaign warned that within the next decade, the suburbs of Chicago might find their supply running dry [4]. The drought in the Southeast during 2007 threatened many cities' water use habits with reports of only a few months' supply in some reservoirs.

Because issues about water tend to be emotional ones, the costs both financial and otherwise that might be associated with new technologies will tend to be overshadowed by the powerful claims of clean drinking water. However, we may be approaching a tipping point when it comes to public engagement. As science and technology continue to make inroads into the day-to-day operation of our lives, the public may be becoming more concerned about choices they make and those foisted upon them. Having observed the failure of government to protect public safety, citizens may be beginning to accept more responsibility when it comes to determining their own public safety.

Although an argument can be made, the public act more like sheep than wolves when it comes to being governed; history bespeaks there comes a time when the public has had enough and rises to challenge the regulators. Both the BSE (bovine spongiform encephalopathy or "mad cow disease") and GMO (genetically modified organisms, esp. seeds) controversies in the United Kingdom and Western Europe offer powerful lessons. Once empowered having had their appetites whetted, the public tends to exercise prerogatives of engagement more regularly. This has clearly been the case in France, the United Kingdom, and some other developed countries. Environmental disasters in Seveso, Italy, and Bhopal, India, were aggravated due to the absence of a *bona fide* engagement strategy between industry and the public and both the Roche Group and Union Carbide have approached engagement as a necessary tool in doing business ever since. Nanotechnology may offer an instance when we can turn the tides of engagement to produce processes that benefit all stakeholders.

Since other chapters examine issues such as specific technologies and applications, environmental health and safety, and so forth, the following will concentrate on public engagement concerns as they may relate to treatment approaches. How can we best involve the public in the decision-making process to maximize the effectiveness of nano-based water treatment strategies?

The public interfaces with the adoption of new technologies of this sort in three ways: management, adoption, and maintenance. Communities of people need to purchase and install the technology, the people need to use it, and the technology needs to be maintained by people and this is especially true for point-of-use approaches.

38.3 Nanotechnology treatment strategies

Potable water comes primarily from surface- and groundwater. They are treated on-site by a community or municipal authority or at the point of use in the village or the home. Freedonia reported world demands for treatment will increase 6% per year through 2009 to more than $35 billion [5].

Indeed, when it comes to cleaning and filtering water in the traditional way, "we've gotten as far as we can go on the larger scale," says Sustich [4]. "It's not black and white" says Mamadou S. Diallo from Cal Tech's Molecular Environmental Technology program. "No one wants to drink nanoparticles with their water" [4]. The perceived risks are especially problematic for this industry. As was somewhat evident from the Samsung washer issue, the water industry is typically conservative and risk adverse. Since most water companies are publicly owned, they aren't allowed to make a profit. And if something went wrong, the water company could be held responsible for a public health crisis, says Sustich [4].

Various nanotechnologies are being studied including carbon nanotubes (CNTs), nanoclays and zeolites, dendrimers, nanoscale metals, nanofibers, and membranes. The following lists of applications and developments are meant to be neither comprehensive nor exhaustive. For a much more complete analysis, see the $5000 Frost & Sullivan report—"Impact of Nanotechnology in Water and Wastewater Treatment [6]—and the three free Meridian Institute reports—"Nanotechnology, Water & Development" [2], "Overview and Comparison of Conventional Water Treatment Technology-Nano-Based Treatment Technologies" [7], and "Workshop on Nanotechnology, Water & Development" [8].

In general, nanofilter technology takes advantages of the higher surface area and throughput and observers, especially from the industry, hail the superiority of nanofilters. It has been reported that nano-based filters are able to achieve 99.5% efficiency when compared with conventional technologies removing protozoan cysts, oocysts, and helminth ova, in some cases bacteria and viruses, and provide effective treatment of contaminants such as mercury, arsenic, and perchlorate [6].

Unsurprisingly, CNTs are receiving a lot of attention. Because of their high flow rates and high selectivity to filter out very small impurities and other organic materials, carbon membranes offer much promise and research is ongoing. A Meridian Report noted CNT filters could remove 25-nm-sized polio viruses from water as well as larger pathogens such as *Escherichia coli* and *Staphylococcus aureus* bacteria [2]. For example, a team, led by Olgica Bakajin from Lawrence Livermore National Laboratory (see Chapter 11), has developed a CNT membrane with high selectivity and high flow rate. Based in principle on aquaporins, the water channels in cells, this team has reported promising results [9,10]. Seldon Laboratories has developed its nanomesh filter media. They claim to be able to remove more than 99.99% of bacteria, viruses, cysts, molds, coliform, parasites, and fungi and also significantly reduce lead and arsenic [7].

CNT membrane technologies are in advanced stages of development. For example, the Pacific Northwest National Lab developed a polypyrrole–CNT nanocomposite. This membrane is made with a thin film of absorbent polymer called polypyrrole on a matrix of CNTs. Other researchers at Rensselaer Polytechnic Institute may have solved for the hydrophobic nature of CNT arrays such that CNT membranes become practical [11]. Membrane nanotechnologies are undergoing varied tests. For example, Nanyang Technological University and the Public Utilities Board in Singapore announced the results of tests that they describe as promising. A pilot plant at Chua Chu Kang Waterworks using nanotechnology to remove dissolved salts and chemical compounds will be up in 2 years [12].

Nanoporous ceramics have garnered a lot of press as well. For example, Porous Ceramic Shapes acquired by MetaMateria Partners offers a line of lightweight ceramic products with controlled porosity called Cell-Pore™. Their ceramic filter hosts aerobic bacteria that convert different pollutants into nontoxic substances [7]. NanoDynamics has introduced cell-pore ceramic filters with highly absorbent nanocrystals [5]. Nanovation AG has its Nanopore® ceramic membrane filters made from ceramic nanopowders on a support material such as alumina. Nanovation claims they effectively remove bacteria, viruses, and fungi from water [7]. Argonide's NanoCeram uses aluminum oxide nanofibers on a glass filter substrate and claims over 99.99% effectiveness over viruses, bacteria, parasites, natural organic matter and 99.9% of salt, radioactive materials, and heavy metals such as chromium, arsenic, and lead [7]. Finally, Steward Environmental Solutions is bringing to market the Pacific Northwest National Laboratory SAMMS™ technology made from ceramic materials with nanoscale pores to which a monolayer of molecules can be attached. Both the monolayer and the mesoporous support can be functionalized to remove specific pollutants including mercury, lead, chromium, arsenic, radionuclides, cadmium, and other metal toxins [7].

38.4 Modalities

There are at least two primary modes by which nanotechnologies used for water purification will interface with the public. They include large-scale centralized community treatment plants as well as diffused treatment facilities and point-of use including end-of-faucet or spigot applications. In addition, there are targeted remediation treatment technologies that can be used across both these modalities.

38.4.1 Municipal systems

Major municipal water treatment technologies need large capital investments, management systems, and governance structures. As the worldwide market for water will exceed $400 billion by 2010, many entrants are likely. As such, the market seems to be getting increasingly global as well as increasingly consolidated. "Large companies like GE, Pentair, and ITT are pursuing both industrial and residential/commercial sectors throughout the world" [13]. Furthermore, there are dozens of start-ups. For example, in the San Francisco area alone, Novazone, Pionetics, and Hydropoint are raising venture capital (VC) funding for their novel technologies [14].

Although water utilities tend to be slow-moving leviathans, there are a few interesting large-scale applications. For example, there is a pilot effort by Ondeo, which has installed an ultrapurification system involving pores of 0.1 μm size in one of its plants outside Paris [15]. Generale des Eaux is collaborating with the Dow Chemical subsidiary Filmtec to produce a nanofiltration system as well. Finally, the Long Beach Water Department in Long Beach, CA, has installed and tested a pilot-scale, dual-stage nanofiltration process (see Chapter 13).

38.4.2 Point-of-use systems

It might be possible to build nanomembrane plants as portable units that can be assembled in the major centers and then transported to outlying areas where they are needed [2]. Point-of-use treatment technologies seem to be poised to make significant contributions to water use needs in developed and especially developing countries. For example, granular media filters using charged metal oxides and hydroxides of iron, aluminum, calcium, and magnesium are under development. Disk filters with colloidal silver are relatively inexpensive and up to 100% effective against bacteria [7]. Some are already on the market. IIT-Madras and Mumbai-based Eureka Forbes claim to have marketed the first nanotechnology-based filter with the first 1000 units in place. It uses silver nanoparticles to remove pesticides such as endosulfan, malathion, and chlorpyrifos from drinking water [3,16].

38.4.3 Targeted systems

Some developments target specific pollutants and other hazards. For example, Japan's Royal Electric Co. released the RVK–Ni oxygen/ozone micro-nano bubble water sterilizer. It mixes ozone nano-bubbles with ozone/oxygen micro-bubbles and proved an effective treatment against the norovirus [17]. SolmeteX has ArsenX™, a resin made of hydrous iron oxide nanoparticles on a polymer substrate that has been shown to remove arsenic, vanadium, uranium, chromium, antimony and molybdenum [7]. Arsenic is receiving special attention. For example, Houston's Rice University's Center for Biological and Environmental Nanotechnology has a nanorust project to clean arsenic from drinking water. Due to its simplicity and low cost, it offers hope for millions of people in developing countries where thousands of cases of arsenic poisoning each year are linked to contaminated wells [18]. As many as half the wells drilled in the late 1960s to counter Bangladesh's severe surface water pollution have been found to be contaminated by arsenic [2]. Rice researcher Mason Tomson warns, however, "no one knows the risks of the arsenic residue being consumed by accident or leaching from landfills back into water supplies" [19].

38.5 Water and public engagement

Generally, the assumption held by the expert community has been "if we build it, they will buy it," and for a large proportion of the population that may be valid. Indeed, given the exclusivity awarded to water utilities, there does not seem to be a realistic alternative for most of the consuming public. However, even those utility contracts need to be awarded and renewed and a poor record of public participation can make this process troublesome for a water provider.

In general, public participation broadens social development ideals enabling the public to participate fully in the decision-making process, and ordinary people experience fulfillment, which contributes to a heightened sense of community and a strengthening of community needs [20]. Beyond these more abstract values, there are advantages from engagement that can contribute to the overall success of a treatment strategy, especially in situations when public ownership is important to management and maintenance.

With nanotechnology poised to make significant inroads in water quality sensing as well as treatment technologies, public participation may become critical. Of course, the approach taken for a treatment strategy will affect the process of engagement. Adopting a strategy for a point-of-use system involves some different variables than would apply to a large municipal system.

Engagement takes many forms: public meetings, public hearings, open houses, workshops, citizen advisory committees, social surveys (such as consensus conferences), focus groups, newsletters, and reports. The forms of engagement to pursue are affected by the experience, if any, the public has with engagement, the amount of information the public has about the technology, how accurate this information may be, the level of comprehension the public possesses, and the context for the exercise. Meetings and hearings can be highly intimidating to the public who often have little experience with advocacy. Consensus conferences can be equally foreboding to some. One of the reasons election caucuses are attended by the same people cycle after cycle is simply a function of familiarity; newcomers confront a high entry barrier. Unfortunately, some of the more meaningful engagement exercises are the more active and demanding ones. An experience at a well-orchestrated public hearing is less easily discarded than a newsletter.

An understanding of how informed the public is about a new technology is challenging as well. Opining does not require information and survey data about advanced technologies is highly suspect. Many surveys of this ilk are closer to push polling than opinion sampling, often incorporating narratives or clever manipulation of phraseology and the order of questions in their technique. Nonetheless, we need some indication of what is known to decide how much time is spent educating and informing the public against time spent in more persuasive appeals.

Unfortunately, public information on advanced technologies is inaccurate, having been gleaned from popular culture and anecdotes. Inaccurate information needs to be debunked and expunged before accurate information is offered and that takes time and expertise. It is insufficient to present competing information. New competing information must be presented using the same or similar warrants that incorporated the original inaccurate information. Given the diversity of warrants a group of the public might have used, the challenge is learning why the inaccurate information was incorporated into their understanding.

The understandability factor is critical when designing an engagement exercise of any sort. For years, it was believed that by improving the science education of the public we could improve their opinions about technology. This deficit theory of scientific education has been a dismal failure. Although there are many benefits to improving science education, persuading the public that scientists are correct is not one of them. By and large, the public selected against an education in science as much as a scientist selected otherwise. Any engagement exercise must speak in a public argot and address issues without deferring to parochial metaphors.

Context is the last major variable. The public engages new information with notions and biases. Generally, the public prefers information consistent with previously held beliefs. In addition, the public searches for stories with a high level of fidelity (they need to ring true to the world around them). Context can modify these sensibilities. If all is generally well, then a new treatment technology is viewed as an expense. Under conditions of an outbreak of waterborne diseases, the new technology will be viewed as an opportunity. If the media has been amplifying fear mongering on the new technology, then the public will reflect apprehensiveness. On the other hand, if the media has been attenuating the same, then the public may be sanguine.

38.5.1 Municipal systems

A Meridian Institute report claims there are multiple requirements for implementing new technologies for water treatment, especially in rural communities; one of the most critical is worth repeating here. The community must be exposed to a comprehensive education program that will inform and educate them about the methodology and benefits of the water treatment project [2].

It is safe to assume there are few people who can summarize how their drinking water is treated. Indeed, most are unable to distinguish between water and waste treatment. This would seem to be true across cultures. As a result, if it weren't for the expense involved, a utility might be willing to forego any engagement with the public altogether. As large systems of this sort are expensive and often involve budgetary trade-offs, municipalities should do what they can to educate the public about the treatment strategy as well as allaying as many of their apprehensions as practicable.

Indeed, in an atmosphere where the public knows little about nanotechnology and there have been no seriously amplified reports of environmental health and safety issues associated with nanotechnology, the claims of safe drinking water should trump reservations especially if the claims are linked to the prevention of waterborne diseases. However, retrofitting or upgrading an existing system or building a new system based on nanotechnologies might be troublesome under a different set of conditions.

The energy industry faces similar financial incentives and their approach has been to call public meetings to solicit public sentiment and support. Unfortunately, these public gatherings are often more proforma than anything else with the decision having already been made and with the gathering used to cement support rather than to engage in dialogue. As such, given a crisis situation, energy consumers become aggrieved antagonists rather than advocates of the industry.

Although nanotechnologies may make inroads into commercial treatment facilities, there seems to be more interests in point-of-use applications at this time and these demand a different engagement strategy altogether.

38.5.2 Point-of-use strategies

The previously mentioned Meridian Report added a second concern: The community must be involved at all stages of the project such as being trained in the operation and maintenance of the new technology to facilitate a sense of community ownership [2]. The same Meridian Report cited earlier suggested heightened community ownership can even reduce vandalism and theft though there are no examples cited in their report.

Nonetheless, it is very important to understand that Western conceptions of hierarchical governance may not be shared in many different cultural settings. Just as group opinion leaders in Western organizations are not necessarily the elected or appointed managers of the organizations, the leader in a non-Western setting might be a tribal leader or chief or a patriarch or matriarch of a clan rather than a government administrator. Getting the government representative to allow distribution of technologies to a community may simply be insufficient when the goals are use, maintenance, and ownership. For decades, if not centuries, clan members may have gathered at wells to get water and to share the news. Public spheres in developing cultures tend to be less formalized and less dependent on public structures or institutions, such as libraries and newspapers. Adding new technologies especially by a third party regardless of intentions risks contamination of a different sort altogether—damage to a public forum.

Point-of-use technologies will demand a more diffused or localized form of public engagement. Whereas in the West, filter technologies that can be used in the home are marketed just like any other product, commercial marketing may be wholly inappropriate for some developing cultures. It is likely advanced technologies for point-of-use water treatment will be perceived with some suspicion.

In addition to many of the variables mentioned earlier (experience, familiarity, comprehension, etc.), there are some special demands when the technology must be situated in a culturally important public setting such as a public well, and home use presents special demands as well. First and foremost, any new technology must be sufficiently well tested such that it not only meets the specific needs of the community but also will not need to be removed, upgraded, or retrofitted. One may get just one shot. Installing a technology that fails may damn subsequent attempts to adopt preferable technologies.

Any new technology introduced into the public sphere will need to be introduced by the public themselves. It has to be viewed as their technology and demands some level of ownership to the extent that it may be desirable to charge the public some costs whether pecuniary or in kind. The technology will need to accommodate both the safe drinking water needs of the community as well as their public sphere needs. Put simply, it must not disrupt the culturally significant activities associated with drawing water from a public well. A technology perceived as foreboding by some members of the community may dissuade them from going to the well altogether. Sending a different member from the clan, more traditional members may isolate themselves from a vibrant public experience, with both them and the community suffering as a result.

Bringing a new technology into the home may be even more challenging. The new technology would need to be introduced by a family member and the family must perceive a level of ownership. Installation and maintenance must be done by a family member as well; hence it must be simple enough that it can be done with minimal training. Point-of-use purification systems require maintenance and this task will need to be done by a family member as well.

Minimizing technological fixes is preferable as it simplifies the maintenance process for families. Durability and ease of use are critical variables. Any technology that disrupts the day-to-day operation of the family should be avoided. Less intrusion is always more desirable.

A serious and sustained educational campaign will need to be mounted. The group opinion leaders of the community should not only participate but also lead the campaign. There should be testimonials presented as narratives as well as demonstrations. Community members should have the opportunity to handle the new technology and participate in a mock installation as well as a mock maintenance exercise. They should name the technology and participate in a ritual whereby they contract for the technology in exchange for some expenses on their part and the expense needs to be meaningful.

Any outbreak of disease subsequent to installation of the treatment technology might be associated with the new technology notwithstanding its falsifiable cause. Responding to this type of misinformation is much easier with the trust that comes with an engagement plan already in place. Beginning a dialogue in the midst of a controversial and damaging event simply suffers from too much mistrust to be productive without a substantial expense on the part of the provider. Rumors can be as disruptive as the truth and they are more easily debunked if a dialogue is ongoing.

38.6 Conclusions

Engagement has intrinsic values, especially the heightened sense of community and a strengthening of community needs that it sustains. In addition, there are highly pragmatic reasons to engage the public in a meaningful way. The public become advocates, in a sense, for the providers and in situations of hazard the dialogic system with attendant trust is already in place. Since so much of the rhetoric over nanotechnology and water seems to address pollution faced by communities in developing economies, we may need to design models of engagement appropriate to the tasks at hand, and we may need to design them very soon.

Acknowledgments

This work was supported by a grant from the National Science Foundation, NSF 06-595, Nanotechnology Interdisciplinary Research Team (NIRT): Intuitive Toxicology and Public Engagement. All opinions expressed within are the author's and do not necessarily reflect those of the National Science Foundation, the University of South Carolina, North Carolina State University (NCSU), and the International Council on Nanotechnology. NCSU communication student Nick Temple assisted in the final disposition of this chapter.

Notes

The Public Communication of Science and Technology (PCOST) Project was conceived at North Carolina State to design and evaluate efforts to improve the public communication of science and technology. It is an informal group, resides in the College of Humanities and Social Sciences, and includes faculty from different departments and schools on the NCSU campus in Raleigh, NC. In addition, PCOST includes a small number of associated members from non-NCSU campuses.

References

[1] M.B. Schiffer (Ed.), Anthropological Perspectives on Technology, University of New Mexico Press, Albuquerque, New Mexico, 2001.

[2] T. Hilie, M. Munasinghe, Y. Deraniyagala, Global dialogue on nanotechnology and the poor: international workshop on nanotechnology. Information Resources Multi- Problem Solving, 2006–2007.

[3] T. Predeep, World's first nano-material based water filter, April 6, 2007, <http://www.thehindubusinessline.com/blnus/34065012.htm> (accessed 09.11.07).

[4] S. Cosier, Big problems, little solutions, September 22, 2006, <http://scienceline.org/2006/09/22/env-cosier-nanotech/> (accessed 07.08.07).

[5] M. Haiken, Nanotech takes on water pollution, July 1, 2007, <http://money.cnn.com/magazines/business2/business2.archive/2007/07/01/100117050/index.htm> (accessed 29.11.07).

[6] Anonymous, Multiple benefits of nanotechnology encourages widespread uptake in water and water treatment, 2007, <http://nanotechwire.com/news.asp?nid=4282&ntid=&pg=43> (accessed 07.09.07).

[7] Meridian Institute, Overview and comparison of convention water nano-based treatment technologies, October 11–12, 2006, <http://www.merid.org/nano/watertechpaper/watertechpaper.pdf> (accessed 26.10.07).

[8] Meridian Institute, Workshop on nanotechnology water & development, 2007, <http://www.merid.org/nano/waterworkshop/assets/workshopsummary.pdf> (accessed 07.09.07).

[9] Anonymous, Carbon nanotubes may offer a cheap technique for desalination, May 22, 2006, <http://www.azonano.com/news.asp?newsID=2322> (accessed 09.11.07).

[10] L. Gallagher, Carbon nanotubes membrane filters fast, May 18, 2006, <http://nanotechweb.org/cws/article/tech/24922> (accessed 11.11.07).

[11] Anonymous, Controlling the movement of water through nanotube membranes, February 13, 2007, <http://www.nanowerk.com/news/newsid=1448.php> (accessed 09.11.07).

[12] Anonymous, Nano-material filters could cut cost of cleaning water, June 4, 2007, <http://www.channelnewsasia.com/stories/singaporelocalnews/view/280171/1/.html> (accessed 11.11.07).

[13] Anonymous, At $19 billion by 2010, filter cartridge market makeup to be one third industrial, July 27, 2006, <http://www.pennnet.com/Articles/Article.Display.cfm?Section=ARTCL&SubSection=Display&PUBLICATION.ID=41&ARTICLE.ID=261077> (accessed 09.11.07).

[14] M. Marshall, Turning to tech for cleaner water: Investors see big growth for firms with new ways to purify what we drink, May 30, 2006, <http://pionetics.com/news-charleston1.htm> (accessed 09.11.07).

[15] Anonymous, Canadian invites the world to pool its resources in clean water, November 18, 2003, <http://www.smalltimes.com/Articles/Article.Display.cfm?ARTICLE.ID=269104&P=109> (accessed 11.11.07).

[16] K. Jayaraman, Pesticide filter debuts in India, April 20, 2007, <http://www.rsc.org/chemistryworld/News/2007/April/20040701.asp> (accessed 11.11.07).

[17] A. Tsukioka, Royal Electric releases ozone micro-nano bubble water sterilizer, February 5, 2007, <http://www.japancorp.net/Article.Asp?Art.ID=14115> (accessed 09.11.07).

[18] Anonymous, Nanorust cleans arsenic from drinking water, November 13, 2006, <http://www.photonics.com/content/news/2006/November/13/85089.aspx> (accessed 09.11.07).

[19] B. Feder, Researchers find method for reducing arsenic levels, November 9, 2006, <http://www.nytimes.com/2006/11/09/science/09cnd-rust.html?ex=1320728400&en=58c07eb3d6c1803&ei=5088&partner=rssnyt&emc=rss> (accessed 29.11.07).

[20] M. Mathabatha, D. Naidoo, A review of public participation in the rural water and sanitation setting, January 1, 2004, <http://www.fwr.org/wrcsa/1381104.htm> (accessed 12.11.07).

CHAPTER

How Can Nanotechnologies Fulfill the Needs of Developing Countries?

39

David J. Grimshaw[1], Lawrence D. Gudza[2] and Jack Stilgoe[3]

[1]*Royal Holloway, University of London, Egham Hill, UK* [2]*Practical Action, Bourton-on-Dunsmore, Rugby, UK* [3]*University College London, London, UK*

39.1 Nanotechnologies and developing countries

We live in a rapidly changing world. Technological advances are increasing productivity and income, quality of life, and life expectancy, in the developed world, that is. The truth is that technological development is focused on meeting the wants of rich consumers. Scant attention is paid to the vital needs of people in the developing world. Each new technology that comes along tends to result in a wider gap between the rich and the poor in the world. Yet some innovations fail to be applied in developing countries where there is the need. The founder of the Intermediate Technology Development Group (now known as Practical Action) observed that: "new technologies are developed only when people of power and wealth back the development" [1]. The challenge is to ensure that nanotechnologies are applied to areas of need in developing countries.

Porritt [2] has argued that to enable sustainable development we need to work with the market system and not against it. This means understanding the market mechanisms, understanding the innovation processes, and then working with the

Street, Sustich, Duncan and Savage. Nanotechnology Applications for Clean Water, 2nd Edition.
 DOI: http://dx.doi.org/10.1016/B978-1-4557-3116-9.00039-1

key stakeholders to enable business models that will deliver on human need rather than on consumer wants. With existing technologies this becomes a challenge because the business models, including the supply chain logistics, are already well established. In the case of new technologies there is a window of opportunity before products are released into the market to negotiate new business models.

In a global economy, many topical issues—for example, sustainable development, climate change, and democracy—are all influenced by the role of science and technology in society. A major challenge is to release public value from science and technology and to channel that public value into developing countries to help reduce poverty [3]. The concept of public value used here refers to value generated by science and technology that is not solely reaped by the market. The central topic of releasing public value from science in a global context is one of the most significant and challenging issues facing societies throughout the world today.

Low-income countries are not only poor in terms of measures of human well-being but are also poor in terms of indicators of technology. They spend a small proportion of GDP on research and development: less than 1 percent compared to high-income countries that spend around 2.5 percent. The number of scientists in low-income countries is less than 50 per 100,000 people compared to over 3,000 in high-income countries. Technology has failed to meet the needs of the poor, with 1.2 billion people living on less than 1 US$ per day. At the centre of these deliberations is the essence pointed out by Sachs [4] that "the single most important reason why prosperity spread, and why it continues to spread is the transmission of technologies and the ideas underlying them."

The challenge faced might be reframed as being one of "how do we enable nanotechnologies to deliver products which fulfil human needs rather than consumer wants?"

39.2 How can nanotechnologies deliver public value?

The role of technology in development is perhaps even more important in the new century than it was in the last. In the era of globalisation, new technologies are rapidly reshaping the livelihoods and lifestyles of people throughout the world. The pace of technological change is increasing, and is beyond the capacity of society to understand and regulate its impacts—even when the implications are profound and far reaching, as is the case with nanotechnologies.

Most scientific and technological research is now in the private sector, producing research for Northern wants rather than Southern needs. Small-scale farmers and the informal sector give little attention to small-scale technological innovation.

Knowledge and communication-based industries are rapidly reshaping the global economy. Many believe that these trends are contributing to a new "knowledge divide" between the information-rich and the information-poor. There is an increasing sense of urgency—in the North and in the South—over the need to regain control over the ways nanotechnologies are developed and used. It is not recognised widely enough that the poor are able to innovate themselves, and innovations arising from developing countries need to be increasingly recognised and supported.

Traditional views of technology that rely on a linear model of innovation and diffusion are not appropriate to programmes that aim to respond to new technologies [5]. The predominant traditional view has been based on technological determinism. As Winner [6] suggested, "the adoption of a particular technical system requires the creation of a particular set of social conditions as the operating environment of that system." Such thinking leads to a technological push philosophy as embodied in the motto of the 1933 Chicago World Fair, "Science explores, technology executes, man conforms" [7]. The worldview on which this philosophy is based is predominantly "Northern," where the power is vested in global enterprises with large research and development budgets and where markets have developed to approximate to monopoly conditions. An example of this is the domination of Microsoft in the market for software. Practical Action views technology as not only meaning the hardware or technical infrastructure, but also the information, knowledge, and skills that surround it, and the capacity to organise and use these.

Thinking from science and technology and the international development traditions were brought together in a recent book edited by Leach, Scoones, and Wynne [8]. The traditional deficit model of public engagement was criticised and a number of themes were articulated, including the issue of how risk is framed and communicated. Wilsdon et al. [3] call for the direction of science to be built on notions of public value. The notion of public value raises issues about equity, efficiency, and the very purpose of science. Those concerned with the ethics of development [9] and the philosophy of science are also making valuable contributions to this debate.

An alternative view of technology is required. Grove-White et al. [10] have suggested that technologies need to be seen as social processes. This alternative view must recognise the role of the user (Southern poor) and the context provided by the cultural and political environment in which the user is based. The distinction being made here has been labelled "technology in use" by Edgerton [11] who argues that the historical emphasis on technology innovation is misleading. Much technology that is in use in the world is adapted or imitated rather than innovated.

The quest to ensure that all people have access to clean drinking water is now enshrined in the Millennium Development Goals. Often approaches to providing water for poor communities have been driven either by economics or by technology. The economics route might typically centre on the importance of regulations, institutions, and open markets whereas the technology approach might focus on designing a water pump, filter system, or novel application of nanotechnology. Yet we know that the technology for providing clean water has been known about and in use for thousands of years (e.g., the Romans around 300 BC). Failure to solve the issue might also be seen as a cultural or indeed political or managerial problem.

39.3 Nanodialogues in Zimbabwe

In 2006, researchers from Demos, Practical Action, and the University of Lancaster collaborated on a process designed to engage Zimbabwean community groups and scientists from both the North and South in debates about new (nano) technologies [12].

The dialogue was one of four experiments, collectively referred to as the nano-dialogues, in public engagement with nanotechnologies, funded by the Office of Science and Technology's "Sciencewise" programme. Sciencewise was created to foster interaction between scientists, government, and the public on impacts of science and technology.

Governments, companies, and NGOs are all talking about nanotechnology as "The Next Big Thing." Alongside the promise of new worthwhile opportunities comes uncertainty about risks, ethics, and the benefits to those people who are too often left out of conversations about the ends of technology—the poor. The potential benefits of the applications of nanotechnologies in developing countries are exciting. But the conversation linking the needs of people in developing countries to the resources and scientific knowledge of researchers around the world needs to be nurtured.

Epworth is a suburb of Harare, but it feels rural. It is just outside the Harare city limits, which means it is cut loose from the support of the city. In 2005 it was the scene of some of the harshest of the slum clearances that formed Robert Mugabe's "Operation Murambatsvina" ("Drive Out Trash"), which left thousands homeless. It is framed by outcrops of rock that have been worn away to resemble meticulously stacked balls. The balancing rocks are famous—they appear on the 10,000 dollar banknote. In the distance, you can see the electricity pylons of Harare's suburbs. But the telegraph poles around Epworth carry no cables. Plans for electricity and telephone lines were abandoned before completion.

Epworth gets its water from a combination of shallow wells and springs (Figure 39.1). The water brought up from the well looks clean enough, but with the pollution from the city, it's impossible to tell what it contains. "We're supposed to check," shrugs our guide, who acts as one of the community leaders. Nearby, a new well is being created. At the bottom of a six-metre pit, a man is filling a bucket with wet sand. His colleagues pull up the bucket and pile the sand around the pit's edge. It has taken two days so far, and will take another three. Then they need to seal it and put a lid on it. The well is next door to a pit latrine. It is far from ideal, which is why new sanitation methods are so important. Though Epworth is cut off, it is near enough to the city to be cramped. There is little space, and the well needs to be dug where there is water.

Any conversation about technology in Epworth has to start from here. In Zimbabwe, there is a headline context—a failing state and an economy that is both shrinking and sliding out of control—and there is an everyday context. In this everyday context, the idea of nanotechnology on its own is not likely to

FIGURE 39.1

Spring water, Epworth, Zimbabwe.

generate excitement. Ask what technologies people would like to see to help them get clean water and they mention rope-and-washer pumps, which replace disease-ridden open wells, and can be made and fixed using old tyres.

People in the developing world don't have much of a voice in science and technology. They are less likely to enjoy the benefits of new technologies and more likely to suffer from their downsides. The Royal Society and Royal Academy of Engineering [13] took issue with the sweeteners often offered to the developing world by nano-marketeers:

> *Much of the "visionary" literature ... contains repeated claims about the major long-term impacts of nanotechnologies upon global society: for example, that it will provide cheap sustainable energy, environmental remediation, radical advances in medical diagnosis and treatment, more powerful IT capabilities, and improved consumer products ... However, it is equally legitimate to ask who will benefit and, more crucially, who might lose out? ... Concerns have been raised over the potential for nanotechnologies to intensify the gap between rich and poor countries because of their different capacities to develop and exploit nanotechnologies, leading to a so-called "nano-divide."*

Other contributions, such as the Meridian Institute's "Global dialogue on nanotechnology and the poor," have stimulated wider discussion about possible benefits. One academic study, collecting the insights of people thinking about nano and development, concluded that the top three applications are energy, agriculture, and water. For our second experiment, we chose to explore the relevance of nanotechnology in the provision of clean water. Demos worked with Practical Action, the development NGO, which for the past 40 years (under its former name of the Intermediate Technology Development Group) has been making technology work for people in poor countries. Its vision is of appropriate, usable, sustainable technologies, driven by human needs rather than markets.

In Harare, we put together a three-day workshop with local mushroom farmers, brick makers, and water scientists. The nonscientists were representatives of communities that work with Practical Action. Three were from Epworth and three from Chakohwa, a rural community near Chimanimani, in the mountains of eastern Zimbabwe. The scientists were from government agencies, universities, and charities. The participants named our workshop Nanokutaurirana, a Shona neologism meaning "Nanodialogue." But for the first day and a half, the word nanotechnology was not mentioned. We wanted people first to define what the problem was.

Their description of the problem had multiple roots. Water is a market commodity, it is unaffordable, it is scarce, it is a long way away, and the responsibility for collecting it normally falls to women and girls. Where wells exist, they are crammed next to latrines and difficult to seal off from contamination. Near Harare, in addition to a recent cholera outbreak, there is chemical pollution from factories.

Away from the city, the rural community reported that water was contaminated by natural salt deposits. By the end of day one, we had a rough map of the issues and the connections between their social, technical, and political dimensions. The more the problem came into view, the further removed nanotechnology seemed as a solution. The community representatives had been let down in the past by well-intentioned technologies. Water pumps had arrived with instructions in English or German. When handles had broken or filters had clogged, they had been unable to find the parts or the expertise to fix them. As one of the community representatives asked, "When the NGO goes away, who has the knowledge to run and maintain their technology?"

In recognition of these characteristics of the problem domain we took a systemic approach. Many complex problems in science, engineering, or indeed other fields have some characteristics in common. Hard systems approaches have sometimes failed, for example, in the case of the Challenger disaster in 1986 when the space shuttle exploded moments after take-off killing all seven crew. Was this an engineering failure or one of managerial or political failure? McConnell [14] says the emphasis at NASA had shifted from technological considerations to managerial, commercial, and political ones. This is a good illustration of how the way we frame problems affects the outcome in terms of the activities that take place to solve the problem situation. Two lessons are taken from this story: first that in complex problem situations a systemic approach has proved worthwhile; and second that "what in fact made the situations ill-defined was that objectives were unclear and that both what to do and how to do it were problematical" [15]. The dialogue took a soft systems approach, which can be depicted at its simplest level as shown in Fig. 39.2. The essence of the soft systems approach is that it allows a natural dialogue to take place with the facilitators using the methodology to capture and keep in a systematic way the outputs of each session.

The problem situation was captured during the workshop held in Zimbabwe. Before the workshops a root definition and CATWOE were conceptualised, ready to be tested with the real dialogue during the first two days of the workshops. Figure 39.3 gives an example of this kind of output, for reference.

Our approach was to build on Practical Action's experience of engaging people in developing countries in debates about new technologies.

Figure 39.4 depicts the problem situation in the form of a rich picture. During the first day of the workshop this rich picture was drawn by the organisers as a reflection of the problem presentation. The idea of the rich picture very simply is that it can convey relationships and connections much more clearly than prose.

In the problem situation identified there were several subsystems. The model in Fig. 39.5 illustrates the three subsystems. Figure 39.6 shows some possible interactions between these three subsystems, with each subsystem being shown in a different colour. The conceptual model shows a set of activities that would realise the root definition.

Academics might ponder such questions as: Is the methodology any good? Does it work? But in the "real world" most people recognise the need to find a methodology that works for them and get on with it.

If a reader tells the author, "I have used your methodology and it works," the author would have to reply "How do you know that better results might have

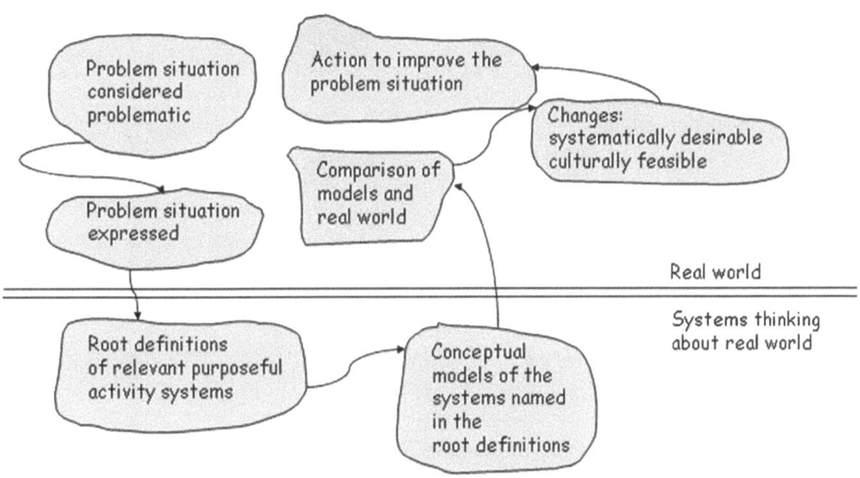

FIGURE 39.2

Soft systems methodology: overview.

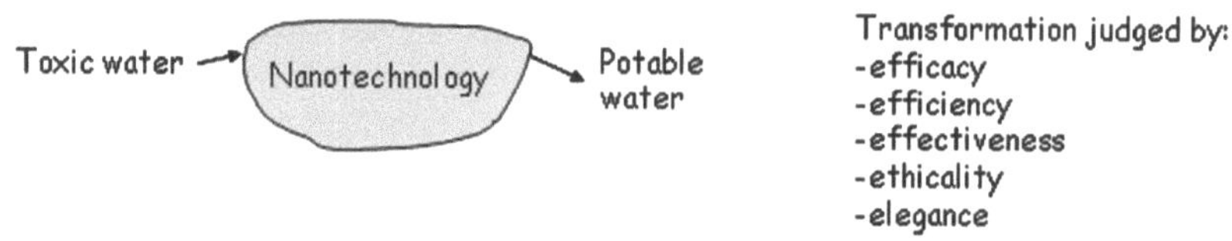

C	customers	The poor rural community
A	actors	Practical Action NGO
T	transformation	From toxic water to potable water
W	worldview	Technology can help to reduce poverty
O	owner	Northern scientists
E	environmental constraints	Low cost solution

Root definition:
The provision of clean drinking water for a poor rural community by Practical Action using a low cost product of nanotechnology that has been developed by Northern Scientists.

FIGURE 39.3

Capturing the problem situation.

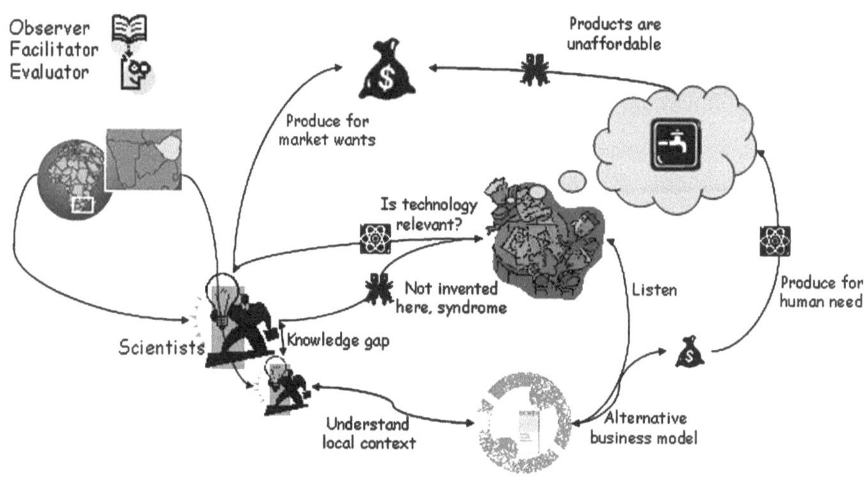

FIGURE 39.4

Rich picture of the problem situation.

been obtained by an ad hoc approach?" If the assertion is: "The methodology does not work," the author can reply, ungraciously but with logic, "How do you know the poor results were not due simply to your incompetence in using the methodology?" [16]

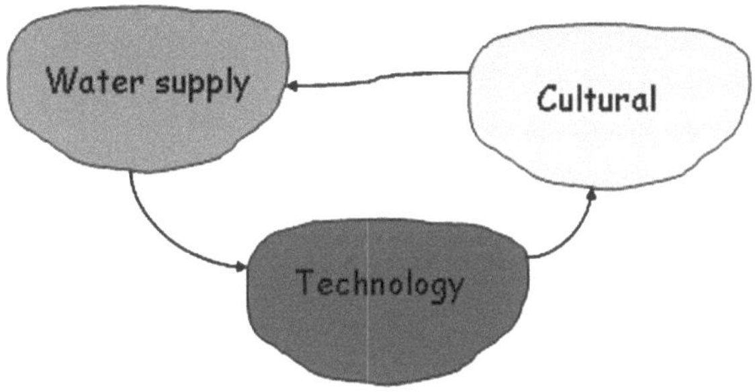

FIGURE 39.5

Subsystems of the problem situation.

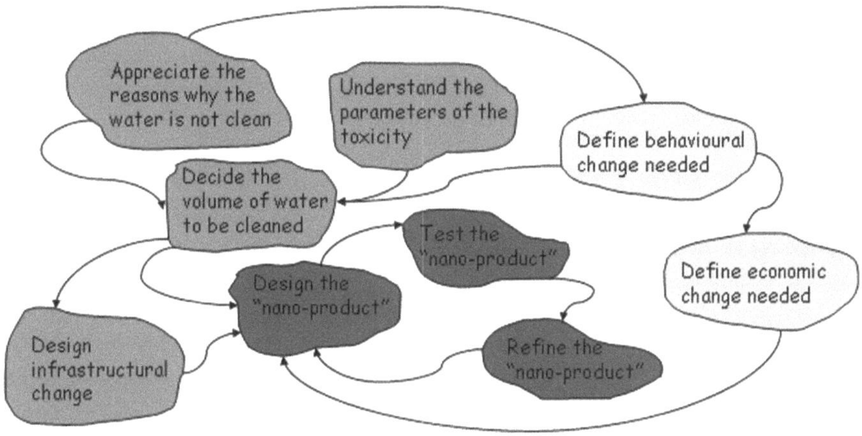

FIGURE 39.6

Conceptual model.

For these communities, local technology was not a matter of pride; it was a matter of what worked. The system that shapes the problem needs to map onto the system that provides the solution. So the rope-and-washer pump makes sense. It is not so much a thing as a system. It is not owned or sold by any one company and it is flexible enough to fit different societies. The participants were well aware that, as one put it, "All these new technologies are old in other countries."

As the historian David Edgerton [11] describes, whereas the West obsesses about the increasing "pace of innovation ... most change is taking place by the transfer of techniques from place to place." Technological systems—the way things are used, abused, and controlled—are political. There are reasons why they end up the way they do, and there are ways in which we can talk about better or worse technologies. We can judge new technologies according to the extent to which they lock people into certain systems (as, e.g., GM crops and centralised nuclear power do) or provide an open platform for new sorts of use (e.g., Linux or micro-renewable energy).

In our first experiment, with the Environment Agency, our participants exposed the politics of technology by talking about issues of detectability (whether we will be able to find nanoparticles once we release them) and reversibility (whether we will be able to backtrack). They realised that, even after we understand the effects of nanoparticles in a lab, when we release them, we will know much less about their impacts. So innovation becomes experimentation as technological systems become bigger and more complicated.

Technologies carry with them some definition of social need and some promise of a technical fix. They define both a problem and a solution. And the systems of research, innovation, and regulation of which they are a part can harden this definition. So whereas in the United Kingdom we may take the system—transport, maintenance, markets, and a stable economy—for granted, in Epworth, this needs close scrutiny. Rather than starting from the technology, we need to start from the local context and think about alternatives.

Edgerton [11] argues that the politics of new technologies have tended to narrow down consideration of alternatives. "Alternatives are everywhere, though they are often invisible." Public discussion reveals these alternatives. Technologies do not force people to do things, but as they open new doors, there is a danger that old ones can close. Whereas good intentions are focused on nano and development, they may lose sight of what else can more easily benefit poor people. In the course of our public engagement work, as we have reflected the public context of nanotechnology back to institutions, we are often asked whether public concerns are specific to nanotechnology or more general. Our response has usually been that nanotechnology is currently a good place to start the conversation, but that the sentiments speak way beyond this. In Zimbabwe, the issue of whether we should be talking about nano at all never seemed more pressing.

Halfway through our workshop's second day, we introduced nanotechnology. Luckily, a few weeks before, Hille et al. [17] had provided some examples of nano-products working in developing countries. Their report was careful to point out that the diffusion of these technologies was some way off, but it provided some examples of nanoscale water filters working in South Africa.

Our participants understandably shared little of the West's excitement for nanotechnology. Even for the scientists there was little prospect of riding nanotechnology's funding wave. So the group asked about applicability, alternatives, environmental impact, cost, maintenance, and the capacity to manufacture and maintain the technology locally. They asked whether these technologies were fixed or adaptable for local needs, whether they would mean an increase in employment for Zimbabwean scientists, or greater reliance on the West. And they asked at what scale these could be used. Were they the sorts of filters that would be used centrally, at government treatment plants, or could they be put in schools and controlled by communities? The experiment revealed the huge gulf between research and diffusion. We began to see the steps that need to be taken to connect innovation to human need in a place like Epworth.

The temptation is to see the problem as intractable, to say that science has nothing to offer and that Zimbabwe needs to provide solutions for its problems. But this would deny the huge potential that exists for constructive collaboration. Around the world, there are efforts under way to direct emerging technologies towards pressing human needs. A more positive approach might ask how these efforts might yield greater benefits.

As a step towards this, we asked our Zimbabwean participants to produce a set of recommendations for U.K. scientists. They concluded that innovation does need to point in a different direction, but collaboration can be hugely positive "when there is a story to tell"—that is, when it starts with some concrete benefit in mind. The Zimbabwean scientists recognised that in many cases, given the asymmetry of resources, Western scientists would have to lead research, but this research should recognise the value of local knowledge and work to empower Southern scientists and build their capacity. They also recommended immediate steps that could be taken, such as opening up access to all scientific journals.

Back in the United Kingdom, we went to visit Mark Welland in Cambridge. We were keen to see what lines could be drawn between the needs of people and science, to stretch the connection back to the research base. Welland runs Cambridge University's Nanoscience Centre, but he is also codirector of the Yousef Jameel Science and Technology Research Center at the American University in Cairo. His research team is driven by scientific curiosity, but he encourages his colleagues to reflect on public value as part of their work. At one of his centre's nanoscience seminars, we told the scientists about our time in Zimbabwe and asked for their thoughts and questions.

In Zimbabwe, scientists saw community participation as a vital, if hugely complicated, part of what it means to do good science and engineering. In the United Kingdom, systems work against community or public engagement. Talking to the young researchers at the Interdisciplinary Research Collaboration (IRC) in Cambridge about our experiment, it became clear that many of them have an appetite to use their skills to contribute to human needs. But advancing in their scientific career often feels like a routine progression through certain stages in which they have "no control over their own research projects, social impacts or otherwise."

At the moment, the gravitational pull for these scientists is towards certain sorts of innovation—marketable technologies or a narrow definition of world-beating basic research. We need a broader understanding of innovation, which places greater value on the needs of people in the developing world. The young scientists in Cambridge recognised the scale of the challenge that this poses to established systems, but were unsure how to continue the conversation and change things from inside.

39.4 Balancing risk and opportunity

Bürgi and Pradeep [18] observed that "the convergence of the newly emerging technologies of the twenty-first century has the potential to revolutionize social and economic development and may offer innovative and viable solutions for the most pressing problems of the world community and its habitat. However, a better understanding of the potential benefits and hazards of nano-scale science and technology is essential because it will provide policymakers with better tools to take responsible choices."

Research on the ethical, legal, and social implications is, according to Mnyusiwalla et al. [19], lagging behind the science. As evidence for this view they quote the low number of citations in the literature and the fact that in the United States, there are research funds available that are not being used. For example, the National Nanotechnology Initiative allocated US$16–28 million to social implications but spent less than half that amount. One of the main reasons quoted for the lack of awards was the paucity of good quality proposals.

In our society, where there is a risk, the insurance industry will be key players in identifying and analysing that risk. The Benfield Group (Benfield is the world's leading independent reinsurance intermediary and risk advisory business) concluded their assessment by stating that "the industry's current focus is on risk management and containment, for example in the manufacture and transportation of nanotechnology" [20]. A study by Munich Re (the largest re-insurance company) echoing the emphasis on risk management states, "up to now, losses involving dangerous products were on a relatively manageable scale whereas, taken to extremes, nanotechnology products can even cause ecological damage which is permanent and difficult to contain" [21].

An increased risk potential of nanotechnology, according to Schmid [21], can arise from:

- New types of loss arising from new material properties such as magnetic fluids.
- Increase in major claims.
- Liability cases arising out of changing legislation designed to protect the wider environment.
- Adverse sociopolitical effects and irreversible ecological damage.

Materials fabricated on the nanoscale have properties that are different from those that are manufactured at a normal scale. For example, the precise way in which the atoms are arranged often leads to unusual optical and electrical properties. Carbon at the nanoscale can conduct electricity better than copper. In other cases the small size may have the effect of being more toxic than normal. A distinction can be made, in terms of risk assessment, between active and passive nanoparticles. Passive particles, such as a coating, are likely to present no more or less a risk than other manufacturing processes according to French [20]. However, she goes on to assert that in the case of active nanoparticles, their ability to move around the environment leads to risks associated with control and containment.

In the United Kingdom, nanotechnology is being seen as an opportunity to have an earlier and more open debate about emerging technologies, to avoid the antagonism and distrust generated with genetically modified (GM) foods. The government is supporting the Royal Society and Royal Academy of Engineering's call for "a constructive and proactive debate about the future of nanotechnologies . . . at a stage when it can inform key decisions about their development and before deeply entrenched or polarized positions appear" [13]. The nano-dialogues are a set of opportunities for early public debate. One of these aims to engage communities in Zimbabwe in discussions about emerging technology.

Views about the relevance of application areas for poor people converge on two sectors, namely, water and energy. These were the sectors, according to an international group of experts convened by the Meridian Institute to advise a Rockefeller project, thought to be where applications of nanotechnology are likely to bring potentially beneficial products that could offer solutions for poor people. According to one recent study, the top three applications that would help developing countries are energy storage, production, and conversion; agricultural productivity enhancement; and water treatment [22].

We chose water treatment as a focus for our dialogue. First, in development terms it is a well-established priority. Second, technology is at a stage where it may be able to make a significant contribution to filtration and decontamination. The Millennium Development Goal is to halve the proportion of people without sustainable access to safe drinking water and basic sanitation by 2015. Our dialogue sought to introduce the views and values of people for whom clean water is an everyday problem into debates about possible technical solutions. By involving scientists who are engaging in leading research, we can move the debate upstream. We hope one of the outcomes will be a sustained dialogue between scientists and end users that enables new technology to deliver on human needs rather than be driven by market wants.

39.5 Future directions

The emergence of nanotechnology has coincided with greater openness in science and innovation policy. For government, public engagement has become a way of avoiding a repeat of past mistakes. Depending on who you ask, nanotechnology might be the Next Big Thing, the Next Asbestos, or the Next GM. But before its impacts have been felt, nanotechnology has become a test case for a new sort of governance. It is an opportunity to reimagine the relationship between science and democracy. For public engagement to matter it must go beyond risk management. New conversations with the public do not provide easy answers. They ask difficult but important questions, and take us into a vital discussion of the politics of science.

The concept of new technology presents many challenges to those concerned with how it can be used to reduce poverty in the world. The promise of many new technologies has been high yet their ability to deliver sustainable change in the lives of poor people has been limited. At the same time the very models and assumptions underpinning much of international development have been economic growth. The essay presented a case for using a new paradigm based on enabling choices to be made that fulfil the needs of people. This requires a move away from the old paradigm, which is supply driven, delivering products to a market at a price that will maximise profits for the owners of the intellectual capital.

The dialogues held in Zimbabwe are one small step towards this new paradigm. They connect the needs of poor people with scientists who are in the process of developing new applications of nanotechnologies. The next step will be to move beyond dialogue towards the engagement of the scientist with relevant stakeholders in developing countries.

References

[1] E.F. Schumacher, Good Work, Jonathan Cape, London, 1979.

[2] J. Porritt, Capitalism as if the World Matters, Earthscan, London, 2006.

[3] J. Wilsdon, B. Wynne, J. Stilgoe, The Public Value of Science: or How to Ensure that Science Really Matters, Demos, London, 2005.

[4] J.D. Sachs, The End of Poverty, Penguin Books, London, 2005.

[5] E. Dantas, The system of innovation approach and its relevance to developing countries, SciDevNet [Internet] Policy Briefs, April 2005 [cited April 26, 2005], Available at <http://www.scidev.net/dossiers/index.cfm?fuseaction=printarticle&dossier=13&policy=61>.

[6] L. Winner, The Whale and the Reactor, University of Chicago Press, Chicago, 1986.

[7] N. Fox, Against the Machine: The Hidden Luddite Tradition in Literature Art and Individual Lives, Shearwater Books, Washington D.C., 2002.

[8] M. Leach, I. Scoones, B. Wynne, Science and Citizens: Globalisation and the Challenge of Engagement, Zed Books, London, 2005.

[9] D. Gasper, The Ethics of Development, Edinburgh University Press, Edinburgh, 2004.

[10] R. Grove-White, P. Macnaghten, B. Wynne, Wising Up: The Public and New Technologies, Research Report, Centre for the Study of Environmental Change, Lancaster University, Lancaster, 2000.

[11] D. Edgerton, The Shock of the Old: Technology and Global History Since 1900, Profile Books, London, 2006.

[12] D.J. Grimshaw, J. Stilgoe, L. Gudza, The role of new technologies in potable water provision: a stakeholder workshop approach, Rugby: Practical Action [Internet] 2006, Available at <http://practicalaction.org/docs/ia4/nano-dialogues-2006-report.pdf> [cited August 15, 2007].

[13] Royal Society and Royal Academy of Engineering, Nanoscience and Nanotechnologies: Opportunities and Uncertainties, London, Royal Society, 2004.
[14] M. McConnell, Challenger: A Major Malfunction, Unwin Hyman, London, 1988.
[15] P.B. Checkland, J. Scholes, Soft Systems Methodology in Action, John Wiley, Chichester, 1990.
[16] P.B. Checkland, Towards a systems based methodology for real world problem solving, J. Syst. Eng. 3 (2) (1972) 87–116.
[17] T. Hille, M. Munasinghe, M. Hlope, Y. Deraniyagala, Nanotechnology, water and development, Meridian Inst. (2006) [Internet]; available at <http://www.merid.org/nano/waterpaper> [cited July 5 2006].
[18] B.R. Bürgi, T. Pradeep, Societal implications of nanoscience and nanotechnology in developing countries, Curr. Sci. 90 (5) (2006) 645–658.
[19] A. Mnyusiwalla, A.S. Daar, P.A. Singer, Mind the gap: science and ethics in nanotechnology, Nanotechnology 14 (2003) 9–13.
[20] A. French, Nanotechnology: New Opportunities, New Risks, Benford Group, London, 2004.
[21] G. Schmid, Nanotechnology: What is in Store for Us? Munich, Munich Re Group, 2002.
[22] F. Salamanca-Buentello, D.L. Persad, E.B. Court, D.K. Martin, A.S. Daar, P.A. Singer, Nanotechnology and the developing world, PLoS Med. 2 (4) (2005) 300–303.

CHAPTER

40 Challenges to Implementing Nanotechnology Solutions to Water Issues in Africa

Mbhuti Hlophe[1] and Thembela Hillie[2,3]

[1]Department of Chemistry, North-West University (Mafikeng Campus), Mmabatho, South Africa [2]National Centre for Nano-Structured Materials, DST/CSIR Nanotechnology Innovation Centre, Pretoria, South Africa [3]Physics Department, University of the Free State, Bloemfontein, South Africa

40.1 Introduction

Africa is the most water-stressed continent in the world and it has already been predicted that there will be a water crisis by 2025 on the continent [1]. Two-thirds of its surface area is affected by aridity to some degree [2]. The scarcity of the water will lead to competition for this finite resource which will in turn result in conflict or cooperation among the various stakeholders. The water situation is so bad that only about 60% of Africans have access to safe drinking water [3]. Moreover, 60% of the people in two of Africa's large cities (Lagos in Nigeria and Nairobi in Kenya) have no running water. The scarcity of water in the continent can be attributed to, among other factors, desert encroachment, recurrent drought, and high population growth. South Africa is confronted with similar problems,

Street, Sustich, Duncan and Savage. Nanotechnology Applications for Clean Water, 2nd Edition.
 DOI: http://dx.doi.org/10.1016/B978-1-4557-3116-9.00040-8

namely water quantity and also quality. The water quantity problem is a direct result of accelerated population growth, industrialization, and the consequent urbanization. The net effect is an ever-increasing demand for treated water. The effluent from the treatment of municipal, industrial, and mine wastewater is discharged into raw water sources, for instance, rivers and dams. The quality of the raw water deteriorates progressively and this impacts negatively on the quality of the treated water. The need for more treated water can only be met by expanding the existing conventional water treatment works. However, the removal efficiency of some organic pollutants (for instance, pesticides, endocrine disruptors, and solvents) by the conventional water treatment method is not satisfactory. This then necessitates the application of an alternative technology, nanotechnology, to solve the problem [4]. The challenges which are encountered in the implementation of water treatment by nanotechnology will be considered.

40.2 Community involvement or ownership

The implementation of nanotechnology in the improvement of water quality will generally be based on a study that was carried out in South Africa by Hlophe and Venter [5]. The researchers used nanostructured membranes whose pore sizes were less than two nanometers. They tested a nanomembrane technology unit for the removal of nitrate, chloride, fluoride, sulphate, calcium, and magnesium ion pollutants from groundwater, and monitored rural consumer knowledge and attitude to water purification.

The success in the implementation of a nanotechnology water treatment project depends on the attitude that the researchers adopt toward the recipient community. The community members have to be involved in all the stages of the project, i.e., from planning up to the implementation stage [6]. The local government or implementing agency should therefore ensure that the community members are part of the decision-making process since they are the end users or consumers of the service. It was found that when the community members played an active role in the implementation of the nanotechnology, the sustainability of the project was assured. On the contrary, other researchers discovered that where the community members were passive stakeholders, the service delivery was not sustainable [5].

Furthermore, the successful implementation of the application of nanotechnology for the improvement of water requires a change of the mindset or attitude of the community members or consumers toward the new technology. The consumers or end users have to be made aware that nanotechnology is a relatively new technology which has wide-ranging benefits for society. However, they should also be aware that there are concerns with respect to its impact on human health and the environment. This implies that the implementation of nanotechnology for water treatment should be accompanied by a well-designed educational program that will educate and inform the consumers about the operation and maintenance of the technology [6]. Providing the consumers with the necessary knowledge results in the consumers taking ownership of the technology [7]. The implementation of the nanotechnology is a joint effort by various stakeholders (for instance, the consumers, researchers, and engineers) for the realization of a common goal. The constant interaction among the stakeholders in their quest to solve the common problem results in attitude change or transformative learning [6]. The latter leads to the development of a permanent change that would ensure the acceptance of the nanotechnology for a sustainable service delivery.

40.3 Community need for the technology

A community can only accept the implementation of a technology if there is a need for it. Hlophe and Venter [5] successfully implemented a nanomembrane brackish groundwater treatment project in the North West Province of South Africa at Madibogo village (a settlement where there is a very poor infrastructure and very little economic activity). This technology was chosen because of its low cost, and because it is easy to operate and maintain. The province is situated in a semiarid region. The Madibogo village community depends solely on untreated groundwater for their livelihood. The groundwater is polluted with nitrate, chloride, calcium, and magnesium ion pollutants. Nitrate ion can be reduced to the toxic nitrite ion which causes methaemoglobinemia in infants of 0–6 months. Nitrite ion can react with amino compounds to form nitrosoamines which are strongly carcinogenic [8]. Calcium and magnesium ions cause water hardness and also pose health risks to body organs such as kidneys, liver, and eyes. The chloride ion is corrosive to metal pipes and harmful to agricultural plants. It imparts a salty taste to water above 300 ppm. A stakeholder meeting was convened at which the health and economic implications of the polluted water was discussed. It was agreed that nanomembrane water treatment was the appropriate technology for the treatment of the water source.

Three nanofiltration (NF) membranes (Desal-DL, NF 90, and NF 270) and two reverse osmosis (RO) membranes (BW 30 and S5) were tested for the removal of the pollutants from the brackish groundwater using a cross-flow module water treatment plant. The South African National Standard (SANS-241) for drinking water was used to determine the water quality of both the raw and treated waters and is given in Table 40.1 for the determinants in the study. The SANS-241 Class 1 water quality specifications were used for assessing water quality in the study. The composition of the brackish groundwater (raw water) or feedwater at Madibogo village is given in Table 40.2 and the composition of the water after treatment by or permeation through a nanomembrane (the permeate) is given in Table 40.3.

Table 40.1 SANS-241 Specification for Determinants in the Study

Determinand	SANS-241 Specification (ppm)
NO_3^-	10
Cl^-	200
F^-	1
Ca^{2+}	150
Mg^{2+}	70

Table 40.2 Composition of Brackish Groundwater at Madibogo Village

Determinand	Concentration (ppm)
NO_3^-	23.6
Cl^-	63.7
Ca^{2+}	176
Mg^{2+}	102

Table 40.3 Composition of Nanomembrane Permeate

Nanomembrane	NO_3^- (ppm)	Cl^- (ppm)	Ca^{2+} (ppm)	Mg^{2+} (ppm)
Desal-DL	9.00	235	65	33
NF 90	3.71	101	25	13
NF 270	10.8	187	61	26
BW 30	0.44	83	9	7
S5	0.21	14	3	1

The results show that the most optimal nanomembrane for the improvement of the water quality at Madibogo village is NF membrane NF 90. The two RO membranes (BW 30 and S5) are not suitable because they essentially remove all the calcium and magnesium cations (nutrients) which are required for the normal development and functioning of the human body. The consumers accepted nanotechnology because it addressed their need (Figures 40.1–40.3).

FIGURE 40.1

A picture of the NF pilot water treatment plant at Madibogo village in North West Province of South Africa. The feedwater tank is mounted on a stand and one of the remaining two tanks contains brine and the other purified water. The NF plant is housed in the facebrick shelter in the center of the three tanks.

FIGURE 40.2

Madibogo primary school learners observing the operation of the NF pilot water treatment plant.

FIGURE 40.3

Primary school learners at Madibogo village sharing purified water from the NF pilot water purification plant.

40.4 Community water quality monitoring

The successful implementation of a nanotechnology water purification project can be further enhanced if the community can monitor the quality of their water supply. One of the most important parameters or indicators of water quality is pH. The significance of water pH cannot be overestimated. It affects both animal and plant life. A pH of less than 3 and greater than 11 is generally fatal to aquatic life which finally affects the whole ecosystem. The variation of pH changes water chemistry with disastrous consequences. pH is one of the 11 parameters of the water quality index (WQI) [9]. This index only gives a general picture of the water quality. Water pH is influenced by effluents that are derived from, among other sources, agriculture, mining, and industry. The application of fertilizers and pesticides for agricultural production, the use of chemicals in industry, and mining sectors introduce a range of pollutants into the environment. The fate of the species derived from these inputs has to be monitored to ensure that the environment continues to support the ecosystem. The monitoring program for this diverse range of pollutants is obviously very expensive as it has to be conducted in well-equipped, specialized laboratories manned by skilled personnel.

Nanotechnology was investigated for the development of a cheap and affordable optical pH nanosensor for the monitoring of water quality. An organic conducting polymer (OCP), polyaniline (PANI), was selected for the study. The advantages of a PANI sensor include ease of preparation, high conductivity, and stability [10,11]. The feature which makes it particularly suitable for the detection of pH is its reversible color change. PANI is dark green in acidic solution (emeraldine salt) and dark blue in basic solution (emeraldine base). A PANI pH nanosensor is reusable as opposed to comparable methods such as a universal indicator and pH papers and has a life span of approximately 5 months [11].

40.5 Infrastructure

The success of technology transfer or implementation of any technology, nanotechnology in this case, depends on the availability of the collateral infrastructure. The infrastructural requirements can be diverse depending on the location where the technology is required. Some factors that should be considered when the implementation will be carried out in remote rural areas (in developing countries) where potable water is required could include such rudimentary basics as electricity and accessibility. A significant part of rural Africa does not have grid electricity and there are no alternative energy sources such as solar and wind power. The fact that most of the advanced technologies (for instance, nanotechnology) are developed in the developed countries assumes that they will mostly be operated on electricity. This alone severely limits the choice of technologies that can be utilized. The approach and the choice of technology strongly depend on the energy infrastructure that is locally available.

Technology transfer can only take place if the necessary equipment is available and for which technical support is required. The equipment is invariably sophisticated and costly, and can only be operated and maintained by highly skilled scientists, technicians, and engineers. The technical support in the water treatment project that was conducted by Hlophe and Venter [5] was provided by the research group from North-West University. The research group consisted of three chemists and a chemical engineer (professors), engineering technician, and three master's students. The technical staff (the professors and the engineering technician) were responsible for teaching the students and community members the operation and maintenance of the nanotechnology water treatment pilot plant. In fact, two community members, a teacher and a security guard at the school, were trained in the operation and maintenance of the nanotechnology water treatment pilot plant.

Finally, collateral on communications has to be considered for the successful implementation and adoption of a nanotechnology for improving water quality. Collateral communications is through established rural authorities, community societies, schools, and churches. The rural authority (the chief and his councilors) serves as the gatekeeper through whom the researcher is introduced to the community [6]. The protocol is thus to first obtain permission from the chief to hold meetings in his community. The information is then disseminated to the different sectors or forums of the community. These forums form lines of communication and are critical in informing the community about the proposed solutions to the water quality problem. This collateral communication ensures the buy-in by the community which should be achieved before the project can start in earnest.

40.6 Capacity development

The community at Madibogo village at which the water treatment nanotechnology was implemented has a very low literacy rate. The proportions of people who have secondary and tertiary education are, respectively, 19% and 2% [12]. It was therefore essential to develop capacity for ensuring the sustainability of the nanotechnology.

The students who provided technical support in the operation and maintenance of the nanotechnology water treatment pilot plant at Madibogo village [5] were developed into human capital, water technologists. They were trained in the principles of the nanotechnology for the treatment of brackish groundwater. The different aspects of the water treatment nanotechnology project were allocated to three master's students to investigate as part of their academic studies. Master's dissertations of two students were based on the removal of pollutant concentrations of some anions and cations from brackish groundwater. The third master's dissertation was based on the monitoring of rural consumer knowledge and attitude to water purification. The net result was the empowerment of the three students on the practice or knowledge of a given technology who would then subsequently apply it to solve similar water quality problems.

40.7 Improvements in quality of life

The implementation of the nanotechnology for water quality improvement results in the improvement in the quality of life of the end users with respect to social image, health, and economic issues. Rural communities in general, and the Madibogo community in particular, had some misconceptions about their water source [6]. They thought that the water caused patches on the face and also discolored the hair. The community was aware of the hardness of their water and used to soften it through the addition of paraffin, cement, or foam bath. The results of the implementation of the nanotechnology were that the community was educated about water quality and treatment or purification. The empowerment of the community with this information restored their self-esteem and thus their quality of life improved.

Health-related chemical determinants that are associated with groundwater in the South African provinces of North West and Limpopo are nitrate and fluoride ions. Pollutant concentrations of nitrate ions were readily removed by nanomembrane treatment (Table 40.3). Schoeman and Steyn [13] also used nanomembrane treatment for the removal of excess concentrations of nitrate ions and hardness from groundwater in a rural village in Zava in Limpopo Province (which shares borders with Botswana, Zimbabwe, and Mozambique). Fluoride ion pollution is very high in groundwater sources around Sun City (the Las Vegas of Africa) in North West Province of South Africa [14]. Fluoride ion concentrations that are greater than 1 ppm cause mottled teeth ("chocolate teeth") and concentrations that are greater than 4 ppm result in fluorosis [15,16]. Herbert [17] constructed and installed nanomembrane treatment plants at Western and Northern Cape provinces in South Africa, Botswana, and Namibia for the removal of pollutant concentrations of, mainly, sodium, calcium, magnesium, chloride, fluoride, and nitrate ions from brackish groundwater. South Africa has also experienced several outbreaks of microbial-related pollution in two of its provinces in the past 3 years. Two cholera outbreaks occurred in Mpumalanga Province (at Delmas), one in 2006 and the other in 2007. There has been at least one outbreak of diarrhea in North West Province at Bloemhof. These microbe pollutants can be readily removed by nanomembrane treatment which therefore assures an improved quality of life.

Nanomembrane treatment of brackish groundwater in implementation areas decreases costs that are caused by polluted water or water of poor quality. The water has to be boiled if it has to be used for drinking and for taking medicine for disinfection purposes. This contributes in increasing the energy bill for the economically challenged communities. Additives such as chlorine compounds (for instance, jik) are also used for disinfection of water of poor microbiological quality. The removal of water hardness means that there will be no longer any scale formation on elements of electrical devices such as kettles and geysers and this will lead to a reduction in energy consumption. Furthermore, the addition of extra soap or detergent to compensate for water hardness in laundering is not required when the water quality is good. The untreated water caused clothes to be stained and thus their quality deteriorated and had to be replaced relatively regularly. The hardships which were due to the hard water were readily removed by nanomembrane treatment and this resulted in an improved quality of life for the community.

40.8 Commercialization of nanotechnologies

The commercialization of nanotechnology is a huge problem in Africa and other parts of the developing world as much as it is in the developed world, due to the general issues associated with commercializing new technologies and more especially to the uncertainties linked with nanotechnology. The general issues include the market uptake, differentiation of the solution, and its existing cheap substitutes in the market and intellectual property (IP) positioning. To execute a winning commercialization plan, access to market research data is required, which is not readily available. This data would provide the size of the addressable market, the growth trends, competitive technologies, and potential substitutes of the technology. Technology development requires a local know-how and infrastructure to meet the demands of upscaling to address the available market. There is an obvious lack of these facilities in Africa and other parts of the developing world and it is expensive to outsource this task to foreign countries. Most of the developing countries do not have IP laws to protect the inventors and guard against illegal exploitation of the IP. Investors are very cautious and do not invest in unprotected technologies.

There is also a predicament in finding viable business models to capture value in technologies that are created from local research. The models include IP licensing, establishment of start-ups, and partnership with established companies for joint ventures. These models have key advantages and disadvantages regarding control, financial and collateral resources, and infrastructure [18]. IP licensing is relatively easy to execute if it is a compelling technology as this is usually taken up by reputable companies that have infrastructure and established distribution channels to manufacture, upscale, and dispense the technology. There is a lack of innovation-based enterprises (IBEs) in Africa and other parts of the developing world to take up sophisticated technologies to the market; therefore, commercialization still remains a huge challenge in implementing water-based nanotechnology solutions.

40.9 Conclusions

The successful implementation of a nanotechnology for the improvement of water quality can be achieved by addressing four factors: community involvement, community need, infrastructure, and capacity development. The active participation of the community in all the stages of project implementation results in the community taking ownership of the project. This is very important because it discourages vandalism and ensures the community operates and maintains the nanotechnology water treatment plant. The existence of a problem or need in water quality guarantees that any nanotechnology which has a potential to address the need will be favorably considered. The implementation of nanotechnology for water quality improvement can only become a reality when the necessary infrastructure is available. For example, in the South African case study, power supply, groundwater, equipment and skilled researchers, and technical backup were available. Finally, capacity development is vital for the implementation of nanotechnology for water quality improvement. The buy-in by academia, the custodians of the nanotechnology, is crucial as they transfer or impart it to the community.

References

[1] K.B. Showers, Water scarcity and urban Africa: an overview of urban-rural water linkages, World Development 30 (4) (2002) 621–648.

[2] P.A. Shaw, Geomorphology of the world's arid zones: Africa and Europe, in: Thomas (Ed.), Arid zone geomorphology, John Wiley & Sons, UK, 1977.

[3] J.C. Glen, T.J. Gordon, Global challenges (Chapter I), Millennium Project's 2002, State of the Future, American Council for the United Nations University, Washington, DC, 2002.
[4] C.F. Schutte and W. Focke, Evaluation of nanotechnology for application in water and wastewater treatment and related aspects in South Africa. Water Research Commission (WRC) Report no. KV 195/07. Pretoria, South Africa (2007).
[5] M. Hlophe and M. Venter, The testing of a membrane technology unit for the removal of nitrate, chloride, fluoride, sulphate, calcium and magnesium ion pollutants from groundwater, and the monitoring of rural consumer knowledge and attitude to water purification. Water Research Commission project K5/1529. Pretoria, South Africa (2008).
[6] U. Kolanisi, A South African Study of Consumers Perception and Household Utilization of a Rural Water Service, Master's thesis, North-West University, Potchefstroom campus, South Africa, 2005.
[7] T. Hillie, M. Munasinghe, M. Hlophe, Y. Deraniyagala, Nanotechnology, Water and Development, Meridian Institute, Washington, U.S.A, 2006.
[8] B.C. Challis, J.A. Challis, in: S. Patai (Ed.), Chemistry of Functional Groups. Supplement F. The Chemistry of Amino, Nitroso and Nitro Compounds and their Derivatives. Part 2, Chapter 26, Wiley & Sons, New York, 1982, pp. 1151–1223.
[9] P.S.P. Herrmann, Nanosensors applied to water quality: developing a low-cost sensor for natural water, and applications of other techniques, Int. J. Nuclear Desalination 4 (3) (2011) 218–229.
[10] E.C. Venancio, L.H.C. Mattoso, P.S.P. Herrmann, A.G. MacDiarmid, Line patterning of graphite and the fabrication of cheap, inexpensive, "throw-away" sensors, Sens. Actuators, B 130 (2008) 723–729.
[11] P.T. Sotomayor, I.M. Raimundo Jr., A.J.G. Zarbin, J.J.R. Rohwedder, G.O. Neto, O.L. Alves, Construction and evaluation of an optical pH sensor based on polyaniline-porous Vycor glass nanocomposite, Sens. Actuators, B 74 (2001) 157–162.
[12] Integrated Development Plan (2007) for Ratlou Municipality in the Central District Municipality of Ditsobotla District of North West Province of South Africa.
[13] J.J. Schoeman and A. Steyn, Defluoridation, denitrification and desalination of water using ion-exchange and reverse osmosis. Water Research Commission (WRC), Report No.: TT 123/00, (2000) ISBN: 1 86845 590 4.
[14] S.J. Modise and H.M. Krieg. Evaluation of nanofiltration for the treatment of rural groundwater for potable water use. WRC Report No.: 1230/04 (2004).
[15] Department of Water Affairs and Forestry, and Department of Health. A guide for the health-related assessment of the quality of water supplies (1996).
[16] S.K.Sharma, High Fluoride in Groundwater Cripples Life in Parts of India, Diffuse pollution conference in Dublin, 2003.
[17] N. Herbert, Malutsa (PTY) Limited, Paarl, South Africa, 2008.
[18] K.T. Hillie, MBA Thesis, Massachusetts Institute of Technology, Cambridge, USA, 2012.

CHAPTER

41 Life Cycle Inventory of Semiconductor Cadmium Selenide Quantum Dots for Environmental Applications

Hatice Sengül[1] and Thomas L. Theis[2]

[1]*Department of Environmental Engineering, Hacettepe University, Ankara, Turkey*

[2]*Institute for Environmental Science and Policy, University of Illinois at Chicago, Chicago, IL, USA*

41.1 Introduction

Advanced technologies to improve the quality of the environment require manufacturing and employment of a wide range of materials, chemicals, and equipment. Such improvements can be direct, for example, removal or transformation of one or more contaminants from air, land, or water, or indirect, in which contaminant emissions are reduced as new nano-based technologies are implemented, for example, more efficient energy production, transmission, or use. Yet, life cycle environmental impacts of these technologies may sometimes be overlooked. The same holds true for environmental applications of nanotechnology and determining the tradeoffs of adopting advanced technologies represents a significant challenge. Life cycle assessment is an effective method to compile and evaluate inputs, outputs, and potential environmental impacts of a system through its life cycle, carried out in four consecutive steps: goal and scope definition, inventory analysis, impact analysis, and interpretation [1].

Street, Sustich, Duncan and Savage. Nanotechnology Applications for Clean Water, 2nd Edition.
 DOI: http://dx.doi.org/10.1016/B978-1-4557-3116-9.00041-X

Current research regarding nanotechnologies applied to improve environmental quality is mainly dominated by studies focused on understanding the potential benefits and impacts of applications. The integration of environmental impact assessment into decision-making at the inception of a technology—during research, development, and design phases—presents significant opportunities for reducing wider life cycle impacts. To date, however, such research is almost nonexistent and the potential benefits that may be gained from such an approach remain largely unexplored. This is especially true for nano-based manufacturing methods, for example, there are many alternative methods for synthesizing nanostructured materials and products with nanoscale features. The range of alternative precursors and solvents that can be employed for nanomanufacturing is also extensive (see, e.g., Cushing et al. [2] for a review of alternative liquid phase synthesis methods for nanostructured materials). The environmental impacts of these alternative pathways are likely to vary considerably. There is also a need to address comparative assessment of manufacturing of alternative products (both nano and micro/macro) on an application basis.

Nanostructured materials and products with nanoscale features can be produced from macroscopic scale materials using top–down manufacturing methods such as lithography, etching, and milling, or from lower (atomic) scale materials using bottom–up methods such as sol–gel processing, chemical vapor deposition, arc discharge, and electrostatic self-assembly or a combination of both [3]. Both top–down and bottom–up methods can have significant environmental impacts for several reasons: (i) strict purity requirements and less tolerance for contamination during processing than more conventional manufacturing processes, (ii) low process yields or material efficiencies, (iii) repeated processing, postprocessing, or reprocessing steps of a single product or batch during manufacturing, (iv) use of toxic/basic/acidic chemicals and organic solvents, (v) need for moderate-to-high vacuum and other specialized environments such as high heat or cryogenic processing, (vi) use of or generation of greenhouse gases, (vii) high water consumption, and (viii) chemical exposure potential in the workplace and through technological/natural disasters [4].

For bottom–up techniques, which are usually the preferred methods for the synthesis of most nanoparticles and nanostructured materials, the literature contain a wealth of empirical data and analysis about process design, material precursors and solvents, and operating parameters [2,5]; however, only a small percentage takes into account environmental impacts. One class of nanostructured materials that have considerable potential for a wide variety of environmental applications is quantum dots, crystalline semiconductors of small enough radius to confine the motion of conduction band electrons, valence band holes, or excitons. Figure 41.1 compares number of sources over time in the scientific literature on the investigation of quantum dots, number of studies on a specific type of quantum dot (cadmium selenide, CdSe), and the number of sources that refer to environmental aspects of CdSe quantum dot manufacturing. There are more than twenty-five thousand sources in the literature investigating synthesis

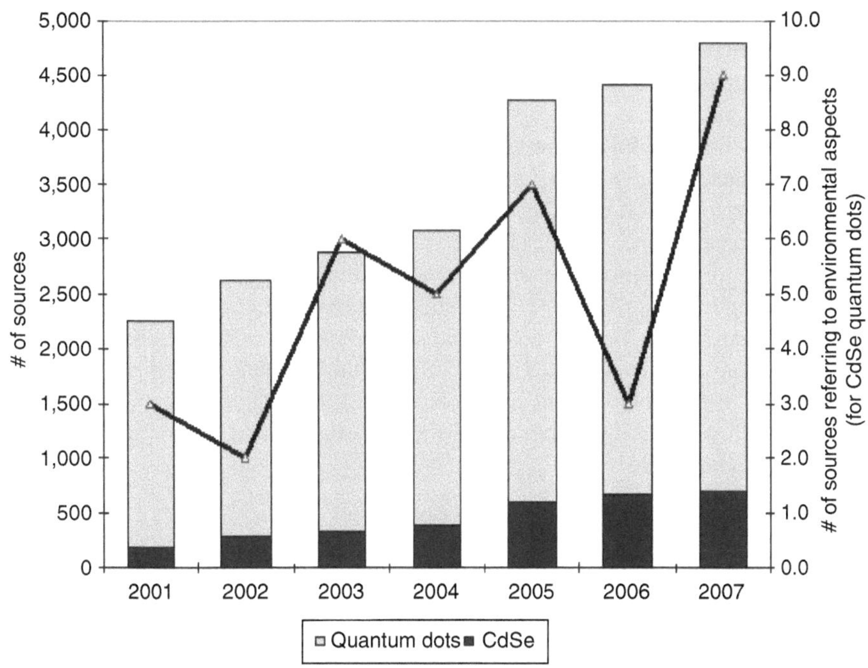

FIGURE 41.1

Comparison of literature sources of quantum dots, cadmium selenide (CdSe) quantum dots, and sources referring to environmental aspects of CdSe quantum dots. *Note*: For quantum dot related sources query keywords were quantum dot* or semiconductor nanoparticle* or semiconductor nanocrystal* at ISI Web of Science; for CdSe, quantum dot keywords and CdSe were combined; for sources referring to environmental aspects, after a query at ISI Web, Google Scholar was also queried to search in full texts rather than limiting the search to abstracts only. Keywords used were environmentally, green, clean, nontoxic; individual records were then scanned to exclude irrelevant records.

and applications of quantum dots (three thousand related to CdSe quantum dots) but only thirty-five refer to greener, cleaner, or environment-friendly synthesis, and only one [6] solely addresses environmentally benign synthesis of CdSe quantum dots. Whereas the number of quantum dot sources has increased markedly over time, there is no clear trend on environmental aspects of manufacturing.

Candidate quantum dot materials are usually compound semiconductors (e.g., CdSe, PbTe, ZnS, InAs) made up of elements of group II and VI, IV and VI, or III and V. Unfortunately, many of these substances are highly toxic and nonrenewable elements, negative attributes that may be exacerbated by their nanoscale dimensions, which may significantly increase their toxicity and environmental impacts [7]. The synthesis of quantum dots also requires highly toxic precursors and solvents with significant upstream and downstream environmental impacts.

41.2 Applications and synthesis of quantum dots

For bulk semiconductors, the carriers (electrons and holes) are free in all three dimensions and have a continuous valence band. When the size of a material becomes smaller than its exciton (electron–hole pair) radius, carriers are confined in space. Quantum dots are a unique group of semiconductor particles that are small enough to exhibit quantum confinement effects. As a result, the density of states that carriers can occupy becomes quantized. Due to quantum confinement, electronic and optical properties of materials dramatically change [8].

Some applications require a core-shell structure (e.g., for sharper luminescence) in which case the core compound semiconductor is surrounded by another compound semiconductor (e.g., CdSe/ZnS). Colloidal quantum dots are bound to stabilization agents (also called passivation or capping agents) that give the dot its stability in solution. Quantum dots are being investigated for a wide variety of applications, including:

- solar cells,
- displays,
- bioimaging,
- targeted drug delivery,
- polymer nanocomposites,
- single-electron transistors,
- lasers,
- nanosensors,
- infrared/near infrared photodetectors,
- light emitting diodes (LEDs),
- optical fibers,
- optical amplifiers,
- optical memory, and
- quantum computing.

Some of these applications have already been commercialized or are in the process of being introduced in the market [9]. Enhanced luminescence and band gap tunability of quantum dots enable them to be applied as nanosensors for environmental analysis and monitoring (screening, diagnostic applications, and monitoring). Several pollutants and pathogens in a sample can be detected simultaneously with high sensitivity [10]. Recent applications include detection of single cells of *E. coli* using CdSe/ZnS quantum dots [11], detection of *Cryptosporidium parvum* and *Giardia lamblia* quantum dot–antibody conjugates [12], and detection of *E. coli* O157:H7 and *S. typhimurium* [13]. Goldman et al. have performed multiplexed sandwich immunoassays by conjugating CdSe/ZnS quantum dots to antibodies to simultaneously detect four toxins that eliminate the need for numerous excitation sources/emission windows and complex processing [14].

Chemical sensors based on quantum dots are used to detect ions in aqueous solutions. Chen and Rosenzweig [15] reported the analysis of Cu(II) and Zn(II) ions by CdSe quantum dots capped with polyphosphate, *L*-cysteine, and thioglycerol in water samples. Jin et al. [16] reported detection of cyanide ions in water samples by a CdSe quantum dot nanosensor. Konishi and Hiratani [17] used an oligo (ethylene glycol) capped cadmium sulfide quantum dot nanosensor to detect copper ions (Cu[II] and Cu[I]). Gattas-Asfura and Leblanc [18] reported the detection of Cu(II) and Ag(I) with a peptide-coated cadmium sulfide quantum dot nanosensor. Sirinakis et al. [19] used a CdSe/ZnS quantum dot nanosensor to detect aromatic hydrocarbons.

Indirect application of quantum dots toward improving environmental quality includes the use of quantum dot films as active layers for solar cells. As part of the novel so-called third generation photovoltaics that aim to eliminate the shortcomings of conventional solar cells, quantum dot solar cells offer a dual solution for advancing the solar technology by increasing the efficiency and allowing roll-to-roll production and thus increasing energy production and throughput and lowering manufacturing costs [20,21]. Quantum dot solar cells can absorb nearly all of the incident solar radiation for wavelengths above their absorption onset with a film of only 200 nm thickness [22]. Candidate materials as quantum dots for solar cells are mostly compound semiconductors of group II–VI, IV–VI, or III–V of the periodic table, including CdSe, CdTe, CdS, InP, InAs, InSb, ZnS, ZnO, ZnSe, ZnTe, PbSe, PbTe, PbS, HgTe, GaN, GaP, GaAs, GaSb, Si, Ge, AlAs, and AlSb.

Quantum dots can be synthesized through two major routes: vapor-phase and liquid-phase deposition, or colloidal synthesis. In vapor-phase synthesis, quantum dots are grown through epitaxial self-assembly by deposition on the surface of a semiconductor layer that has a lattice structure compatible with the quantum dot-compound semiconductor. In colloidal synthesis, precursors of groups II and VI are separately dissolved in organophosphorus solvents such as trioctyl phosphine (TOP), tributyl phosphine (TBP), or triisopropyl phosphine (i-TPP) and injected into a solution of heated solvent (usually a coordinating solvent such as trioctyl phosphine oxide [TOPO]) or a solvent mixture. Quantum dots nucleate and grow instantaneously upon injection of both precursors—or only group V or VI precursors—into a flask containing the solvent—or a mixture of solvent plus group II or III precursors. Further growth occurs through Ostwald ripening (i.e., small particles are absorbed by bigger ones).

Colloidal synthesis of CdSe is the most well-known method for the synthesis of quantum dots, and has been extensively researched [23–27]. It has become “a model system” to study colloidal synthesis of quantum dots in general [28]. Figure 41.2 shows the flow for synthesizing CdSe quantum dots using the conventional and most widespread method pioneered by Murray et al. [23].

Colloidal synthesis is a batch process with low energy and process requirements. The vast majority of environmental impacts are likely to occur in connection with raw material acquisition; thus the choice of precursors and solvents is quite important in reducing the cumulative impact. For the dual precursor route, the source of selenium is selenium powder—there are no alternatives; however, it can be dissolved in different solvents before injection. For cadmium, alternative cadmium compounds can be employed. Dimethyl cadmium, an organometallic compound, is used as the cadmium precursor in conventional synthesis; however, dimethyl cadmium is an extremely toxic, expensive, and unstable solvent that limits its suitability for large-scale synthesis [29]. This has led researchers to search for

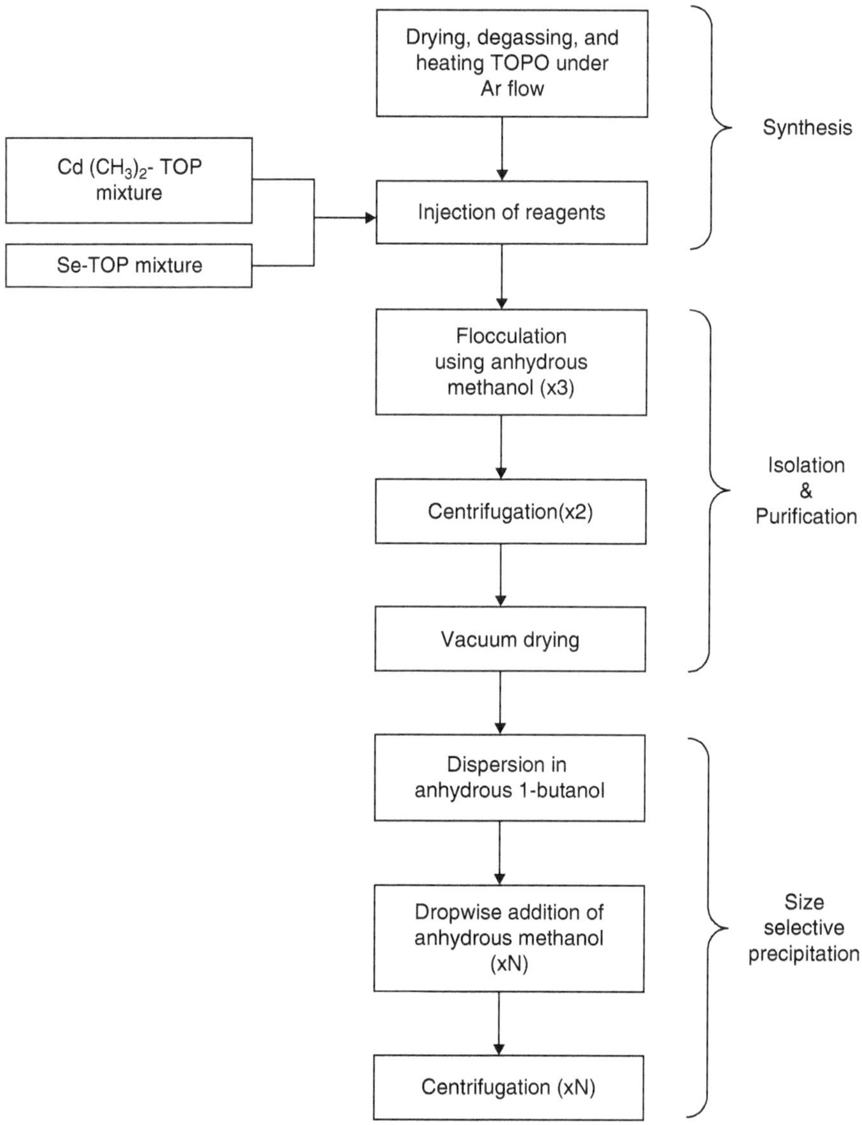

FIGURE 41.2

Liquid-phase synthesis of cadmium selenide (CdSe) quantum dots.

alternative cadmium compounds including other organometallics such as alkyl/alkoxy cadmium compounds (e.g., dineopentylcadmium), cadmium salts of fatty acids (cadmium acetate, cadmium oleate, cadmium laurate), and inorganic forms (cadmium oxide, cadmium carbonate). Instead of using dual precursor sources, single source precursors may also be used although they are less commonly known and applied [30–32].

During CdSe quantum dot synthesis the cadmium and selenium precursors are dissolved in TOP, TBP, or i-TPP and form complexes with cadmium and selenium. The solvent precursors are injected in TOPO (termed "the hot matrix") at high temperatures (250–300°C) [33]. Multicomponent solvents where phosphonic acids act as cosurfactants may also be used. Addition of tetradecylphosphonic acid (TDPA) to a hexadecyl amine (HDA)-TOPO-TOP stabilizing mixture slows nanocrystal growth resulting in good crystallinity and improved size distribution [34]. Addition of phosphonic acids also enables morphological control of quantum dots and leads to synthesis of rod- or tetrapod-shapes that show better performance for specific applications (e.g., solar cells) [35]. Trioctyl phosphine oxide surfactant also acts as a capping agent and ensures the solubility of quantum dots in nonpolar solvents, that is, quantum dots do not exist as stand-alone particles, TOPO molecules are attached to CdSe in solution [36]. Methods are also available to render CdSe soluble in aqueous solvents [37,38].

The need to search for alternative solvents stems from the cost of organophosphorus solvents. The price of quantum dots per gram is currently about $2,000 with solvents accounting for up to 90 percent of the cost [39,40]. Recent trends are to use saturated and unsaturated fatty acids (e.g., oleic acid, stearic acid, lauric acid) [41] and heat transfer fluids based on phenyls [33]. The variety of capping agents is far more extensive and application-dependent, including alkylthiols, alkylamines, peptides, and carboxylic acids in addition to organophosphorus ones. Figure 41.3 presents alternative precursors and solvent pathways for the synthesis of CdSe quantum dots. Table 41.1 provides a listing of sources with varying precursors and solvents/ligands.

41.3 Methodology

This chapter addresses the environmental impacts of nanostructured semiconductor quantum dots as applied to the production of solar energy, using CdSe quantum dots as an example. A life cycle approach is used, facilitated by application of the life cycle assessment software SimaPro 7.1.4 [42] and data from the Swiss national LCI database Ecoinvent [43]. Materials were modeled in SimaPro using information and data from the Ecoinvent database and other sources referenced in the following paragraphs.

For infrastructure and transport of materials, generic data are employed. Infrastructure use was approximated by the "chemical plant, organics, RER (Country code for Europe in the database)" module and the transport was

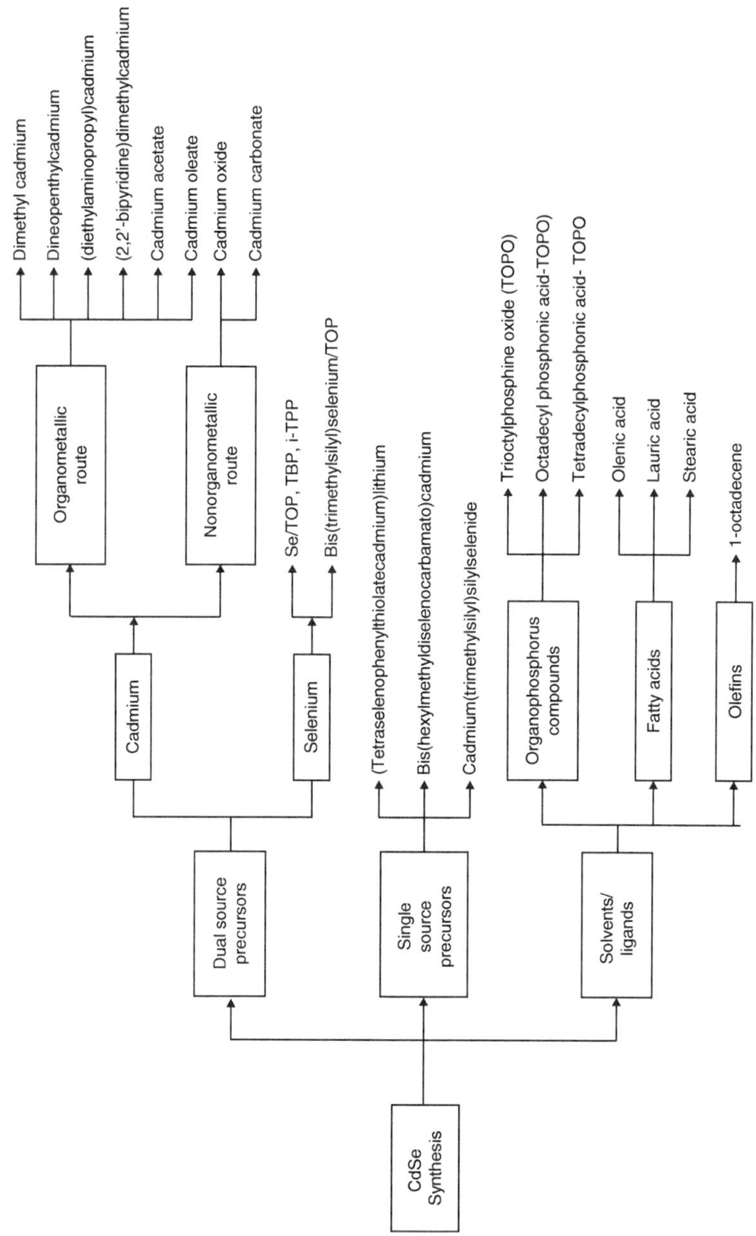

FIGURE 41.3

Most common precursor and solvent pathways for the synthesis of cadmium selenide (CdSe) quantum dots.

Table 41.1 Alternative Precursors and Solvent Systems for the CdSe Synthesis-Dual Source Precursors

Precursors		
Cadmium Precursor/Solvent	**Selenium/ Solvent**	**Solvent(s)**
Dimethyl cadmium/TOP/HPA/ TDPA [23,35]	Bis(trimethylsilyl) selenium/TOP [90]	Dichloromethane [90]
Cadmium oxide/TDPA-TOPO/ ODPA-TOPO* [29]	Se/TOP [23]	TOPO [23]
Cadmium methoxide/ HAD-TOPO* [85]	Se/TBP [71]	TOPO/TBPA [35]
Cadmium oleate/Squalane-Oleyl amine-TOP [41]	Se/i-TPP [71]	1-octadecene [40]
Cadmium 2,4-pentanedionate/ Squalane-Oleyl amine-Oleic acid [76]	Se/TOP-squalane [76]	Dowtherm A, Therminol 66 (T66)-Oleic acid [90]
Cadmium acetate, cadmium carbonate/SA/ SA-TOPO/Lauric acid/Technical TOPO/ TDPA-TOPO* [86]	Sodium selenosulfate [89]	
Dineopentylcadmium, bis (3-diethylaminopropyl) cadmium and (2,2-bipyridine)dimethylcadmium/TBP [87]	Se/TOP-CTAB/ TOP-DDAB [90]	
Cadmium laurate, Cadmium myristate, Cadmium palmitate [88]		
Cadmium oxalate/ Ethylenediamine* [89]		

Notes*: (1) Syntheses designated with "*" are one-pot approaches meaning only selenium solution is injected to the solution for the growth of quantum dots, cadmium precursors are already dissolved in the solvent prior to the injection of selenium containing solution. (2) Dowtherm A and Therminol 66 are commercial names for two heat transfer fluids containing phenyls.*
Abbreviations*: CTAB: Cetyltrimethylammonium bromide; DDAB: Didodecyldimethylammonium bromide; HPA: Hexylphosphonic acid; Me: Methyl group; ODPA: Octadecyl phosphonic acid; SA: Stearic acid; TBP: Tri-n-butylphosphine; TDPA: Tetradecylphosphonic acid; TOP: Trioctylphoshine; TOPO: Trioctylphosphine oxide; i-TPP: Triisopropylphosphine.*

approximated by the "transport, lorry 16t" module in the database assuming an average distance of 100 km for transportation of quantum dots from the production site to the use site. The waste disposal option is selected as hazardous waste incineration for which a life cycle inventory (LCI) already exists. Energy consumption for some materials is based on estimates for similar chemicals in the database. Primary energy consumption for TOP is based on unit process inventory for organophosphorus compounds in the Ecoinvent report No. 15 [44]. The energy consumption during synthesis (drying and heating TOPO before and during synthesis) is estimated assuming 3 J/g K is a representative value for heat capacity of TOPO and the reaction mixture [45]. A similar figure is obtained using power consumed by a heating mantle for the average reaction duration. Since LCI of materials are based on electricity mix for Europe, emission values reported here are conservative as the share of coal in the electricity mix for Europe has a much lower value than that of United States and share of nuclear energy is much higher [46].

41.4 Life cycle inventory of synthesis of CdSe quantum dots

As expressed earlier, the acquisition of raw materials for quantum dot synthesis is likely to be a major contributor to environmental impacts. Twenty-two materials are required for the synthesis of CdSe quantum dots. The LCI data for six materials are not available in the Ecoinvent database. These material processes were modeled in SimaPro using LCI data for other materials available in the Ecoinvent database and other sources referenced in the following paragraphs. Figure 41.4 presents the material flows for the synthesis of CdSe quantum dots. Gray filled boxes represent materials modeled in SimaPro using the Ecoinvent database and the literature/patents; other boxes represent materials for which data is already available in the Ecoinvent database.

Life cycle inventory for the synthesis of CdSe quantum dots is based on the original and widely used method of Murray et al. [23]. Cadmium precursor is dimethyl cadmium [Cd(CH3)2]. Cadmium compounds are included in the list of 189 chemicals listed as hazardous air pollutants under the 1990 Clean Air Act. Cadmium is also on the list of chemicals appearing in the Emergency Planning and Community Right-To-Know Act of 1986 [47]. Cadmium is produced as a by-product of zinc (80 percent) and lead (20 percent) mining and refining. Sources include:

- purification sludge from electrolytic zinc plants;
- fumes and dusts collected in ESPs (electrostatic precipitators) from zinc, lead, lead–zinc, or copper–lead–zinc ore processing;
- recycled zinc metal containing cadmium [48];
- recycled nickel–cadmium batteries [49];
- recycling cadmium from municipal solid waste (MSW) has also been proposed [50].

Approximately 3 kg of cadmium is produced for each ton of zinc produced. Environmental impacts from zinc processing allocated to cadmium production are small, 0.5 percent and 0.58 percent of environmental impacts are contributed by cadmium based on mass and economic value allocation [51]. Cadmium is refined in four major steps: leaching, precipitation, reduction or cadmium plating, and casting. During cadmium refining, cadmium is released from melting furnaces, retorting, casting and tapping, and packaging [48].

Cadmium precursors can be produced through a number of pathways and from different starting materials [48,52,53]. Figure 41.5 presents the network for the

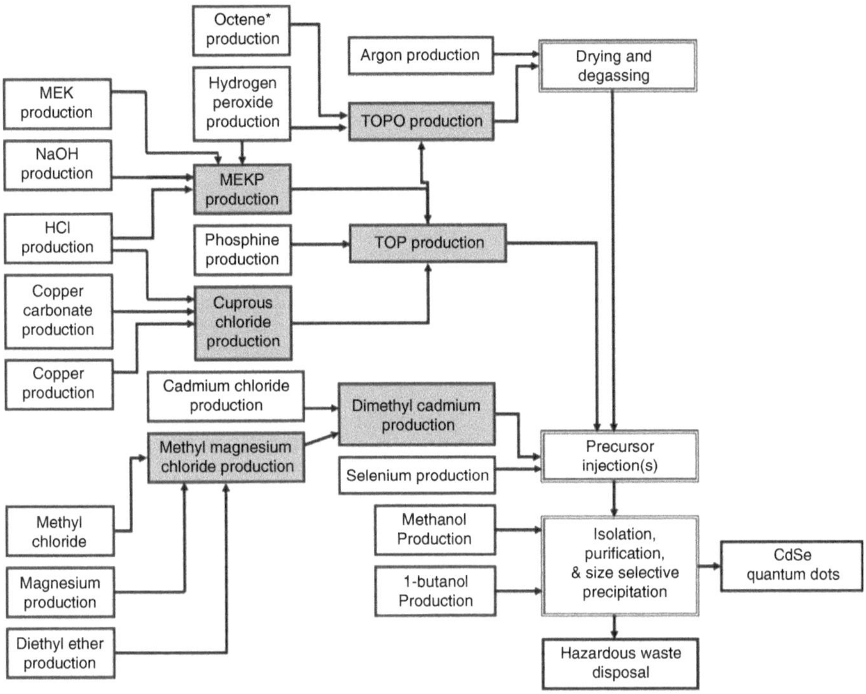

FIGURE 41.4

Material flows for the synthesis of cadmium selenide (CdSe) quantum dots. *Note*: *Life cycle inventory (LCI) for octene is derived by mass allocation from LCI of n-olefins in the Ecoinvent database.

synthesis of alternative cadmium compounds used as precursors for the synthesis of CdSe quantum dots. Cadmium oxide has the shortest route and is probably the one with the lowest environmental impact in terms of energy and materials consumption as cadmium can be easily oxidized upon exposure to air in furnaces. But cadmium oxide vapors are more toxic than cadmium nitrate, cadmium acetate, or cadmium chloride. Therefore, the toxicities of alternative cadmium compounds must be evaluated for a comparative assessment. In the case of dimethyl cadmium, as can be seen in Fig. 41.5, it can be synthesized either from cadmium acetate or cadmium chloride. Its synthesis from cadmium acetate avoids the extra steps of synthesizing cadmium chloride, which is more toxic than cadmium acetate, using Grignard reagents (CH3MgI). However, it involves the use of another chlorinated substance, acetyl chloride (CH3COCl). This is the case if cadmium chloride is to be produced from cadmium acetate. If cadmium chloride is produced directly from cadmium using hydrochloric acid or chlorine gas, the relative impacts of synthesizing dimethyl cadmium using cadmium acetate or cadmium chloride as the starting material are difficult to compare.

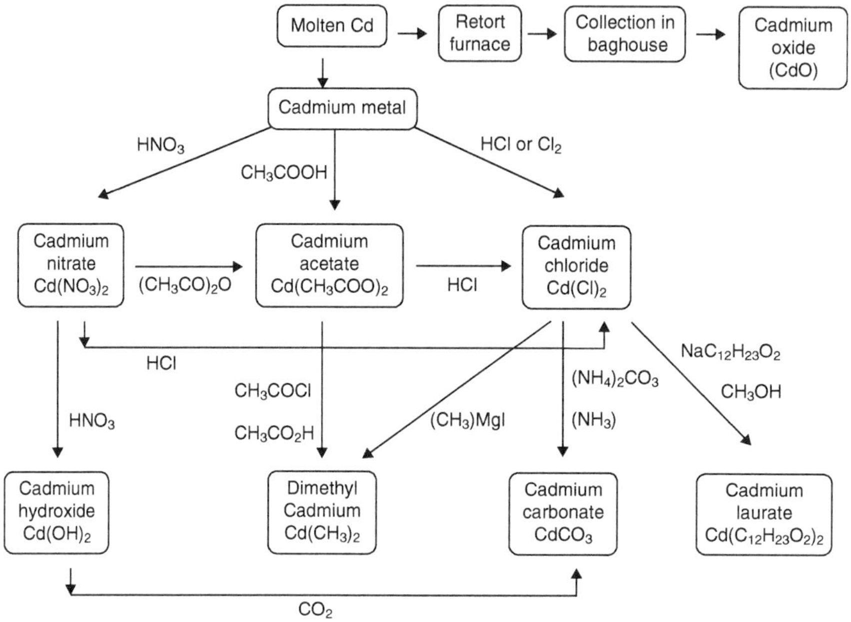

FIGURE 41.5

Cadmium flows and material inputs for the synthesis of cadmium precursors.

Further, cadmium chloride is usually the commercially available compound for the synthesis of other cadmium compounds. No information exists for the toxicity of dimethyl cadmium but parallels can be drawn with other organometallic compounds, for example, organocadmium compounds show similar toxicities to organomercury compounds [54].

Life cycle inventory of dimethyl cadmium is not available in the Ecoinvent database. The synthesis of dimethyl cadmium from methyl magnesium chloride and cadmium chloride were modeled in SimaPro with data and information from the literature and patents. Synthesis of Grignard reagent, methyl magnesium chloride (CH3MgCl), required to produce dimethyl cadmium is also modeled in Simapro [55–61].

Selenium powder is used as selenium precursor. Usually a solution of selenium dissolved in TOP is prepared and used as a stock solution. Selenium is an element regulated under the Safe Drinking Water Act. It is a widely distributed mineral produced as a by-product of the electrolytic refining of copper [62]. Selenium can be extracted from copper refinery slime by roasting with soda ash or sulfuric acid. Other methods include wet chlorination, oxidative leaching with sodium hydroxide solution under pressure, and the hydrometallurgical process of chlorination in hydrochloric acid [63]. Production of selenium based on roasting process with sodium carbonate uses hydrochloric acid, soda, and sulphur dioxide releasing hydrogen chloride, sulfur dioxide, chloride, sodium, and sulfate [64] and inventory data for its production is already available in the Ecoinvent database.

The organophosphorus surfactants TOP and TOPO can be produced from elemental phosphorus, phosphines, or halogenophosphines [65]. Trioctyl phosphine is manufactured by the radical-catalyzed addition of 1-octene to phosphine. Trioctyl phosphine oxide can then be produced from TOP by peroxide oxidation, according to Equation 41.1 [54]:

$$(C_8H_{17})_3P + H_2O_2 \rightarrow (C_8H_{17})_3P = O + H_2O \tag{41.1}$$

The catalytic synthesis of TOP and TOPO (via cuprous chloride) is initiated with methylethylketone peroxide [MEKP, (C6H5CO)2O2] generated from the oxidation of methylethylketone (MEK) by hydrogen peroxide (H2O2). The complete synthesis was modeled in Simapro using data and information from the literature and patents [54,66–70].

The reaction chemistry and stoichiometry of CdSe synthesis is one of the poorly understood parts of CdSe quantum dot synthesis [71]. Also, the thermal decomposition products of dimethyl cadmium in TOPO are currently unknown. Thermal decomposition releases free methyl radicals that recombine and decompose to form hydrocarbons. Gas-phase decomposition yields methane, ethane, propane, ethylene gases [72]. In the TOPO solvent, ethers may form as the methyl radicals combine with the oxygen in TOPO [73]. Ethers may subsequently form complexes with unreacted selenium. To date, no record exists for mass spectrometric analysis of individual compounds. The reaction yield for the synthesis of quantum dots is somewhat difficult to control, ranging from 25 to 97 percent depending on nanocrystal size, heating time, and precursor [34,74–76]. Murray et al. report that the waste from isolation and purification steps is composed mostly of elemental cadmium and selenium as analyzed by powder X-ray diffraction and energy dispersive X-ray [23]. The waste also includes excess TOPO, TOP, and alcohols used to precipitate quantum dots. The waste from the size selective precipitation step contains primarily methanol and 1-butanol.

The subsequent application of this information to solar collectors was conducted as part of this study. A "functional unit" of 1 watt of peak power was selected. The actual power a solar cell can produce depends on the installation location, the tilt angle, the energy conversion efficiency, and the energy loss in the inverter. Solar spectra are defined by an air mass (AM) value, which is a measure of the length of the path through the earth's atmosphere that the solar radiation travels. The value is calculated as 1/cos z, where z is the zenith angle between a line perpendicular to the earth's surface and a line intersecting the sun. AM 1 describes the case in which the sun is directly overhead. The AM 1.5 (at an angle 42.8°) spectrum is commonly used for testing and reporting solar cell devices meant for terrestrial use [77]. The peak power is calculated from efficiency, solar cell area, and input light irradiance using Equation 41.2:

$$\eta = \frac{P_m}{E^* A_c} \tag{41.2}$$

where η = energy conversion efficiency; $E = 1000\ W/m^2$ for AM 1.5 at 25°C; A_c = surface area of solar cell.

The current performance of quantum dot solar cells does not exceed 5 percent (actually laboratory measurements are less, 5 percent is taken as base scenario) but they are expected to reach a conversion efficiency of 42 percent or higher (best case scenario) [78]. A quantum dot layer requires 120 μl of quantum dot solution for a solar cell area of 220 mm^2 [79–82]. Accordingly, for 1 watt of peak power at standard conditions, the required amount of quantum dots ranges from 920 to 2290 mg for the base scenario, and from 110 to 270 mg for best case scenario. To produce 1 watt of peak power, a minimum of 110 mg of quantum dots needs to be synthesized. Material inputs and emissions to soil, water, and air for the synthesis of 1 gram of CdSe quantum dots are provided in Figs 41.6–41.9. On a mass basis, air emissions contribute the most to the cumulative emissions, followed by water and soil emissions. Overall, approximately 110 g of waste materials are produced for the synthesis of 110 mg quantum dots for 1 watt peak of energy, thus the global waste burden of synthesis of CdSe quantum dots for solar cell films is in the range 110–2,225 kg/kWp, corresponding to a conservative waste-to-product ratio of 972 for quantum dots. Note that the waste-to-product ratio was calculated only for the synthesis of quantum dots, not the whole life cycle (i.e. raw materials acquisition and synthesis) and excludes

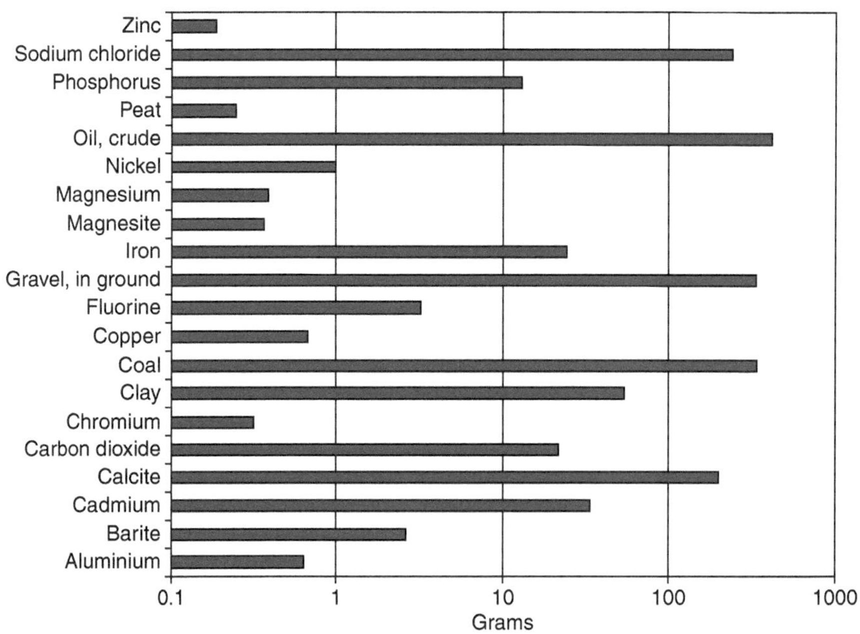

FIGURE 41.6

Raw material use for the synthesis of 1 g of cadmium selenide (CdSe) quantum dots—20 materials with the highest amount of use by mass (except water and other materials reported by volumetric units rather than mass).

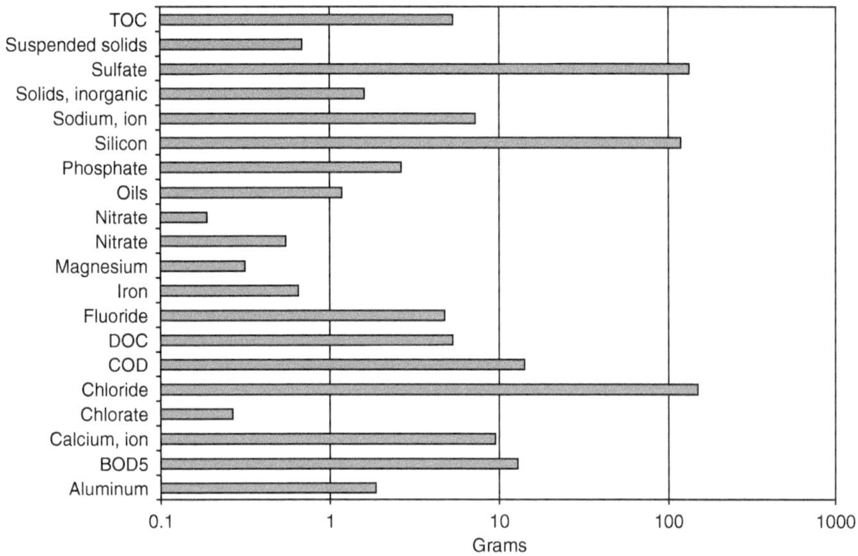

FIGURE 41.7

Emissions to water from the synthesis of 1 g of cadmium selenide (CdSe) quantum dots. BOD5: Biological Oxygen Demand; COD: Chemical Oxygen Demand; DOC: Dissolved Organic Carbon; TOC: Total Organic Carbon.

the waste from energy demand and transport, consistent with literature that reports waste-to-product ratio [83]. The waste-to-product ratio is highly dependent on the amount of solvent used for isolation and purification steps. The number of isolation and size selective precipitation steps depends on the material, surfactants, and target application. The application of three to five isolation and purification steps is accepted as reasonable for quantum dot synthesis [84]. In this chapter, we report LCI results for five repetitions of isolation, and size selective precipitation steps. The total volume of solvents (i.e., methanol and 1-butanol) used is 1.1 liters per gram of quantum dots. The waste-to-product ratio drops to 590 for three repetitions of isolation, purification, and size selective precipitation. For comparison, bulk chemical synthesis ratios are typically less than 10, fine chemicals of less than 50, and many pharmaceuticals of the order of 100 [83].

It must be emphasized that synthesis of quantum dots is based on the original organometallic route, other routes may give different results. It must also be stated that the results are based on laboratory studies and that amount of quantum dot used is for deposition of quantum dot layer by spin-coating, wherein most of the solution becomes waste and cannot be recycled; adoption of other deposition techniques may result in less waste. This analysis does not include the material flow analysis of the manufacture of solar cells.

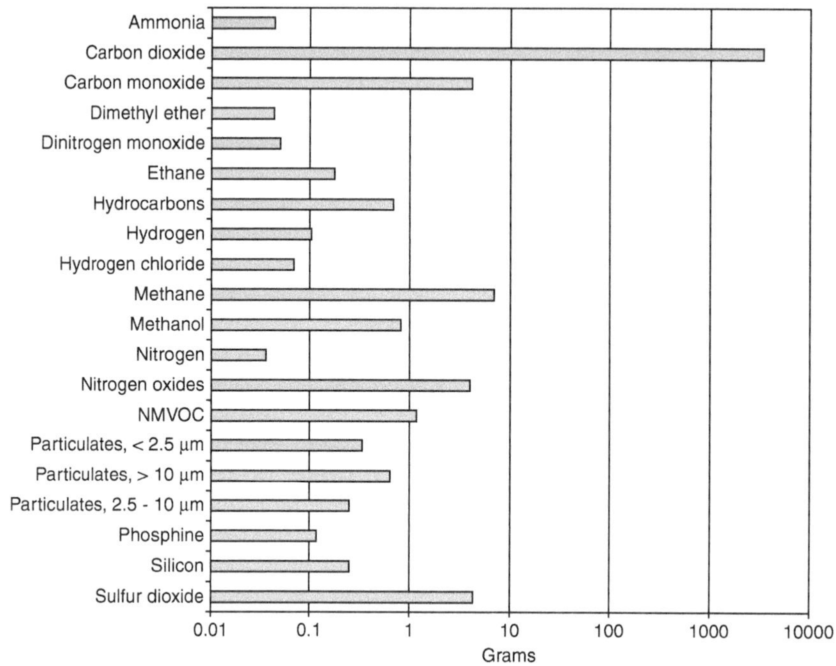

FIGURE 41.8

Emissions to air from the synthesis of 1 g of cadmium selenide (CdSe) quantum dots. NMVOC: Non-Methane Volatile Organic Compounds.

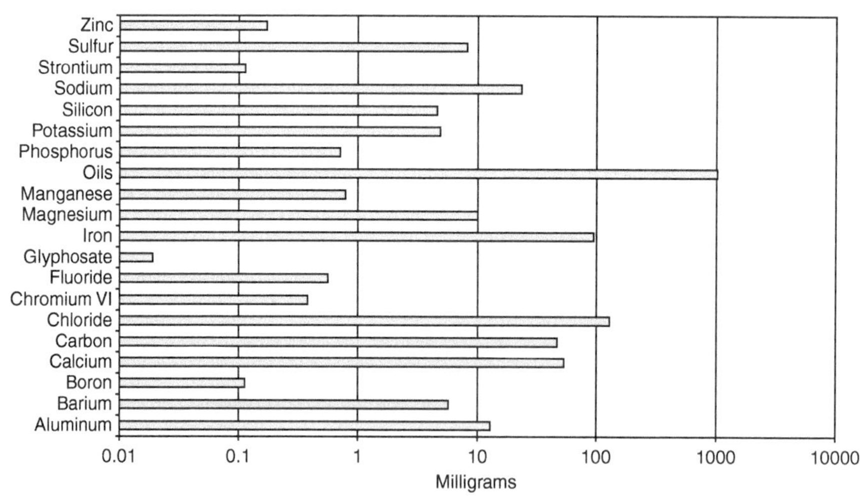

FIGURE 41.9

Emissions to soil from the synthesis of 1 g of cadmium selenide (CdSe) quantum dots.

41.5 Conclusions and future perspective

Life cycle analyses of nano-based technologies are essential to identify environmental advantages and disadvantages of nanoproducts and meet the objective of early diagnosis and treatment of environmental risks of technological developments in this area.

Nanosensors and other applications of quantum dots will be in the market soon as the transition from research to markets has been fast for many nanoproducts. Quantum dot lasers and quantum dot biodetection products have already reached markets. Given the prospective market size of quantum dots, the waste burden and environmental impacts linked to the emissions to the environment caused by the production and use of quantum dots may become significant in the future. In an effort to identify associated risks, we used life cycle assessment software SimaPro, the LCI database Ecoinvent and other data sources to establish a LCI for the synthesis of CdSe quantum dots, which are the most frequently used quantum dots so far. Raw material use, energy use, and emissions to air, water, and soil were quantified. On a mass basis, air emissions contribute the most to the cumulative emissions, followed by water emissions contributing to 12 percent of the total emissions. These emission values are specific for the conventional route of synthesis. Further research is required that applies Life Cycle Assessment (LCA)/LCI to raw materials selection to identify alternative raw materials that are less polluting. Our results indicate that the isolation, purification, and size selective precipitation contribute the most to environmental impacts.

Acknowledgments

This work has been supported by grant number 0646336, "Life Cycle of Nanomanufacturing Technologies" of the Division of Chemical, Bioengineering, Environmental, and Transport Systems, National Science Foundation, Cynthia Eckstein project manager.

References

[1] ISO (International Organization for Standardization), (2006) CAN/CSA ISO 14040-00—Environmental management—Life cycle assessment—Principles and framework. ISBN 1-55324-156-8.16.

[2] B.L. Cushing, V.L. Kolesnichenko, C.J. O'Connor, Recent advances in the liquid-phase syntheses of inorganic nanoparticles, Chem. Rev. 104 (9) (2004) 3893–3946.

[3] The Royal Society and the Royal Academy of Engineering, Nanotechnologies: Opportunities and Uncertainties, The Royal Society, London, 2004.

[4] H. Sengül, T.L. Theis, S. Ghosh, Towards sustainable nanoproducts: an overview of nanomanufacturing methods, J. Ind. Eco. 12 (3) (2008) 329–359.

[5] M.T. Swihart, Vapor-phase synthesis of nanoparticles, Curr. Opin. Colloid Interface Sci. 8 (1) (2003) 127–133.

[6] X. Peng, Green chemical approaches toward high-quality semiconductor nanocrystals, Chem.-Eur. J. 8 (2) (2002).

[7] J.S. Tsuji, A.D. Maynard, P.C. Howard, J.T. James, C.W. Lam, D.B. Warheit, et al., Research strategies for safety evaluation of nanomaterials, Part IV: risk assessment of nanoparticles, Toxicol. Sci. 89 (1) (2006) 42–50.

[8] A.J. Nozik, Third Generation Solar Photon Conversion-Exceeding the 32% Shockley–Queisser Limit, Presentation at the Solar 2006, American Solar Energy Society, accessed online from <http://www.solar2006.org/presentations/forums/f33-nozik.pdf>, 2007.

[9] Willems and van den Wildenberg (W&W), Work document on nanomaterials: state of the art overview and forecasts based on existing information of nanotechnology in the field of nanomaterials, available online at <http://www.nanoroadmap.it/>, 2004 (accessed in 2007).

[10] Science-Metrix Nanotechnology Stewardship, Canadian stewardship practices for environmental nanotechnology, March (2005) accessed online at <http://www.science-metrix.com/pdf/SM_2004_016_EC_Executive_Stewarship_Nanotechnology_Environment.pdf>, 2007.

[11] M.A. Hahn, J.S. Tabb, T.D. Krauss, Detection of single bacterial pathogens with semiconductor quantum dots, Anal. Chem. 77 (15) (2005) 4861–4869.

[12] L. Zhu, S. Ang, W.T. Liu, Quantum dots as a novel immunofluorescent detection system for Cryptosporidium parvum and Giardia lamblia, Appl. Environ. Microbiol. 70 (2004) 597–598.

[13] L.J. Yang, Y.B. Li, Simultaneous detection of Escherichia coli O157: H7 and Salmonella typhimurium using quantum dots as fluorescence labels, Analyst 131 (3) (2006) 394–401.

[14] E.R. Goldman, I.L. Medintz, H. Mattoussi, Luminescent quantum dots in immunoassays, Anal. Bioanal. Chem. 384 (3) (2006) 560–563.

[15] Y.F. Chen, Z. Rosenzweig, Luminescent CdS quantum dots as selective ion probes, Anal. Chem. 74 (19) (2002) 5132–5138.

[16] W.J. Jin, M.T. Fernandez-Arguelles, J.M. Costa-Fernandez, R. Pereiro, A. Sanz-Medel, Photoactivated luminescent CdSe quantum dots as sensitive cyanide probes in aqueous solutions, Chem. Commun. 7 (2005) 883–885.

[17] K. Konishi, T. Hiratani, Turn-on and selective luminescence sensing of copper ions by a water-soluble Cd10S16 molecular cluster Angew, Chem. Int. Ed. 45 (31) (2006) 5191–5194.

[18] K.A. Gattas-Asfura, R.M. Leblanc, Peptide-coated CdS quantum dots for the optical detection of copper(II) and silver(I), Chem. Commun. 21 (2003) 2684–2685.

[19] G. Sirinakis, Z.Y. Zhao, Y. Sevryugina, A. Tayi, and M. Carpenter, Tailored nanomaterials: Selective & sensitive chemical sensors for hydrocarbon analysis, School of NanoSciences and NanoEngineering University at Albany, SUNY, available online at <http://eqs.syr.edu/documents/research/>, 2003.

[20] A.J. Nozik, Quantum dot solar cells, Phys. E 14 (1–2) (2002) 115–120.

[21] A. Slaoui, R.T. Collins, Advanced inorganic materials for photovoltaics, MRS Bull. 32 (3) (2007) 211–218.

[22] H.U. Huynh, J.J. Dittmer, N. Teclemariam, S.J. Milliron, A.P. Alivisatos, K.W.J. Barnham, Charge transport in hybrid nanorod-polymer composite photovoltaic cells, Phys. Rev. B 67 (11) (2003) Art. No. 115326.

[23] C.B. Murray, D.J. Norris, M.G. Bawendi, Synthesis and characterization of nearly monodisperse CdE (E = S, Se, Te) semiconductor nanocrystallites, J. Am. Chem. Soc. 115 (19) (1993) 8706–8715.

[24] C.B. Murray, C.R. Kagan, M.G. Bawendi, Synthesis and characterization of monodisperse nanocrystals and close-packed nanocrystal assemblies, Annu. Rev. Mat. Sci. 30 (2000) 545–610.

[25] Y. Yin, A.P. Alivisatos, Colloidal nanocrystal synthesis and the organic–inorganic interface, Nature 437 (7059) (2005) 664–670.

[26] D.V. Talapin, I. Mekis, S. Gotzinger, A. Kornowski, O. Benson, H. Weller, CdSe/CdS/ZnS and CdSe/ZnSe/ZnS core-shell-shell nanocrystals, J. Phys. Chem. B 108 (49) (2004) 18826–18831.

[27] Z.A. Peng, X.G. Peng, Nearly monodisperse and shape-controlled CdSe nanocrystals via alternative routes: nucleation and growth, J. Am. Chem. Soc. 124 (13) (2002) 3343–3353.

[28] O. Schmelz, A. Mews, T. Basche, A. Herrmann, K. Mullen, Supramolecular complexes from CdSe nanocrystals and organic fluorophors, Langmuir 17 (9) (2001) 2861–2865.

[29] Z.A. Peng, X.G. Peng, Formation of high-quality CdTe, CdSe, and CdS nanocrystals using CdO as precursor, J. Am. Chem. Soc. 123 (1) (2001) 183–184.

[30] T. Trindade, P. O'Brien, A single source approach to the synthesis of CdSe nanocrystallites, Adv. Mat. 8 (2) (1996) 161.

[31] S.L. Cumberland, K.M. Hanif, A. Javier, G.A. Khitrov, G.F. Strouse, S.M. Woessner, et al., Inorganic clusters as single-source precursors for preparation of CdSe, ZnSe, and CdSe/ZnS nanomaterials, Chem. Mat. 14 (4) (2002) 1576–1584.

[32] M.A. Malik, N. Revaprasadu, P. O'Brien, Air-stable single-source precursors for the synthesis of chalcogenide semiconductor nanoparticles, Chem. Mat. 13 (3) (2001) 913–920.

[33] S. Asokan, K.M. Krueger, A. Alkhawaldeh, A.R. Carreon, Z.Z. Mu, V.L. Colvin, et al., The use of heat transfer fluids in the synthesis of high-quality CdSe quantum dots, core/shell quantum dots, and quantum rods, Nanotechnology 16 (10) (2005) 2000–2011.

[34] I. Mekis, D.V. Talapin, A. Kornowski, M. Haase, H. Weller, One-pot synthesis of highly luminescent CdSe/CdS core-shell nanocrystals via organometallic and 'greener' chemical approaches, J. Phys. Chem. B 107 (30) (2003) 7454–7462.

[35] Z.A. Peng, X.G. Peng, Mechanisms of the shape evolution of CdSe nanocrystals, J. Am. Chem. Soc. 123 (7) (2001) 1389–1395.

[36] A.P. Alivisatos, Perspectives on the physical chemistry of semiconductor nanocrystals, J. Phys. Chem. 100 (31) (1996) 13226–13239.

[37] M. Bruchez, M. Moronne, P. Gin, S. Weiss, A.P. Alivisatos, Semiconductor nanocrystals as fluorescent biological labels, Science 281 (5385) (1998) 2013–2016.

[38] W.C.W. Chan, S.M. Nie, Quantum dot bioconjugates for ultrasensitive nonisotopic detection, Science 281 (5385) (1998) 2016–2018.

[39] G.P. Collins, Cheaper dots, Sci. Am. 293 (6) (2005).

[40] W.W. Yu, X.G. Peng, Formation of high-quality CdS and other II-VI semiconductor nanocrystals in noncoordinating solvents: tunable reactivity of monomers, Angew. Chem. Int. Ed. 41 (13) (2002) 2368–2371.

[41] B.K.H. Yen, N.E. Stott, K.F. Jensen, M.G. Bawendi, A continuous-flow microcapillary reactor for the preparation of a size series of CdSe nanocrystals, Adv. Mat. 15 (21) (2003) 1858–1862.
[42] PRe Consultants, SimaPro, Life Cycle Assessment software package, Version 7.1.4 Amersfoort, The Netherlands, 2008.
[43] Ecoinvent Centre, Ecoinvent Data v2.0., Swiss Centre for Life Cycle Inventories, Dubendorf, Switzerland, 2004.
[44] T. Nemecek, A. Heil, O. Huguenin, S. Meier, S. Erzinger, S. Blaser, et al., Life cycle inventories of agricultural production systems, Ecoinvent report No. 15, Dubendorf, April 2004.
[45] N.E. Stott, K.F. Jensen, M.G. Bawendi, B.K.H. Yen, Method of preparing nanocrystals, US Patent 7,229,497, 2007.
[46] S. Pacca, D. Sivaraman, G.A. Keoleian, Parameters affecting the life cycle performance of PV technologies and systems, Energ. Policy 35 (6) (2007) 3316–3326.
[47] ATSDR (Agency for Toxic Substances and Disease Registry), Toxicological Profile for Cadmium, Division of Toxicology-Toxicology Information Branch, Atlanta, Georgia, 1999.
[48] USEPA, Locating and estimating air emissions from sources of cadmium and cadmium compounds, Office of Air Quality Planning and Standards and EPA-454/r-93-040, September 1993.
[49] D.C.R. Espinosa, A.M. Bernardes, J.A.S. Tenorio, An overview on the current processes for the recycling of batteries, J. Power Sources 135 (1–2) (2004) 311–319.
[50] T.R. Hawkins, H.S. Matthews, C. Hendrickson, Closing the loop on cadmium—an assessment of the material cycle of cadmium in the US, Int. J. LCA 11 (1) (2006) 38–48.
[51] V.M. Fthenakis, Life cycle impact analysis of cadmium in CdTePV production, Renew. Sust. Energy Rev. 8 (4) (2004) 303–334.
[52] M. Farnsworth, Cadmium Chemicals, International lead zinc research organization, Inc., 1980.
[53] Kirk-Othmer, Kirk-Othmer Encyclopedia of Chemical Technology, fifth ed., Wiley-Interscience, 2004.
[54] P. Craig, Organometallic Compounds in the Environment, second ed., John Wiley and Sons, West Sussex, UK, 2003.
[55] J.B. Mullin, J.C. Hamilton, E.D. Orrell, P.R. Jacobs, D.V. Shenai-Khatkhate, Preparation of group II metal alkyls, US Patent 4,812,586, 1989.
[56] G. Yordanov, C. Dushkin, B. Bochev, G. Gicheva, E. Adachi, Preparation of dimethyl cadmium precursor for the hot-matrix synthesis of CdSe nanoparticles, Annuaire De L'Universite De Sofia ST, Kliment Ohridski, 2005.
[57] G. Deguest, L. Bischoff, C. Fruit, F. Marsais, Regioselective opening of N-Cbz glutamic and aspartic anhydrides with carbon nucleophiles, Tetrahedron Asymmetr. 17 (14) (2006) 2120–2212.
[58] A.B. Charette, H. Lebel, (2S,3S)-(+)-(3-phenylcyclopropyl)methanol, Org. Synth. Coll. 10 (2004) 613; 76 (1999) 86.
[59] H. Kryk, G. Hessel, W. Schmitt, Improvement of process safety and efficiency of Grignard reactions by real-time monitoring, Org. Process Res. Dev. 11 (6) (2007) 1135–1140.

[60] J.E. Garst, M.R. Soriaga, Grignard reagent formation, Coord. Chem. Rev. 248 (2004) 623–652.

[61] M. Orchin, The Grignard reagent: preparation, structure, and some reactions, J. Chem. Educ. 66 (7) (1989).

[62] USGS (US Geological Survey), Mineral Commodity Summaries, Selenium, February 1997.

[63] V.M. Fthenakis, H.C. Kim, W. Wang, Life cycle inventory analysis in the production of metals used in photovoltaics, Brookhaven National Laboratory BNL-77919-2007, March 2007.

[64] R. Hischier, Life cycle inventories of packaging and graphical paper, Final report Ecoinvent data v2.0 Swiss Centre for LCI, EMPA.

[65] D.G. Gilheany, C.M. Mitchell, Preparation of phoshines, in: F.R. Hartley (Ed.), The Chemistry of Organophosphorus Compounds, vol. I, John Wiley and Sons, West Sussex, UK, 1990.

[66] H.J. Liaw, C.J. Chen, C.C. Yur, The multiple runaway-reaction behavior prediction of MEK-oxidation reactions, J. Loss Prev. Process Ind. 14 (5) (2001) 371–378.

[67] W. Wu, G. Qian, X.G. Zhou, W.K. Yuan, Peroxidization of methyl ethyl ketone in a microchannel reactor, Chem. Eng. Sci. 62 (18–20) (2007) 5127–5132.

[68] C.G. Ferrari, G. Higashiuchi, Methods for preparing oxidatively active compositions, US Patent 3,047,406, 1962.

[69] G.L. Tembe, Process for manufacture of linear alpha-olefins using a titanium component and an organoaluminum halide component, US Patent 6,121,502, September 19, 2000.

[70] P.A.T. Hoye and J.W. Ellis, Process for preparing organophosphines, US Patent 5,284,555, 1994.

[71] H.T. Liu, J.S. Owen, A.P. Alivisatos, Mechanistic study of precursor evolution in colloidal group II-VI semiconductor nanocrystal synthesis, J. Am. Chem. Soc. 29 (2) (2007) 305–312.

[72] L.M. Dyagileva, Y.A. Aleksandrova, The reactivity of organozinc and organocadmium compounds in decomposition reactions, Russ. Chem. Rev. 55 (11) (1986) 1854–1866.

[73] C.B. Murray, E-mail communication, 2007.

[74] C.A. Leatherdale, W.K. Woo, F.V. Mikulec, M.G. Bawendi, On the absorption cross section of CdSe nanocrystal quantum dots, J. Phys. Chem. B 106 (31) (2002) 7619–7622.

[75] A.S. Alkhawaldeh, M. Pasquali, M.S. Wong, Solvents and new method for the synthesis of CdSe semiconductor nanocrystals, US Patent 20,070,204,790, 2007.

[76] B.K.H. Yen, A. Gunther, M.A. Schmidt, K.F. Jensen, M.G. Bawendi, A microfabricated gas-liquid segmented flow reactor for high-temperature synthesis: the case of CdSe quantum dots, Angew. Chem. Int. Ed. 44 (34) (2005) 5447–5451.

[77] S.E. Shaheen, D.S. Ginley, G.E. Jabbour, Organic-based photovoltaics. Toward low-cost power generation, MRS Bull. 30 (1) (2005) 10–19.

[78] V.I. Klimov, Spectral and dynamical properties of multilexcitons in semiconductor nanocrystals, Annu. Rev. Phys. Chem. 58 (2007) 635–673.

[79] E.C. Scher, Nanostructure and nanocomposite based compositions and photovoltaic devices, US Patent 7,087,832, August 8, 2006.

[80] C. Wadia, E-mail Communication, University of California Berkeley Alivisatos group, 2008.

[81] R.A. Sheldon, Catalysis: the key to waste minimization, J. Chem. Tech. Biotech. 68 (4) (1997) 381–388.
[82] X.B. Liu, Y.H. Lin, Y.M. Chen, L.J. An, X.L. Ji, B.Z. Jiang, New organometallic approach to synthesize high-quality CdSe quantum dots, Chem. Lett. 34 (9) (2005) 1284–1285.
[83] L.H. Qu, Z.A. Peng, X.G. Peng, Alternative routes toward high quality CdSe nanocrystals, Nano Lett. 1 (6) (2001) 333–337.
[84] D. Talapin Assistant Professor, Department of Chemistry, University of Chicago, E-mail communication, 2008.
[85] J. Hambrock, A. Birkner, R.A. Fischer, Synthesis of CdSe nanoparticles using various organometallic cadmium precursors, J. Mat. Chem. 11 (12) (2001) 3197–3201.
[86] L.L. Han, D.H. Qin, X. Jiang, Y.S. Liu, L. Wang, J.W. Chen, et al., Synthesis of high quality zinc-blende CdSe nanocrystals and their application in hybrid solar cells, Nanotechnology 17 (18) (2006) 4736–4742.
[87] S.H. Yu, Y.S. Wu, J. Yang, Z.H. Han, Y. Xie, Y.T. Qian, et al., A novel solvothermal synthetic route to nanocrystalline CdE (E = S, Se, Te) and morphological control, Chem. Mat. 10 (9) (1998) 2309.
[88] S.M. Stuczynski, J.G. Brennan, M.L. Steigerwald, Formation of metal chalcogen bonds by the reaction of metal alkyls with silyl chalcogenides, Inorg. Chem. 28 (25) (1989) 4431–4432.
[89] J.H. Li, C.L. Ren, X. Liu, Z. De, Hu, and D.S. Xue, 'Green' synthesis of starch capped CdSe nanoparticles at room temperature, Mat. Science Eng. A. 548 (1–2) (2007) 319.
[90] S. Asokan, K.M. Krueger, V.L. Colvin, M.S. Wong, Shape-controlled synthesis of CdSe tetrapods using cationic surfactant ligands, Small 3 (7) (2007) 1164–1169.

PART

Outlook

6

Nanotechnology Solutions for Improving Water Quality

Mamadou S. Diallo[1,2], Jeremiah S. Duncan[3], Nora Savage[4], Anita Street[5] and Richard Sustich[6]

[1]Materials and Process Simulation Center, Division of Chemistry and Chemical Engineering, California Institute of Technology [2]Department of Civil Engineering, Howard University [3]Department of Atmospheric Science and Chemistry, Plymouth State University, Plymouth, NH, USA [4]Office of Research and Development, US Environmental Protection Agency [5]US Department of Energy, Washington, DC, USA [6]Center of Advanced Materials for Purification of Water with Systems, University of Illinois at Urbana-Champaign (Retired)

The availability of clean water has emerged as one of the most serious problems facing global society in the 21st century. Nanotechnology has great potential for providing efficient, cost effective and environmentally acceptable solutions for improving water quality and for increasing quantities of fresh water. Nanomaterials have a number of key physicochemical properties that make them particularly attractive as separation media for water purification. On a mass basis, they have much larger surface areas than bulk particles. Thus, they are ideal building blocks for developing high capacity sorbents with the ability to be functionalized to enhance their affinity and selectivity. Nanomaterials can serve as high capacity and recyclable ligands for cations, anions, radionuclides, and organic compounds. They provide unprecedented opportunities for developing more efficient water-purification catalysts and redox active media due their large surface areas and their size and shape-dependent optical and electronic properties. Nanomaterials can also be used to develop chlorine-free biocides through functionalization with chemical groups that selectively target key biochemical constituents of water borne bacteria and viruses. The research and commercial examples described in the preceding chapters of this book offer but a small glimpse into the potential benefits of nanotechnology-based solutions for water purification. As discussed in this book, the first generation of "passive" nanomaterials such as metal oxide nanosorbents are being incorporated into point-of-use (POU) and point-of-entry (POE) water purification systems. However, the use of these passive nanomaterials alone will not lead to the revolutionary advances needed to tackle the global water purification challenges facing the world. For example, the development of nanoporous membranes with biofilm-resistant surfaces and embedded sensors/actuators that can automatically adjust membrane rejection, permeability and selectivity could provide a low pressure/energy desalination technology. We anticipate that active nanomaterials and nanosystems [e.g., bioactive nanostructures and 2-D arrays of multifunctional and adaptive nanomaterials] will be key components of such membranes.

Increasingly, water scientists and engineers are questioning the viability of building and operating large and centralized water treatment plants to meet the water demands of users with different water quality requirements. The convergence of nanotechnology and information technology has the potential to accelerate the development of small scale and customized water treatment systems that can either treat a local water source or provide water of a specified quality to meet a local need. For example, "smart" nano-info sensing devices—with the ability to perform a specified action upon detection of a compound—are being developed. Such devices could be placed in surface water systems or subsurface environments to track contaminant migration and and implement preventive measures to keep critical compounds from contaminating local water sources. Reduced size coupled with an accompanying increase in computational power will make these sensors particularly effective. This combined capability allows for ubiquitous placement in areas of need—water, air and soil—thereby effectively increasing spatial coverage. Current disinfectants such as chlorine and ozone have the potential to generate toxic byproducts (e.g. trihalomethanes and bromide) and have limited activeness against emerging microbes and viruses. Thus, we anticipate convergence between nanotechnology and biotechnology to accelerate the development of novel biocides able to selectively deactivate key cellular signaling and metabolic pathways of water-borne microbes without generating disinfection by-products. Ultimately, the convergence of nanotechnology with other existing and emerging technologies may lead to the development of personalized water treatments that could be mass-produced cheaply and distributed throughout the world.

Funding of innovative research will be critical to the development of the next generation of nanotechnology-based solutions for water purification. A major concern is that the current trend of decreasing research budgets will result in a significant decrease in innovative research. Note that many funding agencies are mission-oriented, i.e., they carry out and/or fund research directed toward specific goals. Consequently, when research budgets shrink, available resources become primarily devoted to the agency core mission activities at the expense of funding for exploratory and innovative research. Industry and venture capital firms often prefer to invest in technologies that are far beyond the "idea" or proof-of-concept stage. When resources are unavailable or limited for the development of these ideas into potential commercial products, society bears the burden of not being able to capitalize on the benefits of using nanotechnology to solve the water quality and supply challenges facing the world.

The characterization of the fate, transport and impacts of nanomaterials on humans and ecosystems will determine to a large extent regulatory and public acceptance of nanotechnology-based solutions for water purification. These impacts may be an impairment of human health or ecosystem viability, or they may be of a philosophical, ethical or legal nature. Whatever the possible adverse effects, these must be systematically explored. This requires research into both human and environmental health as well as in to the ethical, legal and social impacts of this technology. Finally, we would like to point out that communication of research results and public engagement will be key factors in the successful development and implementation of nanotechnology solutions to our global water needs. This communication must extend beyond scientific and technical publications. An effort should be made to condense and compile research results into more easily accessible formats for all stakeholders worldwide. Ultimately, this global communication will stimulate the development and deployment of more effective and affordable technologies for solving the water quality and supply challenges facing the world.

Index

Note: Page numbers followed by "*f*", and "*t*" refers to figures and tables respectively.

U

V

W

X

Z